MANY MATHEMATICIANS HAVE CONTRIBUTED IN AREAS OTHER THAN THOSE IN WHICH THEY ARE MENTIONED. \

CHAPTER 3 Logic	CHAPTER 4 Systems of Numeration	CHAPTER 5 Numbers and the Real Number System	
	3000 B.C. Egyptian system, an additive system. Babylonian system, a place-value system (without zero). 2000 B.C. Multiplication by duplation and mediation.	3000 B.C. Egyptians and Babylonians had a knowledge of counting and unit fractions.	3000 B.C. Egyptians and Babylonians had some knowledge of algebra, solved algebraic problems by trial and error rather than formalized procedures. 1850 B.C. Problems of algebraic nature in Moscow Papyrus. 1650 B.C. Rhind Papyrus used legs walking to right to represent $+$, legs walking to left to represent $-$. The word "aha" (heap) was used as the unknown quantity.
380 B.C. Plato stressed reasoning. 340 B.C. Aristotle was the first to use systematized deductive logic.	230 B.C. Eratosthenes developed the prime number sieve.	600 B.C. Pythagoras and followers took first steps in the development of number theory—friendly numbers, perfect numbers, irrational numbers, prime numbers. 75 B.C. Heron—approximations of square roots.	A.D. 250 Diophantus wrote *Arithmetica*, an algebraic treatment of number theory. First to take steps toward algebraic notation. Solved first-, second-, and third-degree equations.
	A.D. 1000 Beginning of development of the symbols we use for numbers, 1–9. Hindus introduced the number 0 before A.D. 800.		820 Arab mathematician al-Khowarizmi wrote *Al-jabr*, an influential book on algebra and Hindu numerals. 1202 Fibonacci wrote *Liber abaci*, a book on arithmetic and elementary algebra that included advanced Hindu Arabic notation, a method of calculating with integers and fractions, solution of equations by algebraic processes, and the Fibonacci sequence, as well as some geometry and trigonometry.
		1491 Calandra first printed process of long division. 1525 Rudolff first used $\sqrt{\ }$ symbol for square root.	1545 Tartaglia solved cubic equation. Cardino wrote *Ara magna*, which included negative roots of an equation, imaginary numbers.
1682 Leibniz's ideas and writings played an important part in the development of symbolic logic. 1854 Boole wrote *The Laws of Thought*, which many feel marks the beginning of symbolic logic; also wrote on differential equations. 1855 DeMorgan's law. 1879 Frege used logic to study the foundations of arithmetic; developed theory of quantifiers. 1910 Russell and Whitehead stressed logical structure in *Principia Mathematica*. 1931 Gödel—logic and axiomatics.	1617 Napier's Rods	1636 Fermat is often credited with founding modern number theory. 1820 Gauss, considered one of the three greatest mathematicians (with Archimedes and Newton), made many contributions to number theory. Proved the Fundamental Theorem of Arithmetic and contributed to many other branches of mathematics. 1872 Dedekind—irrational numbers. 1874 Cantor made many contributions in the area of real numbers. 1889 Peano's axioms for natural numbers. 1910 *Principia Mathematica*—logical structure of arithmetic.	1580 Viète wrote *In Artem* and *Canon*. Did much for the development of symbolic algebra. Used vowels to represent unknown quantities. 1630 Girard—algebra, geometry, and trigonometry. First to use abbreviations sin, cos, tan. Descartes is credited with using the beginning letters of the alphabet for constants and the last letters for variables.

A SURVEY *of* MATHEMATICS WITH APPLICATIONS

4TH EDITION

Allen R. Angel

Monroe Community College

Stuart R. Porter

Emeritus
Monroe Community College

ADDISON-WESLEY PUBLISHING COMPANY

Reading, Massachusetts   Menlo Park, California   New York   Don Mills, Ontario
Wokingham, England   Amsterdam   Bonn   Sydney   Singapore   Tokyo   Madrid
San Juan   Milan   Paris

Sponsoring Editor:	Bill Poole
Development Editor:	Stephanie Botvin
Assistant Development Editor:	Maribeth Jones
Managing Editor:	Karen Guardino
Production Supervisor:	Jack Casteel
Math Marketing Manager	Andy Fisher
Copy Editor:	Cynthia Benn
Proofreader:	Joyce Grandy
Text Design:	DECODE, INC.
Technical Art Supervisor:	Joseph Vetere
Art Coordinator:	Connie Hulse
Illustration:	*Electronic:* Tech Graphics, Network Graphics
	Other: James Bryant
Production Editorial Services:	Barbara Pendergast
Manufacturing Supervisor:	Judy Sullivan
Cover Design:	Peter Blaiwas

Photo Credits may be found on page A-55.

To my wife, Kathy, and my sons, Robert and Steven

A.R.A.

To my family: Joyce, Lisa, Tod, Teri, Adam, Andrew Matthew, Emily

S.R.P.

Library of Congress Cataloging-in-Publication Data

Angel, Allen R. 1942-
  A survey of mathematics with applications/Allen Angel, Stuart R. Porter—4th ed.
    p.   cm.
  Includes index.
  ISBN 0-201-54994-8
  1. Mathematics.   I. Porter, Stuart R., 1932-   II. Title.
QA39.2.A54 1993
510–dc20   92-35566
                                                                     CIP

6 7 8 9 10 DO 9695

PREFACE

The title of this text, *A Survey of Mathematics with Applications*, provides a key to both its content and the type of course for which it is intended. Prospective users include students majoring in liberal arts, elementary education, the social sciences, business, and nursing and the allied health fields.

This text meets the needs of those states, including California, Florida, New Jersey, and Texas, that now require students to obtain a minimum competency in mathematics for graduation or transfer.

The text may be used in either a one-semester or two-semester course. The flexible structure of the material allows instructors to teach the chapters in any order they wish.

Our primary goal in writing this book was to give students a text that they can read, understand, and enjoy while learning mathematics. We hope that as students read through the text they will begin to understand and enjoy mathematics and possibly decide to venture on to another mathematics course.

FEATURES NEW TO THIS EDITION

In this edition we have made a special effort to provide interesting, informative, and motivational marginal material that illustrates the usefulness of mathematics in our world. Through this material, and the practical applications used in examples and exercises, we hope that students will come to appreciate mathematics and its multifaceted role in their everyday lives.

New features in this edition include:

▶ Attractive and functional full-color design that not only enhances the text's pedagogy but also serves to highlight motivational material.

▶ Exciting and motivating photoessays at the beginning of each chapter aimed at showing students the real-world applications of each topic.

▶ Enhanced, colorful **Did You Know** boxes and other illustrative art that highlight the connections of mathematics to history, to the arts and sciences, to both non-Western and Western cultures, and to technology.

▶ Biographies of the men and women who have advanced the study of mathematics and mathematical applications.

► Writing exercises that ask students to explain their answers. These exercises are intended to test student understanding of mathematical concepts and processes, and can be used as discussion exercises. They are indicated with a colored number.

PEDAGOGY

In the fourth edition, we have aimed to provide students with features that will help them improve their skills and acquire an understanding of mathematics. We have also designed the pedagogy to make the book both easy to use and to study. Pedagogical features include the following:

► Clear explanations of topics.

► Short, simple sentences to increase the students' understanding of the material.

► An abundance of detailed, worked-out examples.

► An ample supply of well-graded exercises at the end of each section. The more difficult exercises are indicated with an asterisk, *, and the most difficult are under the heading Problem Solving.

► Important information highlighted in color.

► Practical applications used wherever possible to reinforce the material and motivate the student.

► Application problems, which include names from many cultures and portray men and women in non-gender-specific roles.

► Problem-solving exercises at the end of many exercise sets, which reinforce the techniques of problem solving presented in Chapter 1. Many of these problems are excellent assignments for group or cooperative learning projects.

► Research exercises at the end of many exercise sets, which can be assigned to embellish the material. Many of these exercises are appropriate for extra-credit projects.

► Chapter summaries, which include Key Terms and Important Facts, Review Exercises, and Chapter Tests.

► Answers to odd-numbered exercises and to all Review Exercises and Chapter Tests. These are included at the end of the text.

CONTENT REVISION

Our revisions to the content are based on the advice of a large number of professors who teach the course. At their request, we have carried out a careful revision of the text to improve the clarity and organization of the material. In addition, we have added or increased coverage of selected topics, as follows:

- ► An expanded chapter on critical thinking, including a new section on estimation.
- ► Inclusion of more material on the use of calculators at appropriate locations in the text.
- ► More examples and more exercises than in previous editions.
- ► Reorganization of Chapter 2, "Sets."
- ► Reorganization of Chapter 3, "Logic."
- ► Increased coverage of number theory.
- ► A greatly expanded section on the Fibonacci sequence.
- ► Increased emphasis on formulas and functions in Chapter 6, "Algebra, Graphs, and Functions."
- ► An expanded chapter on geometry, with additional material on graph theory and fractals, as well as a section on Logo.
- ► Updated coverage of consumer mathematics topics for the 1990s.
- ► Reorganization of Chapter 11, "Probability," with a section on conditional probability added.
- ► Integration of the material on computers into other topics as appropriate.
- ► Inclusion of a new appendix on the metric system and dimensional analysis.
- ► Greater emphasis on problem solving throughout the text.

SUPPLEMENTS

A full teaching and learning package accompanies this text and provides resources for both the instructor and the student. Supplements include:

Instructor's Solutions Manual, which contains the worked-out solutions to all exercises in the text.

Omnitest II, an extremely easy-to-use, state-of-the-art computerized testing system. As an algorithm-driven system, it can easily create up to 99 versions of the same test by automatically inserting random numbers into model problems. While the numbers are random, they are constrained to result in reasonable answers. Omnitest II also allows instructors to add and edit questions with full graphics capability, as well as preview and edit items on the screen. It uses pull-down menus for fast access to the functions used most, and utilizes a WYSIWYG ("What You See Is What You Get") interface for perfect match from screen to printout. It runs on IBM PC/compatibles.

Printed Test Bank, which includes three alternative tests per chapter. Instructors can use these items as actual tests or as reference for creating tests with or without the computer.

Student's Solutions Manual, which contains detailed, worked-out solutions to all the odd-numbered exercises in the text. Students will find this manual very helpful.

Guide to CLAST Mathematical Competency (State of Florida), which provides all of the necessary material to help students prepare for the computational portion of the CLAST test. It includes worked-out examples and practice for CLAST skills as well as a practice test. Optional topics in trigonometry are provided for those who wish to brush up in this area as well.

ACKNOWLEDGMENTS

We would like to thank the many students and faculty members who offered suggestions for improving the text. We would also like to thank our colleagues at Monroe Community College for their valued suggestions, in particular, Larry Clar, Christine Dunn, Gary Egan, Annette Leopard, and Robert Nenno. Special thanks go to Robert Nenno for his many interesting suggestions for improving the text and to Frank Swetz, Pennsylvania State University at Harrisburg, for providing as well as checking some of the historical **Did You Knows**. We would also like to thank our students for their input and Judy Conturo for typing part of the manuscript.

Our wives, Kathy and Joyce, and our children, Robert and Steven, and Tod, Teri, Lisa, and Brian deserve our special thanks. Without their support and great sacrifice, this book could not have become a reality.

It is our pleasure to acknowledge the assistance given us by the staff of the Addison-Wesley Publishing Company. In particular, we appreciate the advice and encouragement of our acquisitions editor, William Poole; our development editor, Stephanie Botvin; our production editor, Jack Casteel; our art development editors, Mary Hill and Maribeth Jones; and our illustrator, Jim Bryant. We also thank Barbara Pendergast for editorial production services.

We would like to acknowledge the contribution of our supplement authors: Christine Dunn and Gary Egan, Monroe Community College, for the *Instructor's Solutions Manual* and *Student's Solutions Manual*; Evelyn Woodward, St. Petersburg Junior College, for the *CLAST Supplement*; ips Publishing, Inc., Vancouver, WA, for *OMNITEST II*.

We would also like to thank all the conscientious reviewers who provided us with invaluable suggestions as well as those who responded to questionnaires and surveys.

REVIEWERS

Frank Asta
College of DuPage

Hughette Bach
California State University–Sacramento

Madeline Bates
Bronx Community College, NY

Una Bray
Skidmore College, NY

David Dean
Santa Fe Community College, FL

Karen Estes
St. Petersburg Junior College, FL

Kurtis Fink
Northwest Missouri State University

Mary Lois King
Tallahassee Community College, FL

Edwin Owens
Pennsylvania College of Technology

Joanne Peeples
El Paso Community College, TX

Ronald Ruemmler
Middlesex County College, NJ

Rosa Rusinek
Queensborough Community College, NY

John Samoylo
Delaware County Community College

Richard Schwartz
College of Staten Island, NY

Joyce Wellington
Southeastern Community College, NC

Sue Welsch
Sierra Nevada College

James Wooland
Florida State University

CLAST REVIEWERS CONFERENCE

Karen Estes
St. Petersburg Junior College

Peter Kerwin
Tallahassee Community College

Robert Sonnenberg
Valencia Community College

Evelyn P. Woodward
St. Petersburg Junior College

QUESTIONNAIRE

Jerry Bartolomeo
Nova University, FL

Denis Beveridge
Lincoln Land Community College, IL

Janis M. Cimperman
St. Cloud State University, MN

Virginia Ellen M. Hanks
Western Kentucky University

Edward Heltzel
Luzerne County Community College, IL

Gail Levine
Nova University, FL

Gary L. Long
Tennessee Wesleyan University

Michael Mays
West Virginia University

Lyman Milroy
Pennsylvania College of Technology

Sister Dorothy Regina
Immaculate College, PA

Rebecca W. Stamper
Western Kentucky University

Sherry Tornwall
University of Florida

Lynn R. Williams
Indiana University at South Bend

SURVEY

Carole Bauer
Triton College, IL

Jack Brillhart
Tri-State University, IN

Charles Burrell
Western Carolina University

Elizabeth A. Byrne
Bristol Community College, MA

Mark E. Christie
Louisiana State University

Don Cohen
SUNY - Cobleskill, NY

Benjamin Collins
Saint John's University, MN

Laura Corrigan
Central Florida Community College

Sherralyn Craven
Central Missouri State University

Arlene Dowshen
Widener University, PA

John F. Evans
Peninsula College, WA

Teresa Floyd
Mississippi College

Linda Greico
Hillsborough Community College, FL

William Greiner
McLennan Community College, TX

Ron L. Harris
Lee College, TN

Francis Hildebrand
Western Washington University

Bernard Hoerbelt
Genesee Community College, NY

G. R. Irby
Northwest Missouri State University

Doug Jones
Tallahassee Community College, FL

Maryann Justinger
Erie Community College South, NY

Andrew Kiin
Westfield State College, MA

Vuryl Klassen
California State University, CA

Mark Klespis
St. Xavier College, IL

Sarah Mabrouk
Boston University, MA

Antonio Magliaro
Southern Connecticut State University

Robert D. McMillan
Oklahoma Christian University

Gerald Meike
Wright State University

John Mislom
Butler County Community College, PA

Kanan Mishra
New Community College of Baltimore, MD

Lynda Morton
University of Missouri

Jack D. Murphy
Pennsylvania College of Technology

Carl Myhill
Western Connecticut State University

Michael P. Nestarick
Pennsylvania College of Technology

Dena Jo Perkins
Oklahoma City University

Judith A. Pokrop
Cardinal Stritch College, WI

Mary Rack
Johnson County Community College, KS

Dennis M. Ragan
Kishwaukee College, IL

Fred Rispoli
Dowling College, NY

Eric Robinette
Columbia College, MO

Stewart M. Robinson
Cleveland State University, OH

Howard Rolf
Baylor University, TX

Kenneth Schoen
Worcester State College, MA

Katherine Soderbom
Massasoit Community College, MA

Margaret Stempien
Indiana University of Pennsylvania

David Sutherland
Middle Tennessee State University

Norman Sweet
SUNY-Oneonta, NY

William Tomhave
Concordia College, MN

R. T. Walden
Mid-Continent Baptist College, KY

Mary Wiest
Mankato State University, MN

Jon Wilkin
North Virginia Community College

Charles Ziegenfus
James Madison University, VA

TO THE STUDENT

Mathematics is an exciting, living study. It has applications that shape the world around you and influence your everyday life. We hope that as you read through this book you will realize just how important mathematics is and gain an appreciation of both its usefulness and its beauty. We also hope to teach you some practical mathematics that you can use in your everyday life and that will prepare you for further courses in mathematics.

Our primary purpose in writing this text was to provide material that you could read, understand, and enjoy. To this end we have used straightforward language and tried to relate mathematical concepts to everyday experiences. We have also provided many detailed examples for you to follow.

The concepts, definitions, and formulas that deserve special attention have been either boxed or set in boldface type. The exercises are graded so that the more difficult problems appear at the end of the exercise set. The problems with numbers set in color are writing exercises. The exercises that are aster-isked, *, are more challenging, and the problem-solving problems are the most challenging.

Each chapter has a summary, review exercises, and a chapter test. When studying for a test, be sure to read the chapter summary, work the review exer-cises, and take the chapter test. The answers to the odd-numbered exercises, all review exercises, and the chapter tests appear in the Answers section in the back of the text. However, you should use the answers only to check your work.

It is difficult to learn mathematics without becoming involved. To be suc-cessful, we suggest you read the text carefully *and work each exercise in each assignment in detail*. Check with your instructor to determine which supplements are available for your use. Also determine whether a calculator may be used on homework and tests.

We welcome your suggestions and your comments. Our address is Monroe Community College, Rochester, NY 14623. Good luck in your adventure in mathematics!

INDEX OF APPLICATIONS

CONTENTS

Chapter 5 NUMBER THEORY & THE REAL NUMBER SYSTEM   168

Chapter 6 ALGEBRA, GRAPHS, & FUNCTIONS   238

Chapter 7 SYSTEMS OF LINEAR EQUATIONS & INEQUALITIES   314

Chapter 8 GEOMETRY 358

Chapter 9 MATHEMATICAL SYSTEMS 426

Chapter 10 CONSUMER MATHEMATICS 454

CRITICAL THINKING SKILLS

Charles Lindbergh, who made the first solo flight across the Atlantic in 1927, had to solve many problems, such as the route over the ocean and the financing of the trip. Most importantly, the aircraft design had to be modified to carry enough fuel for the 3600–mile trip. To keep the load light, Lindbergh flew without a copilot, parachute, or radio and carried only five sandwiches and a quart of water on his 33–hour, 32–minute trip.

It's Friday night and you've just arrived at a party at a friend's apartment. You're the first to arrive, but your friend has invited 15 people, and knows that 8 guests will be bringing dates. How many people must you greet when you walk in the door?

Solving this problem requires critical thinking: considering the information given, deciding what question must be answered, and making and acting on a plan to find the answer. When you reread the question, you see that it asks how many guests are at the party when you arrive. Since the problem states that you are the first guest to arrive, the only person you will have to greet is the host.

Mathematics provides the tools to think critically and solve problems. Every day, you make decisions that require you to use critical thinking skills. Sometimes, the problem may be one that can be solved by computation. For example, in a

Symbol for element (Magnesium) → Mg

Atomic weight → 24

Data from Mendeleyev's Periodic Table of Elements

Group I	Group II	Group III	Group IV	Group V	Group VI
Na 23	Mg 24	Al 27.3	Si 2.8	P 31	S 32
K 39	Ca 40	Sc 44	Ti 48	V 51	Cr 52
Cu	Zn 65	Ga 68	Ge 72	As 75	Se 78
Cr	Yt	Zr	Nb		

Seeing patterns often helps in solving problems. In 1869, chemists had isolated 63 of 109 of the chemical elements known today, but had not found any apparent underlying order. In that year, Russian chemist Dmitri Mendeleyev saw a pattern: He proposed a table of the elements organized by increasing atomic weight and grouped according to similar properties. To make this work, he had to predict the existence of three then-unknown elements: scandium, gallium, and germanium.

drugstore you may be choosing between two bottles of shampoo. One is 6 ounces 2.03 49 and costs $2.95, the other is 8 ounces and costs $3.50. Which is the 2.28 .43 better price? The answer to this problem could be found by finding the unit cost: Compute the cost for one ounce of each shampoo and select the bottle that costs the least per ounce.

Sometimes solving a problem may require you to make a reasonable estimate, to look for clues, or to experiment with several possible solutions before choosing the right one. As you saw in the problem about the party, sometimes the most important part of solving a problem is just understanding what question must be answered.

Problem solving can be as satisfying as putting together a puzzle. Using the evidence from the pieces that are available, we can make guesses and test our assumptions. As George Polya said, "A great discovery solves a great problem, but there is a grain of discovery in the solution of any problem."

Life constantly presents new problems. The great inventors, scientists, scholars, politicians, and artists made their contributions to civilization by confronting and solving problems. To learn the techniques for solving problems requires practice and patience. But once the basic principles are understood, they can be applied to each new challenge. The goal of this chapter is to help you master the skills of reasoning, estimating, and problem solving. These skills will aid you in solving the problems in the remainder of this book, as well as problems that you will encounter in your everyday life.

One way of determining the answer to the question "how many" is to estimate, using a small sampling of a larger group. This technique can be used to guess how many jelly beans are in a jar, or how many people are in attendance at a political rally.

1.1 Inductive Reasoning

The goal of this chapter is to help you improve your reasoning and problem-solving skills. This section introduces inductive and deductive reasoning, which are used in problem solving. The next section introduces the concept of estimation. Estimation is a technique that can be used to determine if an answer obtained for a problem or on a calculation is "reasonable." Section 1.3 introduces and applies problem-solving techniques.

Before looking at some examples of inductive reasoning and problem solving, let us first review a few facts about certain numbers. The **natural numbers** or **counting numbers** are the numbers 1, 2, 3, 4, 5, 6, 7, 8, ... The three dots, called an **ellipsis**, mean that 8 is not the last number, but the numbers continue in the same manner. A word that we sometimes use is "divisible." If $a \div b$ has a remainder of zero, then a *is divisible by* b. The counting numbers that are divisible by 2 are 2, 4, 6, 8, ... These are called the *even counting numbers*. The numbers that are not divisible by 2 are 1, 3, 5, 7, 9, ... These are the *odd counting numbers*.

Recognizing patterns is sometimes helpful in solving problems, as Examples 1 and 2 illustrate.

▶ **Example 1**

If two odd counting numbers are multiplied together, will the product always be an odd counting number?

Solution: To answer this question, we will examine the products of several pairs of odd numbers to see if there is a pattern.

$1 \times 3 = 3$	$3 \times 5 = 15$	$5 \times 7 = 35$
$1 \times 5 = 5$	$3 \times 7 = 21$	$5 \times 9 = 45$
$1 \times 7 = 7$	$3 \times 9 = 27$	$5 \times 11 = 55$
$1 \times 9 = 9$	$3 \times 11 = 33$	$5 \times 13 = 65$

None of the products is divisible by 2. Thus we might predict from these examples that the product of any two odd numbers is an odd number.

▶ **Example 2**

If an odd counting number is multiplied by an even counting number, will the product be an odd or an even counting number?

Solution: Let us look at a few examples.

$1 \times 2 = 2$	$3 \times 2 = 6$	$5 \times 2 = 10$
$1 \times 4 = 4$	$3 \times 4 = 12$	$5 \times 4 = 20$
$1 \times 6 = 6$	$3 \times 6 = 18$	$5 \times 6 = 30$
$1 \times 8 = 8$	$3 \times 8 = 24$	$5 \times 8 = 40$

All of these products are divisible by 2. Therefore we might predict that the product of an odd and an even number is an even number.

In Examples 1 and 2 we cannot conclude that the results are true for all counting numbers. However, from the patterns developed, we can make predictions. This type of reasoning process, arriving at a general conclusion from specific observations or examples, is called **inductive reasoning**, or **induction**.

Induction often involves observing a pattern and from that pattern predicting a conclusion. Imagine an endless row of dominoes. You knock down the first, which knocks down the second, which knocks down the third, and so on. Assuming the pattern will continue uninterrupted, you conclude that eventually all the dominoes will fall, even though you may not witness the event.

> **Inductive reasoning** is the process of reasoning to a general conclusion through observations of specific cases.

Inductive reasoning is often used by mathematicians and scientists to predict answers to complicated problems. For this reason, inductive reasoning is part of the **scientific method**. When a scientist or mathematician makes a prediction based on specific observations, it is called a hypothesis or **conjecture**. After looking at the products in Example 1 we might conjecture that the product of two odd counting numbers will be an odd counting number.

Examples 3 and 4 illustrate how we arrive at a conclusion using inductive reasoning.

▶ **Example 3**

What reasoning process has led to the conclusion that no two people have the same fingerprints? This conclusion has resulted in fingerprints being used in courts of law as evidence to convict persons of crimes.

Solution: In millions of tests, no two people have been found to have the same fingerprints. By induction, then, we believe that fingerprints provide a unique identification and can therefore be used in a court of law as evidence. Is it possible that sometime in the future, two people will be found who do have exactly the same fingerprints?

▶ **Example 4**

Consider the conjecture "If the sum of the digits of a number is divisible by 3, then the number is divisible by 3." Test several numbers to determine whether the conjecture appears true or false.

Solution: Let us look at some numbers, the sum of whose digits is divisible by 3.

Number	Sum of the Digits	Sum of the Digits Divided by 3	Number Divided by 3
114	$1 + 1 + 4 = 6$	$6 \div 3 = 2$	$114 \div 3 = 38$
234	$2 + 3 + 4 = 9$	$9 \div 3 = 3$	$234 \div 3 = 78$
7020	$7 + 0 + 2 + 0 = 9$	$9 \div 3 = 3$	$7020 \div 3 = 2340$
2943	$2 + 9 + 4 + 3 = 18$	$18 \div 3 = 6$	$2943 \div 3 = 981$
9873	$9 + 8 + 7 + 3 = 27$	$27 \div 3 = 9$	$9873 \div 3 = 3291$

In each of the examples we find that the sum of the digits is divisible by 3 and the number itself is divisible by 3. From these specific examples we might be tempted to generalize that the conjecture "If the sum of the digits of a number is divisible by 3, then the number is divisible by 3" is true.

▶ **Example 5**

Pick any number, multiply the number by 4, add 6 to the product, divide the sum by 2, and subtract 3 from the quotient. Repeat this procedure for several different numbers and then make a conjecture about the relationship between the original number and the final number.

Solution: Let us go through this one together.

Pick a number:	say, 5
Multiply the number by 4:	$4 \times 5 = 20$
Add 6 to the product:	$20 + 6 = 26$
Divide the sum by 2:	$26 \div 2 = 13$
Subtract 3 from the quotient:	$13 - 3 = 10$

Note that we started with the number 5 and finished with the number 10. If you start with the number 2, you will end with the number 4. Starting with 3 would result in a final number of 6, 4 would result in 8, and so on. Based upon these few examples many of you would conjecture that when you follow the given procedure, the number you end with will always be twice the original number.

The result reached by inductive reasoning is often correct for the specific cases studied but not correct for all cases. History has shown that not all conclusions arrived at by inductive reasoning are correct. For example, Aristotle (384–322 B.C.) reasoned inductively that heavy objects fall at a faster rate than light objects. About 2000 years later, Galileo dropped two pieces of metal—one 10 times heavier than the other—from the Leaning Tower of Pisa in Italy. He found that both hit the ground at exactly the same moment, so they must have traveled at the same rate.

When forming a general conclusion using inductive reasoning, you should test it with several special cases to see whether the conclusion appears correct. A **counterexample** is any special case that shows that a conclusion is not

An Experiment Revisited

Apollo 15 Astronaut David Scott used the moon as his laboratory to show that a heavy object (a hammer) does indeed fall at the same rate as a light object (a feather). Had Galileo dropped a hammer and feather from the tower of Pisa, the hammer would most certainly have fallen more quickly to the ground and he still would have concluded that a heavy object falls faster than a lighter one. If it is not the object's mass that is affecting the outcome, then what is it? The answer is air resistance or friction: The earth has an atmosphere; the moon does not.

correct. Galileo's counterexample disproved Aristotle's conjecture. If a counterexample cannot be found, the conjecture is neither proven nor disproven.

A second type of reasoning process is called **deductive reasoning**, or **deduction**. Mathematicians use deductive reasoning to *prove* conjectures true or false.

> **Deductive reasoning** is the process of reasoning to a specific conclusion from a general statement.

▶ **Example 6**

Prove, using deductive reasoning, that the procedure in Example 5 will always result in twice the original number selected.

Solution: To use deductive reasoning we begin with the *general* case rather than specific examples, as was illustrated in Example 5. Let us select the letter n to represent *any number*.

Pick any number: n
Multiply the number by 4: $4n$ (4n means 4 times n)
Add 6 to the product: $4n + 6$

Divide the sum by 2: $\dfrac{4n + 6}{2} = \dfrac{\overset{2}{4n} + \overset{3}{6}}{\underset{1}{2}} = 2n + 3$

Subtract 3 from the quotient: $2n + 3 - 3 = 2n$

Notice that for any number n selected, the result is $2n$, or twice the original number selected.

In Example 5 you may have conjectured, using specific examples and inductive reasoning, that the result would be twice the original number selected. In Example 6 we proved, using deductive reasoning, that the result will always be twice the original number selected.

 Section 1.1 Exercises

1. a) List the natural numbers.
  b) What is another name for the natural numbers?
2. a) What does it mean to say, "a is divisible by b," where a and b represent natural numbers?
  b) List three natural numbers that are divisible by 5.
  c) List three natural numbers that are divisible by 12.

In Ex. 3–6, explain your answer using a sentence or sentences.
3. What is a conjecture?
4. What is inductive reasoning?
5. What is deductive reasoning?
6. What is a counterexample?

In Ex. 7–11, use inductive reasoning to predict the next line in the pattern.

7. $(1 \times 9) + 2 = 11$
$(12 \times 9) + 3 = 111$
$(123 \times 9) + 4 = 1111$

8. $1 = 1$
$1 + 2 = 3$
$1 + 2 + 3 = 6$
$1 + 2 + 3 + 4 = 10$
$1 + 2 + 3 + 4 + 5 = 15$

9.
$$1 = 1$$
$$1 + 3 = 4$$
$$1 + 3 + 5 = 9$$
$$1 + 3 + 5 + 7 = 16$$

10.

$$\begin{array}{r} 1 \\ + 10 \\ \hline 11 \end{array} \qquad \begin{array}{r} 1 \\ 10 \\ + 100 \\ \hline 111 \end{array} \qquad \begin{array}{r} 1 \\ 10 \\ 100 \\ + 1000 \\ \hline 1111 \end{array} \qquad \begin{array}{r} 1 \\ 10 \\ 100 \\ 1000 \\ + 10000 \\ \hline ? \end{array}$$

11.
$$\begin{array}{ccccccc} & & & 1 \\ & & 1 & & 1 \\ & 1 & & 2 & & 1 \\ 1 & & 3 & & 3 & & 1 \\ 1 & 4 & & 6 & & 4 & 1 \end{array}$$

In Ex. 12–20, use inductive reasoning to predict the next three numbers in the pattern (or sequence).

12. $1, 3, 5, 7, \ldots$

13. $12, 10, 8, 6, \ldots$

14. $5, 3, 1, -1, -3, \ldots$

15. $1, -1, 1, -1, 1, \ldots$

16. $1, -3, 9, -27, \ldots$

17. $1, 1/2, 1/4, 1/8, \ldots$

18. $1, 4, 9, 16, 25, \ldots$

19. $1, 1, 2, 3, 5, 8, 13, 21, \ldots$

20. $0, 3, 8, 15, 24, 35, 48, 63, \ldots$

In Ex. 21–24, look for a pattern in the first three products and use it to find the fourth product.

21.
$$\begin{array}{r} 1 \\ \times 1 \\ \hline 1 \end{array} \qquad \begin{array}{r} 11 \\ \times 11 \\ \hline 121 \end{array} \qquad \begin{array}{r} 111 \\ \times 111 \\ \hline 12321 \end{array} \qquad \begin{array}{r} 1111 \\ \times 1111 \\ \hline ? \end{array}$$

22.
$$\begin{array}{r} 9 \\ \times 9 \\ \hline 81 \end{array} \qquad \begin{array}{r} 99 \\ \times 9 \\ \hline 891 \end{array} \qquad \begin{array}{r} 999 \\ \times 9 \\ \hline 8991 \end{array} \qquad \begin{array}{r} 9999 \\ \times 9 \\ \hline ? \end{array}$$

23.
$$\begin{array}{r} 9 \\ \times 9 \\ \hline 81 \end{array} \qquad \begin{array}{r} 909 \\ \times 9 \\ \hline 8181 \end{array} \qquad \begin{array}{r} 90909 \\ \times 9 \\ \hline 818181 \end{array} \qquad \begin{array}{r} 9090909 \\ \times 9 \\ \hline ? \end{array}$$

24. Study the entries that follow and use the pattern they exhibit to complete the last two rows:
$1 + 3 = 4$ or 2^2
$1 + 3 + 5 = 9$ or 3^2
$1 + 3 + 5 + 7 = 16$ or 4^2
$1 + 3 + 5 + 7 + 9 = ?$
$1 + 3 + 5 + 7 + 9 + 11 = ?$

25. Consider the following products. What patterns can you detect?

$1 \times 9 = 9$	$4 \times 9 = 36$	$7 \times 9 = 63$
$2 \times 9 = 18$	$5 \times 9 = 45$	$8 \times 9 = 72$
$3 \times 9 = 27$	$6 \times 9 = 54$	$9 \times 9 = 81$

26. Consider the number 142,857 and its first four multiples:

$$\begin{array}{r} 142857 \\ \times \quad 1 \\ \hline 142857 \end{array} \qquad \begin{array}{r} 142857 \\ \times \quad 2 \\ \hline 285714 \end{array} \qquad \begin{array}{r} 142857 \\ \times \quad 3 \\ \hline 428571 \end{array} \qquad \begin{array}{r} 142857 \\ \times \quad 4 \\ \hline 571428 \end{array}$$

a) Observe the digits in the product and use inductive reasoning to make a conjecture about the digits that will appear in the product $142,857 \times 5$.

b) Multiply 142,857 by 5 to see whether your conjecture appears to be correct.

c) Can you make a more general conjecture about the digits in the product of a multiplication problem where 142,857 is multiplied by a one-digit positive number?

d) Multiply 142,857 by the digits 6 through 8 and see whether your conjecture appears to be correct.

27. The ancient Greeks labeled certain numbers as triangular numbers. The numbers 1, 3, 6, 10, 15, 21, and so on are triangular numbers.

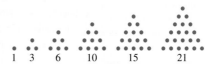

a) Can you determine the next two triangular numbers?

b) Describe a procedure to determine the next five triangular numbers without drawing the figures.

c) Is 72 a triangular number? Explain how you determined your answer.

28. Just as there are triangular numbers, there are also square numbers. The numbers 1, 4, 9, 16, 25, and so on are square numbers.

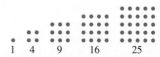

a) Determine the next three square numbers.

b) Describe a procedure to determine the next five square numbers without drawing the figures.

c) Is 72 a square number? Explain how you determined your answer.

29. The pattern shown here is taken from a quilt design known as a Triple Irish Chain. Complete the color pattern by indicating the color assigned to each square.

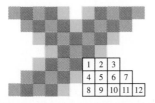

30. Pick a number, multiply the number by 2, add 4 to the product, divide the sum by 2, and subtract 2 from the quotient. See Example 5.

a) What is the relationship between the number you started with and the final number?

b) Arbitrarily select some different numbers and repeat the given process, recording the original number and the result.

c) Can you make a conjecture about the relationship between the original number and the final number?

d) Prove, using deductive reasoning, the conjecture you made in part (c). See Example 6.

31. Pick any number, multiply the number by 6. Add 3 to the product. Divide the sum by 3, and subtract 1 from the quotient.

a) What is the relationship between the number you started with and the final answer?

b) Arbitrarily select some different numbers and repeat the given process, recording the original number and the results.

c) Can you make a conjecture about the relationship between the original number and the final number?

d) Try to prove, using deductive reasoning, the conjecture you made in part (c).

32. Pick any number and add 1 to it. Find the sum of the new number and the original number. Add 9 to the sum. Divide the sum by 2 and subtract the original number from the quotient.

a) What is the final number?

b) Arbitrarily select some different numbers and repeat the given process. Record the results.

c) Can you make a conjecture about the final number?

d) Try to prove, using deductive reasoning, the conjecture you made in part (c).

33. Pick a number, add 5 to the number, divide the sum by 5, subtract 1 from the quotient, and multiply the result by 5.

a) What is the relationship between the number you started with and the final number?

b) Arbitrarily select some different numbers and repeat the given process, recording the original number and the result.

c) Can you make a conjecture about the relationship between the original number and the final number?

d) Try to prove, using deductive reasoning, the conjecture you made in part (c).

In Ex. 34–38, find a counterexample to show that each of the statements is incorrect.

34. The product of any two counting numbers is divisible by 2.

35. Every counting number greater than 5 is the sum of either two or three consecutive counting numbers. For example, 9 = 4 + 5 and 17 = 9 + 8.

36. The product of a number multiplied by itself is even.

37. The sum of any two odd numbers is divisible by 3.

38. When a counting number is added to 3 and the sum is divided by 2, the quotient will be an even number.

39. a) Construct a triangle and measure the three interior angles with a protractor. What is the sum of the measures?

b) Construct three other triangles, measure the angles, and record the sums. Are your answers the same?

c) Make a conjecture about the sum of the measures of the three interior angles of a triangle.

40. a) Construct a quadrilateral (a four-sided figure) and measure the four interior angles with a protractor. What is the sum of their angle measures?

b) Construct three other quadrilaterals, measure the angles, and record the sums. Are your answers the same?

c) Make a conjecture about the sum of the measures of the four interior angles of a quadrilateral.

41. a) Select one- and two-digit numbers and multiply each by 99. Record your results.

b) Find the sum of the digits in each of your products in part (a).

c) Make a conjecture about the sum of the digits when a one- or two-digit number is multiplied by 99.

42. a) Calculate the squares of 15, 25, 35, and 45 and record the results. (Note that the square of 15 is 15 × 15 = 225.) Examine the products and see whether you can establish a pattern in relation to the numbers being multiplied.

b) Make a conjecture about how to mentally calculate the square of any number whose unit digit is 5.

c) Calculate the squares of 55, 75, 95, and 105 using your conjecture.

Research Activity

43. a) Using newspapers, magazines, and other sources, find examples of where conclusions are arrived at using inductive reasoning.

b) Explain how inductive reasoning was used in arriving at the conclusion.

1.2 Estimation

An important step in solving mathematical problems—or, in fact, *any* problem—is to make sure that the answer you've arrived at makes sense. One technique for determining whether an answer is reasonable is to estimate. **Estimation** is the process of arriving at an approximate answer to a question. This section demonstrates several estimation methods.

To estimate, or approximate, an answer we often round off numbers as illustrated in the following examples.

▶ **Example 1**

Estimate the cost of 22 poster boards at $0.89 each.

Solution: We may round off the amounts as follows to obtain an estimate.

$$
\begin{array}{r}
22 \rightarrow 20 \\
\times\, 0.89 \rightarrow \times\, 0.90 \\
\hline
18.00
\end{array}
$$

Thus the 22 poster boards will cost about $18.00.

In Example 1 we could have rounded the 22 to 20 and the $0.89 to $1.00, which would result in an estimate of $20. The true cost is $0.89 × 22, or $19.58. Estimates are not meant to give exact values for answers but are a means of determining if your answer is reasonable. If you calculated an answer of $19.58 and then did a quick estimate to check it, you would know that the answer is reasonable because it is close to your estimated answer.

▶ **Example 2**

The number of bushels of grapes produced at a vineyard are 62,408 Cabernet Sauvignon, 118,916 French Colombard, 106,490 Chenin Blanc, 5960 Charbono, and 12,104 Chardonnay. Select the best estimates of the total number of bushels produced by the vineyard.

a) 500,000 **b)** 30,000 **c)** 300,000 **d)** 5,000,000

Solution: Following are suggested roundings. On the left, the numbers are rounded to thousands. For a less close estimate, round to ten thousands, as

Little Mistake, Big Discovery

The World, 1492

It was a mistake in estimation that brought Christopher Columbus to our part of the world in 1492. He believed the world to be round and thought of distance from one place to another in terms of degrees, with 360° longitude representing a "round trip." He estimated the distance from Spain over land (east) to the far coast of "India" (Asia) to be 282°, and hence very long. That meant the distance west, by sea, was only 78°, or very short. He assumed 56.6 miles to a degree, allowing roughly 4848 feet per mile (compared to our 5280 feet per mile). The outcome of his miscalculations put "India" about 4000 miles west of Spain, more or less where the Americas happen to be. Columbus died in 1506 still unaware that he had discovered a new continent.

illustrated on the right.

62,408 → 62,000	62,408 → 60,000
118,916 → 119,000	118,916 → 120,000
106,490 → 106,000	106,490 → 110,000
5,960 → 6,000	5,960 → 10,000
12,104 → 12,000	12,104 → 10,000
305,000	310,000

Either rounding procedure indicates that the best estimate is (c), 300,000.

▶ **Example 3**

Which of the answers (a–d) is the best estimate of the number of heartbeats in a lifetime? Assume 78 heartbeats per minute and an average life expectancy of 76 years.

a) 3,000,000,000   **b)** 3,000,000
c) 300,000,000   **d)** 30,000,000

Solution: Begin by setting up the calculations to determine the heartbeats in a lifetime.

78×60	number of heartbeats per hour
$78 \times 60 \times 24$	number of heartbeats per day
$78 \times 60 \times 24 \times 365$	number of heartbeats per year
$78 \times 60 \times 24 \times 365 \times 76$	number of heartbeats per lifetime

Round the multipliers to obtain an estimate of the number of heartbeats in a lifetime.

$$78 \times 60 \times 24 \times 365 \times 76$$
$$\downarrow \quad \downarrow \quad \downarrow \quad \downarrow \quad \downarrow$$
$$80 \times 60 \times 20 \times 400 \times 80$$

Now move the decimals in each number until each number is a number greater than or equal to 1 and less than 10.

$$80 \times 60 \times 20 \times 400 \times 80$$

Multiply the nonzero digits.

$$8 \times 6 \times 2 \times 4 \times 8 = 3072$$

Since each number originally multiplied is greater than 10 (arrows all moved to the left) we add the total number of zeros in the six numbers multiplied, 6 zeros, back to the product 3072 to obtain an estimate of 3,072,000,000.

Therefore the best estimate is (a) 3,000,000,000, three billion.

▶ **Example 4**

The odometer of an automobile reads 54,186.2 mi. If the automobile averaged 23.6 miles per gallon (mpg) for the given mileage, (a) estimate the number of gallons of gasoline used. (b) If the cost of the gasoline averaged $1.04 per gallon, estimate the total cost of the gasoline.

Solution:

a) To estimate the number of gallons divide the mileage by the number of miles per gallon.

$$\frac{54{,}186.62}{23.6}$$

Round these numbers to obtain an estimate.

$$\frac{50{,}000}{20} = 2500$$

Therefore the car used approximately 2500 gallons (gal) of gasoline.

b) Rounding the price of the gasoline to $1.00 per gallon, the cost of the gasoline is 2500 × 1, or $2500.

Did You

Know...

ESTIMATING TECHNIQUES IN MEDICINE

Estimating is one of the diagnostic tools used by the medical profession. Physicians take a small sample of blood, tissue, or body fluids to be representative of the body as a whole. Human blood contains different types of white blood cells that fight infection. When a bacterium or virus gets into the blood, the body responds by producing more of the type of white blood cell whose job it is to destroy that particular invader. So an increased level of white blood cells in a sample of blood not only indicates the presence of an infection but also helps identify its type. A trained medical technician estimates the relative number of each kind of white blood cell found in a count of 100 white blood cells. An increase in any one kind indicates the type of infection present. The accuracy of this diagnostic tool is impressive when you consider that there are normally 5 to 9 thousand white blood cells in a dropful of blood (1 cubic millimeter).

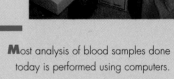

Most analysis of blood samples done today is performed using computers.

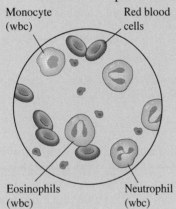

Monocyte (wbc) Red blood cells Eosinophils (wbc) Neutrophil (wbc)

A high count of esinophils (a particular kind of white blood cell, or wbc) can be an indicator of an allergic reaction.

Now let us look at some different types of estimation problems.

▶ **Example 5**

The scale on the accompanying map is $\frac{1}{4}$ inch = 70 miles.
a) Estimate the distance from Los Angeles to San Francisco, via Route 101 (the colored route).
b) If your average speed during the trip is likely to be 60 miles per hour (mph), estimate how long the trip will take.

Solution:

a) Begin by measuring the map distance in inches from Los Angeles to San Francisco. You should obtain a length of about $1\frac{1}{2}$ in. (depending upon how you estimate the curves). One and one half inches equals 6 one quarter inches. Thus the distance is approximately 6×70, or 420 mi.

b) The average speed is found by dividing the distance by the rate. Thus the trip will take about 420 mi ÷ 60 mph, or 7 hr.

▶ **Example 6**

Utility bills sometimes contain graphs illustrating the amount of electricity and gas used. The following graphs show gas and electric use at a specific residence for a period of one year, as well as the current month's use by the average residential customer. Using these graphs, can you answer the questions?

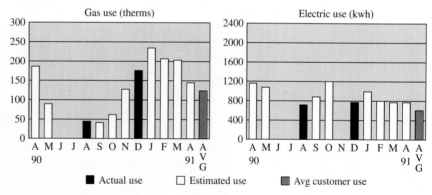

a) How often was an actual gas and electric reading made?
b) Estimate the number of therms of gas used by the average residential customer in the month of April 1991.
c) Estimate the gas used by the resident in April 1991.
d) If the cost of gas is 58.6151 cents per therm, estimate the gas bill in November.
e) In which month was the most electricity used? How many kilowatt hours (kwh) were used in this month?
f) If the cost of electricity is 8.8499 cents per kilowatt hour, estimate the cost of electricity in the month of February.

Solution:
a) Actual readings were made in only two months, August and December.
b) Approximately 125 therms were used, as shown by the height of the red bar.
c) Approximately 145 therms were used (slightly less than 150).
d) In November about 125 therms were used. The rate, 58.6151 cents per therm, is the same as $0.586151 per therm. To get a rough approximation round off the rate to $0.60 per therm.

$$0.60 \times 125 = 75*$$

Thus basic cost for gas was about $75.
e) The most electricity (approximately 1200 kwh) was used in October.
f) In February about 800 kwh were used. Write 8.8499 cents as $0.08849. Rounding the rate to $0.09 per kilowatt hour and multiplying by 800 yields an estimate of $72.

$$0.09 \times 800 = \$72$$

▶ Example 7

Scientists are concerned that large-scale fishing in the Antarctic may dangerously deplete the food supply of the Emperor penguins that inhabit Antarctica. The effect will ultimately be measured by any significant change in their number. Estimate the number of penguins in the accompanying photograph.

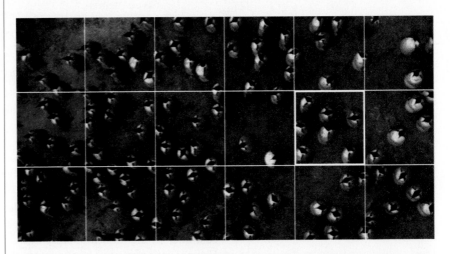

Solution: To estimate the number of penguins, we can divide the photograph into rectangles with equal areas, then select one area that appears to be representative of all the areas. Estimate (or count) the number of birds in this single area, and then multiply this number by the number of equal areas.

---

* These amounts do not include the basic monthly charge, taxes, and other extra charges often included on utility bills.

A team of researchers with the
National Wildlife Federation
used a remote-control camera
attached to a helium balloon to
estimate the Emperor penguins'
number. To be accurate in
counting the sample region, the
researchers put a pinhole
through the image of each
penguin pictured and then turned
over the photos and actually
counted the holes. One colony
was found to have as many as
10,960 penguins.

Let's divide the photo into 18 approximately equal areas. We select the
region indicated in the accompanying enlargement as our representative
region, and count the penguins in this sample. If half a bird is in the region,
we count it (see enlargement of the region). There are 7 penguins in this
region. Multiplying this value by 18 gives $18 \times 7 = 126$. Thus there are
about 126 penguins in the photo.

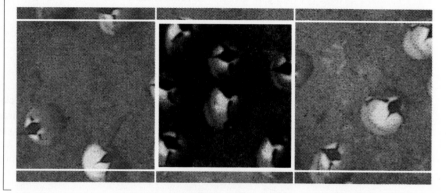

In problems similar to Example 7, the number of regions or areas into
which you choose to divide the total area is arbitrary. Generally the more regions the better the approximation, as long as the region selected is representative of the other regions in the map, diagram, or photo.

When you estimate an answer, the amount that your approximation differs from the actual answer will depend upon how you round off the numbers.
Thus in estimating the product of $196,000 \times 0.02520$, using the rounded values
$195,000 \times 0.025$ would yield an estimate much closer to the true answer than
using the rounded values $200,000 \times 0.03$. However, without a calculator, the
product of $195,000 \times 0.025$ might be more difficult to find than $200,000 \times 0.03$.
When estimating, you will need to determine the accuracy desired in your estimate and round the numbers accordingly.

Section 1.2 Exercises

In Ex. 1–22, estimate the answer. In this exercise set,
your answers may vary from the answers given in the
back of the text, depending upon how you round your
answers.

1. $1.43 + 100.6 + 156.9 + 179 + 0.23 + 416$

2. $192,600 \times 4120$ **3.** 1854×0.0096

4. $\dfrac{315}{0.062}$ **5.** $196.43 - 85.964$

6. 0.048×1964 **7.** 9% of 2164

8. $41,640 \times 89,264$ **9.** $\dfrac{0.0498}{0.00052}$

10. $592 \times 2070 \times 992.62$

11. The cost of twenty-two 29 cent postage stamps.

12. The distance traveled when driving 58 mph for 6 hr and
5 minutes (min).

13. One fourth of 82 pounds (lb).

14. The total weight of three steaks if their individual weights
are 1.96 lb, 2.21 lb, and 0.82 lb.

15. Your salary if you work for 42.8 hr at $6.85 per hour.

16. The total cost of six grocery items if their prices are $2.49,
$3.79, $9.99, $3.21, $2.59, and $13.85.

17. A 6 percent sales tax on a car that sells for $8,164.

18. A 3.84 lb package of ground beef divided into five approximately equal parts.

19. The gas mileage of a car that traveled 235.6 mi on 9.16 gal of gasoline.

20. One-third of a profit of $83,197.

21. The total cost of 12 packs of gum that cost 55 cents each.

22. The cost of one bottle of soda if a six-pack costs $3.59.

23. Using the scale on the map of the Orlando, Florida, area, estimate (a) in miles and (b) in kilometers the distance via the route indicated in color, from the Gatorland Zoo to Sea World.

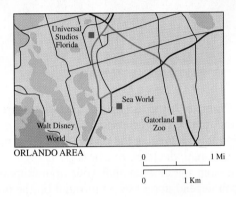

24. Using the scale on the map of Houston, Texas, estimate (a) in miles and (b) in kilometers the distance of the colored loop, Route 610, that goes around the city.

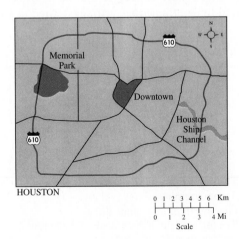

25. Using the scale on the map of Washington, D.C., estimate the distance, in miles, tourists walk if they follow the path indicated in color.

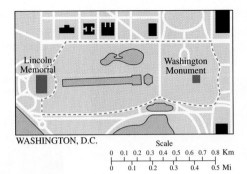

26. Consider the circle graph before answering the question. Ms. Weiss, a retiree, has a total yearly retirement income of $36,000 from various sources, as depicted in the circle graph. Estimate her income from (a) investment savings and (b) her pension.

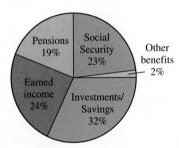

Sources of Retirement Income (for Retirees with at Least $20,000 in Annual Income in 1988)

27. Consider the graph on pet ownership before answering the questions.

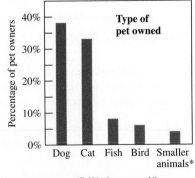

a) Estimate the percentage of pet owners who own dogs.

b) Estimate the percentage of pet owners listed in the graph.

c) Estimate the percentage of pet owners who own a pet other than those listed.

28. Use the three graphs developed from information provided by the Department of Labor to answer the questions.

U.S. Consumer Price Inflation

(a)

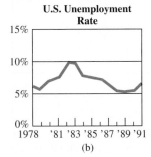

U.S. Unemployment Rate

(b)

Quarterly Unemployment Rate

Canada

U.S.

(c)

a) Estimate the U.S. consumer price inflation for 1991.
b) Describe briefly what happened to the unemployment rate from 1981 to 1990. When did the greatest increase occur? The greatest decrease?
c) Estimate the difference between the unemployment rate in Canada and the rate in the United States in the third quarter of 1990.
d) Estimate the difference between the 1980 rate of inflation and 1990 rate.

29. Based on the information in the chart, estimate the number of female lawyers and judges in a city where the total number of lawyers and judges is 480.

Occupation	% Women
Lawyers and judges	19.5
Teachers in higher education	38.5
Administrators in education	48.9
Secondary teachers	51.0
Elementary teachers	64.0
Kindergarten and pre-K teachers	88.2

Source: Statistical Abstract of the United States, 1990.

30. Studies show that in many occupations, men and women doing the same jobs are not paid the same wage. Use the following chart to estimate the annual income of a male manager if the annual income of a female manager is $35,000.

Occupation	Amount Earned by a Woman for Each $1.00 Earned by a Man
Elementary school teacher	94.3¢
Cashier	79.9
Manager	51.7
Accountant/auditor	59.8
Janitor/cleaner	82.6
Sales/personal service	59.8

31. The Uhligs are planning a vacation in Yosemite National Park. Their air fare from Memphis, Tennessee, to San Francisco, California, totals $792. Car rental is $39 per day. Lodging is a total of $79 per day, and they estimate a total of $70 per day for food, gas, and other miscellaneous items. If they are planning to stay 6 full days and nights estimate their total expenses.

32. Gary, Sue, Andy, and Megan Gilligan are planning a skiing vacation in the Rockies. Their airfare from Raleigh, North Carolina, to Denver, Colorado, totals $892. Car rental is $29 per day, ski lift tickets for each person cost $25 per day, lodging is a total of $89 per night, and food and miscellaneous items are estimated to total $80 per day. They plan to fly into Denver on Friday morning, drive to Aspen that same day, begin skiing on Saturday, ski up to and including Wednesday, drive back to Denver on Thursday, and leave from Denver Thursday evening. Estimate the total cost of the vacation.

In Ex. 33 and 34, estimate the maximum number of smaller figures (at left) that can be placed in the larger figure (at right) without the smaller figures overlapping.

33.

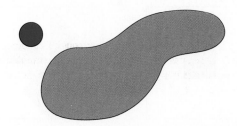

34.

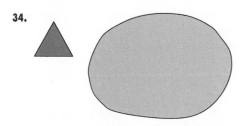

35. Estimate the number of ears of corn shown in the photo.

In Ex. 36–38, estimate, in degrees, the measure of the angles depicted. For comparison purposes a right angle, ⌐ , measures 90°.

36.

37.

38.

Estimate the percent of area that is shaded in the following figures.

39.

40.

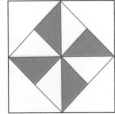

If each square represents one square unit, estimate the area of the shaded figure in square units.

41.

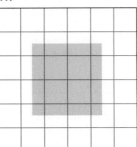

42.

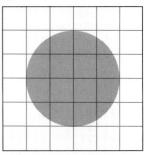

43. If the base of the statue that Buddha is seated upon is 8 ft high, estimate the total height of the statue including the base.

44. Estimate the number of berries in the square container if the container is 3 in. deep.

45. If the hand, which covers the end of the large leaf, is 7 in. long, estimate the length of the leaf.

46. If the diameter of Saturn is 75,100 mi, estimate the diameter of Saturn's largest ring.

47. Estimate, without a ruler, a distance of 12 in. Measure the distance. How good was your estimate?

48. In a bag place objects that you feel have a total weight of 10 lb. Weigh the bag to determine the accuracy of your estimate.

49. Estimate the number of times the phone will ring in 1 min if unanswered. Have a classmate phone you so you can count the rings and thus test your estimate.

50. Fill a glass with water and estimate the water's temperature. Then use a thermometer to measure the temperature and check your estimate.

51. Estimate the ratio of your height to your neck size. Then, have a friend measure you to determine this ratio and check the accuracy of your estimate.

52. Estimate the number of pennies that will fill a 3 ounce (oz) paper cup. Then actually fill a 3 oz paper cup with pennies, counting them to determine the accuracy of your answer.

53. Estimate how long it will take you to walk 60 ft. Then mark off a distance of 60 ft, and use a watch with a second hand to time how long it takes you to walk it.

54. Make a shopping list of 20 items you use regularly that can be purchased at a supermarket. Beside each item write down what you estimate to be its price. Add these price guesses to estimate the total cost of the 20 items. Next, make a trip to your local supermarket and record the actual price of each item. Add these prices to determine the actual total cost. How close was your estimate? (Don't forget to add tax on the taxable items.)

1.3 Problem Solving

Solving mathematical puzzles and real-life mathematical problems can be enjoyable. You should work as many exercises in this section as possible. By doing so, you will sample a variety of problem-solving techniques.

We can approach any problem by using a general procedure developed by George Polya. Before learning Polya's problem-solving procedure, let us consider an example.

▶ **Example 1**

Mike has just returned from a trip to the Grand Canyon and Monument Valley. During the trip he used 20 rolls of slide film, 12 at the Grand Canyon and 8 at Monument Valley. Of the 20 rolls of film, 13 are rolls of 36 exposures and 7 are rolls of 24 exposures. The speed of all the film was 64.

Mike wants to store his slides in slide trays and asks your advice. Slide trays can be purchased in two different sizes. The size that holds 140 slides costs $12.99. The other size that holds 80 slides costs $8.99. How many of

each type tray should Mike purchase to accommodate all his slides yet keep the cost of the trays to a minimum?

Solution: The first thing to do to solve the problem is to read it carefully. Read the problem at least twice and make sure you understand what facts are given and what you are being asked to find.

Next make a list of the given facts and determine which are relevant to answering the question asked.

Given Information
20 rolls of film
12 rolls of Grand Canyon and 8 rolls of Monument Valley
13 rolls of 36 exposures and 7 rolls of 24 exposures
64 is the speed of the film.
Cost of a slide tray that holds 140 slides is $12.99.
Cost of a slide tray that holds 80 slides is $8.99.

Finding the most economical way to store the slides requires knowing the total number of slides, the number of slides held by each type of tray, and the cost of each type of tray. All other given information is irrelevant to finding the answer.

Relevant Information
13 rolls of 36 exposures and 7 rolls of 24 exposures
Cost of a slide tray that holds 140 slides is $12.99.
Cost of a slide tray that holds 80 slides is $8.99.

The next step is to determine the total number of slides. The total is the sum of the slides from the rolls of 36 exposures and from rolls of 24 exposures.

Number of slides from 36-exposure rolls = 36 × 13 rolls = 468
Number of slides from 24-exposure rolls = 24 × 7 rolls = 168
636

At this point we can rephrase the question in simpler form: If slide trays that hold 140 slides cost $12.99, and slide trays that hold 80 slides cost $8.99, what is the least expensive way to store 636 slides in trays?

A plan for solving the problem is needed now. One method is to set up a table or chart to compare the costs of different combinations of trays of the two sizes. Mike needs enough trays to hold a total of 636 slides. Start by using all 140-slide trays and work down to all 80-slide trays. The following table illustrates the possible combinations of trays.

Combination of Trays to Hold 636 Slides	*Cost of Trays*
5 trays of 140 and 0 trays of 80 (700 slides)	$64.95
4 trays of 140 and 1 tray of 80 (640 slides)	$60.95
3 trays of 140 and 3 trays of 80 (660 slides)	$65.94
2 trays of 140 and 5 trays of 80 (680 slides)	$70.93
1 tray of 140 and 7 trays of 80 (700 slides)	$75.92
0 trays of 140 and 8 trays of 80 (640 slides)	$71.92

This chart shows that the least expensive way to store the slides is to purchase four trays of 140-slide size and one tray of 80-slide size.

Following is a general procedure for problem solving as given by George Polya. Notice that Example 1 demonstrates many of these guidelines.

George Polya *(1887–1985) was educated in Europe and taught at Stanford University. In his book* How to Solve It, *Polya outlines four steps in problem solving. We will use Polya's four steps as guidelines for problem solving.*

Guidelines for Problem Solving

1. *Understand the problem.*
   - Read the problem *carefully* at least twice. In the first reading get a general overview of the problem. In the second reading determine (a) exactly what you are being asked to find, and (b) what information the problem provides.
   - Try to make a sketch to illustrate the problem. Label the information given.
   - Make a list of the given facts. Are they all pertinent to the problem?
   - Determine if the information you are given is sufficient to solve the problem.
2. *Devise a plan to solve the problem.*
   - Have you seen the problem or a similar problem before? Are the procedures you used to solve the similar problem applicable to the new problem?
   - Can you express the problem in terms of an algebraic equation? (We will explain how to write algebraic equations in Chapter 6.)
   - Look for patterns or relationships in the problem that may help in solving it.
   - Can you express the problem more simply?
   - Can you substitute smaller or simpler numbers to make the problem more understandable?
   - Will listing the information in a table help in solving the problem?
   - Can you make an educated guess at the solution? Sometimes if you know an approximate solution you can work backward and eventually determine the correct procedure to solve the problem.
3. *Carry out the plan.*
   Use the plan you devised in step 2 to solve the problem.
4. *Check the results.*
   - Ask yourself, "Does the answer make sense?" and "Is the answer reasonable?" If the answer is not reasonable, recheck your method for solving the problem and your calculations.
   - Can you check the solution using the original statement?
   - Is there an alternative method to arrive at the same conclusion?
   - Can the results of this problem be used to solve other problems?

The following examples show how to apply the guidelines for problem solving.

▶ **Example 2**

A bus operates between John Fitzgerald Kennedy airport and downtown Manhattan, 25 miles away. It makes 10 round trips per day carrying an average of 42 passengers per trip. The fare each way is $4.50. What are the receipts from one day's operation?

Solution: A careful reading of the problem shows that the task is to find the total receipts from one day's operation. Make a list of all the information given and determine if it is all pertinent to the problem. Following are the facts given.

Distance from airport to city = 25 miles
Number of round trips daily = 10
Average number of passengers per trip = 42
Fare each way = $4.50

What information do you need to determine the total receipts for the day? Is all the information given needed in solving the problem? Some thought should reveal that the distance between the airport and the city is not needed to determine the answer. You should realize that the total receipts are dependent upon (a) the number of one-way trips per day, (b) the average number of passengers per trip, and (c) the cost per passenger each way. The product of these three numbers will result in the total daily receipts.

$$\begin{pmatrix} \text{Receipts} \\ \text{for one} \\ \text{day} \end{pmatrix} = \begin{pmatrix} \text{Number of} \\ \text{one-way} \\ \text{trips per day} \end{pmatrix} \times \begin{pmatrix} \text{Number of} \\ \text{passengers} \\ \text{per trip} \end{pmatrix} \times \begin{pmatrix} \text{Cost per} \\ \text{passenger} \\ \text{each way} \end{pmatrix}$$

$$= (2 \times 10) \times 42 \times 4.50 = \$3780.$$

Is the answer obtained in Example 2 reasonable for the information given? A quick estimate may be obtained as follows:

Cost of one round trip for 1 person $9
Cost of one round trip for 40 people $9 \times 40 = \$360$
Cost of 10 round trips for 40 people $360 \times 10 = \$3600$

With an estimate of $3600, the answer $3780 seems reasonable.

▶ **Example 3**

The cost of Heather's meal before tax is $13.60. A $7\frac{1}{2}\%$ sales tax of $1.02 is then added to bring the check to a total of $14.62.

a) If Heather wishes to leave a 10% tip on the *pretax* cost of the meal, how much should she leave?

b) If she wishes to leave a 15% tip on the pretax cost of the meal, how much should she leave?

Solution:

a) To find 10% of any number simply move its decimal point one place to the left. Thus, a 10% tip would be $1.36, which rounds off to $1.40. (When we tip we rarely give an exact percentage since we generally round the answer.)

b) One method for finding a 15% tip is to find 10% of the cost, as in part (a), then add half that amount. Thus, a 15% tip would be

$$\$1.40 + (\$1.40/2) = \$1.40 + \$0.70 = \$2.10$$

In Example 3(b) can you find another, and perhaps easier, method to determine a 15% tip from the information given?

▶ **Example 4**

The following chart shows the amount of each ingredient recommended to make two, four, and eight servings of Betty Crocker Potato Buds. Determine the amount of each ingredient necessary to make six servings of Potato Buds.

Servings	2	4	8
Water	$\frac{2}{3}$ cup	$1\frac{1}{3}$ cups	$2\frac{2}{3}$ cups
Milk	2 tbsp	$\frac{1}{3}$ cup	$\frac{2}{3}$ cup
Butter or margarine	1 tbsp	2 tbsp	4 tbsp
Salt*	$\frac{1}{4}$ tsp	$\frac{1}{2}$ tsp	1 tsp
Potato Buds®	$\frac{2}{3}$ cup	$1\frac{1}{3}$ cups	$2\frac{2}{3}$ cups

* Less salt can be used if desired.

Solution: As with Example 3(b), there are various ways this problem may be solved. Before you begin solving the problem, spend some time considering the possible alternatives, then select the plan that appears easiest. One method to determine the solution is to add the amounts for two servings with the amounts for four servings. A second method is to average the amounts for 4 and 8 servings. That is, add the amounts for 4 and 8 servings together, then divide the sum by 2. A third method involves multiplying the quantities given for 2 servings by 3. Which method appears easiest? Many of you would agree the third method is easiest. Below we calculate the amount of each ingredient needed to make 6 servings by multiplying the amount for each ingredient for 2 servings by 3.*

Water: $3(\frac{2}{3}) = 2$ cups
Milk: $3(2) = 6$ tablespoons (tbsp)
Butter or margarine: $3(1) = 3$ tbsp
Salt: $3(\frac{1}{4}) = \frac{3}{4}$ teaspoons (tsp)
Potato Buds®: $3(\frac{2}{3}) = 2$ cups

Example 4 illustrates that there is often more than one method that can be used to solve problems. If you solve the problem by either of the other

* Addition, subtraction, multiplication, and division of fractions are discussed in detail in Section 5.3.

methods you will obtain a slightly different answer for the amount of milk. Can you explain why?

Many real-life problems, such as Example 5, can be solved using proportions.* A proportion is a statement of equality between two ratios (or fractions).

▶ **Example 5**

The instructions on a bottle of insecticide indicate that 1.5 oz of insecticide should be mixed with 5 gal of water. Mr. Wildas wishes to spray his garden. How much insecticide should he mix with 8 gal of water to get the proper strength solution?

Solution: The fact is given that 1.5 oz of insecticide is to be mixed with 5 gal of water. Use this given ratio to set up a proportion.

$$\text{Given ratio} \begin{cases} \dfrac{1.5 \text{ oz}}{5 \text{ gal water}} = \dfrac{? \text{ oz}}{8 \text{ gal}} \end{cases}$$

Item to be found

Other information given

Notice in the proportion that ounces and gallons are placed in the same relative positions. Often the unknown quantity is replaced by an x. The proportion may be written as follows and solved using cross multiplication.

$$\frac{1.5}{5} = \frac{x}{8}$$

$$1.5(8) = 5x$$

$$12.0 = 5x$$

$$\frac{12.0}{5} = \frac{\cancel{5}x}{\cancel{5}} \qquad \text{Divide both sides of the equation by 5 to solve for } x.$$

$$2.4 = x$$

Thus, the gardener must mix 2.4 oz insecticide with 8 gal water to get the proper strength solution.

Most of the problems solved so far have been practical ones. However, many people enjoy solving brainteasers. Two examples of such puzzles follow.

▶ **Example 6**

The odometer of a motor home showed 14,941 miles. The driver said that this number was *palindromic*; that is, it reads the same backward as forward. "Look at this," the driver said to the passengers, "it will be a long time before this happens again." But after another day's drive, the odometer showed five new palindromic numbers. Can you find the numbers?

Solution: The problem is to find five palindromic numbers larger than 14,941. The numbers must be of the form △□◇□△. The number we use in place

---

* Proportions are discussed in greater detail in Section 6.2.

Did You

Know...

East Meets West: Magic Squares

A Chinese myth says that in about 2200 B.C., a divine tortoise emerged from the Yellow River. On his back was a special diagram of numbers from which all of mathematics was derived. The Chinese called this diagram Lo Shu, and represented the numbers by knots tied in white and black cords. The Lo Shu diagram is the first known magic square.

Arab traders brought the Chinese magic square to Europe during the Middle Ages, when the Black Death plague was killing millions of people. Magic squares were considered strong talismans against evil, and possession of a magic square was thought to insure health and wealth.

of the triangles must be a 1. If the triangles were replaced with a 2, the motor home would have had to travel over 5000 miles in a day. This is impossible. Next we replace the squares and the diamond with numbers. Since the number of miles increased, the next number that could replace the squares is 5. The number now looks like this: 15 ◇ 51. The diamond could be replaced with any number and the number formed would be palindromic. If the diamond is replaced with a 0, the result, 15051, is the smallest palindromic number that is larger than 14941. The remaining four numbers desired are easily found. All that needs to be done is to replace the diamond with 1, 2, 3, and 4. Thus the next five palindromic numbers are 15051, 15151, 15251, 15351, and 15451.

▶ Example 7

A magic square is a square array of numbers such that the numbers in all rows, columns, and diagonals have the same sum. Using the digits 1, 2, 3, 4, 5, 6, 7, 8, and 9, construct a magic square.

Solution: The first step is to create a figure with nine cells as in Fig. 1.1(a). The problem is to place the nine numbers in the cells so that the same sum is obtained in each row, column, and diagonal. Common sense tells us that 7, 8, and 9 cannot be in the same row, column, or diagonal. We need some small and large numbers in the same row, column, and diagonal. To see a relationship, list the numbers in order:

$$1, 2, 3, 4, 5, 6, 7, 8, 9.$$

Note that the middle number is 5 and the smallest and largest numbers are 1 and 9, respectively. The sum of 1, 5, and 9 is 15. If the sum of 2 and 8 is added to 5 the sum is 15. Likewise 3, 5, 7 and 4, 5, 6 have sums of 15. We see that in each group of three numbers the sum is 15 and 5 is a member of the group.

Figure 1.1

		8
9	5	1
(a)

		8
9	5	1
2		
(b)

4		8
9	5	1
2		6
(c)

4	3	8
9	5	1
2	7	6
(d)

Since 5 is the middle number in the list of numbers, place 5 in the center square. Place 9 and 1 to the left and right of 5 as in Fig. 1.1(a). Now we place the 2 and the 8. The 8 cannot be placed next to 9 since $8 + 9 = 17$, which is greater than 15. Place the smaller number 2 next to the larger number 9. We elected to place the 2 in the lower left-hand cell, and the 8 in the upper right-hand cell as in Fig. 1.1(b). The sum of 8 and 1 is 9. To arrive at a sum of 15, place 6 in the lower right-hand cell as in Fig. 1.1(c). The sum of 9 and 2 is 11. To arrive at a sum of 15, place 4 in the upper left-hand cell as in Fig. 1.1(c). Now the diagonals 2, 5, 8 and 4, 5, 6 have sums of 15. The numbers that remain to be placed in the empty cells are 3 and 7. Using arithmetic we can see that 3 goes in the top middle cell and 7 goes in the bottom middle cell as in Fig. 1.1(d). A check shows that the sum in all the rows, columns, and diagonals is 15.

The solution to Example 7 is not unique. There are other arrangements of the nine numbers in the cells that will produce a magic square. Also, there are other techniques of arriving at a solution for a magic square. In fact, the process described will not work if the number of squares is even—for example, 16 instead of 9. Magic squares are not limited to the operation of addition or to the set of counting numbers.

Section 1.3 Exercises

Solve the following real-life problems.

1. Ships crossing the Panama Canal have to travel 48 mi from the Atlantic end of the canal to the Pacific end. The tides have an average height of 2.04 ft per day at the Atlantic end of the canal and 12.54 ft per day at the Pacific end. How many more feet do the tides average on the Pacific end?

2. A ship took 10.5 hr to travel through the Albert Canal in Belgium. The Albert Canal is 128.1 km long and 5.03 m deep. At what rate, in kilometers per hour, did the ship travel through the canal?

3. How long will it take to cut 10 ft of plastic pipe into five 2-ft lengths if each cut takes 3 min?

4. Bernardo wants to purchase a fax machine that sells for $420. He can either pay the total amount at the time of purchase, or he can agree to pay the store $120 down and $30 a month for 12 months. How much money can he save by paying the total amount at the time of purchase?

5. A person-to-person call from New York City to Denver costs $3.25 for the first 3 min and $0.42 for each additional minute. How much would a 23-min phone call cost?

6. Sarah placed a 30-min phone call from Sioux City to Detroit. She paid $0.64 for the first minute and $0.42 for each additional minute. Dan placed a similar call, but he called person-to-person. Dan's call cost $1.75 for the first minute and $0.48 for each additional minute. How much more did Dan's phone call cost than Sarah's?

7. An office manager purchased 60 boxes of paper clips at $0.28 a box and a carton of file folders for $15.25. He gave the salesperson $40.00. How much change did he receive?

8. The stage crew built a 4.8 yd by 6.2 yd garden for outdoor scenes in a community theater play. The stage manager bought artificial grass at $3.15 a square yard to cover the garden area. She gave the salesperson a $100 bill. How much change did she receive?

9. The stage crew had 20.2 m of chicken wire. They used some of it to fence in a 2.6 m by 1.75 m rectangular garden. How many meters of chicken wire did they have left?

10. Jessica makes and sells pillows. On August 4 she bought three large bags of foam rubber. Each bag contained 20 lb of foam rubber. She used 4.25 lb to make each of three large pillows, 10.2 lb to make small pillows, and 9.6 lb to make a mat. How many pounds of foam rubber did she have left?

11. The Main Street Garage charges $1.50 for the first hour of parking and $1.00 for each additional hour. Ron parks his car in the garage from 9 A.M. to 5 P.M., 5 days a week. How much money does he save by paying a weekly parking rate of $30.00?

12. Mr. Greene parked his car in a garage that charged $1.25 for the first hour and $0.80 for each additional hour. When he returned from shopping, he gave the attendant $10.00. The attendant gave him $3.95 change. How many hours did he park his car in the garage?

13. South Street Parking lot charges $1.00 for the first half hour, $0.75 for the second half hour, and $0.50 for each additional hour or part of an hour. Roberta paid $4.25 to park her car. For how long was her car parked in the lot?

14. A taxicab charges $1.20 for the first 1/8 mi and 10 cents for each additional 1/8 mi. Determine the cost of a 5-mi trip.

15. A landscape designer is preparing the plan for a town park using a scale of 1 in. = 2.5 yards (yd). He draws a 50.4-in. line to represent the northern boundary of the playground. What actual distance (in yards) does this boundary line represent?

16. An architect is designing a shopping mall. The scale of her plan is 1 in. = 12 ft. If one store in the mall is to have a frontage of 82 ft, how long will the line representing that store's frontage be on the blueprint?

17. If three boxes of pencils cost a total of $1.05, find the cost of eight boxes of pencils.

18. If 20 lb of fertilizer covers an area of 5000 square feet (ft^2), how much fertilizer is needed to cover an area of 12,000 ft^2?

19. The instructions on a bottle of concentrated liquid cleaner reads "Mix 3 ounces with a gallon of water." If a building custodian wishes to mix the solution in a $2\frac{1}{2}$ gal bucket of water, how much concentrate should he use to obtain the proper strength mixture?

20. If three cans of juice cost $1.20, how many cans of juice can be purchased for $10.00? (Only whole cans of juice may be purchased.)

21. At a given time of day the ratio of the height of a tree to its shadow is the same for all trees. If a 3-ft stick in the ground casts a shadow of 1.2 ft, how tall is a tree that casts a shadow of 48.4 ft?

22. The charts give the approximate energy values of some foods, in kilojoules (kJ), and the energy requirements of some activities. How long would it take to use up the energy from
   a) a fried egg by swimming?
   b) a hamburger by walking?
   c) a piece of strawberry shortcake by cycling?
   d) a hamburger and a chocolate milkshake by walking?

Food	Energy Value (kJ)	Activity	Energy Consumption (kJ/min)
Chocolate milkshake	2200	Walking	25
		Cycling	35
Fried egg	460	Swimming	50
Hamburger	1550	Running	80
Strawberry shortcake	1400		
Glass of skim milk	350		

23. A plane is flying at an altitude of 15,000 ft, where temperature is −6 degrees Fahrenheit (°F). The nearby airport where the pilot intends to land is at an altitude of 3000 ft, and the control tower there reports precipitation. If the temperature increases 2.5°F for every 1000 ft decrease in altitude, will the precipitation be rain or snow? Assume rain changes to snow at 32°F.

24. Twelve square posts, 6 in. on a side, are placed in a straight line at 4-ft intervals to construct a fence. How far is it from the beginning of the first post to the end of the last post?

25. A student started out poorly on his first mathematics test. However, he doubled his score on each of the next two tests. The third test grade was 96. What was the grade on the first test?

26. A faucet is leaking at a rate of one drop of water per second. If the volume of one drop of water is 0.1 cubic centimeter (0.1 cm^3), find
   a) the volume of water in cubic centimeters lost in a year,
   b) how long it would take, in days, to fill a rectangular basin 30 cm by 20 cm by 20 cm.

27. A faucet leaks 1 oz water per minute.
   a) How many gallons of water are wasted in a year? (a gallon contains 128 oz)
   b) If water costs $4.20 per 1000 gal, how much additional money is being spent on the water bill?

28. When a car's tire pressure is 28 pounds per square inch (psi) it averages 15.8 miles per gallon (mpg) of gasoline. If the tire pressure is increased to 30 psi the car averages 16.4 mpg of gasoline.
   a) If a commuter drives an average of 15,000 mi per year, how many gallons of gasoline will he save in a year by increasing his tire pressure from 28 to 30 psi?
   b) If gasoline costs $1.20 per gallon, how much will he save in a year?
   c) If we assume there are about 140 million cars in the United States, and these changes are typical of each car, how many gallons of gasoline would be saved if all drivers increased their tire pressure?

29. A gymnasium floor has an area of 2400 square yards (yd^2). Each gallon of floor sealant covers an area of 350 square feet (ft^2). How many gallons of sealant are needed to cover the gymnasium floor?

In Ex. 30–38, explain how you determined each answer.

30. A large drum filled with water is to be drained using a small opening at the top. One 1-in. diameter hose, or two 1/2-in. hoses, to be used in siphoning out the water, can be placed in the small opening. Would it be faster to use the one 1-in. hose or the two 1/2-in. hoses?

31. Assume that the rate of inflation is 6% for the next two years. What will be the cost of goods two years from now, adjusted for inflation, if the goods cost $450.00 today?

32. The cost of a car increases by 20% and then decreases by 20%. Is the resulting price of the car greater than, less than, or equal to the original price of the car?

33. Consider the news brief below, from the November 25, 1991, issue of *Marketing News*.

> **Pizza puts bite on burger biz**
>
> Pizza is poised to take a bite out of the burger business. The number of hamburger orders decreased by 7% between 1987 and 1990, while pizza orders advanced by 12%, according to the National Restaurant Association.

Does this mean that more people are ordering pizza than hamburgers these days?

34. A new high school graduate receives two job offers: Company A offers a starting salary of $15,000 a year with a $600 raise every 6 months. Company B offers $15,000 a year with a $1200 raise every 12 months. Which offer will provide the most income?

35. How many square inches, 1 in. by 1 in., fit in an area of 1 square foot, 1 ft by 1 ft?

36. How many cubic inches fit in 1 cubic foot (ft³)?

37. If the length and width of a rectangle each double, what happens to the area of the rectangle?

38. If the length, width, and height of a cube all double, what happens to the volume of the cube?

Exercises 39–50 are recreational problems. Often answers to such problems can be found by more than one method.

39. Fill in the three boxes using the symbols +, −, ×, and ÷ to make a true statement of equality:

$$7 \,\square\, 7 \,\square\, (7 \,\square\, 7) = 13$$

40. While visiting a friend's home, I saw kittens and children playing in the backyard. Counting heads, I got 18. Counting feet, I got 60. How many kittens and how many children were in the backyard?

41. A 24-ft-by-24-ft carpet is partitioned into 4-ft-by-4-ft squares. How many squares will there be?

42. Find the next two palindromic numbers after 37,473. See Example 6.

43. Using a balance scale and only the four weights 1 gram (g), 3 g, 9 g, and 27 g, explain how you could show that an object had the following weights.

a) 5 g   **b)** 16 g

(*Hint:* Weights must be added to both sides of the balance scale.)

44. A woman purchased a dress that cost $45 and gave the merchant a $100 bill. After the woman had gone with her dress and her change, the merchant took the $100 bill to the bank. The bank clerk informed him that the $100 bill was counterfeit. What was the financial loss to the merchant?

45. The Sunday morning chef is stuck with a pan that holds only two slices of bread for French toast. It takes 30 sec to brown one side of a piece of toast. How can the chef brown both sides of three slices in $1\frac{1}{2}$ min instead of 2 min? (*Hint:* Partially finish two slices and then start the third slice.)

46. Create a magic square using the numbers 1, 3, 5, 7, 9, 11, 13, 15, and 17. The sum of the numbers in every column, row, and diagonal must be 27.

47. Create a magic square using the numbers 2, 4, 6, 8, 10, 12, 14, 16, and 18. The sum of the numbers in every column, row, and diagonal must be 30.

In Ex. 48–50, use the three magic squares illustrated to answer.

6	5	10
11	7	3
4	9	8

3	2	7
8	4	0
1	6	5

10	9	14
15	11	7
8	13	12

48. Examine the 3-by-3 magic squares and find the sum of the four corner entries of each magic square. How can the sum be found using a key number in the magic square?

49. For a 3-by-3 magic square, how can the sum of the numbers in any particular row, column, or diagonal be determined by using a key value in the magic square?

50. For a 3-by-3 magic square, how can the sum of all the numbers in the square be determined using a key value in the magic square?

51. Here is a flat pattern for a cube to be formed by folding. The sides of each square are 6 cm. Find the volume of the cube.

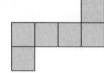

52. Consider a domino with six dots as illustrated on the next page. Two ways of connecting the three dots on the left with the three dots on the right are shown. In how many ways can the three dots on the left be connected with the three dots on the right?

53. Five salespeople gather for a sales meeting. How many handshakes will each person make if each must shake hands with each of the four others?

Problem Solving

54. Find the values of H, E, and A from this addition problem.

$$\begin{array}{r} H\ E \\ H\ E \\ H\ E \\ H\ E \\ \hline A\ H \end{array}$$

55. Find the value of each letter.

$$\begin{array}{r} U \\ C\ A\ N \\ D\ O \\ T\ H\ I\ S \\ \hline 10,\ 4\ 4\ 0 \end{array}$$

56. Rectangle ABCD is made up entirely of squares. The black square has a side of 1 unit. Find the area of ABCD.

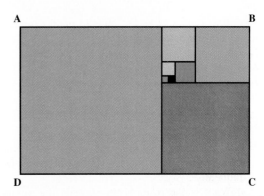

57. Peter, Paul, and Mary are three sports professionals. One is a tennis player, one is a golfer, and one is a skier. They live in three adjacent houses on City View Drive. From the information below determine which is the professional skier. (A table may be helpful.)

Mary does not play tennis.
Peter skis and plays tennis, but does not like golf.
The golfer and the skier live beside each other.
Three years ago, Paul broke his leg skiing and has not tried it since.
Mary lives in the last house.
The golfer and the tennis player share a common backyard swimming pool.

CHAPTER 1 SUMMARY

Key Terms

1.1
conjecture
counterexample
deductive reasoning (deduction)
divisible
inductive reasoning (induction)
scientific method

1.2
estimation

1.3
problem solving

Important Facts
Natural or counting numbers 1, 2, 3, 4, . . .

Guidelines for Problem Solving
1. Understand the problem.
2. Devise a plan to solve the problem.
3. Carry out the plan.
4. Check the results.

CHAPTER 1 REVIEW EXERCISES

1.1*

In Ex. 1–6, use inductive reasoning to predict the next three numbers in the pattern.

1. 3, 5, 7, 9, . . .

2. 1, 4, 9, 16, . . .

3. 4, −8, 16, −32, . . .

4. 5, 7, 10, 14, 19, . . .

5. 25, 24, 22, 19, 15, . . .

6. 1, 1, 2, 3, 5, 8, 13, . . .

7. Pick any number, multiply the number by 2. Add 10 to the product. Divide the sum by 2. Subtract 5 from the quotient.
  a) What is the relationship between the number you started with and the final number?
  b) Arbitrarily select some different numbers and repeat the given process, recording the original number and the results.
  c) Can you make a conjecture about the original number and the final number?
  d) Prove, using deductive reasoning, the conjecture you made in part (c).

8. Choose a number between 1 and 20. Add 5 to the number. Multiply the sum by 6. Subtract 12 from the product. Divide the difference by 2. Divide the quotient by 3. Subtract the number you started with from the quotient. What is your answer? Try this process with a different number. Can you make a conjecture as to what your final answer will always be?

9. Find a counterexample to the statement "The difference between two squares is an odd number."

1.2

In Ex. 10–18, estimate the answer. Your answers may vary from those given in the back of the book, depending upon how you round to arrive at the answer.

10. 194,600 × 2024

11. $\dfrac{18,254.5}{624.3}$

12. 146.2 + 96.402 + 1.04 + 897 + 421

13. 21% of 985

14. Estimate the distance from your wrist to your elbow and estimate the length of your foot. Which do you think is greater? With the help of a friend measure both lengths to determine which is larger.

15. Estimate the cost of eight pairs of socks if each pair cost $1.09.

16. Estimate a 6% sales tax on an item that cost $192.

17. Estimate your average walking speed in miles per hour if you walked 1.1 mi in 22 min.

18. Estimate the total cost of six grocery items that cost $2.49, $0.79, $1.89, $0.10, $2.19, $6.75.

19. The scale of the map is $\frac{1}{4}$ in. = 0.1 mi. Estimate the distance of the walking path indicated in red.

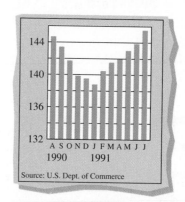

In Ex. 20 and 21, consider the graph before answering.

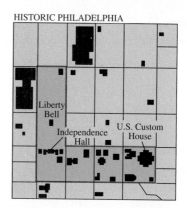

Index of Leading Indicators (seasonally adjusted index, 1982 = 100)

20. How much greater was the index in July 1991 than in October 1990?

21. Describe the trend of the indicators from August 1990 to July 1991.

---

* The number in color indicates the section in which the material is covered.

22. If each square represents one square unit, estimate the shaded area.

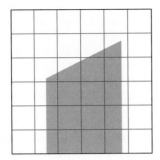

23. The scale of a model railroad is 1 in. = 12.5 ft. Estimate the size of an actual box car if this drawing is the same size as the model box car.

1.3

Solve the following problems.

24. An operator-assisted telephone call to another state costs $10.15 for 15 min. The same call dialed directly costs $0.60 for the first minute and $0.40 for each additional minute. How much money is saved by dialing direct?

25. Mr. Egan parked his car in a lot that charged $2.00 for the first hour and $0.90 for each additional hour. He left the car in the lot for 8 hr. How much change did he receive from a ten-dollar bill?

26. The rates at a parking lot in the city are $0.95 for the first half hour, $0.75 for the second half hour, and $0.40 for each additional hour or part of an hour. Ms. Agnew parked her car for 5 hr. How much did she pay to park her car?

27. A six-pack of cola costs $3.45. A carton of 4 six-packs costs $12.60. How much will be saved by purchasing the carton rather than 4 individual six-packs?

28. The rental cost of a jet ski from vendor A is $10 per 15 min, and the cost from vendor B is $25 per half hour. If you plan to rent the jet ski for 2 hr, which is the better deal, and by how much?

29. A taxicab charges $1.10 for the first 1/5 mi and 15 cents for each additional 1/5 mi. Determine the cost of a 10-mi trip.

30. If the cost of three cans of corn is $1.29, find the cost of eight cans of corn.

31. If 1.5 milligrams (mg) of a medicine is to be given for 10 lb of body weight, how many milligrams should be given to a child who weighs 47 lb?

32. The banks will grant an applicant a mortgage if the monthly payments are not greater than 25 percent of the person's take-home pay. What is the maximum monthly mortage payment you can make if your gross salary is $3800 a month and your payroll deductions are 30 percent of your gross salary?

33. New York City is on eastern standard time, St. Louis is on central standard time (1 hr earlier than eastern standard time), and Las Vegas is on Pacific standard time (3 hr earlier than eastern standard time). A flight leaves New York City at 9 A.M. eastern standard time, stops for 50 min in St. Louis, then arrives in Las Vegas at 1:35 P.M. Pacific time. How long is the plane actually flying?

34. The international date line is an imaginary line of longitude (from North to South Pole) on the earth's surface between Japan and Hawaii in the Pacific Ocean. Crossing the line when traveling west, one adds a day to the present date. Crossing the line when traveling east, one subtracts a day. At 3:00 P.M. on July 25 in Hawaii, what is the time and date in Tokyo, Japan, which is four time zones to the west?

35. 1 in. = 2.54 cm.

a) How many square centimeters are in a square inch?

b) How many cubic centimeters are in a cubic inch?

c) How long is a centimeter in terms of inches?

36. If the following pattern is continued, how many dots will be in the hundredth figure?

37. How many different ways can the letters ABC be arranged?

38. How many different ways can the digits of the number 1144 be arranged?

39. a) Find the sum of the first three, four, five, and ten counting numbers.

b) Use these results to develop a procedure that can be used to find the sum of the first n counting numbers. (*Hint:* The sum will equal the product of n and a second number expressed in terms of n, all divided by 2.)

c) Use the procedure to find the sum of the first 20 counting numbers.

d) Use the procedure to find the sum of the first 50 counting numbers.

40. Complete the magic square using each even number 2–18 only once.

16	2	12
		14
8	18	

41. Complete the magic square using the numbers 6–21 exactly once.

21	7		18
10		15	
14	12	11	17
9	19		

42. Create a magic square using the numbers 13, 15, 17, 19, 21, 23, 25, 27, and 29.

43. Create a magic square using the numbers 20, 40, 60, 80, 100, 120, 140, 160, and 180.

44. a) Consider the number 9876. Reverse the order of the digits and find the difference. Reverse the order of the digits in the difference and add this number to the difference. Record your result.

b) Make up several four-digit numbers with digits in descending order and follow the procedure in part (a). Record your results in each case.

c) Write a conjecture that describes your results for the procedure followed in part (a).

45. Three friends check into a single room in a motel and pay $10 apiece. The room costs $25 instead of $30, so a clerk is sent to the room to give $5 back. The friends each take back $1, and the clerk is given $2 for his trouble. Now each of the friends paid $9, a total of $27, and the clerk received $2. What happened to the missing dollar?

46. a) Take any three-digit number with digits A, B, C. Write the digits twice again with the order of the digits changed as below:

$$ABC$$
$$BCA$$
$$CAB$$

Find the sum of the three 3-digit numbers, and divide by the sum of A + B + C. What is the quotient?

b) Try this for any three-digit number. What is the quotient?

c) Will the answer found in part (a) be the quotient for any three-digit number? Explain why or why not.

47. A faucet is leaking 1 quart (qt) of water every $1\frac{3}{4}$ hr. How many quarts of water will it leak in $4\frac{1}{2}$ hr?

48. Would your paycheck be higher at the beginning of the third year if you received a 15% raise for the first year followed by a 15% cut for the second year or if there were no change in salary for the entire two-year period?

49. Arrange six toothpicks to form four triangles so that each edge of the figure is one toothpick long. (*Hint:* Not a two-dimensional figure.)

50. Four women in a room have an average weight of 130 lbs. A fifth woman who weighs 180 lb enters the room. Find the average weight of all five women.

51. A colony of microbes doubles in number every second. A single microbe is placed in a jar, and in an hour the jar is full. When was the jar half full?

52. A red cup contains a pint of water, and a blue cup contains a pint of wine. A teaspoon of water is taken from the red cup and placed in the blue cup, and the contents of the blue cup are mixed thoroughly. A teaspoon of the mixture in the blue cup is returned to the red cup. Is there more wine in the red cup or water in the blue cup?

53. Find the sum of the first 500 counting numbers. (*Hint:* Group in pairs.)

54. Which numbers, when divided by themselves, become larger than when multiplied by themselves?

55. Write the number 24 using only odd digits and the operation of addition.

56. A man who has a garden 10 meters square (10 m by 10 m) wishes to know how many posts will be required to enclose his land. If the posts are placed exactly 1 m apart how many are needed? Disregard the thickness of the posts.

57. Can you place the letters A, B, C, D, and E in a 5-by-5 grid so that no letter appears twice in any row, in any column, or in any diagonal?

58. You have thirteen coins, all of which look alike. Twelve coins weigh exactly the same, but the other one is heavier. You have a pan balance. Tell how to find the odd coin in just three weighings.

59. Often the name of a puzzle provides a clue to its solution that beginners overlook. Below is an *alphamagic square* invented by Lee Swallows, an electrical engineer from

44	61	57
67	54	41
51	47	64

the Netherlands. Every row, column, and diagonal will add up to the same number (162). By taking a clue from the word "alphamagic," can you discover another property that makes this an interesting magic square? (*Hint:* Count the number of letters in each number, when written out, and construct a magic square using those numbers.)

60. In a rectangular room, how can you place ten small tables so that there are three against each wall?

61. In how many ways can
   a) two people stand in a line?
   b) three people stand in a line?
   c) four people stand in a line?
   d) five people stand in a line?
   e) Using the results from parts (a) through (d), can you make a conjecture about the number of ways in which *n* people can stand in a line?

Use inductive reasoning to determine the next three numbers in the pattern.

1. 0, 4, 8, 12, . . . **2.** $1, \frac{1}{3}, \frac{1}{9}, \frac{1}{27}, \ldots$

3. Pick any number, multiply the number by 5, add 10 to the number. Divide the sum by 5. Subtract 1 from the quotient.
  a) What is the relationship between the number you started with and the final answer?
  b) Arbitrarily select some different numbers and repeat the given process. Record the original number and the results.
  c) Can you make a conjecture about the relationship between the original number and the final answer?
  d) Prove, using deductive reasoning, the conjecture made in part (c).

In Ex. 4 and 5, estimate the answers.

4. $0.000417 \times 930{,}000$ **5.** $\dfrac{89000}{0.00302}$

6. If each square represents one square unit, estimate the area of the shaded figure.

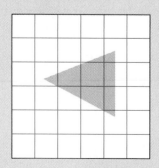

7. Consider the graph before answering the questions that follow.

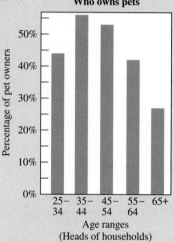

  a) What percent of heads of household in the 55–64 age group own a pet?
  b) What percent of heads of household in the 25–34 age group own a pet?
  c) What percent of heads of household in the 25–34 age group do not own a pet?

8. A gas company charges $6.42 for the basic monthly fee, which includes the first 3 therms of gas used. It charges 59 cents for each additional therm used. If the Smiths' gas bill for December was $71.32, how many therms of gas did they use during that month?

9. At a "20% off sale" John purchased a jacket for $60. What was the original price of the jacket?

10. How long will it take a carpenter to cut a 4 in. by 4 in. by 8 ft piece of treated lumber into four equal pieces if each cut takes $2\frac{1}{2}$ min?

11. In this photo of *Le Moulin de la Galette*, 1876, by Pierre Auguste Renoir, 1 inch equals 24 inches on the actual painting. Find the dimensions of the actual painting.

12. A worker gets $11.25 per hour with time and a half for any time over 44 hr per week. What are his wages when he works a 49-hr week?

13. Create a magic square using the numbers 5, 10, 15, 20, 25, 30, 35, 40, and 45. The sum of the numbers in every row, column, and diagonal must be 75.

14. A square is 2 in. on a side. A second square is 4 in. on a side. How many times larger in area is the second square than the first square?

15. Ilana drove from her home to the beach that is 30 mi from her house. The first 15 mi she drove at 60 mph, and the next 15 mi she drove at 30 mph. Would the trip take more, less, or the same time if she traveled the entire 30 mi at a steady 45 mph?

16. Given the six numbers 2, 6, 8, 9, 11, and 13, pick five that, when multiplied, give 11,232.

17. The period of time the clock takes to strike seven times at 7 o'clock is 7 sec. Assume that the duration of the strike and the time between strikes is the same. How long does it take the same clock to strike twelve times at 12 o'clock? (*Hint:* Make a chart showing times of strikes and times of breaks between strikes.)

SETS

Set building is a fundamental learning tool for even the smallest children. As babies, they learn to distinguish "me" from "mom" and "dad." As toddlers, they learn to distinguish and categorize objects as members of a set according to size, color, or shape. The TV show "Seseme Street" teaches children set building in the game "One of these things is not like the other."

O ne of the most basic human impulses is to sort and classify

Early civilizations looked at the stars and saw images of things that had meaning for them, like the mythical figure Orion shown above. Astronomers studying the skies today categorize the stars and planets they see by more complex characteristics — for example, a star is categorized by its age, magnitude, and temperature.

things. People may be categorized according to where they live: in the mountains or by the water, in igloos or in treehouses, in palaces or in hovels. Or, they may be categorized by other characteristics: by what language they speak, by whether they eat rice or bread, by whether they are brown-haired or blond. Each of these categories allows us to sort a large universe of people or objects into smaller groupings.

Consider yourself, for example. How many different sets are you a member of? You might start with some simple categories, such as whether you are male or female, what age group you belong to, what town, city, or state you live in. Then you might think about what groups your family belongs to: what ethnic group, what socioeconomic group, what nationality. Then you could think about what groups you have chosen to belong to: your political affiliation, your clubs or special-interest groups, your college or university. As you can see, there are many, many ways you could describe yourself to other people.

What use is this activity of categorization? As we will see through the course of this chapter, putting elements into sets helps us order and arrange our world. It allows us to deal with large quantities of information. Set building is a learning tool which helps to answer the question "What are the characteristics of this group?"

Companies often do market research to determine their customer's profile. Such customer surveys often ask the age group, the income range, and how the customer heard about the product.

Sets underlie other mathematical topics, such as logic and abstract algebra. In fact, the book *Eléments de Mathématique*, written by a group of French mathematicians under the pseudonym Nicolas Bourbaki, states, "Nowadays it is possible, logically speaking, to derive the whole of known mathematics from a single source, the theory of sets."

Many games depend on forming sets. One of the oldest games in the world is the African game of mancala, a game where two players distribute seeds into cells and then use strategic moves to capture their opponent's seeds while protecting their own.

2.1 | **Set Concepts**

We encounter sets in many different ways every day of our lives. A **set** is a collection of objects, which are called **elements** or **members** of the set. For example, the United States is a collection or set of 50 states. The 50 individual states are the members or elements of the set that is called the United States.

A set is **well defined** if its contents can be clearly determined. The set of justices presently serving on the U.S. Supreme Court is a well-defined set, since its contents, the justices, can be named. The set of the three best cars is not a well-defined set, since the word "best" is interpreted differently by different people. This text uses only well-defined sets.

Three methods are commonly used to indicate a set: (1) description, (2) roster form, and (3) set-builder notation.

The method of indicating a set by **description** is illustrated in the following example.

▶ **Example 1**

Write a description of the set containing the elements Monday, Tuesday, Wednesday, Thursday, Friday, Saturday, and Sunday.

Solution: The set is the days of the week

Georg Cantor *(1845–1918),
born in St. Petersburg, Russia, is
recognized as the founder of set
theory. Cantor's creative work
in mathematics was nearly lost
when his father insisted that he
become an engineer rather than
a mathematician. His two major
books on set theory,* Foundations
of General Theory of Aggregates
and Contributions to the Founding
of the Theory of Transfinite
Numbers, *were published in
1883 and 1895 respectively.*

Listing the elements of a set inside a pair of braces, { }, is called **roster form**. The braces are an essential part of the notation, since they identify the contents as a set. For example, {1, 2, 3} is notation for the set whose elements are 1, 2, and 3, but (1, 2, 3) and [1, 2, 3] are not sets because parentheses and brackets do not indicate a set. For a set written in roster form, commas separate the elements of the set.

Sets are generally named with capital letters. For example, the name commonly selected for the set of **natural numbers** or **counting numbers** is N.

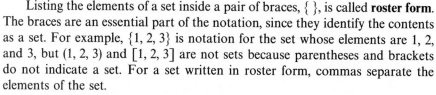

Natural numbers
$N = \{1, 2, 3, 4, 5, \ldots\}$

The three dots after the 5, called an **ellipsis**, indicate that the elements in the set continue in the same manner. An ellipsis followed by a last element indicates that the elements continue in the same manner up to and including the last element. This is illustrated in Example 2(b).

▶ **Example 2**

Express the following in roster form.
a) Set A is the set of natural numbers less than 4.
b) Set B is the set of natural numbers less than or equal to 80.
c) Set P is the set of planets in the earth's solar system.

The planets of the earth's solar system.

Solution:

a) The natural numbers less than 4 are 1, 2, and 3. Thus set A in roster form is $A = \{1, 2, 3\}$.

b) $B = \{1, 2, 3, 4, \ldots, 80\}$. The 80 after the ellipses indicates that the set continues in the same manner until the number 80.

c) $P = \{$Mercury, Venus, Earth, Mars, Jupiter, Saturn, Uranus, Neptune, Pluto$\}$

▶ **Example 3**

Express the following in roster form.

a) The set of natural numbers between 5 and 8.

b) The set of natural numbers between 5 and 8 inclusive.

Solution:

a) $A = \{6, 7\}$

b) $B = \{5, 6, 7, 8\}$. Note that the word "inclusive" indicates that the values of 5 and 8 are included in the set.

The symbol $\in$ is used to indicate membership in a set. In Example 3, since 6 is an element of set A, we write $6 \in A$. This may also be written $6 \in \{6, 7\}$. We may also write $8 \notin A$, meaning 8 is not an element of set A.

Set-builder notation (sometimes called set-generator notation) may be used to symbolize a set. Set-builder notation is frequently used in algebra. The following example illustrates its form.

$$
\begin{array}{cccccc}
D & = & \{ & x & | & \text{Condition(s)}\} \\
\uparrow & \uparrow & \uparrow & \uparrow & \uparrow & \uparrow \\
\text{Set } D & \text{is} & \text{the} & \text{all} & \text{such} & \text{the condition(s)} \\
& & \text{set of} & \text{elements} & \text{that} & x \text{ must meet in} \\
& & & x & & \text{order to be a} \\
& & & & & \text{member of the set.}
\end{array}
$$

Consider $E = \{x \mid x \in N \text{ and } x > 10\}$. The statement is read: "Set E is the set of all the elements x such that x is a natural number and x is greater than 10." The conditions x must meet in order to be a member of the set are $x \in N$, which means that x must be a natural number, and $x > 10$, which means x must be greater than 10. The numbers that meet both conditions are the set of natural numbers greater than 10. The set in roster form is $E = \{11, 12, 13, 14, \ldots\}$.

▶ **Example 4**

a) Write set $B = \{1, 2, 3\}$ using set-builder notation.

b) Write, in words, how you would read set B in set-builder notation.

Solution:

a) Since set B consists of the natural numbers less than 4 we write

$$B = \{x \mid x \in N \quad \text{and} \quad x < 4\}.$$

Another acceptable answer is $B = \{x \mid x \in N \quad \text{and} \quad x \leq 3\}$.

b) Set B is the set of all elements x such that x is a natural number and x is less than 4.

▶ **Example 5**

a) Write set $D = \{$Buick, Cadillac, Chevrolet, Oldsmobile, Pontiac, Saturn$\}$ using set-builder notation.

b) Write in words how you would read set D in set-builder notation.

Solution:

a) $D = \{x \mid x$ is a car produced by General Motors Corporation$\}$

b) Set D is the set of all elements x such that x is a car produced by General Motors Corporation.

▶ **Example 6**

Given set $A = \{x \mid x \in N$ and $3 \le x < 9\}$, write set A in roster form.

Solution: $A = \{3, 4, 5, 6, 7, 8\}$

▶ **Example 7**

Given set $C = \{x \mid x$ is one of the five most populated cities in the United States in 1991$\}$, use the following diagram to write set C in roster form.

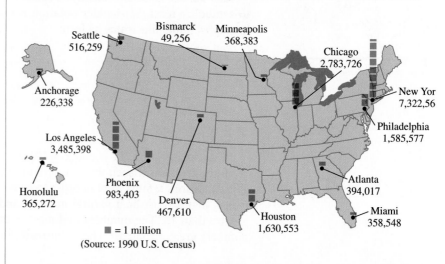

Seattle
516,259

Bismarck
49,256

Minneapolis
368,383

Chicago
2,783,726

Anchorage
226,338

New Yor
7,322,56

Los Angeles
3,485,398

Philadelphia
1,585,577

Honolulu
365,272

Phoenix
983,403

Denver
467,610

Atlanta
394,017

Houston
1,630,553

Miami
358,548

■ = 1 million
(Source: 1990 U.S. Census)

Solution: $C = \{$New York, Los Angeles, Chicago, Houston, Philadelphia$\}$

A set is said to be **finite** if it either contains no elements, or the number of elements in the set is a natural number. The set $B = \{2, 4, 6, 8, 10\}$ is a finite set because the number of elements in the set is 5, and 5 is a natural number. A set that is not finite is said to be **infinite**. The set of counting numbers is one example of an infinite set. Infinite sets are discussed in more detail in Section 2.6.

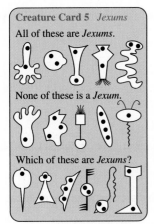

We learn to group objects according to what we see as the relevant distinguishing characteristics. One way used by educators to measure this ability is through visual cues. An example can be seen in this test, called "Creature Cards" offered by the Education Development Center. How would you describe membership in the set of Jexums?

Another important concept is equality of sets.

> Set A is **equal** to set B, symbolized by $A = B$, if and only if they contain exactly the same elements.

For example, if set $A = \{1, 2, 3\}$ and set $B = \{3, 1, 2\}$, then $A = B$, since they contain exactly the same elements. The order of the elements in the set is not important. If two sets are equal, both must contain the same number of elements. The number of elements in a set is called its **cardinal number**.

> The **cardinal number** of set A, symbolized by $n(A)$, is the number of elements in set A.

Both set $A = \{1, 2, 3\}$ and set $B = \{$England, France, Japan$\}$ have a cardinal number of 3; that is $n(A) = 3$, and $n(B) = 3$. We can say that set A and set B both have a cardinality of 3.

Two sets are said to be **equivalent** if they contain the same number of elements.

> Set A is **equivalent** to set B if and only if $n(A) = n(B)$.

Any sets that are equal must also be equivalent. However, not all sets that are equivalent are equal. The sets $D = \{$a, b, c$\}$ and $E = \{$apple, orange, pear$\}$ are equivalent, since both have the same cardinal number, 3. However, since the elements differ, the sets are not equal.

Two sets that are equivalent or have the same cardinality can be placed in **one-to-one correspondence**. Set A and set B can be placed in one-to-one correspondence if every element of set A can be matched with exactly one element of set B and every element of set B can be matched with exactly one element of set A. For example there is a one-to-one correspondence between the student names on a class list and the student identification numbers, since we can match each name with a student ID number.

Consider set C, cowboys, and set H, horses:

$$C = \{\text{Roy Rogers, Gene Autry, Lone Ranger}\},$$
$$H = \{\text{Trigger, Silver, Champion}\}.$$

Two different one-to-one correspondences for sets C and H follow.

$$C = \{\text{Roy Rogers, Gene Autry, Lone Ranger}\},$$
$$H = \{\text{Trigger, Silver, Champion}\}.$$

$$C = \{\text{Roy Rogers, Gene Autry, Lone Ranger}\},$$
$$H = \{\text{Trigger, Silver, Champion}\}.$$

Other one-to-one correspondences between sets C and H are possible. Do you know which cowboy rode which horse?

Null or Empty Set

There are some sets that do not contain any elements—for example, the set of zebras that are in this room.

> The set that contains no elements is called the **empty set** or **null set**, and is symbolized by $\{\ \}$ or $\emptyset$.

Please note that $\{\emptyset\}$ is not the empty set, since this set contains the element $\emptyset$.

▶ **Example 8**

Indicate the set of natural numbers that satisfies the equation $x + 2 = 0$.

Solution: The values that satisfy the equation are those that make the equation a true statement. Only the number -2 satisfies this equation. Since -2 is not a natural number, the solution set of this equation is $\{\ \}$ or $\emptyset$.

Universal Set

Another important set is a **universal set**.

> A **universal set**, symbolized by U, is a set that contains all the elements for any specific discussion.

When a universal set is given, only the elements in the universal set may be considered when working the problem. If, for example, the universal set for a particular problem is defined as $U = \{1, 2, 3, 4, \ldots, 10\}$, then only the elements 1–10 may be used in that problem.

Section 2.1 Exercises

Determine whether each set is well defined.
1. The set of the best posters in college resident halls.
2. The set of people who have conducted the New York Philharmonic Orchestra from 1985 to 1992.
3. The set of the 10 largest national parks in the United States.
4. The set of the five best country-and-western singers.
5. The set of countries that border France.
6. The set of students in this class that live off campus.

Determine whether each of the following sets is finite or infinite.
7. $\{2, 4, 6, 8, \ldots\}$
8. The set of odd numbers greater than 5.
9. The set of multiples of 3.
10. The set of fractions between 0 and 1.
11. The set of even numbers greater than 5.
12. The set of buildings in Chicago on January 1, 1993.
13. The number of needles (leaves) on the giant redwood trees in California at a particular instant.

14. The set of crickets chirping in the town park on a warm July 4 evening at 10:00 P.M.

Express each set in roster form. You may need to use a world almanac or some other reference source.

15. The set of continents in the world.

16. The set of individuals alive today who served as president of the United States.

17. The set of natural numbers between 165 and 397.

18. $B = \{x | x \in N \text{ and } x \text{ is even}\}$

19. $C = \{x | x - 11 = 2\}$

20. $D = \{x | x \in N \text{ and } 11 - x = 14\}$

21. The set of states that have a common border with Mexico.

22. The set of states west of Minnesota that have a common border with Canada.

23. $E = \{x | x \in N \text{ and } 3 \le x < 99\}$

24. The set of states in the United States whose names begin with the letter N.

Express each set in set-builder notation.

25. $B = \{1, 2, 3, 4, 5, 6\}$

26. $A = \{3, 4, 5, 6, 7\}$

27. $C = \{1, 3, 5, 7, \ldots\}$

28. D is the set of the three major television networks with the largest viewing population.

29. E is the set of even natural numbers.

30. A is the set of national holidays in the United States in the month of July.

31. G is the set of the three automobile manufacturers in the United States that produce the largest number of cars.

32. $F = \{1, 2, 3, 4, \ldots, 45\}$

Write a description of each set.

33. $A = \{1, 2, 3, 4, 5, 6, 7\}$

34. $B = \{3, 6, 9, 12, 15, \ldots\}$

35. $L = \{\text{Superior, Michigan, Huron, Erie, Ontario}\}$

36. $C = \{a, b, c, d\}$

37. $S = \{\text{Bashful, Doc, Dopey, Grumpy, Happy, Sleepy, Sneezy}\}$

38. $D = \{7, 8, 9, 10\}$

39. $E = \{x | x \in N \text{ and } x > 18 \text{ and } x < 7\}$

40. $F = \{x | x \in N \text{ and } x > 4 \text{ and } x < 17\}$

In Ex. 41–48, state whether each is true or false.

41. $\{a\} \in \{a, b, c, d, e, f\}$

42. $g \in \{a, b, c, d, e, f\}$

43. $d \in \{a, b, c, d, e, f\}$

44. Japan $\in \{x | x \text{ is a country in the North American continent}\}$

45. $12 \notin \{x | x \in N \text{ and } x \text{ is odd}\}$

46. IBM $\in \{x | x \text{ is a manufacturer of computers}\}$

47. Peter Pan $\notin \{x | x \text{ is a brand of peanut butter}\}$

48. $2 \in \{x | x \text{ is an odd natural number}\}$

Given set $A = \{2, 4, 6, 8\}$, $B = \{1, 3, 7, 9, 13, 21\}$, $C = \{ \ \}$, and $D = \{\#, \&, \%, \Box, *\}$, find

49. $n(A)$

50. $n(B)$

51. $n(C)$

52. $n(D)$

In Ex. 53–58, determine whether the pairs of sets are equal, equivalent, both, or neither.

53. $A = \{5, 6, 7\}$, $\quad B = \{6, 5, 7\}$

54. $A = \{8, 9, 10\}$, $\quad B = \{8, 7, 6\}$

55. $A = \{\text{Donald Duck, Goofy, Mickey Mouse}\}$
$B = \{\text{Walt Disney, Mickey Mouse, John Wayne, Curly}\}$

56. A is the set of faculty members at Michigan State University.
B is the set of faculty members at Michigan State University teaching freshman English.

57. A is the set of counties in the state of Oklahoma.
B is the set of county seats in the state of Oklahoma.

58. A is the set of dog catchers in the United States.
B is the set of dogs in the United States.

59. Set-builder notation is often more versatile and efficient than listing a set in roster form. This is illustrated with the following two sets:

$$A = \{x | x \in N \text{ and } x > 2\}$$
$$B = \{x | x > 2\}$$

a) Write a description of set A and set B.

b) Explain the difference between set A and set B. (*Hint:* Is $4\frac{1}{2} \in A$? Is $4\frac{1}{2} \in B$?)

c) Write set A in roster form.

d) Can set B be written in roster form? Explain your answer.

60. Given sets $A = \{x | 2 < x \le 5, x \in N\}$ and $B = \{x | 2 < x \le 5\}$:

a) Write a description of set A and set B.

b) Explain the difference between set A and set B.

c) Write set A in roster form.

d) Can set B be written in roster form? Explain your answer.

2.2 Subsets

In our complex world we often break down larger sets into smaller more manageable sets, called subsets. For example, consider the totality of human knowledge as a universal set. Within this universal set are certain broad areas of study: physical science, social science, and liberal arts. Each of these areas may be considered a subset. The subsets can be further separated. For example, the physical sciences can be divided into smaller subsets that include mathematics, biology, physics, chemistry, and geology.

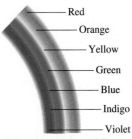

— Red
— Orange
— Yellow
— Green
— Blue
— Indigo
— Violet

Colors of primary rainbow

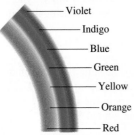

— Violet
— Indigo
— Blue
— Green
— Yellow
— Orange
— Red

Colors of secondary rainbow

Most rainbows we see are primary rainbows, but there are rare moments when a second, fainter rainbow can be seen behind the first. In this secondary rainbow, the light pattern has been reversed. Both rainbows contain the same set of colors, so each set of colors is a subset of the other.

> Set A is a **subset** of set B, symbolized by $A \subseteq B$, if and only if all the elements of set A are also elements of set B.

The symbol $A \subseteq B$ indicates that "Set A is a subset of set B." The symbol $\nsubseteq$ is used to indicate "is not a subset." Thus, $A \nsubseteq B$ indicates that set A is not a subset of set B. *To show that set* A *is not a subset of set* B *it is necessary to find at least one element of set* A *that is not an element of set* B.

▶ **Example 1**

Determine whether set A is a subset of set B.
a) $A = \{$French, German, Russian$\}$,
   $B = \{$Italian, French, Spanish, German, Russian, Arabic$\}$
b) $A = \{2, 3, 4, 5\}$ $B = \{2, 3\}$
c) $A = \{x \mid x$ is a student at the University of Missouri$\}$
   $B = \{x \mid x$ is a student at Clark County College$\}$
d) $A = \{$Snickers, Baby Ruth, Power House$\}$
   $B = \{$Power House, Snickers, Baby Ruth$\}$

Solution:
a) Since all the elements of set A are contained in set B, $A \subseteq B$.
b) Since the elements 4 and 5 are in set A but not in set B, $A \nsubseteq B$ (A is not a subset of B). In this example, however, all the elements of set B are contained in set A; therefore $B \subseteq A$.
c) Since there are students in set A who are not in set B, $A \nsubseteq B$.
d) Since all the elements of set A are contained in set B, $A \subseteq B$. Note that set $A = $ set B.

Proper Subsets

> Set A is a **proper subset** of set B, symbolized by $A \subset B$, if and only if all the elements of set A are elements of set B and set $A \neq$ set B (that is, set B must contain at least one element not in set A).

Consider the sets $A = \{$red, blue, yellow$\}$ and $B = \{$red, orange, yellow, green, blue, violet$\}$. Set A is a *subset* of set B, $A \subseteq B$, since every element of set A is also an element of set B. Set A is also a *proper subset* of set B, $A \subset B$, since sets A and B are not equal. Now consider $C = \{$car, bus, train$\}$ and $D = \{$train, car, bus$\}$. Set C is a subset of set D, $C \subseteq D$, since every element of set C is also an element of set D. However, set C is not a proper subset of set D, $C \not\subset D$, since set C and set D are equal sets.

▶ **Example 2**

Determine whether set A is a proper subset of set B.

a) $A = \{$Ford, Dodge, Buick$\}$
  $B = \{$Ford, Chevrolet, Dodge, Plymouth, Chrysler, Buick$\}$
b) $A = \{a, b, c, d\}$ $B = \{a, c, b, d\}$

Solution:

a) All elements of set A are contained in set B, and sets A and B are not equal; thus $A \subset B$.
b) Since set $A = $ set B, $A \not\subset B$. (However, $A \subseteq B$.)

Every set is a subset of itself, but no set is a proper subset of itself. For all sets A, $A \subseteq A$, but $A \not\subset A$. For example, if $A = \{1, 2, 3\}$ then $A \subseteq A$ since every element of set A is contained in set A. On the other hand $A \not\subset A$ since set $A = $ set A.

Let $A = \{\ \}$ and $B = \{1, 2, 3, 4\}$. Is $A \subseteq B$? In order to show $A \not\subseteq B$, you must find at least one element of set A that is not an element of set B. Since this cannot be done, it must be true that $A \subseteq B$. Using the same reasoning, we can show that the empty set is a subset of every set, including itself.

▶ **Example 3**

Determine whether the following are true or false.

a) $3 \in \{3, 4, 5\}$ **b)** $\{3\} \in \{3, 4, 5\}$
c) $\{3\} \in \{\{3\}, \{4\}, \{5\}\}$ **d)** $\{3\} \subseteq \{3, 4, 5\}$
e) $3 \subseteq \{3, 4, 5\}$ **f)** $\{\ \} \subseteq \{3, 4, 5\}$

Solution:

a) $3 \in \{3, 4, 5\}$ is a true statement since 3 is member of the set $\{3, 4, 5\}$.
b) $\{3\} \in \{3, 4, 5\}$ is a false statement, since $\{3\}$ is a set, and the set $\{3\}$ is not an element of the set $\{3, 4, 5\}$.
c) $\{3\} \in \{\{3\}, \{4\}, \{5\}\}$ is a true statement because $\{3\}$ is an element in the set. The elements of the set $\{\{3\}, \{4\}, \{5\}\}$ are themselves sets.
d) $\{3\} \subseteq \{3, 4, 5\}$ is a true statement because every element of the first set is an element of the second set.
e) $3 \subseteq \{3, 4, 5\}$ is a false statement. Since the 3 is not in braces it is not a set, and thus cannot be a subset. The 3 is an element of the set as was indicated in part (a).
f) $\{\ \} \subseteq \{3, 4, 5\}$ is a true statement since the empty set is a subset of every set.

Number of Subsets

How many distinct subsets can be made from a given finite set? The empty set has no elements and has exactly one subset, the empty set. A set with one element has two subsets. A set with two elements has four subsets. A set with three elements has eight subsets. This information is illustrated in Table 2.1. How many subsets will a set with four elements contain?

Table 2.1

Set	Subsets	Number of Subsets
{}	{}	$1 = 2^0$
{a}	{a}, {}	$2 = 2^1$
{a, b}	{a, b} {a}, {b}, {}	$4 = 2 \times 2 = 2^2$
{a, b, c}	{a, b, c} {a, b}, {a, c}, {b, c} {a}, {b}, {c} {}	$8 = 2 \times 2 \times 2 = 2^3$

By continuing this table with larger and larger sets, we can develop a general formula for finding the number of distinct subsets that can be made from any given set.

> The **number of distinct subsets** of a finite set A is 2^n, where n is the number of elements in set A.

▶ **Example 4**

a) Determine the number of distinct subsets for the set {H, E, A, T}.
b) List all the distinct subsets for the set {H, E, A, T}.
c) How many of the distinct subsets are proper subsets?

Solution:

a) Since the number of elements in the set is 4, the number of distinct subsets is $2^4 = 2 \times 2 \times 2 \times 2 = 16$.

b)

{H, E, A, T}	{H, E, A}	{H, E}	{H}	{}
	{H, E, T}	{H, A}	{E}	
	{H, A, T}	{H, T}	{A}	
	{E, A, T}	{E, A}	{T}	
		{E, T}		
		{A, T}		

c) There are 15 proper subsets. Every subset except {H, E, A, T} is a proper subset.

With seven distinct Scrabble tiles, there are 5040 different ways the tiles can be arranged. The four letters H, E, A, T can be arranged in 24 distinct ways.

▶ **Example 5**

Alvaro is purchasing a new computer. The model that he has selected can be purchased stripped down or with as many as eight optional features. How many different variations of this computer can be made using the various options?

Solution: One technique used in problem solving is to consider similar problems you have solved. If you think about this problem you will realize that this problem is the same as "How many distinct subsets can be made from a set with eight elements?" We know how to answer this question. The computer can have from zero to eight different optional features. The number of different variations of the computer is the same as the number of possible subsets of a set that has eight elements. There are therefore 2^8 or 256 possible distinct variations of the computer.

Did You

Know...

THE LADDER OF LIFE

Scientists use set building to classify and categorize knowledge. In biology, the science of classifying all living things is called taxonomy, and was probably practiced by the earliest cave-dwellers. Over 2000 years ago Aristotle formalized animal classification with his "ladder of life": higher animals, lower animals, higher plants, lower plants.

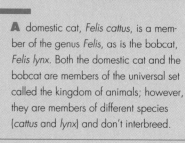

A domestic cat, *Felis cattus*, is a member of the genus *Felis*, as is the bobcat, *Felis lynx*. Both the domestic cat and the bobcat are members of the universal set called the kingdom of animals; however, they are members of different species (*cattus* and *lynx*) and don't interbreed.

Contemporary biologists use a system of classification called the Linnaean system, named after Swedish biologist Carolus Linnaeas (1707–1778). The Linnaean system starts with the smallest unit (member) and assigns it to a specific genus (set) and species (subset).

Even more general groupings of living things are made according to shared characteristics. The groupings, from most general to most specific, are: kingdom, phylum, class, order, family, genus, and species.

Section 2.2 Exercises

Answer true or false.
1. Bach $\subseteq$ {Bach, Beethoven, Brahms}
2. { } $\in$ {red, green, blue}
3. { } $\subseteq$ {Curly, Larry, Moe}
4. red $\subset$ {red, green, blue}
5. 3 $\notin$ {2, 4, 6}
6. {3, 5, 7} $\subseteq$ {5, 6, 7}
7. { } = {$\varnothing$}
8. {5} $\in$ {5, 6, 7}
9. { } = $\varnothing$
10. {1, 5, 9} $\subseteq$ {1, 9, 5}
11. {1} $\subseteq$ {1, 5, 9}
12. {1, 5, 9} $\subset$ {1, 9, 5}
13. { } $\subset$ { }
14. {1} $\in$ {{1}, {2}, {3}}
15. {3, 4, 5} $\nsubseteq$ {5, 4, 3}
16. {1, 3, 7} $\nsubseteq$ {1, 7, 3}

Determine whether $A = B$, $A \subseteq B$, $B \subseteq A$, $A \subset B$, $B \subset A$, or none of these. (There may be more than one answer.)
17. $A = \{a, b, d\}$
    $B = \{b, d, e\}$
18. $A = \{x \mid x$ is a color of the rainbow$\}$
    $B = \{$red, yellow, blue$\}$
19. $A = \{x \mid x \in N$ and $5 < x \leq 9\}$
    $B = \{6, 7, 8, 9\}$
20. $A = \{$records, tapes, compact discs$\}$
    $B = \{$tapes, compact discs$\}$
21. $A = \{1, 3, 5, 7\}$
    $B = \{1, 5\}$
22. $A = \{x \mid x$ is a sport that uses a ball$\}$
    $B = \{$basketball, soccer, tennis$\}$
23. Set A is the set of natural numbers between 3 and 8. Set B is the set of natural numbers greater than 3, but less than 8.
24. Set A is the set of white keys on a piano.
    Set B is the set of keys on a piano.

In Ex. 25–28, list all the subsets of the sets given.
25. $D = \varnothing$
26. $A = \{\triangle\}$
27. $B = \{$pizza, cola$\}$
28. $C = \{$apple, peach, nectarine$\}$

29. Given set $A = \{$a, b, c, d$\}$
    a) List all the subsets of set A.
    b) State which of the subsets in part (a) are not proper subsets of set A.

30. A set contains seven elements.
    a) How many subsets does it contain?
    b) How many proper subsets does it contain?
31. Write a formula for determining the number of proper subsets for a set with n elements.

In Ex. 32–43, if the statement is true for all sets A and B, write the word "true." If it is not true for all sets A and B, write the word "false." Assume that $A \neq \varnothing$, $U \neq \varnothing$, and $A \neq U$.
32. If $A \subseteq B$, then $A \subset B$.
33. If $A \subset B$, then $A \subseteq B$.
34. $A \subset A$
35. $A \subseteq A$
36. $\varnothing \subseteq A$
37. $\varnothing \subset A$
38. $\varnothing \subset \varnothing$
39. $A \subseteq U$
40. $U \subseteq \varnothing$
41. $\varnothing \subset U$
42. $\varnothing \subseteq \varnothing$
43. $A \subset U$

44. If $E \subseteq F$ and $F \subseteq E$, what other relationship exists between E and F? Explain.
45. There are five vocalists in a group. How many different trios can be formed using the five vocalists? (*Hint:* Use subsets.)
46. The Schmidt family is purchasing a new clothes washer. The machine that they are selecting can be purchased with no extra features or with as many as five extra features. For the appliance store to display machines with all the possible combinations of features, how many washers must it have on display?
47. Customers ordering hamburgers at Vic and Irv's Hamburger stand are always asked, "What do you want on it?" The choices are members of the set {catsup, mustard, relish, hot sauce, onions, lettuce, tomato}. How many different variations are there for ordering a hamburger?

Problem Solving

48. The executive committee of the student senate consists of Ashton, Bailey, Katz, and Snyder. Each member of the executive committee has exactly one vote (no abstentions), and a simple majority of the committee is needed to pass or reject a motion. If a motion is neither passed nor rejected, then it is considered blocked and will be considered again. Determine the number of specific ways the members can vote so that a motion can be (a) passed, (b) rejected, or (c) blocked.

2.3 Venn Diagrams and Set Operations

A useful technique for picturing set relationships is the Venn diagram, named for the English mathematician John Venn (1834–1923). Venn invented the diagrams and used them to illustrate ideas in his text on symbolic logic, published in 1881.

Figure 2.1

In a Venn diagram, a rectangle usually represents the universal set, *U*. The set of points inside the rectangle may be divided into subsets of the universal set. The subsets are usually represented by circles. In Fig. 2.1 the circle labeled *A* represents set *A*, which is a subset of the universal set.

Two sets may be represented in a Venn diagram in any of four different ways (see Fig. 2.2). Two sets *A* and *B* are **disjoint**, when they have no elements in common. Two disjoint sets *A* and *B* are illustrated in Fig. 2.2(a). If set *A* is a proper subset of set *B*, $A \subset B$, the two sets may be illustrated as in Fig. 2.2(b). If set *A* contains exactly the same elements as set *B*, that is, $A = B$, the two sets may be illustrated as in Fig. 2.2(c). Two sets *A* and *B* with some elements in common are shown in Fig. 2.2(d), which is regarded as the most general form of the diagram.

If we label the regions of the diagram in Fig. 2.2(d) using I, II, III, and IV, then we can illustrate the four possible cases with this one diagram, Fig. 2.3.

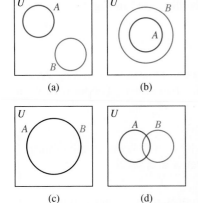

Figure 2.2

Case 1: Disjoint Sets When sets *A* and *B* are disjoint, they have no elements in common. Therefore region II of Fig. 2.3 is empty.

Case 2: Subsets When $A \subseteq B$, every element of set *A* is also an element of set *B*. Thus there can be no elements in region I of Fig. 2.3. On the other hand, if $B \subseteq A$, then region III of Fig. 2.3 is empty.

Case 3: Equal Sets When set *A* = set *B*, all the elements of set *A* are elements of set *B*, and all the elements of set *B* are elements of set *A*. Thus regions I and III of Fig. 2.3 are empty.

Case 4: Overlapping Sets When sets *A* and *B* have elements in common, those elements are in region II of Fig. 2.3. The elements that belong to set *A* but not to set *B* are in region I. The elements that belong to set *B* but not to set *A* are in region III.

Figure 2.3

In each of the four cases, any element not belonging to set *A* or set *B* is placed in region IV.

Venn diagrams will be helpful in understanding set operations. The operations of arithmetic are $+$, $-$, $\times$, and $\div$. When we see these symbols, we know what procedure to follow to determine the answer. Some of the operations in set theory are $'$, $\cup$, $\cap$. They represent complement, union, and intersection, respectively.

Figure 2.4

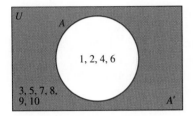

Figure 2.5

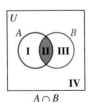

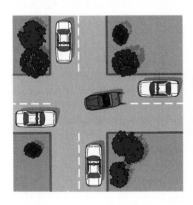

Figure 2.6

Complement

> The **complement** of set A, symbolized by A', is the set of all the elements in the universal set that are not in set A.

In Fig. 2.4 the shaded region outside of set A within the universal set represents the complement of set A, or A'.

▶ **Example 1**

Given
$$U = \{1, 2, 3, 4, 5, 6, 7, 8, 9, 10\} \quad \text{and}$$
$$A = \{1, 2, 4, 6\},$$

find A' and illustrate the relationship among sets U, A, and A' in a Venn diagram.

Solution: The elements in U that are not in set A are 3, 5, 7, 8, 9, 10. Thus $A' = \{3, 5, 7, 8, 9, 10\}$. The Venn diagram is illustrated in Fig. 2.5.

Intersection

The term "intersection" brings to mind the area common to two crossing streets. The car in the figure is in the intersection of the two streets. The set operation is defined as follows.

> The **intersection** of sets A and B, symbolized by $A \cap B$, is the set containing all the elements that are common to both set A and set B.

The shaded region, region II, in Fig. 2.6 represents the intersection of sets A and B.

▶ **Example 2**

Draw a Venn diagram illustrating the following sets.

Set U is the set of the 50 states in the United States.
Set A is the set of the six most populous states in the United States in 1990.
Set B is the set of the six states in the United States with the most marriages in 1990.

Solution: The 1990 Census lists California, New York, Texas, Florida, Pennsylvania, and Illinois as the most populous states (set A) and California, Texas, New York, Florida, Nevada, and Ohio as the states with the most marriages (set B). First determine the intersection of sets A and B. Since California, New York, Texas, and Florida are common to both sets

$$A \cap B = \{\text{CA, NY, TX, FL}\}.$$

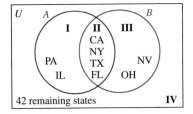

Figure 2.7

Place these elements in region II of Fig. 2.7. Now place the elements in set *A* that have not been placed in region II, in region I. Therefore we place PA and IL in region I. Complete region III by determining the elements in set *B* that have not been placed in region II. Thus NV and OH go in region III. Finally, place those elements in *U* that are not in either set outside both circles. This includes the remaining 42 states, which we write in region IV.

▶ **Example 3**

Given

$$U = \{1, 2, 3, 4, 5, 6, 7, 8, 9, 10\}$$
$$A = \{1, 2, 4, 6\}$$
$$B = \{1, 3, 6, 7, 9\}$$
$$C = \{\ \}$$

Find

a) $A \cap B$,　　**b)** $A \cap C$,
c) $A' \cap B$,　　**d)** $(A \cap B)'$

Solution:

a) $A \cap B = \{1, 2, 4, 6\} \cap \{1, 3, 6, 7, 9\} = \{1, 6\}$. The elements common to both set *A* and set *B* are 1 and 6.

b) $A \cap C = \{1, 2, 4, 6\} \cap \{\ \} = \{\ \}$. There are no elements common to both set *A* and set *C*.

c) 　　$A' = \{3, 5, 7, 8, 9, 10\}$
　　$A' \cap B = \{3, 5, 7, 8, 9, 10\} \cap \{1, 3, 6, 7, 9\}$
　　　　　 $= \{3, 7, 9\}$

d) To find $(A \cap B)'$, first determine $A \cap B$.

$$A \cap B = \{1, 6\} \text{ from part (a)}$$
$$(A \cap B)' = \{1, 6\}' = \{2, 3, 4, 5, 7, 8, 9, 10\}$$

Union

The word "union" means to unite or join together, as in marriage, and that is exactly what is done when performing the operation of union.

> The **union** of sets *A* and *B*, symbolized by $A \cup B$, is the set containing all the elements that are members of set *A* or of set *B* (or of both sets).

$A \cup B$

Figure 2.8

The three shaded regions of Fig. 2.8, regions I, II, and III together represent the union of sets *A* and *B*. If an element is common to both sets, it is listed only once in the union of the sets.

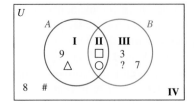

Figure 2.9

▶ **Example 4**

Given the Venn diagram in Fig. 2.9, determine the following sets.

a) U **b)** A

c) B' **d)** $A \cap B$

e) $A \cup B$ **f)** $(A \cup B)'$

g) $n(A \cup B)$

Solution:

a) The universal set consists of all the elements within the rectangle. Thus $U = \{9, \triangle, \square, \bigcirc, 3, 7, ?, \#, 8\}$.

b) Set A consists of the elements in regions I and II. $A = \{9, \triangle, \square, \bigcirc\}$.

c) B' consists of the elements outside set B, or the elements in regions I and IV. $B' = \{9, \triangle, \#, 8\}$.

d) $A \cap B$ consists of elements that belong to both set A and set B (region II). Thus $A \cap B = \{\square, \bigcirc\}$.

e) $A \cup B$ consists of the elements that belong to set A or set B (regions I, II, or III). $A \cup B = \{9, \triangle, \square, \bigcirc, 3, 7, ?\}$.

f) $(A \cup B)'$ consists of the elements in U that are not in $A \cup B$. $(A \cup B)' = \{\#, 8\}$.

g) $n(A \cup B)$ represents the number of elements in the union of sets A and B. $n(A \cup B) = 7$, since there are seven elements in the union of sets A and B.

▶ **Example 5**

Given

$$U = \{1, 2, 3, 4, 5, 6, 7, 8, 9, 10\},$$
$$A = \{1, 2, 4, 6\},$$
$$B = \{1, 3, 6, 7, 9\},$$
$$C = \{\ \}.$$

Find

a) $A \cup B$ **b)** $A \cup C$

c) $A' \cup B$ **d)** $(A \cup B)'$

Solution:

a) $A \cup B = \{1, 2, 4, 6\} \cup \{1, 3, 6, 7, 9\} = \{1, 2, 3, 4, 6, 7, 9\}$

b) $A \cup C = \{1, 2, 4, 6\} \cup \{\ \} = \{1, 2, 4, 6\}$

c) In order to determine $A' \cup B$, it is necessary to determine A'.

$$A' = \{3, 5, 7, 8, 9, 10\}$$
$$A' \cup B = \{3, 5, 7, 8, 9, 10\} \cup \{1, 3, 6, 7, 9\}$$
$$= \{1, 3, 5, 6, 7, 8, 9, 10\}$$

d) Find $(A \cup B)'$ by first determining $A \cup B$, and then find the complement of $A \cup B$.

$$A \cup B = \{1, 2, 3, 4, 6, 7, 9\} \text{ from part (a)}$$
$$(A \cup B)' = \{1, 2, 3, 4, 6, 7, 9\}' = \{5, 8, 10\}$$

▶ **Example 6**

Given

$$U = \{a, b, c, d, e, f, g\},$$
$$A = \{a, b, e, g\},$$
$$B = \{a, c, d, e\},$$
$$C = \{b, e, f\}.$$

Find

a) $(A \cup B) \cap (A \cup C)$ **b)** $(A \cup B) \cap C'$ **c)** $A' \cap B'$.

Solution:

a) $(A \cup B) \cap (A \cup C) = \{a, b, c, d, e, g\} \cap \{a, b, e, f, g\}$
$$= \{a, b, e, g\}$$

b) $(A \cup B) \cap C' = \{a, b, c, d, e, g\} \cap \{a, c, d, g\}$
$$= \{a, c, d, g\}$$

c) $A' \cap B' = \{c, d, f\} \cap \{b, f, g\}$
$$= \{f\}$$

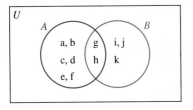

Figure 2.10

Having looked at unions and intersections we can determine a relationship between $n(A \cup B)$, $n(A)$, $n(B)$, and $n(A \cap B)$. Suppose set A has eight elements, set B has five elements, and $A \cap B$ has two elements: How many elements are in $A \cup B$? Let us make up some arbitrary sets that meet the criteria specified and draw a Venn diagram. If we let $A = \{a, b, c, d, e, f, g, h\}$, then set B must contain five elements, two of which are also in set A. Let $B = \{g, h, i, j, k\}$. Construct a Venn diagram by filling in the intersection first, Fig. 2.10. The number of elements in $A \cup B$ is 11. The elements g and h are in both sets and if we add $n(A) + n(B)$ we are counting these elements twice.

To find the number of elements in the union of sets A and B we can add the number of elements in sets A and B and then subtract the number of elements common to both sets.

For any finite sets A and B

$$n(A \cup B) = n(A) + n(B) - n(A \cap B)$$

▶ **Example 7**

Set A contains 20 elements, set B contains 16 elements, and 4 elements are common to sets A and B. How many elements are in $A \cup B$?

Solution:

$$n(A \cup B) = n(A) + n(B) - n(A \cap B)$$
$$= 20 + 16 - 4$$
$$= 32$$

The Meaning of "and" and "or"

The words "and" and "or" are very important in many areas of mathematics. You will see these words used in a number of different chapters in this book, including the probability chapter. The word **or** is generally interpreted to mean **union**, while **and** is generally interpreted to mean **intersection**. Suppose $A = \{1, 2, 3, 5, 6, 8\}$ and $B = \{1, 3, 4, 7, 9, 10\}$. Then the elements that belong to set A *or* set B are 1, 2, 3, 4, 5, 6, 7, 8, 9, and 10. These are the elements in the union of the sets. The elements that belong to set A *and* set B are 1 and 3. These are the elements in the intersection of the sets.

▶ **Example 8**

Set A contains six letters and five numbers. Set B contains four letters and nine numbers. Two letters and one number are common to both sets A and B. Find the number of elements in set A or set B.

Solution: You are asked to find the number of elements in set A or set B, which is $n(A \cup B)$. Since $n(A \cup B) = n(A) + n(B) - n(A \cap B)$, if you can determine $n(A)$, $n(B)$, and $n(A \cap B)$ you can solve the problem. Since set A contains 6 letters and 5 numbers, $n(A) = 11$. Since set B contains 4 letters and 9 numbers, $n(B) = 13$. Since 2 letters and 1 number are common to both sets, $n(A \cap B) = 3$.

$$n(A \cup B) = n(A) + n(B) - n(A \cap B)$$
$$= 11 + 13 - 3 = 21$$

Thus the number of elements in set A or set B is 21.

Section 2.3 Exercises

1. Given the sets U, A, and B, construct a Venn diagram and place the elements in the proper regions.

   $U = \{a, b, c, d, e, f, g, h, i, j\}$
   $A = \{a, b, c, d, f, h\}$
   $B = \{b, c, e, f, i\}$

2. Given the sets U, A, and B, construct a Venn diagram and place the elements in the proper regions.

   $U = \{$rhombus, square, rectangle, parallelogram, trapezoid$\}$
   $A = \{$rhombus, square, rectangle$\}$
   $B = \{$square, trapezoid$\}$

3. Let U represent the set of all married people in the state of Texas in 1993. Let A represent the set of people in the state of Texas in 1993 who have been married less than 25 years. Describe A'.

4. Let U represent the set of all the rock bands in the United States during the period from 1983 to 1993. Let set A be the set of all the rock bands that have recorded exactly three albums during that period of time. Describe A'.

In Ex. 5–10, use the following information to describe the sets in words.

U is the set of college students,
A is the set of college students with a major in English,
B is the set of college students with a major in history,
C is the set of college students with a major in biology.

5. $A \cap B$
6. $A \cup B$
7. $A' \cup C$
8. $A' \cup B$
9. $A \cap B \cap C$
10. $A' \cap B'$

In Ex. 11–16, use the following information to describe the set in words.

U is the set of cities in the State of North Carolina,
A is the set of cities in the State of North Carolina with a
population over 50,000,
B is the set of cities in the State of North Carolina with a
professional sports team,
C is the set of cities in the State of North Carolina with a
philharmonic orchestra.

11. $A \cup C$ **12.** $B \cap C$ **13.** $B' \cap C$
14. $A \cup B \cup C$ **15.** $A' \cup B'$ **16.** $A \cap B \cap C$

In Ex. 17–22, use the following information to describe the
sets in words.

$U = \{x \mid x$ is a student at the University of Iowa$\}$,
$A = \{x \mid x$ has a major in agriculture$\}$,
$B = \{x \mid x$ grew up on a farm$\}$,
$C = \{x \mid x$ is not a resident of the state of Iowa$\}$.

17. $A \cup B$ **18.** $A \cap B$
19. $A' \cap B'$ **20.** $A \cap B \cap C$
21. $A \cup B \cup C$ **22.** $A \cup B' \cup C$

In Ex. 23–30, use the Venn diagram in Fig. 2.11 to list the
set of elements in roster form.

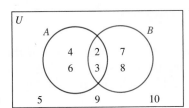

Figure 2.11

23. A **24.** B **25.** U
26. $A \cup B$ **27.** $A \cap B$ **28.** $A \cap B'$
29. $A' \cap B$ **30.** $(A \cup B)'$

In Ex. 31–38, use the Venn diagram in Fig 2.12 to list the
set of elements in roster form. The universal set is the set of
14 states with the largest harvested acreage of principal crops
in 1989. The elements of set A are the five states with the

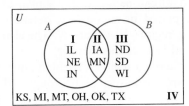

Figure 2.12

largest production of corn in 1989. The elements of set B are
the five states with the largest production of oats in 1989.

31. A **32.** B **33.** U
34. $A \cup B$ **35.** $A \cap B$ **36.** $(A \cap B)'$
37. $(A \cup B)'$ **38.** $A' \cup B$

Given $U = \{1, 2, 3, 4, 5, 6, 7, 8\}$,
$A = \{1, 2, 4, 5, 8\}$,
$B = \{2, 3, 4, 6\}$,
find the following.

39. $A \cup B$ **40.** $A \cap B$
41. B' **42.** $A \cup B'$
43. $(A \cup B)'$ **44.** $A' \cap B'$
45. $(A \cup B)' \cap B$ **46.** $(A \cup B) \cap (A \cup B)'$
47. $(B \cup A)' \cap (B' \cup A')$ **48.** $A' \cup (A \cap B)$

Given $U = \{0, \square, \triangle, \#, \alpha, \beta\}$,
$A = \{0, \square, \triangle, \#\}$,
$B = \{\#, \alpha, \beta,\}$,
$C = \{0, \square, \triangle\}$,
find the following.

49. $A \cup B$ **50.** $A \cap B$
51. $A' \cup B$ **52.** $(B \cup C)'$
53. $A \cap B'$ **54.** $A \cap C'$
55. $(B \cap C)'$ **56.** $(A \cup B) \cap C$
57. $(C \cap B) \cup A$ **58.** $(C \cup A) \cap B$
59. $(A' \cup C) \cap B$ **60.** $(A \cap B') \cup C$
61. $(A \cup B)' \cap C$ **62.** $(A \cap C)' \cap B$

Given $U = \{a, b, c, d, e, f, g, h, i, j, k\}$,
$A = \{a, c, f, g, i\}$,
$B = \{b, c, d, f, g\}$,
$C = \{a, b, f, i, j\}$,
find the following.

63. A' **64.** $B \cup C$
65. $A \cap C$ **66.** $A' \cup B$
67. $(A \cap C) \cup B$ **68.** $(A \cap C)'$
69. $A \cup (C \cap B)'$ **70.** $A' \cap (B \cap C)$
71. $(C \cap B) \cap (A' \cap B)$ **72.** $A \cup (C' \cup B')$

73. Consider the formula

$$n(A \cup B) = n(A) + n(B) - n(A \cap B).$$

a) Show that this relation holds for $A = \{a, b, c, d\}$ and
$B = \{b, d, e, f, g, h\}$.
b) Make up your own sets A and B each consisting of
at least six elements. Using these sets, show that the
relation holds.
c) Using a Venn diagram, can you explain why the relation holds for any two sets A and B?

74. The Venn diagram given in Fig. 2.13 shows a technique of labeling the regions to indicate membership of elements in a particular region. Define each of the four regions with a set statement. (*Hint: A ∩ B′* defines region I.)

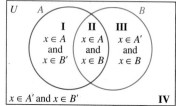

Figure 2.13

Given $U = \{0, 1, 2, 3, 4, 5, \ldots\}$,
  $A = \{1, 2, 3, 4, \ldots\}$,
  $B = \{4, 8, 12, 16, \ldots\}$,
  $C = \{2, 4, 6, 8, \ldots\}$,
find the following.

75. $A \cup B$ **76.** $A \cap B$ **77.** $B \cap C$

78. $B \cup C$ **79.** $A \cap C$ **80.** $A' \cap C$
81. $B' \cap C$ **82.** $(B \cup C)' \cup C$ **83.** $(A \cap C) \cap B'$
84. $U' \cap (A \cup B)$

Problem Solving

For each of the following, determine whether the answer is ∅, *A*, or *U*. (Assume that $A \neq \emptyset$, $A \neq U$.)

85. $A \cup A'$ **86.** $A \cap A'$ **87.** $A \cup \emptyset$
88. $A' \cup U$ **89.** $A \cap \emptyset$ **90.** $A \cup U$
91. $A \cap U$ **92.** $A \cap U'$

Determine the relationship between set *A* and set *B* if
93. $A \cap B = B$ **94.** $A \cup B = B$ **95.** $A \cap B = \emptyset$
96. $A \cup B = A$ **97.** $A \cap B = A$

Draw a Venn diagram as in Fig. 2.2 that illustrates the relationship between set *A* and set *B* (assume $A \neq B$) if
98. $A \subset B$ **99.** $A \cup B = B$
100. $A \cap B = B$ **101.** $A \cap B = \emptyset$

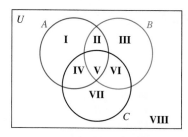

Figure 2.14

2.4

Venn Diagrams with Three Sets and Verification of Set Statements

Venn diagrams can be used to illustrate three or more sets. For three sets, *A*, *B*, and *C*, the diagram is drawn so the three sets overlap (Fig. 2.14), creating eight regions. The diagrams in Fig. 2.15 emphasize selected regions of three intersecting sets.

> **General Procedure for Constructing Venn Diagrams with Three Sets, A, B, and C**
>
> **1.** Determine the elements to be placed in region V by finding the elements that are common to all three sets, $A \cap B \cap C$.
> **2.** Determine the elements to be placed in region II. Find the elements in $(A \cap B)$. The elements in this set belong in regions II and V. Place the elements in the set $A \cap B$ that are not listed in region V in region II. The elements in regions IV and VI can be found in a similar manner.
> **3.** Determine the elements to be placed in region I by determining the elements in set *A* that are not in regions II, IV, and V. The elements in regions III and VII can be found in a similar manner.
> **4.** Determine the elements to be placed in region VIII by finding the elements in the universal set that are not in regions I through VII.

Example 1 illustrates the general procedure.

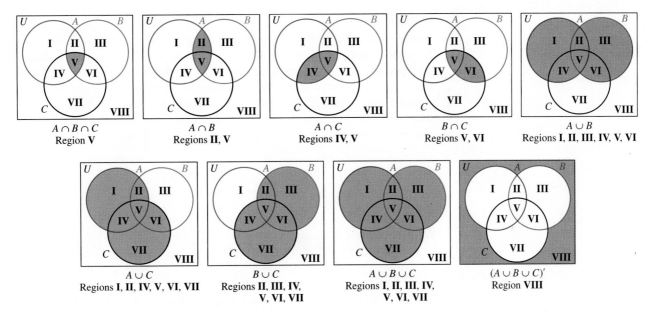

Figure 2.15

▶ **Example 1**

Construct a Venn diagram illustrating the following sets.

$$U = \{1, 2, 3, 4, 5, 6, 7, 8, 9, 10, 11, 12, 13, 14, 15\}$$
$$A = \{1, 2, 3, 4, 7, 9, 11\}$$
$$B = \{2, 3, 4, 5, 10, 12, 14\}$$
$$C = \{1, 2, 4, 8, 9\}$$

Solution: First find the intersection of all three sets. Since the elements 2 and 4 are in all three sets, $A \cap B \cap C = \{2, 4\}$. The elements 2 and 4 are placed in region V (Fig. 2.16). Next complete region II by determining the intersection of sets A and B.

$$(A \cap B) = \{2, 3, 4\}$$

$A \cap B$ consists of regions II and V. Since 2 and 4 have already been placed in region V, the element 3 must be placed in region II. In a similar manner, complete regions IV and VI. The only elements of set A that have not previously been placed in regions II, IV, and V are 7 and 11. Therefore place the elements 7 and 11 in region I. The elements in region I are only in set A. Complete regions III and VII in a similar manner. To determine the elements in region VIII, find the elements in U that have not been placed in regions I–VII. The elements 6, 13, and 15 have not been placed in regions I–VII, so place them in region VIII.

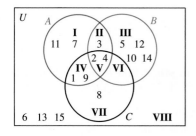

Figure 2.16

Venn diagrams can be used to illustrate and analyze many everyday problems. One example follows.

▶ Example 2

Human blood is classified (typed) according to the presence or absence of the specific antigens A, B, and Rh in the red blood cells. Antigens are highly specified proteins and carbohydrates that will trigger the production of antibodies in the blood to fight infection. Blood lacking the Rh antigen is labeled negative and blood lacking both A and B antigens is type O. Sketch a Venn diagram with three sets A, B, and Rh and place each type of blood listed in the proper region. A person has only one type of blood.

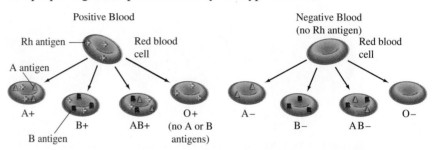

Solution: As illustrated in Chapter 1, the first thing to do is to read the question carefully and make sure you understand what is given and what you are asked to find. From the information given, there are three antigens A, B, and Rh. Therefore begin by naming the three circles in a Venn diagram with the three antigens, Fig. 2.17.

Since any blood containing the Rh antigen is positive, all blood within the Rh circle is positive, and all blood outside the Rh circle is negative. The intersection of all three sets, region V, is AB+. Region II contains only antigens A and B and is therefore AB−. Region I is A−, since it contains only antigen A. Region III is B−, region IV is A+, and region VI is B+. Region VII is O+, containing only the Rh antigen. Region VIII, which lacks all three antigens, is O−.

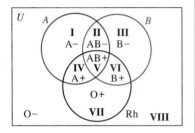

Figure 2.17

Verification of Set Statements

Consider the question, Is $A' \cup B = A' \cap B$ for all sets A and B? For the specific sets $U = \{1, 2, 3, 4, 5\}$, $A = \{1, 3\}$ and $B = \{2, 4, 5\}$ is $A' \cup B = A' \cap B$? To answer the question, do the following.

Find $A' \cup B$.	Find $A' \cap B$.
$A' = \{2, 4, 5\}$	$A' = \{2, 4, 5\}$
$A' \cup B = \{2, 4, 5\}$	$A' \cap B = \{2, 4, 5\}$

For these sets $A' \cup B = A' \cap B$, since both sets are equal to $\{2, 4, 5\}$. At this point you may believe that $A' \cup B = A' \cap B$ for all sets A and B.

If we select the sets $U = \{1, 2, 3, 4, 5\}$, $A = \{1, 3, 5\}$, and $B = \{2, 3\}$, we see that $A' \cup B = \{2, 3, 4\}$ and $A' \cap B = \{2\}$. For this case $A' \cup B \neq A' \cap B$. Thus we have proved that $A' \cup B \neq A' \cap B$ for all sets A and B by using a **counterexample**. A counterexample, as explained in Chapter 1, is an example that shows a statement is not true.

In Chapter 1 it was explained that to prove a statement we use deductive reasoning. Recall that deductive reasoning begins with a general statement and works to a specific conclusion. To verify, or determine whether set statements are equal for any two sets selected, we will use deductive reasoning with Venn diagrams. Venn diagrams are used because they can illustrate general cases.

Two **set statements are equal** when they represent the same set of elements for any given sets. We will use Venn diagrams to determine whether two set statements are equal.

▶ **Example 3**

Determine whether $(A \cup B)' = A' \cap B'$ for all sets A and B.

Solution: Draw a Venn diagram with two sets A and B, as in Fig. 2.18. Label the regions as indicated.

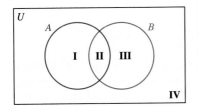

Figure 2.18

Find $(A \cup B)'$.

Set	Corresponding Regions
A	I, II
B	II, III
$A \cup B$	I, II, III
$(A \cup B)'$	IV

Find $A' \cap B'$.

Set	Corresponding Regions
A'	III, IV
B'	I, IV
$A' \cap B'$	IV

Since both set statements are represented by the same regoon, IV, of the Venn diagram, $(A \cup B)' = A' \cap B'$ for all sets A and B.

In proving that $(A \cup B)' = A' \cap B'$ we started with two general sets and worked to the specific conclusion that both statements represented the same regions of the Venn diagram.

▶ **Example 4**

Determine whether $A \cap (B \cup C) = (A \cap B) \cup (A \cap C)$ for all sets, A, B, and C.

Solution: Because the statements include three sets, A, B, and C, three circles must be used. The Venn diagram illustrating the eight regions is shown in Fig. 2.19.

First we will find the regions that correspond to $A \cap (B \cup C)$; then we will find the regions that correspond to $(A \cap B) \cup (A \cap C)$. If both answers are the same, the set statements are equal.

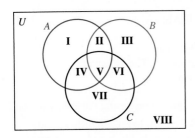

Figure 2.19

Find $A \cap (B \cup C)$.

Set	Corresponding Regions
A	I, II, IV, V
$B \cup C$	II, III, IV, V, VI, VII
$A \cap (B \cup C)$	II, IV, V

Find $(A \cap B) \cup (A \cap C)$

Set	Corresponding Regions
$A \cap B$	II, V
$A \cap C$	IV, V
$(A \cap B) \cup (A \cap C)$	II, IV, V

The regions that correspond to $A \cap (B \cup C)$ are II, IV, and V, and the regions that correspond to $(A \cap B) \cup (A \cap C)$ are also II, IV, and V.

The results show that both set statements are represented by the same regions, namely, II, IV, and V, and therefore $A \cap (B \cup C) = (A \cap B) \cup (A \cap C)$ for all sets A, B, and C.

De Morgan's Laws

In set theory, logic, and other branches of mathematics a pair of related theorems known as De Morgan's laws make it possible to transform statements and formulas into alternative and often more convenient forms. In set theory, **De Morgan's laws** are symbolized as

1. $(A \cup B)' = A' \cap B'$

2. $(A \cap B)' = A' \cup B'$.

Law 1 was verified in Example 3. We suggest that you verify law 2 at this time. The laws were expressed verbally by William of Ockham in the fourteenth century. In the nineteenth century, Augustus De Morgan expressed them mathematically. De Morgan's laws will be discussed more thoroughly in the chapter on logic.

Section 2.4 Exercises

1. Construct a Venn diagram illustrating the following sets.

$U = \{a, b, c, d, e, f, g, h, i, j, k\}$
$A = \{b, c, e, g, h, k\}$
$B = \{a, b, c, d, g, i\}$
$C = \{a, b, c, d, g, k\}$

2. Construct a Venn diagram illustrating the following sets. The elements of the sets are the days Amy, Peter, and Carlos work at Al's Sandwich Shop.

$U = \{$Sunday, Monday, Tuesday, Wednesday, Thursday, Friday, Saturday$\}$
Amy = $\{$Monday, Tuesday, Wednesday, Thursday$\}$
Peter = $\{$Sunday, Monday, Tuesday, Wednesday, Thursday, Friday$\}$
Carlos = $\{$Tuesday, Sunday$\}$

3. Construct a Venn diagram illustrating the following sets.

$U = \{$Jan, Feb, March, April, May, June, July, Aug, Sept, Oct, Nov, Dec$\}$
$A = \{$Jan, April, Aug, Sept, Oct, Dec$\}$
$B = \{$Feb, Aug, Oct, Nov, Dec$\}$
$C = \{$Feb, March, June, Dec$\}$

4. Given the sets U, A, B, and C, construct a Venn diagram and place the elements in the proper regions.

$U = \{$trout, bass, salmon, shark, whale, carp, catfish, whitefish, bluefish, eel$\}$
$A = \{$salmon, shark, whale, eel, whitefish, bluefish$\}$
$B = \{$trout, salmon, eel, bass, carp$\}$
$C = \{$catfish, eel, carp$\}$

5. Given the sets U, A, B, and C, construct a Venn diagram and place the elements in the proper regions.

$U = \{$Joyce, Darnell, Mia, Allen, Ayanna, Steven, Tod, Stuart$\}$
$A = \{$Joyce, Mia, Ayanna$\}$
$B = \{$Darnell, Mia, Steven, Tod, Ayanna$\}$
$C = \{$Mia, Tod, Allen$\}$

6. Construct a Venn diagram illustrating the following sets.

$U = \{$hickory, black walnut, hazel, English walnut, pecan, beech, maple, ash, birch, red oak, white oak, cedar, white pine, yellow pine, blue spruce, aspen, red wood$\}$
$A = \{$hickory, black walnut, hazel, English walnut, pecan, beech$\}$

$B = \{$maple, beech, black walnut, ash, birch, white oak, cedar, white pine, yellow pine$\}$
$C = \{$blue spruce, aspen, beech, white pine, yellow pine, hazel$\}$

Listed below the figure are the top five female (LPGA) and top five male (PGA) golfers as of mid-August 1991. The amount of money each has won on tour and the amount of career winnings are also listed. For each player, indicate the region in Fig. 2.20 in which that player would be placed.

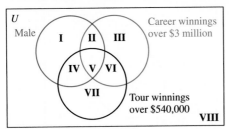

Figure 2.20

Female	Tour Winnings	Career Winnings
7. Pat Bradley	$542,234	$3,888,381
8. Meg Mallon	447,704	646,233
9. Beth Daniel	428,736	3,322,219
10. Deb Richard	342,030	893,411
11. Ayako Okamoto	329,029	2,371,496

Male	Tour Winnings	Career Winnings
12. Corey Pavin	$844,996	$3,138,684
13. Steve Pate	633,887	2,458,605
14. Lanny Wadkins	606,378	5,220,759
15. Billy Andrade	535,419	1,043,974
16. Rocco Mediate	505,482	1,141,206

In Ex. 17–28, indicate in Fig. 2.21 the region in which each of the figures would be placed.

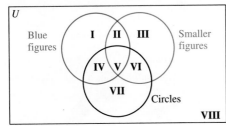

Figure 2.21

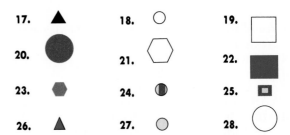

In Ex. 29–42, use the Venn diagram in Fig. 2.22 to list the sets in roster form.

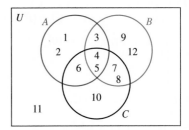

Figure 2.22

29. A **30.** B **31.** C
32. U **33.** $A \cap B$ **34.** $A \cap C$
35. $(B \cap C)'$ **36.** $A \cap B \cap C$ **37.** $A \cup B$
38. $B \cup C$ **39.** $(A \cup C)'$ **40.** $A \cup B \cup C$
41. A' **42.** $(A \cup B \cup C)'$

43. A hematology text gives the following information on percentages of the different types of blood worldwide.

Type	Positive Blood, %	Negative Blood, %
A	37	6
O	32	6.5
B	11	2
AB	5	0.5

Construct a Venn diagram similar to the one in Example 2 and place the correct percent in each of the eight regions.

Using Venn diagrams, determine whether the following statements are equal for all sets A and B.
44. $(A \cup B)'$, $A' \cap B'$
45. $(A \cup B)'$, $A' \cap B$
46. $A' \cup B'$, $A \cap B$
47. $(A \cup B)'$, $(A \cap B)'$
48. $A' \cup B'$, $(A \cup B)'$
49. $A \cap B'$, $A' \cup B$
50. $(A \cap B')'$, $A' \cup B$
51. $A' \cap B'$, $(A' \cap B')'$

Using Venn diagrams, determine whether the following statements are equal for all sets *A*, *B*, and *C*.

52. $A \cup (B \cap C)$, $\qquad$ $(A \cup B) \cap C$
53. $A \cup (B \cap C)$, $\qquad$ $(B \cap C) \cup A$
54. $A \cap (B \cup C)$, $\qquad$ $(B \cup C) \cap A$
55. $A' \cup (B \cap C)$, $\qquad$ $A \cap (B \cup C)'$
56. $A \cap (B \cup C)$, $\qquad$ $(A \cap B) \cup (A \cap C)$
57. $A \cup (B \cap C)$, $\qquad$ $(A \cup B) \cap (A \cup C)$
58. $A \cap (B \cup C)'$, $\qquad$ $A \cap (B' \cap C')$
59. $(A \cup B) \cap (B \cup C)$, $\qquad$ $B \cup (A \cap C)$
60. $(C \cap B)' \cup (A \cap B)'$, $\qquad$ $A \cap (B \cap C)$
61. $(A \cup B)' \cap C$, $\qquad$ $(A' \cup C) \cap (B' \cup C)$

62. Let $U = \{1, 2, 3, 4, 5, 6, 7, 8, 9, 10\}$,
$\qquad A = \{1, 2, 3, 4\}$,
$\qquad B = \{3, 6, 7\}$,
$\qquad C = \{6, 7, 9\}$.
a) Show that $(A \cup B) \cap C = (A \cap C) \cup (B \cap C)$ for these sets.
b) Make up your own sets *A*, *B*, and *C*. Verify that $(A \cup B) \cap C = (A \cap C) \cup (B \cap C)$ for your sets *A*, *B*, and *C*.
c) Using Venn diagrams, verify that $(A \cup B) \cap C = (A \cap C) \cup (B \cap C)$ for all sets *A*, *B*, and *C*.

63. Let $U = \{a, b, c, d, e, f, g, h, i\}$,
$\qquad A = \{a, c, d, e, f\}$,
$\qquad B = \{c, d\}$,
$\qquad C = \{a, b, c, d, e\}$.
a) Determine whether $(A \cup C)' \cap B = (A \cap C)' \cap B$ for the sets given.
b) Make up your own sets, *A*, *B*, and *C*. Determine whether $(A \cup C)' \cap B = (A \cap C)' \cap B$ for your sets.
c) Determine whether $(A \cup C)' \cap B = (A \cap C)' \cap B$ for all sets *A*, *B*, and *C*.

64. Given sets *U*, *A*, *B*, and *C*, construct a Venn diagram and place the elements in the proper regions.

$$U = \{1, 2, 3, 4, 5, 6, 7, 8, 9, 10, 11, 12\}$$

Set *A* is the set of even natural numbers less than or equal to 12.
Set *B* is the set of odd natural numbers less than or equal to 12.
Set *C* is the set of natural numbers less than or equal to 12 that are multiples of 3. (A multiple of 3 is any number that is divisible by 3.)

65. Given sets *U*, *A*, *B*, and *C*, construct a Venn diagram and place the elements in the proper regions.

$$U = \{1, 2, 3, 4, 5, 6, 7, 8, 9, 10, 11, 12\}$$

Set *A* is the set of even natural numbers less than or equal to 12.
Set *B* is the set of natural numbers less than or equal to 12 that are multiples of 3.
Set *C* is the set of natural numbers that are factors of 12. (A factor of 12 is a number that divides 12.)

66. Define each of the eight regions in Fig. 2.23 with a set statement. (*Hint:* $A \cap B' \cap C'$ defines region I.)

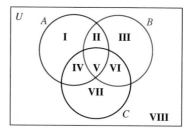

Figure 2.23

Problem Solving

67. a) Construct a Venn diagram illustrating four sets, *A*, *B*, *C*, and *D*. (*Hint:* Four circles cannot be used, and you should end up with 16 *distinct* regions.) Have fun!
b) Label each region with a set statement (see Ex. 66). Check all 16 regions to make sure that *each is distinct*.

Research Activity

68. The two Venn diagrams illustrate what happens when colors are added or subtracted. Do research in an art text, an encyclopedia, or another source, and write a report explaining the creation of the colors in the Venn diagrams using such terms as union of colors and subtraction (or difference) of colors.

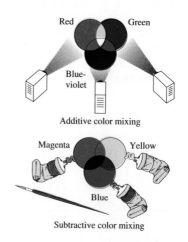

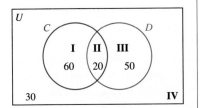

Figure 2.24

2.5 Applications of Sets

We can solve application problems involving sets by using the problem-solving process discussed in Chapter 1: Understand the problem, devise a plan, carry out the plan, and then examine the results. First determine: What is the problem? or What am I looking for? To devise the plan list all the facts that are given and how they are related. Look for key words or phrases like: "only set A," "set A and set B," "set A or set B," "set A and set B and not set C." Remember that "and" means intersection, "or" means union, and "not" means complement. The problems we will solve in this section contain two or three sets of elements, which can be represented in a Venn diagram. Our plan will generally include drawing a Venn diagram, labeling the diagram, and filling in the regions of the diagram. Whenever possible follow the procedure in Section 2.4 for completing the Venn diagram, then answer the questions.

▶ **Example 1**

At the Summit Sports Shop 160 customers were selected at random and asked whether they enjoy cross-country or downhill skiing. The results of the survey follow.

80 enjoy cross-country.
70 enjoy downhill.
20 enjoy both cross-country and downhill.

a) Of those surveyed, how many people do not enjoy downhill or cross-country skiing?
b) How many people enjoy cross-country skiing but not downhill?
c) How many people enjoy downhill skiing but not cross-country?
d) How many people enjoy downhill or cross-country skiing?

Solution: The problem provides us with the following information.
The number of people surveyed is 160: $n(U) = 160$.
The number of people who enjoy cross-country skiing is 80: $n(C) = 80$.
The number of people who enjoy downhill skiing is 70: $n(D) = 70$.
The number of people who enjoy both cross-country and downhill skiing is 20: $n(C \cap D) = 20$.
We can illustrate this information with a Venn diagram, as in Fig. 2.24. From previous sections we know that $C \cap D$ corresponds to region II. Since $n(C \cap D) = 20$, we write 20 in region II. Set C consists of regions I and II. From statement 2 we know that set C contains 80 people. Therefore region I contains $80 - 20$, or 60 people, and we write the number 60 in region I. Set D consists of regions II and III. Since $n(D) = 70$, there must be a total of 70 in these two regions. Region II contains 20, leaving 50 for region III.
The total number of people who enjoy cross-country or downhill skiing is $n(C \cup D) = 60 + 20 + 50 = 130$. The number of people in region IV is the

difference between $n(U)$ and $n(C \cup D)$. There are $160 - 130$, or 30 people in region IV.

a) The people who do not enjoy downhill or cross-country skiing are those members of the universal set who are not contained in set C or set D. The 30 people in region IV do not enjoy downhill or cross-country skiing.

b) The 60 people in region I are those who enjoy cross-country skiing but not downhill.

c) The 50 people in region III are those who enjoy downhill skiing but not cross-country.

d) The people in regions I, II, or III enjoy cross-country or downhill skiing. Thus there are $60 + 20 + 50$ or 130 people who enjoy one or both types of skiing.

Similar problems can be done with three sets, as illustrated in the next example.

▶ Example 2

Concerning the first 41 presidents of the United States we know the following facts.

8 held cabinet posts.
14 served as vice-president.
15 served in the U.S. Senate.
2 served in cabinet posts and as vice-president.
4 served in cabinet posts and in the U.S. Senate.
6 served in the U.S. Senate and as vice-president.
1 served in all three positions.

How many presidents served in
a) none of these positions?
b) only the U.S. Senate?
c) at least one of the three positions?
d) exactly two positions?

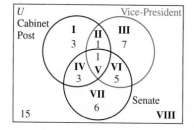

Figure 2.25

Solution: Construct a Venn diagram with three circles: cabinet post, vice-president, and Senate (Fig. 2.25). Label the eight regions.

Whenever possible, work from the center outward. First find the number of presidents that appear in all three sets. Since one president served in all three positions, place a 1 in region V. Next determine the number to be placed in region II. Two presidents served both in cabinet posts and as vice-president. Regions II and V represent the intersection of presidents who served both in cabinet posts and as vice-president. Therefore the sum of the elements in these two regions must be 2. Since a 1 has previously been placed in region V, this leaves a 1 for region II. Four presidents served in cabinet posts and in the Senate. The sum of regions IV and V must therefore be 4. Since there is already a 1 in region V, place a 3 in region IV. Using similar reasoning, place a 5 in region VI.

Now find the numbers in regions I, III, and VII. A total of 8 presidents served in cabinet posts. The sum of regions II, IV, and V is 5, so there are 3 in region I. The total number of presidents who served as vice-president was 14. The sum of the numbers in regions II, V, and VI is 7, so there must be 7 in region III. Similarly, the number in region VII must be 15 − 9, or 6. So far we have accounted for a total of 26 presidents in regions I through VII. There must be 41 − 26, or 15 presidents who did not serve in any of these capacities. Place the number 15 in region VIII. Now use the completed Venn diagram to determine the answers to questions (a) through (d).

a) Fifteen presidents did not serve in any of the three positions. This number is found in region VIII.

b) Region VII represents the presidents who served only in the Senate. There were six.

c) "At least one" means one or more. Thus the number of presidents who served in at least one of the three positions is found by summing the numbers in regions I through VII. Twenty-six presidents served in at least one of these positions.

d) The presidents represented in regions II, IV, and VI served in exactly two positions. Summing the numbers in the three regions shows that nine presidents served in exactly two positions.

When you are solving problems of this type, it is a good idea to check your Venn diagram carefully. The most common error made by students is forgetting to subtract the number in region V from the respective values in determining the numbers to place in regions II, IV, and VI.

▶ Example 3

Common ailments of senior citizens are arthritis, arteriosclerosis, and loss of hearing. A survey of 200 senior citizens found that 70 had arthritis, 60 had arteriosclerosis, 80 had loss of hearing, 35 had arthritis and arteriosclerosis, 33 had arthritis and loss of hearing, 31 had arteriosclerosis and loss of hearing, and 15 had all three. How many of those surveyed had

a) none of these ailments?

b) arthritis but neither of the other two ailments?

c) exactly one of these ailments?

d) arteriosclerosis and loss of hearing, but not arthritis?

e) arteriosclerosis or loss of hearing, but not arthritis?

f) exactly two of these ailments?

Solution: Construct a Venn diagram with three sets as in Example 2. You should go through the process of developing the Venn diagram before consulting Fig. 2.26.

a) The 74 senior citizens without any of these ailments are found in region VIII.

b) The 17 people in region I had only arthritis.

c) Those listed in regions I, III, and VII had only one of the ailments. The

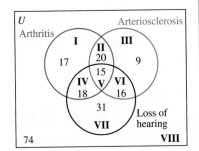

Figure 2.26

sum of the numbers in these regions, $17 + 9 + 31$, or 57, is the number of senior citizens with exactly one of the ailments.

d) The 16 people in region VI had arteriosclerosis and loss of hearing but not arthritis.

e) The word "or" in this type of problem means one or the other or both. All the people in regions II, III, IV, V, VI, and VII had either arteriosclerosis or loss of hearing or had both. Those in regions II, IV, and V also had arthritis. The 56 senior citizens with either arteriosclerosis or loss of hearing but not arthritis are found by adding the numbers in regions III, VI, and VII.

f) The people in regions II, IV, and VI, a total of 54, had exactly two of the three ailments.

Section 2.5 Exercises

1. A financial analyst interviewed 150 people to determine whether they had invested in stocks or bonds. The results follow.

   80 had invested in stocks.
   40 had invested in bonds.
   23 had invested in both stocks and bonds.

   Of those interviewed,
   a) How many had invested only in stocks?
   b) How many had invested only in bonds?
   c) How many had not invested in stocks or bonds?

2. Last year, at East High School, a survey showed that
   65 students participated in soccer or basketball.
   50 participated in soccer.
   10 participated in both sports.

   How many played basketball?

3. A congressional representative polled her constituents regarding two bills. The results showed that of 200 persons in the sample:

   55 favored bill I.
   110 favored bill II.
   40 favored both bills.

   a) How many favored only bill I?
   b) How many were not in favor of either bill?
   c) Did the majority of those surveyed favor bill II?

4. The results of a survey of 50 students showed the following preferences in writing instruments.

   16 liked pencils.
   18 liked felt-tipped pens.
   24 liked ballpoint pens.
   4 liked pencils and felt-tipped pens.
   5 liked felt-tipped pens and ballpoint pens.
   8 liked pencils and ballpoint pens.
   3 liked all three.

   Find the number of students that liked
   a) none of the three writing instruments.
   b) pencils and felt-tipped pens but did not like ballpoint pens.
   c) only pencils.
   d) only ballpoint pens.
   e) felt-tipped pens or ballpoint pens.

5. The results of a survey of 875 students at Youngstown State University were as follows.

   213 were wearing blue jeans.
   217 were freshmen.
   316 were taking an economics course.
   92 were freshmen and wearing blue jeans.
   110 were freshmen and taking an economics course.
   100 were wearing blue jeans and taking an economics course.
   70 were freshmen, wearing jeans, and taking an economics course.

   How many were
   a) doing none of these things?
   b) taking only an economics course?
   c) taking an economics course and wearing blue jeans but were not freshmen?
   d) taking an economics course or wearing blue jeans but were not freshmen?
   e) doing at least one of the three?

6. A market researcher surveyed 35 men to determine whether they like lemon-scented, menthol, or regular shaving cream.

20 like lemon-scented.
20 like menthol.
19 like regular.
12 like lemon and menthol.
11 like lemon and regular.
10 like regular and menthol.
 8 like all three.

How many of those surveyed:
a) like only lemon?
b) like lemon or regular?
c) like only one of the three?
d) like lemon and regular but not menthol?
e) like lemon or regular but not menthol?
f) do not like any of the three?

7. In a recent blood drive the following data on donors were recorded (see Example 2 in Section 2.4):

243 had the A antigen
 93 had the B antigen
 28 had the A and the B antigens
 80 had the B and the Rh antigens
325 had the Rh antigen
 33 had none of the antigens
210 had the A and the Rh antigens
 25 had all three antigens

How many donors had
a) A-positive blood?
b) B-negative blood?
c) AB-positive blood?
d) How many donors are represented here?

8. In a poll, 300 randomly selected voters are surveyed to determine candidate preferences. The three candidates are Professor Plum, Colonel Mustard, and Miss Scarlet. The results of the survey were summarized as follows:

130 choose Professor Plum
110 choose Colonel Mustard
160 choose Miss Scarlet
 80 choose Professor Plum and Colonel Mustard
 65 choose Professor Plum and Miss Scarlet
 40 choose Colonel Mustard and Miss Scarlet
 27 choose all three candidates.

How many voters
a) prefer only Professor Plum?
b) prefer none of the candidates?
c) prefer Miss Scarlet and Colonel Mustard but not Professor Plum?
d) prefer Professor Plum or Colonel Mustard?
e) prefer Professor Plum or Colonel Mustard but not Miss Scarlet?

9. A statistician was commissioned to determine the newspapers read daily by New York City residents. She randomly selected an appropriate sample and had them complete the following form.

I read the following papers daily.
☐ *USA Today*
☐ *New York Times*
☐ *New York Daily News*
The results were summarized as follows:

 80 read all three newspapers
138 read the *Times* and *USA Today*
170 read the *Times* and the *News*
320 read the *USA Today* and the *News*
500 read the *News*
540 read the *USA Today*
700 read the *Times*
208 read none of the three papers

a) How many read only the *New York Times*?
b) How many read at least one of the three papers?
c) How many read at most one of the three papers?
d) How many read the *Times* and *USA Today* but not the *News*?
e) How many read the *Times* or the *USA Today* but not the *News*?

10. A company launching a new toothpaste took a survey to determine how many people preferred wintergreen, peppermint, or cherry flavors of the new toothpaste. One hundred people were questioned. The results were summarized as follows:

44 preferred cherry.
40 preferred wintergreen.
37 preferred peppermint.
16 preferred peppermint and cherry.
15 preferred wintergreen and peppermint.
11 preferred wintergreen and cherry.
 4 preferred all three flavors.

a) How many people in the survey did not prefer any of the flavors of the new toothpaste?
b) How many people prefer only wintergreen flavored toothpaste?
c) How many people prefer cherry or wintergreen flavored toothpaste?
d) How many people prefer cherry or wintergreen flavored but not peppermint flavored toothpaste?

11. Dreamy Cream Ice Cream hired Molly to find out what kind of ice cream people liked. She surveyed 100 people, with the following results: 78 liked hard ice cream, 61 liked soft ice cream, and 40 liked both hard ice cream and soft ice cream. Every person interviewed liked one or the other or both kinds of ice cream. Does this seem right? Explain your answer.

12. An immigration agent samples cars going from the United States into Canada. In his report he indicates of the 85 cars sampled:

    35 cars are driven by women.
    53 cars are driven by U.S. citizens.
    43 cars have two or more passengers.
    27 cars are driven by women who are U.S. citizens.
    25 cars are driven by women and have two or more passengers.
    20 cars are driven by U.S. citizens and have two or more passengers.
    15 cars are driven by women who are U.S. citizens and have two or more passengers.

    After his supervisor reads the report, she explains to the agent that he made a mistake. Explain how his supervisor knew that the agent's report contained an error.

Problem Solving

13. At a rodeo, 24 of the contestants entered into three events—calf roping, steer wrestling, bronco riding.

    8 who roped calves did not wrestle steers.
    8 who wrestled steers did not rope calves.
    3 of the 8 who entered the bronco-riding event entered only that event.
    6 entered only the calf-roping contest.
    1 entered all three events.

    a) How many contestants entered only the steer-wrestling event?
    b) How many contestants entered only one event?
    c) How many contestants entered the steer-wrestling or the calf-roping or the bronco-riding events?
    d) How many contestants entered the calf-roping and the bronco-riding events, or the steer-wrestling event?

14. At an airplane service center 60 planes were inspected. Assume that 33 planes needed new tires and 44 planes needed new brakes.
    a) What is the least number of planes that could have needed both tires and brakes?
    b) What is the greatest number of planes that could have needed both?
    c) What is the greatest number of planes that could have needed neither?

15. A survey of 500 farmers in a midwestern state showed the following.

    125 grew only wheat.
    110 grew only corn.
     90 grew only oats.
    200 grew wheat.
     60 grew wheat and corn.
     50 grew wheat and oats.
    180 grew corn.

    Find the number who
    a) grew at least one of the three.
    b) grew all three.
    c) did not grow any of the three.
    d) grew exactly two of the three.

16. Brothers Tom and Rob Reardon and their respective families vacation together annually. When the two families discussed where to go this year they found that of all their children

    8 refused to go to the family cottage.
    7 refused to go to the ocean.
    9 refused to go camping.
    3 will go neither to the family cottage nor to the ocean.
    4 will go neither to the ocean nor camping.
    6 will go neither to the family cottage nor camping.
    2 will not go to the family cottage or the ocean or camping.
    no children will agree to go to all three places.

    What is the total number of children in the two families?

17. On Diplomat row, a suburb of Washington, D.C., there are five houses. Each owner is a different nationality, each has a different pet, each has a favorite food and a different favorite drink, and each house is painted a different color.

    The green house is directly to the right of the ivory house.
    The Senegalese has the red house.
    The dog belongs to the Spaniard.
    The Afghanistani drinks tea.
    The person who eats cheese lives next door to the fox.
    The Japanese eats fish.
    Milk is drunk in the middle house.
    Apples are eaten in the house next to the horse.
    Ale is drunk in the green house.
    The Norwegian lives in the first house.
    The peach eater drinks whiskey.
    Apples are eaten in the yellow house.
    The banana eater owns a snail.
    The Norwegian lives next door to the blue house.

For each house find
a) the color.
b) the nationality of the occupant.
c) the type of food eaten.

d) the owner's favorite drink.
e) the owner's pet.
f) Finally, the crucial question is: Does the zebra's owner drink vodka or ale?

Infinite Sets

In Section 2.1 we state that a set is finite if it either contains no elements or the number of elements in the set is a natural number. To determine the number of elements in a finite set we can place it in a one-to-one correspondence with a subset of the set of counting numbers. For example, the set $A = \{\#, ?, \$\}$ can be placed in one-to-one correspondence with set $B = \{1, 2, 3\}$, a subset of the set of countering numbers.

$$A = \{\#, ?, \$\}$$
$$\downarrow \; \downarrow \; \downarrow$$
$$B = \{1, 2, 3\}$$

Since the cardinal number of set B is 3, the cardinal number of set A is also 3. Any two sets such as A and B that can be placed in a one-to-one correspondence must have the same number of elements (therefore the same cardinality) and must be equivalent sets. Note that $n(A)$ and $n(B)$ both equal 3.

The German mathematician Georg Cantor (1845–1918), known as the father of set theory, thought about sets that were not bounded. An unbounded set he called an "infinite set" and provided the following definition.

> An **infinite set** is a set that can be placed in one-to-one correspondence with a proper subset of itself.

In Example 1 we will use Cantor's definition of infinite sets to show that the set of counting numbers is infinite.

▶ **Example 1**

Show that $N = \{1, 2, 3, 4, 5, \ldots, n, \ldots\}$ is an infinite set.

Solution: To show that the set N is infinite we will establish a one-to-one correspondence between the counting numbers and a proper subset of itself. By removing the first element from the set of counting numbers, we get the set $\{2, 3, 4, 5, \ldots\}$, which is a proper subset of the set of counting numbers. Now we establish the one-to-one correspondence.

$$\text{Counting numbers} = \{1, 2, 3, 4, 5, \ldots, \quad n \quad , \ldots\}$$
$$\downarrow \downarrow \downarrow \downarrow \downarrow \qquad \downarrow$$
$$\text{Proper subset} \quad = \{2, 3, 4, 5, 6, \ldots, n + 1, \ldots\}$$

Notice for any number, n, in the set of counting numbers, its corresponding number in the proper subset is one greater, or $n + 1$. We have now shown the desired one-to-one correspondence and thus the set of counting numbers is infinite.

In the set of counting numbers n represents the general term. For any other set of numbers the general term will be different. A general term should be written in terms of n such that when 1 is substituted for n in the general term we get the first number in the set; when 2 is substituted for n in the general term we get the second number in the set; when 6 is substituted for n in the general term we get the sixth number in the set; and so on.

Consider the set $\{4, 9, 14, 19, \ldots\}$. Suppose we wish to write a general term for this set (or sequence) of numbers. What would the general term be? Since the numbers differ by 5, the general term will be of the form $5n$ plus or minus some number. Substituting 1 for n yields $5(1)$, or 5. Since the first number in the set is 4, we need to subtract 1 from the 5. Thus the general term is $5n - 1$. Notice when $n = 1$, the value is $5(1) - 1$ or 4, when $n = 2$ the value is $5(2) - 1$ or 9, when $n = 3$ the value is $5(3) - 1$ or 14, and so on. Therefore we write the set of numbers with a general term as

$$\{4, 9, 14, 19, \ldots, 5n - 1, \ldots\}.$$

Now that you are aware of how to determine the general term of a set of numbers, we can do some more problems involving sets.

▶ **Example 2**

Show that the set of even counting numbers $\{2, 4, 6, \ldots, 2n, \ldots\}$ is an infinite set.

Solution: First, create a proper subset of the set of even counting numbers by removing the first number from the set. Then establish the one-to-one correspondence.

Even counting numbers: $\{2, 4, 6, 8, \ldots, \quad 2n \quad, \ldots\}$
$\qquad\qquad\qquad\qquad\qquad\quad \downarrow \downarrow \downarrow \downarrow \qquad\quad \downarrow$
Proper subset: $\{4, 6, 8, 10, \ldots, 2n + 2, \ldots\}$

Since a one-to-one correspondence exists between the two sets, the set of even counting numbers is infinite.

▶ **Example 3**

Show that the set $\{5, 10, 15, 20, \ldots, 5n, \ldots\}$ is an infinite set.

Solution:

Given set: $\{\, 5\,, 10, 15, 20, 25, \ldots, \quad 5n \quad, \ldots\}$
$\qquad\qquad\quad\;\; \downarrow \;\; \downarrow \;\; \downarrow \;\; \downarrow \;\; \downarrow \qquad\qquad \downarrow$
Proper subset: $\{10, 15, 20, 25, 30, \ldots, 5n + 5, \ldots\}$

Therefore the given set is an infinite set.

Countable Sets

In his work with infinite sets, Cantor developed ideas on how to determine the cardinal number of an infinite set. He called the cardinal number of infinite sets "transfinite cardinal numbers" or "transfinite powers." He defined a set as **countable** if it is finite or if it can be placed in one-to-one correspondence with the set of counting numbers. All infinite sets that can be placed in one-to-one correspondence with the set of counting numbers have cardinal number, **aleph-null**, symbolized $\aleph_0$ (the first Hebrew letter, aleph, with a zero subscript, read "null").

The mathematician **Leopold Kronecker**, Cantor's former mentor, ridiculed Cantor's theories and prevented Cantor from gaining a position at the University of Berlin. Although Cantor's work on infinite sets is now considered a masterpiece, it generated heated controversy when originally published. Cantor's claim that the infinite set was unbounded offended the religious views of the time that God had created a complete universe, which could not be wholly comprehended by man. Eventually Cantor was given the recognition due to him, but by then the criticism had taken its toll on his health. He had several nervous breakdowns and spent his last days in a mental hospital.

▶ **Example 4**

Show that the set of even counting numbers has cardinal number $\aleph_0$.

Solution: In Example 2 we showed that a set of even counting numbers is infinite by setting up a one-to-one correspondence between the set and a proper subset of itself.

Now we will show that it is countable and has cardinality $\aleph_0$ by setting up a one-to-one correspondence between the set of counting numbers and the set of even counting numbers.

Counting numbers: $N = \{1, 2, 3, 4, \ldots, n, \ldots\}$

Even counting numbers: $E = \{2, 4, 6, 8, \ldots, 2n, \ldots\}$

We see that for each number n in the set of counting numbers, its corresponding number is $2n$. Since we found a one-to-one correspondence, the set of even counting numbers is not only infinite, it is also countable. Thus the cardinal number of the set of even counting numbers is $\aleph_0$, that is $n(E) = \aleph_0$.

> Any set that can be placed in a one-to-one correspondence with the set of counting numbers has cardinality $\aleph_0$ and is countable.

▶ **Example 5**

Show that the set of odd counting numbers has cardinality $\aleph_0$.

Solution: To show that the set of odd counting numbers has cardinality $\aleph_0$, we need to show a one-to-one correspondence between the counting numbers and the odd counting numbers.

Counting numbers: $N = \{1, 2, 3, 4, 5, \ldots, n, \ldots\}$

Odd counting numbers: $O = \{1, 3, 5, 7, 9, \ldots, 2n - 1, \ldots\}$

Since there is a one-to-one correspondence, the odd counting numbers have cardinality $\aleph_0$; that is to say, $n(O) = \aleph_0$.

We have shown that both the odd and even counting numbers have cardinality $\aleph_0$. Since the odd counting numbers pooled together with the even counting numbers gives the set of counting numbers, we may reason that

$$\aleph_0 + \aleph_0 = \aleph_0.$$

This may seem strange, but it is true. However, what could such a statement mean? Well, consider a hotel with an infinite number of rooms. If all the rooms are occupied then the hotel is, of course, full. If more guests appear wanting accommodations, will they have to be turned away? The answer is no, for if the room clerk were to reassign each guest to a new room with a room number twice that of the present room, then all the odd-numbered rooms would become unoccupied and there would be space for more guests!

In Cantor's work he showed that there are different orders of infinity. Sets that are countable and have cardinal number $\aleph_0$ are the lowest order of infinity. Cantor showed that the set of integers and the set of rational numbers (fractions of the form p/q, where $q \neq 0$) are infinite sets with cardinality of $\aleph_0$. He also showed the set of real numbers (to be discussed in Chapter 5) could not be placed in a one-to-one correspondence with the set of counting numbers, and they have a higher order of infinity, aleph-one, $\aleph_1$.

Welcome to

HOTEL

INFINITY

... where there's always room for one more ...

Section 2.6 Exercises

For each of the following sets, show that the set is infinite by placing it in a one-to-one correspondence with a proper subset of itself.

1. $\{3, 4, 5, 6, 7, \ldots\}$

2. $\{4, 8, 12, 16, 20, \ldots\}$

3. $\{2, 3, 4, 5, 6, \ldots\}$

4. $\{3, 5, 7, 9, 11, \ldots\}$

5. $\{4, 7, 10, 13, 16, \ldots\}$

6. $\{6, 11, 16, 21, 26, \ldots\}$

7. $\{8, 10, 12, 14, 16, \ldots\}$

8. $\{1, \frac{1}{2}, \frac{1}{3}, \frac{1}{4}, \frac{1}{5}, \ldots\}$

9. $\{1, \frac{1}{3}, \frac{1}{5}, \frac{1}{7}, \frac{1}{9}, \ldots\}$

10. $\{\frac{5}{8}, \frac{6}{8}, \frac{7}{8}, \frac{8}{8}, \frac{9}{8}, \ldots\}$

For each of the following sets show that the set has cardinal number $\aleph_0$ by establishing a one-to-one correspondence between the set of counting numbers and the given set.

11. $\{3, 6, 9, 12, 15, \ldots\}$

12. $\{100, 101, 102, 103, 104, \ldots\}$

13. $\{4, 6, 8, 10, 12, \ldots\}$

14. $\{0, 2, 4, 6, 8, \ldots\}$

15. $\{2, 5, 8, 11, 14, \ldots\}$

16. $\{4, 9, 14, 19, 24, \ldots\}$

17. $\{5, 8, 11, 14, 17, \ldots\}$

18. $\{\frac{1}{2}, \frac{1}{4}, \frac{1}{6}, \frac{1}{8}, \ldots\}$

19. $\{\frac{1}{3}, \frac{1}{4}, \frac{1}{5}, \frac{1}{6}, \frac{1}{7}, \ldots\}$

20. $\{\frac{1}{2}, \frac{2}{3}, \frac{3}{4}, \frac{4}{5}, \frac{5}{6}, \ldots\}$

Problem Solving

Show that the following sets have cardinality $\aleph_0$ by establishing a one-to-one correspondence between the set of counting numbers and the given set.

21. $\{1, 4, 9, 16, 25, 36, \ldots\}$

22. $\{2, 4, 8, 16, 32, \ldots\}$

23. $\{3, 9, 27, 81, 243, \ldots\}$

24. $\{\frac{1}{3}, \frac{1}{6}, \frac{1}{12}, \frac{1}{24}, \frac{1}{48}, \ldots\}$

Research Activities

25. Do research to explain how Cantor proved that the set of rational numbers has cardinal number $\aleph_0$.

26. Do research to explain how it can be shown that the real numbers do not have cardinal number $\aleph_0$.

CHAPTER 2 SUMMARY

Key Terms

2.1
cardinal number
counting number
description of a set
element
empty set
equal sets
equivalent sets
finite set
infinite set
member of a set
natural number
null set
one-to-one correspondence
roster form
set
set-builder notation
universal set
well-defined set

2.2
proper subset
subset

2.3
complement
disjoint sets
intersection
union

2.4
De Morgan's laws
verification of set statements

2.6
aleph-null
countable set
infinite set

Important Facts
For any sets A and B:
$n(A \cup B) = n(A) + n(B) - n(A \cap B)$.

Number of distinct subsets of a finite set $= 2^n$.

CHAPTER 2 REVIEW EXERCISES

2.1–2.2, 2.6
Answer true or false.
1. The set of chickens that lay green eggs on St. Patrick's day is a well-defined set.
2. The set of students who ate too much pizza last Friday night is a well-defined set.
3. fish $\in$ {water, rod, reel, net, worms, sinker, hook, net}
4. $\{\ \} \subset \varnothing$
5. $\{3, 6, 9, 12, \ldots\}$ and $\{2, 4, 6, 8, \ldots\}$ are disjoint sets.
6. $\{23, 33, 43, 53, \ldots\}$ is an example of a set in roster form.
7. {bird, plane, ship} = {ship, plane, gull}
8. {ant, bee, wasp, cricket} is equivalent to {horse, sheep, cow, pig}
9. If $A = \{1, 2, 3, \ldots, 14\}$, then $n(A) = 14$.
10. $A = \{1, 4, 9, 16, \ldots\}$ is a countable set.
11. $A = \{1, 4, 7, 10, \ldots, 31\}$ is a finite set.

12. $\{3, 6, 7\} \subseteq \{7, 6, 3, 5\}$.
13. $\{x \mid x \in N \text{ and } 3 < x \leq 5\}$ is a set in set-builder form.
14. $\{x \mid x \in N \text{ and } 5 < x \leq 15\} \subseteq \{1, 2, 3, 4, 5, \ldots, 20\}$.

Express each set using roster form.
15. The set of counting numbers between 6 and 13.
16. $A = \{x \mid x \in N \text{ and } x > 13\}$.
17. The set of the lower 48 states that border the Pacific Ocean.
18. $B = \{x \mid x \in N \text{ and } 357 \leq x < 753\}$

Express each set using set-builder notation.
19. $\{13, 14, 15, 16, 17, 18\}$
20. The set of natural numbers greater than 38.
21. The set of all natural numbers beween 17 and 25 inclusive.

22. {bicycle, car, bus, train, airplane, taxi, motorcycle, scooter, boat, truck, trolley, horse and carriage, . . .}

Express each using a description.
23. $A = \{x \mid x$ is a letter of the English alphabet from E through M inclusive$\}$
24. $B = \{$penny, nickel, dime, quarter, half dollar$\}$
25. $C = \{$ABC, CBS, NBC$\}$
26. $D = \{x \mid x$ is an English teacher who owns a red car and does not live in Nebraska$\}$

2.2–2.3
Let $U = \{1, 2, 3, 4, 5, 6, 7, 8, 9\}$,
  $A = \{1, 2, 3, 4, 5\}$,
  $B = \{4, 6, 8, 9\}$,
  $C = \{2, 4, 6\}$,
Find the following.
27. $A \cap B$ **28.** $A \cup B'$
29. $A' \cap B$ **30.** $(A \cup B)' \cup C$
31. The number of subsets of set B.
32. The number of proper subsets of set A.

33. Given the following sets, construct a Venn diagram and place the elements in the proper regions.
  $U = \{1, 2, 3, 4, 5, 6, 7, 8, 9, 10\}$
  $A = \{1, 2, 3, 4, 5\}$
  $B = \{3, 4, 5, 6, 7\}$

Use Fig. 2.27 to find the following.

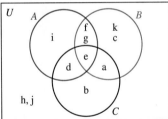

Figure 2.27

34. $A \cup B$ **35.** $A \cap B'$ **36.** $A \cup B \cup C$
37. $A \cap B \cap C$ **38.** $(A \cup B) \cap C$ **39.** $(A \cap B) \cup C$

2.4
Construct a Venn diagram to determine whether the following statements are true.
40. $(A' \cup B')' = A \cap B$
41. $(A \cup B') \cup (A \cup C') = A \cup (B \cap C)'$

2.5
42. A pizza chain was willing to pay $1 to each person interviewed about his or her likes and dislikes of types of pizza crust. Of the people interviewed, 200 liked thin crust, 270 liked thick crust, 70 liked both, and 50 did not like pizza at all. What was the total cost of the survey?

The Cookie Shoppe conducted a survey to determine its customers preferences. The results follow.
200 people like chocolate chip cookies.
190 people like peanut butter cookies.
210 people like sugar cookies.
100 people like chocolate chip and peanut butter cookies.
150 people like peanut butter and sugar cookies.
110 people like chocolate chip and sugar cookies.
 70 people like all three.
  5 people like none of these cookies.
How many people did the following?
43. Completed the survey?
44. Liked only peanut butter cookies?
45. Liked peanut butter and chocolate chip but not sugar cookies?
46. Liked peanut butter cookies or sugar cookies but not chocolate chip cookies?

A survey was taken of 62 people who live around a lake. Its aim was to determine their daily involvement in water activities in the summer months. The survey found the following results.
15 only go swimming.
19 only go sailing.
10 water ski.
 7 swim and water ski.
 5 sail and water ski.
 8 swim and sail.
 3 swim, sail, and water ski.
Find the number of people who do the following.
47. None of the water sports
48. Exactly two of the water sports
49. Exactly one of the water sports
50. Swimming and water skiing but not sailing.
51. Water skiing but no swimming or sailing

2.6
In Ex. 52 and 53, show that the sets are infinite by placing each set in a one-to-one correspondence with a proper subset of itself.
52. $\{2, 4, 6, 8, 10, \ldots\}$ **53.** $\{3, 5, 7, 9, 11, \ldots\}$

In Ex. 54 and 55, show that each set has cardinal number $\aleph_0$ by setting up a one-to-one correspondence between the set of counting numbers and the given set.
54. $\{5, 8, 11, 14, 17, \ldots\}$ **55.** $\{4, 9, 14, 19, 24, \ldots\}$

CHAPTER TEST

Chapter 2

State whether each is true or false.

1. $\{1, 3, 5, 7\} = \{7, 5, 3, 1\}$

2. $\{a, c, d, e\}$ is equivalent to $\{e, c, a\}$

3. $\{q, w, e\} \subset \{q, w, e, r, t\}$ **4.** $6 \in \{2, 4, 6, 8\}$

5. $\{\ \} \subseteq \{1\}$ **6.** $\{8\} \subseteq \{x \mid x \in N$ and $x > 3\}$

7. If $A \cap B = \varnothing$ then A and B are disjoint sets.

8. $\{a, g, k\}$ has seven proper subsets.

9. For any set A, $A \cap A' = U$.

10. For any set A, $A \cap U = A$.

In Ex. 11 and 12, use set $A = \{1, 2, 3, 4, 5\}$.

11. Write set A in set-builder notation.

12. Write a description of set A.

In Ex. 13–17, use the following information.
$U = \{2, 4, 6, 8, 10, 12, 14, 16\}$
$A = \{2, 4, 6, 10\}$
$B = \{4, 8, 10, 12\}$
$C = \{2, 4, 12, 14\}$

Find the following.

13. $A \cup B$ **14.** $A \cap C'$

15. $A \cap (B \cup C)'$ **16.** $n(A \cap C)$

17. Draw a diagram illustrating the relationship among sets U, A, B, and C.

18. Use a Venn diagram to determine whether $A \cup (B \cap C)' = (A \cup B') \cap (A \cup C')$ for all sets A, B, and C.

In a class of 35 college students the results of a questionnaire determined that this group's most popular methods of exercise were aerobics, swimming, and tennis. The results showed further that

3 students used all three activities for exercise.
5 students used aerobics and swimming.
4 students used aerobics and tennis.
7 students used swimming and tennis.
12 students used aerobics.
18 students used swimming.
13 students used tennis.

19. Construct a Venn diagram, and record the number of students in each region.

20. How many students used only swimming to exercise?

21. How many students do not use aerobics, swimming, or tennis to exercise?

22. How many students used swimming or aerobics to exercise?

23. How many students used swimming and aerobics but not tennis to exercise?

24. Show that the set is infinite by setting up a one-to-one correspondence between the set and a proper subset of itself.

$$\{3, 4, 5, 6, \ldots\}$$

25. Show that the set has cardinal number $\aleph_0$ by setting up a one-to-one correspondence between the set of counting numbers and the given set.

$$\{2, 4, 6, 8, \ldots\}$$

LOGIC

Logical reasoning can tell us whether a conclusion follows from a set of given premises, but not whether those premises are true. For example, Greek astronomers, using the assumption that the planets revolved around the earth, correctly predicted the positions of the planets even though their premise was false.

T hat's highly illogical, Captain," says the Vulcan, Mr. Spock, on the TV reruns of "Star Trek." His statement is full of scorn for the often illogical and emotional workings of human interaction.

P.E.B. JOURDAIN
The statement on the other side of this card is true.

.B. JOURDAIN
The statement on the other side of this card is false.

The French mathematician P. E. B. Jourdain had an interesting calling card. On one side was written "The statement on the other side of this card is true." On the other side was written "The statement on the other side of this card is false." This is the liar's paradox first formulated by Eubulides, a Greek philosopher who asked, "Does the liar speak the truth when he says he is lying?"

However, by using deductive reasoning we too can analyze complicated situations and come to a reasonable conclusion from a given set of information.

The ancient Greeks were the first people to analyze systematically the way people think and arrive at a conclusion. Aristotle, whose study of logic is captured in a work called *Organon*, is called the "father of logic." Since Aristotle's time, the study of logic has been continued by other great mathematicians, including Wilhelm Leibniz and George Boole. Perhaps the most popular logician of all time was Charles Dodgson, better known by his pen name, Lewis Carroll, author of *Alice's Adventures in Wonderland* and *Through the Looking Glass*. Though written for children, Carroll's books incorporate many puzzles and references to concepts of logic.

Though most people believe that logic deals with the way people think, it is important to realize that logic does not, in fact, describe the actual way in which the human thought process works. Even the most rational people often think irrationally, using false information or confused facts. Would we want to live in a world where everyone's thoughts followed the strict rules of logical thinking? It is unlikely, since some of the world's greatest inventions, discoveries, and works of art have resulted from creative, nonrational inspiration. For example, consider the German chemist Friedrich Kekulé, who reported that his discovery of the ring structure of benzene came to him in a dream in which dancing snakes formed a ring by taking one another's tails in their mouths.

'And how do you know that you're mad?'

'To begin with,' said the Cat, 'a dog's not mad. You grant that?'

'I suppose so,' said Alice.

'Well then,' the Cat went on, 'you see a dog growls when it's angry, and wags its tail when it's pleased. Now I growl when I'm pleased, and wag my tail when I'm angry. Therefore, I'm mad.'

If human thought does not always follow the rules of logic, then why do we study it? For centuries, logic has been studied because it provides a procedure to determine the validity of an argument. Logic enables you to communicate effectively, to make more convincing arguments, and to develop patterns of reasoning that will help you in making decisions. The study of logic also prepares an individual for other areas of mathematics.

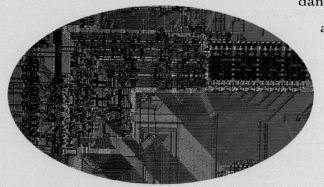

Every day, trillions of microscopic switches turn off and on as computers everywhere process and store information. Here is a microscopic view of only a few of the over 4 million switches found in a computer chip. The fact that data can be encoded and manipulated by the simple on/off of a switch has its beginning in the work of English mathematician George Boole.

Statements and Logical Connectives

History

The ancient Greeks were the first people to analyze systematically the way humans think and arrive at conclusions. Aristotle (384–322 B.C.) organized the study of logic for the first time in a work called *Organon*. As a result of his work in logic, Aristotle is called the Father of Logic. The logic from this period, called **Aristotelian logic**, has been taught and studied for more than 2000 years.

Since Aristotle's time the study of logic has been continued by other great philosophers and mathematicians. Gottfried Wilhelm Leibniz (1646–1716) had a deep conviction that all mathematical and scientific concepts could be derived from logic. As a result he became the first serious student of **symbolic logic**. One difference between symbolic logic and Aristotelian logic is that in symbolic logic, as its name implies, symbols (usually letters) represent written statements. The forms of the statements in the two types of logic are different. The self-educated English mathematician George Boole (1815–1864) is considered to be the founder of symbolic logic because of his impressive work in this area. Among Boole's publications are *The Mathematical Analysis of Logic* (1847) and *An Investigation of the Law of Thought* (1854).

British philosophers Alfred North Whitehead (1861–1947) and Bertrand Russell (1872–1970) wrote a detailed development of arithmetic starting with only undefined concepts and assumptions of logic. Their work appeared in three large volumes called *Principia Mathematica*, published between 1910 and 1913. Charles Dodgson, better known as Lewis Carroll, incorporated many interesting ideas from logic into his books *Alice's Adventures in Wonderland* and *Through the Looking Glass* and his other children's stories.

Logic has been studied through the ages to exercise the mind's ability to reason. Understanding logic will enable you to think clearly, communicate effectively, make more convincing arguments, and develop patterns of reasoning that will help you in making decisions. It will also help you to detect the fallacies in the reasoning or arguments of others. Studying logic has practical applications as well, such as helping you to understand legal documents such as wills and contracts.

The study of logic is also good preparation for other areas of mathematics. If you preview Chapter 11, on probability, you will see formulas for the probability of *a* or *b* and the probability of *a* and *b*, symbolized as $P(A$ or $B)$ and $P(A$ and $B)$ respectively. Special meanings of common words such as "or" and "and" apply to all areas of mathematics. The meaning of these and other special words will be discussed in this chapter.

Logic and the English Language

In reading, writing, and speaking we use many words such as "*and*," "*or*," and "*if . . . then . . .*" to connect thoughts. In logic we call these words **connectives**. How are these words interpreted in daily communication? A judge announces

to a convicted offender: "I hereby sentence you to five months of community service *and* a fine of $100." In this case we normally interpret the word "and" to indicate that *both* events will take place. That is, the person must do community service and must also pay a fine.

Now suppose a judge states: "I sentence you to six months in prison *or* 10 months of community service." In this case we interpret the connective "or" as meaning the convicted person must either spend time in jail or do community service, but not both. The word "or" in this case is the **exclusive "or."** It excludes one of the possibilities named; one or the other event takes place, but *not both*.

In a restaurant a waiter asks: "Can I interest you in a cup of soup or a salad?" This question offers three possibilities: You may order soup, you may order salad, or you may order both soup and a salad. The "or" in this case is the **inclusive "or."** It includes the possibilities named individually or in combination; one or the other *or both* events can take place.

If-then statements are often used to relate two ideas, as in the bank policy statement: "If the average daily balance is greater than $500 then there will be no service charge." If-then statements are also used to emphasize a point or add humor, as in the statement "If the Cubs win then I will be a monkey's uncle."

Now we will look at logic from a mathematical point of view.

Statements and Logical Connectives

A sentence that can be judged either true or false is called a **statement**. Labeling a statement true or false is called **assigning a truth value**. Here are some examples of statements.

1. The famous Gateway Arch in St. Louis is exactly as wide as it is tall.
2. The Apple Macintosh SE is a computer.
3. New Orleans is on the Ohio River.

In each case we can say that the sentence is either true or false.

The three sentences are examples of **simple statements** because they convey one idea. Sentences combining two or more ideas that can be assigned a truth value are called **compound statements**. Compound statements will be discussed shortly.

Quantifiers

Sometimes it is necessary to change a statement to its opposite meaning. To accomplish this we use the **negation** of a statement. For example the negation of the statement "Emily is at home" is "Emily is not at home." The negation of a true statement is always a false statement and the negation of a false statement is always a true statement. We must use special caution when negating statements containing the words **all**, **none** (or **no**), and **some**. These are referred to as **quantifiers**.

Consider the statement "All teachers are rich." We know that this statement is false because a teacher we know, call him Professor John Doe, is not rich. Its negation must therefore be true. We may be tempted to write its negation as "No teachers are rich," but this statement is also false because one teacher we know, call him Professor Bill Smith, *is* rich. Therefore, "No teachers are rich" is not the negation of "All teachers are rich." The correct negation of "All teachers are rich" is "Not all teachers are rich" or "At least one teacher is not rich" or "Some teachers are not rich." These statements all imply that there is at least one teacher who is not rich.

Now consider the statement "No birds can swim." This statement is false, since there is at least one type of bird, the penguin, that can swim. Therefore the negation of this statement must be true. We may be tempted to write the negation as "All birds can swim," but since this is also false it cannot be the negation. The correct negation of the statement is the true statement, "Some birds can swim" or "At least one bird can swim."

Now let's consider statements involving the quantifier "some." Consider, for example, "Some students have a driver's license." This is a true statement since it means "There is at least one student who has a driver's license." The negation of this statement must therefore be false. The negation is "No student has a driver's license."

Consider the statement "Some students do not ride motorcycles." This is a true statement since it means "There is at least one student who does not ride a motorcycle." The negation of this statement must therefore be false. The negation is "All students ride motorcycles."

The negation of quantified statements is summarized as follows:

Form of Statement	Form of Negation
All are.	Some are not.
None are.	Some are.
Some are.	None are.
Some are not.	All are.

This diagram might help you to remember the statements and their negations:

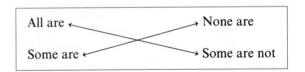

The quantifiers diagonally opposite each other are the negations of each other.

▶ **Example 1**

Write the negation of each statement.

a) Some dogs have short tails.

b) All books have pictures.

Solution:

a) Since "some" means "at least one," the statement "Some dogs have short tails" is the same as "At least one dog has a short tail." Since this is a true statement, its negation must be false. The negation is "No dogs have short tails."

b) The statement "All books have pictures" is false. Its negation must therefore be true. The negation may be written as "Some books do not have pictures," or "Not all books have pictures," or "At least one book does not have pictures."

Compound Statements

Statements consisting of two or more simple statements are compound statements. The connectives often used to join two simple statements are "and," "or," "if . . . then . . . ," and "if and only if." In addition, we consider a simple statement that has been negated to be a compound statement.

To reduce the amount of writing in logic it is common to represent each simple statement with a lowercase letter and each connective with a symbol. It is customary to use the letters p, q, r, and s to represent simple statements. However, other letters may be used instead. Now we will study the connectives used to make compound statements.

Not Statements

The **negation** is symbolized by $\sim$ and read "not." For example, the negation of the statement "Steve is a college student" is "Steve is not a college student." If p represents the simple statement "Steve is a college student," then $\sim p$ represents the compound statement "Steve is not a college student." For any statement p, $\sim(\sim p) = p$. For example, the negation of the statement "Steve is not a college student" is "Steve is a college student."

And Statements

The **conjunction** is symbolized by $\wedge$ and read "and." Let p and q represent the simple statements

> p: You will perform five months of community service.
> q: You will pay a $100 fine.

Then the following is the conjunction written in symbolic form.

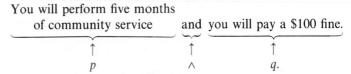

You will perform five months of community service $\underbrace{\qquad}_{p}$ and $\underbrace{\qquad}_{\wedge}$ you will pay a $100 fine. $\underbrace{\qquad}_{q.}$

The conjunction is generally expressed as "and." Other words sometimes used to express a conjunction are "but," "however," or "nevertheless."

▶ **Example 2**
Write the following conjunction in symbolic form. The horse is gray, but its mane is not white.

> **Solution:** Let *h* and *m* represent the simple statements.
>
> h: The horse is gray.
> m: Its mane is white.
>
> In symbolic form the compound statement is $h \wedge \sim m$.

Or Statements

The **disjunction** is symbolized by $\vee$ and read "or." The "or" we will use in this book (except where indicated in the exercise sets) is the inclusive or as described earlier.

▶ **Example 3**

Let

p: Joyce will go to the movie.
q: Joyce will eat popcorn.

Write the following statements in symbolic form.
a) Joyce will go to the movie or Joyce will eat popcorn.
b) Joyce will eat popcorn or Joyce will go to the movie.
c) Joyce will not eat popcorn or Joyce will go to the movie.

Solution: **a)** $p \vee q$ **b)** $q \vee p$ **c)** $\sim q \vee p$

When a compound statement contains more than one connective, a comma can be used to indicate which simple statements are to be grouped together. When writing the statement symbolically, the simple statements on the same side of the comma are to be grouped together within parentheses.

For example, "Long Shot is a horse (*h*) or Patti is a pilot (*p*), and Carl owns a car (*c*)" is written $(h \vee p) \wedge c$. Notice that *h* and *p* are both on the same side of the comma. The statement "Long Shot is a horse, or Patti is a pilot and Carl owns a car" is written $h \vee (p \wedge c)$. In this case *p* and *c* are on the same side of the comma.

A compound statement containing only all conjunctions, or all disjunctions, may have the parentheses placed around any two statements. For example $p \wedge q \wedge r$ may be interpreted as either $(p \wedge q) \wedge r$ or $p \wedge (q \wedge r)$. In this book when we place parentheses around statements like $p \wedge q \wedge r$ we generally place parentheses around the first two statements on the left.

▶ **Example 4**

Let

p: Dinner includes soup.
q: Dinner includes salad.
r: Dinner includes the vegetable of the day.

Write the following statements in symbolic form.
a) Dinner includes soup, and salad or the vegetable of the day.
b) Dinner includes soup and salad, or the vegetable of the day.

Solution:

a) The comma tells us to group the statement "Dinner includes salad" with the statement "Dinner includes the vegetable of the day." Note that both these statements are on the same side of the comma. The statement in symbolic form is $p \wedge (q \vee r)$. In mathematics we always evaluate the information within the parentheses first. Since the conjunction, $\wedge$, is outside the parentheses and is evaluated last, this statement is considered a conjunction.

b) The comma tells us to group the statement "Dinner includes soup" with the statement "Dinner includes salad." Note that both these statements are on the same side of the comma. The statement in symbolic form is $(p \wedge q) \vee r$. Since the disjunction, $\vee$, is outside the parentheses, this statement is considered a disjunction.

An important point to remember is that a negation has the effect of negating only the statement that directly follows it. To negate a compound statement, we must use parentheses. When a negation symbol is placed in front of a statement in parentheses, it negates the entire statement in parentheses. The negation symbol in this case is read, "It is not true that . . ." or "It is false that . . . ."

▶ **Example 5**

Let   *p:*   Tanya is the star in the home movies.
            q: Matthew is operating the camcorder.

Write the following symbolic statements in words.
a) $\sim p \wedge \sim q$ **b)** $\sim (p \vee q)$

Solution:

a) Tanya is not the star in the home movies and Matthew is not operating the camcorder.

b) It is false that Tanya is the star in the home movies or Matthew is operating the camcorder.

Part (a) of Example 5 is a conjunction, since it can be written $(\sim p) \wedge (\sim q)$. Part (b) is a negation, since it negates the entire statement. The similarity of these two statements is discussed in Section 3.4.

Occasionally, we come across a **neither-nor** statement. For example, "John is *neither* handsome *nor* rich." This means that John is not handsome *and* John is not rich. If *p* represents "John is handsome" and *q* represents "John is rich," this statement is symbolized $\sim p \wedge \sim q$.

If-Then Statements

The **conditional statement** is symbolized by $\rightarrow$ and is read "if-then." The statement $p \rightarrow q$ is read "If *p*, then *q*."* The conditional statement consists of two

*The self-taught English mathematician **George Boole** (1815–1864) took the operations of algebra and used them to extend Aristotelian logic. He used symbols like x and y to represent particular qualities or objects in question. For example, if x represents all butterflies, then 1 − x represents everything else except butterflies. If y represents the color yellow, then (1 − x) (1 − y) represents everything except butterflies and things that are yellow, or yellow butterflies. This added a computational dimension to logic that provided a basis for twentieth-century work in the field of computing.*

---

* Some books indicate $p \rightarrow q$ may also be read "*p* implies *q*." However, many higher level mathematics books indicate that $p \rightarrow q$ may be read "*p* implies *q*" only under certain conditions. Implications are discussed in Section 3.3.

parts; the part that precedes the arrow is the **antecedent**, and the part that follows the arrow is the **consequent**.* In the statement $p \rightarrow q$, p is the antecedent and q is the consequent. An example of a conditional statement is "If you drink your milk, then you will grow up to be healthy." A conditional symbol may be placed between any two statements even if the statements are not related.

▶ **Example 6**

Let
| p: | I go into the bookstore. |
| q: | I will spend money. |

Write the following statements symbolically.
a) If I go into the bookstore, then I will spend money.
b) If I do not go into the bookstore, then I will not spend money.
c) It is false that if I go into the bookstore then I will spend money.

Solution: **a)** $p \rightarrow q$ **b)** $\sim p \rightarrow \sim q$ **c)** $\sim(p \rightarrow q)$

▶ **Example 7**

Let
p:	Sally is enrolled in calculus.
q:	Sally's major is nursing.
r:	Sally's major is engineering.

Write the following symbolic statements in words and indicate whether the statement is a negation, conjunction, disjunction, or conditional.
a) $(q \rightarrow \sim p) \vee r$ **b)** $q \rightarrow (\sim p \vee r)$

Solution: The parentheses tell where to place the commas in the sentences.
a) "If Sally's major is nursing then Sally is not enrolled in calculus, or Sally's major is engineering." This statement is a disjunction since $\vee$ is outside the parentheses.
b) "If Sally's major is nursing, then Sally is not enrolled in calculus, or Sally's major is engineering." This is a conditional statement since $\rightarrow$ is outside the parentheses.

If and Only If Statements

The **biconditional** is symbolized by $\leftrightarrow$ and is read "if and only if." The phrase "if and only if" is sometimes abbreviated as "iff." The statement $p \leftrightarrow q$ is read "p if and only if q."

▶ **Example 8**

Let
| p: | The cow is brown. |
| q: | The milk is chocolate. |

---

* Some books refer to the antecedent as the hypotheses or premise, and the consequent as the conclusion.

Write the following symbolic statements in words.

a) $q \leftrightarrow p$ **b)** $\sim(p \leftrightarrow \sim q)$

Solution:

a) The milk is chocolate if and only if the cow is brown.

b) It is false that the cow is brown if and only if the milk is not chocolate.

You will learn later that $p \leftrightarrow q$ means the same as $(p \rightarrow q) \wedge (q \rightarrow p)$. Therefore the statement "I will go to college if and only if I can pay the tuition" has the same logical meaning as "If I go to college then I will pay the tuition and If I pay the tuition then I will go to college."

Here is a summary of the connectives discussed in this section.

Formal Name	Symbol	Read	Symbolic Form
Negation	$\sim$	not	$\sim p$
Conjunction	$\wedge$	and	$p \wedge q$
Disjunction	$\vee$	or	$p \vee q$
Conditional	$\rightarrow$	if-then	$p \rightarrow q$
Biconditional	$\leftrightarrow$	if and only if	$p \leftrightarrow q$

Section 3.1 Exercises

Indicate whether each statement is a simple or a compound statement. If it is a compound statement, indicate whether it is a negation, conjunction, disjunction, conditional, or biconditional using both the word and its appropriate symbol (for example "a negation," $\sim$).

1. The charcoals are hot and the chicken is on the grill.
2. Today is not Tuesday.
3. The dog will bite if and only if it is cornered.
4. If your father is a teacher, then you are a teacher.
5. The farm is on Willow Street or on Birch Street.
6. The jacket is neither red nor tan.
7. The report is 20 pages long.
8. Mike will pay the bill on the due date or he will be assessed an interest charge.
9. It is false that the sofa will be delivered by Friday or we will not pay for it.
10. If shopkeepers closed their stores early, then they wanted to get home.
11. We decided not to go to class, but we planned to get the notes.
12. Everyone had the same plan, nevertheless most of us ended up failing the quiz.
13. If the door is closed, then class has started or the class is not meeting.

14. It is false that if the computer is a PC then there is only one piece of software available.
15. If you are planning to work past age 62 and expect to make more than the earnings limit for Social Security recipients, then you're better off delaying benefits.
16. If you are a full-time college student and decide to drop out after the fourth week, then you will lose the total tuition payment.

In Ex. 17–30, write the negation of each statement.

17. Some spiders are black.
18. No dogs have fleas.
19. Some llamas do not have tails.
20. No mail will be delivered on Sunday.
21. All plants are living things.
22. Some students will pass this course.
23. Some symphony conductors do not use a baton.
24. All people are friendly.
25. No one likes squash.
26. Some soda contains sugar.
27. Some soda does not contain sugar.
28. All people who work pay taxes.
29. All nurses have a degree.
30. Some people who work do not pay taxes.

In Ex. 31–36, write each statement in symbolic form.

Let   *p*: The class is 50 minutes.
          q: The teacher lectures for 40 minutes.

31. The class is 50 minutes and the teacher lectures for 40 minutes.
32. The class is not 50 minutes.
33. The class is 50 minutes or the teacher does not lecture for 40 minutes.
34. The teacher lectures for 40 minutes if and only if the class is not 50 minutes.
35. The class is 50 minutes, nevertheless the teacher does not lecture for 40 minutes.
36. The teacher does not lecture for 40 minutes, however the class is 50 minutes.

In Ex. 37–40, write each statement in symbolic form.

Let   *p*: Kate is a member of the marching band.
          q: Luke is a member of the jazz band.

37. If Kate is a member of the marching band, then Luke is a member of the jazz band.
38. Luke is a member of the jazz band but Kate is not a member of the marching band.
39. Neither is Kate a member of the marching band nor is Luke a member of the jazz band.
40. It is false that Kate is a member of the marching band and Luke is a member of the jazz band.

In Ex. 41–50, write each compound statement in words.

Let   *p*: The horse is five years old.
          q: The horse won the race.

41. $\sim p$ **42.** $\sim q$
43. $p \wedge q$ **44.** $q \vee p$
45. $\sim p \rightarrow q$ **46.** $\sim p \leftrightarrow \sim q$
47. $\sim (q \vee p)$ **48.** $\sim p \wedge q$
49. $\sim p \vee \sim q$ **50.** $\sim (p \wedge q)$

In Ex. 51–60, write each statement in symbolic form.

Let   *p*: The temperature is 90°.
          q: The air conditioner is working.
          r: The apartment is hot.

51. The temperature is 90° and the air conditioner is working, or the apartment is hot.
52. If the temperature is 90° or the air conditioner is not working, then the apartment is hot.

53. If the temperature is 90°, then the air conditioner is working or the apartment is not hot.
54. The apartment is hot if and only if the temperature is not 90°, or the apartment is hot.
55. The temperature is not 90° if and only if the air conditioner is not working, or the apartment is not hot.
56. If the apartment is hot and the air conditioner is working, then the temperature is 90°.
57. The apartment is hot if and only if the air conditioner is working, and the temperature is 90°.
58. It is false that if the apartment is hot then the air conditioner is not working.
59. The apartment is hot or the air conditioner is not working, if and only if the temperature is 90°.
60. If the air conditioner is working, then the temperature is 90° if and only if the apartment is hot.

In Ex. 61–70, write each symbolic statement in words.

Let   *p*: The sun is shining.
          q: We will go to the lake.
          r: We will go swimming.

61. $(p \vee q) \wedge r$ **62.** $(\sim p \vee q) \wedge \sim r$
63. $(q \rightarrow p) \vee r$ **64.** $\sim p \wedge (q \vee r)$
65. $\sim r \rightarrow (q \wedge p)$ **66.** $(q \wedge r) \rightarrow p$
67. $(r \rightarrow q) \wedge p$ **68.** $\sim p \rightarrow (q \vee r)$
69. $(q \leftrightarrow p) \wedge r$ **70.** $q \rightarrow (p \leftrightarrow r)$

Translate the following statements into symbolic form. Indicate the letters you use to represent each simple statement.
71. "Each language has a beauty of its own and forms of expression which are duplicated nowhere else." (Margaret Mead)
72. "Our constitution is in actual operation, everything appears to promise that it will last, but in this world nothing is certain but death and taxes." (Benjamin Franklin)
73. The Constitution requires that the president be at least 35 years old, a natural-born citizen of the United States, and a resident of the country for 14 years.
74. "You can fool all of the people some of the time and you can fool some of the people all of the time, but you can't fool all of the people all of the time." (Abraham Lincoln)
75. "If you want to kill an idea in the world today, then get a committee working on it." (C. F. Kettering)
76. "Happiness is beneficial for the body but it is grief that develops the powers of the mind." (Marcel Proust)
77. "I did not know the dignity of their birth, but I do know the glory of their death." (Douglas MacArthur)

3.2 Truth Tables for Negation, Conjunction, and Disjunction

A truth table is a device used to determine when a compound statement is true or false. There are five basic truth tables that are used in constructing other truth tables. Three will be discussed in this section and two will be discussed in the next section. Section 3.6 uses truth tables in determining whether a logical argument is valid or invalid.

Negation

The first truth table is for **negation**. If p is a true statement, then the negation of p, "not p," is a false statement. If p is a false statement, then "not p" is a true statement. For example, if the statement "The shirt is blue" is true, then the statement "The shirt is not blue" is false. These relationships are summarized in Table 3.1. For a simple statement there are exactly two true-false cases, as illustrated in Table 3.1.

If a compound statement consists of two simple statements p and q there are four possible cases, as illustrated in Table 3.2. Consider the statement "The test is today and the test covers Chapter 5." The simple statement "The test is today" has two possible truth values, true or false. The simple statement "The test covers Chapter 5" also has two truth values, true or false. Thus for these two simple statements there are four distinct possible true-false arrangements. Whenever we construct a truth table for a compound statement that consists of two simple statements we will begin by listing the four true-false cases.

Table 3.1 Negation

	p	$\sim p$
Case 1	T	F
Case 2	F	T

Table 3.2

	p	q
Case 1	T	T
Case 2	T	F
Case 3	F	T
Case 4	F	F

Conjunction

To illustrate the conjunction, consider the following situation. You have recently purchased a new house. To decorate the house, you ordered a new carpet and new furniture. You explain to the salesperson that the carpet must be delivered before the furniture. He promises that the carpet will be delivered on Thursday and the furniture will be delivered on Friday.

To help determine whether the salesperson kept his promise we assign letters to each simple statement. Let p be "The carpet will be delivered on Thursday" and q be "The furniture will be delivered on Friday." The salesperson's statement written in symbolic form is $p \wedge q$. There are four possible true-false situations to be considered (Table 3.3).

Table 3.3 Conjunction

	p	q	$p \wedge q$
Case 1	T	T	T
Case 2	T	F	F
Case 3	F	T	F
Case 4	F	F	F

Case 1: p is true and q is true. The carpet is delivered on Thursday and the furniture is delivered on Friday. The salesperson has kept his promise and the compound statement is true.

Case 2: p is true and q is false. The carpet is delivered on Thursday but the furniture is not delivered on Friday. Since the furniture was not delivered as promised, the compound statement is false.

Case 3: p is false and q is true. The carpet is not delivered on Thursday but the furniture is delivered on Friday. Since the carpet was not delivered on Thursday as promised, the compound statement is false.

Case 4: p is false and q is false. The carpet is not delivered on Thursday and the furniture is not delivered on Friday. Since the carpet and furniture were not delivered as promised, the compound statement is false.

Examining the four cases, we see there is only one case where the salesperson has kept his promise, that is in case 1. Therefore case 1 (T, T) is true. In cases 2, 3, and 4 the salesperson has not kept his promise and the compound statement is false. The results are summarized in Table 3.3, the truth table for the conjunction.

> The **conjunction** $p \wedge q$ is true only when both p and q are true.

▶ **Example 1**

Construct a truth table for $p \wedge \sim q$.

Solution: Since there are two statements, p and q, construct a truth table with four cases (Table 3.4a). Then write the truth values under the p in the compound statement and label this column 1 (Table 3.4b). These truth values are copied directly from the p column on the left. Write the corresponding truth values under the q in the compound statement and call this column 2 (Table 3.4c). The truth values for column 2 are copied directly from the q column on the left. Now find the truth values of $\sim q$ by negating the truth

Table 3.4

(a)

	p	q	$p \wedge \sim q$
Case 1	T	T	
Case 2	T	F	
Case 3	F	T	
Case 4	F	F	

(b)

p	q	$p \wedge \sim q$
T	T	T
T	F	T
F	T	F
F	F	F
		1

(c)

p	q	$p \wedge \sim q$
T	T	T T
T	F	T F
F	T	F T
F	F	F F
		1 2

(d)

p	q	$p \wedge \sim q$
T	T	T F T
T	F	T T F
F	T	F F T
F	F	F T F
		1 3 2

(e)

p	q	p	$\wedge$	$\sim$	q
T	T	T	F	F	T
T	F	T	T	T	F
F	T	F	F	F	T
F	F	F	F	T	F
		1	4	3	2

values in column 2 and call this column 3 (Table 3.4d). Using the conjunction table, Table 3.3, and the entries in columns 1 and 3, complete column 4 (Table 3.4e). The results in column 4 are obtained as follows:

row 1: T ∧ F is F
row 2: T ∧ T is T
row 3: F ∧ F is F
row 4: F ∧ T is F

The answer is always the last column completed. Columns 1, 2, and 3 are only aids in arriving at the answer in column 4.

The statement $p \wedge \sim q$ actually means $p \wedge (\sim q)$. In the future, instead of listing a column for q and a separate column for its negation, we will make one column for $\sim q$, which will have the opposite values of those in the q column on the left.

▶ Example 2

a) Construct a truth table for the following statement.

The furnace is not on and we are not wasting energy.

b) Under which conditions will the compound statement be true?
c) Suppose "The furnace is on" is a false statement, and "We are wasting energy" is a true statement. Is the compound statement given in part (a) true or false?

Solution:

a) First write the simple statements in symbolic form using simple non-negated statements.

Let
p: The furnace is on.
q: We are wasting energy.

Therefore the compound statement may be written $\sim p \wedge \sim q$. Now construct a truth table with four cases, Table 3.5.

Fill in the table column labeled 1 by negating the truth values that are under p on the far left. Fill in the column labeled 2 by negating the values that are under q in the second column from the left. Fill in the column labeled 3 using the columns labeled 1 and 2 and the definition of conjunction.

b) The compound statement in part (a) will be true only in case 4 (circled in blue) when both simple statements, p and q, are false. That is when the furnace is not on and we are not wasting energy.
c) "The furnace is on," p, is a false statement. "We are wasting energy," q, is a true statement. From the truth table (Table 3.5) we can determine that when p is false and q is true, case 3, the compound statement, is false (circled in red).

Table 3.5

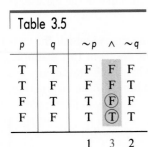

p	q	$\sim p$	$\wedge$	$\sim q$
T	T	F	F	F
T	F	F	F	T
F	T	T	(F)	F
F	F	T	(T)	T
		1	3	2

Disjunction

Consider the following situation. A job description contains the following requirements:

> **Civil Technician**
>
> Municipal program for redevelopment seeks on-site technician. **The applicant must have a two-year college degree in civil technology or five years of experience in the field.** Interested candidates please call 555-1234.

Table 3.6 Disjunction

p	q	p ∨ q
T	T	T
T	F	T
F	T	T
F	F	F

Who qualifies for the job? To help analyze the statement translate it into symbolic form. Let p be "A requirement for the job is a two-year college degree in civil technology" and q be "A requirement for the job is five years of experience in the field." The statement in symbolic form is $p \vee q$. Since there are two simple statements there are four distinct cases. See Table 3.6.

Case 1: p is true and q is true. A candidate has a two-year college degree in civil technology and five years of experience in the field. The candidate has both requirements, and qualifies for the job. Consider qualifying for the job as a true statement and not qualifying as a false statement.

Case 2: p is true and q is false. A candidate has a two-year college degree in civil technology but does not have five years of experience in the field. The candidate still qualifies for the job with the two-year college degree.

Case 3: p is false and q is true. The candidate does not have a two-year college degree in civil technology but does have five years of experience in the field. The candidate qualifies for the job with the five years of experience in the field.

Case 4: p is false and q is false. The candidate does not have a two-year college degree in civil technology and does not have five years of experience in the field. Since the candidate does not meet either of the two requirements the candidate does not qualify for the job.

In examining the cases we see that there is only one case in which the candidate does not qualify for the job, case 4. As we can see from this example, an "or" statement will be true in every case, except when both statements are false. The results are summarized in Table 3.6, the truth table for the disjunction.

> The **disjunction**, $p \vee q$, is true when either p is true, q is true, or both p and q are true.

The disjunction $p \vee q$ will be true except when p and q are both false.

▶ **Example 3**

Construct a truth table for $\sim(q \vee \sim p)$.

Table 3.7

p	q	~	(q	∨	~p)
T	T	F	T	T	F
T	F	T	F	F	F
F	T	F	T	T	T
F	F	F	F	T	T
		4	1	3	2

Solution: First construct the standard truth table listing the four cases. Then determine the truth values for the statement within the parentheses. The order to be followed is indicated by the numbers below the columns (see Table 3.7). In column 1, copy the values from the q column on the left. Under $\sim p$, column 2, write the negation of the p column on the left. Next complete the "or" column, column 3, using columns 1 and 2 and the truth table for the disjunction. The "or" column is false only when both statements are false, as in case 2. Finally negate the values in the "or" column, and place these negated values in column 4. By examining the truth table you can see that the compound statement $\sim(q \vee \sim p)$ is true only in case 2, that is, when p is true and q is false.

A General Procedure to Construct Truth Tables

1. Complete the columns under the simple statements, p, q, r, and their negations, $\sim p$, $\sim q$, $\sim r$, within parentheses. If there are nested parentheses (one pair of parentheses within another pair) work with the innermost pair first.
2. Complete the column under the connective within the parentheses. We will use these truth values in determining the final answer in step 4.
3. Complete the columns under any remaining statements and their negations.
4. Complete the column under any remaining connective using the truth values in the last column completed on the left and right sides of the connective.

In Section 3.1 we explained how to classify a compound statement as a negation, conjunction, disjunction, conditional, or biconditional. The final column completed, and the answer to the truth table, will be under the symbol that determines the classification of the statement. For example, if the statement is a conjunction the final column completed will be under the symbol ∧.

Table 3.8

p	q	(~p	∨	q)	∧	~p
T	T	F	T	T	F	F
T	F	F	F	F	F	F
F	T	T	T	T	T	T
F	F	T	T	F	T	T
		1	3	2	5	4

▶ **Example 4**

Construct a truth table for the statement $(\sim p \vee q) \wedge \sim p$.

Solution: We will follow the procedure outlined in the box. This statement is a conjunction and so the answer will be under the conjunction symbol. Complete columns under $\sim p$ and q within the parentheses and call these columns 1 and 2 respectively (see Table 3.8). Complete the column under the disjunction, ∨, using the values in columns 1 and 2, and call this column 3. Next, complete the column under $\sim p$, and call this column 4. The answer column, 5, is determined using the definition of the conjunction and the truth values in columns 3 and 4.

Table 3.9

	p	q	r
Case 1	T	T	T
Case 2	T	T	F
Case 3	T	F	T
Case 4	T	F	F
Case 5	F	T	T
Case 6	F	T	F
Case 7	F	F	T
Case 8	F	F	F

So far all the truth tables we have constructed have contained at most two simple statements. Now we will explain how to construct a truth table that consists of three simple statements, such as $(p \wedge q) \wedge r$. When a compound statement consists of three simple statements there are eight different true-false possibilities, as illustrated in Table 3.9. To begin such a truth table write four Ts and four Fs in the column under p. Under the second statement, q, pairs of Ts alternate with pairs of Fs. Under the third statement, r, T alternates with F. This technique ensures that each case is unique and no cases are omitted.

▶ **Example 5**

a) Construct a truth table for the statement "Matthew is cutting the grass and he will not water the lawn, or he will weed the flower beds."

b) Suppose "Matthew is cutting the grass" is a false statement, "Matthew will water the lawn" is a true statement, and "Matthew will weed the flower beds" is a true statement. Is the compound statement in part (a) true or false?

Solution:

a) First we will translate the statement into symbolic form.

Let
p: Matthew is cutting the grass.
q: Matthew will water the lawn.
r: Matthew will weed the flower beds.

Thus in symbolic form the statement is $(p \wedge \sim q) \vee r$.

Since this statement is composed of three simple statements, there are eight cases (Table 3.10). Fill out the truth table by working in parentheses first. Place values under p, column 1, and $\sim q$, column 2. Then find the conjunctions of columns 1 and 2 to obtain column 3. Place the values of r in column 4. To obtain the answer, column 5, use columns 3 and 4 and your knowledge of the disjunction.

Table 3.10

p	q	r	(p	∧	~q)	∨	r
T	T	T	T	F	F	T	T
T	T	F	T	F	F	F	F
T	F	T	T	T	T	T	T
T	F	F	T	T	T	T	F
F	T	T	F	F	F	(T)	T
F	T	F	F	F	F	F	F
F	F	T	F	F	T	T	T
F	F	F	F	F	T	F	F

| | | | 1 | 3 | 2 | 5 | 4 |

b) In case 5 of the truth table p, q, and r are respectively F, T, and T. Therefore under these conditions this statement is true (as circled in the table).

The number of distinct cases in a truth table with n distinct simple statements is 2^n. The compound statement $(p \lor q) \lor (r \land {\sim}s)$ has four simple statements p, q, r, s. Thus a truth table for this compound statement would have 2^4, or 16, distinct cases.

When we construct a truth table we determine the truth values of a compound statement for every possible case. If we wish to find the truth value of the compound statement for any one specific case where we know the truth values of the simple statements, we do not have to develop the entire table. For example, to determine the truth value for the statement

$$2 + 3 = 5 \quad \text{and} \quad 1 + 1 = 3$$

we let p be $2 + 3 = 5$ and q be $1 + 1 = 3$. Now we can write the compound statement as $p \land q$. We know that p is a true statement and q is a false statement. Thus we can substitute T for p and F for q and evaluate the statement.

$$p \land q$$
$$\text{T} \land \text{F}$$
$$\text{F}$$

Therefore this compound statement is false.

▶ **Example 6**

Determine the truth value for each simple statement. Then, using these truth values, determine the truth value of the compound statement.

a) 3 is greater than or equal to 2 ($3 \geq 2$).

b) Canada is south of Mexico and New York City is east of the Mississippi River, or Dallas is not a city in Texas.

Solution:

a) Let

p: 3 is greater than 2
q: 3 is equal to 2.

The statement "3 is greater than or equal to 2" means 3 is greater than 2 or 3 is equal to 2. The compound statement can be expressed as $p \lor q$. We know that p is a true statement and q is a false statement. Substitute T for p and F for q and evaluate the statement.

$$p \lor q$$
$$\text{T} \lor \text{F}$$
$$\text{T}$$

Therefore the compound statement "3 is greater than or equal to 2" is a true statement.

b) Let

p: Canada is south of Mexico.
q: New York City is east of the Mississippi River.
r: Dallas is a city in Texas.

The compound statement can be written in symbolic form as $(p \land q) \lor {\sim}r$. We know that p is a false statement since Canada is north of Mexico and

q is a true statement since New York City is east of the Mississippi River. Since Dallas is a city in Texas r is a true statement. Its negation, $\sim r$, is therefore a false statement. Substitute F for p, T for q, and F for $\sim r$ and then evaluate the statement.

$$(p \land q) \lor \sim r$$
$$(F \land T) \lor F$$
$$F \quad \lor F$$
$$F$$

Therefore the compound statement is a false statement.

Di∂ You

Know...

APPLICATIONS OF LOGIC

George Boole provided the key that would unlock the door to modern computing. However, it wasn't until 1938 that Claude Shannon, in his master's thesis for MIT, proposed uniting the on–off capability of electrical switches with Boole's two-value system of 0's and 1's. The operations AND, OR, NOT, and the rules of logic laid the foundation for computer gates. Such gates determine whether or not the current will pass. The closed switch (current flow) is represented as 1 and the open switch (no current flow) is represented by 0.

Early computers were large, noisy, and fairly slow, and used actual relay switches that mechanically clicked open or shut, often causing breakdowns. During Hopper's work on Mach II, one such breakdown was found to be caused by a moth caught in a relay. She used the word "bug" in describing the problem. The term "bug" is now generally used to describe many computer problems.

AND Gate

0, 0 → 0
1, 0 → 0
0, 1 → 0
1, 1 → 1

OR Gate

0, 0 → 0
1, 0 → 1
0, 1 → 1
1, 1 → 1

NOT Gate

0 → 1   1 → 0

Rear Admiral Grace Hopper (1906–1992), a pioneer of computer science and the inventor of the computer language COBOL, worked on one of the earliest computer projects, Harvard's Mach II.

As you can see from the accompanying diagrams, the gates function in essentially the same way as a truth table. The gates shown here are the simplest ones, representing simple statements. There are other gates like the NAND and NOR gates that are combinations of NOT, AND, and OR gates. The microprocessing unit of a computer uses thousands of these switches.

Section 3.2 Exercises

Construct a truth table for each of the following statements.

1. $p \wedge \sim p$

2. $p \vee \sim p$

3. $q \vee \sim p$

4. $p \wedge \sim q$

5. $\sim (p \wedge \sim q)$

6. $\sim p \vee \sim q$

7. $\sim (p \vee \sim q)$

8. $\sim (\sim p \wedge \sim q)$

9. $(p \wedge q) \vee \sim q$

10. $(p \vee \sim q) \wedge r$

11. $q \vee (p \vee \sim r)$

12. $(r \wedge q) \wedge \sim p$

13. $p \wedge (q \wedge r)$

14. $p \wedge (q \vee r)$

15. $r \vee (p \wedge q)$

16. $\sim r \vee (\sim p \wedge q)$

In Ex. 17–26, write each statement in symbolic form and construct a truth table.

17. The book is small but it is difficult reading.

18. The cake is round and the cookie is not square.

19. Tom is a member of the glee club and Karen is not a member of the chess club.

20. It is false that the pen will not write or the paper is poor quality.

21. Brian will go water skiing and he will go swimming, but he will not go sailing.

22. The dress is expensive and the shoes are not exactly what I wanted, however I need them for a meeting on Monday.

23. The necktie is stained, but it is the only one that is not too wide and it goes with my brown suit.

24. The desk is too high for typing or the chair is too low, but my arms are not too short.

25. The football game is not too long and my team is winning, or my team is not winning.

26. The words are not too small, or my glasses need new lenses or my glasses are not clean.

In Ex. 27–32, if p is true, q is false, and r is true, find the truth value of the statements.

27. $(\sim p \wedge r) \wedge q$

28. $\sim p \vee (q \wedge r)$

29. $(\sim q \wedge \sim p) \vee \sim r$

30. $(\sim p \vee \sim q) \vee \sim r$

31. $(p \wedge \sim q) \vee r$

32. $(p \vee \sim q) \wedge \sim (p \wedge \sim r)$

In Ex. 33–38, if p is false, q is true, and r is false, find the truth value of the statements.

33. $(\sim r \wedge p) \vee q$

34. $\sim q \vee (r \wedge p)$

35. $(\sim q \vee \sim p) \wedge r$

36. $(\sim r \vee \sim p) \vee \sim q$

37. $(\sim p \vee \sim q) \vee (\sim r \vee q)$

38. $(\sim r \wedge \sim q) \wedge (\sim r \vee \sim p)$

In Ex. 39–46, determine the truth value for each simple statement. Then using these truth values, determine the truth value of the compound statement. (You may have to use a reference source such as an almanac.)

39. $2 + 3 = 7$ or $6 + 6 = 11$

40. $9 - 6 = 15$ and $7 - 4 = 3$

41. There are five Great Lakes in the United States or the country of Brazil is in Africa.

42. The capital of New York State is Albany or Ohio is west of the Mississippi River, and Los Angeles is not the capital of California.

43. Seventeen and 19 are prime numbers, but 2 is the only even prime number. (A prime number is a natural number that has exactly two distinct divisors: 1 and itself.)

44. Paris is in France or Rome is not in Germany, and London is in England.

45. The cheetah is the fastest land animal or the giraffe is the tallest land animal, and the elephant is the largest land animal.

46. Mars is a planet and the sun is a star, or the moon is not a star.

In Ex. 47–50,

let p: Omar passed English.

            q: Elaine passed English.

Translate the following statements into symbols. Then construct a truth table for each, and indicate under what conditions the compound statement is true.

47. Omar did not pass English but Elaine passed English.

48. Omar passed English and Elaine did not pass English.

49. Omar passed English or Elaine did not pass English.

50. Omar did not pass English or Elaine did not pass English.

In Ex. 51–54,

let p: The house is owned by an engineer.

            q: The heat is solar generated.

            r: The car is run by electric power.

Translate the following statements into symbols. Then construct a truth table for each, and indicate under what conditions the compound statement is true.

51. The house is owned by an engineer and the heat is solar generated, or the car is run by electric power.

52. The car is run by electric power or the heat is solar generated, but the house is owned by an engineer.

53. The heat is solar generated, or the house is owned by an engineer and the car is not run by electric power.

54. The house is not owned by an engineer, and the car is not run by electric power and the heat is solar generated.

In Ex. 55 and 56 read the requirements and each applicant's qualifications for obtaining a loan.

a) Identify which of the applicants would qualify for the loan.

b) For the applicants who do not qualify for the loan, explain why.

55. To qualify for a loan of $30,000 an applicant must have a gross income of $25,000 if single, $40,000 combined income if married, and assets of at least $5,000.

Mr. Vapmek, married with three children, makes $37,000 on his job. Mrs. Vapmek does not have an income. The Vapmeks have assets of $50,000.

Ms. Bell is not married, works in sales, and earns $29,000. She has assets of $8,000.

Mrs. Spatz and her husband have total assets of $35,000. One earns $25,000 and the other earns $23,500.

56. To qualify for a loan of $45,000 an applicant must have a gross income of $30,000 if single, $50,000 combined income if married, and assets of at least $10,000.

Mr. Argento, married with two children, makes $37,000 on his job. Mrs. Argento earns $15,000 at a part-time job. The Argentos have assets of $25,000.

Tina McVey, single, has assets of $19,000. She works in a store and earns $25,000.

Mr. Henke earns $24,000 and Mrs. Henke earns $28,000. Their assets total $8,000.

57. An airline advertisement states, "To get the special fare you must purchase your tickets between January 1 and February 15 and fly round trip between March 1 and April 1. You must depart on a Monday, Tuesday, or Wednesday, and return on a Tuesday, Wednesday, or Thursday, and stay over at least one Saturday.

a) Determine which of the following individuals will qualify for the special fare.

b) If the person does not qualify for the special fare, explain why.

Mr. James plans to purchase his ticket on January 15, depart on Monday, March 3, and return on Tuesday, March 18.

Ms. Vela plans to purchase her ticket on February 1, depart on Wednesday, March 10, and return on Thursday, April 2.

Mr. Davis plans to purchase his ticket on February 14, depart on Tuesday, March 5, and return on Monday, March 18.

Ms. Chang plans to purchase her ticket on January 4, depart on Monday, March 8, and return on Thursday, March 11.

Ms. Ghandi plans to purchase her ticket on January 1, depart on Monday, March 3, and return on Monday, March 10.

Research Activities

58. How does the computer make use of logic? Visit the computer center at your school, and ask the person in charge for specific references that can assist you in answering this question.

59. Digital computers use gates that work like switches to perform calculations. Information is fed into the gates and information leaves the gates, according to the type of gate. The three basic gates used in computers are the NOT gate, the AND gate, and the OR gate. Do research on the three types of gates.

a) Explain how each gate works.

b) Explain the relationship between each gate and the corresponding logic connectives *not*, *and*, and *or*.

c) Illustrate how two or more gates can be combined to form a more complex gate.

60. In addition to the NOT, AND, and OR gates, there are also NOR gates and NAND gates.

a) Do the necessary research, then explain how each gate works.

b) Explain the relationship between the NOR gate and the OR gate.

c) Explain the relationship between the NAND gate and the AND gate.

3.3 Truth Tables for the Conditional and Biconditional

Conditional

In Section 3.1 we mentioned that the statement preceding the conditional symbol is called the antecedent. The statement following the conditional symbol is called the **consequent**. For example, consider $(p \vee q) \rightarrow [\sim(q \wedge r)]$. In this statement, $(p \vee q)$ is the antecedent and $[\sim(q \wedge r)]$ is the consequent.

Now we will look at the truth table for the conditional. Consider the statement "If you get an A in class, then I will buy you a car." Assume that this statement is true except when I have actually broken my promise to you.

Let
p: You get an A.
q: I buy you a car.

Translated into symbolic form, the statement becomes $p \rightarrow q$. Let's examine the four cases shown in Table 3.11.

Case 1: (T, T) You get an A, and I buy a car for you. I have met my commitment, and the statement is true.

Case 2: (T, F) You get an A, and I do not buy a car for you. I have broken my promise, and the statement is false.

What happens if you don't get an A? If you don't get an A, I no longer have a commitment to you, and therefore I cannot break my promise.

Case 3: (F, T) You do not get an A, and I buy you a car. I have not broken my promise, and therefore the statement is true.

Case 4: (F, F) You do not get an A, and I don't buy you a car. I have not broken my promise, and therefore the statement is true.

The conditional statement is false when the antecedent is true and the consequent is false. In every other case the conditional statement is true.

> The **conditional statement** $p \rightarrow q$ is true in every case except when p is a true statement and q is a false statement.

Table 3.11 Conditional

p	q	$p \rightarrow q$
T	T	T
T	F	F
F	T	T
F	F	T

▶ **Example 1**

Construct a truth table for the statement $p \rightarrow \sim q$.

Solution: Fill out the truth table by placing the appropriate values under p, column 1, and under $\sim q$, column 2 (Table 3.12). Then, using the information given in the truth table for the conditional and the truth values in columns 1 and 2, determine the solution, column 3. In row 1 the antecedent, p, is true and the consequent, $\sim q$, is false. Row 1 is T → F, which is F. Row 2 is T → T, which is T. Row 3 is F → F, which is T. Row 4 is F → T, which is T.

Table 3.12

p	q	p	$\rightarrow$	$\sim q$
T	T	T	F	F
T	F	T	T	T
F	T	F	T	F
F	F	F	T	T
		1	3	2

▶ **Example 2**

Construct a truth table for the statement $p \rightarrow (q \wedge \sim r)$.

Solution: Work within the parentheses first. Place truth values under q, column 1, and $\sim r$, column 2 (Table 3.13). Then take the conjunction of columns 1 and 2 to obtain column 3. Place the values of p in column 4. To obtain the answer, column 5, use columns 3 and 4 and your knowledge of the conditional

Table 3.13

p	q	r	p	→	(q	∧	~r)
T	T	T	T	F	T	F	F
T	T	F	T	T	T	T	T
T	F	T	T	F	F	F	F
T	F	F	T	F	F	F	T
F	T	T	F	T	T	F	F
F	T	F	F	T	T	T	T
F	F	T	F	T	F	F	F
F	F	F	F	T	F	F	T
			4	5	1	3	2

statement. Column 4 represents the truth values of the antecedent, and column 3 represents the truth values of the consequent. Remember that the conditional is false only when the antecedent is true and the consequent is false, as in cases 1, 3, and 4 of column 5.

Every day you make decisions using logical reasoning about how to spend your time, energy, and money. The decisions you make result in series of trade-offs: If you spend $12 on a new CD, then you have $12 less to spend on your other needs; if you go to your early morning exercise class then you forfeit an hour that could be spent studying or sleeping.

▶ Example 3

An advertisement in a magazine makes the following claim: "If you use Sunshine Face Lotion, then your skin will be radiant and have no wrinkles." Translate the statement into symbolic form and construct a truth table.

Solution:

Let

        p: You use Sunshine Face Lotion
        q: Your skin will be radiant
        r: Your skin will have wrinkles.

In symbolic form the statement is

$$p \to (q \land \sim r)$$

This symbolic statement is identical to the statement in Table 3.13, and the truth tables are the same. In the table $(q \land \sim r)$, which represents "Your skin will be radiant and have no wrinkles," is true in cases 2 and 6 of column 3. In case 2, since *p* is true, Sunshine Face Lotion is used. But in case 6, since *p* is false, Sunshine Face Lotion is not used. From this information we can conclude that it is possible to get skin that is radiant and has no wrinkles without using Sunshine Face Lotion (case 6). Also by examining cases 1, 3, and 4 we see it is possible to use Sunshine Face Lotion and not have skin that is radiant and has no wrinkles.

A truth table alone cannot tell us if a statement is true or false. However, it can be used to examine the various possibilities.

Biconditional

The **biconditional statement**, $p \leftrightarrow q$, means $p \rightarrow q$ and $q \rightarrow p$. Symbolically, this is $(p \rightarrow q) \wedge (q \rightarrow p)$. To determine the truth table for $p \leftrightarrow q$, we will construct the truth table for $(p \rightarrow q) \wedge (q \rightarrow p)$ (Table 3.14). Table 3.15 shows the truth values for the biconditional statement.

Table 3.14

p	q	(p	$\rightarrow$	q)	$\wedge$	(q	$\rightarrow$	p)
T	T	T	T	T	T	T	T	T
T	F	T	F	F	F	F	T	T
F	T	F	T	T	F	T	F	F
F	F	F	T	F	T	F	T	F
		1	3	2	7	4	6	5

Table 3.15 Biconditional

p	q	$p \leftrightarrow q$
T	T	T
T	F	F
F	T	F
F	F	T

> The **biconditional statement**, $p \leftrightarrow q$, is true only when both p and q have the same truth value—that is, when both are true or both are false.

▶ Example 4

Construct a truth table for the statement $p \leftrightarrow (q \rightarrow \sim r)$.

Solution: Since there are three letters, there must be eight cases. The parentheses indicate that the answer must be under the biconditional (Table 3.16). Use columns 3 and 4 to obtain the answer, column 5. When columns 3 and 4 have the same truth values, place a T in column 5. When columns 3 and 4 have different truth values, places an F in column 5.

Table 3.16

p	q	r	p	$\leftrightarrow$	(q	$\rightarrow$	$\sim r$)
T	T	T	T	F	T	F	F
T	T	F	T	T	T	T	T
T	F	T	T	T	F	T	F
T	F	F	T	T	F	T	T
F	T	T	F	T	T	F	F
F	T	F	F	F	T	T	T
F	F	T	F	F	F	T	F
F	F	F	F	F	F	T	T
			4	5	1	3	2

In the previous section we illustrated that finding the truth value of a compound statement for a specific case does not require constructing an entire truth

"If you build a field, they will come. People will come, Ray. They'll come to Iowa for reasons they cannot fathom." So said James Earl Jones to Kevin Costner, who played Ray Kinsella in the film Field of Dreams. *He was right. As many as 1000 tourists a week have come to Dyersville, Iowa, to the ball field used in the making of the movie.*

table. Example 5 illustrates this technique using the conditional and the biconditional.

▶ **Example 5**

Determine the truth value of the statement $(q \leftrightarrow r) \to (\sim p \wedge r)$ when p is true, q is false, and r is true.

Solution: Substitute the truth value for each simple statement.

$$
\begin{array}{ccc}
(q \leftrightarrow r) & \to & (\sim p \wedge r) \\
(F \leftrightarrow T) & \to & (F \wedge T) \\
F & \to & F \\
& T &
\end{array}
$$

For this specific case the statement is true.

▶ **Example 6**

Determine the truth value for each simple statement. Then using the truth values determine the truth value of the compound statement.
a) If $3 + 3 = 8$, then $5 + 2 = 11$
b) George Washington was the first president of the United States and George Washington was born in Texas, if and only if Thomas Jefferson is credited with designing our monetary system.

Solution:
a) Let p: $3 + 3 = 8$
   q: $5 + 2 = 11$.

Then the statement "If $3 + 3 = 8$, then $5 + 2 = 11$" can be written $p \to q$. We know that p is a false statement and q is a false statement. Substitute F for p and F for q and evaluate the statement.

$$
\begin{array}{c}
p \to q \\
F \to F \\
T
\end{array}
$$

Therefore, "If $3 + 3 = 8$, then $5 + 2 = 11$" is a true statement.
b) Let p be "George Washington was the first president of the United States," q be "George Washington was born in Texas," and r be "Thomas Jefferson is credited with designing our monetary system." The compound statement can be written $(p \wedge q) \leftrightarrow r$. We know that p is a true statement. Statement q is false since Washington was born in Virginia. Statement r is true. Substitute T for p, F for q, and T for r and evaluate.

$$
\begin{array}{ccc}
(p \wedge q) & \leftrightarrow & r \\
(T \wedge F) & \leftrightarrow & T \\
F & \leftrightarrow & T \\
& F &
\end{array}
$$

Therefore the given compound statement is false.

Self-contradictions, Tautologies, and Implications

Two special situations can occur in the truth table of a compound statement: The statement may always be true or the statement may always be false. We give such statements special names.

> A **self-contradiction** is a compound statement that is always false.

When every truth value in the answer column of the truth table is false, then the statement is a self-contradiction.

▶ Example 7

Construct a truth table for the statement $(p \leftrightarrow q) \wedge (p \leftrightarrow \sim q)$.

Solution: See Table 3.17. In this example, the truth values are false in each case of column 5. This is an example of a self-contradiction or a *logically false statement*.

Table 3.17

p	q	$(p \leftrightarrow q)$	$\wedge$	$(p$	$\leftrightarrow$	$\sim q)$
T	T	T	F	T	F	F
T	F	F	F	T	T	T
F	T	F	F	F	T	F
F	F	T	F	F	F	T
		1	5	2	4	3

> A **tautology** is a compound statement that is always true.

When every truth value in the answer column of the truth table is true, then the statement is a tautology.

▶ Example 8

Construct a truth table for the statement $(p \wedge q) \rightarrow (p \vee r)$.

Solution: The answer is given in column 3 of truth Table 3.18. The truth values are true in every case. Thus the statement is an example of a tautology or a *logically true statement*.

Table 3.18

p	q	r	$(p \wedge q)$	$\rightarrow$	$(p \vee r)$
T	T	T	T	T	T
T	T	F	T	T	T
T	F	T	F	T	T
T	F	F	F	T	T
F	T	T	F	T	T
F	T	F	F	T	F
F	F	T	F	T	T
F	F	F	F	T	F
			1	3	2

"Heads I win, tails you lose." Do you think that this statement is a tautology, self-contradiction, or neither?

The conditional statement $(p \wedge q) \rightarrow (p \vee q)$ is a tautology. Conditional statements that are tautologies are called *implications*. In Example 8 we can say that $p \wedge q$ implies $p \vee r$.

> An **implication** is a conditional statement that is a tautology.

In any implication the antecedent of the conditional statement implies the consequent. This means that if the antecedent is true, then the consequent must also be true. That is, the consequent will be true whenever the antecedent is true.

▶ **Example 9**

Determine whether the conditional statement $[(p \wedge q) \wedge p] \to q$ is an implication.

Solution: If the truth table of the conditional statement is a tautology, the conditional statement is an implication. Since the truth table is a tautology (see Table 3.19), the conditional statement is an implication. The antecedent $[(p \wedge q) \wedge p]$ implies the consequent q. Note that the antecedent is true only in case 1 and the consequent is also true in case 1.

Table 3.19

p	q	$[(p \wedge q)$	$\wedge$	$p]$	$\to$	q
T	T	T	T	T	T	T
T	F	F	F	T	T	F
F	T	F	F	F	T	T
F	F	F	F	F	T	F
		1	3	2	5	4

Section 3.3 Exercises

Construct a truth table for each of the following statements.
1. $\sim p \to q$
2. $\sim q \to \sim p$
3. $\sim(q \leftrightarrow p)$
4. $\sim(p \to q)$
5. $\sim p \leftrightarrow q$
6. $(p \leftrightarrow q) \to p$
7. $p \leftrightarrow (q \vee p)$
8. $(\sim p \wedge q) \to \sim q$
9. $q \to (p \to \sim q)$
10. $(p \vee q) \leftrightarrow (p \wedge q)$

Construct a truth table for each of the following statements.
11. $p \to (q \vee r)$
12. $r \wedge (\sim p \to q)$
13. $r \leftrightarrow (q \wedge p)$
14. $(q \leftrightarrow p) \wedge \sim r$
15. $(p \vee \sim r) \leftrightarrow q$
16. $(p \vee r) \to (q \wedge r)$
17. $(\sim p \vee \sim q) \to r$
18. $[p \wedge (q \vee \sim r)] \leftrightarrow \sim p$
19. $(p \to q) \leftrightarrow (\sim q \to \sim r)$
20. $(\sim p \leftrightarrow \sim q) \to (\sim q \leftrightarrow r)$

In Ex. 21–26, write the statements in symbolic form. Then construct a truth table for each.
21. If the clouds fill the sky, then the stars are not visible and the moon is hidden.
22. If the class is interesting, then the time passes rapidly and it is fun to be in class.
23. The pizza will be delivered if and only if the driver finds the house, or the pizza will not be hot.
24. If it rains then the roof will leak, and if the sun shines then the roof will not leak.
25. If the bookstore is out of textbooks then I will use my friend's copy, or I will not read the assignment.
26. I will read the assignment if and only if it is less than 10 pages, or there is no good movie on television.

27. What is a tautology?

28. What is a self-contradiction?

29. What is an implication?

In Ex. 30–35, determine whether the statement is a tautology, a self-contradiction, or neither.

30. $\sim p \to q$

31. $(p \vee q) \vee \sim p$

32. $(p \wedge q) \wedge \sim p$

33. $(\sim p \to q) \vee \sim p$

34. $(\sim p \vee \sim q) \to p$

35. $[(p \wedge q) \wedge \sim r] \leftrightarrow [(p \vee q) \wedge r]$

In Ex. 36–41, determine if the statement is an implication.

36. $(p \vee q) \to q$

37. $q \to (p \vee q)$

38. $(p \wedge q) \to (q \wedge p)$

39. $(p \vee q) \to (p \vee \sim q)$

40. $[(p \to q) \wedge (q \to p)] \to (p \leftrightarrow q)$

41. $[(p \vee q) \wedge r] \to (p \vee q)$

In Ex. 42–51, if p is true, q is false, and r is true, find the truth value of the statement.

42. $\sim p \to (q \vee \sim r)$

43. $\sim p \to (q \wedge \sim r)$

44. $p \leftrightarrow (\sim q \wedge r)$

45. $(q \wedge \sim p) \to \sim r$

46. $(p \wedge \sim q) \wedge r$

47. $\sim [p \to (q \wedge r)]$

48. $(p \vee r) \leftrightarrow (p \wedge \sim q)$

49. $(\sim p \vee q) \to \sim r$

50. $(\sim p \leftrightarrow r) \vee (\sim q \leftrightarrow r)$

51. $(r \to \sim p) \wedge (q \to \sim r)$

In Ex. 52–59, determine the truth value for each simple statement. Then using the truth values determine the truth value of the compound statement.

52. If horses have tails, then ducks do not have wings.

53. If $6 + 13 = 18$, then $5 + 6 = 11$

54. 7 and 11 are odd numbers, if and only if $7 + 9 = 14$

55. If Paul McCartney is president of the United States, then Margaret Thatcher is the leader of Cuba or Florida is north of Georgia.

56. If the British Isles are in the Pacific Ocean and Hawaii is in the Atlantic Ocean, then $4^2 = 16$.

57. If John Adams was the second president of the United States and Thomas Jefferson was the third president of the United States, then Peter Pan was the sixteenth president of the United States.

58. The Empire State Building is in Lincoln, Nebraska, if and only if Alberta is one of the 48 states and the United Nations is the ruling body of Spain.

59. If a dollar has the same value as 100 pennies and 100 pennies has the same value as 21 nickels, then a dollar has the same value as 21 nickels.

In Ex. 60–65, suppose that both of these statements are true:

p: Amy received a grade of A on her final examination.

q: Amy passed the course.

Find the truth values of each of the following compound statements:

60. If Amy received a grade of A on the final examination, then she passed the course.

61. If Amy did not receive a grade of A on the final examination, then she did not pass the course.

62. If Amy did not receive a grade of A on the final examination, then she passed the course.

63. Amy passed the course if and only if she received a grade of A on the final examination.

64. Amy did not pass the course if and only if she received a grade of A on the final examination.

65. If Amy did not pass the course, then she did not receive a grade of A on the final examination.

66. Your father makes the statement: "If I get a raise in salary, then I will buy a new car." At the end of the semester you come home and there is a new car in the driveway. Can you conclude that your father got a raise in salary? Explain.

67. Consider the statement: "If your interview goes well, then you will be offered the job." If you are interviewed and then offered the job, can you conclude that your interview went well? Explain.

Problem Solving

68. The Barr triplets have an annoying habit: Whenever a question is asked of the three of them, two tell the truth and the third lies. When I asked them which of them was born last, they replied as follows:

Mary: Katie was born last.

Katie: I am the youngest.

Annie: Mary is the youngest.

Which of the Barr triplets was born last?

69. Helen Hurry shares a faculty office with three men: Sam Swift, Nick Nickelson, and Peter Perkins. One of the four teachers is a historian, another is a biologist, a third is a mathematician, and the fourth is a chemist.

According to the rumors that I heard when I was a student in the college: Professor Swift, who is the historian, and the biologist were fraternity brothers in college. The historian and the biologist were once legally married to each other. Professor Nickelson and the chemist are engaged to be married to each other. Helen and her boyfriend play tennis each week with the biologist and his wife. If exactly three of these four statements are true, what is each person's subject matter?

70. On this page is a photograph of a logic game at the Ontario Science Centre. There are 12 balls on top of the game board, numbered from left to right, with ball 1 on the extreme left and ball 12 on the extreme right. On

the platform in front of the players are 12 buttons, one corresponding to each of the balls. When six buttons are pushed, the six respective balls are released. When one or two balls reach an "and" or "or" gate, a single ball may, or may not, pass through the gate. The object of the game is to select a proper combination of six buttons that will allow one ball to reach the bottom. Using your knowledge of "and" and "or," select a combination of six buttons that will result in a win. Explain your answer.

71. Construct a truth table for (a) $(p \vee q) \to (r \wedge s)$ and (b) $(q \to \sim p) \vee (r \leftrightarrow s)$.

Research Activity

72. Select an advertisement from a newspaper or magazine that makes or implies a conditional statement. Analyze the advertisement to determine whether the consequent necessarily follows from the antecedent. Explain your answer. (See Example 3.)

3.4 Equivalent Statements and De Morgan's Laws

Equivalent statements are an important concept in the study of logic.

> Two statements are **equivalent** (symbolized ⇔*) if both statements have exactly the same truth values in the answer columns of the truth tables.

Sometimes the words **logically equivalent** are used in place of the word equivalent.

To determine whether two statements are equivalent, construct a truth table for each statement and compare the answer columns of the truth tables. If the answer columns are identical, then the statements are equivalent. If the answer columns are not identical, then the statements are not equivalent.

▶ **Example 1**

Show that the following two statements are equivalent.

$$[p \vee (q \vee r)] \qquad [(p \vee q) \vee r]$$

Solution: Construct a truth table for each statement (see Table 3.20). Since the truth tables have the same answer (column 3 for both tables), the statements are equivalent. Thus we can write

$$[p \vee (q \vee r)] \Leftrightarrow [(p \vee q) \vee r].$$

---

* The symbol ≡ is also used to indicate equivalent statements.

Table 3.20

p	q	r	[p	∨	(q ∨ r)]	[(p ∨ q)	∨	r]
T	T	T	T	T	T	T	T	T
T	T	F	T	T	T	T	T	F
T	F	T	T	T	T	T	T	T
T	F	F	T	T	F	T	T	F
F	T	T	F	T	T	T	T	T
F	T	F	F	T	T	T	T	F
F	F	T	F	T	T	F	T	T
F	F	F	F	F	F	F	F	F
			1	3	2	1	3	2

▶ **Example 2**

Determine whether the following statements are equivalent.

a) If you study hard and get a good night's sleep, then you will do well on the test.

b) If you do not study hard or do not get a good night's sleep, then you will not do well on the test.

Solution: Write each statement in symbolic form; then construct a truth table for each statement (Tables 3.21 and 3.22). If the answer columns of both tables are identical, then the statements are equivalent. If the answer columns are not identical, then the statements are not equivalent.

Let

    p: You study hard.
    q: You get a good night's sleep.
    r: You will do well on the test.

In symbolic form the statements are
a) $(p \land q) \rightarrow r$ and   **b)** $(\sim p \lor \sim q) \rightarrow \sim r$

Table 3.21

p	q	r	(p	∧	q)	→	r
T	T	T	T	T	T	T	T
T	T	F	T	T	T	F	F
T	F	T	T	F	F	T	T
T	F	F	T	F	F	T	F
F	T	T	F	F	T	T	T
F	T	F	F	F	T	T	F
F	F	T	F	F	F	T	T
F	F	F	F	F	F	T	F
			1	3	2	5	4

Table 3.22

p	q	r	(~p	∨	~q)	→	~r
T	T	T	F	F	F	T	F
T	T	F	F	F	F	T	T
T	F	T	F	T	T	F	F
T	F	F	F	T	T	T	T
F	T	T	T	T	F	F	F
F	T	F	T	T	F	T	T
F	F	T	T	T	T	F	F
F	F	F	T	T	T	T	T
			1	3	2	5	4

Since the answers in the columns labeled 5 are not identical, the statements are not equivalent.

▶ **Example 3**

Select the statement that is logically equivalent to "It is not true that the plane is both overbooked and departing late."
a) If the plane is not departing late, then the plane is not overbooked.
b) The plane is not overbooked or the plane is not departing late.
c) The plane is not departing late and the plane is not overbooked.
d) If the plane is not overbooked, then the plane is not departing late.

Did You

Know...

ADVENTURES IN LOGIC

One of the more interesting and well-known students of logic was Charles Dodgson (1832–1898), better known to us as Lewis Carroll, the author of *Alice's Adventures in Wonderland* and *Through the Looking-Glass*.

Although the books have a child's point of view, many would argue that the audience best equipped to enjoy them is an adult one. Dodgson, a mathematician, logician, and photographer (among other things), uses the naïveté of a seven-year-old girl to show what can happen when the rules of logic are taken to absurd extremes.

You should say what you mean," the March Hare went on.
"I do," Alice hastily replied; "at least—at least I mean what I say—that's the same thing, you know."
"Not the same thing a bit!" said the Hatter. "You might as well say that 'I see what I eat' is the same thing as 'I eat what I see'!"

Table 3.23(a)

p	q	~	(p	∧	q)
T	T	F	T	T	T
T	F	T	T	F	F
F	T	T	F	F	T
F	F	T	F	F	F

Solution: To determine if any of the choices are equivalent to the given statement write the given statement, and the choices in symbolic form. Then construct and compare their truth tables.

Let p: The plane is overbooked.
                          q: The plane is departing late.

The given statement may be written "It is not true that the plane is overbooked and the plane is departing late." The statement is expressed in symbolic form as $\sim(p \wedge q)$. Using p and q as indicated, choices (a–d) may be expressed symbolically as

a) $\sim q \to \sim p$ **b)** $\sim p \vee \sim q$ **c)** $\sim q \wedge \sim p$ **d)** $\sim p \to \sim q$.

Now construct a truth table for the given statement (Table 3.23a) and each possible choice (Table 3.23b–d).

Table 3.23(b)

p	q	(a) $\sim q$	$\to$	$\sim p$	(b) $\sim p$	$\vee$	$\sim q$	(c) $\sim q$	$\wedge$	$\sim p$	(d) $\sim p$	$\to$	$\sim q$
T	T	F	T	F	F	F	F	F	F	F	F	T	F
T	F	T	F	F	F	T	T	T	F	F	F	T	T
F	T	F	T	T	T	T	F	F	F	T	T	F	F
F	F	T	T	T	T	T	T	T	T	T	T	T	T

Examining the truth tables we see that the given statement, $\sim(p \wedge q)$, is logically equivalent to choice (b), $\sim p \vee \sim q$. Therefore the correct answer is "The plane is not overbooked or the plane is not departing late." This is logically equivalent to "It is not true that the plane is both overbooked and departing late."

De Morgan's Laws

Example 3 showed that a statement of the form $\sim(p \wedge q)$ is equivalent to a statement of the form $\sim p \vee \sim q$. Thus we may write, $\sim(p \wedge q) \Leftrightarrow \sim p \vee \sim q$. This equivalent statement is one of two special laws called De Morgan's laws. The laws, named after Augustus De Morgan, an English mathematician, were first introduced in Chapter 2, where they applied to sets.

> **De Morgan's Laws**
> **1.** $\sim(p \wedge q) \Leftrightarrow \sim p \vee \sim q$
> **2.** $\sim(p \vee q) \Leftrightarrow \sim p \wedge \sim q$

You can demonstrate that De Morgan's second law is true by constructing and comparing truth tables for $\sim(p \vee q)$ and $\sim p \wedge \sim q$.

Augustus De Morgan
(1806–1871), the son of a member of the East India Company, was born in India and educated at Trinity College, Cambridge. One of the great reformers of logic in the nineteenth century, De Morgan made his greatest contribution to the subject by realizing that logic as it had come down from Aristotle was narrow in scope and could be applied to a wider range of arguments. His work laid the foundation for modern, symbolic logic.

When using De Morgan's Laws, if it becomes necessary to negate an already negated statement, make use of the fact that $\sim(\sim p) = p$. For example the negation of the statement "Today is not Monday" is "Today is Monday."

▶ **Example 4**

Using De Morgan's laws, write an equivalent statement for "The car is too long or the garage is too short."

Solution: The first step is to write the statement in symbolic form.

Let
p: The car is too long.
q: The garage is too short.

The statement in symbolic form is $p \lor q$. To create an equivalent statement using De Morgan's laws, we do the following: (1) Negate the complete statement, $\sim(p \lor q)$. (2) Negate each statement inside the parentheses, $\sim(\sim p \lor \sim q)$. (3) Change the $\lor$ to an $\land$. Thus $p \lor q \Leftrightarrow \sim(\sim p \land \sim q)$. The equivalent statement reads: "It is false that the car is not too long and the garage is not too short."

▶ **Example 5**

Select the statement that is logically equivalent to "The sun is not shining but it is not raining."
a) It is not true that the sun is shining and it is raining.
b) It is not raining or the sun is not shining.
c) The sun is shining or it is raining.
d) It is not true that the sun is shining or it is raining.

Solution: To determine which statement is equivalent, write each statement in symbolic form.

Let
p: The sun is shining.
q: It is raining.

The statement "The sun is not shining but it is not raining" written symbolically is $\sim p \land \sim q$. Remember, the word "but" means the same as the word "and." Write parts (a) through (d) symbolically.
a) $\sim(p \land q)$ **b)** $\sim q \lor \sim p$ **c)** $p \lor q$ **d)** $\sim(p \lor q)$
De Morgan's law shows that $\sim p \land \sim q$ is equivalent to $\sim(p \lor q)$. Therefore, the answer is (d), "It is not true that the sun is shining or it is raining."

▶ **Example 6**

Write a statement that is logically equivalent to "It is not true that the microwave must be set at level 6 or you will burn the food."

Solution: This statement is of the form $\sim(p \lor q)$. An equivalent statement using De Morgan's laws is $\sim p \land \sim q$. Therefore, an equivalent statement is "The microwave must not be set at level 6 and you will not burn the food."

There are strong similarities between the topics of sets and logic. We can see them by examining De Morgan's laws for sets and logic.

De Morgan's Laws: Set Theory	*De Morgan's Laws: Logic*
$(A \cap B)' = A' \cup B'$	$\sim(p \wedge q) \Leftrightarrow \sim p \vee \sim q$
$(A \cup B)' = A' \cap B'$	$\sim(p \vee q) \Leftrightarrow \sim p \wedge \sim q$

The complement in set theory, $'$, is similar to the negation, $\sim$, in logic. The intersection, $\cap$, is similar to the conjunction, $\wedge$; and the union, $\cup$, is similar to the disjunction, $\vee$. If we were to interchange the set symbols with the logic symbols, De Morgan's laws would remain, but in a different form.

Both $'$ and $\sim$ can be interpreted as "not."
Both $\cap$ and $\wedge$ can be interpreted as "and."
Both $\cup$ and $\vee$ can be interpreted as "or."

For example, the set statement $A' \cup B$ can be written as a statement in logic as $\sim a \vee b$.

Statements containing connectives other than "and" and "or" may have equivalent statements. For example, if you construct truth tables for $p \rightarrow q$ and $\sim p \vee q$ you will see that the truth tables have the same answer columns and therefore the statements are equivalent. That is,

$$p \rightarrow q \Leftrightarrow \sim p \vee q.$$

With these equivalent statements we can write a conditional statement as a disjunction or a disjunction as a conditional statement. For example, the statement "If the game is polo then you ride a horse" can be equivalently stated as "The game is not polo or you ride a horse."

To change a conditional statement to a disjunction: Negate the antecedent, change the conditional symbol to a disjunction symbol, and keep the consequent the same. To change a disjunction statement to a conditional statement: Negate the first statement, change the disjunction symbol to a conditional symbol, and keep the second statement the same.

▶ **Example 7**

Write a conditional statement that is logically equivalent to "The Eagles will win or the Lions will lose." Assume the negation of winning is losing.

Solution:

Let

p: The Eagles will win.
q: The Lions will win.

The statement may be written symbolically as $p \vee \sim q$. To write an equivalent statement negate the first statement, p, and change the disjunction symbol to a conditional symbol. Symbolically the statement is $\sim p \rightarrow \sim q$. The equivalent statement is "If the Eagles lose then the Lions will lose."

In the previous section we showed that $p \leftrightarrow q$ has the same truth table as $(p \rightarrow q) \wedge (q \rightarrow p)$. Therefore these statements are equivalent, a useful fact for Example 8.

▶ **Example 8**

Write the following statement as an equivalent biconditional statement. "If the figure is a triangle then the figure has three angles and if the figure has three angles then the figure is a triangle."

Solution: An equivalent statement is "A figure is a triangle if and only if the figure has three angles."

Section 3.4 Exercises

1. What are equivalent statements?
2. Explain how you can determine whether two statements are equivalent.
3. Suppose two statements are connected with the biconditional, and the truth table is constructed. If the answer column of the truth table has all trues, what must be true about these two statements? Explain.
4. Write De Morgan's laws for logic.

In Ex. 5–14, use De Morgan's laws to determine whether the two statements are equivalent.

5. $\sim p \wedge \sim q$, $\sim(p \wedge q)$
6. $\sim(p \vee q)$, $\sim p \wedge \sim q$
7. $\sim(p \wedge q)$, $\sim p \wedge q$
8. $\sim p \wedge q$, $\sim(\sim p \vee q)$
9. $\sim(p \wedge q)$, $\sim p \vee \sim q$
10. $\sim(p \wedge \sim q)$, $\sim p \vee q$
11. $\sim p \wedge q$, $\sim(q \vee \sim p)$
12. $\sim(p \wedge q) \rightarrow r$, $(\sim p \vee \sim q) \rightarrow r$
13. $q \rightarrow \sim(p \wedge \sim r)$, $q \rightarrow \sim p \vee r$
14. $p \leftrightarrow (q \vee \sim r)$, $p \leftrightarrow \sim(\sim q \wedge r)$

In Ex. 15–24, use a truth table to determine whether the two statements are equivalent.

15. $p \rightarrow q$, $\sim p \vee q$
16. $\sim p \rightarrow q$, $p \wedge q$

17. $p \rightarrow q$, $\sim q \rightarrow \sim p$
18. $(p \vee q) \vee r$, $p \vee (q \vee r)$
19. $(p \wedge q) \wedge r$, $p \wedge (q \wedge r)$
20. $\sim p \rightarrow (q \wedge r)$, $p \vee (q \wedge r)$
21. $p \leftrightarrow (q \wedge \sim r)$, $p \rightarrow (q \vee r)$
22. $(p \rightarrow q) \wedge (q \rightarrow r)$, $(p \rightarrow q) \rightarrow r$
23. $\sim(q \rightarrow p) \vee r$, $(p \vee q) \wedge \sim r$
24. $\sim q \rightarrow (p \wedge r)$, $\sim(p \vee r) \rightarrow q$

25. Show that $(p \rightarrow q) \wedge (q \rightarrow p) \Leftrightarrow (p \leftrightarrow q)$.
26. Determine whether or not $[\sim(p \rightarrow q)] \wedge [\sim(q \rightarrow p)] \Leftrightarrow \sim(p \leftrightarrow q)$.

In Ex. 27–38, determine whether the pairs of statements are equivalent. You may use De Morgan's laws, the fact that $p \rightarrow q \Leftrightarrow \sim p \vee q$, or truth tables.

27. **a)** It is not true that the song is a hit or the rap group is Salt-N-Pepa.
    b) The song is not a hit and the rap group is not Salt-N-Pepa
28. **a)** It is false that the light bulb uses a lot of energy and the toaster does not use a lot of energy.
    b) The light bulb does not use a lot of energy or the toaster uses a lot of energy.
29. **a)** The Atlanta Braves are not in first place or the Kansas City Royals are in first place.

b) It is not true that the Atlanta Braves are in first place and the Kansas City Royals are not in first place.

30. a) If it is raining, then the flag must not be flying.

b) It is not raining or the flag must not be flying.

31. a) If you pay your credit card bill within 30 days, then you will not be assessed a penalty.

b) Pay your credit card bill within 30 days or you will not be assessed a penalty

32. a) If *Spy* magazine is at the newsstand, then I will buy the magazine.

b) *Spy* magazine is at the newsstand or I will buy the magazine.

33. a) The sales tax is 7% if and only if the tax on a $200 purchase is $14.

b) If the tax on a $200 purchase is $14, then the sales tax is 7%.

34. a) If the razor blade is sharp, then you will get a good shave or you will get cut.

b) If the razor blade is not sharp, then you will not get a good shave and you may not get cut.

35. a) If the grass is fertilized or it rains, then the grass will grow.

b) If the grass grows, then the grass is fertilized or it has rained.

36. a) If you are 18 years of age and a citizen of the United States, then you can vote in the presidential election.

b) If you vote in the presidential election, then it is not true that you are not 18 years of age or not a citizen of the United States.

37. a) You can receive a scholarship, and if you go to college then you will get a degree.

b) You can receive a scholarship, or if you go to college then you will not get a degree.

38. a) The television is not broken, but if we lose power then we will not be able to watch television.

b) If we are not able to watch television, then it is not true that the television is broken and we lost power.

In Ex. 39–46, use De Morgan's laws to write an equivalent statement for each of the sentences.

39. It is not true that both Neil Diamond and AC/DC sang "Cherry Cherry."

40. It is not true that the fire is too low and the water will not boil.

41. The new carpet is treated with Scotch Guard or it is not a good carpet.

42. I did not study and I did not get a good night's sleep.

43. The figure is neither a triangle nor a square.

44. If the treadmill is new, then it is not true that the treadmill is not under warranty or it will not work for years.

45. If the house does not need painting, then the siding does not need repair or we will be able to sell the house.

46. We will travel to Washington, D.C., or we will not visit the Smithsonian Institution and we will not visit the White House.

In Ex. 47–52, use the fact that $p \to q$ is equivalent to $\sim p \lor q$ to write each statement in an equivalent form.

47. If the wheat is yellow, then it is ready to be cut.

48. The construction crew is working or it is snowing.

49. If the frog has legs, then it is no longer a tadpole.

50. Ray will watch either the CNN or the ABC newscast.

51. The mail will not be delivered today or today is not Sunday.

52. The fish are biting or I will not go fishing.

In Ex. 53–56, write an equivalent statement.

53. If a figure is a quadrilateral then it has four sides, and if a figure has four sides then it is a quadrilateral.

54. If you drive fast then you will use more gas and if you use more gas then you drive fast.

55. The bird will fly if and only if it has wings.

56. You need to pay taxes if and only if you receive income.

3.5 The Conditional

Forms of the Conditional Statement

The conditional may be stated in many different ways both in logic and in everyday life. Some of the different forms of the conditional statement are given in the list on the following page. Each of the forms listed means $p \to q$.

Impossible Connections

When we look at the work of the Dutch graphic artist M. C. Escher (1902–1972), we can see that Escher takes great pleasure in defying visual logic. He creates paradoxical worlds in which up is down and front is back. In Waterfall, the water drops into a channel that appears to move down and away from us, only to end up where it began. In creating the illusion of a three-dimensional image in just two dimensions, Escher made two impossible connections. If you allow connection 1, as shown in the second diagram, connections 2 and 3 cannot follow—at least not in a three-dimensional world.

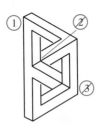

Forms of the Conditional Statement

Statement	*Symbolic Form*
If p, then q	$p \to q$
p only if q	$p \to q$
q if p	$p \to q$
p is a sufficient condition for q	$p \to q$
q is a necessary condition for p	$p \to q$

Let's examine some illustrations of the different forms of the conditional. Consider these two statements:

p: You live in Los Angeles.
q: You live in California.

We can join these two statements together in many different ways.

1. If you live in Los Angeles, then you live in California.
2. You live in Los Angeles only if you live in California.
3. You live in California if you live in Los Angeles.
4. You live in Los Angeles is a sufficient condition for you to live in California.
5. You live in California is a necessary condition for you to live in Los Angeles.

Each of these statements can be represented as $p \to q$.

▶ **Example 1**

To be declared legally intoxicated in New York State it is necessary for your blood alcohol level be at least 0.10. Write this statement in the if-then form.

Solution: If your blood alcohol level is at least 0.10, then you can be declared legally intoxicated in New York State.

▶ **Example 2**

Being the youngest of four children is sufficient for you to be the recipient of many jokes. Write this statement in the if-then form.

Solution: If you are the youngest of four children, then you will be the recipient of many jokes.

Variations of the Conditional

Having looked at the alternative ways to express a conditional, let us now consider *variations of the conditional statement*. The variations are the **converse** of the conditional, the **inverse** of the conditional, and the **contrapositive** of the conditional. Do not confuse the *variations* of the conditional statement with the *forms* of the conditional statement just discussed. The variations of the conditional are made by switching and/or negating the antecedent and the consequent of a conditional statement.

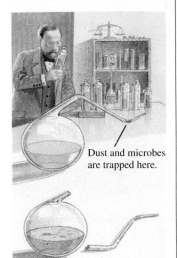

Dust and microbes
are trapped here.

Logic is often used by scientists
when developing theories. Sometimes their theories are supported
by experimental evidence and
thus accepted; other times their
theories are contradicted by the
evidence and rejected. For
example, Aristotle observed that
bugs appeared in spoiling meat
and reasoned that life arose
spontaneously from nonliving
matter. With advances in technology, scientists have had more
means at their disposal to test
their theories. Logical reasoning
often plays a part in developing
procedures to test theories. To
refute Aristotle's claim of spon-
taneous generation of life, Louis
Pasteur in 1862 conducted an
experiment by isolating some
meat broth in a sterile flask in
order to demonstrate that the
bugs Aristotle observed grew
from microscopic life forms too
small to be seen.

Listed here are the variations of the conditional with their symbolic form and the words we say to read each one.

Variations of the Conditional Statement

Name	*Symbolic Form*	*Read*
Conditional	$p \rightarrow q$	If p, then q.
Converse of the conditional	$q \rightarrow p$	If q, then p.
Inverse of the conditional	$\sim p \rightarrow \sim q$	If not p, then not q.
Contrapositive of the conditional	$\sim q \rightarrow \sim p$	If not q, then not p.

From the chart we see that to write the converse of the conditional state-ment, switch the order of the antecedent and the consequent. To write the inverse, negate both the antecedent and the consequent. To write the contrapositive, switch the order of the antecedent and the consequent and then negate both of them.

Are any of the variations of the conditional statement equivalent? To deter-mine the answer we can construct a truth table for each variation (Table 3.24). Examining Table 3.24 reveals that the conditional statement is equivalent to the contrapositive statement, and that the converse statement is equivalent to the inverse statement.

Table 3.24

p	q	Conditional $p \rightarrow q$	Contrapositive $\sim q \rightarrow \sim p$	Converse $q \rightarrow p$	Inverse $\sim p \rightarrow \sim q$
T	T	T	T	T	T
T	F	F	F	T	T
F	T	T	T	F	F
F	F	T	T	T	T

▶ **Example 3**

For the conditional statement "If there is a space station, then we will explore outer space." Write the
a) converse   **b)** inverse   **c)** contrapositive.

Solution:
a) If we explore outer space, then there is a space station.
b) If there is no space station, then we will not explore outer space.
c) If we do not explore outer space, then there is no space station.

▶ **Example 4**

For p and q as follows, write statements (a–d) and determine which are true.
a) the conditional statement, $p \rightarrow q$ **b)** the converse of $p \rightarrow q$

c) the inverse of $p \to q$ **d)** the contrapositive of $p \to q$

     p: The number is divisible by 9.
     q: The number is divisible by 3.

Solution:

a) *Conditional statement:* $(p \to q)$
If the number is divisible by 9, then the number is divisible by 3. This statement is true. A number divisible by 9 must also be divisible by 3, since 3 is a divisor of 9.

b) *Converse of the conditional:* $(q \to p)$
If the number is divisible by 3, then the number is divisible by 9. This statement is false. For instance: 6 is divisible by 3, but 6 is not divisible by 9.

c) *Inverse of the conditional:* $(\sim p \to \sim q)$
If the number is not divisible by 9, then the number is not divisible by 3. This statement is false. For instance: 6 is not divisible by 9, but 6 is divisible by 3.

d) *Contrapositive of the conditional:* $(\sim q \to \sim p)$
If the number is not divisible by 3, then the number is not divisible by 9. The statement is true, since any number that is divisible by 9 must be divisible by 3.

▶ **Example 5**

Use the contrapositive to write a statement logically equivalent to "If the boat is 24 ft long, then it will not fit into the boathouse."

Solution:

Let p: The boat is 24 ft long.
                 q: The boat will fit into the boathouse.

The given statement written symbolically is

$$p \to \sim q.$$

The contrapositive of the statement is

$$q \to \sim p.$$

Therefore, an equivalent statement is "If the boat will fit into the boathouse, then the boat is not 24 ft long.

     The contrapositive of the conditional is very important in mathematics. Consider the statement "If a^2 is not a whole number, then a is not a whole number." Is this statement true? You may find this question difficult to answer. Writing the statement's contrapositive may enable you to answer the question. The contrapositive is "If a is a whole number, then a^2 is a whole number." Since the contrapositive is a true statement, the original statement must also be true.

Section 3.5 Exercises

1. For a statement of the form if *p*, then *q*, symbolically indicate the form of the
   a) converse   b) inverse   c) contrapositive.
2. Which of the following are equivalent statements?
   a) the converse   b) the contrapositive
   c) the inverse   d) the conditional

In Ex. 3–10 determine which, if any, of the three statements are equivalent.

3. a) If it is raining, then I will use an umbrella.
   b) I will use an umbrella if it is raining.
   c) I will use an umbrella only if it is raining.
4. a) The signal light being red is sufficient for you to stop the car.
   b) You stop the car if the signal light is red.
   c) The signal light being red is a necessary condition for you to stop the car.
5. a) You will get to your destination on time only if you leave by 9:00 A.M.
   b) You have left by 9:00 A.M. if you get to your destination on time.
   c) Leaving by 9:00 A.M. is necessary to get to your destination on time.
6. a) You need to pay a luxury tax if the car costs over $30,000.
   b) If the car costs over $30,000, then you will need to pay a luxury tax.
   c) You need to pay a luxury tax only if the car costs over $30,000.
7. a) Less than three days having passed is necessary for cancelling the contract without penalty.
   b) A contract may be cancelled without penalty if less than three days have passed.
   c) Less than three days having passed is sufficient for cancelling the contract without penalty.
8. a) I will go swimming if the water temperature is at least 80° Fahrenheit.
   b) I will go swimming only if the water temperature is at least 80° Fahrenheit.
   c) If the water temperature is at least 80° Fahrenheit, then I will go swimming.
9. a) The album being *The Wall* is necessary for the group to be Pink Floyd.
   b) If the album is *The Wall*, then the group is Pink Floyd.
   c) The album is *The Wall* if the group is Pink Floyd.
10. a) We will use the microwave only if the food is frozen.
    b) Using the microwave is necessary when the food is frozen.
    c) Using the microwave is sufficient when the food is frozen.

In Ex. 11–22, rewrite the statements in if-then form.

11. You can buy the paper bag only if you have 50 cents.
12. I will go skiing if at least three trails are open.
13. I will go hunting if it does not snow.
14. I will go shopping only if I have been paid.
15. Renting an apartment is sufficient for buying rental insurance.
16. Starting a fire in the wood-burning stove is sufficient for staying warm.
17. Buying a parachute is necessary for hang gliding.
18. Planting the seed is necessary for digging a garden.
19. I will eat my hat if you get an A.
20. Chewing food properly is sufficient for a healthy life.
21. Getting paid is necessary for going to work.
22. I will go to the movie only if the movie is a thriller.

In Ex. 23–32, write the converse, inverse, and contrapositive of the given statement. (For Ex. 30–32, use De Morgan's laws.)

23. If today is Friday, then we will party.
24. If the car is red, then the owner has a lot of money.
25. If the score on the test is greater than 70, then I will pass the course.
26. If the paper is blank, then I froze during the test.
27. If the pear tree bears fruit, then I'll be a monkey's uncle.
28. If there is a full moon, then the werewolves will howl.
29. If the restaurant is clean, then I will enjoy the food.
30. If Carmen is a police officer or a race car driver, then Carmen is in a dangerous occupation.
31. If the figure is a rhombus, then the opposite sides are parallel and equal.
32. If I buy a house or rent an apartment, then I will make monthly payments.

In Ex. 33–39, write the contrapositive of the statement. Use the contrapositive to determine whether the conditional statement is true or false.

33. If the number is not divisible by 3, then the sum of the digits is not divisible by 3.
34. If the triangle is not isosceles, then two angles of the triangle are not equal.

35. If 2 does not divide the counting number, then 2 does not divide the units digit of the counting number.

36. If $1/n$ is not a natural number, then n is not a natural number.

37. If two lines do not intersect in at least one point, then the two lines are parallel.

38. If $\dfrac{m \cdot a}{m \cdot b} \neq \dfrac{a}{b}$, then m is not a counting number.

39. If a and b are not both even counting numbers, then the product of a and b is not an even counting number.

40. Can a conditional statement and its converse both be false statements? Explain your answer with the use of an example.

41. Can a conditional statement be true if its converse is false? Explain your answer with the use of an example.

42. Can a conditional statement and its contrapositive both be false? Explain your answer with the use of an example.

43. Can a conditional statement be true if its contrapositive is false? Explain.

3.6 Symbolic Arguments

In previous sections in this chapter symbolic logic was used to determine the truth value of a compound statement. We will now extend those basic ideas to determine whether symbolic arguments are valid or invalid.

Consider the following statements.

> If Jason is a singer, then he is well known.
> Jason is a singer.

If you accept these two statements as true, then a conclusion that necessarily follows is

> Jason is well known.

The three statements in the following form constitute a **symbolic argument**.

A symbolic argument consists of a set of **premises** and a **conclusion**. It is called a symbolic argument since we generally write the argument in symbolic form to determine its validity.

Premise 1:	If Jason is a singer, then he is well known.
Premise 2:	Jason is a singer.
Conclusion:	Therefore Jason is well known.

> An **argument is valid** when its conclusion necessarily follows from a given set of premises.
>
> An **argument is invalid** or a **fallacy** when the conclusion does not necessarily follow from the given set of premises.

An argument that is not valid is invalid. The argument just presented is an example of a valid argument, since the conclusion necessarily follows from the premises. Now we will discuss a procedure to determine if an argument is valid or invalid. We begin by writing the argument in symbolic form. To write the argument in symbolic form we let p and q be as follows:

p: Jason is a singer.

q: Jason is well known.

Symbolically, the argument is written

Premise 1: $p \rightarrow q$
Premise 2: $\underline{p}$
Conclusion: $\therefore q$ (The three-dot triangle is read "therefore.")

Write the argument in the following form.

$$\text{If } [\text{ premise 1 \textbf{and} premise 2}] \textbf{ then } \text{conclusion}$$
$$[\ (p \rightarrow q)\ \wedge\ p\ ]\ \rightarrow\ q$$

Then construct a truth table for the statement $[(p \rightarrow q) \wedge p] \rightarrow q$ (Table 3.25). *If the truth table answer column is true in every case, then the statement is a tautology, and the argument is valid. If the truth table is not a tautology, then the argument is invalid.* Since the statement is a tautology the conclusion necessarily follows from the premises and the argument is valid.

In the case of the disappearance of the racehorse Silver Blaze, Sherlock Holmes demonstrated that sometimes the absence of a clue is itself a clue. The local police inspector asked him, "Is there any point to which you would wish to draw my attention?" Holmes replied, "To the curious incident of the dog in the nighttime." The inspector, confused, asked: "The dog did nothing in the nighttime." "That was the curious incident," remarked Sherlock Holmes. From the lack of the dog's bark, Holmes concluded that the horse had been "stolen" by a stable-hand. How did Holmes reach his conclusion?

Table 3.25

p	q	$[(p$	$\rightarrow$	$q)$	$\wedge$	$p]$	$\rightarrow$	q
T	T		T		T	T	T	T
T	F		F		F	T	T	F
F	T		T		F	F	T	T
F	F		T		F	F	T	F
		1			3	2	5	4

Once we have demonstrated that an argument in a particular form is valid, all arguments with exactly the same form will also be valid. In fact, many of these forms have been assigned names. The argument form just discussed,

$$p \rightarrow q$$
$$\underline{p}$$
$$\therefore q$$

is called the **law of detachment** (or *modus ponens*).

Consider the following argument.

If flowers smell sweet, then the honey is bitter.
Flowers smell sweet.
$\overline{\therefore \text{ Honey is bitter.}}$

Now translate the argument into symbolic form.

Let
s: Flowers smell sweet.
b: Honey is bitter.

In symbolic form the argument is

$$s \rightarrow b$$
$$\underline{s}$$
$$\therefore b$$

This argument is also the law of detachment, and therefore it is a valid argument.

Notice that the argument is valid even though the conclusion, "Honey is bitter," is a false statement. It is also possible to have an invalid argument in which the conclusion is a true statement. When an argument is valid, it means that the conclusion necessarily follows from the premises. It is not necessary for the premises or the conclusion to be true statements in an argument.

From ancient times well into the Renaissance, 1 was not considered a number, but rather the builder of numbers. Two was considered to be the first real number. The mathematician-engineer Simon Stevin (1548—1620) offered the following counterargument:

"If from a number is subtracted no number, the number remains; but if from 3 we take 1, 3 does not remain; hence, 1 is not no number."

Procedure to Determine Whether an Argument Is Valid

1. Write the argument in symbolic form.
2. Compare the form of the argument with forms that are known to be valid or invalid. If there are no known forms to compare then go on to step 3.
3. Write a conditional statement of the form

(conjunction of the premises) → conclusion.

4. Construct a truth table for the statement in step 3.
5. If the answer column of the truth table has all trues, then the statement is a tautology, and the argument is valid. If the answer column does not have all trues, then the argument is invalid.

▶ **Example 1**

Determine whether the following argument is valid or invalid.

If Tim drives the truck, then the shipment will be delivered.
The shipment will not be delivered.
∴ Tim does not drive the truck.

Solution: We first write the argument in symbolic form.

Let p: Tim drives the truck.
   q: The shipment will be delivered.

In symbolic form the argument is

$$p \rightarrow q$$
$$\underline{\sim q}$$
$$\therefore \sim p$$

Since we have not tested an argument in this form, we will construct a truth table to determine whether it is valid or invalid.

Write the argument in the form $[(p \rightarrow q) \wedge \sim q] \rightarrow \sim p$ and construct a truth table (Table 3.26). Since the answer, column 5, has all Ts, the argument is valid.

Table 3.26

p	q	[(p → q)	∧	~q]	→	~p
T	T	T	F	F	T	F
T	F	F	F	T	T	F
F	T	T	F	F	T	T
F	F	T	T	T	T	T
		1	3	2	5	4

The argument form in Example 1 is an example of the **law of contraposition** (or *modus tollens*)

▶ **Example 2**

Determine whether the following argument is valid or invalid.

> The car is red or the car is new.
> The car is new.
> ∴ The car is red.

Solution:

Let
    p: The car is red.
    q: The car is new.

In symbolic form the argument is

$$p \vee q$$
$$\underline{q}$$
$$\therefore p$$

Since this is not one of the forms we are familiar with we will construct a truth table. Write the argument in the form $[(p \vee q) \wedge q] \to p$. Next construct a truth table, as shown in Table 3.27. The answer to the truth table, column 5, is not true in *every case*. Therefore the statement is not a tautology, and the argument is invalid, or is a fallacy.

Table 3.27

p	q	[(p ∨ q)	∧	q]	→	p
T	T	T	T	T	T	T
T	F	T	F	F	T	T
F	T	T	T	T	F	F
F	F	F	F	F	T	F
		1	3	2	5	4

Standard forms of commonly used arguments are given in the following chart.

Standard Forms of Arguments

Valid Arguments	*Law of Detachment*	*Law of Contraposition*	*Law of Syllogism*	*Disjunctive Syllogism*
	$p \rightarrow q$	$p \rightarrow q$	$p \rightarrow q$	$p \vee q$
	p	$\sim q$	$q \rightarrow r$	$\sim p$
	$\therefore q$	$\therefore \sim p$	$\therefore p \rightarrow r$	$\therefore q$

Invalid Arguments	*Fallacy of the Converse*	*Fallacy of the Inverse*
	$p \rightarrow q$	$p \rightarrow q$
	q	$\sim p$
	$\therefore p$	$\therefore \sim q$

▶ **Example 3**

Determine whether the argument is valid or invalid.

> If I drink hot milk, then I can sleep.
> If I can sleep, then I can dream.
> $\therefore$ If I drink hot milk, then I can dream.

Solution:

Let

	p:	I drink hot milk.
	q:	I can sleep.
	r:	I can dream.

In symbolic form the argument is

$$
\begin{array}{c}
p \rightarrow q \\
q \rightarrow r \\
\hline
\therefore p \rightarrow r
\end{array}
$$

The argument is in the form of the law of syllogism. Therefore the argument is valid and there is no need to construct a truth table.

Now we consider an argument that has more than two premises.

▶ **Example 4**

Use a truth table to determine whether the following argument is valid or invalid.

> If the ice cream is soft, then it tastes better.
> The ice cream tastes better or it is fruit flavored.
> The ice cream is fruit flavored or the ice cream is soft.
> $\therefore$ The ice cream is soft.

Solution:

Let

p: The ice cream is soft.
q: The ice cream tastes better.
r: The ice cream is fruit flavored.

In symbolic form the argument is

$$p \to q$$
$$q \lor r$$
$$\underline{r \lor p}$$
$$\therefore p$$

Write the argument in the form

$$[(p \to q) \land (q \lor r) \land (r \lor p)] \to p.$$

Then construct the truth table (Table 3.28). The answer, column 7, is not true in every case. Thus the argument is a fallacy.

Table 3.28

p	q	r	[(p → q)	∧	(q ∨ r)	∧	(r ∨ p)]	→	p
T	T	T	T	T	T	T	T	T	T
T	T	F	T	T	T	T	T	T	T
T	F	T	F	F	T	F	T	T	T
T	F	F	F	F	F	F	T	T	T
F	T	T	T	T	T	T	T	F	F
F	T	F	T	T	T	F	F	T	F
F	F	T	T	T	T	T	T	F	F
F	F	F	T	F	F	F	F	T	F
			1	3	2	5	4	7	6

We will now investigate how we can arrive at a valid conclusion from a given set of premises.

▶ **Example 5**

Determine a logical conclusion that follows from the given statements. "If you own a house, then you will pay property tax. You own a house. Therefore, . . . "

Solution: If you recognize a specific form of an argument, you can use your knowledge of that form to draw a logical conclusion.

Let

p: You own a house.
q: You will pay property tax.

$$p \to q$$
$$\underline{p}$$
$$\therefore ?$$

Too many artificial colors in your dog's food?

Advertisements often use humorous twists of logic in order to grab the reader's interest. For example, consider this ad. Can you write a conditional statement that is implied by the ad?

If the question mark is replaced with a q, this argument is of the form of the law of detachment. A logical conclusion is "Therefore, you will pay property tax."

Section 3.6 Exercises

1. What does it mean when an argument is valid?
2. Is it possible for an argument to be valid if its conclusion is false? Explain your answer.
3. Explain how to determine if an argument with premises p_1 and p_2 and conclusion c is a valid or invalid argument.

In Ex. 4–7, (a) list the form of the following valid arguments, and (b) write an original argument in words for each form.
4. Law of detachment
5. Law of syllogism
6. Law of contraposition
7. Disjunctive syllogism

In Ex. 8 and 9, (a) list the form of the following fallacies, and (b) write an original argument in words for each form.
8. Fallacy of the converse
9. Fallacy of the inverse

In Ex. 10–28, determine whether the argument is valid or invalid. You may compare the argument to a standard form, or use a truth table.

10. $p \rightarrow q$
$\dfrac{\sim p}{\therefore q}$

11. $p \wedge \sim q$
$\dfrac{q}{\therefore \sim p}$

12. $\sim p \vee q$
$\dfrac{q}{\therefore p}$

13. $p \rightarrow q$
$\dfrac{p}{\therefore q}$

14. $p \rightarrow q$
$\dfrac{\sim q}{\therefore \sim p}$

15. $p \vee q$
$\dfrac{\sim p}{\therefore \sim q}$

16. $p \vee q$
$\dfrac{\sim q}{\therefore p}$

17. $p \rightarrow q$
$\dfrac{q}{\therefore p}$

18. $\sim p \rightarrow q$
$\dfrac{\sim q}{\therefore \sim p}$

19. $q \wedge \sim p$
$\dfrac{\sim p}{\therefore q}$

20. $q \wedge p$
$\dfrac{q}{\therefore \sim p}$

21. $p \rightarrow q$
$\dfrac{q \rightarrow r}{\therefore p \rightarrow r}$

22. $p \leftrightarrow q$
$\dfrac{q \rightarrow r}{\therefore \sim r \rightarrow \sim p}$

23. $p \rightarrow q$
$\dfrac{q \wedge r}{\therefore p \vee r}$

24. $r \leftrightarrow p$
$\dfrac{\sim p \wedge q}{\therefore p \wedge r}$

25. $p \vee q$
$\dfrac{r \wedge p}{\therefore q}$

*26. $p \rightarrow q$
$q \vee r$
$\dfrac{r \vee p}{\therefore p}$

*27. $p \rightarrow q$
$r \rightarrow \sim p$
$\dfrac{p \vee r}{\therefore q \vee \sim p}$

*28. $p \rightarrow q$
$q \rightarrow r$
$r \rightarrow p$
$\dfrac{}{\therefore q \rightarrow p}$

In Ex. 29–40, translate the argument into symbolic form. Determine whether each argument is valid or invalid. You may compare the argument to a standard form, or use a truth table.

29. If it is cold, then my feet are cold.
It is cold.
$\overline{\therefore \text{ My feet are cold.}}$

30. I like broiled fish or I like baked chicken.
I do not like broiled fish.
$\overline{\therefore \text{ I like baked chicken.}}$

31. If the story is interesting, then the people will listen.
The people will listen.
$\overline{\therefore \text{ The story is interesting.}}$

32. The dog is a collie and the cat is a Siamese.
If the cat is Siamese, then the bird is a parrot.
$\overline{\therefore \text{ If the bird is not a parrot, then the dog is not a collie.}}$

33. The painting is a Rembrandt or the painting is a Picasso.
If the painting is a Picasso, then it is not a Rembrandt.
$\overline{\therefore \text{ The painting is not a Picasso.}}$

34. If you cook the meal, then I will vacuum the rug.
I will not vacuum the rug.
$\overline{\therefore \text{ You will not cook the meal.}}$

35. The watch is not expensive.
If the watch is expensive, then the watch cost a lot of money.
$\overline{\therefore \text{ The watch cost a lot of money.}}$

36. It is snowing and I am going skiing.
If I am going skiing, then I will wear a coat.
∴ If it is snowing, then I will wear a coat.

37. It is Bart's house or the house is made of brick.
If the house is not made of brick, then it is Bart's house.
∴ The house is made of brick or the house is Bart's house.

38. If the house has three bedrooms, then the Cleavers will buy the house.
If the price is not less than $90,000, then the Cleavers will not buy the house.
∴ If the house has three bedrooms, then the price is less than 90,000.

39. If the main dish is fish, then Teri will come to dinner.
If Teri comes to dinner, then we will eat at 6:00 P.M.
∴ If the main dish is fish, then we will eat at 6:00 P.M.

40. If there is an atmosphere, then there is gravity.
If an object has weight, then there is gravity.
∴ If there is an atmosphere, then an object has weight.

In Ex. 41–50, translate the argument into symbolic form. Then determine whether the argument is valid or invalid.

41. There are no good programs on T.V. or it is raining. There are good programs on T.V. Therefore it is raining.

42. If all truck drivers are frustrated little boys, then I am a truck driver. I am not a truck driver. Therefore not all truck drivers are frustrated little boys.

43. If Bonnie passes the bar exam, then she will practice law. Bonnie will not practice law. Therefore Bonnie did not pass the bar exam.

44. The music is soft and soothing. The music is not soothing or it is soft. Therefore the music is not soft.

45. The birds are singing and the cat is listening. The cat is not listening or the birds are singing. Therefore the birds are not singing.

46. If the car is new, then the car has air conditioning. The car is not new and the car has air conditioning. Therefore the car is not new.

47. If the football team wins the game, then Dave played quarterback. If Dave played quarterback, then the team is not in second place. Therefore, if the football team wins the game, then the team is in second place.

48. The engineering courses are difficult and the chemistry labs are long. If the chemistry labs are long, then the art tests are easy. Therefore the engineering courses are difficult and the art tests are not easy.

49. If the lights are on then we can play ball. If the umpires are present, then we can play ball. Therefore if the lights are on, then the umpires are present.

50. Money is the root of all evil or the salvation of the poor. If money is the salvation of the poor, then it is not the root of all evil. Money is the root of all evil. Therefore money is not the salvation of the poor.

In Ex. 51–59, using the standard forms of arguments and other information you have learned, supply what you believe is a logical conclusion to the argument. Verify that the argument is valid for the conclusion you supplied.

51. If you drink orange juice, then you will be healthy.
You drink orange juice.
Therefore . . .

52. If you have lunch, then you must pay the bill.
You did not have lunch.
Therefore . . .

53. John reads the history assignment or he fixes the car.
John does not read the history assignment.
Therefore . . .

54. If the electric bill is too high, then I will not be able to pay the bill.
I will not be able to pay the bill.
Therefore . . .

55. If the dog is brown, then the dog is a cocker spaniel.
The dog is not a cocker spaniel.
Therefore . . .

56. If you do not read a lot, then you will not gain knowledge.
You do not read a lot.
Therefore . . .

57. If you overdraw your checking account, then the bank will send you a notice.
If the bank sends you a notice, then you will have to pay a fee.
Therefore . . .

58. If you are a waiter, then you must take orders.
If you do not get tips, then you do not take orders.
Therefore . . .

*59. If the fire will burn, then the logs are dry.
If the fire will not burn, then you will have to dry the logs.
Therefore . . .

Problem Solving

60. Determine whether the argument is valid or invalid.

If Lynn wins the contest or strikes oil, then she will be rich.
If Lynn is rich, then she will stop working.
∴ If Lynn does not stop working, she did not win the contest.

Research Activities

61. Show how logic is used in advertising. Discuss several advertisements, and show how logic is used to persuade the reader.

62. Show how logic is used in politics. Collect articles on a particular political campaign, and show how logic is used by the candidates to gain votes.

3.7 Euler Circles and Syllogistic Arguments

The Swiss mathematician **Leonhard Euler** *(1707–1783) produced such a large volume of work that all of it has not yet been compiled. It has been estimated that a complete edition of his work would comprise over 60 volumes. While his greatest contribution was made in the field of modern analysis, his work also explored physics, astronomy, trigonometry, and calculus. Euler also had a great influence on the notation and symbolism of mathematics, and it was through his work that the symbols e and π came into common use. His facility for mental calculation and his uncommon memory allowed him to continue his prolific career after he lost his vision in both eyes.*

In the previous section we showed how to determine the validity of *symbolic arguments* using truth tables and comparing the arguments to standard forms. This section presents another form of argument called a **syllogistic argument**, better known by the short name **syllogism**. The validity of a syllogistic argument is determined by using Euler circles as will be explained shortly.

Syllogistic logic, a deductive process of arriving at a conclusion, was developed by Aristotle around 350 B.C. Aristotle considered the relationships among the four types of statements that follow.

All _____ are _____.
No _____ are _____.
Some _____ are _____.
Some _____ are not _____.

An example of such statements follows. *All doctors are tall. No doctors are tall. Some doctors are tall. Some doctors are not tall.* Since Aristotle's time, other types of statements have been added to the study of syllogistic logic, two of which are

_____ is a _____.
_____ is not a _____.

An example of such statements follows. *Maria is a doctor. Maria is not a doctor.*

The difference between a symbolic argument and a syllogistic argument can be seen in the following chart. Symbolic arguments use the connectives "and," "or," "not," "but," "if-then," and "if and only if." Syllogistic arguments use the quantifiers "all," "some," and "none," which were discussed in Section 3.1.

Symbolic Arguments vs. Syllogistic Arguments		
	Words or Phrases Used	*Method of Determining Validity*
Symbolic argument	and, or, not, if-then, if and only if	truth tables or by comparison with standard forms of arguments
Syllogistic argument	all are, some are, none are, some are not	Euler diagrams

As with symbolic logic the premises and the conclusion together form an argument. An example of a syllogistic argument is

All German shepherds are dogs.
All dogs bark.
∴ All German shepherds bark.

This is an example of a valid argument. Recall from the previous section that an argument is *valid* when its conclusion necessarily follows from a given set of premises. An argument in which the conclusion does not necessarily follow from the given premises is said to be an **invalid argument** or **fallacy**.

Before we give another example of a syllogism, let us review the Venn diagrams discussed in Section 2.3 in relationship with Aristotle's four statements.

All *A*s are *B*s	No *A*s are *B*s	Some *A*s are *B*s	Some *A*s are not *B*s
If an element is in set *A* then it is in set *B*.	If an element is in set *A* then it is not in set *B*.	There is at least one element that is in both set *A* and set *B*.	There is at least one element that is in set *A* that is not in set *B*.

One method used to determine whether an argument is valid or is a fallacy is by means of **Euler diagrams**, named after **Leonhard Euler** (1707–1783) who used circles to represent sets in syllogistic arguments. The technique of using Euler diagrams is illustrated in Example 1.

▶ **Example 1**

Determine whether the following syllogism is valid or a fallacy.

All teachers are rich.
All rich people are doctors.
∴ All teachers are doctors.

Solution: The statement "All teachers are rich" may be interpreted as "If a person is a teacher, then the person is rich." Construct the diagram and represent the first premise, "All teachers are rich," as in Fig. 3.1. The outer circle represents all rich people and the inner circle represents all teachers. Now illustrate the second premise, "All rich people are doctors," as in Fig. 3.2. The set containing all doctors must contain all the rich people, as illustrated in the diagram. Note that the premises force the set of teachers to be within the set of doctors. Therefore the argument is valid, since the conclusion "All teachers are doctors" necessarily follows from the set of premises.

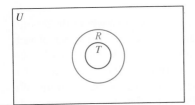

Figure 3.1

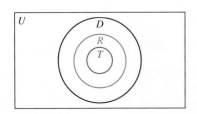

Figure 3.2

The argument in Example 1 is valid even though the conclusion, "All teachers are doctors," is obviously a false statement. Similarly, an argument can be a fallacy even if the conclusion is a true statement.

When we determine the validity of an argument, we are determining whether the conclusion necessarily follows from the premises. When we say an argument is valid, we are saying that if all the premises are true statements, then the conclusion must also be a true statement.

It is the form of the argument that determines its validity, not the particular statements. For example, consider the syllogism

> All earth people have two heads.
> All people with two heads can fly.
> ∴ All earth people can fly.

The form of this argument is the same as that of the previous valid argument. Therefore this argument is also valid.

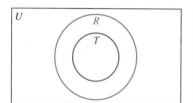

Figure 3.3

▶ **Example 2**

Determine whether the following syllogism is valid or is a fallacy.

> All teachers are rich.
> Pete is a teacher.
> ∴ Pete is rich.

Solution: The statement "All teachers are rich" is illustrated in Fig. 3.3. The second premise, "Pete is a teacher," tells us that Pete must be placed in the inner circle (see Fig. 3.4). The Euler diagram illustrates that we must accept the conclusion "Pete is rich" as true (when we accept the premises as true). Therefore the argument is valid.

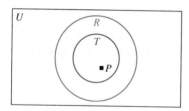

Figure 3.4

In both Example 1 and Example 2 we had no choice as to where the second premise was to be placed in the Euler diagram. In Example 1, the set of rich people had to be placed inside the set of doctors. In Example 2 Pete had to be placed inside the set of teachers. Often when determining the truth value of a syllogism a premise can be placed in more than one area in the diagram. **We always try to draw the Euler diagram so that the conclusion does not necessarily follow from the premises. If this can be done then the conclusion does not necessarily follow from the premises and the argument is invalid. If we cannot show that the argument is invalid, only then do we accept the argument as valid.** This is illustrated in Example 3.

▶ **Example 3**

Determine whether the following syllogism is valid or is a fallacy.

> All football players are strong.
> Christine is strong.
> ∴ Christine is a football player.

Solution: The statement "All football players are strong" is illustrated in Fig.

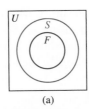
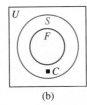
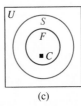

Figure 3.5   (a)   (b)   (c)

3.5(a). The next premise, "Christine is strong," tells us that Christine must be placed in the set of strong people. Two diagrams in which both premises are satisfied are shown in Fig. 3.5(b) and 3.5(c). By examining Fig. 3.5(b), however, we see that Christine is not a football player. Therefore the conclusion "Christine is a football player" does not necessarily follow from the set of premises. Thus the argument is a fallacy.

▶ **Example 4**

Determine whether the following syllogism is valid or is a fallacy.

> No airplane pilots eat spinach.
> Janet does not eat spinach.
> ∴ Janet is an airplane pilot.

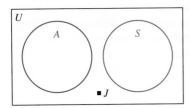

Figure 3.6

Solution: The diagram in Fig. 3.6 satisfies the two given premises and also shows that Janet is not an airplane pilot. Therefore the argument is invalid, or a fallacy.

Note that in Example 4 if we placed Janet in circle *A*, the argument would appear to be valid. Remember, *whenever testing the validity of an argument, always try to show that the argument is invalid.* If there is any way of showing that the conclusion does not necessarily follow from the premises, then the argument is invalid.

▶ **Example 5**

Determine whether the following syllogism is valid or invalid.

> All *A*s are *B*s
> Some *B*s are *C*s.
> ∴ Some *A*s are *C*s

Figure 3.7

Solution: The statement "All *A*s are *B*s" is illustrated in Fig. 3.7. The statement "Some *B*s are *C*s" means that there is at least one *B* that is a *C*. We can illustrate this set of premises in four ways as illustrated in Fig. 3.8.

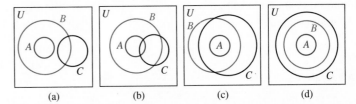

Figure 3.8   (a)   (b)   (c)   (d)

In all four illustrations we see that (1) all *A*s are *B*s and (2) some *B*s are *C*s. The conclusion is "Some *A*s are *C*s." Since at least one of the illustrations (Fig. 3.8a) shows that the conclusion does not necessarily follow from the given premises, the argument is invalid.

▶ **Example 6**

Determine whether the following syllogism is valid or invalid.

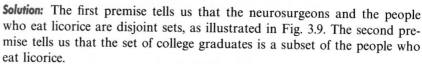

No neurosurgeons eat licorice.
All college graduates eat licorice.
_____
∴ No neurosurgeons are college graduates.

Solution: The first premise tells us that the neurosurgeons and the people who eat licorice are disjoint sets, as illustrated in Fig. 3.9. The second premise tells us that the set of college graduates is a subset of the people who eat licorice.

The set of college graduates and neurosurgeons cannot be made to intersect without violating a premise. Thus no neurosurgeons can be college graduates, and the syllogism is valid. Note that we did not say that this conclusion was true—only that the argument was valid.

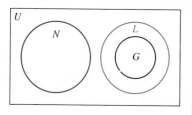

Figure 3.9

Section 3.7 Exercises

1. What does it mean when we determine that an argument is valid?

2. Explain the differences between a symbolic argument and a syllogistic argument.

In Ex. 3–25, use Euler diagrams to determine whether the syllogism is valid or is a fallacy.

3. All freshmen study English.
Irvin is a freshman.
_____
∴ Irvin studies English.

4. No candy is sweet.
All Tootsie Rolls are sweet.
_____
∴ Tootsie Rolls are not candy.

5. All grass is green.
All things that are green will grow.
_____
∴ All grass will grow.

6. All *A*s are *B*s.
All *B*s are *C*s.
_____
∴ All *A*s are *C*s.

7. All healthy people are happy.
Jane Edgar is happy.
_____
∴ Jane Edgar is healthy.

8. All physicists are intelligent.
Madame Curie was intelligent.
_____
∴ Madame Curie was a physicist.

9. All football players have agents.
Morris Jones is a football player.
_____
∴ Morris Jones has an agent.

10. All numbers that are divisible by 2 are even.
The number 7 is divisible by 2.
_____
∴ The number 7 is even.

11. Some swans are graceful swimmers.
Andy is a swan.
_____
∴ Andy is a graceful swimmer.

12. Some swans are not graceful swimmers.
Candy is a swan.
_____
∴ Candy is not a graceful swimmer.

13. Some medical researchers have made important discoveries.
Some people who have made important discoveries are famous.

    ∴ Some medical researchers are famous.

14. All hummingbirds are small.
Some small birds can hover.

    ∴ All hummingbirds can hover.

15. No prime numbers are divisible by 2.
The number 10 is not divisible by 2.

    ∴ The number 10 is a prime number.

16. Some cars can float.
All things that float are light.

    ∴ Some cars are light.

17. Some people love mathematics.
All people who love mathematics love physics.

    ∴ Some people love physics.

18. Some desks are made of wood.
All paper is made of wood.

    ∴ Some desks are made of paper.

19. No xs are ys.   20. All pilots can fly.
    No ys are zs.   All astronauts can fly.

    ∴ No xs are zs.   ∴ Some pilots are astronauts.

21. Some tall people wear glasses.
Maria wears glasses.

    ∴ Maria is not tall.

22. All rainy days are cloudy.
Today it is cloudy.

    ∴ Today is a rainy day.

23. All sweet things taste good.
All things that taste good are fattening.
All things that are fattening put on pounds.

    ∴ All sweet things put on pounds.

24. All Addison-Wesley books have red covers.
All books that have red covers are interesting.
Some books that are interesting are lengthy.

    ∴ All interesting books are lengthy.

25. All squares are rectangles.
All rectangles are quadrilaterals.
All quadrilaterals are polygons.

    ∴ All squares are polygons.

Problem Solving

26. Statements in logic can be translated into set statements—for example, $p \wedge q$ is similar to $P \cap Q$; $p \vee q$ is similar to $P \cup Q$; $p \rightarrow q$ is equivalent to $\sim p \vee q$, which is similar to $P' \cup Q$. Euler circles can also be used to show that arguments similar to those discussed in this section are valid or invalid. Use Euler circles to show that the symbolic argument is invalid.

$$p \rightarrow q$$
$$\underline{p \vee q}$$
$$\therefore \sim p$$

CHAPTER 3 SUMMARY

Key Terms

3.1
antecedent	compound statement	connectives	negation (not)	statement
Aristotelian logic	conditional (if . . . then . . .)	consequent	quantifier	symbolic logic
biconditional (if and only if)	conjunction (and)	disjunction (or)	simple statement	

3.2
truth table

3.3
implication
self-contradiction
tautology

3.4
De Morgan's laws
equivalent statements
logically equivalent

3.5
contrapositive
converse
inverse

3.6
fallacy
invalid argument
symbolic argument
valid argument

3.7
Euler diagram
syllogism
syllogistic argument

Important Facts

Quantifiers

Form of Statement	Form of Negation
All are.	Some are not.
None are.	Some are.
Some are.	None are.
Some are not.	All are.

Summary of Connectives

Formal Name	Symbol	Read	Symbolic Form
Negation	$\sim$	not	$\sim p$
Conjunction	$\wedge$	and	$p \wedge q$
Disjunction	$\vee$	or	$p \vee q$
Conditional	$\rightarrow$	if-then	$p \rightarrow q$
Biconditional	$\leftrightarrow$	if and only if	$p \leftrightarrow q$

Basic Truth Tables

Negation

p	$\sim p$
T	F
F	T

p	q	Conjunction $p \wedge q$	Disjunction $p \vee q$	Conditional $p \rightarrow q$	Biconditional $p \leftrightarrow q$
T	T	T	T	T	T
T	F	F	T	F	F
F	T	F	T	T	F
F	F	F	F	T	T

De Morgan's Laws

$\sim(p \wedge q) \Leftrightarrow \sim p \vee \sim q$

$\sim(p \vee q) \Leftrightarrow \sim p \wedge \sim q$

Other Equivalent Forms

$p \rightarrow q \Leftrightarrow \sim p \vee q$

$p \leftrightarrow q \Leftrightarrow [(p \rightarrow q) \wedge (q \rightarrow p)]$

Forms of the Conditional Statement

Statement	Symbolic Form
If p, then q	$p \rightarrow q$
p only if q	$p \rightarrow q$
q if p	$p \rightarrow q$
p is a sufficient condition for q	$p \rightarrow q$
q is a necessary condition for p	$p \rightarrow q$

Variations of the Conditional Statement

Name	Symbolic Form	Read
Conditional	$p \rightarrow q$	If p, then q.
Converse of the conditional	$q \rightarrow p$	If q, then p.
Inverse of the conditional	$\sim p \rightarrow \sim q$	If not p, then not q.
Contrapositive of the conditional	$\sim q \rightarrow \sim p$	If not q, then not p.

Standard Forms of Arguments

Valid Arguments

Law of Detachment	*Law of Contraposition*	*Law of Syllogism*	*Disjunctive Syllogism*
$p \rightarrow q$	$p \rightarrow q$	$p \rightarrow q$	$p \vee q$
p	$\sim q$	$q \rightarrow r$	$\sim p$
$\therefore q$	$\therefore \sim p$	$\therefore p \rightarrow r$	$\therefore q$

Invalid Arguments

Fallacy of the Converse	*Fallacy of the Inverse*
$p \rightarrow q$	$p \rightarrow q$
q	$\sim p$
$\therefore p$	$\therefore \sim q$

Symbolic Argument vs. Syllogistic Argument

	Words or Phrases Used	*Method of Determining Validity*
Symbolic argument	and, or, not, if-then, if and only if	truth tables or by comparison with standard forms of arguments
Syllogistic argument	all are, some are, none are, some are not	Euler diagrams

CHAPTER 3 REVIEW EXERCISES

3.1
Write the negation of the following statements.

1. All wood is hard

2. No birds are yellow.

3. Some apples are not red.

4. Some squirrels fly.

5. Not all people wear glasses.

6. No rabbits wear glasses.

In Ex. 7–12, write each compound statement in words.

p: I opened a can of soda.
q: I spilled some soda.
r: The soda is red.

7. $p \wedge q$ **8.** $\sim p \vee r$

9. $p \rightarrow (q \wedge \sim r)$ **10.** $p \leftrightarrow q$

11. $(p \vee \sim q) \wedge \sim r$ **12.** $\sim p \leftrightarrow (r \wedge \sim q)$

3.2
In Ex. 13–18, use the statements for p, q, and r as in Ex. 7–12 to write the statement in symbolic form.

13. I spilled some soda or the soda is not red.

14. I opened a can of soda and the soda is red.

15. If the soda is red, then I did not open a can of soda or I spilled some soda.

16. I spilled some soda if and only if I opened a can of soda, and the soda is not red.

17. The soda is red and I opened a can of soda, or I spilled some soda.

18. It is false that I spilled some soda or the soda is red.

Construct a truth table for each of the following statements.

19. $(p \vee \sim q) \wedge p$

20. $\sim p \wedge \sim q$

21. $\sim p \rightarrow (q \wedge \sim p)$

22. $\sim p \leftrightarrow q$

23. $p \wedge (\sim q \vee r)$

24. $p \rightarrow (q \wedge \sim r)$

25. $(p \vee q) \leftrightarrow (p \vee r)$

26. $(p \wedge q) \rightarrow \sim r$

3.2, 3.3
Find the truth value of the following statements.

27. If the moon is square, then Texas is in the United States.

28. George Boole was a mathematician or Charles Dodgson wrote children's stories.

29. If Aristotle is known for his automobile designs, then George Bush was a dance teacher.

30. $3 + 7 = 11$ or $6 + 5 = 11$, and $7 \cdot 6 = 42$.

31. French is a language, if and only if $2 + 2 = 7$ or $3 + 5 = 8$.

3.3
Find the truth values of the following statements when p is T, q is F, and r is F.

32. $(p \vee q) \rightarrow (\sim r \wedge p)$

33. $(q \rightarrow \sim r) \vee (p \wedge q)$

34. $\sim r \leftrightarrow [(p \vee q) \leftrightarrow \sim p]$

35. $\sim[(q \wedge r) \rightarrow (\sim p \vee r)]$

36. $[\sim(q \wedge p)] \rightarrow \sim(\sim p \vee r)$

3.4, 3.5
In Ex. 37–42, determine whether the pairs of statements are equivalent. You may use De Morgan's laws, the fact that $(p \rightarrow q) \Leftrightarrow (\sim p \vee q)$, truth tables, or equivalent forms of the conditional statement.

37. $\sim p \rightarrow \sim q$ $p \vee \sim q$

38. $\sim p \vee \sim q$ $\sim p \leftrightarrow q$

39. $\sim p \vee (q \wedge r)$ $(\sim p \vee q) \wedge (\sim p \vee r)$

40. $(\sim q \rightarrow p) \wedge p$ $\sim(\sim p \leftrightarrow q) \vee p$

41. a) If it is snowing, then I will go skiing.

b) It is not snowing or I will not go skiing.

42. a) The *Workbench* magazine is on the desk or the magazine is in my workshop.

b) If the *Workbench* magazine is not in my workshop, then the magazine is on my desk.

Determine whether any of the statements are equivalent.

43. a) If there is school today, then I will ride the bus.

b) I will ride the bus if there is school today.

c) I will ride the bus only if there is school today.

44. a) The apples are sweet is sufficient for Saul to eat three apples.

b) Saul eats three applies if the apples are sweet.

c) The apples are sweet is necessary for Saul to eat three apples.

Use De Morgan's laws or the fact that $(p \rightarrow q) \Leftrightarrow (\sim p \vee q)$ to write an equivalent statement for each of the following.

45. The pencil is not yellow or the desk is orange.

46. The Flyers score goals and the Flyers win games.

47. It is not true that New Orleans is in California or Chicago is not a city.

48. If the water is blue, then the fish will not bite.

49. I did not go to the party and I did not finish my special report.

Write the contrapositive for each of the following statements.

50. If the railroad crossing light is flashing red, then you must stop.

51. If it is an expensive car, then the repair costs are high.

52. If today is not a holiday, then I will be at work.

53. If the computer has a color monitor and an extended key board, then it is Sam's computer.

54. Write the converse, inverse, and contrapositive of the conditional statement "If I study, then I will get a good grade."

3.6, 3.7

Determine whether the following arguments are valid.

55. Nicole is in the hot tub or she is in the shower.
Nicole is in the hot tub.
_____

∴ Nicole is not in the shower.

56. If the car has a sound system, then Rick will buy the car. If the price is not less than $18,000, then Rick will not buy the car. Therefore if the car has a sound system, then the price is less than $18,000.

57. All grasshoppers are green.
Some crickets are green.
_____

∴ Some crickets are grasshoppers.

58. No mathematics books are dull.
All textbooks are dull.
_____

∴ Some mathematics books are textbooks.

Chapter 3

CHAPTER TEST

In Ex. 1–3, write the statement in symbolic form.

p: The house is red
q: The car is blue
r: The owner is a lawyer.

1. The car is blue or the house is red, and the owner is a lawyer.

2. If the car is blue and the owner is a lawyer, then the house is red.

3. It is false that the owner is a lawyer if and only if the house is red.

In Ex. 4 and 5, use p, q, and r as above to write an English sentence for each symbolic statement.

4. $\sim(p \rightarrow \sim r)$ **5.** $p \leftrightarrow (q \wedge r)$

In Ex. 6–8, construct a truth table for the given statement.

6. $\sim p \vee q$ **7.** $\left[\sim(p \rightarrow r)\right] \wedge q$

8. $(q \leftrightarrow \sim r) \vee p$

In Ex. 9 and 10, find the truth value of the statement.

9. $2 + 6 = 8$ or $7 - 12 = 5$.

10. If Germany is a country in Europe and India is a country in Asia, then Mexico is a country in Canada.

Given that p is true, q is false, and r is true, find the truth value of the following statements.

11. $\left[\sim(r \rightarrow \sim p)\right] \wedge (q \rightarrow p)$ **12.** $(r \vee q) \leftrightarrow (p \wedge \sim q)$

In Ex. 13–15, determine whether the pairs of statements are equivalent. You may use De Morgan's laws, the fact that $(p \rightarrow q) \Leftrightarrow (\sim p \vee q)$, truth tables, or equivalent forms of the conditional.

13. a) $\sim p \vee q$,   **b)** $\sim(p \wedge \sim q)$

14. a) It is not true that the test is today or the concert is tonight.
    b) The test is not today and the concert is not tonight.

15. a) If it is snowing, then the game will not be played.
    b) It is not snowing or the game will not be played.

16. Translate the following argument into symbolic form. Determine whether the argument is valid or invalid by comparing the argument to a recognized form or by using a truth table.

If the soccer team wins the game, then Sue played fullback. If Sue played fullback, then the team is in second place. Therefore, if the soccer team wins the game, then the team is in second place.

17. Use an Euler diagram to determine whether the following syllogistic argument is valid or invalid.

All numbers divisible by 7 are odd.
The number 28 is divisible by 7.
∴ The number 28 is odd.

In Ex. 18 and 19, write the negation of the statement.

18. All birds are black.

19. Some people are funny.

20. The conditional statement "If the apple is red, then it is a delicious apple" is given. Write the inverse, converse, and contrapositive of the conditional statement.

SYSTEMS OF NUMERATION

One of the earliest reasons humans needed numbers was for reckoning time, marking off days in the lunar month, so the seasonal changes that dictated human activity could be anticipated. The Mayans, the Egyptians, and the ancient Britons constructed monumental stone observatories that enabled them to mark the passage of the seasons, especially the summer solstice, using the alignment of the sun as a guide.

We are so accustomed to reading and writing that it is hard to imagine what our lives would be like without them. The number system we use — called the Hindu-Arabic system — seems to be a permanent, unchanging means of communicating quantities. However, just as languages evolve over time, so do numerical symbols which represent numbers.

Mathematics began with the practical problem of counting and record keeping. This happened so far back in human history that no one can point to a time or place and say, "It all started here." People had to count their herds, the passage of days, and objects of barter. They used physical objects — stones, shells, fingers — to represent the objects counted. For example, shepherds counted their flocks by moving one pebble from the "out" pile to the "in" pile as each animal passed by, making sure all were accounted for.

Archaeologist Denise Schmandt-Besserat made a breakthrough discovery about early systems of numeration. She realized that the little clay geometric objects that had been found in many archaeological sites had actually been used by people to account for their goods. Later in history, these tokens were impressed on a clay tablet to represent quantities — the beginning of writing.

As primitive cultures grew from villages to cities, the complexity of human activities increased. Now people needed better ways of recording and communicating. It was a revolutionary step when people started using physical objects to represent not only specific objects like sheep and grain, but also the concept of pure quantity. The later invention of devices for counting — counting boards, the abacus, wooden or ivory rods — made it even easier to work with large numbers. At the same time, new writing materials expanded the amount of record keeping that could be done.

Through the course of human history, the evolution of numeration systems has expanded our knowledge and abilities for record keeping,

This antique print shows one person calculating with Hindu-Arabic numerals, the other with a counting board. Looking over their shoulders is the spirit Arithmeticae.

communication, and computation. As a society's numeration system changes, so do the capabilities of that society. For example, without an understanding of the binary number system, the computer as we know it today could not exist. And we can ask, without the computer, would we have had the ability to explore space or track the activities of a global marketplace? Studying the evolution of numbers, counting, and calculation methods allows us to glimpse a fascinating aspect of our history as well as our future.

Because of its flexibility, it is the Hindu-Arabic system that has become the universal language of mathematics.

4.1 Additive, Multiplicative, and Ciphered Systems of Numeration

Just as the first attempts to write were made long after the development of speech, the first representation of numbers by symbols came long after people had learned to count. A tally system using physical objects, such as scratch marks in the soil or on a stone, notches on a stick, pebbles, or knots on a vine, was probably the earliest method of recording numbers.

In primitive societies such a tally system adequately served the limited need for recording livestock, agriculture, or whatever was counted. As civilization developed, however, more efficient and accurate methods of calculating and keeping records were needed. Because tally systems are impractical and inefficient, societies developed symbols to replace them. For example, the Egyptians used the symbol ∩ and the Babylonians used the symbol ◀ to represent the number we symbolize by 10.

A **number** is a quantity, and it answers the question "How many?" A **numeral** is a symbol such as ∩, ◀, and 10 used to represent the number. One thinks a number but writes a numeral. The distinction between number and numeral will be made here only if it is helpful to the discussion.

In language, relatively few letters of the alphabet are used to construct a large number of words. Similarly, in arithmetic a small variety of numerals can be used to represent all numbers. One of the greatest accomplishments of humankind has been the development of systems of numeration, whereby all numbers are "created" from a few symbols. Without such systems, mathematics would not have developed to its present level.

> A **system of numeration** consists of a set of numerals and a scheme or rule for combining the numerals to represent numbers.

Four types of numeration systems used by different cultures are the topic of this section. They are additive (or repetitive), multiplicative, ciphered, and place-value systems. It is not necessary to memorize all the symbols, but you should understand the principles behind each system. By the end of this chapter, we hope you better understand the system we use, the Hindu-Arabic system, and its relationship to other types of systems.

Additive Systems

An additive system is one in which the number represented by a particular set of numerals is simply the sum of the values of the numerals. The additive system of numeration is one of the oldest and most primitive types of numeration systems. One of the first additive systems, the Egyptian hieroglyphic system, dates back to about 3000 B.C. The Egyptians used symbols for the powers of 10: 10^0 or 1; 10^1 or 10; 10^2 or $10 \cdot 10$; 10^3 or $10 \cdot 10 \cdot 10$; and so on. Table 4.1

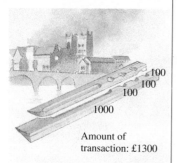

Table 4.1 Egyptian Hieroglyphics		
Hindu-Arabic Numerals	Egyptian Numerals	Description
1	ǀ	Staff (vertical stroke)
10	∩	Heel bone (arch)
100	৭	Scroll (coiled rope)
1,000	⚶	Lotus flower
10,000	⟨	Pointing finger
100,000	∝	Tadpole (or whale)
1,000,000	ⱷ	Astonished person

lists the Egyptian hieroglyphic numerals with the equivalent Hindu-Arabic numerals.

In order to write the number 600 in Egyptian hieroglyphics, one writes the numeral for 100 six times: ৭৭৭৭৭৭

▶ **Example 1**

Write the following numeral as a Hindu-Arabic numeral.

⟨⟨৭৭৭ǀ

Solution: $10,000 + 10,000 + 100 + 100 + 100 + 1 = 20,301$

▶ **Example 2**

Write 43,628 as an Egyptian numeral.

Solution:

$$43,628 = 40,000 + 3,000 + 600 + 20 + 8$$

⟨⟨⟨⟨⚶⚶⚶৭৭৭৭৭৭∩∩ǀǀǀǀǀǀǀǀ

In this system the order of the symbols is not important. For example, ৭৭∩ ∝ǀǀ and ǀǀ৭৭∝∩ both represent 100,212.

Users of additive systems easily accomplished addition and subtraction by combining or removing symbols. Multiplication and division were more difficult; they were performed by a process called **duplation and mediation** (see Section 4.5). The Egyptians had no symbol for zero, but they did have an understanding of fractions. The symbol ⌣ was used to take the reciprocal of a number; thus ⌣∩ meant $\frac{1}{3}$ and ⌣∩ was $\frac{1}{11}$. Writing large numbers in the Egyptian system would have taken longer than in other systems because so many symbols had to be listed. For example, 45 symbols are needed to represent the number 99,999.

The Roman numeration system, a second example of an additive system, was developed later than the Egyptian system. Roman numerals (Table 4.2 on page 138) were used in most European countries until the eighteenth century.

Table 4.2

Roman numerals	I	V	X	L	C	D	M
Hindu-Arabic numerals	1	5	10	50	100	500	1000

They are still commonly seen on buildings, on clocks, and in books. Roman numerals are selected letters of the Roman alphabet.

The Roman system has two advantages over the Egyptian system. The first is that it uses the subtraction principle as well as the addition principle. Starting from the left, we add each numeral unless its value is smaller than the value of the numeral to its right. In that case, we subtract it from that numeral. Only the numbers 1, 10, 100, . . . can be subtracted, and only from the next two higher numbers. For example, C (100) can be subtracted only from D (500) or M (1000). The symbol DC represents $500 + 100$, or 600, and CD represents $500 - 100$, or 400. Similarly, MC represents $1000 + 100$, or 1100, and CM represents $1000 - 100$, or 900.

▶ **Example 3**

Write CLXII as a Hindu-Arabic numeral.

Solution: Since each numeral is larger than the one on its right, no subtraction is necessary.

$$CLXII = 100 + 50 + 10 + 1 + 1 = 162$$

▶ **Example 4**

Write DCXLVI as a Hindu-Arabic numeral.

Solution: Checking from left to right, we see that X (10) has a smaller value than L (50). Therefore XL represents $50 - 10$, or 40.

$$DCXLVI = 500 + 100 + (50 - 10) + 5 + 1 = 646$$

▶ **Example 5**

Write 289 as a Roman numeral.

Solution:

$$289 = 200 + 80 + 9 = 100 + 100 + 50 + 10 + 10 + 10 + 9$$

(Nine is treated as $10 - 1$.)

$$289 = CCLXXXIX$$

The second advantage of the Roman numeration system over the Egyptian system is that it makes use of the multiplication principle for numbers over 1000. A bar above a symbol or a group of symbols indicates that the symbol or symbols are to be multiplied by 1000. Thus $\bar{V} = 5 \times 1000 = 5000$, $\bar{X} = 10 \times 1000 = 10,000$, and $\overline{CD} = 400 \times 1000 = 400,000$. This greatly reduces the number of symbols needed to write large numbers. Still, it requires 19 symbols, including the bar, to write the number 33,888.

Roman numerals remained popular on large clock faces long after their disappearance from daily transactions because they are easier to read from a distance than Hindu-Arabic numerals.

Table 4.3													
Traditional Chinese numerals	一	二	三	四	五	六	七	八	九	十	百	千	
Hindu-Arabic numerals	1	2	3	4	5	6	7	8	9	10	100	1000	

Multiplicative Systems

Multiplicative numeration systems are more similar than additive systems to our Hindu-Arabic system. The number 642 in a multiplicative system might be written (6) (100) (4) (10) (2) or

$$
\begin{array}{c}
6 \\
100 \\
4 \\
10 \\
2
\end{array}
$$

Note that no addition signs are needed to represent the number. From this illustration, try to formulate a rule explaining how multiplicative systems work.

The principal example of a multiplicative system is the traditional Chinese system. The numerals used in this system are given in Table 4.3.

Chinese numerals are always written vertically. The number on top will be a number from 1 to 9 inclusive. This number is to be multiplied by the power of 10 below it. The number 20 is written

$$
\left.\begin{array}{c} 二 \\ 十 \end{array}\right\} 2 \times 10 = 20
$$

The number 400 is written

$$
\left.\begin{array}{c} 四 \\ 百 \end{array}\right\} 4 \times 100 = 400
$$

▶ **Example 6**

Write 428 as a Chinese numeral.

Solution:

$$
428 = \begin{cases} 400 = \begin{cases} 4 & 四 \\ 100 & 百 \end{cases} \\[2ex] 20 = \begin{cases} 2 & 二 \\ 10 & 十 \end{cases} \\[2ex] 8 = \quad\; 8 \quad 八 \end{cases}
$$

Notice in Example 6 that the units digit, the 8, is not multiplied by a power of the base.

Have you noticed that there is no symbol for zero in the Chinese system? Why is a symbol for zero not needed?

The number system used today in China is different from the traditional system. The present-day system is a positional-value system rather than a multiplicative system and uses the symbol 0 for zero.

Ciphered Systems

Ciphered numeration systems require the memorization of many different symbols but have the advantage that numbers can be written in a compact form. The ciphered numeration system that we will discuss is the Ionic Greek (Table 4.4). The Ionic Greek system was developed around 3000 B.C. and used letters of their alphabet for numerals. Other ciphered systems include the Hebrew, Coptic, Hindu, Brahmin, Syrian, Egyptian Hieratic, and early Arabic.

Since the Greek alphabet contains 24 letters but 27 symbols were needed, the Greeks borrowed the symbols ζ, Q, π from the Phoenician alphabet.

Table 4.4 Ionic Greek Numerals

1	α	alpha	60	ξ	xi
2	β	beta	70	o	omicron
3	γ	gamma	80	π	pi
4	δ	delta	90	Q	koph*
5	ϵ	epsilon	100	ρ	rho
6	ζ	vau*	200	σ	sigma
7	ζ	zeta	300	τ	tau
8	η	eta	400	υ	upsilon
9	θ	theta	500	ϕ	phi
10	ι	iota	600	χ	chi
20	κ	kappa	700	ψ	psi
30	λ	lambda	800	ω	omega
40	μ	mu	900	π	sampi*
50	ν	nu			

* Taken from the Phoenician alphabet.

The number $24 = 20 + 4$. When 24 is written as a Greek numeral, the plus sign is omitted:

$$24 = \kappa\delta.$$

The number 996 written as a Greek numeral is $\pi\,Q\,\zeta$.

When a prime (') is placed above a number, it multiplies that number by 1000. For example,

$$\beta' = 2 \times 1000 = 2000,$$
$$\sigma' = 200 \times 1000 = 200{,}000.$$

▶ **Example 7**

Write $\chi\nu\gamma$ as a Hindu-Arabic numeral.

Solution: $\chi = 600$, $\nu = 50$, and $\gamma = 3$. Adding these numbers gives 653.

▶ **Example 8**

Write 8652 as an Ionic Greek numeral.

Solution:
$$8652 = 8000 + 600 + 50 + 2$$
$$= (8 \times 1000) + 600 + 50 + 2$$
$$= \eta' \qquad \chi \qquad \nu \qquad \beta$$
$$= \eta'\chi\nu\beta$$

Section 4.1 Exercises

Write each of the following as a Hindu-Arabic numeral.

1. ⟨⟨⟨∩IIII

2. ∩⟨II

3. ⟨⟨⟨⟨⟨⟨∩II

4. ⟨⟨⟨⟨⟨∩IIII

5. ⟨⟨⟨⟨⟨⟨⟨⟨⟨⟨∩IIII

6. ⟨⟨⟨⟨⟨⟨⟨⟨⟨∩∩∩I

Write each of the following as an Egyptian numeral.

7. 246   **8.** 315   **9.** 1687

10. 1492   **11.** 162,451   **12.** 1,342,567

Write each of the following as a Hindu-Arabic numeral.

13. LXIV   **14.** XLIX

15. CXVII   **16.** DCLXIV

17. MDCCCXCII   **18.** MCMLXVI

19. MCMXCIX   **20.** MMCDXLIV

21. V̄MMDCII   **22.** ĪV̄DLV

23. X̄DCLXXIX   **24.** ĪX̄DCCCXLII

Write each of the following as a Roman numeral.

25. 38   **26.** 95   **27.** 148

28. 397   **29.** 1978   **30.** 2435

31. 2606   **32.** 9436   **33.** 4321

34. 15,954

Write each of the following as a Hindu-Arabic numeral.

35. 五十七

36. 二百八十

37. 千九百七十六

38. 三千一十七

39. 二千六百五

40. 三千四百八十七

Write each of the following as a traditional Chinese numeral.

41. 42   **42.** 156   **43.** 248

44. 2483   **45.** 9007   **46.** 1999

Write each of the following as a Hindu-Arabic numeral.

47. $\tau\xi\delta$ **48.** $\chi o\eta$ **49.** $\mu'\beta'\phi\,\epsilon$

50. $\rho'\nu'\omega\iota\gamma$ **51.** $\theta'\zeta$ **52.** $\alpha'\,\pi\,Q\,\theta$

Write each of the following as an Ionic Greek numeral.

53. 3   **54.** 78   **55.** 123

56. 2076   **57.** 35,412   **58.** 102,685

59. Construct your own additive numeration system. You can use the subtraction and multiplication properties if you wish. Write 1992 in your system.

60. Construct your own multiplicative numeration system. Write 1992 in your system.

61. Construct your own ciphered numeration system. Write 1992 in your system.

List the advantages and disadvantages of a ciphered system of numeration compared with each of the following systems.

62. An additive system

63. A multiplicative system

64. The Hindu-Arabic system

In Ex. 65–68, write each of the following as numerals in the indicated systems of numeration.

65. ⌜∩∩∩⌝ in Hindu-Arabic, Roman, Chinese, and Greek.

66. MCMXXXVI in Hindu-Arabic, Egyptian, Greek, and Chinese.

67. in Hindu-Arabic, Egyptian, Roman, and Greek.

68. νκβ in Hindu-Arabic, Egyptian, Roman, and Chinese.

Research Activity

69. Use books on the history of mathematics, an encyclopedia, and other sources to write a report on the Rhind Papyrus. Indicate what information archaeologists learned from the papyrus.

4.2 Place-Value or Positional-Value Numeration Systems

The eighteenth-century mathematician Pierre Simon, Marquis de Laplore, speaking of the positional principle, said: "The idea is so simple that this very simplicity is the reason for our not being sufficiently aware of how much attention it deserves."

Today the most common type of numeration system is the place-value system. The Hindu-Arabic numeration system, used in the United States and many other countries, is an example of a place-value system. In a **place-value system** the value of the symbol depends on its position in the representation of the number. For example, the 2 in 20 represents 2 tens, and the 2 in 200 represents 2 hundreds. A true positional-value system requires a **base** and a set of symbols, including a symbol for zero, and one for each counting number less than the base. Although any number can be written in any base, the most common positional system is the base 10 system (the decimal number system).

The Hindus in India are credited with the invention of zero and the other symbols used in our system. The Arabs, who traded regularly with the Hindus, also adopted the system—thus the name Hindu-Arabic. However, it was not until the middle of the fifteenth century that the Hindu-Arabic numerals took the form we know today.

The Hindu-Arabic numerals and the positional system of numeration revolutionized mathematics by making addition, subtraction, multiplication, and division much easier to learn and very practical to use. Merchants and traders no longer had to depend on the counting board, or abacus. The first group of mathematicians, who computed with the Hindu-Arabic system rather than with pebbles or beads on a wire, were known as the "algorists."

In the Hindu-Arabic system the symbols 0, 1, 2, 3, 4, 5, 6, 7, 8, and 9 are called **digits**. The base 10 system developed from counting on fingers, and the word "digit" comes from the Latin word for fingers.

The positional values in the Hindu-Arabic system are

$$\ldots, (10)^5, (10)^4, (10)^3, (10)^2, 10, 1.$$

To evaluate a number in the Hindu-Arabic system, the first digit on the right is multiplied by 1. The second digit from the right is multiplied by the base 10. The third digit from the right is multiplied by the base squared, 10^2 or 100. The fourth digit from the right is multiplied by the based cubed, 10^3 or 1000, and so on. In general, the digit that is n places from the right is multiplied by 10^{n-1}. Therefore the digit eight places from the right is multiplied by 10^7. Using the place-value rule, we can write a number in **expanded form**. The number 1234 written in expanded form is

$$1234 = (1 \times 10^3) + (2 \times 10^2) + (3 \times 10^1) + (4 \times 1)$$
$$(1 \times 1000) + (2 \times 100) + (3 \times 10) + 4$$

The oldest known numeration system that resembled a place-value system was developed by the Babylonians around 2500 B.C. Their system resembled a place-value system with a base of 60, a sexagesimal system. It was not a true place-value system because it lacked a symbol for zero. The lack of a symbol for zero led to a great deal of ambiguity and confusion. Table 4.5 gives the Babylonian numerals.

The positional values in the Babylonian system are

$$\ldots (60)^3, (60)^2, 60, 1.$$

In a Babylonian numeral a gap is left between the characters to distinguish between the various place values. In reading from right to left, the sum of the first group of numerals is multiplied by 1. The sum of the second group is multiplied by 60. The sum of the third group is multiplied by $(60)^2$, and so on.

Table 4.5		
Babylonian numerals	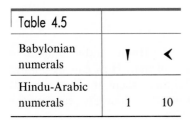	
Hindu-Arabic numerals	1	10

▶ Example 1

Write 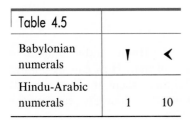 as a Hindu-Arabic numeral.

Solution:

$$\underbrace{\text{<<}}_{60\text{'s}} \qquad \underbrace{\text{<<IIII}}_{\text{units}}$$

$$\underbrace{10 + 10}_{60\text{'s}} \qquad \underbrace{10 + 10 + 1 + 1 + 1 + 1}_{\text{units}}$$

$$(20 \times 60) + (24 \times 1)$$
$$1200 + 24 = 1224$$

The Babylonians used the symbol to indicate subtraction. The numeral 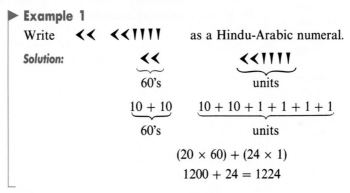 represents $10 - 2$, or 8. The numeral represents $35 - 12$, or 23 in decimal notation.

▶ Example 2

Write as a Hindu-Arabic numeral.

Solution: The place value of these three groups of numerals from left to right is

$$(60)^2, \quad 60, \quad 1$$
$$\text{or} \quad 3600, \quad 60, \quad 1$$

Counting Boards

One of the earliest counting devices, found in most ancient civilizations, was the counting board. On such a board each column represents a positional value. The number of times a value occurs is represented by markers (beads, stones, sticks) in the column. An empty column signifies "no value." The widespread use of counting boards meant that Europeans were already long accustomed to working with positional values when they were introduced to Hindu-Arabic numerals in the fifteenth century. People in China, Japan, the former Soviet Union, and Eastern Europe still commonly use a type of counting board known as the abacus to perform routine computations.

The numeral in the group on the right has a value of $20 - 2$, or 18. The numeral in the center has a value of $10 + 1$, or 11. The numeral on the left represents 1. Multiplying each group by its positional value gives

$$(1 \times 60^2) + (11 \times 60) + (18 \times 1)$$
$$= (1 \times 3600) + (11 \times 60) + (18 \times 1)$$
$$= 3600 + 660 + 18$$
$$= 4278$$

To explain the procedure used to convert from a Hindu-Arabic numeral to a Babylonian numeral, we will consider a length of time. How can we change 9820 seconds into hours, minutes, and seconds? Since there are 3600 seconds in an hour (60 seconds to a minute and 60 minutes to an hour), we can find the number of hours by dividing 9820 by 60^2, or 3600.

$$
\begin{array}{r}
2 \leftarrow \text{hours} \\
3600 \overline{) 9820} \\
7200 \\
\hline
2620 \leftarrow \text{remaining seconds}
\end{array}
$$

Now we can determine the number of minutes by dividing the remaining seconds by 60, the number of seconds in a minute.

$$
\begin{array}{r}
43 \leftarrow \text{minutes} \\
60 \overline{) 2620} \\
2400 \\
\hline
220 \\
180 \\
\hline
40 \leftarrow \text{remaining seconds}
\end{array}
$$

Since the remaining number of seconds, 40, is less than the number of seconds in a minute, our task is complete.

$$9820 \text{ sec} = 2 \text{ hr}, 43 \text{ min}, \text{ and } 40 \text{ sec}$$

The same procedure is used to convert a decimal (base 10) number to a Babylonian number or any number in a different base.

▶ **Example 3**

Write 1602 as a Babylonian numeral.

Solution: The Babylonian numeration system has positional values of

$$\ldots, 60^3, 60^2, 60, 1,$$

which can be expressed as

$$\ldots, 216000, 3600, 60, 1.$$

The largest positional value less than or equal to 1602 is 60. To determine how many groups of 60 are in 1602, divide 1602 by 60:

$$
\begin{array}{r}
26 \\
60 \overline{)\ 1602} \\
120 \\
\hline
402 \\
360 \\
\hline
42
\end{array}
$$

Thus, $1602 \div 60 = 26$ with remainder 42. There are 26 groups of 60 and 42 units remaining. Since the remainder, 42, is less than the base, 60, no further division is necessary. The remainder represents the number of units when the number is written in expanded form. Therefore $1602 = (26 \times 60) + (42 \times 1)$. When written as a Babylonian numeral, 1602 is

$$\text{<<????}?? \ \ \text{<<<<??}$$

▶ **Example 4**

Write 6270 as a Babylonian numeral.

Solution: Divide 6270 by the largest positional value less than or equal to 6270. That value is 3600.

$$6270 \div 3600 = 1 \text{ with remainder } 2670$$

There is one group of 3600 in 6270. Next divide the remainder 2670 by 60 to determine the number of groups of 60 in 2670:

$$2670 \div 60 = 44 \text{ with remainder } 30$$

There are 44 groups of 60 and 30 units remaining.

$$6270 = (1 \times 60^2) + (44 \times 60) + (30 \times 1)$$

Thus 6270 written as a Babylonian numeral is

$$? \ \ \text{<<<<????} \ \ \text{<<<}$$

Another place-value system is the Mayan numeration system. The Mayans, who lived on the Yucatan Peninsula, developed a sophisticated numeration system based upon their religious and agricultural calendar. The numbers in this system are written vertically rather than horizontally, with the units position on the bottom. In the Mayan system the number in the bottom row is to be multiplied by 1. The number in the second row from the bottom is to be multiplied by 20. The number in the third row is to be multiplied by 18×20, or 360. You probably expected the third row to be multiplied by 20^2 rather than 18×20. It is believed that the Mayans used 18×20 so that their numeration system would conform to their calendar of 360 days. The positional values

Did You Know...

Sacred Mayan Glyphs

3

4

In addition to their base 20 numerals the Mayans had a holy numeration system used by priests to create and maintain calendars. They used a special set of hieroglyphs that consisted of pictograms of Mayan gods. For example, the number 3 was represented by the god of wind and rain, the number 4 by the god of the sun.

above 18×20 are 18×20^2, 18×20^3, and so on.

Positional Values in the Mayan System

$\ldots 18 \times (20)^3$,	$18 \times (20)^2$,	18×20,	20,	1
or $\ldots 144{,}000$,	7200,	360,	20,	1

The digits $0, 1, 2, 3, \ldots, 19$ of the Mayan system are formed by a simple grouping of dots and lines, as shown in Table 4.6.

Table 4.6 Mayan Numerals

0	1	2	3	4	5	6	7	8	9

10	11	12	13	14	15	16	17	18	19

▶ **Example 5**

Write ⠶ as a Hindu-Arabic numeral.

Solution: In the Mayan numeration system the first three positional values are

$$18 \times 20$$
$$20$$
$$1$$

$$= 7 \times (18 \times 20) = 2520$$
$$= 2 \times 20 \quad\quad = 40$$
$$= 13 \times 1 \quad\quad = \underline{13}$$
$$2573$$

▶ **Example 6**

Write as a Hindu-Arabic numeral.

Solution:

$$= 8 \times (18 \times 20) = 2880$$
$$= 11 \times 20 \quad\quad = 220$$
$$= 4 \times 1 \quad\quad\quad = \underline{4}$$
$$3104$$

▶ **Example 7**

Write 1023 as a Mayan numeral.

Solution: To convert from a Hindu-Arabic to a Mayan numeral, we use a procedure similar to the one used to convert to a Babylonian numeral. The Mayan positional values are ..., 7200, 360, 20, 1. The greatest positional value less than or equal to 1023 is 360. Divide 1023 by 360:

$$1023 \div 360 = 2 \text{ with remainder } 303.$$

There are two groups of 360 in 1023. Next divide the remainder, 303, by 20:

$$303 \div 20 = 15 \text{ with remainder } 3.$$

There are 15 groups of 20 with three units remaining.

$$1023 = (2 \times 360) + (15 \times 20) + (3 \times 1)$$

1023 written as a Mayan numeral is

$$\begin{cases} 2 \times 360 \\ 15 \times 20 \\ 3 \times 1 \end{cases} = \begin{matrix} \bullet\bullet \\ \equiv \\ \bullet\bullet\bullet \end{matrix}$$

Section 4.2 Exercises

Write the Hindu-Arabic numeral in expanded form.
1. 48   **2.** 96   **3.** 942
4. 765   **5.** 3,452   **6.** 1,265
7. 64,521   **8.** 32,687   **9.** 245,672
10. 148,562

Write each of the following as a Hindu-Arabic numeral.

11. ⟨⟨⟨▾▾

12. ⟨⟨⟨⟨▾▾▾▾▾

13. ⟨▾▾ ▾▾▾▾

14. ▾⟨ ⟨⟨▾▾▾

15. ▾ ⟨⟨⟨⟨▾▾ ⟨▾▾▾

16. ⟨ ⟨⟨▾▾▾▾ ▾▾

Write each of the following as a Babylonian numeral.
17. 70   **18.** 95   **19.** 121
20. 512   **21.** 3878   **22.** 3030

Write each of the following as a Hindu-Arabic numeral.

23. •••
       ≡

24. ≡
       —

25. ••••
       ≡
       ⬭
       •

26. ••
       ••
       ••
       —

27. •
       ≡
       ••
       ⬭

28. ••••
       ≡
       —

Write each of the following as a Mayan numeral.
29. 15   **30.** 227

31. 300   **32.** 406
33. 3060   **34.** 1978
35. The Babylonians lacked a symbol for zero. Why did this fact lead to confusion?
36. What are the advantages and disadvantages of a place-value system compared with (a) additive numeration systems, (b) multiplicative numeration systems, (c) ciphered numeration systems?
37. Create your own place-value system. Write 1992 in your system.

In Ex. 38 and 39, write the numeral in the indicated systems of numeration.
38. ⟨⟨⟨▾▾▾ in Hindu-Arabic and Mayan.
39. —— in Hindu-Arabic and Babylonian.
       ••

       ••••

Research Activity

40. Investigate and write a report on the development of the Hindu-Arabic system of numeration. Start with the earliest records of this system in India. Possible sources: *World of Mathematics* by J. R. Newman; *A History of Mathematics* by Carl B. Boyer; *An Introduction to the History of Mathematics* by Howard Eves.

4.3 Other Bases

The positional values in the Hindu-Arabic numeration system are

$$\ldots, (10)^4, (10)^3, (10)^2, 10, 1.$$

The positional values in the Babylonian numeration system are

$$\ldots, (60)^4, (60)^3, (60)^2, 60, 1.$$

The numbers 10 and 60 are called the **bases** in the Hindu-Arabic and Babylonian systems respectively.

Any counting number greater than 1 may be used as a base for a positional-value numeration system. If a positional-value system has a base b, then its positional values will be

$$\ldots, b^4, b^3, b^2, b, 1.$$

The positional values in a base 8 system are

$$\ldots, 8^4, 8^3, 8^2, 8, 1$$

and the positional values in a base 2 system are

$$\ldots, 2^4, 2^3, 2^2, 2, 1.$$

The Kewa people of Papua, New Guinea, have gone well beyond counting on their fingers—they use the entire upper body. Going from the little finger of one hand, down the elbow, to shoulder to head to shoulder, down the other elbow, to the little finger of the opposite hand provides them with a count of 68.

As we indicated earlier, the Mayan numeration system is based on the number 20. However, it is not a true base 20 positional-value system. Why?

The fact that human beings have ten fingers is believed to be responsible for the almost universal acceptance of base 10 numeration systems. Even so, there are still some positional-value numeration systems that use bases other than 10. Some societies are still using a base 2 numeration system. They include some groups of people in Australia, New Guinea, Africa, and South America. Bases 3 and 4 are also used in some areas of South America. The only base 5 system in pure form at present seems to be the one used in Saraveca, a South American Arawakan language. Elsewhere, base 5 systems are combined with base 10 or base 20 systems. The pure base 6 system occurs only sparsely in Northwest Africa. Base 6 also occurs in other systems in combination with base 12, the *duodecimal system.*

We continue to see remains of other base systems in many countries. For example, there are 12 inches in a foot, 12 months in a year. Base 12 is also evident in the dozen, the 24-hour day, and the gross (12×12). English uses the word "score" to mean 20, as in "Four score and seven years ago." Other traces are found in pre-English Celtic, Gaelic, Danish, and Welsh. Remains of base 60 are found in measurements of time (60 seconds to a minute, 60 minutes to an hour) and angles (60 seconds to one minute, 60 minutes to one degree).

The base 2, or binary, number system has become very important because it is the internal language of the computer. For example, when the grocery store's cash register computer records the price of your groceries using a scanning device, the bar codes it scans on the packages are in binary form. Computers

The *I ching*

One of the most influential books in Chinese history is the *I ching,* or *Book of Changes,* said to have been written by the emperor Fu Hsi in the 29th century B.C. The Chinese used the 64 hexagrams of the *I ching* to make predictions. These were formed from eight trigrams said to encompass all that happens in heaven and on earth: the creative, the receptive, the arousing, the abysmal, keeping still, the gentle, the clinging, the joyous. The German mathematician Gottfried Leibniz (1646–1716) recognized in the *I ching's* patterns of broken and unbroken lines a system similar to binary arithmetic. If you imagine the broken line to be a zero and the unbroken line to be a one, each hexagram can be interpreted as a binary number.

use a two-digit "alphabet" that consists of the numerals 0 and 1. Every character on a standard keyboard can be represented by a combination of those two numerals. A single numeral such as 0 or 1 is called a **bit**. Other bases that computers make use of are base 8 and base 16. A group of four bits is called a **nibble**, a group of eight bits is called a **byte**. In the ASCII code (American Standard Code for Information Interchange), used in most computers, the byte 01000001 represents the character A, 01100001 represents the character a, 00110000 represents the character 0, and 00110001 represents the character 1.

A place-value system with base b must have b distinct symbols, one for zero and one for each number less than the base. A base 6 system must have symbols for the numbers 0, 1, 2, 3, 4, 5. A base 8 system must have symbols for 0, 1, 2, 3, . . . , 7. A number in a base other than 10 will be indicated by a subscript to the right of the number. Thus 123_5 represents a number in base 5. The value of 123_5 is not the same as the value of 123_{10}. A base 10 number may be written without a subscript. For clarity, in certain problems we will use the subscript 10 to indicate a number in base 10. Remember, there is never a single symbol for the base number itself. For example, in base 5 the only symbols are those for the numbers 0, 1, 2, 3, and 4, and all other numbers are constructed from these five. The symbol that we will use for the base, in base 5, is 10_5, which means one group of five and no units.

To change a number in a base other than 10 to a base 10 number, we follow the same procedure we used in Section 4.2 to change the Babylonian and Mayan numbers to base 10 numbers. Multiply each digit in the number by its respective positional value. Then find the sum of the products.

▶ **Example 1**

Convert 234_6 to base 10.

Solution: In base 6 the positional values are . . . $6^3, 6^2, 6, 1$. In expanded form,

$$234_6 = (2 \times 6^2) + (3 \times 6) + (4 \times 1)$$
$$= (2 \times 36) + (3 \times 6) + (4 \times 1)$$
$$= \quad 72 \quad + \quad 18 \quad + \quad 4$$
$$= 94$$

▶ **Example 2**

Convert 3615_8 to base 10.

Solution: $\quad 3615_8 = (3 \times 8^3) \ + (6 \times 8^2) + (1 \times 8) + (5 \times 1)$
$$= (3 \times 512) + (6 \times 64) + (1 \times 8) + (5 \times 1)$$
$$= \quad 1536 \quad + \quad 384 \quad + \quad 8 \quad + \quad 5$$
$$= 1933$$

A base 12 system must have 12 distinct symbols. This text will use the symbols 0, 1, 2, 3, 4, 5, 6, 7, 8, 9, T, E, where T represents ten and E represents eleven. Why will the numerals 10_{12} and 11_{12} have different meanings than 10 and 11?

▶ **Example 3**

Convert $12T6_{12}$ to base 10.

Solution:
$$\begin{aligned}
12T6_{12} &= (1 \times 12^3) + (2 \times 12^2) + (T \times 12) + (6 \times 1) \\
&= (1 \times 1728) + (2 \times 144) + (10 \times 12) + (6 \times 1) \\
&= \quad 1728 \quad + \quad 288 \quad + \quad 120 \quad + \quad 6 \\
&= 2142
\end{aligned}$$

▶ **Example 4**

Convert 101101_2 to base 10.

Solution:

$$\begin{aligned}
101101_2 &= (1 \times 2^5) + (0 \times 2^4) + (1 \times 2^3) + (1 \times 2^2) + (0 \times 2) + (1 \times 1) \\
&= \quad 32 \quad + \quad 0 \quad + \quad 8 \quad + \quad 4 \quad + \quad 0 \quad + \quad 1 \\
&= 45
\end{aligned}$$

To change a number from a base 10 system to a different base, we will use the procedure explained in Section 4.2. Divide the number by the highest power of the base less than or equal to the given number. Record this quotient. Then divide the remainder by the next smaller power of the base and record this quotient. Repeat this procedure until the remainder is a number less than the base. The answer is the set of quotients listed from left to right, with the remainder on the far right. This procedure is illustrated in the following examples.

▶ **Example 5**

Convert 406 to base 8.

Solution: The positional values in the base 8 system are ..., 8^3, 8^2, 8, 1, or ..., 512, 64, 8, 1. The highest power of 8 that is less than or equal to 406 is 8^2, or 64. Divide 406 by 64.

first digit in answer

$$406 \div 64 = 6 \text{ with remainder } 22$$

Therefore there are six groups of 8^2 in 406. Next divide the remainder, 22, by 8.

second digit in answer

$$22 \div 8 = 2 \text{ with remainder } 6$$

third digit in answer

There are two groups of 8 in 22 and 6 units remaining. Since the remainder, 6, is less than the base, 8, no further division is required.

$$\begin{aligned}
406 &= (6 \times 64) + (2 \times 8) + (6 \times 1) \\
&= (6 \times 8^2) + (2 \times 8) + (6 \times 1) \\
&= 626_8
\end{aligned}$$

▶ **Example 6**

Convert 273 to base 3.

Solution: The place values in the base 3 system are . . . , 3^6, 3^5, 3^4, 3^3, 3^2, 3, 1, or . . . , 729, 243, 81, 27, 9, 3, 1. The highest power of the base that is less than 273 is 3^5, or 243. Successive divisions by the powers of the base give

$$273 \div 243 = 1 \text{ with remainder of } 30$$
$$30 \div 81 = 0 \text{ with remainder } 30$$
$$30 \div 27 = 1 \text{ with remainder } 3$$
$$3 \div 9 = 0 \text{ with remainder } 3$$
$$3 \div 3 = 1 \text{ with remainder } 0$$

Since the remainder, 0, is less than the base, 3, no further division is necessary. To obtain the answer, list the quotients from top to bottom followed by the remainder in the last division.

The number 273 can be represented as one group of 243, no groups of 81, one group of 27, no groups of 9, one group of 3, and no units.

$$273 = (1 \times 243) + (0 \times 81) + (1 \times 27) + (0 \times 9) + (1 \times 3) + (0 \times 1)$$
$$= (1 \times 3^5) + (0 \times 3^4) + (1 \times 3^3) + (0 \times 3^2) + (1 \times 3) + (0 \times 1)$$
$$= 101010_3$$

▶ **Example 7**

Convert 558 to base 12.

Solution: The place values in base 12 are . . . , 12^3, 12^2, 12, 1, or . . . , 1728, 144, 12, 1.

$$558 \div 144 = 3 \text{ with remainder } 126$$
$$126 \div 12 = T \text{ with remainder } 6$$

(Remember that T is used to represent ten in base 12.)

$$558 = (3 \times 12^2) + (T \times 12) + (6 \times 1) = 3T6_{12}$$

Section 4.3 Exercises

In Ex. 1–18, convert the number to a number in base 10.

1. 5_7	**2.** 40_8	**3.** 12_5
4. 101_2	**5.** 1011_2	**6.** 1101_2
7. 67_{12}	**8.** 20221_3	**9.** 674_9
10. 654_7	**11.** 20432_5	**12.** 101111_2
13. 3001_4	**14.** $123E_{12}$	**15.** 123_8
16. 1023_8	**17.** 10047_8	**18.** 84721_9

Convert the base 10 number to a number in the base indicated.

19. 8 to base 2	**20.** 16 to base 2
21. 23 to base 2	**22.** 46 to base 5
23. 406 to base 8	**24.** 809 to base 4
25. 1695 to base 12	**26.** 100 to base 3
27. 230 to base 6	**28.** 64 to base 2

29. 286 to base 12   **30.** 1234 to base 5
31. 1011 to base 2   **32.** 1589 to base 7
33. 2408 to base 8   **34.** 13,469 to base 8

Assume that a base 16 positional-value system uses the numerals 0, 1, 2, 3, 4, 5, 6, 7, 8, 9, A, B, C, D, E, F where A through F represent 10 through 15 respectively. Convert each of the following to base 10.

35. 734_{16} **36.** 285_{16}
37. $6D3B_{16}$ **38.** $24FE_{16}$

Convert the following to base 16.

39. 307   **40.** 349
41. 5478   **42.** 34,721

Convert 1992 to each of the following bases.

43. 2   **44.** 3   **45.** 5
46. 7   **47.** 8   **48.** 16

If any of the following numerals is written incorrectly, explain why.

49. 4063_5 **50.** 1203_3
51. 674_8 **52.** 1206_{12}

53. There is an alternative method for changing a number in base 10 to a different base. This method will be used to convert 328 to base 5. Dividing 328 by 5 gives a quotient of 65 and a remainder of 3. Write the quotient below the dividend and the remainder on the right, as illustrated below.

Continue this process of division by 5.

```
5 | 328   remainder
5 |  65   3
5 |  13   0     answer
5 |   2   3
      0   2
```

(Since the dividend, 2, is smaller than the divisor, 5, the quotient is 0 and the remainder is 2.)

Note that the division continues until the quotient is zero. The answer is read from the bottom number to the top number in the remainder column. Thus $328 = 2303_5$.

a) Explain why this procedure results in the proper answer.

b) Convert 683 to base 5 by this method.

c) Convert 763 to base 8 by this method.

Problem Solving

54. The American Standard Code for Information Interchange (ASCII), employed by most computers, uses the last seven positions of an eight-bit byte to represent all the characters on a standard keyboard. How many different orderings of 0s and 1s (or how many different characters) can be made using the last seven positions of an eight-bit byte?

55. Find b if $111_b = 43$.

Research Activity

56. Investigate and write a report on how digital computers use the binary number system.

4.4 Computation in Other Bases

Addition

When computers perform calculations, they do so in base 2, the binary system. In this section we explain how to perform calculations in base 2 and other bases.

In a base 2 system the only digits are 0 and 1, and the place values are

$$\ldots, 2^4, 2^3, 2^2, 2, 1$$
$$\text{or} \ldots, 16, 8, \ 4, \ 2, 1.$$

Suppose we wish to add $1_2 + 1_2$. The subscript 2 indicates that we are adding in base 2. Remember the answer to $1_2 + 1_2$ must be written using only the digits 0 and 1. The sum of $1_2 + 1_2$ is 10_2 since $1_2 + 1_2$ is one group of two and no units. Recall that 10_2 means $1(2) + 0(1)$.

If we wished to find the sum of $10_2 + 1_2$ we would add the digits in the right-hand, or units, column. Since $0_2 + 1_2 = 1_2$, the sum of $10_2 + 1_2 = 11_2$.

Since we are going to perform additional examples and exercises in base 2, rather than performing individual calculations in every problem, we can construct and use an addition table, Table 4.7, for base 2 (just as we use an addition table in base 10 when we are first learning to add in base 10).

Table 4.7

+	0	1
0	0	1
1	1	10

▶ **Example 1**

Add 1101_2
111_2

Solution: Begin by adding the numbers in the right-hand or units column. From our previous discussion, and as can be seen in Table 4.7, $1_2 + 1_2 = 10_2$. Place the 0 under the units column and carry the 1 to the twos column on the left, as follows:

$$\begin{array}{cccc} 1 & 1 & {}^1 0 & 1 \\ & 1 & 1 & 1 \\ \hline & & & 0_2 \end{array}$$

Now add the three digits in the twos column, $1_2 + 0_2 + 1_2$. Treat this as $(1_2 + 0_2) + 1_2$. Therefore we add $1_2 + 0_2$ to get 1_2, then add $1_2 + 1_2$ to get 10_2. Place the 0 under the twos column and carry the 1 to the 2^2 column (the third column from the right).

$$\begin{array}{cccc} 1 & {}^1 1 & {}^1 0 & 1 \\ & 1 & 1 & 1 \\ \hline & & 0 & 0_2 \end{array}$$

Now add the three 1s in the 2^2 column to get $(1_2 + 1_2) + 1_2 = 10_2 + 1_2 = 11_2$. Place the 1 under the 2^2 column and carry the 1 to the 2^3 column (the fourth column from the right).

$$\begin{array}{cccc} {}^1 1 & {}^1 1 & {}^1 0 & 1 \\ & 1 & 1 & 1 \\ \hline 1 & 0 & 0_2 \end{array}$$

Now add the two 1s in the 2^3 column, $1_2 + 1_2 = 10_2$. Place the 10 as follows:

$$\begin{array}{ccccc} {}^1 1 & {}^1 1 & {}^1 0 & 1 \\ & 1 & 1 & 1 \\ \hline 1 & 0 & 1 & 0 & 0_2 \end{array}$$

Therefore the sum is 10100_2.

Let us now look at addition in a base 5 system. In base 5 the only digits

Table 4.8

+	0	1	2	3	4
0	0	1	2	3	4
1	1	2	3	4	10
2	2	3	4	10	11
3	3	4	10	11	12
4	4	10	11	⑫	13

are 0, 1, 2, 3, and 4, and the positional values are

$$\ldots, 5^4, \quad 5^3, 5^2, 5, 1$$
$$\text{or} \quad \ldots, 625, 125, 25, 5, 1.$$

What is the sum of $4_5 + 3_5$? We can consider this to mean $(1 + 1 + 1 + 1) + (1 + 1 + 1)$. We can regroup the seven 1s into one group of five and two units as follows: $(1 + 1 + 1 + 1 + 1) + (1 + 1)$. Thus the sum of $4_5 + 3_5 = 12_5$ (circled in Table 4.8). Recall that 12_5 means $1(5) + 2(1)$. We can use this same procedure in obtaining the values in the base 5 addition table.

▶ **Example 2**

Add 32_5.
$\underline{33_5}$

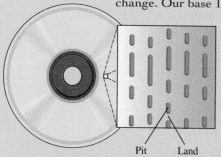

Solution: First determine that $2_5 + 3_5$ is 10_5 from Table 4.8. Record the 0 and carry the 1 to the fives column.

$$\begin{array}{r} ^13\ 2_5 \\ 3\ 3_5 \\ \hline 0_5 \end{array}$$

Add the numbers in the second column, $(1_5 + 3_5) + 3_5 = 4_5 + 3_5 = 12_5$. Record the 12.

$$\begin{array}{r} ^13\ 2_5 \\ 3\ 3_5 \\ \hline 1\ 2\ 0_5 \end{array}$$

The sum is 120_5.

▶ **Example 3**

Add 1234_5.
$\quad\ 2042_5$

Solution:

$$\begin{array}{r} 1\ {}^12\ {}^13\ 4_5 \\ 2\ 0\ 4\ 2_5 \\ \hline 3\ 3\ 3\ 1_5 \end{array}$$

You can develop an addition table for any base and use it to add in that base. However, as you get more comfortable with addition in other bases, you may prefer to add numbers in other bases using mental arithmetic. To do so, convert the sum of the numbers being added from the given base to base 10, and then convert the base 10 number back into the given base. You must clearly understand how to convert from base 10 to the given base, as was discussed in Section 4.3. As an example, to add $7_9 + 8_9$ add $7 + 8$ in base 10 to get 15_{10}, and then mentally convert 15_{10} to 16_9 using the procedure given earlier. Remember, 16_9 when converted to base 10 becomes $1(9) + 6(1)$, or 15. Addition using this procedure is illustrated in Examples 4 and 5.

▶ **Example 4**

Add 1022_3.
$\quad\ 2121_3$

Solution: To solve this problem make the necessary conversions by using mental arithmetic. $2 + 1 = 3_{10} = 10_3$. Record the 0 and carry the 1.

$$\begin{array}{r} 1\ 0\ {}^12\ 2_3 \\ 2\ 1\ 2\ 1_3 \\ \hline 0_3 \end{array}$$

$1 + 2 + 2 = 5_{10} = 12_3$. Record the 2 and carry the 1.

$$\begin{array}{r} 1\;{}^{1}0\;{}^{1}2\;2_3 \\ 2\;\;1\;\;2\;1_3 \\ \hline 2\;\;0_3 \end{array}$$

$1 + 0 + 1 = 2_{10} = 2_3$. Record the 2.

$$\begin{array}{r} 1\;{}^{1}0\;{}^{1}2\;2_3 \\ 2\;\;1\;\;2\;1_3 \\ \hline 2\;\;2\;\;0_3 \end{array}$$

$1 + 2 = 3_{10} = 10_3$. Record the 10.

$$\begin{array}{r} 1022_3 \\ 2121_3 \\ \hline 10220_3 \end{array}$$

▶ **Example 5**

Add 444_5.
$$\begin{array}{r} 444_5 \\ 244_5 \\ 143_5 \\ 214_5 \\ \hline \end{array}$$

Solution: Adding the digits in the right-hand column, $4 + 4 + 3 + 4 = 15_{10} = 30_5$. Record the 0 and carry the 3. Adding the 3 with the digits in the next column yields $3 + 4 + 4 + 4 + 1 = 16_{10} = 31_5$. Record the 1 and carry the 3. Adding the 3 with the digits in the left-hand column gives $3 + 4 + 2 + 1 + 2 = 12_{10} = 22_5$. Record both digits. The sum of these four numbers is 2210_5.

$$\begin{array}{r} {}^{3}4\;{}^{3}4\;4_5 \\ 2\;\;4\;4_5 \\ 1\;\;4\;3_5 \\ 2\;\;1\;4_5 \\ \hline 2\;\;2\;\;1\;\;0_5 \end{array}$$

Subtraction

Subtraction can also be performed in other bases. It is important to remember that when you "borrow," you borrow the amount of the base. In the following two examples, we will perform the subtraction in base 10 when convenient and convert the results to the given base.

▶ **Example 6**

Subtract $\quad 3032_5$.
$$\underline{-\;1004_5}$$

Solution: Since 4 is greater than 2, we must borrow one group of 5 from the preceding column. This gives a sum of $5 + 2$, or 7 in base 10. Now subtract 4 from 7; the difference is 3. We complete the problem in the usual manner. The 3 in the second column becomes a 2, $2 - 0 = 2, 0 - 0 = 0$, and $3 - 1 = 2$.

$$
\begin{array}{r}
3032_5 \\
-\ 1004_5 \\
\hline
2023_5
\end{array}
$$

▶ **Example 7**

Subtract $\quad 468_{12}.$
$\qquad -\ 295_{12}$

Solution: $8 - 5 = 3$. Next we must subtract 9 from 6. Since 9 is greater than 6, borrowing is necessary. We must borrow one group of 12 from the preceding column. We then have a sum of $12 + 6 = 18$ in base 10. Now we subtract 9 from 18, and the difference is 9. The 4 in the left column becomes a 3, and $3 - 2 = 1$.

$$
\begin{array}{r}
468_{12} \\
-\ 295_{12} \\
\hline
193_{12}
\end{array}
$$

Multiplication

Multiplication can also be performed in other bases. Doing so is helped by forming a multiplication table for the base desired. Suppose we wish to determine the product of $4_5 \times 3_5$. In base 10, 4×3 means there are four groups of three units. Similarly in a base 5 system, $4_5 \times 3_5$ means there are four groups of three units, or

$$(1 + 1 + 1) + (1 + 1 + 1) + (1 + 1 + 1) + (1 + 1 + 1).$$

Regrouping the 12 units into groups of five gives

$$(1 + 1 + 1 + 1 + 1) + (1 + 1 + 1 + 1 + 1) + (1 + 1),$$

or two groups of five, and two units. Thus, $4_5 \times 3_5 = 22_5$.

We can construct other values in the base 5 multiplication table in the same way. However, many people find it easier to multiply the values in the base 10 system, and then change the product to base 5 using the procedure discussed in Section 4.3. Multiplying 4×3 in base 10 gives 12, and converting 12 from base 10 to base 5 gives 22_5.

The product of $4_5 \times 3_5$ is circled in Table 4.9, the base 5 multiplication table. The other values in the table may be found by either method discussed.

Table 4.9

×	0	1	2	3	4
0	0	0	0	0	0
1	0	1	2	3	4
2	0	2	4	11	13
3	0	3	11	14	22
4	0	4	13	(22)	31

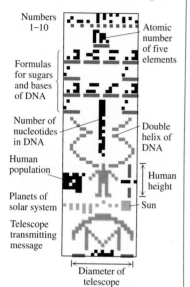
▶ **Example 8**

Multiply $\begin{array}{r} 13_5 \\ \times\ 3_5 \end{array}$ using the base 5 multiplication table.

Solution: Multiply as you would in base 10, but use the base 5 multiplication table to find the products. When the product consists of two digits, record the right digit and carry the left digit. Multiplying $3_5 \times 3_5 = 14_5$. Record the 4 and carry the 1.

$$\begin{array}{r} {}^1 13_5 \\ \times\ 3_5 \\ \hline 4 \end{array}$$

$(3_5 \times 1_5) + 1_5 = 4_5$. Record the 4.

$$\begin{array}{r} {}^1 13_5 \\ \times\ 3_5 \\ \hline 44_5 \end{array}$$

The product is 44_5.

It is often a tedious task to construct a multiplication table, especially when the base is large. To multiply in a given base without the use of a table, multiply in base 10 and convert the products to the appropriate base number before recording them. This procedure is illustrated in Example 9.

▶ **Example 9**

Multiply $\begin{array}{r} 43_7 \\ \times\ 25_7 \end{array}$

Solution: $5 \times 3 = 15_{10} = 2(7) + 1(1) = 21_7$. Record the 1 and carry the 2.

$$\begin{array}{r} {}^2 43_7 \\ \times\ 25_7 \\ \hline 1 \end{array}$$

$(5 \times 4) + 2 = 20 + 2 = 22_{10} = 3(7) + 1(1) = 31_7$. Record the 31.

$$\begin{array}{r} {}^2 43_7 \\ \times\ 25_7 \\ \hline 311 \end{array}$$

$2 \times 3 = 6$. Record the 6.

$$\begin{array}{r} {}^2 43_7 \\ \times\ 25_7 \\ \hline 311 \\ 6 \end{array}$$

$2 \times 4 = 8_{10} = 1(7) + 1(1) = 11_7$. Record the 11. Now add in base 7 to determine the answer. Remember, in base 7 there are no digits greater than 6.

$$\begin{array}{r} ^{2}43_7 \\ 25_7 \\ \hline 311 \\ 116 \\ \hline 1501_7 \end{array}$$

Division

Division is performed in much the same manner as long division in base 10. A detailed example of a division in base 5 is illustrated below. The same procedure is used for division in any other base.

▶ **Example 10**

Divide $2_5 \overline{\smash{\big)}\, 143_5}$.

Solution: Using Table 4.9, the multiplication table for base 5, we list the multiples of the divisor, 2.

$$2_5 \times 1_5 = 2_5$$
$$2_5 \times 2_5 = 4_5$$
$$2_5 \times 3_5 = 11_5$$
$$2_5 \times 4_5 = 13_5$$

Since $2_5 \times 4_5 = 13_5$, which is less than 14_5, 2_5 goes into 14_5 four times:

$$\begin{array}{r} 4 \\ 2_5 \overline{\smash{\big)}\, 143_5} \\ 13 \\ \hline 1 \end{array}$$

Now bring down the 3 as when dividing in base 10:

$$\begin{array}{r} 4 \\ 2_5 \overline{\smash{\big)}\, 143_5} \\ 13 \\ \hline 13 \end{array}$$

We see that $2_5 \times 4_5 = 13_5$. Use this information to complete the problem:

$$\begin{array}{r} 44_5 \\ 2_5 \overline{\smash{\big)}\, 143} \\ 13 \\ \hline 13 \\ 13 \\ \hline 0 \end{array}$$

Thus $143_5 \div 2_5 = 44_5$ with a remainder of zero.

A division problem can be checked by multiplication. If the division was performed correctly, (quotient × divisor) + remainder = dividend. We can check Example 10 as follows:

$$(44_5 \times 2_5) + 0_5 = 143_5$$

$$\begin{array}{r} 44_5 \\ \times\ 2_5 \\ \hline 143_5 \end{array} \quad \text{(check)}$$

▶ **Example 11**

Divide $4_6\overline{)\,2430_6}$.

Solution: The multiples of 4 in base 6 are

$$4_6 \times 1_6 = 4_6$$
$$4_6 \times 2_6 = 12_6$$
$$4_6 \times 3_6 = 20_6$$
$$4_6 \times 4_6 = 24_6$$
$$4_6 \times 5_6 = 32_6.$$

$$\begin{array}{r} 404_6 \\ 4_6\overline{)\,2430_6} \\ 24 \\ \hline 03 \\ 00 \\ \hline 30 \\ 24 \\ \hline 2 \end{array}$$

Thus the quotient is 404_6, with a remainder of 2_6.

Be careful when subtracting! When you borrow, remember that you borrow 10_6, which is the same as 6 in base 10.

Check: Does $(404_6 \times 4_6) + 2_6 = 2430_6$?

$$\begin{array}{r} 404_6 \\ \times\quad 4_6 \\ \hline 2424_6 + 2_6 = 2430_6 \end{array} \quad \text{(check)}$$

Section 4.4 Exercises

Add the following in the indicated base.

1. $\begin{array}{r} 13_4 \\ 103_4 \\ \hline \end{array}$ **2.** $\begin{array}{r} 20_7 \\ 65_7 \\ \hline \end{array}$ **3.** $\begin{array}{r} 4023_5 \\ 2334_5 \\ \hline \end{array}$ **4.** $\begin{array}{r} 101_2 \\ 11_2 \\ \hline \end{array}$ **5.** $\begin{array}{r} 467_{12} \\ 238_{12} \\ \hline \end{array}$ **6.** $\begin{array}{r} 222_3 \\ 22_3 \\ \hline \end{array}$

7. 1012_3
$\underline{1011_3}$

8. 470_{12}
$\underline{347_{12}}$

9. 14631_7
$\underline{6040_7}$

27. 234_9
$\underline{\times\ \ 23_9}$

28. $6T3_{12}$
$\underline{\times\ \ 24_{12}}$

29. 111_2
$\underline{\times\ 111_2}$

*10. $43A_{16}$
$\underline{496_{16}}$

Divide the following in the indicated base.

30. $1_2\lceil\overline{110_2}$

31. $5_6\lceil\overline{342_6}$

32. $3_5\lceil\overline{143_5}$

Subtract the following in the indicated base.

11. 203_4
$\underline{-\ 103_4}$

12. 463_7
$\underline{-\ 124_7}$

13. 2334_5
$\underline{-\ 1243_5}$

33. $6_8\lceil\overline{466_8}$

34. $2_4\lceil\overline{312_4}$

35. $6_{12}\lceil\overline{431_{12}}$

36. $3_7\lceil\overline{2101_7}$

14. 1011_2
$\underline{-\ 101_2}$

15. 463_{12}
$\underline{-\ 13T_{12}}$

16. 1221_3
$\underline{-\ 202_3}$

Problem Solving

Divide the following in the indicated base.

17. 1001_2
$\underline{-\ 110_2}$

18. 1453_{12}
$\underline{-\ 245_{12}}$

19. 4223_7
$\underline{-\ 304_7}$

37. $14_5\lceil\overline{301_5}$

38. $20_4\lceil\overline{223_4}$

*20. $4E7_{16}$
$\underline{-189_{16}}$

Research Activities

39. Investigate and write a report on the use of the duodecimal (base 12) system as a system of numeration. You might wish to contact the Duodecimal Society. The address is Duodecimal Society, Nassau Community College, Garden City, NY 11530.

Multiply the following in the indicated base.

21. 42_5
$\underline{\times\ \ 3_5}$

22. 123_5
$\underline{\times\ \ \ 4_5}$

23. 423_7
$\underline{\times\ \ \ 6_7}$

40. One method used by computers to perform subtraction is the "end around carry method." Do research in books on computers, in encyclopedias, or in other sources, and write a report explaining, with specific examples, how a computer performs subtraction using the end around carry method.

24. 101_2
$\underline{\times\ 11_2}$

25. 302_4
$\underline{\times\ 23_4}$

26. 124_{12}
$\underline{\times\ \ 6_{12}}$

---

* See Ex. 35–38 in Section 4.3.

4.5 Early Computational Methods

Early civilizations used various methods for multiplying and dividing. Multiplication was performed by *duplation and mediation*, by the *galley method*, and by *Napier rods*. Following is an explanation of each method.

Duplation and Mediation

▶ **Example 1**

Multiply 17 × 30 using duplation and mediation.

Solution: Write 17 and 30 with a dash between to separate them. Divide the number on the left, 17, in half, drop the remainder, and place the quotient, 8, under the 17. Double the number on the right, 30, obtaining 60, and place it under the 30. You will then have the following paired lines:

$$17 - 30$$
$$8 - 60$$

Continue this process, taking one-half the number in the left-hand column, disregarding the remainder, and doubling the number in the right-hand column, as shown below. When a 1 appears in the left-hand column, stop.

$$17 - 30$$
$$8 - 60$$
$$4 - 120$$
$$2 - 240$$
$$1 - 480$$

Cross out all the numbers in the left-hand column that are even and the corresponding numbers in the right-hand column.

$$17 - 30$$
$$\cancel{8 - 60}$$
$$\cancel{4 - 120}$$
$$\cancel{2 - 240}$$
$$1 - 480$$

Now add the remaining numbers in the right-hand column, obtaining $30 + 480 = 510$, which is the product you want. If you check, you will find that $17 \times 30 = 510$.

The Galley Method

The galley method (sometimes referred to as the Gelosia method) was developed after duplation and mediation. To multiply 312×75 using the galley method, you must construct a rectangle consisting of three columns (one for each digit of 312) and two rows (one for each digit of 75).

Place the digits 3, 1, 2 above the boxes and the digits 7, 5 on the right of the boxes, as shown in Fig. 4.1. Then place a diagonal in each box.

Complete each box by multiplying the number on top of the box by the number to the right of the box (Fig. 4.2). Place the units below the diagonal and the tens above.

Add the numbers along the diagonals, as shown in Fig. 4.3, starting with the bottom right diagonal. If the sum in a diagonal is 10 or greater, record the

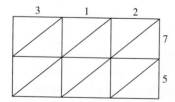

Figure 4.1

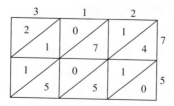

Figure 4.2

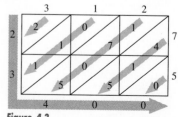

Figure 4.3

*During the seventeenth century, the growth of scientific fields such as astronomy required the ability to perform often unwieldy calculations. The English mathematician **John Napier** (1550–1617) made great contributions toward solving the problem of computing these numbers. His inventions include simple calculating machines and a device for performing multiplication and division known as Napier bones. Napier also developed the theory of logarithms.*

units digit below the rectangle and carry the tens digit to the next diagonal to the left.

The answer is read down the left-hand column and along the bottom, as shown by the arrow in Fig. 4.3. The answer is 23,400.

Napier Rods

The third method was developed from the galley method by John Napier in the seventeenth century. His method of multiplication, known as Napier rods, proved to be one of the forerunners of the modern-day computer. Napier developed a system using separate rods numbered from 0 through 9 and an additional strip for an index, numbered vertically 1 through 9; see Fig. 4.4. Each rod is divided into 10 blocks, and each block contains a multiple of the top number. Units are placed to the right, and tens to the left. Example 2 explains how Napier rods are used to multiply numbers.

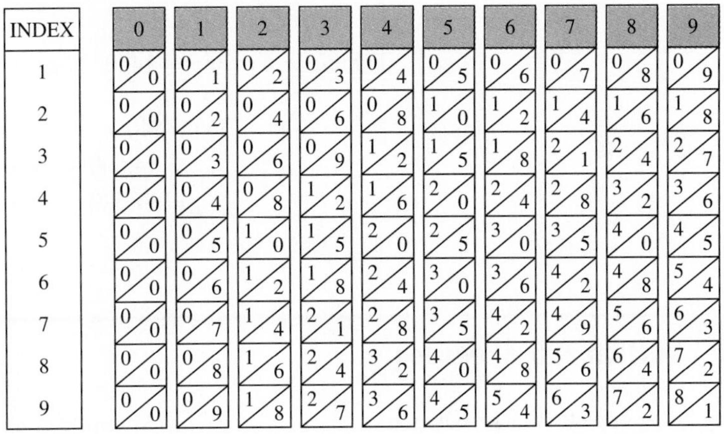

INDEX	0	1	2	3	4	5	6	7	8	9
1	0/0	0/1	0/2	0/3	0/4	0/5	0/6	0/7	0/8	0/9
2	0/0	0/2	0/4	0/6	0/8	1/0	1/2	1/4	1/6	1/8
3	0/0	0/3	0/6	0/9	1/2	1/5	1/8	2/1	2/4	2/7
4	0/0	0/4	0/8	1/2	1/6	2/0	2/4	2/8	3/2	3/6
5	0/0	0/5	1/0	1/5	2/0	2/5	3/0	3/5	4/0	4/5
6	0/0	0/6	1/2	1/8	2/4	3/0	3/6	4/2	4/8	5/4
7	0/0	0/7	1/4	2/1	2/8	3/5	4/2	4/9	5/6	6/3
8	0/0	0/8	1/6	2/4	3/2	4/0	4/8	5/6	6/4	7/2
9	0/0	0/9	1/8	2/7	3/6	4/5	5/4	6/3	7/2	8/1

Figure 4.4

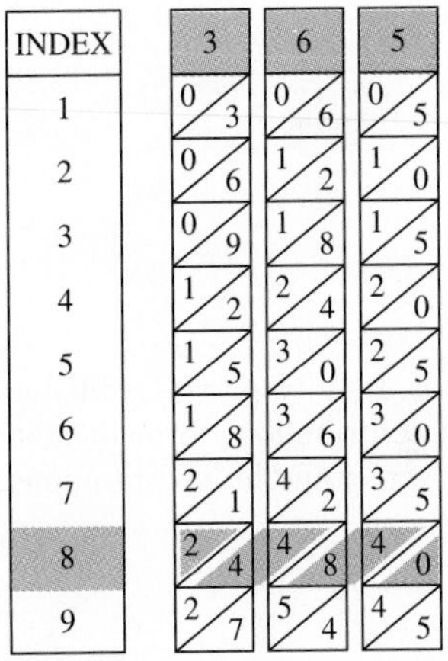

INDEX	3	6	5
1	0/3	0/6	0/5
2	0/6	1/2	1/0
3	0/9	1/8	1/5
4	1/2	2/4	2/0
5	1/5	3/0	2/5
6	1/8	3/6	3/0
7	2/1	4/2	3/5
8	2/4	4/8	4/0
9	2/7	5/4	4/5

Figure 4.5

▶ **Example 2**

Multiply 8 × 365 using Napier rods.

Solution: To multiply 8 × 365 line up the rods 3, 6, and 5 opposite the index, as shown in Fig. 4.5. Add along the diagonals as in the galley method.

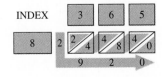

Thus 8 × 365 = 2920.

Example 3 illustrates the procedure to follow to multiply numbers containing more than one digit using Napier rods.

▶ **Example 3**

Multiply 48 × 365, using Napier rods.

Solution: 48 × 365 = (40 + 8) × 365

We can write $(40 + 8) \times 365 = (40 \times 365) + (8 \times 365)$. To find 40×365, determine 4×365, and multiply the product by 10. To evaluate 4×365, set up Napier rods for 3, 6, and 5 with index 4, and then evaluate along the diagonals, as indicated:

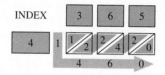

Therefore $4 \times 365 = 1460$. Then $40 \times 365 = 10 \times 1460 = 14{,}600$.

$$48 \times 365 = (40 \times 365) + (8 \times 365)$$
$$= 14{,}600 + 2920$$
$$= 17{,}520$$

(8 × 365 = 2920 from the (preceding example)

Section 4.5 Exercises

Multiply, using duplation and mediation.
1. 18×30
2. 35×23
3. 8×145
4. 120×90
5. 43×221
6. 96×53
7. 75×82
8. 49×124

Multiply, using the galley method.
9. 5×365
10. 6×365
11. 56×365
12. 7×12
13. 75×12
14. 17×256
15. 376×431
16. 525×698

Multiply, using Napier rods.
17. 4×28
18. 7×28
19. 47×28
20. 7×125
21. 5×125
22. 75×125
23. 8×2345
24. 7×3456

Research Activity

25. In addition to Napier rods, John Napier is credited with making other important contributions to mathematics. Write a report on John Napier and his contributions to mathematics.

CHAPTER 4 SUMMARY

Key Terms

4.1
number
numeral
system of numeration

4.2
base
digit
expanded form

4.3
binary system
bit
byte

4.5
duplation and mediation
galley method
Napier rods

Important Facts

Types of numeration systems
Additive (Egyptian hieroglyphics, Roman)
Multiplicative (traditional Chinese)
Ciphered (Ionic Greek)
Place-value (Babylonian, Mayan, Hindu-Arabic)

Early computational methods
Duplation and mediation
The galley method
Napier rods

CHAPTER 4 REVIEW EXERCISES

4.1, 4.2
In Ex. 1–6, assume an additive numeration system in which $a = 1$, $b = 10$, $c = 100$, and $d = 1000$. Find the value of the following numerals.

1. ddca
2. bdccda
3. ddbbaa
4. cbdadaaa
5. dddccbaaaa
6. ccbaddac

In Ex. 7–12, assume the same additive numeration system as in Ex. 1–6. Write the following in terms of a, b, c, and d.
7. 43
8. 167
9. 389
10. 1978
11. 6004
12. 3421

In Ex. 13–18, assume a multiplicative numeration system in which $a = 1$, $b = 2$, $c = 3$, $d = 4$, $e = 5$, $f = 6$, $g = 7$, $h = 8$, $i = 9$, $x = 10$, $y = 100$, and $z = 1000$. Find the value of the following numerals.

13. ixc
14. bxg
15. fydxh
16. bygxc
17. gzdyhxb
18. izeygxd

In Ex. 19–24, assume the same multiplicative numeration system as in Ex. 13–18. Write the following Hindu-Arabic numerals in that system.
19. 43
20. 167
21. 489
22. 1978
23. 6004
24. 2001

In Ex. 25–36, use the ciphered numeration system indicated below.

Decimal	1	2	3	4	5	6	7	8	9
Units	a	b	c	d	e	f	g	h	i
Tens	j	k	l	m	n	o	p	q	r
Hundreds	s	t	u	v	w	x	y	z	A
Thousands	B	C	D	E	F	G	H	I	J
Ten thousands	K	L	M	N	O	P	Q	R	S

Using the table above, convert the following to Hindu-Arabic numerals.
25. rc
26. vg
27. Itrh
28. NGzqc
29. Qqb
30. Pwki

Using the table above, write the following as a numeral in the ciphered numeration system.
31. 43
32. 167
33. 489
34. 1978
35. 23,685
36. 75,496

In Ex. 37–42, convert 1462 to a numeral in the indicated numeration systems.

37. Egyptian   **38.** Roman   **39.** Chinese

40. Ionic Greek   **41.** Babylonian   **42.** Mayan

In Ex. 43–48, convert the numeral to a Hindu-Arabic numeral.

43. ⊲𝕄𝕄𝕤𝕤∩∩|||||

44. 八千二百五十四

45. $\phi\pi\epsilon$

46. MCMXCI

47. ⟨⟨❘   ⟨⟨❮❘❘❘❘

48. ——
    •••
    ══
    •

4.3

Convert each of the following to a Hindu-Arabic numeral.

49. 46_7 **50.** 101_2 **51.** 304_5

52. 2663_8 **53.** $T0E_{12}$ **54.** 20220_3

Convert 463 to a numeral in the base indicated.

55. base 5   **56.** base 6   **57.** base 2

58. base 4   **59.** base 12   **60.** base 8

4.4

Add in the base indicated.

61. 43_8
74_8

62. 10110_2
1101_2

63. TE_{12}
46_{12}

64. 234_7
456_7

65. 4023_5
4023_5

66. 1407_8
7014_8

Subtract in the base indicated.

67. 4032_7
$-\ 121_7$

68. 1001_2
$-\ 101_2$

69. $4TE_{12}$
$-\ E7_{12}$

70. 4321_5
$-\ 442_5$

71. 2473_8
$-\ 567_8$

72. 2021_3
$-\ 222_3$

Multiply in the base indicated.

73. 23_5
$\times\ 4_5$

74. 23_4
$\times\ 22_4$

75. 126_{12}
$\times\ 42_{12}$

76. 202_3
$\times\ 22_3$

77. 1011_2
$\times\ 101_2$

78. 476_8
$\times\ 23_8$

Divide in the base indicated.

79. $1_2\overline{)1011_2}$ **80.** $2_4\overline{)230_4}$ **81.** $3_5\overline{)140_5}$

82. $4_6\overline{)3020_6}$ **83.** $5_6\overline{)2034_6}$ **84.** $6_8\overline{)5072_8}$

4.5

85. Multiply 142 × 24, using the duplation and mediation method.

86. Multiply 142 × 24, using the galley method.

87. Multiply 142 × 24, using Napier rods.

CHAPTER TEST

1. Explain the difference between a numeral and a number.

Convert each of the following to a Hindu-Arabic numeral.

2. ⲋ99∩∩ı

3. MCMXCIX

4. <<<ꀤꀤꀤ <ı

5. 九千八百三十五

In Ex. 6–9, convert the numeral to a numeral in the system indicated:

6. 427 to Egyptian

7. 1820 to Mayan

8. 3563 to Babylonian

9. 875 to Ionic Greek

10. Describe briefly an additive system, multiplicative system, ciphered system, and place-value system of numeration.

In Ex. 11–14, convert the number to a number in base 10.

11. 24_5

12. 63_8

13. 742_9

14. 10101_2

In Ex. 15–18, convert the base 10 number to a number in the base indicated.

15. 38 to base 2

16. 73 to base 5

17. 1347 to base 12

18. 3642 to base 8

Perform the indicated operations.

19. $\begin{array}{r} 121_3 \\ + 212_3 \\ \hline \end{array}$

20. $\begin{array}{r} 576_8 \\ - 347_8 \\ \hline \end{array}$

21. $\begin{array}{r} 45_7 \\ \times 36_7 \\ \hline \end{array}$

22. Multiply 15×17 using duplation and mediation.

23. Multiply 187×26 using the galley method.

5

NUMBER THEORY & THE REAL NUMBER SYSTEM

The strength of the traditional Japanese samurai sword is legendary. The master sword maker prepares the blade by heating a bar of iron until it is white hot, then folding it over and pounding it smooth. He does this 15 times. Each time the metal is folded, the layers of steel are doubled (a geometric sequence). For a sword of 15 folds, the blade contains 2^{15}, or 32,768 layers of steel!

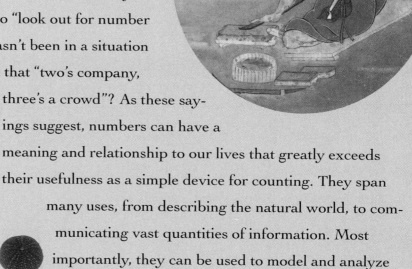

Numbers find their way into every aspect of life. For instance, doesn't everyone want someone to tell them "you're the only one for me"? Has anyone ever warned you to "look out for number one"? And who hasn't been in a situation where they realize that "two's company, three's a crowd"? As these sayings suggest, numbers can have a meaning and relationship to our lives that greatly exceeds their usefulness as a simple device for counting. They span many uses, from describing the natural world, to communicating vast quantities of information. Most importantly, they can be used to model and analyze almost any phenomenon of daily life.

The ancient Greeks were the first to think of numbers in a systematic way and to study their properties. The Greek philosopher Pythagoras believed that the universe revealed itself in the language of numbers and looked at the shapes of nature to find the

Natural objects often exhibit regular patterns that can be expressed in the language of mathematics.

Numbers are a fact of life.

numerical relations in them. The full force of the power of numbers to describe natural phenomena would not be realized until Sir Isaac Newton (1642–1727) and Gottfried Leibniz (1646–1716) developed a branch of mathematics known to us as calculus. Calculus can describe natural phenomena, such as the motion of planets, the swing of a pendulum, and the recoil of a spring.

Interest in the mathematical description of our world continues today. Within the last 20 years, the field of chaos theory has grown in importance because it provides a model for natural phenomena, such as storms and earthquakes, which previously could not be described mathematically. Still, the words of Galileo ring true: Speaking of the universe, he said, "It is written in the language of mathematics … without which it is humanly impossible to understand a single word of it."

The language of mathematics now permeates almost every part of our lives. In this age of information, we are constantly being assaulted by facts, figures, and numbers big and small. In recent years, the news media have increasingly focused on a phenomenon called "innumeracy," the inability to use and understand numerical information. Literacy of numbers, or numeracy, will be important to your career in the twenty-first century.

The artist George Seurat (1859–1891) believed that certain "natural proportions" exist in nature and that those proportions are the ones most likely to please the human eye. You can see this within his painting *La Parade*, which is composed of rectangles within rectangles. The sides of each rectangle are proportioned with a ratio of about 1 to 1.618.

5.1 Number Theory

This chapter introduces **number theory**, the study of numbers and their properties. The numbers we use to count are called the **counting numbers** or **natural numbers**. Since we begin counting with the number 1, the set of natural numbers begins with 1. The set of natural numbers is frequently denoted by N.

$$N = \{1, 2, 3, 4, 5, \ldots\}$$

Any natural number can be expressed as a product of two or more natural numbers. For example $8 = 2 \times 4$, $16 = 4 \times 4$, and $19 = 1 \times 19$. The natural numbers that are multiplied together are called **factors** of the product. For example

$$\underset{\underset{\text{factors}}{\uparrow \quad \uparrow}}{2 \times 4} = 8$$

A given natural number may have many factors. For example, what pairs of numbers have a product of 12?

$$\begin{aligned} 12 \cdot 1 &= 12 \\ 6 \cdot 2 &= 12 \\ \underset{\underset{\text{factors}}{\uparrow \quad \uparrow}}{4 \cdot 3} &= 12 \end{aligned}$$

The numbers 1, 2, 3, 4, 6, and 12 are all factors of 12. Each of these numbers divides 12 without a remainder.

If a and b are natural numbers we say that a is **a divisor** of b or a **divides** b, symbolized $a|b$, if the quotient of b divided by a has a remainder of 0. If a divides b, then b is **divisible** by a. For example, 4 divides 12, symbolized $4|12$, since the quotient of 12 divided by 4 has a remainder of zero. Note that 12 is divisible by 4. The notation $7\nmid12$ means 7 does not divide 12. Note that every factor of a natural number is also a divisor of the natural number. *Caution:* Do not confuse the symbols, $a|b$ and a/b. $a|b$ means "a divides b" and a/b means "a divided by b" $(a \div b)$. The symbols a/b and $a \div b$ indicate the operation of division is to be performed, and b may or may not be a divisor of a.

Prime and Composite Numbers

Every natural number greater than 1 can be classified as either a prime number or a composite number.

> A **prime number** is a natural number greater than 1 that has exactly two factors (or divisors), itself and 1.

The number 5 is a prime number since it is divisible only by the factors 1 and 5. The first eight prime numbers are 2, 3, 5, 7, 11, 13, 17, and 19. The number 2 is the only even prime number. All other even numbers have at least three divisors: 1, 2, and the number itself.

> A **composite number** is a natural number that is divisible by a number other than itself and 1.

Any natural number greater than 1 that is not prime is composite. The first eight composite numbers are 4, 6, 8, 9, 10, 12, 14, and 15.

The number 1 is neither prime nor composite; it is called a **unit**. The number 38 has at least three divisors, 1, 2, and 38, and, hence, is a composite number. In contrast, the number 23 is a prime number since its only divisors are 1 and 23.

The ancient Greeks, more than 2000 years ago, developed a technique for determining which numbers are prime numbers and which are not. This technique is named the **sieve of Eratosthenes**, for the Greek mathematician Eratosthenes of Cyrene who first used it.

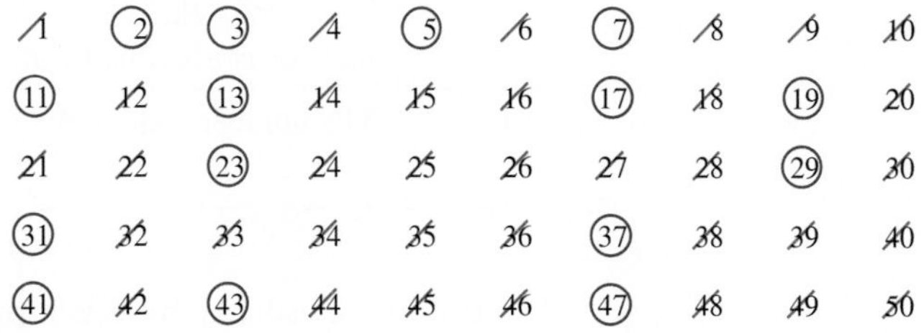

$$
\begin{array}{cccccccccc}
\not1 & ② & ③ & \not4 & \not5 & \not6 & ⑦ & \not8 & \not9 & \not{10} \\
⑪ & \not{12} & ⑬ & \not{14} & \not{15} & \not{16} & ⑰ & \not{18} & ⑲ & \not{20} \\
\not{21} & \not{22} & ㉓ & \not{24} & \not{25} & \not{26} & \not{27} & \not{28} & ㉙ & \not{30} \\
㉛ & \not{32} & \not{33} & \not{34} & \not{35} & \not{36} & ㊲ & \not{38} & \not{39} & \not{40} \\
㊶ & \not{42} & ㊸ & \not{44} & \not{45} & \not{46} & ㊼ & \not{48} & \not{49} & \not{50}
\end{array}
$$

Figure 5.1

To find the prime numbers less than or equal to any natural number, say 50, using this method, list the first 50 counting numbers (Fig. 5.1). Cross out 1 since it is not a prime number. Circle 2, the first prime number. Then cross out all the multiples of 2; that is, 2, 4, 6, 8, . . . , 50. Circle the next prime number, 3. Cross out all multiples of 3 that are not already crossed out. Continue this process until you reach the prime number p, such that $p \cdot p$ or p^2, is greater than the last number listed, in this case 50. Thus, 5 is next circled and its multiples are crossed out. Then 7 is circled and its multiples are crossed out. Since the next prime number is 11 and $11 \cdot 11$, or 121, is greater than 50 we are done. At this point, circle all the remaining numbers to obtain the prime numbers less than 50. The prime numbers less than or equal to 50 are 2, 3, 5, 7, 11, 13, 17, 19, 23, 29, 31, 37, 41, 43, and 47.

Now we turn our attention to composite numbers and their factors. The rules of divisibility given in the following chart are helpful in finding divisors (or factors) of composite numbers.

Rules of Divisibility

Divisible by	Test	Example
2	The number is even.	924 is divisible by 2, since 924 is even.
3	The sum of the digits of the number is divisible by 3.	924 is divisible by 3, since the sum of the digits, $9 + 2 + 4 = 15$, and 15 is divisible by 3.
4	The number formed by last two digits of the number is divisible by 4.	924 is divisible by 4, since the number formed by the last two digits, 24, is divisible by 4.
5	The number ends in 0 or 5.	265 is divisible by 5, since the number ends in 5.
6	The number is divisible by both 2 and 3.	924 is divisible by 6, since it is divisible by both 2 and 3.
8	The number formed by last three digits of the number is divisible by 8.	5824 is divisible by 8, since the number formed by the last three digits, 824, is divisible by 8.
9	The sum of the digits of the number is divisible by 9.	837 is divisible by 9, since the sum of the digits, 18, is divisible by 9.
10	The number ends in 0.	290 is divisible by 10, since the number ends in 0.

The test for divisibility by 6 is a particular case of the general statement that the product of two prime divisors of a number is a divisor of the number. Thus for example if both 3 and 7 divide a number then 21 will also divide the number.

Note that the chart does not list rules of divisibility for the numbers 7 and 11. There are rules for these numbers (see the exercise set at the end of this section), but the rules are difficult to remember. To check divisibility by 7 or 11 it is simpler just to perform the division.

▶ **Example 1**

Determine whether the number 374,832 is divisible by

a) 2   **b)** 3   **c)** 4   **d)** 5   **e)** 6   **f)** 8   **g)** 9   **h)** 10

Solution:

a) Since the number is even, it is divisible by 2.

b) Since the sum of the digits, 27, is divisible by 3, the number is divisible by 3.

c) Since the number formed by the last two digits, 32, is divisible by 4, the number is divisible by 4.

d) Since the last digit is not a 0 or a 5, the number is not divisible by 5.

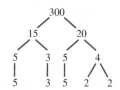

Figure 5.2

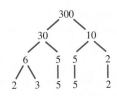

Figure 5.3

e) Since the number is divisible by both 2 and 3, the number is divisible by 6.

f) Since the number formed by the last three digits, 832, is divisible by 8, the number is divisible by 8.

g) Since the sum of the digits, 27, is divisible by 9, the number is divisible by 9.

h) Since the number does not end in 0, the number is not divisible by 10.

Every composite number can be expressed as a product of prime numbers. The process of breaking a given number down into a product of prime numbers is called **prime factorization**. The prime factorization of 18 is $3 \times 3 \times 2$. No other natural number listed as a product of primes will have the same prime factorization as 18. The *fundamental theorem of arithmetic* states this concept formally. (A **theorem** is a statement or proposition that can be proven true.)

> ### The Fundamental Theorem of Arithmetic
> Every composite number can be expressed as a **unique** product of prime numbers.

In writing the prime factorization of a number, the order of the factors is immaterial. For example, to list the prime factors of 18, we may write either $3 \times 3 \times 2$ or $2 \times 3 \times 3$ or $3 \times 2 \times 3$.

A number of techniques can be used to find the prime factorization of a number. Two methods are illustrated.

Method 1: Branching

To find the prime factorization of a number, select any two numbers whose product is the number to be factored. If the factors are not prime numbers, then continue factoring each composite number until all numbers are prime.

▶ ### Example 2

Write 300 as a product of primes.

Solution: Select any two numbers whose product is 300. Among the many choices, two possibilities are $10 \cdot 30$ and $15 \cdot 20$. We will consider $15 \cdot 20$. Now find any two numbers whose product is 15 and any two numbers whose product is 20. Continue branching as shown in Fig. 5.2 until the numbers in the last row are all prime numbers. The branching diagram is sometimes referred to as a **factor tree**.

$$300 = 2 \cdot 2 \cdot 3 \cdot 5 \cdot 5$$
$$= 2^2 \cdot 3 \cdot 5^2$$

Thus the prime factorization of 300 is $2^2 \cdot 3 \cdot 5^2$. Note that if we had selected 30 and 10 as the initial factors, the final results would be the same (Fig. 5.3).

*One of the most interesting mathematicians of modern times was **Srinivasa Ramanujan** (1887–1920). Born to an impoverished middle-class family in India, he virtually taught himself higher mathematics. He went to England to study with the number theorist G.H. Hardy. Hardy tells the story of a taxicab ride he took to visit Ramanujan. The cab had the license plate number 1729 and he defied the young Indian to find anything interesting in that. Without hesitating Ramanujan pointed out that it was the smallest positive integer that could be represented in two different ways as the sum of two cubes: $1^3 + 12^3$ and $9^3 + 10^3$.*

Method 2: Division

To obtain the prime factorization of a number by this method, divide the given number by the smallest prime number by which it is divisible. Place the quotient under the given number. Then divide the quotient by the smallest prime number by which it is divisible, and again record the quotient. Repeat this process until the quotient is a prime number. The prime factorization is the product of all the prime divisors and the prime (or last) quotient. This procedure is illustrated in Example 3.

▶ ### Example 3

Write 300 as a product of prime numbers.

Solution: Since 300 is an even number, the smallest prime number that divides it is 2. Divide 300 by 2. Place the quotient, 150, below the 300. Repeat this process of dividing each quotient by the smallest prime number that divides it:

$$
\begin{array}{r|r}
2 & 300 \\ \hline
2 & 150 \\ \hline
3 & 75 \\ \hline
5 & 25 \\ \hline
  & 5
\end{array}
$$

Since the final quotient, 5, is a prime number, we stop. The prime factorization is

$$300 = 2 \cdot 2 \cdot 3 \cdot 5 \cdot 5 = 2^2 \cdot 3 \cdot 5^2.$$

Notice that despite the two different methods in Example 2 and Example 3, the answer is the same.

Greatest Common Divisor

The discussion, in Section 5.3, of how to reduce fractions makes use of the greatest common divisor (GCD). One technique of finding the GCD is to use prime factorization.

> The **greatest common divisor (GCD)** of a set of natural numbers is the largest natural number that divides (without remainder) every number in that set.

What is the GCD of 12 and 18? One way to determine this is to make a list of the divisors (or factors) of 12 and 18, as illustrated below.

Divisors of 12: {**1, 2, 3**, 4, **6**, 12}
Divisors of 18: {**1, 2, 3, 6**, 9, 18}

The common divisors are 1, 2, 3, and 6, and therefore the greatest common divisor is 6.

If the numbers are large, this method of finding the GCD is not practical. The GCD can be found more efficiently by using prime factorization.

> **To Find the Greatest Common Divisor of Two or More Numbers**
> **1.** Determine the prime factorization of each number.
> **2.** Find each prime factor with the smallest exponent that appears in each of the prime factorizations.
> **3.** Find the product of the factors found in step 2.

Example 4 illustrates this procedure.

▶ **Example 4**

Find the GCD of 24 and 60.

Solution: The branching method of finding the prime factors of 24 and 60 is illustrated in Fig. 5.4.

Figure 5.4

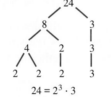
$$24 = 2^3 \cdot 3$$

$$60 = 2^2 \cdot 3 \cdot 5$$

1. The prime factorization of 24 is $2^3 \cdot 3$ and the prime factorization of 60 is $2^2 \cdot 3 \cdot 5$.
2. The prime factors with the smallest exponents that appear in each of the factorizations of 24 and 60 are 2^2 and 3.
3. The product of the factors found in step 2 is $2^2 \cdot 3 = 4 \cdot 3 = 12$. The GCD of 24 and 60 is 12. Twelve is the largest number that divides both 24 and 60.

▶ **Example 5**

Find the GCD of 36 and 150.

Solution:
1. The prime factorization of 36 is $2^2 \cdot 3^2$ and the prime factorization of 150 is $2 \cdot 3 \cdot 5^2$.
2. The prime factors with the smallest exponents that appear in each of the factorizations of 36 and 150 are 2 and 3.
3. The product of the factors found in step 2 is $2 \cdot 3 = 6$. The GCD of 36 and 150 is 6.

Two numbers with a GCD of 1 are said to be **relatively prime**. The numbers 9 and 14 are relatively prime, since the GCD is 1. This can be expressed in a concise form with the notation GCD(9, 14) = 1.

Least Common Multiple

To perform addition and subtraction of fractions (Section 5.3) we will use the least common multiple (LCM). One technique of finding the LCM is to use prime factorization.

> The **least common multiple (LCM)** of a set of natural numbers is the smallest natural number that is divisible (without remainder) by each element of the set.

What is the least common multiple of 12 and 18? One way to determine the LCM is to make a list of the multiples of each number, as illustrated below.

Multiples of 12: {12, 24, **36**, 48, 60, **72**, 84, 96, **108**, 120, 132, **144**, ...}
Multiples of 18: {18, **36**, 54, **72**, 90, **108**, 126, **144**, 162, ...}

Some common multiples of 12 and 18 are 36, 72, 108, and 144. The least common multiple, 36, is the smallest number that is divisible by both 12 and 18. Usually the most efficient method of finding the LCM is to use prime factorization.

> ### To Find the Least Common Multiple of Two or More Numbers
> **1.** Determine the prime factorization of each number.
> **2.** List each prime factor with the greatest exponent that appears in any of the prime factorizations.
> **3.** Find the product of the factors found in step 2.

Example 6 illustrates this procedure.

▶ **Example 6**
Find the LCM of 24 and 60.

Solution:
1. Find the prime factors of each number. In Example 4 we determined that

$$24 = 2^3 \cdot 3 \quad \text{and} \quad 60 = 2^2 \cdot 3 \cdot 5.$$

2. List each prime factor with the greatest exponent that appears in either of the prime factorizations: 2^3, 3, 5.
3. Find the product of the factors found in step 2:

$$2^3 \cdot 3 \cdot 5 = 8 \cdot 3 \cdot 5 = 120$$

Thus 120 is the LCM of 24 and 60. It is the smallest number that is divisible by both 24 and 60.

▶ **Example 7**

Find the LCM of 36 and 150.

Solution:

1. Find the prime factors of each number. In Example 5 we determined that

$$36 = 2^2 \cdot 3^2 \quad \text{and} \quad 150 = 2 \cdot 3 \cdot 5^2$$

2. List each prime factor with the greatest exponent that appears in either of the prime factorizations: 2^2, 3^2, 5^2.

3. Find the product of the factors found in step 2:

$$2^2 \cdot 3^2 \cdot 5^2 = 4 \cdot 9 \cdot 25 = 900$$

Thus 900 is the least common multiple of 36 and 150. It is the smallest number divisible by both 36 and 150.

The Search for Larger Prime Numbers

More than 2000 years ago, the Greek mathematician Euclid proved that there is no largest prime number. However, mathematicians continue to strive to find larger and larger prime numbers.

Marin Mersenne (1588–1648), a seventeenth-century monk, found that numbers of the form $2^n - 1$ are often prime numbers when n is a prime number. For example,

$$2^2 - 1 = 4 - 1 = 3 \qquad 2^3 - 1 = 8 - 1 = 7$$
$$2^5 - 1 = 32 - 1 = 31 \qquad 2^7 - 1 = 128 - 1 = 127$$

Numbers of the form $2^n - 1$ that are prime are referred to as **Mersenne primes**. The first 10 Mersenne primes occur when $n = 2, 3, 5, 7, 13, 17, 19, 31, 61, 89$. The first time the expression $2^n - 1$ does not generate a prime number, for prime number n, is when n is 11. The number $2^{11} - 1$ is a composite number (see Ex. 90). Scientists frequently use Mersenne primes in their search for larger and larger primes.

The largest prime found to date was discovered in March 1992 by a group of British mathematicians at the ABA Technology Hartwell Laboratory in Hartwell, England. The number consists of 227,832 digits and is the 32nd and largest Mersenne prime found to date. This new record eclipses the previous record holder which had a mere 65,050 digits. The new champion, equal to $2^{756,839} - 1$, was found after 19 hours of searching on a Cray 2 Super Computer. However, because no one has yet checked all the Mersenne primes having smaller exponents, mathematicians can't be sure that no Mersenne prime lurks in the vast expanse between the record holder and the second-place Mersenne, which has an exponent of 216,091. It is very difficult to comprehend the magnitude of this

largest prime number. If you stacked $2^{756,839} - 1$ sheets of paper one on top of the other, the height of the stack would be many times the size of our universe.

More about Prime Numbers

Another mathematician who studied prime numbers was Pierre de Fermat (1601–1665). A lawyer by profession, Fermat became interested in mathematics as a hobby. He became one of the finest mathematicians of the seventeenth century. Fermat conjectured that each number of the form $2^{2^n} + 1$, now referred to as the **Fermat number**, was prime for each natural number n. Recall that a **conjecture** is a supposition that has not been proved nor disproved. In 1732 Leonhard Euler proved that for $n = 5$, $2^{32} + 1$ was a composite number, thus disproving Fermat's conjecture.

In the past 250 years mathematicians have only been able to evaluate the sixth, seventh, eighth, and ninth Fermat numbers to determine whether they are prime or composite. Each of these numbers has been shown to be composite. Fermat numbers are very difficult to test due to the sheer magnitude of the computations involved.

In 1772 Euler devised the formula $n^2 - n + 41$, which yields a prime number for any natural number up to 40 but fails for $n = 41$. In 1879 E. B. Escott devised the formula $n^2 - 79n + 1601$, which yields a prime number for any natural number up to 79 but fails for $n = 80$.

In 1742 Christian Goldbach conjectured in a letter sent to Euler that every even number greater than or equal to 6 can be represented as a sum of two prime numbers (*examples:* $6 = 3 + 3$; $8 = 3 + 5$) and that every odd number greater than or equal to 9 can be represented as the sum of three odd primes (*examples:* $9 = 3 + 3 + 3$; $11 = 3 + 3 + 5$). This is still unproven.

Another conjecture is that there are infinitely many pairs of **twin primes** of the form p, $p + 2$ (*examples:* 3, 5; 5, 7). Can you find the next two pairs of twin primes?

Section 5.1 Exercises

1. Define prime number.
2. Define composite number.
3. Define conjecture.
4. What does "a and b are factors of c" mean?
5. What does "a is a divisor of b" mean?
6. Use the sieve of Eratosthenes to find the prime numbers up to and including 100.

In Ex. 7–16, determine whether the statement is true or false.

7. Seven is a factor 42.
8. $7 | 42$
9. Forty-two is a multiple of 7.
10. Seven is a divisor of 42.
11. Forty-two is divisible by 7.
12. Seven is a multiple of 42.
13. If a number is divisible by 3, then every digit of the number is divisible by 3.
14. If every digit of a number is divisible by 3, then the number itself is divisible by 3.
15. If a number is divisible by 2 and 3, then the number is divisible by 6.
16. If a number is divisible by 2 and 4, then the number is divisible by 8.

In Ex. 17–22, determine whether the number is divisible by each of the following numbers: 2, 3, 4, 5, 6, 8, 9, and 10.

17. 48,324

18. 529,200

19. 2,763,105

20. 3,126,120

21. 1,882,320

22. 3,941,221

23. Determine a number that is divisible by 2, 3, 4, 5, and 6.

24. Determine a number that is divisible by 3, 4, 5, 9, and 10.

In Ex. 25–36, find the prime factorization of the number.

25. 39   **26.** 48   **27.** 57

28. 180   **29.** 312   **30.** 500

31. 1026   **32.** 2530   **33.** 485

34. 999   **35.** 1992   **36.** 1168

In Ex. 37–46, find the greatest common divisor of the numbers.

37. 15 and 180

38. 12 and 77

39. 42 and 56

40. 52 and 65

41. 90 and 600

42. 180 and 360

43. 96 and 212

44. 240 and 285

45. 24, 48, 128

46. 18, 78, 198

In Ex. 47–56, find the least common multiple of the numbers.

47. 15 and 180

48. 12 and 77

49. 42 and 56

50. 52 and 65

51. 90 and 600

52. 180 and 360

53. 96 and 212

54. 240 and 285

55. 24, 48, 128

56. 18, 78, 198

In Ex. 57 and 58, are the statements true or false? Explain.

57. If a number is not divisible by 5, then it is not divisible by 10.

58. If a number is not divisible by 10, then it is not divisible by 5.

59. Find the next two sets of twin primes that follow the set 5, 7.

60. The primes 2 and 3 are consecutive natural numbers. Is there another pair of consecutive natural numbers both of which are prime? Explain.

61. Show that Goldbach's conjecture is true for the even numbers 6 through 12 and the odd numbers 9 through 15.

62. Find the fifth Mersenne prime number.

63. Show that each number of the form $2^{2^n} + 1$ is prime for $n = 1, 2,$ and 3.

64. Suppose that the 100 members of the U.S. Senate are placed on committees consisting of more than two members but fewer than 50 members. Each committee is to have an equal number of members and each member is to be on one and only one committee.

a) What size committees are possible?

b) How many committees are there of each size?

65. Sara and Harry both work the 3:00 P.M. to 11:00 P.M. shift. Sara has every fifth night off and Harry has every sixth night off. If they both have tonight off, how many days will pass before they have the same night off again?

66. Sears has a sale on mechanics tool chests every 40 days and a sale on mechanics tool sets every 60 days. If both are on sale today, how long will it be before they are on sale together again?

67. Consider the first eight prime numbers greater than 3. The numbers are 5, 7, 11, 13, 17, 19, 23, and 29.

a) Determine which of these prime numbers differs by 1 from a multiple of the number 6.

b) Use inductive reasoning and the results obtained in part (a) to make a conjecture regarding prime numbers.

c) Select a few more prime numbers and determine whether your conjecture appears to be correct.

68. State a procedure that defines a divisibility test for 15.

Another method that can be used to find the greatest common divisor is known as the **Euclidean algorithm**. We will illustrate this procedure by finding the GCD of 60 and 220.

First divide 220 by 60. Disregard the quotient 3. Then divide 60 by the remainder 40. Continue this process of dividing the divisors by the remainders until you obtain a remainder of 0. The divisor in the last division, in which the remainder is 0, is the GCD.

$$
\begin{array}{ccc}
3 & 1 & 2 \\
60\,\overline{)\,220} & 40\,\overline{)\,60} & 20\,\overline{)\,40} \\
\underline{180} & \underline{40} & \underline{40} \\
40 & 20 & 0
\end{array}
$$

Since 40/20 had a remainder of 0, the GCD is 20.

In Ex. 69–74, use the Euclidean algorithm to find the GCD.

69. 35, 75

70. 20, 160

71. 18, 112

72. 96, 115

73. 150, 180

74. 210, 560

A number whose **proper factors** (factors other than the number itself) add up to the number is called a **perfect number**. For example, 6 is a perfect number because its proper factors are 1, 2, and 3, and $1 + 2 + 3 = 6$. Determine which, if any, of the following numbers are perfect.

75. 12

76. 28

77. 496

78. 48

The rule for divisibility by 7 follows. If the number without its unit digit, minus twice the unit digit of the original number, is divisible by 7, then the original number is divisible by 7. For example, consider the number 2996. Its unit digit is 6. Begin by dropping the unit digit. Then subtract twice the unit digit as follows:

$$299 - 2(6) = 299 - 12 = 287$$

Now divide, $287 \div 7 = 41$. Since 7 divides 287, the original number 2996 is divisible by 7.

In Ex. 79–82, use the stated procedure to determine whether the number is divisible by 7.

79. 3661   **80.** 4375

81. 39,837   **82.** 56,678

The rule for divisibility by 11 is as follows. Find the sum of the digits in the odd positions (from the right) and the sum of the digits in the even positions. Then find the difference between the sums. If the difference is 0 or a number divisible by 11, then the original number is divisible by 11. For example, to determine whether 136,059 is divisible by 11, we do the following:

$$9 + 0 + 3 = 12 \qquad 5 + 6 + 1 = 12$$

Now subtract: $12 - 12 = 0$. Since the difference is 0, the number 136,059 is divisible by 11.

In Ex. 83–86, use the above procedure to determine whether the number is divisible by 11.

83. 65,855   **84.** 134,431

85. 6,185,854   **86.** 672,541

Problem Solving

87. A number in which each digit except 0 appears exactly three times is divisible by 3. For example, 888,444,555 and 714,714,714 are both divisible by 3. Explain why this must be true.

88. A **palindromic** number is a number that reads the same forward and backward.

    a) Check the following palindromic numbers for divisibility by 11 (see Ex. 83–86).

        i) 3553   **ii)** 6116   **iii)** 9229

    b) Explain why a four-digit palindromic number must be divisible by 11.

    c) Is every five-digit palindromic number divisible by 11? Give an example to support your answer.

    d) Is every six-digit palindromic number divisible by 11?

    e) Do you believe that every palindromic number that contains an even number of digits is divisible by 11? Explain your answer.

89. Use the fact that, if $a|b$ and $a|c$ then $a|(b + c)$, to determine whether 36,018 is divisible by 18. (*Hint:* Write 36,018 as $36,000 + 18$.)

90. Show that $2^{11} - 1$ is a composite number.

Research Activity

91. Do the necessary research and write a report on how the ancient Greeks contributed to the development of mathematics.

5.2 The Integers

In Section 5.1 we introduced the natural or counting numbers:

$$N = \{1, 2, 3, 4, \ldots\}$$

Another important set of numbers, the **whole numbers**, help to answer the question "How many?"

$$\text{Whole numbers} = \{0, 1, 2, 3, 4, \ldots\}$$

Note that the set of whole numbers contains the number 0, while the set of counting numbers does not. If a farmer were asked how many chickens were in a coop the answer would be a whole number. If the farmer had no chickens, he or she could answer zero. Although we use the number 0 daily and take it

The poet Virgil was born in 70 B.C. In 1930 some highly respected scholars decided to celebrate his 2000th birthday. The only problem was that the accounting of time does not include a year 0 between B.C. and A.D. So Virgil would have turned 70 in our year 1 A.D. and, had he lived so long, 2000 in 1931. This fact was pointed out to the celebrants, but only after the party was well underway.

for granted, the number 0 as we know it was not used and accepted until the sixteenth century.

If the temperature is 12°F and drops 20°, the resulting temperature is −8°F. This type of problem shows the need for negative numbers. The set of **integers** consists of the negative integers, 0, and the positive integers.

$$\text{Integers} = \{\ldots, \underbrace{-4, -3, -2, -1}_{\text{Negative integers}}, 0, \underbrace{1, 2, 3, \ldots}_{\text{Positive integers}}\}$$

The term "positive integers" is yet another name for the natural numbers or counting numbers.

An understanding of addition, subtraction, multiplication, and division of the integers is essential in understanding algebra (Chapter 6). To aid in our explanation of addition and subtraction of integers, we will introduce the real number line (Fig. 5.5). To construct the real number line, arbitrarily select a point for zero to serve as the starting point. Place the positive integers to the right of 0, equally spaced from one another. Place the negative integers to the left of 0, using the same spacing. The real number line contains the integers and all of the other real numbers that are not integers. Some examples of real numbers that are not integers are indicated in Fig. 5.5. We will discuss real numbers that are not integers in the next two sections.

Figure 5.5

$$-\frac{5}{2} \qquad \frac{1}{2} \quad \sqrt{2} \qquad \pi$$

$$\xleftarrow{\quad} \; \overset{\bullet}{\underset{-5}{|}} \; \overset{}{\underset{-4}{|}} \; \overset{\bullet}{\underset{-3}{|}} \; \overset{}{\underset{-2}{|}} \; \overset{}{\underset{-1}{|}} \; \overset{}{\underset{0}{|}} \; \overset{\bullet}{\underset{1}{|}} \; \overset{\bullet}{\underset{2}{|}} \; \overset{}{\underset{3}{|}} \; \overset{\bullet}{\underset{4}{|}} \; \overset{}{\underset{5}{|}} \; \xrightarrow{\quad}$$

The arrows at the ends of the real number line indicate that the line continues indefinitely in both directions. Note that for any natural number, n, on the number line, the opposite of that number, $-n$, is also on the number line. This real number line was drawn horizontally, but it could just as well have been drawn vertically. In fact, in the next chapter you will see that the axes of a graph are the union of two number lines, one horizontal and the other vertical.

The number line can be used to determine the order of two integers. On the number line, the numbers increase from left to right. The number 3 is greater than 2, written $3 > 2$. Observe that 3 is to the right of 2. Similarly, we can see that $0 > -1$ by observing that 0 is to the right of -1 on the number line.

Instead of stating that 3 is greater than 2, we could have stated that 2 is less than 3, written $2 < 3$. Observe that 2 is to the left of 3 on the number line. Similarly, we can see that $-1 < 0$ by observing that -1 is to the left of 0. The inequality symbol always points to the smaller of the two numbers when the inequality statement is true.

▶ **Example 1**

Insert either $>$ or $<$ in the shaded area between the paired numbers to make the statement correct.

a) $-4 \quad \blacksquare \quad 2$ **b)** $-2 \quad \blacksquare \quad -4$ **c)** $-5 \quad \blacksquare \quad -3$ **d)** $0 \quad \blacksquare \quad -2$

Solution:

a) $-4 < 2$, since -4 is to the left of 2.

b) $-2 > -4$, since -2 is to the right of -4.

c) $-5 < -3$, since -5 is to the left of -3.

d) $0 > -2$, since 0 is to the right of -2.

Addition of Integers

Addition of integers can be represented geometrically with a number line. To accomplish this, begin at 0 on the number line. Represent the first addend by an arrow starting at 0. The arrow will be drawn to the right if the addend is positive. If the addend is negative, the arrow will be drawn to the left. From the tip of the first arrow, draw a second arrow to represent the second addend. The second arrow is drawn to the right or left, as explained above. The sum of the two integers is found at the tip of the second arrow.

"Can you do addition?" the White Queen asked. "What's one and one and one and one and one and one and one and one and one and one?" "I don't know," said Alice, "I lost count." "She can't do addition," the Red Queen interrupted. (Lewis Carrol, Alice's Adventures in Wonderland)

▶ **Example 2**

Evaluate $2 + (-4)$ using the number line.

Solution:

$$2 + (-4) = -2$$

▶ **Example 3**

Evaluate **a)** $-2 + (-5)$ and **b)** $-5 + 2$ using a number line.

Solution:

a)

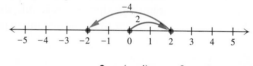

$$-2 + (-5) = -7$$

b)

$$-5 + 2 = -3$$

▶ **Example 4**

Evaluate $4 + (-4)$ using the number line.

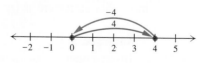

Solution:

$$4 + (-4) = 0$$

The number -4 is said to be the additive inverse of 4, and 4 is the additive inverse of -4, because their sum is 0. In general the **additive inverse** of the number n is $-n$, since $n + (-n) = 0$. Inverses will be discussed more formally in Chapter 9.

Subtraction of Integers

Any subtraction problem can be rewritten as an addition problem. To do so, we use the following definition of subtraction.

Subtraction

$$a - b = a + (-b)$$

The rule for subtraction indicates that to subtract b from a, add the additive inverse of b to a. For example,

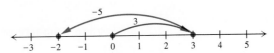

$$3 - 5 = 3 + (-5)$$

subtraction   addition   additive inverse of 5

Now we can determine the value of $3 + (-5)$.

Thus $3 - 5 = 3 + (-5) = -2$.

▶ **Example 5**

Evaluate $-3 - (-2)$ using the number line.

We are subtracting -2 from -3. The additive inverse of -2 is 2; therefore we add 2 to -3.

$$-3 - (-2) = -3 + 2$$

We now add $-3 + 2$ on the number line to obtain the answer -1.

Thus $-3 - (-2) = -3 + 2 = -1$.

As you get more proficient in working with integers you should be able to answer questions involving them without drawing a number line.

▶ **Example 6**

Evaluate   **a)** $-3 - 5$ **b)** $4 - (-3)$

Solution:

a) $-3 - 5 = -3 + (-5) = -8$ **b)** $4 - (-3) = 4 + 3 = 7$

More people have made it into outer space than have stood on the top of Nepal's Mt. Everest. Fewer people still have visited the Pacific Ocean's Mariana Trench. In 1960 Jacques Piccard and Don Walsh were the first to make the journey in a vessel designed to withstand the immense pressure: 7 tons per square inch. The diving time was $8\frac{1}{2}$ hours, of which only 20 minutes were spent on the ocean floor.

▶ **Example 7**

The highest point on earth is Mount Everest, in the Himalayas, at a height of 29,028 ft above sea level. The lowest point on earth is the Mariana Trench, in the Pacific Ocean, at a depth of 36,198 feet below sea level ($-36,198$ ft). Find the vertical height difference between Mount Everest and the Mariana Trench.

Solution: We obtain the vertical difference by subtracting the lower elevation from the higher elevation.

$$29,028 - (-36,198) = 29,028 + 36,198 = 65,226$$

The vertical difference is 65,226 ft.

Multiplication of Integers

The multiplication property of zero is important in our discussion of multiplication of integers. It indicates that the product of 0 and any number is 0.

Multiplication Property of Zero

$$a \cdot 0 = 0 \cdot a = 0$$

We will develop the rules for multiplication of integers using number patterns. The four possible cases are

1. Positive integer × positive integer,
2. Positive integer × negative integer,
3. Negative integer × positive integer,
4. Negative integer × negative integer.

Case 1: Positive integer × positive integer The product of two positive integers can be defined as repeated addition of a positive integer. Thus $3 \cdot 2$ means $2 + 2 + 2$. This sum will always be positive. Thus a positive integer times a positive integer is a positive integer.

Case 2: Positive integer × negative integer Consider the following patterns.

$$4(3) = 12$$
$$4(2) = 8$$
$$4(1) = 4$$

Note that each time the second factor is reduced by 1, the product is reduced by 4. Continuing the process gives

$$4(0) = 0.$$

What comes next?

$$4(-1) = -4$$
$$4(-2) = -8$$

The pattern indicates that a positive integer times a negative integer is a negative integer.

We can confirm this by using the number line. The expression $3(-2)$ means $(-2) + (-2) + (-2)$. Adding $(-2) + (-2) + (-2)$ on the number line, we obtain a sum of -6.

Case 3: Negative integer × positive integer A procedure similar to that used in Case 2 will indicate that a negative integer times a positive integer is a negative integer.

Case 4: Negative integer × negative integer We have illustrated that a positive integer times a negative integer is a negative integer. We will make use of this fact in the following pattern.

$$4(-3) = -12$$
$$3(-3) = -9$$
$$2(-3) = -6$$
$$1(-3) = -3$$

In this pattern, each time the first term is decreased by 1, the product is increased by 3. Continuing this process gives

$$0(-3) = 0,$$
$$(-1)(-3) = 3,$$
$$(-2)(-3) = 6.$$

This pattern illustrates that a negative integer times a negative integer is a positive integer.

The examples were restricted to integers. However, the rules for multiplication can be used for any numbers. The following box summarizes them.

Rules for Multiplication

1. The product of two numbers with **like signs** (positive × positive or negative × negative) is a **positive number**.

2. The product of two numbers with **unlike signs** (positive × negative or negative × positive) is a **negative number**.

▶ **Example 8**

Evaluate   **a)** $4(-7)$ **b)** $(-3)(2)$ **c)** $(-4)(-5)$

Solution:

a) $4(-7) = -28$ **b)** $(-3)(2) = -6$ **c)** $(-4)(-5) = 20$

Division of Integers

You may already realize that there is a relationship between multiplication and division.

$$6 \div 2 = 3 \quad \text{means that} \quad 3 \cdot 2 = 6.$$
$$\frac{20}{10} = 2 \quad \text{means that} \quad 2 \cdot 10 = 20.$$

From the examples above we can see that division is the reverse process of multiplication.

For any a, b, and c where $b \neq 0$, $\dfrac{a}{b} = c$ means that $c \cdot b = a$.

We will discuss the four possible cases for division, which are similar to the ones for multiplication.

Case 1: Positive integer ÷ positive integer A positive integer divided by a positive integer is positive.

$$\frac{6}{2} = 3 \quad \text{since} \quad 3(2) = 6$$

Case 2: Positive integer ÷ negative integer A positive integer divided by a negative integer is negative.

$$\frac{6}{-2} = -3 \quad \text{since} \quad (-3)(-2) = 6$$

Case 3: Negative integer ÷ positive integer A negative integer divided by a positive integer is negative.

$$\frac{-6}{2} = -3 \quad \text{since} \quad (-3)(2) = -6$$

Case 4: Negative integer ÷ negative integer A negative integer divided by a negative integer is positive.

$$\frac{-6}{-2} = 3 \quad \text{since} \quad 3(-2) = -6$$

The examples were restricted to integers. However, the rules for division can be used for any numbers. They are summarized as follows.

Rules for Division
1. The quotient of two numbers with **like signs** (positive ÷ positive or negative ÷ negative) is a **positive number**.
2. The quotient of two numbers with **unlike signs** (positive ÷ negative or negative ÷ positive) is a **negative number**.

▶ **Example 9**

Evaluate   **a)** $\dfrac{20}{-5}$ **b)** $\dfrac{-60}{4}$ **c)** $\dfrac{-25}{-5}$

Solution:

a) $\dfrac{20}{-5} = -4$ **b)** $\dfrac{-60}{4} = -15$ **c)** $\dfrac{-25}{-5} = 5$

In the definition of division we stated that the denominator could not be 0. Division by 0 is not allowed. The quotient of a number divided by zero is said to be **undefined**. For example $\frac{3}{0}$ is undefined.

Section 5.2 Exercises

In Ex. 1–10, evaluate the expression.
1. $-6 + 8$
2. $6 + (-7)$
3. $(-4) + 13$
4. $(-5) + (-5)$
5. $[9 + (-13)] + 0$
6. $(4 + 6) + (-3)$
7. $[(-4) + (-5)] + 8$
8. $[7 + (-4)] + (-6)$
9. $[(-11) + (-8)] + 12$
10. $(3 - 13) + (-7)$

In Ex. 11–20, evaluate the expression.
11. $5 - 7$
12. $-3 - 6$
13. $-7 - 5$
14. $6 - (-4)$
15. $-7 - (-4)$
16. $4 - (-4)$
17. $4 - 11$
18. $5 - (-12)$
19. $[5 + (-3)] - 4$
20. $6 - (6 + 8)$

In Ex. 21–30, evaluate the expression.
21. $-5 \cdot 7$
22. $3(-6)$
23. $(-7)(-7)$
24. $(-5)(7)$
25. $[(-4)(-3)] \cdot 5$
26. $(-3)(5)(-6)$
27. $(5 \cdot 5)(-3)$
28. $(-7)(-1)(-1)$
29. $[(-3)(-3)] \cdot [(-5)(8)]$
30. $[[(-4)(8)](6)](-2)$

In Ex. 31–40, evaluate the expression.
31. $-12 \div (-6)$
32. $-12 \div 6$
33. $16 \div (-4)$
34. $\dfrac{-15}{-3}$
35. $\dfrac{35}{-7}$
36. $\dfrac{-75}{25}$
37. $\dfrac{33}{-11}$
38. $\dfrac{186}{-6}$
39. $140 \div (-7)$
40. $(-600) \div (-4)$

In Ex. 41–50, determine whether the statement is true or false.
41. Every natural number is an integer.
42. Every integer is a natural number.
43. The sum of any two negative integers is a negative integer.
44. The difference of any two negative integers is a negative integer.
45. The difference of a positive integer and a negative integer is a negative integer.
46. The product of any two positive integers is a positive integer.
47. The quotient of any two negative integers is a negative integer.
48. The quotient of a negative integer and a positive integer is a negative integer.

49. The product of a positive integer and a negative integer is a positive integer.

50. The sum of a positive integer and a negative integer is always a positive integer.

In Ex. 51–60, evaluate the expression.

51. $(5 + 5) \div 2$

52. $[5(-4)] + 3$

53. $[5(-4)] - 3$

54. $[(-5)(-6)] - 3$

55. $(4 - 7)(3)$

56. $[14 \div (-2)](-3)$

57. $[5 + (-14)] \div 3$

58. $(5 - 9) \div (-4)$

59. $[(-22)(-3)] \div (2 - 13)$

60. $[15(-4)] \div (-6)$

In Ex. 61–64, write the numbers in increasing order from left to right.

61. $-4, 3, -2, 0, 5, -1$

62. $5, -3, 2, 0, 3, -2$

63. $-5, -2, -3, -1, -4, -6$

64. $113, 33, -47, -108, 72, -76$

65. Karen picked her favorite basketball team to win by 15 points. The team lost by 8 points. By how much was Karen's guess in error?

66. A helicopter drops a package from a height of 842 feet above sea level. The package lands in the ocean and settles at a point 927 feet below sea level. What was the vertical distance the package traveled?

67. Mount Whitney, in the Sierra Nevada mountains of California, is the highest point in the contiguous United States. It is 14,495 feet above sea level. Death Valley, in California and Nevada, is the lowest point in the United States, 282 feet below sea level. Find the vertical height difference between Mount Whitney and Death Valley.

68. In the first four plays of the game the Eagles gained 8 yd, lost 5 yd, gained 3 yd, and gained 4 yd. What is the total number of yards gained in the first four plays? Did the Eagles make a first down? (10 yd are needed for a first down)

69. Yvette records the closing price of her favorite stock every Friday, and the change in price from the previous Friday. The changes in price for the last five Fridays were: gained 7 points, lost 2 points, lost 3 points, lost 2 points, and gained 1 point. What was the net change in the stock's value for the five Fridays?

70. In a game of pinochle Charlie scores the following points in each hand: gains 33 points, loses 26 points, gains 8 points, gains 15 points, loses 35 points, gains 42 points, and gains 37 points. What is the net gain for the game?

71. Explain why $\dfrac{a}{b} = \dfrac{-a}{-b}$.

72. Shown below is part of a World Standard Time Zones chart used by airlines and the United States Navy. The scale along the bottom is just like a number line with the integers $-12, -11, \ldots, 11, 12$ on it.
  a) Find the difference in time between Amsterdam (zone $+1$) and Los Angeles (zone -8).
  b) Find the difference in time between Boston (zone -5) and Puerto Vallerta (zone -7).

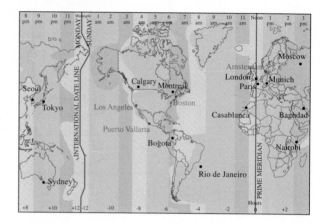

Problem Solving

73. Find the quotient:

$$\frac{-1 + 2 - 3 + 4 - 5 + \cdots - 99 + 100}{1 - 2 + 3 - 4 + 5 - \cdots + 99 - 100}$$

74. Triangular numbers and square numbers were introduced in the Section 1.1 Exercises. There are also **pentagonal numbers**, which were also studied by the Greeks. Four pentagonal numbers are 1, 5, 12, and 22.

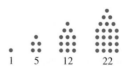

  a) Determine the next three pentagonal numbers.
  b) Describe a procedure to determine the next five pentagonal numbers without drawing the figures.
  c) Is 72 a pentagonal number? Explain how you determined your answer.

75. Place the appropriate plus or minus signs between each digit so that the total will equal 1.

$$0 \quad 1 \quad 2 \quad 3 \quad 4 \quad 5 \quad 6 \quad 7 \quad 8 \quad 9 = 1$$

5.3 The Rational Numbers

We introduced the number line in Section 5.1, and discussed the integers in the previous section. The numbers that fall between the integers on the number line are either rational or irrational numbers. In this section we will discuss the rational numbers and in Section 5.4 we will discuss the irrational numbers.

Any number that can be expressed as a quotient of two integers (denominator not 0) is a rational number.

> The set of **rational numbers**, denoted by Q, is the set of all numbers of the form p/q, where p and q are integers and $q \neq 0$.

The following numbers are examples of rational numbers:

$$\frac{3}{5}, \quad \frac{-2}{7}, \quad \frac{12}{5}, \quad 2, \quad 0$$

The integers 2 and 0 are rational numbers because each can be expressed as the quotient of two integers: $2 = \frac{2}{1}$ and $0 = \frac{0}{1}$. In fact, every integer p is a rational number, since it can be written in the form of $p/1$.

Note the following important property of the rational numbers.

> Every **rational number** when expressed as a decimal will be either a terminating or a repeating decimal.

Examples of terminating decimals are 0.5, 0.75, and 4.65. Examples of repeating decimals are 0.333..., 0.2323..., and 8.456456.... One way to indicate that a number or group of numbers repeat is to place a bar above the number or group of numbers that repeat. Thus 0.333... may be written $0.\overline{3}$, 0.2323... may be written $0.\overline{23}$, and 8.456456... may be written $8.\overline{456}$.

Example 1 uses the words numerator and denominator. The top number in a fraction is called the **numerator**, and the bottom numerator is the **denominator**.

▶ **Example 1**

Show that the following rational numbers are terminating decimals.

a) $\frac{1}{2}$ **b)** $\frac{12}{5}$ **c)** $\frac{3}{8}$

Solution: To express the rational number in decimal form, divide the numerator by the denominator. If you use a calculator, or divide by long division, you will obtain the following results.

a) $1 \div 2 = 0.5$ **b)** $12 \div 5 = 2.4$ **c)** $3 \div 8 = 0.375$

Notice that in each part of Example 1 each quotient has a final nonzero digit. Thus each number is a terminating decimal.

Did You

Know...

Triangular Numbers

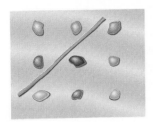

One very simple reason that the ancient Greek mathematicians thought of mathematics in terms of whole numbers and their ratios (the rational numbers) was that they were still working with numbers that were represented by objects, such as squares and triangles, not number symbols. Still, this did not prevent them from drawing some conclusions about number theory. Consider two consecutive triangular numbers, for instance 3 and 6. You can see from the diagram that the sum of the two triangular numbers is equal to the square number 9, the square of a side of the larger triangle.

▶ **Example 2**

Show that the following rational numbers are repeating decimals.

a) $\frac{1}{3}$ **b)** $\frac{23}{99}$ **c)** $\frac{2}{7}$

Solution: If you use a calculator, or divide by long division, you will see that each fraction results in a repeating decimal number.

a) $1 \div 3 = 0.333\ldots$ or $0.\overline{3}$

b) $23 \div 99 = 0.2323\ldots$ or $0.\overline{23}$

c) $2 \div 7 = 0.285714285714\ldots$ or $0.\overline{285714}$

Notice in each part of Example 2 the quotient has no final digit and continues on indefinitely. Each number is a repeating decimal.

When a fraction is converted to a decimal, the maximum number of digits that can repeat is $n - 1$, where n is the denominator of the fraction. For example, when $\frac{2}{7}$ is converted to a decimal, the maximum number of digits that can repeat is $7 - 1$, or 6.

Converting Decimal Numbers to Fractions

We can convert a terminating or repeating decimal into a quotient of integers. The explanation of the procedure will refer to the positional values to the right of the decimal point, as illustrated here:

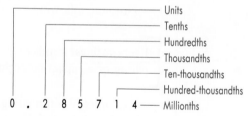

The following examples illustrate how to convert from a decimal to a fraction.

The words *numerator* and *denominator* are derived from the Latin for "the numberer" and "the namer," respectively (*nomin* is Latin for "name"). The ancient Chinese looked at the relationship in a more fanciful way: They called the numerator "the son" and the denominator "the mother."

▶ **Example 3**

Convert the following terminating decimals to a quotient of integers.

a) 0.4   **b)** 0.62   **c)** 0.062

Solution: When converting a terminating decimal to a quotient of integers, we observe the last digit to the right of the decimal point. The position of this digit will indicate the denominator of the quotient of integers.

a) $0.4 = \frac{4}{10}$, since the 4 is in the tenths position.

b) $0.62 = \frac{62}{100}$, since the digit on the right, 2, is in the hundredths position.

c) $0.062 = \frac{62}{1000}$, since the digit on the right, 2, is in the thousandths position.

Converting repeating decimals to a quotient of integers is more difficult. To do so, we must "create" another repeating decimal with the same repeating digits so that when one repeating decimal is subtracted from the other repeating decimal, the difference will be a whole number. To create a number with the same repeating digits, multiply the original repeating decimal by 10 if one digit repeats, by 100 if two digits repeat, by 1000 if three digits repeat, and so on. Examples 4, 5, and 6 illustrate this procedure.

▶ **Example 4**

Convert $0.\overline{3}$ to a quotient of integers.

Solution: $0.\overline{3} = 0.3\overline{3} = 0.333$, and so on.

Let the original repeating decimal be called n; thus $n = 0.\overline{3}$. Since one digit repeats, we multiply both sides of the equation by 10. This gives $10n = 3.\overline{3}$. Then subtract:

$$10n = 3.\overline{3}$$
$$- \quad n = 0.\overline{3}$$
$$9n = 3.0$$

Note that $10n - n = 9n$ and $3.\overline{3} - 0.\overline{3} = 3.0$.

Next, solve for n by dividing both sides by 9:

$$\frac{9n}{9} = \frac{3.0}{9}$$

$$n = \frac{3}{9} = \frac{1}{3}$$

Therefore $0.\overline{3} = \frac{1}{3}$.

▶ **Example 5**

Convert $0.\overline{83}$ to a quotient of integers.

Solution: Let $n = 0.\overline{83}$. Since two digits repeat, multiply both sides of the equation by 100. Thus $100n = 83.\overline{83}$. Now subtract n from $100n$:

$$100n = 83.\overline{83}$$
$$- \quad n = \quad 0.\overline{83}$$
$$99n = 83$$

Finally, divide both sides by 99:

$$\frac{99n}{99} = \frac{83}{99}$$

$$n = \frac{83}{99}$$

Therefore $0.\overline{83} = \frac{83}{99}$.

▶ **Example 6**

Convert $12.14\overline{2}$ to a quotient of integers.

Solution: This problem is different from the two preceding examples in that the repeating digit, 2, is not directly to the right of the decimal point. When this situation arises, move the decimal point to the right until the repeating terms are directly to its right. For each place the decimal point is moved, the number is multiplied by 10. In this example the decimal point must be moved two places to the right. Thus the number must be multiplied by 100.

$$n = 12.14\overline{2}$$
$$100n = 100 \times 12.14\overline{2} = 1214.\overline{2}$$

Now proceed as in the previous two examples. Since one digit repeats, we multiply both sides by 10:

$$100n = 1214.\overline{2}$$
$$10 \times 100n = 10 \times 1214.\overline{2}$$
$$1000n = 12142.\overline{2}$$

Now subtract $100n$ from $1000n$ so that the repeating part will drop out:

$$
\begin{array}{r}
1000n = 12142.\overline{2} \\
-100n = 1214.\overline{2} \\
\hline
900n = 10928
\end{array}
$$

Thus $n = \dfrac{10928}{900}$ or $\dfrac{2732}{225}$.

A set of numbers is said to be **dense** if between any two distinct members of the set there exists a third distinct member of the set. The set of integers is not dense, since between any two consecutive integers there is not another integer. For example, between 1 and 2 there are no other integers. The set of rational numbers is dense because between any two distinct rational numbers there exists a third distinct rational number. Example 7 illustrates this concept.

▶ **Example 7**

Determine a rational number between the two numbers in each of the following pairs.

a) 0.243 and 0.244   **b)** 0.0007 and 0.0008

Solution:

a) The number 0.243 can be written as 0.2430, and 0.244 can be written as 0.2440. There are many numbers between these two. Some of them are 0.2431, 0.2435, and 0.243912. How many numbers are there between these two numbers?

b) 0.0007 can be written as 0.00070, and 0.0008 can be written as 0.00080. One number between them is 0.00075.

Consider the rational number $0.\overline{9}$. What is $0.\overline{9}$ when expressed as a quotient of integers? We will use the procedure we used earlier in this section to answer this question.

Let $n = 0.\overline{9}$.

$$\begin{aligned} 10n &= 9.\overline{9} \\ - \quad n &= 0.\overline{9} \\ \hline 9n &= 9 \end{aligned}$$

$$n = \frac{9}{9} = 1$$

We have just illustrated that $0.\overline{9} = 1$, a statement that you probably question. However, we can show that $0.\overline{9} = 1$, using the denseness property. We stated earlier that the rational numbers are dense. Thus if $0.\overline{9}$ and 1 are two different rational numbers, then we can find a rational number between the two. If we take the average of $0.\overline{9}$ and 1 by adding them and dividing the sum by 2 the result is

$$\frac{0.\overline{9} + 1}{2} = \frac{1.\overline{9}}{2} = 0.\overline{9}$$

Since the average of the two numbers, $0.\overline{9}$, is one of the original numbers, the two numbers must be equal.

Example 7 illustrated a procedure used to find a rational number between any two decimal numbers. It is also possible to find a rational number between any two fractions. One way to do so is to add the fractions and divide the sum by 2. In order to do this we will explain how to add, subtract, multiply, and divide fractions.

Multiplication and Division of Fractions

The product of two fractions is found by multiplying the numerators together and multiplying the denominators together.

Multiplication of Fractions

$$\frac{a}{b} \cdot \frac{c}{d} = \frac{a \cdot c}{b \cdot d} = \frac{ac}{bd} \qquad b \neq 0, \quad d \neq 0$$

▶ **Example 8**

Evaluate the following.

a) $\dfrac{2}{3} \cdot \dfrac{5}{7}$ b) $\left(\dfrac{-2}{3}\right)\left(\dfrac{-4}{9}\right)$ c) $\left(2\dfrac{3}{4}\right)\left(1\dfrac{5}{8}\right)$

Solution:

a) $\dfrac{2}{3} \cdot \dfrac{5}{7} = \dfrac{2 \cdot 5}{3 \cdot 7} = \dfrac{10}{21}$ b) $\left(\dfrac{-2}{3}\right)\left(\dfrac{-4}{9}\right) = \dfrac{(-2)(-4)}{(3)(9)} = \dfrac{8}{27}$

c) First we change each of the mixed numbers to fractions, as follows.

$$2\frac{3}{4} = \frac{(2 \cdot 4) + 3}{4} = \frac{11}{4}$$

$$1\frac{5}{8} = \frac{(1 \cdot 8) + 5}{8} = \frac{13}{8}$$

Now multiply the fractions together:

$$\left(\frac{11}{4}\right)\left(\frac{13}{8}\right) = \frac{143}{32}$$

Note that $\frac{143}{32}$ may be written as a mixed number by dividing the numerator by the denominator.

$$
\begin{array}{r}
4 \\
32 \overline{\smash{)}\,143} \\
128 \\
\hline
15
\end{array}
$$

$$\frac{143}{32} = 4\frac{15}{32} \quad \begin{array}{l} \leftarrow \text{remainder} \\ \leftarrow \text{divisor} \end{array}$$

$$\uparrow$$
$$\text{quotient}$$

The **reciprocal** of a given number is 1 divided by the given number. The product of a number and its reciprocal must equal 1. Examples of some numbers and their reciprocals follow.

Number		Reciprocal		Product
3	$\cdot$	$\frac{1}{3}$	$=$	1
$\frac{3}{5}$	$\cdot$	$\frac{5}{3}$	$=$	1
-6	$\cdot$	$-\frac{1}{6}$	$=$	1

To find the quotient of two fractions, multiply the first fraction by the reciprocal of the second fraction.

Division of Fractions

$$\frac{a}{b} \div \frac{c}{d} = \frac{a}{b} \cdot \frac{d}{c} = \frac{ad}{bc}, \qquad b \neq 0, \quad d \neq 0, \quad c \neq 0$$

▶ **Example 9**

Evaluate the following.

a) $\dfrac{2}{3} \div \dfrac{5}{7}$ b) $\dfrac{\frac{-3}{5}}{\frac{5}{7}}$

Solution:

a) $\dfrac{2}{3} \div \dfrac{5}{7} = \dfrac{2}{3} \cdot \dfrac{7}{5} = \dfrac{2 \cdot 7}{3 \cdot 5} = \dfrac{14}{15}$ b) $\dfrac{\frac{-3}{5}}{\frac{5}{7}} = \dfrac{-3}{5} \cdot \dfrac{7}{5} = \dfrac{-3 \cdot 7}{5 \cdot 5} = \dfrac{-21}{25}$

Addition and Subtraction of Fractions

To add or subtract fractions, it is necessary for the fractions to have a common denominator. A common denominator is another name for a common multiple. The least common multiple (LCM), introduced in Section 5.1, is the lowest common denominator.

To add or subtract two fractions with a common denominator, we add or subtract their numerators and retain the common denominator.

Addition and Subtraction of Fractions

$$\frac{a}{c} + \frac{b}{c} = \frac{a+b}{c}, \quad c \neq 0 \qquad \frac{a}{c} - \frac{b}{c} = \frac{a-b}{c}, \quad c \neq 0$$

▶ **Example 10**

Evaluate the following.

a) $\dfrac{3}{7} + \dfrac{2}{7}$ b) $\dfrac{5}{9} - \dfrac{1}{9}$

Solution:

a) $\dfrac{3}{7} + \dfrac{2}{7} = \dfrac{3+2}{7} = \dfrac{5}{7}$ b) $\dfrac{5}{9} - \dfrac{1}{9} = \dfrac{5-1}{9} = \dfrac{4}{9}$

Notice that in Example 10 the denominators of the fractions being added or subtracted were the same. *When adding or subtracting two fractions with unlike denominators, first rewrite each fraction with a common denominator. Then add or subtract the fractions.*

Writing fractions with a common denominator is accomplished with the *fundamental law of rational numbers.*

> **Fundamental Law of Rational Numbers**
> If a, b, and c are integers, with $b \neq 0$ and $c \neq 0$, then
> $$\frac{a}{b} = \frac{a}{b} \cdot \frac{c}{c} = \frac{a \cdot c}{b \cdot c}$$

Examples 11–13 illustrate the procedure for adding and subtracting fractions with unlike denominators.

▶ **Example 11**

Evaluate $\dfrac{5}{12} - \dfrac{3}{10}$.

Solution: Using prime factorization (Section 5.1), we find that the LCM of 12 and 10 is 60. We will therefore express each fraction with a denominator of 60. Sixty divided by 12 is 5. Therefore the denominator, 12, must be multiplied by 5 to get 60. If the denominator is multiplied by 5, the numerator must also be multiplied by 5 so that the value of the fraction remains unchanged. Multiplying both numerator and denominator by 5 is the same as multiplying by 1.

We follow the same procedure for the other fraction, $\frac{3}{10}$. Sixty divided by 10 is 6. Therefore we multiply both the denominator, 10, and the numerator, 3, by 6 to obtain an equivalent fraction with a denominator of 60.

$$\frac{5}{12} - \frac{3}{10} = \left(\frac{5}{12} \cdot \frac{5}{5}\right) - \left(\frac{3}{10} \cdot \frac{6}{6}\right)$$

$$= \frac{25}{60} - \frac{18}{60}$$

$$= \frac{7}{60}$$

▶ **Example 12**

Add $\dfrac{1}{36} + \dfrac{1}{150}$.

Solution: In Example 7 of Section 5.1 we determined that the LCM of 36 and 150 is 900. We rewrite both fractions using the LCM in the denominator.

$$\frac{1}{36} + \frac{1}{150} = \left(\frac{1}{36} \cdot \frac{25}{25}\right) + \left(\frac{1}{150} \cdot \frac{6}{6}\right)$$

$$= \frac{25}{900} + \frac{6}{900}$$

$$= \frac{31}{900}$$

Now that we know how to add and divide fractions, we can find a fraction between any two given fractions by adding the two given fractions and dividing their sum by 2.

▶ **Example 13**

Find a rational number between $\frac{1}{4}$ and $\frac{1}{3}$.

Solution: First add $\frac{1}{4}$ and $\frac{1}{3}$.

$$\frac{1}{4} + \frac{1}{3} = \frac{3}{12} + \frac{4}{12} = \frac{7}{12}$$

Next divide the sum by 2:

$$\frac{7}{12} \div 2 = \frac{7}{12} \div \frac{2}{1} = \frac{7}{12} \cdot \frac{1}{2} = \frac{7}{24}.$$

Thus $\frac{7}{24}$ is between $\frac{1}{4}$ and $\frac{1}{3}$, written

$$\frac{1}{4} < \frac{7}{24} < \frac{1}{3}.$$

The procedure used in Example 13 will result in a fraction that is exactly halfway between (or the average of) the two given fractions. The result in Example 13 is more obvious if you change all of the denominators to 24. That is,

$$\frac{6}{24} < \frac{7}{24} < \frac{8}{24}.$$

Reducing Fractions

A fraction is said to be in its **lowest terms** (or reduced) when the numerator and denominator are relatively prime (that is, have no common divisors other than 1). To reduce a fraction to its lowest terms, divide both the numerator and the denominator by the greatest common divisor. A procedure for finding the greatest common divisor was discussed in Section 5.1.

▶ **Example 14**

Reduce $\frac{36}{150}$ to its lowest terms.

Solution: In Example 5 of Section 5.1 we determined that the GCD of 36 and 150 is 6. Divide the numerator and the denominator by the GCD, 6.

$$\frac{36 \div 6}{150 \div 6} = \frac{6}{25}$$

Since there are no common divisors of 6 and 25 other than 1, this fraction is in its lowest terms.

▶ **Example 15**

Following are the instructions given on a box of Minute Rice. Determine the amount of (a) rice and water, (b) salt, and (c) butter or margarine needed to make 3 servings of rice.

DIRECTIONS

1. Bring water, salt, and butter (or margarine) to a boil.

2. Stir in rice. Cover: remove from heat. Let stand 5 minutes. Fluff with fork.

TO MAKE	RICE & WATER (equal measures)	SALT	BUTTER OR MARGARINE (if desired)
2 servings	2/3 **cup**	1/4 tsp	1 tsp
4 servings	11/3 **cups**	1/2 tsp	2 tsp

Solution: We can find the amount of each ingredient by finding the average of the amount for 2 and 4 servings.

a) $\dfrac{\dfrac{2}{3} + 1\dfrac{1}{3}}{2} = \dfrac{\dfrac{2}{3} + \dfrac{4}{3}}{2} = \dfrac{\dfrac{6}{3}}{2} = \dfrac{2}{2} = 1$ cup

b) $\dfrac{\dfrac{1}{4} + \dfrac{1}{2}}{2} = \dfrac{\dfrac{1}{4} + \dfrac{2}{4}}{2} = \dfrac{\dfrac{3}{4}}{2} = \dfrac{3}{4} \cdot \dfrac{1}{2} = \dfrac{3}{8}$ tsp

c) $\dfrac{1+2}{2} = \dfrac{3}{2}$ $\left(\text{or } 1\dfrac{1}{2} \text{ tsp}\right)$

The solution to Example 15 can be found in other ways. Can you suggest other procedures for solving the same problem?

Section 5.3 Exercises

Determine whether the following are terminating or repeating decimals.

1. 5

2. 31.9

3. $1.\overline{37}$

4. 0.566566

5. $0.31\overline{31}$

6. 0.7777

7. 3.14151415...

8. 4.14

9. 3.132132132...

10. 0.999...

Express each fraction as a terminating or repeating decimal.

11. $\dfrac{1}{4}$

12. $\dfrac{3}{7}$

13. $\dfrac{5}{7}$

14. $\dfrac{1}{6}$

15. $\dfrac{5}{9}$

16. $\dfrac{22}{7}$

17. $\dfrac{60}{15}$

18. $\dfrac{70}{15}$

19. $\dfrac{80}{15}$

20. $\dfrac{1001}{11}$

Express each terminating decimal as a quotient of two integers.

21. 0.5

22. 0.96

23. 0.345

24. 0.9075

25. 3.006

26. 3.625

27. 1.246

28. 0.56234

29. 2.0678

30. 1.0025

Express each repeating decimal as a quotient of two integers.

31. $0.\overline{4}$ **32.** $0.\overline{7}$ **33.** $1.\overline{9}$

34. $0.\overline{56}$ **35.** $1.\overline{72}$ **36.** $0.\overline{357}$

37. $3.\overline{258}$ **38.** $1.3\overline{4}$ **39.** $1.2\overline{1}$

40. $4.3\overline{73}$

Evaluate each expression.

41. $\dfrac{3}{7} \div \dfrac{5}{8}$ **42.** $\dfrac{3}{7} \cdot \dfrac{5}{8}$

43. $\left(\dfrac{-3}{7}\right)\left(\dfrac{-14}{15}\right)$ **44.** $\left(-\dfrac{3}{5}\right) \div \dfrac{5}{9}$

45. $\dfrac{7}{9} \div \dfrac{9}{7}$ **46.** $\dfrac{7}{9} \div \dfrac{7}{9}$

47. $\left(\dfrac{3}{5} \cdot \dfrac{4}{7}\right) \div \dfrac{1}{3}$ **48.** $\left(\dfrac{4}{7} \div \dfrac{4}{5}\right) \cdot \dfrac{1}{7}$

49. $\left[\left(\dfrac{-3}{4}\right)\left(\dfrac{-2}{7}\right)\right] \div \dfrac{3}{5}$ **50.** $\left(\dfrac{3}{8} \cdot \dfrac{5}{9}\right) \cdot \left(\dfrac{4}{7} \div \dfrac{5}{8}\right)$

Reduce each fraction to lowest terms.

51. $\dfrac{18}{96}$ **52.** $\dfrac{42}{54}$ **53.** $\dfrac{48}{92}$

54. $\dfrac{160}{240}$ **55.** $\dfrac{325}{675}$ **56.** $\dfrac{11}{1001}$

57. $\dfrac{96}{212}$ **58.** $\dfrac{35}{140}$ **59.** $\dfrac{44}{121}$

60. $\dfrac{112}{196}$

Perform the indicated operation, and reduce your answer to lowest terms.

61. $\dfrac{1}{5} + \dfrac{1}{6}$ **62.** $\dfrac{1}{7} - \dfrac{4}{21}$ **63.** $\dfrac{2}{13} + \dfrac{3}{130}$

64. $\dfrac{5}{24} + \dfrac{7}{72}$ **65.** $\dfrac{5}{9} - \dfrac{7}{54}$ **66.** $\dfrac{13}{30} - \dfrac{17}{120}$

67. $\dfrac{1}{12} + \dfrac{1}{48} + \dfrac{1}{72}$ **68.** $\dfrac{3}{5} + \dfrac{7}{15} + \dfrac{9}{75}$

69. $\dfrac{1}{30} - \dfrac{3}{40} - \dfrac{7}{50}$ **70.** $\dfrac{4}{25} - \dfrac{9}{100} - \dfrac{7}{40}$

Alternative methods for adding and subtracting two frac-

tions are given below. These methods may not result in a solution in its lowest terms.

$$\frac{a}{b} + \frac{c}{d} = \frac{ad + bc}{bd} \quad \text{and} \quad \frac{a}{b} - \frac{c}{d} = \frac{ad - bc}{bd}$$

In Ex. 71–76, use one of the above formulas to evaluate the expression.

71. $\dfrac{2}{3} + \dfrac{3}{4}$ **72.** $\dfrac{5}{7} - \dfrac{1}{12}$ **73.** $\dfrac{5}{7} + \dfrac{3}{4}$

74. $\dfrac{7}{3} - \dfrac{5}{12}$ **75.** $\dfrac{3}{8} + \dfrac{5}{12}$ **76.** $\left(\dfrac{2}{3} + \dfrac{1}{4}\right) - \dfrac{3}{5}$

Evaluate each of the following.

77. $\left(\dfrac{1}{5} \cdot \dfrac{1}{4}\right) + \dfrac{1}{3}$ **78.** $\left(\dfrac{3}{5} \div \dfrac{2}{10}\right) - \dfrac{1}{3}$

79. $\left(\dfrac{1}{2} + \dfrac{3}{10}\right) \div \left(\dfrac{1}{5} + 2\right)$ **80.** $\left(\dfrac{1}{9} \cdot \dfrac{3}{5}\right) + \left(\dfrac{2}{3} \cdot \dfrac{1}{5}\right)$

81. $\left(3 - \dfrac{4}{9}\right) \div \left(4 + \dfrac{2}{3}\right)$ **82.** $\left(\dfrac{2}{5} \div \dfrac{4}{9}\right)\left(\dfrac{3}{5} \cdot 6\right)$

Find a rational number between the two numbers in each of the following pairs.

83. 0.47 and 0.48 **84.** 5.3 and 5.003

85. -1.05 and -1.06 **86.** 1.3457 and 1.34571

87. 3.12345 and 3.123451 **88.** 0.4105 and 0.4106

89. 4.872 and 4.873 **90.** -3.7896 and -3.7895

Find a rational number between the two fractions in each of the following pairs.

91. $\dfrac{2}{7}$ and $\dfrac{3}{7}$ **92.** $\dfrac{1}{9}$ and $\dfrac{2}{9}$ **93.** $\dfrac{1}{20}$ and $\dfrac{1}{10}$

94. $\dfrac{5}{13}$ and $\dfrac{6}{13}$ **95.** $\dfrac{1}{4}$ and $\dfrac{1}{5}$ **96.** $\dfrac{1}{3}$ and $\dfrac{2}{3}$

97. $\dfrac{1}{10}$ and $\dfrac{1}{100}$ **98.** $\dfrac{1}{2}$ and $\dfrac{2}{3}$

99. Are the rational numbers dense? Explain.

100. Are the integers dense? Explain.

101. If a person from another planet visited you and asked what Earthlings mean by the symbol $\frac{3}{8}$, how would you explain?

102. Mr. and Mrs. Stulfous have been increasing their children's weekly allowances each year by $\frac{1}{10}$ over the previous year's weekly allowance.

a) If Tom's current weekly allowance is $8.00, what will it be next year?

b) Diana's current weekly allowance is $10.00. What will her weekly allowance be in 2 years?

103. An electrician purchased 310 connectors at a cost of $7\frac{3}{4}$ cents each. What was the total bill if the sales tax is $6\frac{1}{2}\%$?

104. A sociology class is $\frac{2}{5}$ math majors, $\frac{1}{4}$ English majors, and $\frac{1}{10}$ chemistry majors; the remaining members of the class are art majors. What fraction of the class are art majors?

105. Wu, who lives in California, spends $\frac{11}{30}$ of her income on housing, $\frac{2}{5}$ on food and clothing, and $\frac{1}{10}$ on entertainment. If she saves the rest, what fraction of her income does she save?

106. A stairway consists of 12 stairs, each $10\frac{3}{4}$ inches high. What is the vertical height of the stairway?

107. Mary Anne gives Sam the tailor $8\frac{3}{4}$ yards of fabric. She asks him to make a skirt, a pair of slacks, and a jacket. These three items require $1\frac{7}{8}$ yd, $2\frac{5}{8}$ yd, and $2\frac{3}{8}$ yd, respectively, of fabric. How many yards of fabric are left over from the original piece?

108. If a plumber cuts three $2\frac{5}{16}$ ft lengths of pipe from an 8 ft section, how much remains?

109. If a stock price drops from $17\frac{3}{8}$ to $13\frac{7}{8}$, how much did the stock drop?

110. A piece of wood measures $15\frac{3}{8}$ in.

a) How far from one end should you cut the wood if you wish to cut the length in half?

b) What is the length of each piece after the cut? You must allow $\frac{1}{8}$ in. for the saw cut.

111. If a piece of wood is $8\frac{3}{4}$ ft long and is to be cut into four equal pieces, find the length of each piece. (Allow $\frac{1}{4}$ in. for the saw cut.)

112. The instructions for assembling a computer stand include a diagram illustrating its dimensions. Find the total height of the stand.

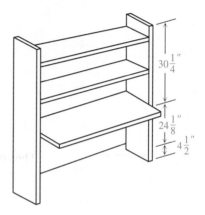

Problem Solving

113. A dress is on sale for $150. What was the original price of the dress if the discount was 25% of the original price?

5.4 | # The Irrational Numbers and the Real Number System

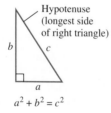

Hypotenuse (longest side of right triangle)

$a^2 + b^2 = c^2$

Figure 5.6

Pythagoras (ca. 585–500 B.C.), a Greek mathematician, provided a proof that in any *right triangle* (a triangle with a 90° angle, see Fig. 5.6) the square of the length of one side (a^2) added to the square of the length of the other side (b^2) equals the square of the length of the hypotenuse (c^2). The formula $a^2 + b^2 = c^2$ is now known as the **Pythagorean theorem**.* Pythagoras found that the solution of the formula, where $a = 1$ and $b = 1$, is not a rational number.

$$a^2 + b^2 = c^2$$
$$1^2 + 1^2 = c^2$$
$$1 + 1 = c^2$$
$$2 = c^2$$

---

* The Pythagorean theorem is discussed in more detail in Section 8.3.

The Pythagoreans were a secret society in southern Italy in the sixth century B.C. These 300 young aristocratic men and women followed their master, Pythagoras. During the first three years of their schooling, the students were asked to listen in silence to the teaching of the master as he spoke to them from behind a curtain about mathematics and philosophy.

There is no rational number that when squared will equal 2. This prompted a need for a new set of numbers, the irrational numbers.

In Section 5.2 we introduced the real number line. The points on the real number line that are not rational numbers are referred to as irrational numbers. Recall that every rational number is either a terminating or a repeating decimal. Therefore irrational numbers, when represented as decimals, will be nonterminating, nonrepeating decimals.

> An **irrational number** is a real number whose decimal representation is a nonterminating, nonrepeating decimal.

Nonrepeating number patterns can be used to indicate irrational numbers. For example, 6.1011011101111 . . . and 0.525225222 . . . are both irrational numbers.

The expression $\sqrt{2}$ is read "the square root of 2" or "radical 2." The symbol $\sqrt{}$ is called the **radical sign** and the number or expression inside the radical sign is called the **radicand**. In $\sqrt{2}$, 2 is the radicand.

The square roots of some numbers are rational while the square roots of other numbers are irrational. The **principal or positive square root** of a number n, written $\sqrt{n}$, is the positive number that when multiplied by itself gives n. Whenever we mention the term "square root" in this text, we mean the principal square root. For example,

$$\sqrt{25} = 5, \quad \text{since} \quad 5 \cdot 5 = 25$$
$$\sqrt{100} = 10, \quad \text{since} \quad 10 \cdot 10 = 100$$

Both $\sqrt{25}$ and $\sqrt{100}$ are examples of numbers that are rational numbers, since their square roots, 5 and 10, are terminating decimals.

Returning to the problem faced by Pythagoras, if $c^2 = 2$ then c has a value of $\sqrt{2}$. But what is $\sqrt{2}$ equal to? The $\sqrt{2}$ is an irrational number and it cannot be expressed as a terminating or repeating decimal. It can only be approximated by a decimal: $\sqrt{2}$ is approximately 1.4142135.

Other irrational numbers include $\sqrt{3}$, $\sqrt{5}$, $\sqrt{37}$. Another important irrational number used to represent the ratio of a circle's circumference to its diameter is pi, symbolized π. Pi is approximately 3.1415926.

We have discussed procedures for performing the arithmetic operations of addition, subtraction, multiplication, and division with rational numbers. We can perform the same operations with the irrational numbers. Before we can proceed, we must have an understanding of numbers that are called perfect squares. Any number that is the square of a natural number is said to be a **perfect square**.

Natural numbers	1,	2,	3,	4,	5,	6, . . .
Squares of the natural numbers	1^2,	2^2,	3^2,	4^2,	5^2,	6^2, . . .
or Perfect squares	1,	4,	9,	16,	25,	36, . . .

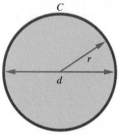
The number which multiplies a radical is called the radical's **coefficient**. For example in $3\sqrt{5}$ the 3 is the coefficient of the radical.

Some irrational numbers can be simplified by determining if there are any perfect square factors in the radicand. If there are, we can use the following rule to simplify the radical.

Product Rule for Radicals

$$\sqrt{a \cdot b} = \sqrt{a} \cdot \sqrt{b} \qquad a \geq 0, \quad b \geq 0$$

For example, $\sqrt{8} = \sqrt{4 \cdot 2} = \sqrt{4} \cdot \sqrt{2} = 2\sqrt{2}$

and $\sqrt{50} = \sqrt{25 \cdot 2} = \sqrt{25} \cdot \sqrt{2} = 5\sqrt{2}$.

▶ **Example 1**

Simplify   **a)** $\sqrt{45}$ **b)** $\sqrt{48}$

Solution:

a) Since 9 is a perfect square factor of 45, write:

$$\sqrt{45} = \sqrt{9 \cdot 5} = \sqrt{9} \cdot \sqrt{5} = 3\sqrt{5}$$

Since 5 has no perfect square factors, $\sqrt{5}$ cannot be simplified.

b) 16 is a perfect square factor of 48.

$$\sqrt{48} = \sqrt{16 \cdot 3} = \sqrt{16} \cdot \sqrt{3} = 4\sqrt{3}$$

In Example 1(b) you can obtain the correct answer if you start out factoring differently.

$$\sqrt{48} = \sqrt{4 \cdot 12} = \sqrt{4} \cdot \sqrt{12} = 2\sqrt{12}$$

Note that 12 has 4 as a perfect square factor.

$$2\sqrt{12} = 2\sqrt{4 \cdot 3} = 2\sqrt{4} \cdot \sqrt{3}$$
$$= 2 \cdot 2\sqrt{3}$$
$$= 4\sqrt{3}$$

We obtain the same answer by doing a little more work.

Addition and Subtraction of Irrational Numbers

To add or subtract two or more square roots with the same radicand, add or subtract their coefficients. The answer is the sum or difference of the coefficients multiplied by the common radical.

▶ **Example 2**

Simplify   **a)** $2\sqrt{7} + 3\sqrt{7}$ **b)** $3\sqrt{5} - 2\sqrt{5} + \sqrt{5}$

Solution:

a) $2\sqrt{7} + 3\sqrt{7} = (2 + 3)\sqrt{7} = 5\sqrt{7}$

b) $3\sqrt{5} - 2\sqrt{5} + \sqrt{5} = (3 - 2 + 1)\sqrt{5} = 2\sqrt{5}.$
(Note that $\sqrt{5} = 1\sqrt{5}.$)

▶ **Example 3**
Simplify $5\sqrt{3} - \sqrt{12}.$

Solution: These radicals cannot be added in their present form because they contain different radicands. When this occurs, determine whether one or more of the radicals can be simplified so that they have the same radicand.

$$5\sqrt{3} - \sqrt{12} = 5\sqrt{3} - \sqrt{4 \cdot 3}$$
$$= 5\sqrt{3} - \sqrt{4} \cdot \sqrt{3}$$
$$= 5\sqrt{3} - 2\sqrt{3}$$
$$= (5 - 2)\sqrt{3} = 3\sqrt{3}$$

Multiplication of Irrational Numbers

When multiplying irrational numbers, we again make use of the product rule for radicals. After the radicands are multiplied, simplify the remaining radical when possible.

▶ **Example 4**
Simplify **a)** $\sqrt{2} \cdot \sqrt{8}$ **b)** $\sqrt{3} \cdot \sqrt{5}$ **c)** $\sqrt{8} \cdot \sqrt{3}$

Solution:
a) $\sqrt{2} \cdot \sqrt{8} = \sqrt{2 \cdot 8} = \sqrt{16} = 4$
b) $\sqrt{3} \cdot \sqrt{5} = \sqrt{3 \cdot 5} = \sqrt{15}$
c) $\sqrt{8} \cdot \sqrt{3} = \sqrt{8 \cdot 3} = \sqrt{24} = \sqrt{4} \cdot \sqrt{6} = 2\sqrt{6}$

Division of Irrational Numbers

To divide irrational numbers, we make use of the following rule. After performing the division, simplify when possible.

Quotient Rule for Radicals

$$\frac{\sqrt{a}}{\sqrt{b}} = \sqrt{\frac{a}{b}} \qquad a \geq 0, \quad b > 0$$

▶ **Example 5**
Divide **a)** $\dfrac{\sqrt{8}}{\sqrt{2}}$ **b)** $\dfrac{\sqrt{96}}{\sqrt{2}}$

Solution:

a) $\dfrac{\sqrt{8}}{\sqrt{2}} = \sqrt{\dfrac{8}{2}} = \sqrt{4} = 2$

b) $\dfrac{\sqrt{96}}{\sqrt{2}} = \sqrt{\dfrac{96}{2}} = \sqrt{48} = \sqrt{16 \cdot 3} = \sqrt{16} \cdot \sqrt{3} = 4\sqrt{3}$

Rationalizing the Denominator

A denominator is **rationalized** when it contains no radical expressions. To rationalize a denominator that contains only a square root, multiply both the numerator and denominator of the fraction by a number that will result in the radicand in the denominator becoming a perfect square. (This is the equivalent of multiplying the fraction by 1 because the value of the fraction does not change.) Then simplify the fractions when possible.

▶ **Example 6**

Rationalize the denominator of

a) $\dfrac{3}{\sqrt{2}}$ **b)** $\dfrac{5}{\sqrt{8}}$ **c)** $\dfrac{\sqrt{3}}{\sqrt{6}}$

Solution:

a) Multiply the numerator and denominator by a number that will make the radicand a perfect square.

$$\frac{3}{\sqrt{2}} = \frac{3}{\sqrt{2}} \cdot \frac{\sqrt{2}}{\sqrt{2}} = \frac{3\sqrt{2}}{\sqrt{4}} = \frac{3\sqrt{2}}{2}$$

Note that the 2s cannot be divided out, since one of the 2s is a radicand and the other is not.

b) $\dfrac{5}{\sqrt{8}} = \dfrac{5}{\sqrt{8}} \cdot \dfrac{\sqrt{2}}{\sqrt{2}} = \dfrac{5\sqrt{2}}{\sqrt{16}} = \dfrac{5\sqrt{2}}{4}$

You could have obtained the same answer to this problem by multiplying both the numerator and denominator by $\sqrt{8}$, and then simplifying.

c) We can write $\dfrac{\sqrt{3}}{\sqrt{6}}$ as $\sqrt{\dfrac{3}{6}}$. Reduce the fraction to obtain $\sqrt{\dfrac{1}{2}}$ or $\dfrac{1}{\sqrt{2}}$. Now rationalize $\dfrac{1}{\sqrt{2}}$.

$$\frac{1}{\sqrt{2}} = \frac{1}{\sqrt{2}} \cdot \frac{\sqrt{2}}{\sqrt{2}} = \frac{\sqrt{2}}{2}$$

Section 5.4 Exercises

1. Explain the difference between a rational number and an irrational number.
2. Explain how to rationalize a denominator that contains only a square root.

In Ex. 3–12, determine whether each number is rational or irrational.

3. $\sqrt{9}$

4. $\sqrt{5}$

5. $\dfrac{3}{5}$

6. $0.303003000\ldots$

7. $5.272272227\ldots$

8. π

9. $\dfrac{22}{7}$

10. 3.14159

11. $3.14159\ldots$

12. $\dfrac{\sqrt{7}}{\sqrt{7}}$

Evaluate the following.

13. $\sqrt{64}$
14. $\sqrt{121}$
15. $\sqrt{49}$
16. $-\sqrt{144}$
17. $-\sqrt{169}$
18. $\sqrt{25}$
19. $-\sqrt{225}$
20. $-\sqrt{36}$
21. $-\sqrt{81}$
22. $\sqrt{256}$

Classify each of the following numbers as belonging to one or more of the following sets: the rational numbers, the integers, the natural numbers, the irrational numbers.

23. 5

24. -7

25. $\sqrt{9}$

26. $\dfrac{3}{7}$

27. $0.131131113\ldots$

28. 1.732

29. $-\dfrac{3}{5}$

30. $0.345345345\ldots$

31. $0.345345\overline{345}$

32. $0.345334533345\ldots$

Simplify each of the radicals.

33. $\sqrt{49}$
34. $\sqrt{20}$
35. $\sqrt{45}$
36. $\sqrt{64}$
37. $\sqrt{125}$
38. $\sqrt{28}$
39. $\sqrt{48}$
40. $\sqrt{50}$
41. $\sqrt{54}$
42. $\sqrt{98}$

Perform the indicated operations.

43. $2\sqrt{5} + 3\sqrt{5}$

44. $\sqrt{7} + 5\sqrt{7}$

45. $3\sqrt{5} - 7\sqrt{5}$

46. $5\sqrt{3} + 4\sqrt{12}$

47. $4\sqrt{12} - 7\sqrt{27}$

48. $2\sqrt{7} + 5\sqrt{28}$

49. $5\sqrt{3} + 7\sqrt{12} - 3\sqrt{75}$

50. $13\sqrt{2} + 2\sqrt{18} - 5\sqrt{32}$

51. $\sqrt{8} - 3\sqrt{50} + 9\sqrt{32}$

52. $\sqrt{63} + 13\sqrt{98} - 5\sqrt{112}$

Perform the indicated operation. Simplify the answer when possible.

53. $\sqrt{3}\sqrt{2}$
54. $\sqrt{6}\sqrt{3}$
55. $\sqrt{4}\sqrt{8}$
56. $\sqrt{5}\sqrt{10}$

57. $\sqrt{6}\sqrt{12}$
58. $\sqrt{13}\sqrt{26}$
59. $\dfrac{\sqrt{8}}{\sqrt{4}}$
60. $\dfrac{\sqrt{125}}{\sqrt{5}}$

61. $\dfrac{\sqrt{72}}{\sqrt{8}}$
62. $\dfrac{\sqrt{136}}{\sqrt{8}}$

Rationalize the denominator.

63. $\dfrac{3}{\sqrt{2}}$
64. $\dfrac{5}{\sqrt{7}}$
65. $\dfrac{\sqrt{7}}{\sqrt{2}}$
66. $\dfrac{\sqrt{7}}{\sqrt{6}}$

67. $\dfrac{\sqrt{12}}{\sqrt{3}}$
68. $\dfrac{\sqrt{32}}{\sqrt{6}}$
69. $\dfrac{\sqrt{9}}{\sqrt{2}}$
70. $\dfrac{\sqrt{15}}{\sqrt{3}}$

71. $\dfrac{\sqrt{10}}{\sqrt{6}}$
72. $\dfrac{7}{\sqrt{7}}$

In Ex. 73–77, determine whether the statement is true or false.

73. $\sqrt{p}$ is irrational for any prime number p.
74. The sum of any two irrational numbers is a rational number.
75. The sum of any two irrational numbers is an irrational number.
76. The product of any two irrational numbers is an irrational number.
77. The product of an irrational and a rational number is always an irrational number.

In Ex. 78–81, give an example to show the stated case can occur.

78. The sum of two irrational numbers may be an irrational number.
79. The sum of two irrational numbers may be a rational number.
80. The product of two irrational numbers may be an irrational number.
81. The product of two irrational numbers may be a rational number.

82. Without doing any calculations, determine whether $\sqrt{3} = 1.732$. Explain your answer.

83. Without doing any calculations, determine whether $\sqrt{11} = 3.3166$. Explain.

84. Give an example to show that $\sqrt{a + b} \neq \sqrt{a} + \sqrt{b}$.

85. The number π is an irrational number. Often the values 3.14 or $\frac{22}{7}$ are used for π. Does π equal either 3.14 or $\frac{22}{7}$? Explain your answer.

86. A carpenter constructing a rectangular door, 3 ft 6 in. by 6 ft 6 in., places a diagonal brace from the upper left corner to the lower right corner. What is the length of the brace to the nearest foot? (*Hint:* Use the Pythagorean theorem.)

87. Students are decorating a rectangular dorm room that measures 12 ft by 18 ft. They want to hang a streamer diagonally across the ceiling of the room from one corner to the other. Determine the length of the streamer needed to go diagonally across the room. Assume there is no slack in the streamer. (*Hint:* Use the Pythagorean theorem.)

88. The time T it takes for a pendulum to swing back and forth may be found by the formula

$$T = 2\pi \sqrt{\frac{l}{g}},$$

where l is the length of the pendulum and g is the acceleration of gravity. Find the time in seconds if $l = 35$ cm and $g = 980 \dfrac{\text{cm}}{\text{sec}^2}$.

Problem Solving

89. One way to find a rational number between two distinct rational numbers is to add the two distinct rational numbers and divide by 2. Do you think that this method will work for finding an irrational number between two distinct irrational numbers? Explain.

90. a) Using the Pythagorean theorem, $a^2 + b^2 = c^2$, construct a right triangle, and then construct a square on each side, as illustrated in Fig. 5.7. Compute the area of the square with side a, the area of the square with side b, and the area of the square with side c. What is the relationship among these areas?

Figure 5.7

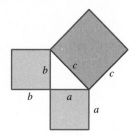

b) Construct a right triangle. Construct a semicircle on each side, with the side as the diameter of the circle. Compute the area of the three semicircles. What is the relationship between these areas?

c) Using parts a) and b) conjecture a rule regarding the areas of similar figures constructed on the sides of a right triangle.

Research Activity

91. Do the necessary research and write a report on the history of the development of the irrational numbers.

5.5 Real Numbers and Their Properties

Now that we have discussed both the rational and irrational numbers, we can discuss the real numbers and the properties of the real number system. The union of the rational numbers and the irrational numbers is the **real numbers**. The set of real numbers is symbolized by $\mathbb{R}$.

Figure 5.8 illustrates the relationship among various sets of numbers. It shows that the natural numbers are a subset of the whole numbers, the integers, the rational numbers, and the real numbers. For example, since the number 3 is a natural or counting number, it is also a whole number, an integer, a rational number and a real number. Since the rational number $\frac{1}{3}$ is outside of the set of integers, it is not an integer or a whole number or a natural number. The number $\frac{1}{3}$ is a real number, however, as is the irrational number $\sqrt{2}$. Notice

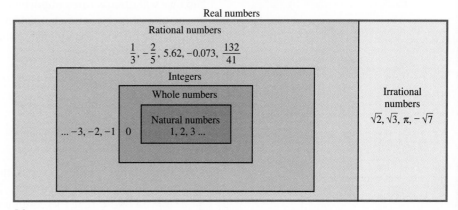

Figure 5.8

that the real numbers are the union of the rational numbers and the irrational numbers.

Properties of the Real Number System

We are now prepared to consider the properties of the real number system. The first property that we will discuss is closure.

> If an operation is performed on any two elements of a set and the result is an element of the set, we say that the set is **closed** under that given operation.

Is the sum of any two natural numbers a natural number? The answer is yes. Thus we say that the natural numbers are closed under the operation of addition.

Are the natural numbers closed under the operation of subtraction? The answer is no. For example, $3 - 5 = -2$, which is not a natural number. Therefore the natural numbers are not closed under the operation of subtraction.

▶ **Example 1**

Determine whether the integers are closed under the operations of (a) multiplication, and (b) division.

Solution:

a) If we multiply any two integers, will the product always be an integer? The answer is yes. Thus the integers are closed under the operation of multiplication.

b) If we divide any two integers, will the quotient be an integer? The answer is no. For example, $6 \div 5 = \frac{6}{5}$, which is not an integer. Therefore the integers are not closed under the operation of division.

Next we will discuss three important properties: the commutative property, the associative property, and the distributive property. A knowledge of these properties is essential for the understanding of algebra. We begin our discussion with the commutative property.

Commutative Property

Addition	Multiplication
$a + b = b + a$	$a \cdot b = b \cdot a$

for any real numbers a and b.

The commutative property states that the **order** in which two numbers are added or multiplied is immaterial. For example $2 + 3 = 3 + 2 = 5$ and $5 \cdot 7 = 7 \cdot 5 = 35$. Note that the commutative property does not hold for the operations of subtraction or division. For example

$$2 - 3 \neq 3 - 2 \quad \text{and} \quad 4 \div 2 \neq 2 \div 4$$

Associative Property

Addition	Multiplication
$(a + b) + c = a + (b + c)$	$(a \cdot b) \cdot c = a \cdot (b \cdot c)$

for any real numbers a, b, and c.

The associative property states that when adding or multiplying three real numbers, parentheses may be placed around any two adjacent numbers. For example

$$
\begin{array}{ll}
(2 + 3) + 4 = 2 + (3 + 4) & (2 \cdot 3) \cdot 4 = 2 \cdot (3 \cdot 4) \\
5 + 4 = 2 + 7 & 6 \cdot 4 = 2 \cdot 12 \\
9 = 9 & 24 = 24
\end{array}
$$

The associative property does not hold for the operations of subtraction and division. For example

$$(4 - 2) - 3 \neq 4 - (2 - 3) \quad \text{and} \quad (8 \div 4) \div 2 \neq 8 \div (4 \div 2)$$

Note the difference between the commutative property and the associative property. The commutative property involves a change in *order*, while the associative property involves a change in *grouping* (or the *association* of numbers that are grouped together).

Another property of the real numbers is the distributive property of multiplication over addition.

Distributive Property of Multiplication over Addition

$$a \cdot (b + c) = a \cdot b + a \cdot c$$

for any real numbers a, b, and c.

For example if $a = 2$, $b = 3$, and $c = 4$, then

$$2 \cdot (3 + 4) = (2 \cdot 3) + (2 \cdot 4),$$
$$2 \cdot 7 = 6 + 8,$$
$$14 = 14.$$

This illustrates that when using the distributive law you may either add first and then multiply or multiply first and then add. Note that the distributive property involves two operations, addition and multiplication. Although positive integers were used in the example, any real numbers could have been used.

We frequently use the commutative, associative, and distributive properties without realizing that we are using them. To add $13 + 4 + 6$, we may add the $4 + 6$ first to get 10. To this sum we then add 13 to get 23. Here we have done the equivalent of placing parentheses around the $4 + 6$. We can do so because of the associative property of addition.

To multiply 102×11 in our heads, we might multiply $100 \times 11 = 1100$ and $2 \times 11 = 22$ and add these two products to get 1122. We are permitted to do this because of the distributive property.

$$102 \times 11 = (100 + 2) \times 11 = (100 \times 11) + (2 \times 11)$$
$$= 1100 + 22 = 1122$$

▶ **Example 2**

Name the property illustrated.

Solution:

a) $4 + 11 = 11 + 4$ Commutative property of addition

b) $(3 + 7) + 13 = 3 + (7 + 13)$ Associative property of addition

c) $(x \cdot y) \cdot z = x \cdot (y \cdot z)$ Associative property of multiplication

d) $13(x + z) = 13 \cdot x + 13 \cdot z$ Distributive property of multiplication over addition

e) $3 + (x + 4) = 3 + (4 + x)$ The only change between left and right sides of the equation is the order of the x and 4. The order is changed from $x + 4$ to $4 + x$ using the commutative property of addition.

f) $5(11 \cdot 9) = (11 \cdot 9)5$ The order of 5 and $(11 \cdot 9)$ is changed by using the commutative property of multiplication.

▶ **Example 3**

Use the distributive property to simplify

a) $3(5 + \sqrt{2})$ **b)** $\sqrt{3}(5 + \sqrt{2})$.

Solution:

a) $3(5 + \sqrt{2}) = (3 \cdot 5) + (3\sqrt{2})$
$$= 15 + 3\sqrt{2}$$

b) $\sqrt{3}(5 + \sqrt{2}) = (\sqrt{3} \cdot 5) + (\sqrt{3} \cdot \sqrt{2})$
$$= 5\sqrt{3} + \sqrt{6}$$

Note that $\sqrt{3} \cdot 5$ is written $5\sqrt{3}$.

▶ **Example 4**

Use the distributive property to multiply $3(x + 2)$. Then simplify the result.

Solution: $3(x + 2) = 3 \cdot x + 3 \cdot 2$
$$= 3x + 6$$

The properties mentioned in this section are summarized below, where a, b, and c are any real numbers.

Commutative property of addition	$a + b = b + a$
Commutative property of multiplication	$a \cdot b = b \cdot a$
Associative property of addition	$(a + b) + c = a + (b + c)$
Associative property of multiplication	$(a \cdot b) \cdot c = a \cdot (b \cdot c)$
Distributive property of multiplication over addition	$a \cdot (b + c) = a \cdot b + a \cdot c$

 Section 5.5 Exercises

Determine whether the natural numbers are closed under the given operation.
1. addition
2. subtraction
3. multiplication
4. division

Determine whether the integers are closed under the given operation.
5. subtraction
6. addition
7. division
8. multiplication

Determine whether the rational numbers are closed under the given operation.
9. addition
10. subtraction
11. multiplication
12. division

Determine whether the irrational numbers are closed under the given operation.
13. addition
14. subtraction
15. multiplication
16. division

Determine whether the real numbers are closed under the given operation.
17. addition
18. subtraction
19. division
20. multiplication

21. Does $3 + (4 + 5) = 3 + (5 + 4)$ illustrate the commutative property or the associative property? Explain your answer.
22. Does $(4 + 5) + 6 = 6 + (4 + 5)$ illustrate the commutative property or the associative property? Explain.
23. Give an example to show that the commutative property of addition may be true for the negative integers.

24. Give an example to show that the commutative property of multiplication may be true for the negative integers.

25. Does the commutative property hold for the integers under the operation of subtraction? Give an example to support your answer.

26. Does the commutative property hold for the real numbers under the operation of subtraction? Give an example to support your answer.

27. Does the commutative property hold for the integers under the operation of division? Give an example to support your answer.

28. Does the commutative property hold for the rational numbers under the operation of division? Give an example to support your answer.

29. Give an example to show that the associative property of addition may be true for the negative integers.

30. Give an example to show that the associative property of multiplication may be true for the negative integers.

31. Does the associative property hold for the integers under the operation of subtraction? Give an example to support your answer.

32. Does the associative property hold for the integers under the operation of division? Give an example to support your answer.

33. Does the associative property hold for the real numbers under the operation of subtraction? Give an example to support your answer.

34. Does the associative property hold for the real numbers under the operation of division? Give an example to support your answer.

35. Does $a + (b \cdot c) = (a + b) \cdot (a + c)$? Give an example to support your answer.

State the name of the property illustrated.

36. $3 + [4 + (-2)] = (3 + 4) + (-2)$

37. $3 \cdot x = x \cdot 3$

38. $(x + 3) + 4 = x + (3 + 4)$

39. $2(3 + 6) = (2 \cdot 3) + (2 \cdot 6)$

40. $(\frac{1}{3} + \frac{1}{4}) + 2 = \frac{1}{3} + (\frac{1}{4} + 2)$

41. $x + y = y + x$

42. $x \cdot y = y \cdot x$

43. $(3 + 4) + (5 + 6) = (5 + 6) + (3 + 4)$

44. $4 + (2 + 3) = (2 + 3) + 4$

45. $2(x + y) = 2x + 2y$

46. $(2 + 3) + 4 = (3 + 2) + 4$

47. $(x + y) + z = z + (x + y)$

48. $(a + b) + c = c + (a + b)$

49. $a \cdot (x + y) = (x + y) \cdot a$

50. $(a + b) + (c + d) = (c + d) + (a + b)$

Use the distributive property to multiply. Then simplify the resulting expression.

51. $4(x + 5)$ **52.** $5(x + 4)$

53. $\sqrt{3}(2 + \sqrt{6})$ **54.** $\sqrt{2}(\sqrt{6} + 3)$

55. $\sqrt{3}(x + \sqrt{3})$ **56.** $x(3 + y)$

Name the property used to go from step to step.

57. $3(x + 2) + 5 = (3x + 3 \cdot 2) + 5$
$$= (3x + 6) + 5$$
58. $\quad = 3x + (6 + 5)$
$$= 3x + 11$$

Name the property used to go from step to step.

59. $5(x + 3) + 4x = (5 \cdot x + 5 \cdot 3) + 4x$
$$= (5x + 15) + 4x$$
60. $\quad = 5x + (15 + 4x)$
61. $\quad = 5x + (4x + 15)$
62. $\quad = (5x + 4x) + 15$
$$= 9x + 15$$

Name the property used to go from step to step.

63. $5 + 3(x + 2) + 4x = 5 + (3 \cdot x + 3 \cdot 2) + 4x$
$$= 5 + (3x + 6) + 4x$$
64. $\quad = 5 + (6 + 3x) + 4x$
65. $\quad = (5 + 6) + 3x + 4x$
$$= 11 + 7x$$
66. $\quad = 7x + 11$

In Ex. 67–74, determine whether each of the following activities can be used to illustrate the property indicated. Assume that for the property to hold the end result must be the same and when using the associative property the items within parentheses must be performed first. Explain your answer.

67. Putting on shoes and socks—commutative property.

68. Putting sugar and cream in coffee—commutative property.

69. Brushing your teeth, washing your face, and combing your hair—associative property.

70. Turning on the lamp and reading a book—commutative property.

71. Cracking an egg, pouring out the egg, and cooking an egg—associative property.

72. Removing the gas cap, putting the nozzle in the tank, and turning on the gas—associative property.

73. Writing on the blackboard and erasing the blackboard—commutative property.

74. A coffee machine dropping the cup, dispensing the coffee, and then adding the sugar—associative property.

75. The man in the photograph is demonstrating that it is possible to remove his sweater without removing his coat. Can the removing of sweater and jacket be used to illustrate the commutative property? Explain your answer.

5.6 | Rules of Exponents and Scientific Notation

An understanding of exponents is important in solving problems in algebra. In the expression 5^2, the 2 is referred to as the **exponent** and the 5 is referred to as **the base**. We read 5^2 as 5 to the second power, or 5 squared. This means

$$5^2 = \underbrace{5 \cdot 5}_{\text{2 factors of 5}}$$

The number 5 to the third power, or 5 cubed, written 5^3, means

$$5^3 = \underbrace{5 \cdot 5 \cdot 5}_{\text{3 factors of 5}}$$

In general, the number b to the nth power, written b^n, means

$$b^n = \underbrace{b \cdot b \cdot b \cdot \cdots \cdot b}_{n \text{ factors of } b}$$

▶ **Example 1**

Evaluate the following.
a) 3^2 **b)** $(-3)^2$ **c)** 4^3 **d)** 1^{100} **e)** 8^1

Solution:
a) $3^2 = 3 \cdot 3 = 9$
b) $(-3)^2 = (-3)(-3) = 9$

c) $4^3 = 4 \cdot 4 \cdot 4 = 64$
d) $1^{100} = 1$. (The number 1 times itself any number of times equals 1.)
e) $8^1 = 8$. (Any number with an exponent of 1 is the number itself.)

Rules of Exponents

Now that we know how to evaluate powers of numbers we can discuss the rules of exponents. Consider

$$2^2 \cdot 2^3 = \underbrace{2 \cdot 2}_{2 \text{ factors}} \cdot \underbrace{2 \cdot 2 \cdot 2}_{3 \text{ factors}} = 2^5$$

This example illustrates the product rule for exponents.

> ### Product Rule
> $$a^m \cdot a^n = a^{m+n}$$

Therefore by using the product rule, $2^2 \cdot 2^3 = 2^{2+3} = 2^5$.

▶ **Example 2**
Use the product rule to simplify.
a) $5^2 \cdot 5^6$ **b)** $8^3 \cdot 8^5$

Solution:
a) $5^2 \cdot 5^6 = 5^{2+6} = 5^8$ **b)** $8^3 \cdot 8^5 = 8^{3+5} = 8^8$

Consider

$$\frac{2^5}{2^2} = \frac{2 \cdot 2 \cdot 2 \cdot \not 2 \cdot \not 2}{\not 2 \cdot \not 2} = 2 \cdot 2 \cdot 2 = 2^3$$

This example illustrates the quotient rule for exponents.

> ### Quotient Rule for Exponents
> $$\frac{a^m}{a^n} = a^{m-n}, \qquad a \neq 0$$

Therefore $\dfrac{2^5}{2^2} = 2^{5-2} = 2^3$.

▶ **Example 3**
Use the quotient rule to simplify.

a) $\dfrac{5^8}{5^5}$ **b)** $\dfrac{8^{12}}{8^5}$

Solution:

a) $\dfrac{5^8}{5^5} = 5^{8-5} = 5^3$ **b)** $\dfrac{8^{12}}{8^5} = 8^{12-5} = 8^7$

Consider $2^3 \div 2^3$. By the quotient rule we see that

$$\frac{2^3}{2^3} = 2^{3-3} = 2^0$$

But $\dfrac{2^3}{2^3} = \dfrac{8}{8} = 1$. Therefore 2^0 must equal 1. This example illustrates the zero exponent rule.

Zero Exponent Rule

$$a^0 = 1, \qquad a \neq 0$$

Note that 0^0 is not defined by the zero exponent rule.

▶ **Example 4**

Use the zero exponent rule to simplify.
a) 5^0 **b)** $(-3)^0$

Solution:
a) $5^0 = 1$ **b)** $(-3)^0 = 1$

Consider $2^3 \div 2^5$. Using the quotient rule, we find that

$$\frac{2^3}{2^5} = 2^{3-5} = 2^{-2}$$

But $\dfrac{2^3}{2^5} = \dfrac{\cancel{2} \cdot \cancel{2} \cdot \cancel{2}}{\cancel{2} \cdot \cancel{2} \cdot \cancel{2} \cdot 2 \cdot 2} = \dfrac{1}{2^2}$. Since $\dfrac{2^3}{2^5}$ equals both 2^{-2} and $\dfrac{1}{2^2}$, then 2^{-2} must equal $\dfrac{1}{2^2}$. This example illustrates the negative exponent rule.

Negative Exponent Rule

$$a^{-m} = \frac{1}{a^m}, \qquad a \neq 0$$

▶ **Example 5**

Use the negative exponent rule to simplify.
a) 5^{-2} **b)** 8^{-1}

Did You
Know...

Large and Small Numbers

1×10^5 light years

Diameter
1×10^{-10} meters

Our everyday activities don't require us to deal with quantities much above those in the thousands: $3.95 for lunch, 100 meters to a lap, a $6000 car loan, and so on. Yet as modern technology has developed, so has our ability to study all aspects of the universe we live in from the very large to the very small. It's not easy to compute values when you have to deal with 10 or 20 digits at a time. Using the rules of exponents and scientific notation allows us to compute with numbers that are out of this world in a very down-to-earth way.

Solution:

a) $5^{-2} = \dfrac{1}{5^2} = \dfrac{1}{25}$ **b)** $8^{-1} = \dfrac{1}{8^1} = \dfrac{1}{8}$

Consider $(2^3)^2$:

$$(2^3)^2 = (2^3)(2^3) = 2^{3+3} = 2^6$$

This example illustrates the power rule for exponents.

Power Rule

$$(a^m)^n = a^{m \cdot n}$$

Thus $(2^3)^2 = 2^{3 \cdot 2} = 2^6$

▶ **Example 6**
Use the power rule to simplify
a) $(3^5)^4$ **b)** $(4^3)^6$

Solution:
a) $(3^5)^4 = 3^{5 \cdot 4} = 3^{20}$ **b)** $(4^3)^6 = 4^{3 \cdot 6} = 4^{18}$

Summary of the Rules of Exponents

$a^m \cdot a^n = a^{m+n}$	Product rule
$\dfrac{a^m}{a^n} = a^{m-n}, \quad a \neq 0$	Quotient rule
$a^0 = 1, \quad a \neq 0$	Zero exponent rule
$a^{-m} = \dfrac{1}{a^m}, \quad a \neq 0$	Negative exponent rule
$(a^m)^n = a^{m \cdot n}$	Power rule

Scientific Notation

Often scientific problems deal with very large and very small numbers. For example, the distance from the earth to the sun is about 93,000,000 miles. The wavelength of a yellow color of light is about 0.0000006 meter. Because it is difficult to work with many zeros, scientists developed a notation that expresses such numbers with exponents. For example consider the distance from the earth to the sun, 93,000,000 miles.

$$93,000,000 = 9.3 \times 10,000,000$$
$$= 9.3 \times 10^7$$

The diameter of a red blood cell may be 0.000004 inch.

$$0.000004 = 4.0 \times 0.000001$$
$$= 4.0 \times 10^{-6}$$

The numbers 9.3×10^7 and 4.0×10^{-6} are written in a form called **scientific notation**. Each number written in scientific notation is written as a number greater than or equal to 1 and less than 10 ($1 \leq a < 10$) multiplied by some power of 10.

Some examples of numbers in scientific notation are the following.

$$3.7 \times 10^3 \qquad 2.05 \times 10^{-3}$$
$$5.6 \times 10^8 \qquad 1.00 \times 10^{-5}$$

The box below shows a simplified procedure for writing a number in scientific notation.

To Write a Number in Scientific Notation

1. Move the decimal point in the original number to the right or left until you obtain a number greater than or equal to 1 and less than 10.
2. Count the number of places you have moved the decimal point to obtain the number in step 1. If the decimal point was moved to the left, the count is to be considered positive. If the decimal point was moved to the right, the count is to be considered negative.
3. Multiply the number obtained in step 1 by 10 raised to the count found in step 2. (Note that the count determined in step 2 is the exponent on the base 10.)

▶ **Example 7**

Write each number in scientific notation.
a) Males have up to 5,800,000 red blood cells per cubic millimeter of blood.
b) The world population was estimated to be approximately 5,333,000,000 in 1990.
c) The probability of winning a lottery may be 0.0000018.
d) The wavelength of an x-ray may be 0.000000000492 meters.

Solution:
a) $5,800,000 = 5.8 \times 10^6$
b) $5,333,000,000 = 5.333 \times 10^9$
c) $0.0000018 = 1.8 \times 10^{-6}$
d) $0.000000000492 = 4.92 \times 10^{-10}$

To convert from a number given in scientific notation to decimal notation we reverse the procedure.

To Change a Number in Scientific Notation to Decimal Notation

1. Observe the exponent of the power of 10.
2. a) If the exponent is positive, move the decimal point in the number to the right the same number of places as the exponent. It might be necessary to add zeros to the number.
   b) If the exponent is negative, move the decimal point in the number to the left the same number of places as the exponent. It might be necessary to add zeros.

▶ **Example 8**

Write each number in decimal notation.
a) The age of the earth is estimated by some scientists to be 4.6×10^9 years.
b) The half-life of plutonium 192 is about 1×10^{15} years.
c) The average grain size in siltstone is 1.35×10^{-3} inches.
d) The quotient of 12 divided by 480,000 is 2.5×10^{-5}.

Solution:
a) $4.6 \times 10^9 = 4,600,000,000$
b) $1 \times 10^{15} = 1,000,000,000,000,000$
c) $1.35 \times 10^{-3} = 0.00135$
d) $2.5 \times 10^{-5} = 0.000025$

In scientific journals and books we occasionally see numbers like 10^{15} and 10^{-6}. These numbers are interpreted as 1×10^{15} and 1×10^{-6} respectively when converting the numbers to decimal form.

A Very Large Number

94,750,475,793,226,037,037,134, 790,392,044,530,094,926,124 is certainly a very large number. But how would we pronounce this value in order to communicate it to someone else? Here goes: "94 tredecillion 750 duodecillion 475 undecillion 793 decillion 226 nonillion 37 octillion 37 septillion 134 sextillion 790 quintillion 392 quadrillion 44 trillion 530 billion 94 million 926 thousand 124." Whew! Let's hope inflation never requires us to carry this many dollars around in our wallets.

▶ **Example 9**

Multiply $(4.3 \times 10^6)(2 \times 10^{-4})$. Write the answer in decimal notation.

Solution: $(4.3 \times 10^6)(2 \times 10^{-4}) = (4.3 \times 2)(10^6 \times 10^{-4})$
$$= 8.6 \times 10^2$$
$$= 860$$

▶ **Example 10**

Divide $\dfrac{0.0000093}{0.003}$. Write the answer in scientific notation.

Solution: First write each number in scientific notation.
$$\frac{0.0000093}{0.003} = \frac{9.3 \times 10^{-6}}{3 \times 10^{-3}} = \left(\frac{9.3}{3}\right)\left(\frac{10^{-6}}{10^{-3}}\right)$$
$$= 3.1 \times 10^{-6-(-3)}$$
$$= 3.1 \times 10^{-6+3}$$
$$= 3.1 \times 10^{-3}$$

▶ **Example 11**

Multiply (42,100,000)(0.008). Write the answer in (a) scientific notation and (b) decimal notation.

Solution:

a) Change each number to scientific notation form:

$$
\begin{aligned}
(42,100,000)(0.008) &= (4.21 \times 10^7)(8 \times 10^{-3}) \\
&= (4.21 \times 8)(10^7 \times 10^{-3}) \\
&= 33.68 \times 10^4 \\
&= 3.368 \times 10^5
\end{aligned}
$$

b) The answer in decimal form is 336,800.

▶ **Example 12**

One of the most powerful computers is the Cray 3 supercomputer. The Cray can perform 16 billion operations per second. How long would it take for the Cray 3 to perform 20 trillion (20,000,000,000,000) operations?

Solution: You may not be sure whether to multiply or divide to determine the solution. One approach to solving problems, as mentioned in Section 1.3, is to substitute simpler numbers in the problem. Suppose the problem stated that the computer performed two operations per second, and you wanted to determine the time needed to perform 10 operations. You would use division to determine that the answer is 5 sec ($10 \div 2 = 5$ sec). Therefore, we must use division to solve this problem. We divide the number of operations desired by the number of operations per second to determine the time required.

$$20 \text{ trillion} = 2.0 \times 10^{13}$$
$$16 \text{ billion operations is } 1.6 \times 10^{10}$$

$$\text{Time to perform the operations} = \frac{2.0 \times 10^{13}}{1.6 \times 10^{10}}$$

$$= 1.25 \times 10^3, \quad \text{or } 1250 \text{ sec (about 21 min)}$$

*The Greek mathematician **Hypatia** (370–415 A.D.) was the first recorded notable woman in mathematics. Daughter of the mathematician and philosopher Theon, she worked in Alexandria (in Egypt), and wrote works on algebra, conic sections, and, it is believed, the construction of scientific instruments. Hypatia actively stood for learning and science at a time in Western history when such learning was associated with paganism.*

She paid the ultimate price for her beliefs—she was brutally murdered by religious zealots. Her death led to the departure of many scholars from Alexandria and marked the beginning of the end of the great age of Greek mathematics.

Scientific Notation on the Calculator

What will your calculator show when you multiply very large or very small numbers? The answer depends on whether your calculator has the ability to display an answer in scientific notation. On calculators without the ability to express numbers in scientific notation you will probably get an error message because the answer will be too large or too small for the display.

For example, on a calculator without scientific notation:

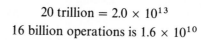

8000000 ☒ 600000 ＝ Error

If your calculator has the ability to give an answer in scientific notation you will probably get the following:

$$8000000 \boxed{\times} 600000 \boxed{=} 4.8 \quad 12$$

This $4.8 \quad 12$ means 4.8×10^{12}.

On a calculator that uses scientific notation:

$$.0000003 \boxed{\times} .004 \boxed{=} 1.2 \quad -9$$

This $1.2 \quad -9$ means 1.2×10^{-9}.

Section 5.6 Exercises

Evaluate each of the following.

1. 4^2

2. 3^3

3. $(-2)^2$

4. $(-2)^3$

5. 5^0

6. $(\frac{1}{2})^2$

7. $(\frac{5}{6})^2$

8. $(-3)^4$

9. $3^2 \cdot 4^3$

10. $\dfrac{6^2}{2^2}$

11. $\dfrac{3^5}{3^2}$

12. $3^2 \cdot 3^3$

13. $\dfrac{2^2}{2^5}$

14. $4^2 \cdot 3^0$

15. $(-5)^0$

16. $(-2)^4$

17. 2^4

18. $2^5 \cdot 4^0$

19. 2^{-2}

20. 2^{-3}

21. $(3^2)^3$

22. $(1^4)^5$

23. $\dfrac{5^7}{5^5}$

24. $4^2 \cdot 4$

25. 6^{-3}

26. 2^{-4}

27. $3^2 \cdot 3^2$

28. $(5^2)^3$

29. $\dfrac{3^5}{3^4}$

30. $2^{-2} \cdot 2$

Express each number in scientific notation.

31. 55,000

32. 4,610,000

33. 900

34. 0.00062

35. 0.053

36. 0.0000561

37. 19,000

38. 5,260,000,000

39. 0.00000186

40. 0.0003

41. 0.00000423

42. 54,000

43. 107

44. 0.02

45. 0.153

46. 416,000

Express each number in decimal notation.

47. 3.1×10^3

48. 1.63×10^{-4}

49. 6×10^7

50. 6.15×10^5

51. 2.13×10^{-5}

52. 2.74×10^{-7}

53. 3.12×10^{-1}

54. 4.6×10^1

55. 9×10^6

56. 7.3×10^4

57. 2.31×10^2

58. 1.04×10^{-2}

59. 3.5×10^4

60. 2.17×10^{-6}

61. 1×10^4

62. 1×10^{-3}

Perform the indicated operation and express each number in decimal notation.

63. $(4 \times 10^2)(3 \times 10^5)$

64. $(2 \times 10^{-3})(3 \times 10^2)$

65. $(5.1 \times 10^1)(3 \times 10^{-4})$

66. $(1.6 \times 10^{-2})(4 \times 10^{-3})$

67. $\dfrac{6.4 \times 10^5}{2 \times 10^3}$

68. $\dfrac{8 \times 10^{-3}}{2 \times 10^1}$

69. $\dfrac{8.4 \times 10^{-6}}{4 \times 10^{-3}}$

70. $\dfrac{25 \times 10^3}{5 \times 10^{-2}}$

71. $\dfrac{4 \times 10^5}{2 \times 10^4}$

72. $\dfrac{16 \times 10^3}{8 \times 10^{-3}}$

Perform the indicated operation by first converting each number to scientific notation. Write the answer in scientific notation.

73. $(700,000)(6,000,000)$

74. $(0.0006)(5,000,000)$

75. $(0.003)(0.00015)$

76. $(230,000)(3000)$

77. $\dfrac{1,400,000}{700}$

78. $\dfrac{20,000}{0.0005}$

79. $\dfrac{0.00004}{200}$

80. $\dfrac{0.0012}{0.000006}$

81. $\dfrac{150,000}{0.0005}$

82. List the numbers from smallest to largest.
4.8×10^5, 3.2×10^{-1}, 4.6, 8.3×10^{-4}

83. List the numbers from smallest to largest.
9.2×10^{-5}, 8.4×10^3, 1.3×10^{-1}, 6.2×10^4

84. The distance from the earth to the planet Jupiter is approximately 4.5×10^8 mi. If a spacecraft traveled at a speed of 25,000 mph, how long, in hours, would it take the spacecraft to travel from the earth to Jupiter. Use distance = rate × time.

85. The distance from the earth to the moon is approximately 239,000 mi. If a spacecraft travels at a speed of 20,000 mph, how long would it take the spacecraft to travel from the earth to the moon? Use distance = rate × time.

86. If a computer can perform 4 million operations per second, how long would it take to perform 10 trillion (10,000,000,000,000) operations?

87. If a computer can do a calculation in 0.000004 sec, how long, in seconds, would it take the computer to do 8 trillion (8,000,000,000,000) calculations?

88. If a cubic millimeter of blood contains 5,800,000 red blood cells, how many red blood cells are contained in 50 cubic millimeters of blood?

89. The half-life of a radioactive isotope is the time required for half the quantity of the isotope to decompose. The half-life of uranium-238 is 4.5×10^9 years, and the half-life of uranium-234 is 2.5×10^5 years. How many times greater is the half-life of uranium-238 than uranium-234?

90. In the metric system 1 meter = 10^3 millimeters. How many times greater is a meter than a millimeter? Explain how you determined your answer.

91. In the metric system 1 gram = 10^3 milligrams and 1 gram = 10^{-3} kilograms. What is the relationship between milligrams and kilograms? Explain how you determined your answer.

92. A treaty between the United States and Canada requires that during the tourist season a minimum of 100,000 cubic feet of water per second flow over Niagara Falls (another 130,000–160,000 cubic feet/sec are diverted for power generation). Find the minimum amount of water that will flow over the falls in a 24-hour period during the tourist season.

Problem Solving

93. a) Light travels at a speed of 1.86×10^5 miles/second. A *light year* is the distance that light travels in one year. Determine the number of miles in a light year.

b) The earth is approximately 93,000,000 mi from the sun. How long does it take light from the sun to reach the earth?

94. The exponential function, $E(t) = 2^{10} \cdot 2^t$ approximates the number of bacteria in a certain culture after t hr.

a) The initial number of bacteria is determined when $t = 0$. What is the initial number of bacteria?

b) How many bacteria are there after $\frac{1}{2}$ hr?

5.7 Arithmetic and Geometric Sequences

Now that you can recognize the various sets of real numbers and know how to add, subtract, multiply, and divide real numbers, we can discuss sequences. A **sequence** is a list of numbers which are related to each other by a given rule. The numbers that form the sequence are called the *terms* of the sequence. If your salary increases or decreases by a fixed amount over a period of time, the listing of the amounts, over time, would form an arithmetic sequence. When interest in a savings account is compounded at regular intervals, the listing of the amounts in the account over time will be a geometric sequence.

Arithmetic Sequences

A sequence in which each term after the first term differs from the preceding term by a constant amount is called an **arithmetic sequence**. The amount by which each pair of successive terms differs is called the **common difference**, d. The common difference can be found by subtracting any term from the term that directly follows it.

Examples of Arithmetic Sequences

1, 5, 9, 13, 17, . . .

$-7, -5, -3, -1, 1, \ldots$

$\dfrac{5}{2}, \dfrac{3}{2}, \dfrac{1}{2}, -\dfrac{1}{2}$

Common Differences

$d = 5 - 1 = 4$

$d = -5 - (-7) = -5 + 7 = 2$

$d = \dfrac{3}{2} - \dfrac{5}{2} = -\dfrac{2}{2} = -1$

Carl Friedrich Gauss *(1777–1855),
often called the "Prince of
Mathematicians," made significant
contributions to the fields of
algebra, geometry, and number
theory, including a proof of the
fundamental theorem of arithmetic. When Gauss was only 10, his
mathematics teacher gave him
the problem of finding the sum
of the first 100 natural numbers,
thinking that this would keep
him busy for a while. Gauss
recognized a pattern in the
sequence of numbers when he
considered the sum of the
following numbers.*

$$\begin{array}{r} 1+\quad 2+\quad 3+\cdots+\quad 99+100 \\ 100+\ 99+\ 98+\cdots+\quad 2+\quad 1 \\ \hline 101+101+101+\cdots+101+101 \end{array}$$

*He had the required answer in no
time at all: (100)(101)/2 = 5050.*

▶ **Example 1**

Write the first five terms of the arithmetic sequence with first term 4 and common difference 3.

Solution: The first term is 4. The second term is $4 + 3$ or 7. The third term is $7 + 3$ or 10, and so on. Thus the five terms are 4, 7, 10, 13, 16.

▶ **Example 2**

Write the first four terms of the arithmetic sequence with first term 3 and common difference -2.

Solution: $3, 1, -1, -3$

When discussing a sequence, we often represent the first term as a_1 (read "a sub 1"), the second term as a_2, the fifteenth term as a_{15}, and so on. We use the notation a_n to represent the general or nth term of a sequence. Thus a sequence may be symbolized as

$$a_1, a_2, a_3, a_4, \ldots, a_n, \ldots$$

When we know the first term of an arithmetic sequence and the common difference, we can find the value of any specific term using the following formula.

General or *n*th Term of an Arithmetic Sequence

$$a_n = a_1 + (n - 1)d$$

▶ **Example 3**

Find the seventh term of the arithmetic sequence whose first term is 3 and whose common difference is -6.

Solution: To find the seventh term, or a_7, replace n in the formula with 7, a_1 with 3, and d with -6.

$$\begin{aligned} a_n &= a_1 + (n - 1)d \\ a_7 &= 3 + (7 - 1)(-6) \\ &= 3 + (6)(-6) \\ &= 3 - 36 \\ &= -33 \end{aligned}$$

The seventh term is -33. As a check we have shown the first seven terms of the sequence:

$$3, -3, -9, -15, -21, -27, -33$$

▶ **Example 4**

Write an expression for the general or nth term, a_n, for the sequence 6, 9, 12, 15, ...

Solution: In this sequence, $a_1 = 6$ and $d = 3$. We substitute these values in $a_n = a_1 + (n-1)d$ to obtain an expression for the nth term, a_n:

$$a_n = a_1 + (n-1)d$$
$$a_n = 6 + (n-1)3$$
$$a_n = 6 + 3n - 3$$
$$a_n = 3 + 3n$$

Note that when $n = 1$, the first term is $3 + 3(1) = 6$. When $n = 2$, the second term is $3 + 3(2) = 9$, and so on.

We can find the sum of the first n terms in an arithmetic sequence by using the following formula.

> **Sum of the First n Terms in an Arithmetic Sequence**
>
> $$S_n = \frac{n(a_1 + a_n)}{2}$$

In this formula, s_n represents the sum of the first n terms a_1 is the first term, a_n is the nth term, and n is the number of terms in the sequence from a_1 to a_n.

▶ **Example 5**

Find the sum of the first 25 natural numbers.

Solution: The sequence we are discussing is

$$1, 2, 3, 4, 5, \ldots, 25.$$

In this sequence, $a_1 = 1$, $a_{25} = 25$, and $n = 25$. Thus the sum of the first 25 terms is

$$S_n = \frac{n(a_1 + a_n)}{2}$$
$$S_{25} = \frac{25(1 + 25)}{2}$$
$$S_{25} = \frac{25(26)}{2} = 325$$

Thus the sum of the terms $1 + 2 + 3 + 4 + \cdots + 25$ is 325.

Geometric Sequences

The next type of sequence we will discuss is the geometric sequence. A **geometric sequence** is one in which the ratio of any term to the term that directly precedes it is a constant. This constant is called the **common ratio**. The common ratio, r, can be found by taking any term except the first and dividing that term by the preceding term.

Examples of Geometric Sequences	*Common Ratios*
$2, 4, 8, 16, 32$	$r = 4 \div 2 = 2$
$-3, 6, -12, 24, -48$	$r = 6 \div (-3) = -2$
$\dfrac{2}{3}, \dfrac{2}{9}, \dfrac{2}{27}, \dfrac{2}{81}$	$r = \dfrac{2}{9} \div \dfrac{2}{3} = \left(\dfrac{2}{9}\right)\left(\dfrac{3}{2}\right) = \dfrac{1}{3}$

To construct a geometric sequence when the first term, a_1, and common ratio are known, multiply the first term by the common ratio to get the second term. Then multiply the second term by the common ratio to get the third term, and so on.

▶ **Example 6**

Write the first five terms of the geometric sequence with first term 5 and common ratio $\frac{1}{2}$.

Solution: The first term is 5. The second term, found by multiplying the preceding term by $\frac{1}{2}$, is $5(\frac{1}{2})$ or $\frac{5}{2}$. The third term is $(\frac{5}{2})(\frac{1}{2})$ or $\frac{5}{4}$, and so on. The sequence is

$$5, \frac{5}{2}, \frac{5}{4}, \frac{5}{8}, \frac{5}{16}, \cdots$$

When we know the first term of a geometric sequence and the common ratio we can find the value of the general or nth term, a_n, by using the following formula.

General or nth Term of a Geometric Sequence

$$a_n = a_1 r^{n-1}$$

▶ **Example 7**

Find the seventh term of the geometric sequence whose first term is -3 and whose common ratio is -2.

Solution: In this sequence, $a_1 = -3$, $r = -2$, and $n = 7$. Substituting the values, we obtain

$$a_n = a_1 r^{n-1}$$
$$a_7 = -3(-2)^{7-1}$$
$$a_7 = -3(-2)^6$$
$$a_7 = -3(64)$$
$$a_7 = -192$$

As a check we have listed the first seven terms of the sequence: $-3, 6, -12, 24, -48, 96, -192$.

▶ Example 8

Write an expression for the general or nth term, a_n, of the sequence 5, 15, 45, 135, ...

Solution: In this sequence, $a_1 = 5$ and $r = 3$. We substitute these values in $a_n = a_1 r^{n-1}$ to obtain an expression for the nth term, a_n:

$$a_n = a_1 r^{n-1}$$
$$a_n = 5(3)^{n-1}$$

Note that when $n = 1$, $a_1 = 5(3)^0 = 5(1) = 5$. When $n = 2$, $a_2 = 5(3)^1 = 15$, and so on.

▶ Example 9

Suppose we form stacks of white chips such that there is one white chip in the first stack and in each successive stack we double the number of chips. Thus we have stacks of 1 chip, 2 chips, 4 chips, 8 chips and so on. We also form stacks of red chips, starting with one red chip then tripling the number of chips in each successive stack. Thus the stacks will contain 1 chip, 3 chips, 9 chips, 27 chips, and so on. How many more chips would there be in the sixth stack of red chips than in the sixth stack of white chips?

Solution: The number of chips in each stack is multiplied by a constant to get the number of chips in the next stack. The number of chips in any stack may be found using the formula $a_n = a_1 r^{n-1}$.

$$a_n = a_1 r^{n-1}$$

$$\text{white:} \quad a_6 = 1(2)^{6-1} = 1 \cdot 32 = 32$$
$$\text{red:} \quad a_6 = 1(3)^{6-1} = 1 \cdot 243 = 243$$

Since the sixth stack of the red chips has 243 chips and the sixth stack of the white chips has 32 chips, there are $243 - 32 = 211$ more chips in the sixth stack of the red chips than in the 6th stack of the white chips.

We can find the sum of the first n terms of a geometric sequence by using the following formula.

Sum of the First n Terms of a Geometric Sequence

$$s_n = \frac{a_1(1 - r^n)}{1 - r}, \qquad r \neq 1$$

▶ **Example 10**

Find the sum of the first five terms in the geometric sequence whose first term is 3 and whose common ratio is 4.

Solution: In this sequence, $a_1 = 3, r = 4,$ and $n = 5$. Substituting the values, we obtain

$$s_n = \frac{a_1(1 - r^n)}{1 - r}$$

$$s_5 = \frac{3[1 - (4)^5]}{1 - 4}$$

$$s_5 = \frac{3(1 - 1024)}{-3}$$

$$s_5 = \frac{1(-1023)}{-1} = \frac{-1023}{-1} = 1023$$

The sum of the first five terms of the sequence is 1023. The first five terms of the sequence are 3, 12, 48, 192, 768. If you add these five numbers, you will obtain the sum of 1023.

▶ **Example 11**

Determine whether the following sequences are arithmetic or geometric, and find the next two terms.
a) $2, 5, 8, 11, \ldots$ **b)** $2, 6, 18, 54, \ldots$
c) $2, -4, 8, -16, \ldots$ **d)** $7, 3, -1, -5, \ldots$

Solution:
a) Each term is 3 more than the preceding term. Therefore this sequence is arithmetic with $d = 3$. The next two terms are 14 and 17.
b) Each term is 3 times the preceding term. Therefore this sequence is geometric with $r = 3$. The next two terms are 162 and 486.

c) Each term is -2 times the preceding term, so the sequence is geometric with $r = -2$. The next two terms are 32 and -64.

d) Each term is 4 less than the preceding term, so the sequence is arithmetic with $d = -4$. The next two terms are -9 and -13.

 ## Section 5.7 Exercises

Write the first five terms of the arithmetic sequence for the given first term, a_1, and common difference, d.

1. $a_1 = 4$, $d = 3$ **2.** $a_1 = 5$, $d = 2$

3. $a_1 = -4$, $d = 5$ **4.** $a_1 = -3$, $d = 1$

5. $a_1 = 6$, $d = -2$ **6.** $a_1 = 8$, $d = -5$

7. $a_1 = 24$, $d = -8$ **8.** $a_1 = \frac{1}{2}$, $d = 2$

9. $a_1 = \frac{1}{2}$, $d = \frac{1}{2}$ **10.** $a_1 = \frac{5}{2}$, $d = -\frac{3}{2}$

Find the indicated term for the arithmetic sequence for the given first term, a_1, and common difference, d.

11. Find a_5, $a_1 = 12$, $d = 6$

12. Find a_9, $a_1 = -8$, $d = -5$

13. Find a_8, $a_1 = -4$, $d = 8$

14. Find a_{12}, $a_1 = 9$, $d = -7$

15. Find a_{10}, $a_1 = -30$, $d = -12$

16. Find a_{20}, $a_1 = \frac{3}{5}$, $d = -2$

17. Find a_{15}, $a_1 = -\frac{1}{2}$, $d = 5$

18. Find a_{11}, $a_1 = -20$, $d = -\frac{1}{2}$

Write an expression for the general or nth term, a_n, for the given arithmetic sequences.

19. $4, 8, 12, 16, \ldots$

20. $5, 2, -1, -4, \ldots$

21. $8, 18, 28, 38, \ldots$

22. $-3, -6, -9, -12, \ldots$

23. $-\frac{5}{2}, -\frac{3}{2}, -\frac{1}{2}, \frac{1}{2}, \ldots$

24. $-18, -9, 0, 9, \ldots$

25. $-5, -\frac{5}{2}, 0, \frac{5}{2}, \ldots$

26. $0, 4, 8, 12, \ldots$

27. $-3, -\frac{7}{2}, -4, -\frac{9}{2}, \ldots$

28. $-100, -95, -90, -85, \ldots$

Find the sum of the terms of the arithmetic sequence.

29. $1, 2, 3, 4, \ldots, 12$

30. $5, 10, 15, \ldots, 45$

31. $20, 18, 16, 14, \ldots, 0$

32. $-1, -4, -7, -10, \ldots, -25$

33. $8, 5, 2, -1, \ldots, -16$

34. $\frac{1}{2}, \frac{5}{2}, \frac{9}{2}, \frac{13}{2}, \ldots, \frac{29}{2}$

35. $-9, -\frac{17}{2}, -8, -\frac{15}{2}, \ldots, -\frac{1}{2}$

36. $100, 110, 120, 130, \ldots, 190$

37. $10, 5, 0, -5, \ldots, -20$

38. $\frac{3}{5}, \frac{4}{5}, \frac{5}{5}, \frac{6}{5}, \ldots, 4$

Write the first five terms of the geometric sequence with the given first term, a_1, and common ratio, r.

39. $a_1 = 3$, $r = 2$

40. $a_1 = 2$, $r = 3$

41. $a_1 = 5$, $r = -2$

42. $a_1 = 3$, $r = \frac{1}{2}$

43. $a_1 = -2$, $r = -1$

44. $a_1 = 6$, $r = -3$

45. $a_1 = -2$, $r = \frac{1}{2}$

46. $a_1 = -10$, $r = -\frac{1}{2}$

47. $a_1 = 5$, $r = \frac{3}{5}$

48. $a_1 = -4$, $r = -\frac{2}{3}$

Find the indicated term for the geometric sequence for the given first term, a_1, and common ratio, r.

49. Find a_6, $a_1 = 2$, $r = 3$

50. Find a_4, $a_1 = 5$, $r = 2$

51. Find a_5, $a_1 = 3$, $r = 2$

52. Find a_6, $a_1 = -2$, $r = -2$

53. Find a_4, $a_1 = -8$, $r = -3$

54. Find a_7, $a_1 = 10$, $r = -3$

55. Find a_3, $a_1 = 3$, $r = \frac{1}{2}$

56. Find a_7, $a_1 = -3$, $r = -3$

57. Find a_5, $a_1 = \frac{1}{2}$, $r = 2$

58. Find a_4, $a_1 = 4$, $r = \frac{1}{3}$

Write an expression for the general or nth term, a_n, for the indicated geometric sequence.

59. $2, 4, 8, 16, \ldots$

60. $3, 12, 48, 192, \ldots$

61. $1, 3, 9, 27, \ldots$

62. $-6, 6, -6, 6, \ldots$

63. $-8, -4, -2, -1, \ldots$

64. $\frac{1}{3}, 1, 3, 9, \ldots$

65. $-6, 18, -54, 162, \ldots$

66. $10, \frac{10}{3}, \frac{10}{9}, \frac{10}{27}, \ldots$

67. $-4, -\frac{8}{3}, -\frac{16}{9}, -\frac{32}{27}, \ldots$

Find the sum of the first n terms of the geometric sequence for the given values of a_1 and r.

68. $n = 3$, $a_1 = 2$, $r = 5$
69. $n = 4$, $a_1 = 2$, $r = 2$
70. $n = 4$, $a_1 = 5$, $r = 3$
71. $n = 5$, $a_1 = 3$, $r = 5$
72. $n = 6$, $a_1 = 3$, $r = 4$
73. $n = 5$, $a_1 = 4$, $r = 3$
74. $n = 4$, $a_1 = -6$, $r = 2$
75. $n = 5$, $a_1 = -3$, $r = -2$

76. Find the sum of the first 50 natural numbers.
77. Find the sum of the first 50 even natural numbers.
78. Find the sum of the first 50 odd natural numbers.
79. Find the sum of the first 20 multiples of 3.
80. Donna is given a starting salary of $20,200 and promised a $1200 raise per year after each of the next eight years. Find her salary during her eight years of work.
81. Each swing of a pendulum (from far left to far right) is 3 in. shorter than its preceding swing. The first swing is 8 ft.
  a) Find the length of the twelfth swing.
  b) Determine the total distance traveled by the pendulum during the first 12 swings.
82. Each time a ball bounces, the height attained by the ball is 6 in. less than previous height attained. If on the first bounce the ball reaches a height of 6 ft, find the height attained on the eleventh bounce.
83. If you are given $1 on January 1, $2 on January 2, $3

on the 3rd, and so on, how much money will you have accumulated during the month of January (31 days)?
84. A certain substance decomposes and loses 20% of its weight each hour. If there are originally 200 grams of the substance, how much remains after six hr?
85. If your salary increases at a rate of 6% per year, find your salary during your fifteenth year if your present salary is $20,000.
86. The population in the United States in 1990 was 248.7 million. If the population grows at a rate of 6% per year, find the population in 12 years.
87. When dropped, a ball rebounds to four-fifths of its original height. How high will the ball rebound after the fourth bounce if it is dropped from a height of 30 ft.
88. A clock strikes once at 1 o'clock, twice at 2 o'clock, and so on. How many times does it strike over a 12-hour period?

Problem Solving

89. A geometric sequence has $a_1 = 82$ and $r = \frac{1}{2}$; find s_6.
90. Determine how many numbers between 7 and 1610 are divisible by 6.
91. Find r and a_1 for the geometric sequence with $a_2 = 24$ and $a_5 = 648$.
92. A ball is dropped from a height of 30 ft. On each bounce it attains a height four-fifths of its original height (or of the previous bounce). Find the total vertical distance traveled by the ball after it has completed its fifth bounce (therefore has hit the ground six times).

5.8   Fibonacci Sequence

Leonardo of Pisa (1170–1250) is credited with introducing the Hindu-Arabic number system into Europe. When he began writing, he referred to himself as **Fibonacci**, or "son of Bonacci," the name by which he is known today.

We cannot leave the topic of sequences without discussing a very interesting and exciting sequence known as the **Fibonacci sequence**.

Leonardo of Pisa, known as Fibonacci, is often considered the most distinguished mathematician of the Middle Ages. He was born in Italy, but was sent by his father to study calculations with an Arab master. His book, *Liber Abacci* (Book of the Abacus), contains interesting problems that Fibonacci liked to invent, such as the following: "A certain man put a pair of rabbits in a place surrounded on all sides by a wall. How many pairs of rabbits can be produced from that pair in a year if it is assumed that every month each pair begets a new pair which from the second month becomes productive?"

The solution to this problem (Fig. 5.9 on page 228) led to the development of the sequence that bears its author's name: the Fibonacci sequence. The sequence is shown in Table 5.1. The numbers in the column titled Pairs of Adults form the Fibonacci sequence.

Table 5.1

Month	Pairs of Adults	Pairs of Babies	Total Pairs
1	1	0	1
2	1	1	2
3	2	1	3
4	3	2	5
5	5	3	8
6	8	5	13
7	13	8	21
8	21	13	34
9	34	21	55
10	55	34	89
11	89	55	144
12	144	89	233

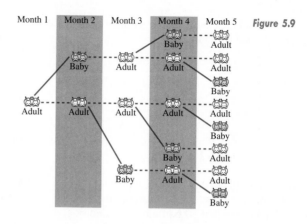

Figure 5.9

Fibonacci Sequence

$$1, 1, 2, 3, 5, 8, 13, 21, \ldots$$

In the Fibonacci sequence the first and second terms are 1. The sum of these two terms is the third term. The sum of the second and third terms is the fourth term, and so on.

In the middle of the nineteenth century, mathematicians made a serious study of this sequence and found strong similarities between it and many natural phenomena. Fibonacci numbers appear in the seed arrangement of many species of plants and in the petal counts of various flowers. For example, when the flowering head of the sunflower matures to seed, the seeds' spiral arrangement becomes clearly visible. A typical count of these spirals may give 89 steeply curving to the right, 55 curving more shallowly to the left, and 34 again shallowly to the right. The largest known specimen to be examined had spiral counts of 144 right, 89 left, and 55 right. These numbers, like the other three mentioned, are consecutive terms of the Fibonacci sequence.

On the heads of many flowers, petals (or florets in composite plants) surrounding the central disk generally yield a Fibonacci number. For example, some daisies contain 21 petals, and others contain 34, 55, or 89 petals. (People who use a daisy to play the "love me, love me not" game will likely pluck 21, 34, 55, or 89 petals before arriving at an answer.)

Fibonacci numbers are also observed in the structure of pinecones and pineapples. The tablike or scalelike structures called bracts that make up the main body of the pinecone form a set of spirals that start from the cone's attachment to the branch. Two sets of oppositely directed spirals can be observed, one steep and the other more gradual. A count on the steep spiral will reveal a Fibonacci number and a count on the gradual one will be the adjacent smaller Fibonacci number, or if not, the next smaller Fibonacci number. In one investigation of 4290 pinecones from 10 species of pine trees found in California, it was determined that only 74 cones, or 1.7%, deviated from this Fibonacci pattern.

Many objects in nature exhibit patterns of Fibonacci numbers.

Like pinecone bracts, pineapple scales are patterned into spirals, and because they are roughly hexagonal in shape, three distinct sets of spirals can be counted.

Fibonacci Numbers and Divine Proportions

Numbers	Ratio
1, 1	$\frac{1}{1} = 1$
1, 2	$\frac{2}{1} = 2$
2, 3	$\frac{3}{2} = 1.5$
3, 5	$\frac{5}{3} = 1.66\ldots$
5, 8	$\frac{8}{5} = 1.6$
8, 13	$\frac{13}{8} = 1.625$

In 1753, while studying the Fibonacci sequence, Robert Simson, a mathematician at the University of Glasgow, noticed that when he took the ratio of any term to the term that immediately preceded it, the value he obtained remained in the vicinity of one specific number. To illustrate this, we indicate in the table to the left the ratio of various pairs of sequential Fibonacci numbers.

The ratio of the 50th term to the 49th term is 1.6180. Simson proved that the ratio of the $(n + 1)$ term to the nth term as n gets larger and larger is the irrational number $(\sqrt{5} + 1)/2$, which begins 1.61803 . . . . This number was already well known to mathematicians at that time as the **golden number**.

Many years earlier the Bavarian astronomer and mathematician Johannes Kepler wrote that for him the golden number symbolized the Creator's intention "to create like from like." The golden number $(\sqrt{5} + 1)/2$ is frequently referred to as "phi," symbolized by the Greek letter Φ.

The ancient Greeks, in about the sixth century B.C., sought unifying principles of beauty and perfection, which they believed could be described using mathematics.

A   C   B

Figure 5.10

In their study of beauty the Greeks used the terminology "*golden ratio.*" In order to understand their golden ideas, consider the line segment AB in Fig. 5.10. When this line segment is divided at a point C, such that the ratio of the whole, AB, to the larger part, AC, is equal to the ratio of the larger part, AC, to the smaller part, CB, then each ratio AB/AC and AC/CB is referred to as a **golden ratio**. And the proportion they form $AB/AC = AC/CB$ is called the **golden proportion**. Furthermore each ratio in the proportion will have a value equal to the golden number, $(\sqrt{5} + 1)/2$.

$$\frac{AB}{AC} = \frac{AC}{CB} = \frac{\sqrt{5} + 1}{2} \approx 1.618$$

The Great Pyramid of Gizeh in Egypt, built about 2600 B.C., is the earliest known example of the golden ratio in architecture. The ratio of the base of one of its square sides (775.75 ft) to its altitude (481.4 ft) is about 1.611. Other evidence of the use of the golden ratio appears in other Egyptian buildings and tombs.

In medieval times people referred to the golden proportion as the **divine proportion**, reflecting their belief in its relationship to the will of God.

The twentieth-century architect Le Corbusier developed a scale of proportions for the human body that he called the Modulor (Fig. 5.11). Notice that the navel separates the entire body into golden proportions, as does the neck and knee.

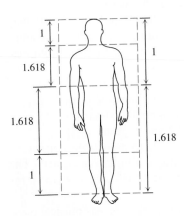

Figure 5.11

Figure 5.12

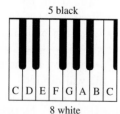

5 black

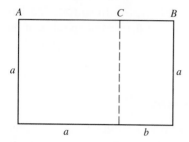

C D E F G A B C

8 white

Figure 5.14   13 total

From the golden proportion the **golden rectangle** can be formed, as shown in Fig. 5.12.

$$\frac{\text{Length}}{\text{Width}} = \frac{a+b}{a} = \frac{a}{b} = \frac{\sqrt{5}+1}{2}$$

Note that when a square is cut off at one end of a golden rectangle, as in Fig. 5.12, the remaining rectangle has the same properties as the original golden rectangle (creating "like from like" as Johannes Kepler had written) and is therefore itself a golden rectangle. It is interesting to note that the curve derived from a succession of diminishing golden rectangles, as shown in Fig. 5.13, is the same as the spiral curve of the chambered nautilus. The same curve appears on the horns of rams and some other animals. It is this curve that is observed in the plant structures arranged in Fibonacci sequence mentioned earlier—in sunflowers and other flower heads, in pinecones and pineapples. The curve closely approximates what mathematicians term a logarithmic spiral.

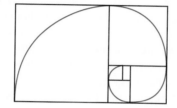

Figure 5.13

Ancient Greek civilization used the golden rectangle in art and architecture. The main measurements of many buildings of antiquity, including the Parthenon in Athens, are governed by golden ratios and rectangles. Greek statues, vases, urns, and so on also exhibit characteristics of the golden ratio. It is for Phidas, considered the greatest of Greek sculptors, that the golden ratio was named "phi." The proportions can be found abundantly in his work.

The proportions of the golden rectangle can be found in the works of many artists, from the old masters to the moderns. For example the golden rectangle can be seen in the painting "La Parade" by George Seurat, a French neoimpressionist artist, in the chapter opening display.

Fibonacci numbers are also found in another form of art, namely music. Perhaps the most obvious link between Fibonacci numbers and music can be found on the piano keyboard. An octave (Fig. 5.14) on a keyboard has 13 keys: 8 white, and 5 black (the 5 are in one group of 2 and a group of 3).

In Western music, the most complete scale, the chromatic scale, consists of 13 notes. Its predecessor, the diatonic scale, contains 8 notes (an octave). The diatonic scale was preceded by a 5-note "pentatonic scale" (*penta* is Greek for "five"). Each number is a Fibonacci number.

The visual arts deal with what is pleasing to the eye, whereas musical composition deals with what is pleasing to the ear. While art achieves some of its goals by using division of planes and area, music achieves some of its goals by a similar division of time, using notes of various duration and spacing. The musical intervals considered by many to be the most pleasing to the ear are the major sixth and minor sixth. A major sixth, for example, consists of the note C, vibrating at about 264* vibrations per second, and note A, vibrating at about 440 vibrations per second. The ratio of 440 to 264 reduces to 5 to 3, or $\frac{5}{3}$, a ratio of two consecutive Fibonacci numbers. An example of a minor sixth would be E (about 330 vibrations per second) and C (about 528 vibrations per second). The ratio 528 to 330 reduces to 8 to 5, or $\frac{8}{5}$, the next ratio of two consecutive Fibonacci numbers. The vibrations of any sixth interval reduce to a similar ratio.

Patterns that can be expressed mathematically in terms of Fibonacci relationships have been found in Gregorian chants and works of many composers, including Bach, Beethoven, and Bartok. A number of twentieth-century works, including Ernst Krened's *Fibonacci Mobile*, have been deliberately structured by using Fibonacci proportions.

A number of studies have tried to explain why the Fibonacci series and related items are linked to so many real-life situations. It appears that the Fibonacci numbers are a part of natural harmony that is pleasing to both the

* Frequencies of notes vary in different parts of the world, and change over time.

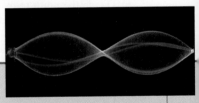

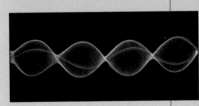

Did You Know...

MATHEMATICAL MUSIC

The roadway in a suspension bridge is not unlike a string stretched between two points and so vibration must be kept to a minimum. This was the lesson learned in 1940, in the state of Washington when high winds set the Tacoma Narrows bridge into a pattern of vibration that ripped the roadway from its supports.

The ancient Greeks are considered by many to have been the first true mathematicians. They studied mathematics not because of its applications, but because of its beauty. They believed that in nature, all harmony and everything of beauty could be explained using ratios of whole numbers (called

When the string of a musical instrument is plucked or bowed, it moves in a wavelike pattern like the strings shown here. The vibration this creates in the surrounding air is what your eardrums detect as sound.

rational numbers). This was reinforced by the discovery that the sound of plucked strings could be quite pleasing if the strings plucked were in the ratio of 1 to 2 (an octave), 2 to 3 (a fifth), 3 to 4 (a fourth), and so on. In other words, the secret of harmony lies in the ratio of whole numbers such as 1/2, 2/3, and 3/4.

The theory of vibrating strings has applications today that go well beyond music. How materials vibrate, and hence the stress they can absorb, is a matter important in the construction of rockets, buildings, and bridges.

eye and the ear. In the nineteenth century the German physicist and psychologist Gustav Fechner tried to determine which dimensions were most pleasing to the eye. Fechner, along with the psychologist Wilhelm Wundt, found that most people do unconsciously favor golden dimensions when purchasing greeting cards, mirrors, and other rectangular objects. This discovery has been heavily used by commercial manufacturers in their packaging and labeling designs, by retailers in their store displays, and in other areas of business and advertising.

Section 5.8 Exercises

1. Explain how to construct the Fibonacci sequence.
2. **a)** Find the eighth, ninth, and tenth terms of the Fibonacci sequence.
   b) Divide the ninth term by the eighth term, rounding to the nearest thousandth.
   c) Divide the tenth term by the ninth term, rounding to the nearest thousandth.
   d) Try a few more divisions and then conjecture what the result will be.
3. The ratio of the a_{n+1}/a_n terms of the Fibonacci sequence approaches 1.6180 (rounded to four decimal places) as n increases. Find the ratio of a_n/a_{n+1} rounded to four decimal places as n increases.
4. What is the golden number? Explain the relationship between the golden number and the golden proportion.
5. Select a piece of art and see if you can determine whether the artist used the golden rectangle. Write a brief description of your findings.
6. Explain what is meant by the divine proportion.
7. **a)** To what decimal value is $(\sqrt{5} + 1)/2$ approximately equal?
   b) To what decimal value is $(\sqrt{5} - 1)/2$ approximately equal?
   c) By how much do the results in part (a) and (b) differ?
8. The eleventh Fibonacci number is 89. Examine the first six terms in the decimal expression of its reciprocal, $\frac{1}{89}$. What do you find?
9. Find the ratio of the second to the first term of the Fibonacci sequence. Then find the ratio of the third to the second term of the sequence and determine whether this ratio was an increase or decrease from the first ratio. Continue this process for 10 ratios and then make a conjecture regarding the increasing or decreasing values in consecutive ratios.
10. A musical composition is indicated in the next column. Explain why this piece is based upon the golden ratio.

Entire Composition

34 measures	55 measures	21 measures	34 measures
Theme	Fast, Loud	Slow	Repeat of theme

11. The greatest common factor of any two consecutive Fibonacci numbers is 1. Show this is true for the first 15 Fibonacci numbers.
12. The sum of any 10 consecutive Fibonacci numbers is always divisible by 11. Select any 10 consecutive Fibonacci numbers and show that for your selection this is true.
13. Twice any Fibonacci number minus the next Fibonacci number equals the second number preceding the original one. Select a number in the Fibonacci sequence and show this is true for the number selected.
14. For any four consecutive Fibonacci numbers, the difference of the squares of the middle two numbers equals the product of the smallest and largest numbers. Select four consecutive Fibonacci numbers and show this is true for the numbers you selected.
15. **a)** What are the length and width of standard size index cards?
    b) Determine the ratio of the length to the width of the index cards and compare the ratio to Φ.
16. Determine the ratio of the length to the width of this textbook, and compare this ratio to Φ.
17. Determine the ratio of the length to the width of your television screen, and compare this ratio to Φ.

Determine whether each of the following is a Fibonacci type sequence (each term after the second term is the sum of the two preceding terms). If it is, determine the next two terms of the sequence.

18. 1, 3, 4, 7, 11, 18, . . .   **19.** 2, 3, 5, 8, 21, . . .
20. 2, 2, 2, 2, . . .   **21.** 15, 30, 45, 60, 75, . . .
22. 15, 30, 45, 75, 120, . . .   **23.** $\frac{1}{2}, \frac{1}{2}, 1, \frac{3}{2}, \frac{5}{2}, 4, \frac{13}{2}$, . . .

24. a) Select any two nonzero digits and add them to obtain a third digit. Continue adding the two previous terms to get a Fibonacci type sequence.
b) Form ratios of successive terms to show how they will eventually approach the golden number.

25. Repeat Ex. 24 for two different nonzero numbers.

26. a) Select any three consecutive terms of a Fibonacci sequence. Subtract the product of the terms on each side of the middle term from the square of the middle term. What is the difference?
b) Repeat part (a) with three different consecutive terms of the sequence.
c) Make a conjecture about what will happen when you repeat this process for any three consecutive terms of a Fibonacci sequence.

27. One of the most famous number patterns involves **Pascal's triangle**. The Fibonacci sequence can be found using Pascal's triangle. Can you explain how this can be done? A hint is shown.

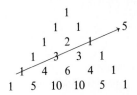

28. a) A sequence related to the Fibonacci sequence is the **Lucas sequence**. The Lucas sequence is formed in a manner similar to the Fibonacci sequence. The first two numbers of the Lucas sequence are 1 and 3. Write the first eight terms of the Lucas sequence.
b) Complete the next two lines of the following chart.

$$1 + 2 = 3$$
$$1 + 3 = 4$$
$$2 + 5 = 7$$
$$3 + 8 = 11$$
$$5 + 13 = 18$$

c) What do you notice about the first column in the chart in part (b)?

29. When two panes of glass are placed face to face, four interior reflective surfaces exist labeled 1, 2, 3, and 4. If light is not reflected, it has just one path through the glass. If it has one reflection, it can be reflected in two ways. If it has two reflections, it can be reflected in three ways. Use this information to answer parts (a–c).

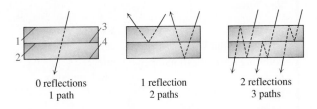

0 reflections   1 reflection   2 reflections
1 path   2 paths   3 paths

a) If a ray is reflected three times, there are five paths it can follow. Show the paths.
b) If a ray is reflected four times there are eight paths it can follow. Show the paths.
c) How many paths can a ray follow if it is reflected five times? Explain how you determined your answer.

Problem Solving

30. A plant grows for two months and then adds a new branch. Each new branch grows for two months, and then adds another branch. After the second month, each branch adds a new branch every month. Assuming that the growth begins in January
a) How many branches will there be in February?
b) How many branches will there be in May?
c) How many branches will there be after 12 months?

31. The divine proportion is $(a + b)/a = a/b$ (Fig. 5.12). This can be written $1 + (b/a) = a/b$. Now let $x = a/b$, which gives $1 + (1/x) = x$. Multiply both sides of this equation by x to get a quadratic equation and then use the quadratic formula (Section 6.8) to show that one answer is $x = (1 + \sqrt{5})/2$ (the golden ratio).

32. Draw a line of length 5 in. Determine and mark the point on the line that will create the golden ratio. Explain how you determined your answer.

Research Activities

33. The digits 1 through 9 have evolved considerably since they first appeared in Fibonacci's book *Liber Abacci*. Write a report tracing the history of the evolution of the digits 1–9 since Fibonacci's time.

34. Write a report on the history and contributions of Fibonacci.

35. Write a report indicating where the golden ratio and golden rectangle have been used in art and architecture. An art teacher or a staff member of an art museum might be able to give you some information and a list of resources. You may also wish to contact the Fibonacci Society.

CHAPTER 5 SUMMARY

Key Terms

5.1
composite number
counting (or natural) number
divisible
divisor
factor
Fermat number
greatest common divisor
least common multiple
Mersenne prime
number theory
prime factorization
prime number
relatively prime

5.2
additive inverse
integer
whole number

5.3
dense
lowest terms
rational number
reciprocal

5.4
irrational number
perfect square
principal square root

radicand
rationalized denominator

5.5
associative property
closure
cummutative property
distributive property
real number

5.6
base
exponent
scientific notation

5.7
arithmetic sequence
common difference
common ratio
geometric sequence
sequence

5.8
divine proportion
Fibonacci sequence
golden number
golden proportion
golden ratio
golden rectangle

Important Facts

Fundamental theorem of arithmetic Every composite number can be expressed as a unique product of prime numbers.

Sets of numbers
Natural or counting numbers: $\{1, 2, 3, 4, \ldots\}$
Whole numbers: $\{0, 1, 2, 3, 4, \ldots\}$
Integers: $\{\ldots, -3, -2, -1, 0, 1, 2, 3, \ldots\}$
Rational numbers: Numbers of the form p/q where p and q are integers, $q \neq 0$. Every rational number when expressed as a decimal will be either a terminating or repeating decimal.
Irrational number: A real number whose representation is a nonterminating, nonrepeating decimal (not a rational number).

Definition of subtraction $a - b = a + (-b)$

Fundamental law of rational numbers

$$\frac{a}{b} = \frac{a}{b} \cdot \frac{c}{c} = \frac{ac}{bc}, \quad b \neq 0, \quad c \neq 0.$$

Rules of radicals
Product rule for radicals:

$$\sqrt{a \cdot b} = \sqrt{a} \cdot \sqrt{b} \quad a \geq 0, \quad b \geq 0.$$

Quotient rule for radicals:

$$\frac{\sqrt{a}}{\sqrt{b}} = \sqrt{\frac{a}{b}}, \quad a \geq 0, \quad b > 0.$$

Properties of real numbers
Commutative property of addition: $a + b = b + a$
Commutative property of multiplication: $a \cdot b = b \cdot a$
Associative property of addition:

$$(a + b) + c = a + (b + c)$$

Associative property of multiplication:

$$(a \cdot b) \cdot c = a \cdot (b \cdot c)$$

Distributive property: $a \cdot (b + c) = ab + ac$

Rules of exponents
Product rule for exponents: $a^m \cdot a^n = a^{m+n}$

Quotient rule for exponents: $\dfrac{a^m}{a^n} = a^{m-n}, \quad a \neq 0$

Zero exponent rule: $a^0 = 1, \quad a \neq 0$

Negative exponent rule: $a^{-m} = \dfrac{1}{a^m}, \quad a \neq 0$

Power rule: $(a^m)^n = a^{mn}$

Arithmetic sequence	Geometric sequence	Fibonacci sequence 1, 1, 2, 3, 5, 8, 13, 21, ...

$$a_n = a_1 + (n-1)d \qquad\qquad a_n = a_1 r^{n-1}$$

Golden number $\dfrac{\sqrt{5}+1}{2}$

$$s_n \doteq \frac{n(a_1 + a_n)}{2} \qquad\qquad s_n = \frac{a_1(1 - r^n)}{1 - r}, \quad r \neq 1$$

Golden proportion $\dfrac{a+b}{a} = \dfrac{a}{b}$

CHAPTER 5 REVIEW EXERCISES

5.1

In Ex. 1 and 2, determine whether or not the number is divisible by each of the following numbers: 2, 3, 4, 5, 6, 8, 9, and 10.

1. 148,632
2. 400,644

Find the prime factorization of the following numbers.
3. 192   **4.** 240   **5.** 180   **6.** 1260   **7.** 960

Find the GCD and the LCM of the following numbers.
8. 32, 20   **9.** 80, 15   **10.** 148, 216
11. 840, 320   **12.** 60, 40, 96   **13.** 36, 108, 144

5.2

Use the number line to evaluate the following.
14. $-4 + 3$ **15.** $4 + (-2)$
16. $4 - 6$ **17.** $-3 + (-2)$
18. $-4 - 3$ **19.** $-4 - (-5)$
20. $(-1 + 9) - 4$ **21.** $-1 + (9 - 4)$
22. $-1 - (9 - 4)$

Evaluate the following.
23. $(-2)(-6)$ **24.** $-5(3)$ **25.** $6(-4)$

26. $\dfrac{-4}{-2}$ **27.** $\dfrac{8}{-2}$ **28.** $[8 \div (-4)](-3)$

29. $[(-4)(-3)] \div 2$ **30.** $[(-30) \div (10)] \div (-1)$

5.3

Express each fraction as a terminating or repeating decimal.
31. $\frac{5}{8}$ **32.** $\frac{8}{10}$ **33.** $\frac{9}{12}$
34. $\frac{15}{4}$ **35.** $\frac{3}{7}$ **36.** $\frac{5}{12}$
37. $\frac{3}{8}$ **38.** $\frac{7}{8}$ **39.** $\frac{2}{7}$

Express each decimal as a quotient of two integers.
40. 0.624   **41.** $0.\overline{6}$ **42.** 2.43
43. $1.\overline{84}$ **44.** 12.083   **45.** 0.0042

46. $2.1\overline{5}$ **47.** $2.\overline{34}$ **48.** $5.0\overline{62}$

Find a rational number between the two numbers in each of the following pairs.
49. 0.0042 and 0.0043   **50.** $\frac{3}{4}$ and $\frac{3}{5}$
51. 2.406 and 2.407   **52.** $\frac{9}{12}$ and $\frac{10}{12}$
53. $\frac{6}{12}$ and 0.51   **54.** 0.2 and $\frac{3}{9}$

Perform the indicated operation, and reduce the answer to lowest terms.

55. $\dfrac{2}{3} + \dfrac{1}{5}$ **56.** $\dfrac{3}{5} - \dfrac{2}{4}$

57. $\dfrac{7}{18} + \dfrac{9}{12}$ **58.** $\dfrac{4}{5} \cdot \dfrac{7}{9}$

59. $\dfrac{5}{9} \div \dfrac{6}{7}$ **60.** $\left(\dfrac{4}{5} + \dfrac{5}{7}\right) \div \dfrac{4}{5}$

61. $\left(\dfrac{2}{3} \cdot \dfrac{1}{7}\right) \div \dfrac{4}{7}$ **62.** $\left(\dfrac{1}{5} + \dfrac{2}{3}\right)\left(\dfrac{3}{8}\right)$

63. $\left(\dfrac{1}{5} \cdot \dfrac{2}{3}\right) + \left(\dfrac{1}{5} \div \dfrac{1}{2}\right)$

5.4

Simplify each of the following. Rationalize the denominator when necessary.
64. $\sqrt{12}$ **65.** $\sqrt{72}$ **66.** $\sqrt{2} + 3\sqrt{2}$
67. $\sqrt{3} - 4\sqrt{3}$ **68.** $\sqrt{8} + 6\sqrt{2}$ **69.** $\sqrt{3} - 7\sqrt{27}$
70. $\sqrt{75} + \sqrt{27}$ **71.** $\sqrt{3} \cdot \sqrt{6}$ **72.** $\sqrt{8} \cdot \sqrt{6}$

73. $\dfrac{\sqrt{18}}{\sqrt{2}}$ **74.** $\dfrac{\sqrt{56}}{\sqrt{2}}$ **75.** $\dfrac{3}{\sqrt{2}}$

76. $\dfrac{\sqrt{3}}{\sqrt{5}}$ **77.** $5(3 + \sqrt{5})$ **78.** $\sqrt{3}(4 + \sqrt{6})$

5.5

State the name of each property illustrated.

79. $x + 2 = 2 + x$

80. $3 \cdot x = x \cdot 3$

81. $(2 + 3) + 4 = 2 + (3 + 4)$

82. $5 \cdot (2 + x) = 5 \cdot 2 + 5 \cdot x$

83. $(6 + 3) + 4 = 4 + (6 + 3)$

84. $(3 + 5) + (4 + 3) = (4 + 3) + (3 + 5)$

85. $(3 \cdot a) \cdot b = 3 \cdot (a \cdot b)$

86. $a \cdot (2 + 3) = (2 + 3) \cdot a$

87. $(x + 3)2 = (x \cdot 2) + (3 \cdot 2)$

88. $x \cdot 2 + 6 = 2 \cdot x + 6$

Determine whether the following sets of numbers are closed under the given operation.

89. Natural numbers, addition

90. Integers, addition

91. Integers, division

92. Real numbers, subtraction

93. Irrational numbers, multiplication

94. Rational numbers, division

5.6

Evaluate each of the following.

95. 2^3

96. 3^{-2}

97. $\dfrac{5^5}{5^4}$

98. $5^2 \cdot 5$

99. 3^0

100. 5^{-3}

101. $(2^3)^2$

102. $(3^2)^2$

Write each number in scientific notation.

103. 3200

104. 0.0000423

105. 0.00168

106. 4,950,000

Express each number in decimal notation.

107. 4.2×10^2

108. 3.87×10^{-5}

109. 1.75×10^{-4}

110. 1×10^5

Perform the indicated operations and express the answer in scientific notation.

111. $(2 \times 10^6)(3.2 \times 10^{-4})$

112. $(3 \times 10^2)(4.6 \times 10^2)$

113. $\dfrac{8.4 \times 10^3}{4 \times 10^2}$

114. $\dfrac{1.5 \times 10^{-3}}{5 \times 10^{-4}}$

Perform the indicated operation by first converting each number to scientific notation. Write the answer in decimal notation.

115. (80,000)(420,000)

116. (75,000)(0.0003)

117. $\dfrac{9,600,000}{3000}$

118. $\dfrac{0.000002}{0.0000004}$

119. At noon there were 12,000 bacteria in the culture. At 6:00 P.M. there were 300,000 bacteria in the culture. How many times greater is the number of bacteria at 6:00 P.M. than at noon?

5.7

Determine whether the following sequences are arithmetic or geometric. Then determine the next two terms of the sequence.

120. $3, 8, 13, 18, \ldots$

121. $-4, 12, -36, 108, \ldots$

122. $0, -4, -8, -12, \ldots$

123. $1, \frac{1}{2}, \frac{1}{4}, \frac{1}{8}, \ldots$

124. $1, 4, 7, 10, 13, \ldots$

125. $2, -2, 2, -2, 2, \ldots$

Write the first five terms of the sequence with the given first term, a_1, and common difference, d, or ratio, r.

126. $a_1 = 4, \ d = 5$

127. $a_1 = -6, \ d = -3$

128. $a_1 = -\frac{1}{2}, \ d = -2$

129. $a_1 = 4, \ r = 2$

130. $a_1 = 16, \ r = \frac{1}{2}$

131. $a_1 = \frac{1}{2}, \ r = -\frac{1}{2}$

Find the indicated term of the sequence with the given first term, a_1, and common difference, d, or ratio, r.

132. Find a_5, given $a_1 = 6, \ d = 2$

133. Find a_7, given $a_1 = -6, \ d = -5$

134. Find a_{10}, given $a_1 = 20, \ d = 8$

135. Find a_4, given $a_1 = 8, \ r = 3$

136. Find a_5, given $a_1 = 4, \ r = \frac{1}{2}$

137. Find a_4, given $a_1 = -6, \ r = 2$

Find the sum of the arithmetic sequence.

138. $4, 2, 0, -2, \ldots, -18$

139. $-4, -3\frac{3}{4}, -3\frac{1}{2}, -3\frac{1}{4}, \ldots, -2\frac{1}{4}$

140. $100, 94, 88, 82, \ldots, 58$

Find the sum of the first n terms of the geometric sequence with the given values of a_1 and r.

141. $n = 3, \ a_1 = 4, \ r = 3$

142. $n = 4, \ a_1 = 2, \ r = 4$

143. $n = 5, \ a_1 = 3, \ r = -2$

First determine whether the sequence is arithmetic or geometric; then write an expression for the general or nth term, a_n.

144. $7, 4, 1, -2, \ldots$

145. $0, 5, 10, 15, \ldots$

146. $4, \frac{5}{2}, 1, -\frac{1}{2}, \ldots$

147. $3, 6, 12, 24, \ldots$

148. $4, -4, 4, -4, \ldots$

149. $5, \frac{5}{3}, \frac{5}{9}, \frac{5}{27}, \ldots$

5.8

Determine whether each of the following is a Fibonacci type sequence. If so determine the next two terms in the sequence.

150. $1, 2, 3, 5, 8, 13, \ldots$

151. $2, 4, 8, 16, 32, \ldots$

152. $3, 8, 13, 18, 23, \ldots$

153. $2, 2, 4, 6, 10, 16, \ldots$

Chapter 5

CHAPTER TEST

1. Which of the numbers 2, 3, 4, 5, 6, 8, 9, and 10, divide 481.248?

2. Find the prime factorization of 360

3. Evaluate $[(-8) + (-5)] + 7$.

4. Evaluate $-7 - 15$.

5. Evaluate $[(-50)(-2)] \div (8 - 10)$.

6. Determine a rational number between 0.435 and 0.436.

7. Determine a rational number between $\frac{1}{8}$ and $\frac{1}{9}$.

8. Write $\frac{3}{8}$ as a terminating or repeating decimal.

9. Express 2.45 as a quotient of two integers.

10. Evaluate $(\frac{5}{16} \div 3) + (\frac{4}{5} \cdot \frac{1}{2})$.

11. Perform the operation and reduce the answer to lowest terms:

$$\frac{17}{24} - \frac{3}{40}$$

12. Simplify $\sqrt{18} + \sqrt{50}$.

13. Rationalize $\dfrac{\sqrt{5}}{\sqrt{3}}$.

14. Determine whether the integers are closed under the operation of multiplication. Explain your answer.

Name the properties illustrated.

15. $(2 + x) + 3 = 2 + (x + 3)$

16. $3(x + y) = 3x + 3y$

Evaluate

17. $\dfrac{4^5}{4^2}$ **18.** $2^3 \cdot 2^2$ **19.** 7^{-2}

20. Perform the operation by first converting the numerator and denominator to scientific notation. Write the answer in scientific notation.

$$\frac{64,000}{0.008}$$

21. Write an expression for the general or nth term, a_n, of the sequence $-4, -8, -12, -16, \ldots$.

22. Find the sum of the terms of the arithmetic sequence $-2, -5, -8, -11, \ldots, -32$.

23. Find a_5 when $a_1 = 3$ and $r = 3$.

24. Find the sum of the first five terms of the sequence when $a_1 = 3$ and $r = 4$.

25. Write an expression for the general or nth term, a_n, of the sequence $3, 6, 12, 24, \ldots$.

26. Write the first eight terms of the Fibonacci sequence.

ALGEBRA, GRAPHS, & FUNCTIONS

Math phobic's nightmare

Gary Larson, creator of *The Far Side* comic strip, touched a common fear in a cartoon titled "Math Phobic's Nightmare." In the cartoon, St. Peter guards the gates to Heaven, admitting only those who can answer an algebraic word problem. Word problems — the very mention of them is enough to frighten most people. And yet, algebra is one of the most practical tools for solving everyday problems. You probably use algebra in your daily life without realizing it.

For example, you use a coordinate system when you consult your car map to find directions to a new destination. You solve simple equations when you change a recipe to increase or decrease the number of servings. To evaluate how much interest you will earn on a savings account, or to figure out how long it will take you to travel a given distance, you use common formulas that are algebraic equations. The list of the applications of algebra could go on and on. For this reason, the French mathematician and encyclopaedist Jean-le-Rond d'Alembert remarked, "Algebra is

In the seventh century, Baghdad became the intellectual center of the world, where the mathematics and science practiced in the ancient civilizations of the East and West came together. It was here that Islamic translators brought together mathematics and the Hindu system of base 10 numeration. Not until the Crusades (1050–1250) would Europeans discover more fully the great intellectual treasures that the Islamic world possessed.

generous, she often gives more than is asked of her." Indeed, the greatest value of algebra is that it provides the means to solve everyday problems.

The symbolic language of algebra makes it an excellent tool for solving problems. Symbolism has three advantages. First, it allows us to write lengthy expressions in compact form: The eye can take in an entire statement and the mind can retain it. Thus "the product of three times a number multiplied by a second number multiplied by itself and added to two times the product of the first, second, and third numbers" can be expressed simply as $3xy^2 + 2xyz$. Second, symbolic language is clear: Each symbol has a precise meaning. Finally, symbolism allows us to consider a large or infinite number of separate cases with a common property. For example, we can use the symbolic representation $ax + b = 0$ to represent all linear equations in one variable.

The English philosopher Alfred North Whitehead explained the power of algebra when he stated, "By relieving the brain of all unnecessary work, a good notation sets the mind free to concentrate on more advanced problems and in effect increases the mental power of the race."

Translating mathematical concepts into symbols allows us to discuss them and solve them more easily.

Algebraic equations can be graphed using computer algebra systems like Mathematica, Maple, or Derive. The results, even for relatively simple equations, are often surprisingly beautiful, and can be applied to study many fields, from physics to biology. The image here is a 3-D depiction of the equation $x^5 + y^5 = z^5$.

6.1 | # Order of Operations

Algebra is a generalized form of arithmetic used to solve problems. It developed from the Babylonian number system as early as 1800 B.C. and spread to Greece and Europe, as well as to India, China, and Japan. The Islamic world became the center of mathematical studies between A.D. 800 and A.D. 1100. The word *algebra* is derived from the Arabic word *al-jabr* (meaning "reunion of broken parts"), which was the title of a book written by the mathematician Muhammed ibn-Musa al Khwarizmi around A.D. 825. Algebra was reintroduced to Europe in the thirteenth and fourteenth centuries.

Why study algebra? Many problems in everyday life can be solved using arithmetic or by trial and error, but with a knowledge of algebra you can find the solutions with less effort. Other problems, like some we will present in this chapter, can only be solved using algebra.

Algebra uses letters of the alphabet called **variables** to represent numbers. Often the letters x and y are used to represent variables. However, any letter may be used as a variable. A symbol that represents a specific quantity is called a **constant**.

Multiplication of numbers and variables may be represented in several different ways in algebra. Since the "times" sign might be confused with the variable x, a dot between two numbers or variables indicates multiplication. Thus $3 \cdot 4$ means 3 times 4, and $x \cdot y$ means x times y. Placing two letters or a number and a letter next to one another, with or without parentheses, also indicates multiplication. Thus $3x$ means 3 times x, xy means x times y, and $(x)(y)$ means x times y.

An **algebraic expression** (or simply **expression**) is a collection of variables, numbers, parentheses, and operation symbols. Here are some examples of algebraic expressions:

$$x; \qquad x + 2; \qquad 3(2x + 3); \qquad \frac{3x + 1}{2x - 3}; \qquad x^2 + 7x + 3$$

Two algebraic expressions joined by an equal sign form an **equation**. Here are some examples of equations:

$$x + 2 = 4; \qquad 3x + 4 = 1; \qquad x + 3 = 2x$$

The **solution** to an equation is the number or numbers that replace the variable to make the equation a true statement. For example, the solution to the equation $x + 3 = 4$ is $x = 1$. When we find the solution to an equation, we **solve the equation**.

The solution to any equation can be **checked** by substituting the value obtained for the variable in the original equation. To check the answer, $x = 1$, in the previous equation, we do the following:

Check:

$$x + 3 = 4$$
$$1 + 3 = 4 \qquad \text{Substitute 1 for } x.$$
$$4 = 4 \qquad \text{True}$$

Since the same number is obtained on both sides of the equal sign, the solution is correct.

To **evaluate an expression** means to find the value of the expression for a given value of the variable. To evaluate expressions and solve equations, you must have an understanding of exponents. Exponents (Section 5.6) are used to abbreviate repeated multiplication. For example, the expression 5^2 means $5 \cdot 5$. The 2 in the expression 5^2 is the **exponent**, and the 5 is the **base**. We read 5^2 as "5 to the second power" or "5 squared." 5^2 means $5 \cdot 5$ or 25.

In general, the number b to the nth power, written b^n, means

$$\overset{\text{exponent}}{\underset{\text{base}}{b^n}} = \underbrace{b \cdot b \cdot b \cdot \cdots \cdot b}_{n \text{ factors of } b}$$

An exponent refers only to its base. In the expression -5^2 the base is 5. In the expression $(-5)^2$, the base is -5.

$$-5^2 = -(5)^2 = -(5)(5) = -25$$
$$(-5)^2 = (-5)(-5) = 25$$

Notice that $-5^2 \neq (-5)^2$ since $-25 \neq 25$.

Order of Operations

To evaluate an expression or to check an equation, we need to know the **order of operations** to follow. For example, suppose we wish to evaluate the expression $2 + 3x$ when $x = 4$. Substituting 4 for x, we obtain $2 + 3 \cdot 4$. What is the value of $2 + 3 \cdot 4$? Does it equal 20, or does it equal 14? Some standard rules called the order of operations have been developed to ensure that there is only one correct answer. In mathematics, unless parentheses indicate otherwise, always perform multiplication before addition. Thus the correct answer is 14:

$$2 + 3 \cdot 4 = 2 + (3 \cdot 4) = 2 + 12 = 14.$$

The order of operations to be followed in evaluating an expression is given in the following chart.

Order of Operations

1. First, perform all operations within parentheses or other grouping symbols (according to the order given below).
2. Next, perform all exponential operations (that is, raising to powers or finding roots).
3. Next, perform all multiplication and division from left to right.
4. Finally, perform all addition and subtraction from left to right.

▶ **Example 1**

Evaluate the expression $-x^2 + 4x + 25$ for $x = 3$.

Solution: Substitute 3 for x and use the order of operations to evaluate the expression.

$$-x^2 + 4x + 25$$
$$= -(3)^2 + 4(3) + 25$$
$$= -9 + 12 + 25$$
$$= 28$$

▶ **Example 2**

The temperature T, in degrees Celsius, in a sauna n minutes after the sauna is turned on can be approximated by the expression $-0.04n^2 + 1.6n + 20$ (assuming n is between 0 and 20). Find the sauna's temperature after it has been on for 12 minutes.

Solution: Substitute 12 for n.

$$-0.04n^2 + 1.6n + 20$$
$$= -0.04(12)^2 + 1.6(12) + 20$$
$$= -0.04(144) + 19.2 + 20$$
$$= -5.76 + 19.2 + 20$$
$$= 33.44$$

The temperature in the sauna is approximately 33°C (or 92°F) after 12 minutes.

This painting by Charles Demuth, called I Saw the Figure 5 in Gold, depicts the abstract nature of numbers. The mathematician and philosopher Bertrand Russell observed in 1919 that it must have required "many ages to discover that a brace of pheasants and a couple of days were both instances of the number 2 . . . ." The discovery that numbers could be used not only to count objects such as the number of birds but also to represent abstract quantities represented a breakthrough in the development of algebra.

▶ **Example 3**

Evaluate $-3x^2 + 4xy - y^2$ when $x = 3$ and $y = 4$.

Solution: Substitute 3 for x and 4 for y; then evaluate using the order of operations.

$$-3x^2 + 4xy - y^2$$
$$= -3(3)^2 + 4(3)(4) - 4^2$$
$$= -3(9) + 12(4) - 16$$
$$= -27 + 48 - 16$$
$$= 21 - 16$$
$$= 5$$

▶ **Example 4**

Determine whether 3 is a solution to the equation $2x^2 + 4x - 9 = 21$.

Solution: To determine whether 3 is a solution to the equation, substitute 3 for each x in the equation. Then evaluate the left side of the equation using the order of operations. If this leads to a 21 on the left side of the equal sign, then both sides of the equation have the same value, and 3 is a solution.

$$2x^2 + 4x - 9 = 21$$
$$2(3)^2 + 4(3) - 9 = 21$$
$$2(9) + 12 - 9 = 21$$
$$18 + 12 - 9 = 21$$
$$30 - 9 = 21$$
$$21 = 21 \quad \text{True}$$

Since 3 makes the equation a true statement, 3 is a solution to the equation.

Section 6.1 Exercises

Evaluate each of the following for the given value(s) of the variable(s).

1. x^2, $x = 4$
2. x^2, $x = -3$
3. $-x^2$, $x = 8$
4. $-x^2$, $x = -5$
5. $-2x^3$, $x = -7$
6. $-x^3$, $x = -3$
7. $x + 7$, $x = 4$
8. $6x - 4$, $x = \frac{3}{2}$
9. $-5x + 8$, $x = -4$
10. $x^2 + 6x - 4$, $x = 5$
11. $-x^2 - 7x + 5$, $x = -2$
12. $5x^2 + 7x - 11$, $x = -1$
13. $\frac{1}{2}x^2 + 4x - 6$, $x = \frac{1}{3}$
14. $x^3 - 3x^2 + 7x + 11$, $x = 2$
15. $4x^3 - 6x + 5$, $x = \frac{1}{2}$
16. $-x^2 + 4xy$, $x = 2$, $y = 3$
17. $x^2 - 2xy + y^2$, $x = -1$, $y = -1$
18. $3x^2 + \frac{2}{5}xy - \frac{1}{2}y^2$, $x = 5$, $y = -2$
19. $4x^2 - 12xy + 9y^2$, $x = 3$, $y = 2$
20. $(x + 3y)^2$, $x = 4$, $y = -1$

Determine whether the given values are a solution to the equation.

21. $5x - 4 = 0$, $x = 2$
22. $3x - 6 = 4$, $x = 3$
23. $2x + y = 0$, $x = 2$, $y = -4$
24. $3x + 2y = 4$, $x = 0$, $y = 2$
25. $x^2 + 3x - 4 = 5$, $x = 2$
26. $2x^2 - x - 5 = 0$, $x = 3$
27. $2x^2 - x = 28$, $x = 4$
28. $y = x^2 + 3x - 5$, $x = 1$, $y = -1$
29. $y = -x^2 - 3x + 2$, $x = 2$, $y = -8$
30. $x^2 + 3y = 1$, $x = -2$, $y = -1$

31. If a car travels at 60 mph for t hours, then the distance traveled can be found using the expression $60t$. Find the distance traveled in (a) 4 hr and (b) 5.6 hr.

32. If the sales tax on an item is 6%, then the sales tax, in dollars, on an item costing d dollars can be found using the expression $0.06d$. Find the sales tax on a washing machine that costs $680.

33. The total cost of an item, including a 7% sales tax, can be found using the expression $d + 0.07d$, where d is the pretax cost of the item. Find the total cost of a boat whose pretax cost is $4000.

34. The number of baskets of oranges that are produced by x trees in a small orchard can be approximated by the expression $25x - 0.2x^2$ (assuming x is no more than 100). Find the number of baskets of oranges produced by 60 trees.

35. The time it takes, in minutes, for clothes hanging on a line outdoors to dry, at a specific temperature and wind speed, depends upon the humidity, h. The time can be approximated by the expression $2h^2 + 80h + 40$, where h is the percent humidity expressed as a decimal number. Find the length of time it will take clothing to dry if there is 60% humidity.

36. The rate of growth of grass in inches per week depends on a number of factors, including rainfall and temperature. For a certain area this can be approximated by the expression $0.2R^2 + 0.003RT + 0.0001T^2$ where R is the weekly rainfall, in inches, and T is the average weekly temperature, in degrees Fahrenheit. Find the growth of grass for a week in which the rainfall is 2 in. and the average temperature is 70°F.

6.2 Linear Equations in One Variable

In Section 6.1 we stated that two algebraic expressions joined by an equal sign form an equation. The solution to some equations, such as $x + 3 = 4$, can be found easily by trial and error. However, solving more complex equations, such as $2x - 3 = 4(x + 3)$, requires understanding the meaning of like terms, and learning four basic properties.

The parts that are added or subtracted in an algebraic expression are called **terms**. The expression $4x - 3y - 5$ contains three terms, namely $4x$, $-3y$, and -5. The $+$ and $-$ signs that break the expression into terms are a part of the terms. When listing the terms of an expression, however, it is not necessary to include the $+$ sign at the beginning of the term.

The numerical part of a term is called its **numerical coefficient** or simply its **coefficient**. In the term $4x$, the 4 is the numerical coefficient. In the term $-4y$, the -4 is the numerical coefficient.

Like terms are terms that have the same variables with the same exponents on the variables. **Unlike terms** have different variables or different exponents on the variables.

Like Terms		Unlike Terms	
$3x$, $-4x$	(Same variable, x)	$3x$, 2	(Only first term has a variable)
$4y$, $6y$	(Same variable, y)	$3x$, $4y$	(Different variables)
5, -6	(Both constants)	x, 3	(Only first term has a variable)
$-6x^2$, $7x^2$	(Same variable with same exponent)	$3x^2$, $5x^3$	(Different exponents)

To **simplify** an expression means to combine like terms by using the commutative, associative, and distributive properties discussed in Chapter 5. For convenience we list the properties here.

Properties of the Real Numbers

$a(b + c) = ab + ac$	Distributive property
$a + b = b + a$	Commutative property of addition
$ab = ba$	Commutative property of multiplication
$(a + b) + c = a + (b + c)$	Associative property of addition
$(ab)c = a(bc)$	Associative property of multiplication

▶ **Example 1**

Combine like terms in each expression.

a) $4x + 3x$ **b)** $5a - 7a$ **c)** $12 + x + 7$

Solution:

a) We use the distributive property (in reverse) to combine like terms:

$$4x + 3x = (4 + 3)x \quad \text{Distributive property}$$
$$= 7x$$

b) $5a - 7a = (5 - 7)a$
$$= -2a$$

c) $12 + x + 7 = x + 12 + 7$
$$= x + 19$$

We are able to rearrange the terms of an expression, as was done in Example 1(c), by the commutative and associative properties that were discussed in Section 5.5.

▶ **Example 2**

Combine like terms in each expression.

a) $3y + 4x - 3 - 2x$ **b)** $-2x + 3y - 4x + 3 - y + 5$

Solution:

a) $4x$ and $-2x$ are the only like terms. Grouping the like terms together gives

$$\underbrace{4x - 2x} + 3y - 3$$
$$\quad 2x \quad + 3y - 3$$

b) Grouping the like terms together gives

$$\underbrace{-2x - 4x} \underbrace{+ 3y - y} \underbrace{+ 3 + 5}$$
$$\quad -6x \quad\quad +2y \quad\quad +8$$

The order of the terms in an expression is not crucial. However, when listing the terms of an expression we generally list the terms in alphabetical order with the constant, the term without a variable, last.

Solving Equations

Recall that to solve an equation means to find the value or values for the variable that make(s) the equation true. In this section we discuss solving **linear (or first degree) equations**. A linear equation in one variable is one in which the exponent on the variable is 1. The following equations are examples of linear equations: $5x - 1 = 3$ and $2x + 4 = 6x - 5$.

To solve any equation, it is necessary to **isolate the variable**. This means getting the variable by itself on one side of the equal sign. The four properties we are about to discuss will be used to isolate the variable. The first is the addition property.

Addition Property

If $a = b$, then $a + c = b + c$ for all real numbers a, b, and c.

The addition property indicates that the same number can be added to both sides of an equation without changing the solution.

▶ **Example 3**

Find the solution to the equation $x - 6 = 4$.

Solution: To isolate the variable, add 6 to both sides of the equation.

$$x - 6 + 6 = 4 + 6$$
$$x + 0 = 10$$
$$x = 10$$

Check: $x - 6 = 4$
  $10 - 6 = 4$ Substitute 10 for *x*.
    $4 = 4$ True

In Example 3 we showed the step $x + 0 = 10$. Generally this step is done mentally and the step is not listed.

Subtraction Property

If $a = b$, then $a - c = b - c$ for all real numbers a, b, and c.

The subtraction property indicates that the same number can be subtracted from both sides of an equation without changing the solution.

▶ **Example 4**

Find the solution to the equation $x + 5 = 7$.

Solution: To isolate the variable, subtract 5 from both sides of the equation.

$$x + 5 - 5 = 7 - 5$$
$$x = 2$$

Note that we did not subtract 7 from both sides of the equation, since this would not result in getting x on one side of the equals sign by itself.

Multiplication Property

If $a = b$, then $a \cdot c = b \cdot c$ for all real numbers a, b, and c where $c \neq 0$.

The multiplication property indicates that both sides of the equation can be multiplied by the same nonzero number without changing the solution.

▶ **Example 5**

Find the solution to the equation $\dfrac{x}{3} = 2$.

Solution: To solve this equation, multiply both sides of the equation by 3:

$$3\left(\frac{x}{3}\right) = 3(2)$$

$$\frac{\overset{1}{\cancel{3}}x}{\underset{1}{\cancel{3}}} = 6$$

$$1x = 6$$

$$x = 6$$

In Example 5 we showed the steps $(3x)/3 = 6$ and $1x = 6$. Generally we will not illustrate these steps.

Division Property

If $a = b$, then $a/c = b/c$ for all real numbers a, b, and c, $c \neq 0$.

The division property indicates that both sides of an equation can be divided by the same nonzero number without changing the solution. Note that the divisor, c, cannot be 0, since division by 0 is not permitted.

▶ **Example 6**

Find the solution to the equation $5x = 35$.

Solution: To solve this equation, divide both sides of the equation by 5:

$$\frac{5x}{5} = \frac{35}{5}$$

$$x = 7$$

An **algorithm** is a general procedure for accomplishing a task. The following is an algorithm for solving linear (or first degree) equations.

Sometimes the solution to an equation may be found more easily by using a variation of the general procedure. Remember that the primary objective in solving any equation is to isolate the variable.

A General Procedure for Solving Linear Equations

1. If the equation contains fractions, multiply both sides of the equation by the lowest common denominator (or least common multiple). This step will eliminate all fractions from the equation.
2. Use the distributive property to remove parentheses when necessary.
3. Combine like terms on the same side of the equals sign when possible.
4. Use the addition or subtraction property to collect all terms with a variable on one side of the equals sign and all constants on the other side of the equals sign. It may be necessary to use the addition or subtraction property more than once. This process will eventually result in an equation of the form $ax = b$, where a and b are real numbers.
5. Solve for the variable using the division or multiplication property. This will result in an answer in the form $x = c$, where c is a real number.

▶ **Example 7**

Solve the equation $3x + 4 = 13$; then check your solution.

Solution: Our goal is to isolate the variable, therefore we start by getting the term $3x$ by itself on one side of the equation.

$$3x + 4 = 13$$
$$3x + 4 - 4 = 13 - 4 \quad \text{Subtract 4 from both sides of the equation (subtraction property) (step 4).}$$
$$3x = 9$$

$$\frac{3x}{3} = \frac{9}{3} \quad \text{Divide both sides of the equation by 3 (division property) (step 5).}$$

$$x = 3$$

A check will show that 3 is the solution to $3x + 4 = 13$.

▶ **Example 8**

Solve the equation $2 = 3 + 5(p + 1)$ for p.

Solution: The goal is to isolate the variable p. To accomplish this we will follow the general procedure for solving equations:

$$2 = 3 + 5(p + 1)$$
$$2 = 3 + 5p + 5 \quad \text{Distributive property (step 2)}$$
$$2 = 5p + 8 \quad \text{Combine like terms (step 3).}$$
$$2 - 8 = 5p + 8 - 8 \quad \text{Subtraction property (step 4)}$$
$$-6 = 5p$$

then

$$-\frac{6}{5} = \frac{5p}{5}$$ Division property (step 5)

$$-\frac{6}{5} = p$$

▶ **Example 9**

Solve the equation: $\dfrac{2x}{3} + \dfrac{1}{3} = \dfrac{3}{4}$.

Solution: When an equation contains fractions we generally begin by multiplying each term of the equation by the lowest common denominator, LCD (see Chapter 5). In this example the LCD is 12, since 12 is the smallest number that is divisible by both 3 and 4.

$$12\left(\frac{2x}{3} + \frac{1}{3}\right) = 12\left(\frac{3}{4}\right)$$ Multiply both sides of the equation by the LCD (step 1).

$$12\left(\frac{2x}{3}\right) + 12\left(\frac{1}{3}\right) = 12\left(\frac{3}{4}\right)$$ Distributive property (step 2)

$$\overset{4}{\cancel{12}}\left(\frac{2x}{\cancel{3}}\right) + \overset{4}{\cancel{12}}\left(\frac{1}{\cancel{3}}\right) = \overset{3}{\cancel{12}}\left(\frac{3}{\cancel{4}}\right)$$ Divide out common factors.

$$8x + 4 = 9$$

$$8x + 4 - 4 = 9 - 4$$ Subtraction property (step 4)

$$8x = 5$$

$$\frac{8x}{8} = \frac{5}{8}$$ Division property (step 5)

$$x = \frac{5}{8}$$

A check will show that $\frac{5}{8}$ is the solution of the equation. You could have worked the problem without first multiplying both sides of the equation by the LCD. Try it!

▶ **Example 10**

Solve the equation $3x + 4 = 5x + 6$.

Solution: Note that the equation has an x on both sides of the equals sign. In equations of this type you might wonder what to do first. It really does not matter as long as you don't forget the goal of isolating the variable x. We will

collect the terms containing a variable on the left-hand side of the equation.

$$3x + 4 = 5x + 6$$
$$3x + 4 - 4 = 5x + 6 - 4 \qquad \text{Subtraction property (step 4)}$$
$$3x = 5x + 2$$
$$3x - 5x = 5x - 5x + 2 \qquad \text{Subtraction property (step 4)}$$
$$-2x = 2$$
$$\frac{-2x}{-2} = \frac{2}{-2} \qquad \text{Division property (step 5)}$$
$$x = -1$$

In the solution to Example 10 the terms containing the variable were collected on the left side of the equals sign. Now work Example 10 collecting the terms with the variable on the right-hand side of the equation. If you do so correctly you should get the same result.

▶ **Example 11**

Solve the equation and check your solution: $4x - 0.48 = 0.8x + 4$.

Solution: This problem may be solved with the decimals, or you may multiply each term by 100 and eliminate the decimals. We will solve the problem with the decimals:

$$4x - 0.48 = 0.8x + 4$$
$$4x - 0.48 + 0.48 = 0.8x + 4 + 0.48 \qquad \text{Addition property}$$
$$4x = 0.8x + 4.48$$
$$4x - 0.8x = 0.8x - 0.8x + 4.48 \qquad \text{Subtraction property}$$
$$3.2x = 4.48$$
$$\frac{3.2x}{3.2} = \frac{4.48}{3.2} \qquad \text{Division property}$$
$$x = 1.4$$

Check:
$$4x - 0.48 = 0.8x + 4$$
$$4(1.4) - 0.48 = 0.8(1.4) + 4 \qquad \text{Substitute 1.4 in place of } x \text{ in the equation.}$$
$$5.6 - 0.48 = 1.12 + 4$$
$$5.12 = 5.12 \qquad \text{True}$$

In Chapter 5 we explained that $a - b$ can be expressed as $a + (-b)$. We will use this principle in the next example.

▶ **Example 12**

Solve $9 = -7 + 8(r - 5)$ for r.

Solution:

$$9 = -7 + 8(r - 5)$$
$$9 = -7 + 8[r + (-5)] \qquad \text{Definition of subtraction of integers}$$
$$9 = -7 + 8(r) + 8(-5) \qquad \text{Distributive property}$$
$$9 = -7 + 8r + (-40)$$
$$9 = -47 + 8r \qquad \text{Combine like terms.}$$
$$9 + 47 = -47 + 47 + 8r \qquad \text{Addition property}$$
$$56 = 8r \qquad \text{Combine like terms.}$$
$$\frac{56}{8} = \frac{8r}{8} \qquad \text{Division property}$$
$$7 = r$$

So far every equation has had exactly one solution. However, some equations have no solution while others have more than one solution. Example 13 illustrates an equation that has no solution, and Example 14 illustrates an equation that has an infinite number of solutions.

▶ **Example 13**

Solve $2(x - 3) + x = 5x - 2(x + 5)$.

Solution:
$$2(x - 3) + x = 5x - 2(x + 5)$$
$$2x - 6 + x = 5x - 2x - 10 \qquad \text{Distributive property}$$
$$3x - 6 = 3x - 10 \qquad \text{Combine like terms.}$$
$$3x - 3x - 6 = 3x - 3x - 10 \qquad \text{Subtraction property}$$
$$-6 = -10 \qquad \text{False}$$

During the process of solving an equation, if you obtain a false statement like $-6 = -10$, or $4 = 0$, then the equation has **no solution**. An equation that has no solution is called an **inconsistent equation**. The equation $2(x - 3) + x = 5x - 2(x + 5)$ is inconsistent and has no solution.

▶ **Example 14**

Solve $3(x + 2) - 5(x - 3) = -2x + 21$.

Solution:
$$3(x + 2) - 5(x - 3) = -2x + 21$$
$$3x + 6 - 5x + 15 = -2x + 21 \qquad \text{Distributive property}$$
$$-2x + 21 = -2x + 21$$

Note that at this point both sides of the equation are the same. Every real number will satisfy this equation. This equation has an infinite number of solutions. An equation of this type is called an **identity**. When solving an equation, if you notice the same expression appears on both sides of the equation, the equation is an identity. The solution to any identity is **all real numbers**. If you continue to solve an equation that is an identity, you will end up with $0 = 0$, as illustrated below:

$$-2x + 21 = -2x + 21$$
$$-2x + 2x + 21 = -2x + 2x + 21 \quad \text{Addition property}$$
$$21 = 21$$
$$21 - 21 = 21 - 21 \quad \text{Subtraction property}$$
$$0 = 0 \quad \text{True for any value of } x$$

Proportions

A **ratio** is a quotient of two quantities. An example is the ratio of 2 to 5, which can be written $2:5$ or $\frac{2}{5}$ or 2/5.

A **proportion** is a statement of equality between two ratios.

An example of a proportion is $\dfrac{a}{b} = \dfrac{c}{d}$. Consider the proportion

$$\frac{x + 2}{5} = \frac{x + 5}{8}.$$

We can solve this proportion by first multiplying both sides of the equation by the least common denominator, 40.

$$\frac{x + 2}{5} = \frac{x + 5}{8}$$
$$\overset{8}{\cancel{40}}\left(\frac{x + 2}{\cancel{5}}\right) = \overset{5}{\cancel{40}}\left(\frac{x + 5}{\cancel{8}}\right)$$
$$8(x + 2) = 5(x + 5)$$
$$8x + 16 = 5x + 25$$
$$3x + 16 = 25$$
$$3x = 9$$
$$x = 3$$

A check will show that 3 is the solution.

Proportions can often be solved more easily using the cross product method.

Cross Product Method to Solve Proportions

If $\dfrac{a}{b} = \dfrac{c}{d}$ then $ad = bc$.

The proportion $(x + 2)/5 = (x + 5)/8$ is solved by the cross product method as follows:

$$\frac{x + 2}{5} = \frac{x + 5}{8}$$

$$8(x + 2) = 5(x + 5)$$
$$8x + 16 = 5x + 25$$
$$3x + 16 = 25$$
$$3x = 9$$
$$x = 3$$

Many practical application problems can be solved using proportions.

To Solve Application Problems Using Proportions

1. Represent the unknown quantity by a variable.
2. Set up the proportion by listing the given ratio on the left-hand side of the equals sign, and the unknown and other given quantity on the right-hand side of the equals sign. When setting up the right-hand side of the proportion, the same respective quantities should occupy the same respective positions on the left and right. For example, an acceptable proportion might be

$$\frac{\text{miles}}{\text{hour}} = \frac{\text{miles}}{\text{hour}}.$$

3. Once the proportion is properly written, drop the units and use the cross product method to solve the equation.
4. Answer the question or questions asked.

▶ **Example 15**

The water rate in Monroe County is $1.04 per 1000 gallons (gal) of water used. What is the water bill if 25,000 gallons are used?

Solution: This problem may be solved by setting up a proportion. One proportion that can be used is

$$\frac{\text{Cost of 1000 gal}}{1000 \text{ gal}} = \frac{\text{Cost of 25,000 gal}}{25,000 \text{ gal}}.$$

Since we wish to find the cost for 25,000 gallons of water, we will call this quantity x. The proportion then becomes

$$\text{Given ratio} \left\{ \frac{1.04}{1000} = \frac{x}{25,000}. \right.$$

Now we solve for x:

$$(1.04)(25{,}000) = 1000x$$
$$1000x = 26{,}000$$

$$x = \frac{26{,}000}{1000} = 26$$

The cost of 25,000 gallons of water is $26.00.

▶ **Example 16**

Insulin comes in 10-cc vials labeled in the number of units of insulin per cubic centimeter (cc) of fluid. A vial of insulin marked U40 has 40 units of insulin per cubic centimeter of fluid. If a patient needs 25 units of insulin, how much fluid should be drawn up into the syringe from the U40 vial?

Solution: The unknown quantity, x, is the number of cubic centimeters of fluid to be drawn into the syringe. Following is one proportion that can be used to find that quantity.

$$\text{Given ratio} \quad \left\{ \frac{40 \text{ units}}{1 \text{ cc}} = \frac{25 \text{ units}}{x \text{ cc}} \right.$$

$$40x = 25(1)$$
$$40x = 25$$

$$x = \frac{25}{40} = 0.625$$

The nurse or doctor putting the insulin in the syringe should draw up 0.625 cc of the fluid.

Section 6.2 Exercises

Combine like terms.

1. $3x - 5x$
2. $-2x - 3x$
3. $2x + 3x - 7$
4. $9 + 2x + 3x$
5. $5x - 3y + 3 + y$
6. $x - 4x + 3$
7. $-3x + 2 - 5x$
8. $-3x + 4x - 2 + 5$
9. $2 - 3x - 2x + 1$
10. $0.4x + 1.1x - 3$
11. $5.2x + 7.4 - 6.3x$
12. $\frac{1}{2}x + \frac{3}{4}x + 5$
13. $\frac{1}{3}x - \frac{1}{4}x - 2$
14. $3x - 4y + 6x + 7y + 12$
15. $4x - 3y - 2y + 6x + 4$
16. $3(p + 2) - 4(p + 3)$
17. $2(r - 3) + 5(r - 2) - 4$
18. $5(w + 5) - 2(w - 12) - 1$
19. $0.1(x + 4) - 2.3(x - 2)$
20. $\frac{1}{2}(x + 1) + \frac{1}{3}x$
21. $\frac{1}{4}x + \frac{4}{5} - \frac{3}{5}x + \frac{1}{3}$
22. $n - \frac{3}{4} + \frac{5}{9}n - \frac{1}{6}$
23. $0.5(2.6x - 4) + 2.3(1.4x - 5)$
24. $\frac{2}{3}(3x + 9) - \frac{1}{4}(2x + 5)$

Solve the following equations.

25. $p + 4 = 10$
26. $5x - 6 = 24$
27. $10 = 6x + 4$
28. $11 = 3 - 4y$
29. $\frac{4}{9} = \frac{x}{11}$
30. $\frac{x + 3}{3} = \frac{x + 7}{5}$
31. $\frac{1}{2}x + \frac{1}{3} = \frac{2}{3}$
32. $\frac{1}{2}y + \frac{1}{3} = \frac{1}{4}$
33. $0.3x + 4.2 = 3.6$
34. $3x - 0.02 = -0.548$
35. $5t + 6 = 2t + 15$
36. $\frac{x}{4} + 2x = -\frac{2}{3}$

37. $\dfrac{x-5}{4} = \dfrac{x-9}{3}$

38. $2r + 8 = 5 + 3r$

39. $\dfrac{x}{15} = 2 + \dfrac{x}{5}$

40. $12x - 1.2 = 3x + 1.5$

41. $4(x + 3) = 36 + 4x$

42. $6y + 3(4 + y) = 8$

43. $6(x + 1) = 4x + 2(x + 3)$

44. $\dfrac{x}{3} + 4 = \dfrac{2x}{5} - 6$

45. $0.34x - 0.15 = 0.23x + 0.41$

46. $\frac{2}{3}(x + 5) = \frac{1}{4}(x + 2)$

47. $2x + 3 - 5x = 11 + 3x - 7x - 5$

48. $3x + 4 - 11x = 14 + 7x - 5 + 6x$

49. $2(x + 3) = 4(x - 5) + 2$

50. $3.4x - 5.2(x + 3) = 6.2$

The water rate in Middlesex County is $1.55 per 1000 gal of water used.

51. What is the water bill if a resident uses 32,500 gal?

52. How many gallons of water can the customer use if the water bill is not to exceed $30.00?

The property tax rate in the town of Brighton is $25.299 per $1000 of assessed value.

53. What is the tax if the property is assessed at $70,000?

54. What can the maximum assessed value of a house be if the taxes are not to exceed $1500?

The cost of electricity in Granville, Ohio, is $0.1475 per kilowatt hour (kwh).

55. What is the monthly electric bill if a customer uses 1675 kwh?

56. Find the maximum amount of electricity that can be used if the total cost of electricity is not to exceed $110 per month.

The cost of gas for heating a house in Reading, Pennsylvania, is $1.62 per cubic foot.

57. What is the monthly gas bill if a customer uses 123 ft^3?

58. Find the maximum amount of gas that can be used if the total cost for gas is not to exceed $90 per month.

How much insulin (in cc) would be given for the following doses? (Refer to Example 16.)

59. 12 units of insulin from a vial marked U40

60. 35 units of insulin from a vial marked U40

61. a) In your own words summarize the procedure to use to solve an equation.
   b) Solve the equation $2(x + 3) = 4x + 3 - 5x$ using the procedure you outlined in part (a).

62. a) What is an identity?
   b) When solving an equation, how will you know if the equation is an identity?

63. a) What is an inconsistent equation?
   b) When solving an equation, how will you know if the equation is inconsistent?

Problem Solving

64. The pressure, P, in pounds per square inch (psi), exerted on an object x feet below the sea is given by the formula $P = 14.70 + 0.43x$. The 14.70 represents the weight in pounds of the column of air (from sea level to the top of the atmosphere) standing over a 1 in. by 1 in. square of seawater. The $0.43x$ represents the weight in pounds of a column of water 1 in. by 1 in. by x ft (see Fig. 6.1).

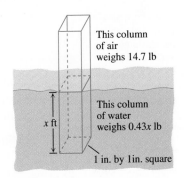

This column of air weighs 14.7 lb

This column of water weighs 0.43x lb

x ft

1 in. by 1 in. square

Figure 6.1

a) A submarine is built to withstand a pressure of 148 psi. How deep can that submarine go?

b) If the pressure gauge in the submarine registers a pressure of 128.65 psi, how deep is the submarine?

Research Activity

65. Do the necessary research and write a report on the historical development of algebra from its inception.

6.3 Formulas

A **formula** is an equation that typically has a real-life application. To **evaluate a formula**, substitute the given values for their respective variables, then evaluate

using the order of operations given in the box in Section 6.1. Many of the formulas given in this section will be discussed in greater detail in other parts of the book.

▶ **Example 1**

The simple interest formula,* interest = principal × rate × time, or $i = prt$, is used to find the interest you must pay on a simple interest loan when you borrow principal, p, at simple interest rate, r, in decimal form, for time, t. Erikka borrows $3000 at a simple interest rate of 9% for 3 years.

a) How much will Erikka pay in interest at the end of 3 years?

b) What is the total amount she will repay the bank at the end of 3 years?

Solution:

a) Substitute the values of p, r, and t into the formula, then evaluate.

$$i = prt$$
$$= 3000(0.09)(3)$$
$$= 810$$

Thus Erikka must pay $810 interest.

b) The total she must pay at the end of 3 years is the principal, $3000, plus the $810 interest, for a total of $3810.

▶ **Example 2**

Use the simple interest formula to find the rate on Ahmed's $6000 loan if the interest is $1080 after 2 years.

Solution: We substitute the appropriate values into the simple interest formula and solve for the desired quantity, r.

$$i = prt$$
$$1080 = (6000)r(2)$$
$$1080 = 12,000r$$
$$\frac{1080}{12,000} = \frac{12,000r}{12,000}$$
$$0.09 = r$$

Therefore the simple interest rate is 9% per year.

Many formulas contain Greek letters, such as μ (mu), σ (sigma), Σ (capital sigma), δ (delta), ϵ (epsilon), π (pi), θ (theta), and λ (lambda). Example 3 makes use of Greek letters.

---

* The simple interest formula will be discussed further in Section 10.2.

▶ **Example 3**

A formula used in the study of statistics to find the standard score (or Z-score) is

$$Z = \frac{\bar{x} - \mu}{\dfrac{\sigma}{\sqrt{n}}}.$$

Find the value of Z when $\bar{x}$ (read "x bar") = 120, μ = 100, σ = 16, and n = 4.

Solution: $Z = \dfrac{\bar{x} - \mu}{\dfrac{\sigma}{\sqrt{n}}} = \dfrac{120 - 100}{\dfrac{16}{\sqrt{4}}} = \dfrac{20}{\dfrac{16}{2}} = \dfrac{20}{8} = 2.5$

Some formulas contain **subscripts**. Subscripts are numbers (or letters) placed below and to the right of variables. They are used to help clarify a formula. For example, if there are two different amounts used in a problem, they may be symbolized as A and A_0, or A_1 and A_2. Subscripts are read using the word "sub"; for example, A_0 is read "A sub zero" and A_1 is read "A sub one."

Many real-life problems, including population growth, growth of bacteria, decay of radioactive substances, and certain other items that increase or decrease at a very rapid rate (exponentially), can be solved using exponential formulas. An **exponential equation** or **exponential formula** is of the form

$$y = a^x, \quad a > 0, \quad a \neq 1.$$

In an exponential formula letters other than x and y may be used to represent the variables. The following equations are examples of exponential formulas: $y = 2^x$, $A = \left(\frac{1}{2}\right)^x$, and $A = 2.3^t$. Notice in the exponential formula the variable is the exponent of some positive constant that is not equal to 1. In many real-life applications the variable t will be used to represent time. Problems involving exponential formulas can be evaluated much more easily if you use a calculator containing a $\boxed{y^x}$ or $\boxed{x^y}$ key.

The following formula, referred to as the **exponential growth** or **decay formula**, is used to solve many real-life problems.

$$P = P_0 a^{kt}, \quad a > 0, \quad a \neq 1$$

In the formula P_0 represents the original amount present, P represents the amount present after t years, and a and k are constants.

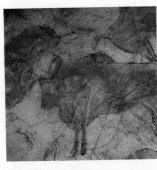

Scientists have estimated that the cave paintings at Lascaux, France, were painted over 15,500 years ago. To determine the age of the painting, they used a formula based on the rate of decay of radioactive carbon, carbon 14. They measured the amount of carbon 14 present and from that computed the age of the painting.

▶ **Example 4**

Carbon dating is used by scientists to find the age of fossils, bones, and other items. The formula used in carbon dating is

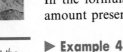

$$P = P_0 2^{-t/5600}$$

where P_0 represents the original amount of carbon 14 (C_{14}) present and P represents the amount of C_{14} present after t years. If 10 milligrams (mg) of C_{14} is present in an animal bone recently excavated, how many milligrams will be present in 5000 years?

Solution: Substituting the values in the formula gives

$$P = P_0 2^{-t/5600}$$
$$P = 10(2)^{-5000/5600}$$
$$P \approx 10(2)^{-0.89} \quad (\approx \text{ means "is approximately equal to"})$$
$$P \approx 10(0.54)$$
$$P \approx 5.4 \text{ mg}$$

Thus in 5000 years approximately 5.4 mg of the original 10 mg of C_{14} will remain.

In Example 4 we used a calculator to evaluate $(2)^{-5000/5600}$. The steps used to find this quantity on a calculator with a $\boxed{y^x}$ key are as follows:

$$2 \boxed{y^x} \boxed{(} 5000 \boxed{+/-} \boxed{\div} 5600 \boxed{)} \boxed{=} 0.538547$$

To evaluate $10(2)^{-5000/5600}$ on a scientific calculator we can press the following keys:

$$10 \boxed{\times} 2 \boxed{y^x} \boxed{(} 5000 \boxed{+/-} \boxed{\div} 5600 \boxed{)} \boxed{=} 5.38547$$

When the a in the formula $P = P_0 a^{kt}$ is replaced with the very special letter e, we get the **natural exponential formula**

$$P = P_0 e^{kt}.$$

The letter e represents an irrational number whose value is approximately 2.7183. The number e plays a very important role in mathematics and is used in finding the solution to many application problems.

▶ **Example 5**

Banks often credit compound interest on a continuous basis. When interest is compounded continuously, the principal amount in the account, P, at any time t can be calculated by the natural exponential formula $P = P_0 e^{kt}$, where P_0 is the initial principal invested, k is the interest rate in decimal form, and t is the time.

Suppose $10,000 is invested in a savings account at an 8% interest rate compounded continuously. What will be the balance (or principal) in the account in 5 years?

Solution: $P = P_0 e^{kt}$

$$= 10,000 e^{(0.08)5}$$
$$= 10,000 e^{0.40}$$
$$\approx 10,000(1.4918)$$
$$\approx 14,918$$

Thus after 5 years the account's value has grown from $10,000 to $14,918, an increase of $4918.

To evaluate $e^{(0.08)5}$ on a calculator (in Example 5) press:*

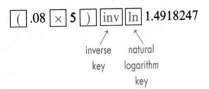

To evaluate $10{,}000e^{(0.08)5}$ on a calculator press the following keys:

10,000 $\boxed{\times}$ $\boxed{(}$ .08 $\boxed{\times}$ 5 $\boxed{)}$ $\boxed{\text{inv}}$ $\boxed{\ln}$ $\boxed{=}$ 14918.247

▶ **Example 6**

The world's population in 1991 was estimated to be 5.384 billion people. It is known that at present, the world's population continues to grow exponentially at the rate of about 1.8% per year. The world's expected population, in billions, in t years is given by the formula $P = 5.384e^{0.018t}$. Find the expected world population in the year 2091, 100 years after 1991.

Solution: $\quad P = 5.384e^{0.018t}$

$\qquad\quad = 5.384e^{0.018(100)}$

$\qquad\quad = 5.384e^{1.8}$

$\qquad\quad \approx 5.384(6.05)$

$\qquad\quad \approx 32.57$

Thus in the year 2091 the world's population is expected to be about 32.6 billion, about 6 times as great as it is today.

Solving for a Variable in a Formula or Equation

Often in mathematics and science courses, you are given a formula or an equation expressed in terms of one variable and asked to express it in terms of a different variable. For example you may be given the formula $P = i^2r$ and asked to solve the formula for r. To do this, treat each of the variables, except the one you are solving for, as if they were constants. Then solve for the variable desired, using the properties previously discussed. Examples 7–9 show how to do this.

When graphing equations in Section 6.6, you will sometimes have to solve the equation for the variable y as is done in Example 7.

▶ **Example 7**

Solve the equation $2x + 3y - 6 = 0$ for y.

Solution: Isolate the term containing the y by moving the constant, -6, and

---

* Keys to be pressed may vary on some calculators.

the term containing the x to the right-hand side of the equation.

$$2x + 3y - 6 = 0$$

$$2x + 3y - 6 + 6 = 0 + 6 \qquad \text{Addition property}$$

$$2x + 3y = 6$$

$$-2x + 2x + 3y = -2x + 6 \qquad \text{Subtraction property}$$

$$3y = -2x + 6$$

$$\frac{3y}{3} = \frac{-2x + 6}{3} \qquad \text{Divison property}$$

$$y = \frac{-2x + 6}{3}$$

$$y = \frac{-2x}{3} + \frac{6}{3}$$

$$y = -\frac{2}{3}x + 2$$

Note that once you have found $y = (-2x + 6)/3$, you have solved the equation for y. The solution can be expressed in the form $y = -\frac{2}{3}x + 2$. This form of the equation is convenient for graphing equations, as will be explained in Section 6.6. Example 7 can also be solved by moving the y term to the right side of the equal sign. Do this now and see that you obtain the same answer.

▶ **Example 8**

An important formula used in statistics is

$$z = \frac{x - \mu}{\sigma}.$$

Solve this formula for x.

Solution: To isolate the term x, use the general procedure for solving linear equations given in the box in Section 6.2. Treat each letter, except x, as if it were a constant.

$$z = \frac{x - \mu}{\sigma}$$

$$z \cdot \sigma = \frac{x - \mu}{\sigma} \cdot \sigma \qquad \text{Multiplication property}$$

$$z\sigma = x - \mu$$

$$z\sigma + \mu = x - \mu + \mu \qquad \text{Add } \mu \text{ to both sides of the equation}$$

$$z\sigma + \mu = x$$

$$\text{or} \quad x = z\sigma + \mu$$

▶ **Example 9**

A formula that may be important to you now or sometime in the future is the tax-free yield formula, $T_f = T_a(1 - F)$. This formula can be used to convert a taxable yield, T_a, into its equivalent tax-free yield, T_f, where F is the federal income tax bracket of the individual. A taxable yield is an interest rate where income tax is paid on the interest made. A tax-free yield is an interest rate where income tax does not have to be paid on the interest made.

a) For someone in a 28% income tax bracket, find the equivalent tax-free yield of an 8% taxable investment.

b) Solve this formula for T_a. That is, write a formula for taxable yield in terms of tax-free yield.

Solution:

a) $T_f = T_a(1 - F)$

$ = 0.08(1 - 0.28) = 0.08(0.72) = 0.0576$ or 5.76%

Did You

Know...

A SUPERIOR GENIUS

Because she was a woman, Sophie Germain (1776-1831) was denied admission to the *École Polytechnic*, the French academy of mathematics and science. Not to be stopped, she obtained lecture notes from courses in which she had an interest, including one taught by Joseph-Louis Lagrange. Under the pen name M. LeBlanc, she submitted a paper on analysis to Lagrange, who was so impressed with the report that he wished to meet the author and personally congratulate "him." When he found out that the author was a woman he became a great help and encouragement to her. Lagrange introduced Germain to many of the French scientists of the time.

In 1801 Germain wrote the great German mathematician Carl Friedrich Gauss to discuss Fermat's equation, $x^n + y^n = z^n$. He commended her for showing "the noblest courage, quite extraordinary talents and a superior genius." Germain's interests included work in number theory and mathematical physics. She would have received an honorary doctorate from the University of Gottingen based upon Gauss's recommendation, but died before the honorary doctorate could be awarded.

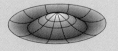

Germain was the first to devise a formula describing elastic motion. The study of the equations for the elasticity of different materials aided the development of acoustical diaphragms in loudspeakers and telephones.

Thus, a taxable investment of 8% is equivalent to a tax-free investment of 5.76% for a person in a 28% income tax bracket.

b) $\quad T_f = T_a(1 - F)$

$$\frac{T_f}{1 - F} = \frac{T_a(\cancel{1 - F})}{\cancel{1 - F}} \qquad \text{Divide both sides of the equation by } 1 - F.$$

$$\frac{T_f}{1 - F} = T_a \quad \text{or} \quad T_a = \frac{T_f}{1 - F}$$

Section 6.3 Exercises

Use the formula to find the value of the indicated variable for the values given. Use a calculator if its use is permitted.

1. $d = rt$; find d when $r = 60$ and $t = 4$ (physics)
2. $P = a + b + c$; find P when $a = 20$, $b = 30$, and $c = 40$ (geometry)
3. $P = 2l + 2w$; find P when $l = 6$ and $w = 10$ (geometry)
4. $E = IR$; find I when $E = 120$ and $R = 720$ (electronics)
5. $C = 2\pi r$; find r when $C = 25.12$ and $\pi = 3.14$ (geometry)
6. $p = i^2 r$; find r when $p = 6000$ and $i = 4$ (electronics)
7. $m = \dfrac{a + b}{2}$;   find b when $m = 40$ and $a = 35$ (statistics)
8. $z = \dfrac{x - \mu}{\sigma}$;   find z when $x = 100$, $\mu = 110$,   and $\sigma = 5$ (statistics)
9. $z = \dfrac{x - \mu}{\sigma}$;   find μ when $z = 3$, $x = 60$,   and $\sigma = 2$ (statistics)
10. $A = \frac{1}{2}h(b_1 + b_2)$;   find h when $A = 60$, $b_1 = 10$, and $b_2 = 30$ (geometry)
11. $T = \dfrac{PV}{k}$;   find P when $T = 80$, $V = 20$,   and $k = 0.5$ (physics)
12. $m = \dfrac{a + b + c}{3}$;   find a when $m = 60$, $b = 80$, and $c = 100$ (statistics)
13. $A = P(1 + rt)$;   find P when $A = 2360$, $r = 0.06$, and $t = 3$ (economics)

14. $K = \frac{1}{2}mv^2$;   find m when $K = 6000$ and $v = 20$ (physics)
15. $V = \pi r^2 h$;   find h when $V = 301.44$, $\pi = 3.14$,   and $r = 4$ (geometry)
16. $F = \frac{9}{5}C + 32$;   find F when $C = 30$ (temperature conversion)
17. $C = \frac{5}{9}(F - 32)$;   find C when $F = 41$ (temperature conversion)
18. $K = \dfrac{F - 32}{1.8} + 273.1$;   find K when $F = 100$ (chemistry)
19. $z = \dfrac{\bar{x} - \mu}{\dfrac{\sigma}{\sqrt{n}}}$;   find z when $\bar{x} = 66$, $\mu = 60$, $\sigma = 15$, and $n = 25$ (statistics)
20. $m = \dfrac{y_2 - y_1}{x_2 - x_1}$;   find m when $y_2 = 4$, $y_1 = -2$, $x_2 = -1$,   and $x_1 = -3$ (mathematics)
21. $C = a + by_D$;   find a when $C = 1100$, $b = 10$,   and $y_D = 60$ (economics)
22. $R_T = \dfrac{R_1 R_2}{R_1 + R_2}$;   find R_T when $R_1 = 100$ and $R_2 = 200$ (electronics)
23. $E = a_1 p_1 + a_2 p_2 + a_3 p_3$;   find E when $a_1 = 10$, $p_1 = 0.3$, $a_2 = 100$, $p_2 = 0.3$, $a_3 = 500$, $p_3 = 0.4$ (probability)
24. $x = \dfrac{-b + \sqrt{b^2 - 4ac}}{2a}$;   find x when $a = 2$, $b = -5$, and $c = -12$ (mathematics)

25. $x = \dfrac{-b - \sqrt{b^2 - 4ac}}{2a}$;   find x when $a = 2$, $b = -5$, and $c = -12$ (mathematics)

26. $R = 0 + (V - D)r$;   find O when $R = 670$, $V = 100$, $D = 10$,   and $r = 4$ (economics)

27. $S = \dfrac{n}{2}(f + l)$;   find n when $S = 52$, $f = 6$,   and $l = 20$ (mathematics)

28. $V = \sqrt{V_x^2 + V_y^2}$;   find V when $V_x = 3$ and $V_y = 4$ (physics)

29. $F = \dfrac{Gm_1m_2}{r^2}$;   find G if $F = 625$, $m_1 = 100$, $m_2 = 200$,   and $r = 4$ (physics)

30. $P = \dfrac{nRT}{V}$;   find V if $P = 12$, $n = 10$, $R = 60$,   and $T = 8$ (chemistry)

31. $S_n = \dfrac{a_1(1 - r^n)}{1 - r}$;   find S_n when $a_1 = 8$, $r = \frac{1}{2}$ and $n = 3$ (mathematics)

32. $A = P\left(1 + \dfrac{r}{n}\right)^{nt}$;   find A when $P = 100$, $r = 6\%$, $n = 1$,   and $t = 3$ (banking)

Solve for y in each of the following equations.

33. $3x + 2y = 5$

34. $2x - 3y = 6$

35. $4x + 7y = 14$

36. $-x + 2y = 16$

37. $2x - 3y + 6 = 0$

38. $3x + 4y = 0$

39. $-4x - 3y = 6$

40. $4x + 3y - 2z = 4$

41. $5x - 3z = 2 + 7y$

42. $2x - 3y + 5z = 0$

Solve for the variable indicated.

43. $d = rt$ for t

44. $E = IR$ for I

45. $p = a + b + c$ for c

46. $p = a + b + s_1 + s_2$ for a

47. $v = lwh$ for l

48. $p = irt$ for t

49. $C = \pi d$ for d

50. $PV = KT$ for T

51. $y = mx + b$ for b

52. $y = mx + b$ for x

53. $P = 2l + 2w$ for l

54. $A = \dfrac{m + d}{2}$ for m

55. $A = \dfrac{a + b + c}{3}$ for c

56. $A = \frac{1}{2}bh$ for b

57. $V = \frac{1}{3}Bh$ for h

58. $y = \dfrac{kX}{Z}$ for Z

59. $F = \frac{9}{5}C + 32$ for C

60. $C = \frac{5}{9}(F - 32)$ for F

61. $y_2 - y_1 = m(x_2 - x_1)$ for m

62. $a_n = a_1 + (n - 1)d$ for a_1

63. Ivan borrowed $3000 from a bank for a 5-year period. The bank is charging 10% simple interest per year for the period of the loan. Determine the interest Ivan will owe the bank at the end of the period.

64. Shakola deposited $10,000 in a savings account, paying simple interest, for 2 years. If the interest at the end of 2 years is $1200, find the rate.

65. Find the volume of a basketball if its diameter is 18 in. Use the formula $V = \frac{4}{3}\pi r^3$, where $\pi = 3.14$.

66. A sail on a sailboat is in the shape of a triangle. If the area of the sail is 160 ft^2 and the height of the sail is 20 ft, find the base of the sail. Use the formula $A = \frac{1}{2}bh$.

67. The number of gametes, g, in a certain species of plant is determined by the equation $g = 2^n$, where n is the number of cells in the species. Determine the number of gametes in a species that has eight cells.

68. The formula $A = P(1 + r)^n$ is the **compound interest formula**. When interest is compounded periodically (yearly, monthly, daily), the formula can be used to find the amount, A, in the account after n periods. In the formula, P is the principal, r is the interest rate per compounding period, and n is the number of compounding periods. Suppose Natasha invests $10,000 at 8% annual interest compounded quarterly for 6 years. Then $P = 10,000$, $r = \dfrac{8\%}{4} = 2\%$, and $n = 6 \cdot 4 = 24$. Find the amount in the account after 6 years.

69. If 20 mg of carbon 14 are originally present in an animal bone, how much will remain at the end of 500 years? (See Example 4.)

70. If a culture has 1000 bacteria initially, and the number of bacteria doubles each hour, the number of bacteria, N, after t hours can be found by the formula $N = 1000(2)^t$. Find the number of bacteria in the culture after 5 hours.

71. Assume that the value of the island of Manhattan has grown at an exponential rate of 8% per year since 1626 when Peter Minuit of the Dutch West India Company purchased the island for $24. Then, the value of the island, V, at any time, t, in years, can be found by the formula $V = 24e^{0.08t}$. What is the value of the island in 1992, 336 years after Minuit purchased the island?

72. Plutonium, a radioactive material used in most nuclear reactors, decays exponentially at a rate of 0.003% per year. The amount of plutonium, P, left after t years can be found by the formula $P = P_0e^{-0.00003t}$ where P_0 is the original amount of plutonium present. If there are originally 1000 grams of plutonium, find the amount of plutonium left after 100 years.

73. For a certain type of soft drink the percentage of a target market, P, that buys the soft drink after t days of advertising can be found by the equation $P = 1 - e^{-0.03t}$. What percentage of the target market buys the soft drink after 30 days of advertising?

Problem Solving

74. On Sunday, December 8, 1991, A. J. Kitt of Rochester, New York, was the first American to win the World Cup men's ski title since Bill Johnson won the title in 1984. Kitt skied the 2.1-mi course in 1 min 55.69 sec. Use the formula distance = rate · time to determine Kitt's average speed (or rate), in miles per hour, for the course.

Research Activity

75. The formula $R = LF\left(\dfrac{1 - C}{2}\right) + M$ may be used to describe an innocent person's reaction to a lie detector test. Write a report describing the variables in this formula. See "Formula for Fibbing" by Margie Cadidy in the March 1975 issue of *Psychology Today*.

6.4 Applications of Linear Equations in One Variable

One of the main reasons for studying algebra is that it can be used to solve everyday problems. In this section we will do two things: (1) show how to translate a written problem into a mathematical equation and (2) show how linear equations can be used in solving everyday problems. We begin by illustrating how English phrases can be written as mathematical expressions. When writing a mathematical expression any letter may be used to represent the variable. In our illustrations below we use the letter x.

Phrase	*Mathematical Expression*
Six more than a number	$x + 6$
A number increased by 3	$x + 3$
Four less than a number	$x - 4$
A number decreased by 9	$x - 9$
Twice a number	$2x$
Four times a number	$4x$

Sometimes the phrase that must be converted to a mathematical expression involves more than one operation, as follows.

Phrase	*Mathematical Expression*
Four less than 3 times a number	$3x - 4$
Ten more than twice a number	$2x + 10$
The sum of 5 times a number and 3	$5x + 3$
Eight times a number decreased by 7	$8x - 7$

The word "is" often represents the equal sign.

Phrase	*Mathematical Equation*
Six more than a number is 10.	$x + 6 = 10$
Five less than a number is 20.	$x - 5 = 20$
Twice a number decreased by 6 is 12.	$2x - 6 = 12$
A number decreased by 13 is 6 times the number.	$x - 13 = 6x$

Below is a general procedure for solving word problems.

To Solve a Word Problem

1. Read the problem carefully at least twice to make sure that you understand it.
2. If possible, draw a sketch to help visualize the problem.
3. Determine which quantity you are being asked to find. Choose a letter to represent this unknown quantity. Write down exactly what this letter represents.
4. Write the word problem as an equation.
5. Solve the equation for the unknown quantity.
6. Answer the question or questions asked.
7. Check the solution.

This general procedure for solving word problems is illustrated in the following examples.

▶ **Example 1**

A car rental agency charges $210 per week plus $0.18 per mile. How far can you travel, to the nearest mile, on a maximum budget of $400?

Solution: In this problem the unknown quantity is the number of miles you can travel. Let us select m to represent the number of miles you can travel. Then we construct an equation using the given information that will allow us to solve for m.

Let $\quad m =$ number of miles you can travel
then $\quad \$0.18m =$ amount spent for m miles traveled at 18¢ per mile

$$\text{rental fee} + \text{mileage charge} = \text{total amount spent}$$
$$\$210 \quad + \quad \$0.18m \quad = \quad \$400$$

Now solve the equation.

$$210 + 0.18m = 400$$
$$210 - 210 + 0.18m = 400 - 210$$
$$0.18m = 190$$
$$\frac{0.18m}{0.18} = \frac{190}{0.18}$$
$$m = 1055.\overline{5}$$
$$m = 1055$$

Therefore you can travel 1055 mi. (If you go 1056 mi, you will have to spend more than $400.)

Check: The check is made with the information given in the original problem.

$$\text{Total amount spent} = \text{rental fee} + \text{mileage charge}$$
$$= 210 + 0.18(1055)$$
$$= 210 + 189.90$$
$$= 399.90$$

The check shows that the solution is correct.

▶ **Example 2**

Forty hours of overtime must be split among three workers. One worker will be assigned twice the number of hours as each of the other two. How many hours of overtime will be assigned to each worker?

Solution: Two workers receive the same amount of overtime, and the third worker receives twice that amount.

Let x = number of hours of overtime for the first worker
       x = number of hours of overtime for the second worker
      $2x$ = number of hours of overtime for the third worker

$$x + x + 2x = \text{total amount of overtime}$$
$$x + x + 2x = 40$$
$$4x = 40$$
$$x = 10$$

Thus two workers are assigned 10 hours, and the third worker is assigned 2(10), or 20, hours of overtime. A check in the original problem will verify that the answer given is correct.

▶ **Example 3**

Mrs. Chang wants to build a sandbox for her daughter. She has only 26 feet of wood to build the perimeter of the sandbox. What should be the dimensions of the sandbox if she wants the length to be 3 feet greater than the width?

Solution: The formula for finding the perimeter of a rectangle is $P = 2l + 2w$, where P is the perimeter, l is the length, and w is the width. A simple diagram such as Fig. 6.2 is often helpful in solving problems of this type.

Let w equal the width of the sandbox. Since the length is 3 feet more than the width, $l = w + 3$. The total distance around the sandbox, P, is 26 feet.

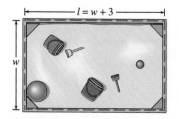

Figure 6.2

$l = w + 3$

w

Substitute the known quantities in the formula:

$$P = 2w + 2l$$
$$26 = 2w + 2(w + 3)$$
$$26 = 2w + 2w + 6$$
$$26 = 4w + 6$$
$$20 = 4w$$
$$5 = w$$

The width of the sandbox is 5 ft, and the length of the sandbox is $5 + 3$ or 8 ft.

In shopping and other daily activities we are occasionally asked to solve problems using percents. The word *percent* means "per hundred." Thus, for example, 7% means 7 per hundred, or $\frac{7}{100}$. When $\frac{7}{100}$ is converted to a decimal number, we obtain 0.07. Thus $7\% = 0.07$.

Let us look at one example involving percent. (See Section 10.1 for a more detailed discussion of percent.)

▶ **Example 4**

Mr. Dobson is planning to open a hot dog stand in New York City. What should be the price of the hot dog before tax if the total cost of the hot dog, including a 7% sales tax, is to be $1.25?

Solution: We are asked to find the cost of the hot dog before sales tax.

Let $\qquad\qquad x =$ the cost of the hot dog before sales tax.
Then $\qquad 0.07x = 7\%$ of the cost of the hot dog (the sales tax).

$$\text{Cost of hot dog before tax} + \text{tax on hot dog} = 1.25$$
$$x + 0.07x = 1.25$$
$$1.07x = 1.25$$
$$\frac{1.07x}{1.07} = \frac{1.25}{1.07}$$
$$x = \frac{1.25}{1.07}$$
$$x = 1.168 \quad \text{or} \quad 1.17$$

Thus the cost of the hot dog before tax is $1.17 to the nearest cent.

Section 6.4 Exercises

Write the following as mathematical expressions.
1. 7 decreased by 3 times x

2. $4r$ increased by 16
3. 8 more than 6 times y

4. 10 decreased by twice *n*

5. 3 times *r* decreased by 10

6. 5 more than *t*

7. 5 decreased by *x*, times 2

8. 5 decreased by two times *x*

9. The sum of 6 and *x*, divided by 7

10. The sum of 6, and *x* divided by 7

11. 5 less than the quantity 7 times *y*, increased by 11

12. *w* more than 11 times *z*, decreased by twice *w*

Write an equation and solve.

13. The sum of a number and 8 is 24.

14. A number multiplied by 4 is 36.

15. A number divided by 7 is 5.

16. A number decreased by 6 is 22.

17. Eleven increased by twice a number is 9.

18. Seven less than 4 times a number is 17.

19. Four more than 3 times a number is 16.

20. Twice a number decreased by 11 is the same as 4 more than the number.

21. A number increased by 11 is the same as 1 more than 3 times the number.

22. A number divided by 3 is the same as 4 less than the number.

23. Three more than a number is 5 times the sum of the number and 7.

24. The product of 3 and a number, decreased by 4, is 4 times the number.

Set up an algebraic equation that can be used to solve each problem. Solve the equation and find the desired value(s).

25. The village of Montgomery, which has a population of 6200, is growing by 500 each year. In how many years will the population reach 12,200?

26. Budget Warehouse has a plan whereby for a yearly fee of $80 you save 8% of the price of all items purchased in the store. What is the total Vito will need to spend during the year for his savings to equal the yearly fee?

27. Delta airlines wishes to keep its airfare between San Francisco and Cleveland at exactly $380 including a 6% sales tax. Find the cost of the ticket before tax.

28. It costs Paula 8 cents to make a copy of a page at a copy shop. She is considering purchasing a photocopy machine that is on sale for $250 including tax. How many copies would Paula have to make in the copy shop for her cost to equal the purchase price of the machine she is considering buying?

29. Eighty-four hours of overtime must be split among three workers. The two younger workers are assigned the same number of hours. The third worker is assigned twice as much time as each of the younger workers. How much overtime is assigned to each worker?

30. Singe budgeted $100 for renting a car. How far can he travel in one day if the charges are $60 a day plus 20¢ per mile?

31. The Connells would like to build a rectangular sandbox for their children. They have a total of 20 ft of wood for the perimeter of the sandbox. What will be the dimensions of the box if the width of the box is to be two-thirds of its length?

32. Mr. Umoja worked a 55-hr week last week. He is paid $1\frac{1}{2}$ times his regular hourly rate for all hours over a 40-hr week. His pay last week was $500. What is his hourly rate?

33. In a day, machine *A* produces twice as many bowls as machine *B*. Machine *C* produces 50 more bowls than machine *B*. If the total production is 6170 bowls, how many bowls does each produce?

34. Stock *A* is selling at $140 per share. Stock *B* is selling at $37 per share. An investor has a total of $10,000 to invest. She wishes to purchase 6 times as many shares of Stock *A* as of Stock *B*.

   a) How many shares of each stock will she purchase?

   b) How much money will be left over? *Note:* Stock may be purchased only in whole shares.

35. At a 20% off sale, a box of floppy diskettes costs $18. Find the regular price of the floppy diskettes.

36. The Peretskys purchased a car. If the total cost including a 5% sales tax was $12,836.25, find the cost of the car before tax.

37. The Goullets will need 186 ft of fence to enclose their rectangular yard. If the length of the yard is 9 ft more than the width what are the dimensions of the yard?

38. A bookcase with 3 shelves is to be built by a woodworking student. If the height of the bookcase is to be 2 ft longer than the width of a shelf, and the total amount of wood to be used is 32 ft, find the dimensions of the bookcase.

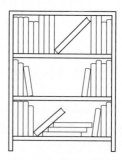

39. The cost of doing the family laundry for a month at a local laundromat is $65. A new washer and dryer cost a total of

$760. How many months would it take for the cost of doing the laundry at the laundromat to equal the cost of a new washer and dryer?

40. Telephones are selling for $34.95. Sandy's charge for renting her telephone from the telephone company is $4.95 per month. If Sandy purchases her phone, how long will it take her to recover the cost of the phone?

41. Jim is about to join a racquet club. He has a choice of two plans. Plan *A* has a flat monthly fee of $22 and no charge for court time. Plan *B* has no monthly fee, but he must pay $6 per hour for court time. How many hours would Jim have to play to make Plan *A* the less expensive plan?

42. Ms. Saplin is being hired as a salesperson. She is given a choice of two salary plans. Plan *A* is $100 a week plus 5% commission on sales. Plan *B* is a straight 15% commission on sales. How many dollars in sales are necessary per week for Plan *B* to result in the same salary as Plan *A*?

43. Florence has been told that with her half-off airfare coupon her airfare from New York to San Diego will be $227.00. The $227.00 includes a 7% tax *on the regular fare*. On the way to the airport, Florence realizes that she has lost her coupon. What will her regular fare be, including tax?

44. The cost of renting a compact car at the Hertz car rental agency is $35 per day plus 8 cents a mile. The cost of renting the same car at the Avis car rental agency is $25 per day plus 12 cents a mile. How far would you have to drive in one day to make the cost of renting from Hertz equal the cost of renting from Avis?

*45. The cost of a telephone call from Denver to St. Louis is 75 cents for the first minute and 50 cents for each additional half minute or any part thereof. If Meg lives in St. Louis and her friend lives in Denver, how long can Meg talk on the telephone if she has only $6.25?

Problem Solving

46. Some states allow a husband and wife to file individual tax returns (on a single form) even though they have filed a joint federal tax return. It is usually to the taxpayers' advantage to do so when both husband and wife work. The smallest amount of tax owed (or the largest refund) will occur when the husband's and wife's taxable incomes are the same.

Mr. Ranieri's 1992 taxable income was $24,200, and Mrs. Ranieri's taxable income for that year was $26,400. The Ranieris' total tax deduction for the year was $3640. This deduction can be divided between Mr. and Mrs. Ranieri any way they wish. How should the $3640 be divided between them to result in each individual's having the same taxable income and therefore the greatest tax refund?

6.5 Linear Inequalities

The first sections of this chapter have dealt with equations. There are many occasions when we may encounter statements of inequality. The symbols of inequality are given here.

Symbols of Inequality

$a < b$ means a is less than b.
$a \leq b$ means a is less than or equal to b.
$a > b$ means a is greater than b.
$a \geq b$ means a is greater than or equal to b.

An **inequality** consists of two (or more) expressions joined by an inequality sign.

Examples of Inequalities

$$3 < 5, \qquad x < 2, \qquad 3x - 2 \geq 5$$

A statement of inequality can be used to indicate a set of real numbers. For example, $x < 2$ represents the set of all real numbers less than 2. It is impossible to list all these numbers. Some numbers less than 2 are $-\frac{1}{2}$, -1, 0, -2, $\frac{97}{163}$, -1.234.

A method of picturing all real numbers less than 2 is to graph the solution on a number line. The number line was discussed in Chapter 5.

To indicate the solution set of $x < 2$ on the number line, we draw an open circle at 2 and a line to the left of 2 with an arrow at its end. This technique indicates that all the points to the left of 2 are part of the solution. The open circle indicates that the solution does not include the number 2.

▶ **Example 1**

Graph the solution set of $x \geq -1$, where x is a real number, on the number line.

Solution: The numbers greater than or equal to -1 are all the points on the number line to the right of -1 and -1 itself. The solid dot at -1 in the accompanying figure shows that -1 is included in the solution.

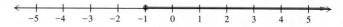

The inequality statements $x < 2$ and $2 > x$ have the same meaning. Note that the inequality symbol points to the x in both cases. Thus one may be written in place of the other. Likewise, $x > 2$ and $2 < x$ have the same meaning. Note that the inequality symbol points to the 2 in both cases. We make use of this fact in the following example.

▶ **Example 2**

Graph $-4 \leq x$, where x is a real number, on the number line.

Solution: We can restate $-4 \leq x$ as $x \geq -4$. Both statements have identical solutions. Any number that is greater than or equal to -4 satisfies the inequality $x \geq -4$. Thus the graph includes -4 and all the points to the right of -4 on the number line.

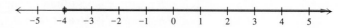

We can find the solution to an inequality by adding, subtracting, multiplying, or dividing both sides of the inequality by the same number or expression. We use the procedure discussed in Section 6.2 to isolate the variable, with one important exception: **When both sides of an inequality are multiplied or divided by a negative number, the order (or sense) of the inequality is reversed.**

▶ **Example 3**

Solve the inequality $-x > 5$ and graph the solution set on the number line.

Solution: To solve this inequality, we must eliminate the negative sign in front of the x. To do this, we can multiply both sides of the inequality by -1 and change the order of the inequality:

$$-x > 5$$
$$-1(-x) < -1(5) \quad \text{Multiply both sides of the inequality by } -1 \text{ and}$$
$$\text{change the order of the inequality.}$$
$$x < -5$$

The solution set is graphed on the number line as follows:

In the production of machine parts, engineers must allow a certain tolerance in the way parts fit together. For example, the boring machines that grind cylindrical openings in an automobile's engine block must create a cylinder that allows the piston to move freely up and down, but still fits tightly enough to ensure that compression and combustion are complete. The allowable tolerance between parts can be expressed as an inequality: For example, the diameter of a cylinder may need to be no less than 3.383 inches and no greater than 3.387 inches.

▶ **Example 4**

Solve the inequality $-2x \geq 6$ and graph the solution set on the number line.

Solution: Solving this inequality requires making the coefficient of the x term 1. To do this, divide both sides of the inequality by -2 and change the order of the inequality:

$$-2x \geq 6$$
$$\frac{-2x}{-2} \leq \frac{6}{-2} \quad \text{Divide both sides of the inequality by } -2 \text{ and change}$$
$$\text{the order of the inequality.}$$
$$x \leq -3$$

The solution set is graphed on the number line as follows:

▶ **Example 5**

Solve the inequality $2x - 4 < 6$ and graph the solution set on the number line.

Solution: To find the solution, isolate x on one side of the inequality symbol:

$$2x - 4 < 6$$
$$2x - 4 + 4 < 6 + 4 \quad \text{Add 4 to both sides of the inequality.}$$
$$2x < 10$$
$$\frac{2x}{2} < \frac{10}{2} \quad \text{Divide both sides of the inequality by 2.}$$
$$x < 5$$

Thus the solution to $2x - 4 < 6$ is all real numbers less than 5.

Notice that in Example 5 the order of the inequality did not change when both sides of the inequality were divided by the positive number 2.

▶ **Example 6**

Solve the inequality $x + 4 < 7$, where x is an integer, and graph the solution set on the number line.

Solution:
$$x + 4 < 7$$
$$x + 4 - 4 < 7 - 4$$
$$x < 3$$

Since x is an integer, and x is less than 3, the solution set is the set of integers less than 3, or $\{\ldots -3, -2, -1, 0, 1, 2\}$. To graph the solution, we make solid dots at the corresponding points on the number line. The three smaller dots to the left of -3 indicate that all the integers to the left of -3 are included.

An inequality of the form $a < x < b$ is called a **compound inequality**. Consider the compound inequality $-3 < x \leq 2$. This means that $-3 < x$ *and* $x \leq 2$.

▶ **Example 7**

Graph the solution set of the inequality $-3 < x \leq 2$,
a) where x is an integer,
b) where x is a real number.

Solution:

a) The solution set is all of the integers between -3 and 2, including the 2 but not including the -3, or $\{-2, -1, 0, 1, 2\}$.

b) The solution set consists of all the real numbers between -3 and 2, including the 2 but not including the -3.

▶ Example 8

Solve the compound inequality for x and graph the solution set.

$$-4 < \frac{x+3}{2} \le 5$$

Solution: To solve a compound inequality, we must isolate the x as the middle term. To do this, we use the same principles used to solve inequalities:

$$-4 < \frac{x+3}{2} \le 5$$

$$2(-4) < 2\left(\frac{x+3}{2}\right) \le 2(5) \qquad \text{Multiply the three terms by 2.}$$

$$-8 < x + 3 \le 10$$
$$-8 - 3 < x + 3 - 3 \le 10 - 3 \qquad \text{Subtract 3 from all three terms.}$$
$$-11 < x \le 7$$

The solution set is graphed on the number line as follows:

▶ Example 9

A student must have an average (the mean) on five tests that is greater than or equal to 80% but less than 90% to receive a final grade of B. Darrell's grades on the first four tests were 98%, 76%, 86%, and 92%. What range of grades on the fifth test would give him a B in the course?

Solution: The unknown quantity is the range of grades on the fifth test. First construct an inequality that can be used to find the range of grades on the fifth exam. The average (mean) is found by adding the grades and dividing the sum by the number of exams.

Let $x =$ the fifth grade:

$$\text{Average} = \frac{98 + 76 + 86 + 92 + x}{5}$$

For Darrell to obtain a B, his average must be greater than or equal to 80 but less than 90:

$$80 \le \frac{98 + 76 + 86 + 92 + x}{5} < 90$$

$$80 \le \frac{352 + x}{5} < 90$$

$$5(80) \le 5\left(\frac{352 + x}{5}\right) < 5(90) \qquad \text{Multiply the three terms of the inequality by 5.}$$

$$400 \leq 352 + x < 450$$
$$400 - 352 \leq 352 - 352 + x < 450 - 352 \qquad \text{Subtract 352 from all three terms.}$$
$$48 \leq x < 98$$

"In mathematics the art of posing problems is easier than that of solving them."

George Cantor

Thus a grade of 48% up to but not including a grade of 98% on the fifth test will result in a grade of B.

Section 6.5 Exercises

1. When solving an inequality, under what conditions is it necessary to change the order of the inequality symbol?
2. Does $x > -3$ have the same meaning as $-3 < x$? Explain.
3. Does $x < 2$ have the same meaning as $2 > x$? Explain.
4. When graphing the solution to an inequality on the number line, when should you use an open circle and when should you use a closed circle?

Graph the solution set of the inequality, where x is a real number, on the real number line.

5. $x < 5$
6. $x \geq -2$
7. $x + 3 < 7$
8. $4x \geq 2$
9. $-5x \leq 15$
10. $-2x < 6$
11. $\dfrac{x}{6} < -2$
12. $\dfrac{x}{2} > 4$
13. $\dfrac{x}{3} \geq -2$
14. $\dfrac{-x}{5} \geq 2$
15. $-3x - 2 \geq 13$
16. $3x + 5 < -7 + 3x$
17. $2(x + 6) \leq 15$
18. $-4(x + 2) - 2x > -6x + 2$
19. $3(x + 4) - 2 < 3x + 10$
20. $-2 \leq x \leq 1$
21. $-1 < x + 3 < 8$
22. $\dfrac{1}{2} < \dfrac{x + 4}{2} \leq 4$

Graph the solution set of the inequality, where x is an integer, on the number line.

23. $x \leq 0$
24. $-2 > x$
25. $4x \geq 16$
26. $-4x \leq 16$
27. $-3x < 24$
28. $4 - x > 5$
29. $\dfrac{x}{6} < -2$
30. $\dfrac{x}{3} > -3$

31. $-\dfrac{x}{6} \geq 3$
32. $\dfrac{2x}{3} \leq 4$
33. $-13 < -7y + 1$
34. $3x + 6 < -2 + 11x$
35. $2(x + 8) \leq 3x + 17$
36. $5(x + 3) < -(6 + x) + 4$
37. $3(x + 4) - 2 \leq 3x + 10$
38. $-2 \leq x \leq 10$
39. $1 > -x > -5$
40. $-2 < 2x + 3 < 6$
41. $5 < 3x - 4 \leq 10$
42. $-3 < \dfrac{x - 3}{2} \leq -2$

43. In Example 9, what range of grades on the fifth test would result in Darrell receiving a grade of B if his grades on the first four tests were 78%, 64%, 88%, and 76%?
44. To obtain a C, a student must have an average (mean) greater than or equal to 70% but less than 80%. Cora's first four test grades are 82%, 68%, 70%, and 85%. What range of grades on Cora's fifth test would result in her having a C average?
45. Two employees are told that the sum of their hours of overtime must be less than 21. Overtime will not be paid for any fractional part of an hour. Employee A is assigned three times as many hours of overtime as Employee B. For how many hours of overtime will each employee be paid? Write a statement of inequality and solve it.
46. For a small plane to take off safely, it must have a maximum load of no more than 1500 lb. The gasoline weighs 400 lb and the passengers weigh a total of 670 lb.
    a) Write an inequality that can be used to determine the maximum weight of the luggage that the plane can safely carry.
    b) Determine the maximum weight of the luggage.
47. The janitor must move a large shipment of books from the first floor to the fifth floor. Each box of books weighs

60 lb, and the janitor weighs 180 lb. The sign on the elevator reads "Maximum weight 1200 lb."

a) Write a statement of inequality to determine the maximum number of boxes of books the janitor can place on the elevator at one time. (The janitor must ride in the elevator with the books.)

b) Determine the maximum number of boxes that can be moved in one trip.

48. Mr. Green wants to invest $10,000, part at 5% and part at 7% simple interest. What is the least he can invest at 7% to earn a minimum return of $616.00?

49. After Mrs. Franklin is seated in a restaurant, she realizes that she has only $19.00. If she must pay 7% tax and wishes to leave a 15% tip, what is the price range of meals she can order?

50. For a business to realize a profit, its revenue, R, must be greater than its costs, C; that is, a profit will result only if $R > C$ (the company breaks even when $R = C$). A record manufacturer has a weekly cost equation $C = 300 + 1.5x$ and a weekly revenue equation $R = 2x$, where x is the number of records produced and sold in a week. How many records must be sold weekly for the company to make a profit?

Problem Solving

51. Teresa's five test grades for the semester are 86%, 74%, 68%, 96%, and 72%. Her final exam counts one-third of her final grade. What range of grades on her final exam would result in Teresa receiving a final grade of **B** in the course? (See Example 9.)

Graphing Linear Equations

In the previous section we solved equations with a single variable. However real-world problems often involve two or more unknowns. For example the profit, p, of a company may depend on the sales, s; or the cost, c, of mailing a package may depend on the weight, w, of the package. Thus it is helpful to be able to work with equations with two variables (for example, $x + 2y = 6$). Doing so requires understanding the **rectangular coordinate system**, also called the **Cartesian coordinate system**, named after the French mathematician René Descartes (1596–1650).

The rectangular coordinate system consists of two perpendicular number lines (Fig. 6.3). The horizontal line is the **x-axis** and the vertical line is the **y-axis**. The point of intersection of the x-axis and y-axis is called the **origin**. Note that the numbers on the axes to the right and above the origin are positive. The numbers on the axes on the left and below the origin are negative. The axes divide the plane into four parts: the first, second, third, and fourth **quadrants**.

We indicate the location of a point in the rectangular coordinate system

Figure 6.3

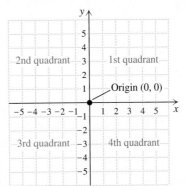

by means of an **ordered pair** of the form (x, y). The x-coordinate is always placed first and the y-coordinate is always placed second in the ordered pair. Consider the point illustrated in Fig. 6.4. Since the x-coordinate of the point is 5 and the y-coordinate is 3, the ordered pair that represents this point is (5, 3).

Figure 6.4

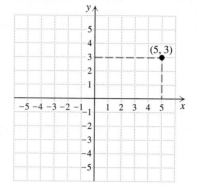

The origin is represented by the ordered pair (0, 0). Every point on the plane can be represented by one and only one ordered pair (x, y), and every ordered pair (x, y) represents one and only one point on the plane.

▶ **Example 1**

Plot the points $A(-2, 4)$, $B(3, -4)$, $C(6, 0)$, $D(4, 1)$,   and $E(0, 3)$.

Solution: Point A has an x-coordinate of -2 and a y-coordinate of 4. Project a vertical line up from -2 on the x-axis and a horizontal line from 4 on the y-axis. The two lines intersect at the point denoted by A (Fig. 6.5). The other points are plotted in a similar manner.

Figure 6.5

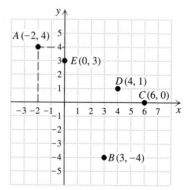

▶ **Example 2**

The points A, B, and C are three vertices of a parallelogram with a side parallel to the x-axis. Plot the three points and find the coordinates of the fourth vertex, D.

$$A(1, 2) \qquad B(2, 4) \qquad C(7, 4)$$

Solution: A parallelogram is a figure that has opposite sides that are of equal length and are parallel. (Parallel lines are two lines in the same plane that do not intersect.) The horizontal distance between points B and C is 5 units (see Fig. 6.6). Therefore the horizontal distance between points A and D must also be 5 units. This problem has two possible solutions, as illustrated in Fig. 6.6(a) and 6.6(b).

Figure 6.6

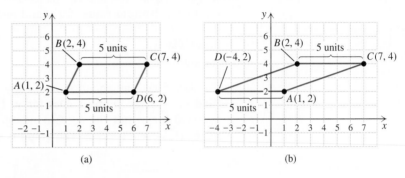

(a)   (b)

The solutions are the points $(6, 2)$ and $(-4, 2)$.

Graphing Linear Equations by Plotting Points

Consider the following equation in two variables: $y = x + 1$. Every ordered pair that makes the equation a true statement is a solution to, or satisfies, the equation. We can mentally find some ordered pairs that satisfy the equation $y = x + 1$ by picking some values of x and solving the equation for y. For example, suppose we let $x = 1$, then $y = 1 + 1 = 2$. The ordered pair $(1, 2)$ is a solution to the equation $y = x + 1$. We can make a chart of other ordered pairs that are solutions to the equation.

x	y	Ordered Pair
1	2	(1, 2)
2	3	(2, 3)
3	4	(3, 4)
4.5	5.5	(4.5, 5.5)
−3	−2	(−3, −2)

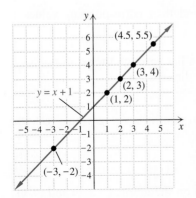

Figure 6.7

How many other ordered pairs satisfy the equation? There are infinitely many ordered pairs that satisfy the equation. Since we cannot list all of the solutions, we illustrate them by means of a graph. A **graph** is an illustration of all of the points whose coordinates satisfy an equation.

The points $(1, 2)$, $(2, 3)$, $(3, 4)$, $(4.5, 5.5)$, and $(-3, -2)$ are plotted in Fig. 6.7. With a straightedge we can draw one line that contains all these points. This line, when extended indefinitely in either direction, passes through all the points in the plane that satisfy the equation $y = x + 1$. The arrows on the ends of the line indicate that the line extends indefinitely.

All equations of the form $ax + by = c$, $a \neq 0$, $b \neq 0$, will be straight lines when graphed. Thus such equations are called **linear equations in two variables**. Since only two points are needed to draw a line, only two points are needed to graph a linear equation. It is always a good idea to graph a third point as a checkpoint. If no error has been made, all three points will be in a line, or **collinear**. One method that can be used to obtain points is to solve the equation for y, then substitute values for x and find the corresponding values of y.

▶ **Example 3**

Graph $y = 3x + 6$.

Solution: Since the equation is already solved for y, select values for x and find the corresponding values for y. The chart indicates values arbitrarily selected for x and the corresponding values for y. The ordered pairs are $(0, 6)$, $(1, 9)$, and $(-2, 0)$. The graph is shown in Fig. 6.8.

x	y
0	6
1	9
-2	0

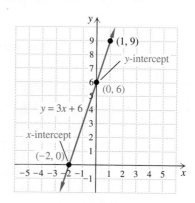

Figure 6.8

To Graph Equations by Plotting Points

1. Solve the equation for y.
2. Select at least three values for x and find their corresponding values of y.
3. Plot the points.
4. If the points are in a straight line, draw a line through the set of points and place arrow tips at both ends of the line.

In step 4 of the procedure, if the points are not in a straight line, recheck your calculations and find your error.

Graphing by Using Intercepts

In Example 3 take note of two special points on the graph, $(-2, 0)$ and $(0, 6)$. It is at these points that the line crosses the x-axis and the y-axis, respectively.

The ordered pairs $(-2, 0)$ and $(0, 6)$ represent the **x-intercept** and the **y-intercept**, respectively. Another method that can be used to graph linear equations is to find the x- and y-intercepts of the graph.

Finding the x- and y- Intercepts

To find the y-intercept, set $x = 0$ and solve the equation for y.
To find the x-intercept, set $y = 0$ and solve the equation for x.

An equation may be graphed by finding the x- and y-intercepts, then plotting the intercepts and drawing a straight line through the intercepts. When graphing by this method, it is always a good idea to plot a checkpoint before drawing your graph. To obtain a checkpoint, select a nonzero value for x and find the corresponding value of y. The checkpoint should be collinear with the x- and y-intercepts.

▶ **Example 4**

Graph $2x + 3y = 6$ using the x- and y-intercepts.

Solution: To find the x-intercept, set $y = 0$ and solve for x:

$$2x + 3(0) = 6$$
$$2x = 6$$
$$x = 3$$

The x-intercept is $(3, 0)$. To find the y-intercept, set $x = 0$ and solve for y:

$$2(0) + 3y = 6$$
$$3y = 6$$
$$y = 2$$

The y-intercept is $(0, 2)$. As a checkpoint try $x = 1$ and find the corresponding value for y:

$$2x + 3y = 6$$
$$2(1) + 3y = 6$$
$$2 + 3y = 6$$
$$3y = 4$$
$$y = \tfrac{4}{3}$$

The checkpoint is the ordered pair $(1, \tfrac{4}{3})$ or $(1, 1\tfrac{1}{3})$.

Since all three points are collinear in Fig. 6.9, draw a line through the three points to obtain the graph.

Grids have long been used in mapping. In archaeological digs, a rectangular coordinate system is used to chart the location of each find.

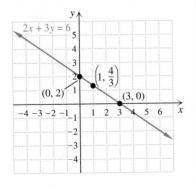

Figure 6.9

Slope

Another useful concept when working with straight lines is slope, which is a measure of the "steepness" of a line. The **slope of a line** is a ratio of the vertical change to the horizontal change for any two points on the line. Consider

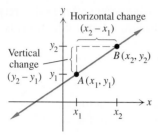

Figure 6.10

Fig. 6.10. Point A has coordinates (x_1, y_1), and point B has coordinates (x_2, y_2). We see that the vertical change between points A and B is $y_2 - y_1$, and the horizontal change between points A and B is $x_2 - x_1$. Thus the slope, which is often symbolized with the letter m, can be found as follows:

$$\textbf{Slope} = \frac{\text{vertical change}}{\text{horizontal change}}$$

$$m = \frac{y_2 - y_1}{x_2 - x_1}$$

The Greek letter delta, Δ, is used to represent the words "the change in." Therefore slope may be defined as

$$m = \frac{\Delta y}{\Delta x}$$

A line may have a positive slope, or a negative slope, or zero slope, as indicated in Fig. 6.11. A line with a positive slope rises from left to right (Fig. 6.11a). A line with a negative slope falls from left to right (Fig. 6.11b). A horizontal line, which neither rises nor falls, has a slope of zero (Fig. 6.11c). Since a vertical line does not have any horizontal change (the x value remains constant), and since we cannot divide by 0, the slope of a vertical line is undefined (Fig. 6.11d).

Figure 6.11

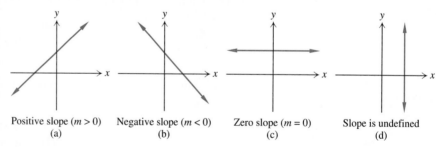

| Positive slope ($m > 0$) | Negative slope ($m < 0$) | Zero slope ($m = 0$) | Slope is undefined |
| (a) | (b) | (c) | (d) |

Although we may not think much about it, the slope of a line is something we are altogether familiar with. You confront it every time you run up the stairs, late for class, moving 8 inches horizontally for every 6 inches up. The 1992 Olympic gold medalist Toni Nieminen is familiar with the concept of slope: He speeds down 120 meters of a ski ramp at speeds of over 60 miles per hour before he takes flight.

▶ **Example 5**

Determine the slope of the line that passes through the points $(-1, -3)$ and $(1, 5)$.

Solution: Let us begin by drawing a sketch, illustrating the points and the line (Fig. 6.12a).

We will let (x_1, y_1) be $(-1, -3)$ and (x_2, y_2) be $(1, 5)$.

$$\text{Slope} = \frac{y_2 - y_1}{x_2 - x_1} = \frac{5 - (-3)}{1 - (-1)} = \frac{5 + 3}{1 + 1} = \frac{8}{2} = \frac{4}{1} \quad \text{or} \quad 4$$

Thus the slope is 4. This means that there is a vertical change of 4 units for each horizontal change of 1 unit (see Fig. 6.12b). The slope is positive, and the line rises from left to right. Note that we would have obtained the same results if we let (x_1, y_1) be $(1, 5)$ and (x_2, y_2) be $(-1, -3)$. Try this now and see.

Figure 6.12

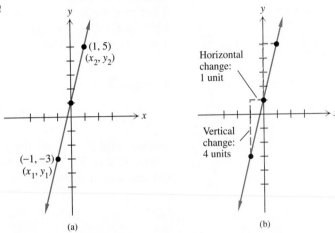

(a)   (b)

Graphing Equations Using the Slope and y-Intercept

An equation given in the form $y = mx + b$ is said to be in slope-intercept form.

> **Slope-Intercept Form of a Line**
>
> $$y = mx + b$$
>
> where m is the slope of the line and b is the y-intercept of the line.

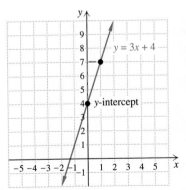

Figure 6.13

Consider the graph of the equation $y = 3x + 4$, which appears in Fig. 6.13. By examining the graph we can see that the y-intercept is 4. We can also see that the graph has a positive slope, since it rises from left to right. Since the vertical change is 3 units for every 1 unit of horizontal change, the slope must be $\frac{3}{1}$ or 3.

We could graph this equation by marking the y-intercept at 4 and then moving *up* 3 units and to the *right* 1 unit to get another point. If the slope were -3, we could start at the y-intercept and move *down* 3 units and to the *right* 1 unit. Thus, if we know the slope and y-intercept of a line we can graph the line.

> **To Graph Equations Using the Slope and y-Intercept**
>
> **1.** Solve the equation for y to place the equation in slope-intercept form.
> **2.** Determine the slope and y-intercept from the equation.
> **3.** Plot the y-intercept.
> **4.** Obtain a second point using the slope.
> **5.** Draw a straight line through the points.

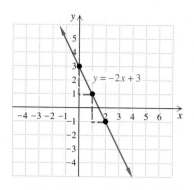

Figure 6.14

▶ **Example 6**

Graph $y = -2x + 3$ using the slope and y-intercept.

Solution: The slope is -2 and the y-intercept is 3. Plot (0, 3) on the y-axis. Then plot the next point by moving *down* two units and to the *right* 1 unit (see Fig. 6.14). A third point has been plotted in the same way. The graph of $y = -2x + 3$ is the line drawn through these three points.

▶ **Example 7**

Graph $y = \dfrac{5}{3}x - 2$ using the slope and y-intercept.

Solution: Plot (0, −2) on the y-axis. Then plot the next point by moving *up* 5 units and to the *right* 3 units (see Fig. 6.15). The graph of $y = \frac{5}{3}x - 2$ is the line drawn through these two points.

Figure 6.15

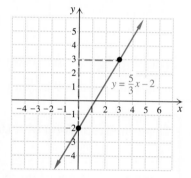

▶ **Example 8**

a) Write $3x + 4y = 4$ in slope-intercept form.
b) Graph the equation.

Solution:
a) Solve the given equation for y:

$$3x + 4y = 4$$
$$3x - 3x + 4y = -3x + 4$$
$$4y = -3x + 4$$
$$\frac{4y}{4} = \frac{-3x + 4}{4}$$
$$y = -\frac{3x}{4} + \frac{4}{4}$$
$$y = -\frac{3}{4}x + 1$$

Thus in slope-intercept form the equation is $y = -\frac{3}{4}x + 1$.

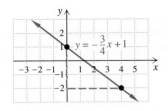

Figure 6.16

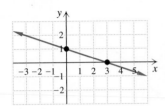

Figure 6.17

b) The y-intercept is 1, and the slope is $-\frac{3}{4}$. Place a dot at $(0, 1)$ on the y-axis then move *down* 3 units and to the *right* 4 units to obtain the second point (see Fig. 6.16). Draw a line through the two points.

▶ Example 9

Determine the equation of the line in Fig. 6.17.

Solution: If we determine the slope and the y-intercept of the line, then we can write the equation using slope-intercept form, $y = mx + b$. We see from the graph that the y-intercept is 1; thus $b = 1$. The slope of the line is negative, since the graph falls from left to right. The change in y is one unit for each three-unit change in x. Thus m, the slope of the line, is $-\frac{1}{3}$:

$$y = mx + b$$

$$y = -\frac{1}{3}x + 1$$

The equation of the line is $y = -\frac{1}{3}x + 1$.

▶ Example 10

In the Cartesian coordinate system, graph (a) $y = 2$ and (b) $x = -3$.

Solution:
a) For any value of x, the value of y is 2. Therefore the graph will be a horizontal line through $y = 2$ (Fig. 6.18).
b) For any value of y, the value of x is -3. Therefore the graph will be a vertical line through $x = -3$ (Fig. 6.19).

Figure 6.18

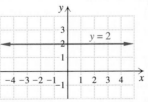

Figure 6.19

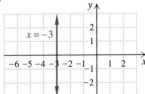

Note that the graph of $y = 2$ has a slope of 0. The slope of $x = -3$ is undefined.

Now let us look at a couple of applications of graphing.

▶ Example 11

Mihály Bencze owns a construction company. He has a contract from the state to pave 20 miles of highway. The distance, d, in miles he can pave in t hours can be approximated by the formula $d = 0.2t$.
a) Make a graph of distance he can pave for up to 300 hours.
b) Estimate the distance paved in 150 hours.

Solution:

a) Since $d = 0.2t$ is a linear equation, its graph will be a straight line. Select three values for t, find the corresponding values for d, and then draw the graph (Fig. 6.20).

$$d = 0.2t$$

$$\text{Let } t = 0, \quad d = 0.2(0) = 0$$
$$\text{Let } t = 100, \quad d = 0.2(100) = 20$$
$$\text{Let } t = 300, \quad d = 0.2(300) = 60$$

t	d
0	0
100	20
300	60

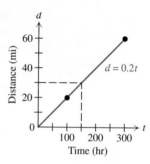

Figure 6.20

b) By drawing a vertical line up to the graph at $t = 150$ and a horizontal line across to the distance axis, we can determine that the distance covered is about 30 miles.

▶ Example 12

Tishia Jones owns a small business that manufactures compact discs. She estimates that the profit (or loss) from each CD produced can be estimated by the formula $P = 3.5S - 200{,}000$, where S is the number of copies of the CD sold.

a) Draw a graph of profit versus sales, for sales up to 500,000 copies.
b) From the graph estimate the number of copies that must be sold for the company to break even on a CD.
c) If the profit from a CD is $1 million, estimate the number of copies sold.

Solution:

a) Select values for S and find the corresponding values of P.

S	P
0	$-200{,}000$
100,000	150,000
500,000	1,550,000

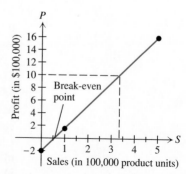

Figure 6.21

b) On the graph (Fig. 6.21) note that the break-even point is about 0.6 or 60,000 copies.

c) We can obtain the answer by drawing a horizontal line from 10 on the profit axis. Since the horizontal line cuts the graph at about 3.4 on the *S* axis, approximately 340,000 copies were sold.

Section 6.6 Exercises

1. Describe the three methods used to graph a linear equation in this section.

2. a) Explain (in your own words) how to find the slope of a line between two points.

b) Using your explanation find the slope of the line through the points (6, 2) and (−3, 5).

Plot the following points on the same set of axes.

3. (2, 3)

4. (−4, 1)

5. (2, −3)

6. (0, 0)

7. (4, 0)

8. (0, 4)

9. (0, 7)

10. $(3\frac{1}{2}, 4\frac{1}{2})$

Plot the following points on the same set of axes.

11. (1, 3)

12. (0, 3)

13. (−2, −3)

14. (0, −4)

15. (−3, −1)

16. (−3, 0)

17. (4, −1)

18. (4.5, 3.5)

Exercises 19–28 are indicated on Fig. 6.22. For each exercise, write the coordinates of the corresponding point.

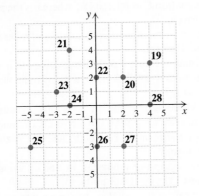

Figure 6.22

The points *A*, *B*, and *C* are three vertices (the points where two lines meet) of a rectangle. Plot the three points. (a) Find the coordinates of the fourth point, *D*, to complete the rectangle. (b) Find the area of the rectangle; use $A = lw$.

29. *A*(2, 3), *B*(7, 3), *C*(7, 5)

30. *A*(−4, 2), *B*(7, 2), *C*(7, 8)

The points *A*, *B*, and *C* are three vertices of a parallelogram with sides parallel to the *x*-axis. Plot the three points. Find the coordinates of the fourth point, *D*, to complete the parallelogram. *Note:* There are two possible answers for point *D*.

31. *A*(3, 2), *B*(5, 5), *C*(9, 5)

32. *A*(1, 6), *B*(−5, 6), *C*(−3, −1)

For what value of *b* will the line joining the points *P* and *Q* be parallel to the indicated axis?

33. *P*(−4, 3), *Q*(*b*, 1); *y*-axis

34. *P*(−5, 2), *Q*(7, *b*); *x*-axis

35. *P*(3*b* − 1, 5), *Q*(8, 4); *y*-axis

36. *P*(−6, 2*b* + 1), *Q*(2, 7); *x*-axis

In Ex. 37–44, determine which ordered pairs satisfy the given equation.

37. $5x + y = 0$ (1, −3), (2, −10), (−1, 5)

38. $x − 4y = 6$ (2, −3), $(0, −\frac{3}{2})$, (6, 0)

39. $4x − 2y = −10$ (0, 5), $(−\frac{5}{2}, 0)$, (2, −3)

40. $2x = 6y + 3$ $(0, −\frac{1}{2})$, (2, −1), $(−\frac{3}{2}, 0)$

41. $\dfrac{x}{2} + 3y = 4$ $(0, \frac{4}{3})$, (8, 0), (10, −2)

42. $7y = 3x − 5$ (1, −1), (−3, −2), (2, 5)

43. $3x − 4y = −4$ (0, 1), (−4, −2), $(2, −\frac{1}{2})$

44. $\dfrac{x}{2} + \dfrac{3y}{4} = 2$ $(0, \frac{8}{3})$, $(1, \frac{11}{4})$, (4, 0)

Graph each equation and state the slope if it exists (see Example 10).

45. $x = 4$

46. $x = −2$

47. $y = 3$

48. $y = -5$

Graph each equation by plotting points as in Example 3.

49. $y = x + 4$
50. $y = x - 1$
51. $y = 4x - 3$
52. $y = -x + 5$
53. $y + 2x = 4$
54. $y = 3x - 3$
55. $y = \frac{1}{2}x + 4$
56. $3y = 2x - 3$
57. $2y = -x + 6$
58. $y = -\frac{3}{4}x$

Graph each equation using the *x*- and *y*-intercepts as in Example 4.

59. $x + y = 3$
60. $x - y = 4$
61. $2x + y = 8$
62. $3x - 2y = 6$
63. $4x = 3y - 12$
64. $y = 6x + 6$
65. $y = -2x - 5$
66. $2x + 4y = 6$
67. $-3x + y = 4$
68. $5y = 3x + 10$

Find the slope of the line through the given points. If the slope is undefined, so state.

69. (5, 2) and (2, 3)
70. (3, 5) and (6, −1)
71. (−2, −4) and (3, 2)
72. (2, −2) and (−3, 5)
73. (5, 2) and (−3, 2)
74. (−3, −5) and (−1, −2)
75. (8, −3) and (−8, 3)
76. (2, 6) and (2, −3)
77. (−2, 3) and (1, −1)
78. (−7, −5) and (5, −6)

Graph each equation using the slope and *y*-intercept as in Examples 6 through 8.

79. $y = x + 4$
80. $y = 2x + 1$
81. $y = -x - 2$
82. $y = -3x + 1$
83. $y = -\frac{2}{3}x + 3$
84. $y = -x - 2$
85. $7y = 4x - 7$
86. $3x + 2y = 6$
87. $3x - 2y + 6 = 0$
88. $3x + 4y - 8 = 0$

Determine the equation of the graph.

89.

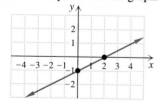

90.

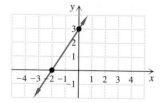

91.

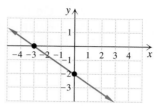

92.

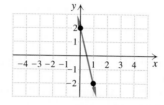

93. The monthly profit at a TV repair shop can be estimated by the equation $p = 30n - 400$, where *n* is the number of TV sets repaired in a week.
  a) Draw a graph of profit versus TV sets repaired for up to 100 TV sets repaired.
  b) Estimate the weekly profit in a week that 70 TV sets are repaired.
  c) How many TV sets would need to be repaired in a week for the shop to break even?

94. Sigfried is an owner of a newly formed magic store. His monthly salary is $1000 plus 20% of the company's net revenues for that month.
  a) Draw a graph of his monthly salary versus net monthly revenue for revenue up to $30,000.
  b) Estimate his monthly salary if the monthly revenue is $12,000.
  c) If his monthly salary is $3000 estimate the monthly revenue.

95. The weekly cost of operating a taxi can be approximated by the equation $C = 60 + 0.18n$, where *n* is the number of miles driven.
  a) Draw a graph of cost versus number of miles driven for up to 1000 mi.
  b) Estimate the cost if 600 mi are driven in a week.
  c) If the weekly cost is $132 how many miles were driven in a week?

96. When $1000 is invested in a savings account giving simple interest for a year, the interest, *i*, obtained can be found by the formula $i = 1000r$, where *r* is the rate in decimal form.
  a) Draw a graph of interest, *i*, in dollars versus simple interest rates for rates up to 15%.
  b) If the rate is 6%, what is the simple interest?
  c) If the rate is 12%, what is the simple interest?

Problem Solving

97. a) Two lines are parallel when they do not intersect no matter how far they are extended. Explain how you can determine, without graphing the equations, whether two equations will be parallel lines when graphed.

b) Determine whether the graphs of the equations $2x - 3y = 6$ and $4x = 6y + 6$ are parallel lines.

6.7 Linear Inequalities in Two Variables

In Section 6.5 we introduced linear inequalities in one variable. Now we will introduce linear inequalities in two variables. Some examples of linear inequalities in two variables are: $2x + 3y \leq 7$, $x + 7y \geq 5$,   and $x - 3y < 6$.

The solution set of a linear inequality in one variable may be indicated on a number line. The solution set of a linear inequality in two variables is indicated on a coordinate plane.

An inequality that is strictly less than ($<$) or greater than ($>$) will have as its solution set a **half-plane**. A half-plane is the set of all the points on one side of a line. An inequality that is less than or equal to ($\leq$) or greater than or equal to ($\geq$) will have as its solution set the set of points that consists of a half-plane and a line. To indicate that the line is part of the solution set, we draw a solid line. To indicate that the line is not part of the solution set, we draw a dashed line.

To Graph Inequalities in Two Variables

1. Mentally substitute the equals sign for the inequality sign and plot points as if you were graphing the equation.

2. If the inequality is $\langle$ or $\rangle$ draw a dashed line through the points. If the inequality is $\leq$ or $\geq$ draw a solid line through the points.

3. Select a test point not on the line and substitute the x- and y-coordinates into the inequality. If the substitution results in a true statement, shade in the area on the same side of the line as the test point. If the test point results in a false statement, shade in the area on the opposite side of the line as the test point.

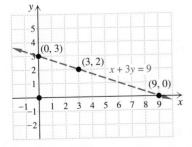

Figure 6.23

▶ **Example 1**

Draw the graph of $x + 3y < 9$.

Solution: To obtain the solution set, start by graphing $x + 3y = 9$. Since the original inequality is strictly "less than," draw a dashed line (Fig. 6.23). The dashed line indicates that the points on the line are not part of the solution set.

The line $x + 3y = 9$ divides the plane into three parts—the line itself and two half-planes. The line is the boundary between the two half-planes. The points in one half-plane will satisfy the inequality $x + 3y < 9$. The points in the other half-plane will satisfy the inequality $x + 3y > 9$.

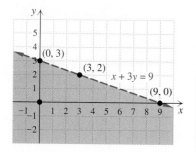

Figure 6.24

To determine the solution set of the inequality $x + 3y < 9$, pick any point on the plane that is not on the line. The simplest point to work with is the origin, $(0, 0)$. Substitute $x = 0$ and $y = 0$ in $x + 3y < 9$.

$$x + 3y < 9$$
$$\text{Is } 0 + 3(0) < 9?$$
$$0 + 0 < 9$$
$$0 < 9 \quad \text{True}$$

Since 0 is less than 9, the point $(0, 0)$ is part of the solution set. All the points on the same side of the line $x + 3y = 9$ as the point $(0, 0)$ are members of the solution set. Indicate this by shading the half-plane that contains $(0, 0)$. The graph is shown in Fig. 6.24.

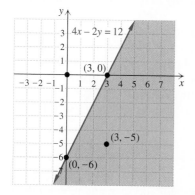

Figure 6.25

▶ **Example 2**

Draw the graph of $4x - 2y \geq 12$.

Solution: First draw the graph of the equation $4x - 2y = 12$. Use a solid line, since the points on the boundary are included in the solution set. Now pick a point that is not on the line. Take $(0, 0)$ as the test point.

$$4x - 2y \geq 12$$
$$\text{Is } 4(0) - 2(0) \geq 12?$$
$$0 \ngeq 12$$

Since zero is not greater than or equal to 12 $(0 \ngeq 12)$, the solution set is the line and the half-plane that does not contain the point $(0, 0)$. The graph is shown in Fig. 6.25.

If you had arbitrarily selected the test point $(3, -5)$, you would have found that the inequality would be true: $4(3) - 2(-5) \geq 12$, or $22 \geq 12$. Thus the point $(3, -5)$ would be in the half-plane containing the solution set.

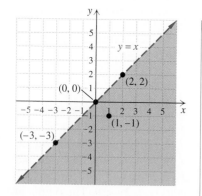

Figure 6.26

▶ **Example 3**

Draw the graph of $y < x$.

Solution: The inequality is strictly "less than," so the boundary is not part of the solution set. In graphing the equation $y = x$, draw a dashed line (Fig. 6.26). Since $(0, 0)$ is *on* the line, it cannot serve as a test point. Let us pick the point $(1, -1)$.

$$y < x$$
$$-1 < 1$$

Since $-1 < 1$ is true, the solution set is the half-plane containing the point $(1, -1)$.

Section 6.7 Exercises

Draw the graph for each of the following.

1. $x > 3$

2. $y \le 2$

3. $y > x + 2$

4. $y \le x - 4$

5. $y \ge 2x - 6$

6. $y < -2x + 2$

7. $3x - 4y > 12$

8. $x + 2y > 4$

9. $5x - 2y \le 10$

10. $2y - 5x \ge 10$

11. $3x + 2y < 6$

12. $-x + 2y < 2$

13. $x + y > 0$

14. $x + 2y \le 0$

15. $3y - 2x \le 0$

16. $y \ge 2x + 7$

17. $3x + 2y > 12$

18. $y \le 3x - 4$

19. $\frac{2}{5}x - \frac{1}{2}y \le 1$

20. $0.1x + 0.3y \le 0.4$

21. $0.2x + 0.5y \le 0.3$

22. Ann is allowed to talk a maximum of 15 min on the telephone. She has two friends (x and y) with whom she wishes to share the time.

a) State this problem as an inequality in two variables.

b) Graph the inequality.

23. Jose has 30 ft of fencing to place around a sandbox he plans to build for his children.

a) Write an inequality illustrating all possible dimensions of the rectangular sandbox. ($P = 2l + 2w$ is the formula for the perimeter of a rectangle.)

b) Graph the inequality.

6.8 Solving Quadratic Equations Using Factoring and Using the Quadratic Formula

In Section 6.2 we solved linear, or first degree, equations. In those equations the exponent on all variables was 1. Now we deal with the **quadratic equation**. The standard form of a quadratic equation in one variable is shown in the box.

Standard Form of a Quadratic Equation

$$ax^2 + bx + c = 0, \qquad a \ne 0$$

The series of steps taken to solve a certain type of equation is called an **algorithm**. The fact that mathematical procedures to solve equations can be generalized makes it possible to program computers to solve problems. For example, the equation $4x^2 + 7x + 3 = 0$ is an equation of the form $ax^2 + bx + c = 0$. If you can find a procedure to solve an equation of the general form, then it can be used to solve all equations of that form. One of the first computer languages, FORTRAN (for FORMula TRANslation), was developed specifically to handle scientific and mathematical applications.

Note that in the standard form of a quadratic equation, the greatest exponent on x is 2 and the right side of the equation is equal to zero. To solve a quadratic equation means to find the value or values that make the equation true. In this section we will solve quadratic equations by factoring and by the quadratic formula.

Before we examine factoring, we will look at the **FOIL method** of multiplying two binomials. A **binomial** is an expression that contains two terms, in which each exponent that appears on a variable is a whole number.

Examples of binomials

$x + 3 \qquad x - 5$

$3x + 5 \qquad 4x - 2$

To multiply two binomials, we can use the FOIL method. The name of the method, FOIL, is an acronym to help its users remember it as a method that obtains the products of the **First, Outer, Inner,** and **Last** terms of the binomials.

$$(a + b)(c + d) = a \cdot c + a \cdot d + b \cdot c + b \cdot d$$

First   Outer   Inner   Last

After multiplying the first, outer, inner, and last terms, combine all like terms.

▶ **Example 1**

Multiply $(x + 2)(x + 4)$.

Solution: The FOIL method of multiplication yields

$$(x + 2)(x + 4) = x \cdot x + (x)(4) + 2 \cdot x + 2 \cdot 4$$
$$= x^2 + 4x + 2x + 8$$
$$= x^2 + 6x + 8$$

Note that $4x$ and $2x$ were combined to get $6x$.

▶ **Example 2**

Multiply $(3x + 2)(x - 5)$.

Solution: $\quad (3x + 2)(x - 5) = 3x \cdot x + 3x(-5) + 2 \cdot x + 2(-5)$
$$= 3x^2 - 15x + 2x - 10$$
$$= 3x^2 - 13x - 10$$

Factoring Trinomials of the Form $x^2 + bx + c$

The expression $x^2 + 6x + 8$ is an example of a trinomial. A **trinomial** is an expression containing three terms in which each exponent that appears on a variable is a whole number.

Example 1 showed that

$$(x + 2)(x + 4) = x^2 + 6x + 8.$$

Since the product of $x + 2$ and $x + 4$ is $x^2 + 6x + 8$, we say that $x + 2$ and $x + 4$ are **factors** of $x^2 + 6x + 8$. **To factor an expression** means to write the expression as a product of its factors. For example, to factor $x^2 + 6x + 8$, we would write

$$x^2 + 6x + 8 = (x + 2)(x + 4).$$

Let us look at the factors more closely:

$$2 + 4 = 6$$
$$2 \cdot 4 = 8$$
$$x^2 + 6x + 8 (x + 2)(x + 4)$$

Note that the sum of the two numbers in the factors is $2 + 4$ or 6. The 6 is the coefficient of the x-term. Also note that the product of the numbers in the two factors is $2 \cdot 4$ or 8. The 8 is the constant. In general, when factoring an expression of the form $x^2 + bx + c$, we need to find two numbers whose product is c and whose sum is b. When we determine the two numbers, the factors will be of the form

$$(x + \boxed{})(x + \boxed{})$$

one   other
number   number

▶ **Example 3**
Factor $x^2 + 5x + 6$.

Solution: We need to find two numbers whose product is 6 and whose sum is 5. Since the product is $+6$, the two numbers must both be positive or both be negative. Since the coefficient of the x-term is positive, only the positive factors of 6 need to be considered. Can you explain why? We begin by listing the positive numbers whose product is 6.

Factors of 6	Sum of Factors
1(6)	$1 + 6 = 7$
2(3)	$2 + 3 = 5$

Since $2 \cdot 3 = 6$ and $2 + 3 = 5$, 2 and 3 are the numbers we are seeking. Thus we write

$$x^2 + 5x + 6 = (x + 2)(x + 3)$$

Note $(x + 3)(x + 2)$ is also an acceptable answer.

To Factor Trinomial Expressions of the Form $x^2 + bx + c$
1. Find two numbers whose product is c and whose sum is b.
2. Write factors in the form

$$(x + \boxed{})(x + \boxed{}).$$

one number   other number
from step 1   from step 1

If, for example, the numbers found in step 1 were 6 and -4, the factors would be written $(x + 6)(x - 4)$.

▶ **Example 4**
Factor $x^2 - x - 12$.

Solution: We must find two numbers whose product is -12 and whose sum is -1. (Remember: $-x$ means $-1x$.) Begin by listing the factors of -12:

Factors of -12	Sum of Factors
$-12(1)$	$-12 + 1 = -11$
$-6(2)$	$-6 + 2 = -4$
$-4(3)$	$-4 + 3 = -1$
$-3(4)$	$-3 + 4 = 1$
$-2(6)$	$-2 + 6 = 4$
$-1(12)$	$-1 + 12 = 11$

The chart lists all the factors of -12. The only numbers with a product of -12 and a sum of -1 are -4 and 3. We listed all factors in this example so that you would see that, for example, $-4(3)$ is a different set of factors than $-3(4)$. Once you find the factors you are seeking, there is no reason to go any further. The trinomial can be written in factored form as

$$x^2 - x - 12 = (x - 4)(x + 3).$$

Factoring Trinomials of the Form $ax^2 + bx + c$, $a \neq 1$

Now we discuss how to factor an expression of the form $ax^2 + bx + c$ where a, the coefficient of the squared term, is not equal to 1.

Consider the multiplication problem $(2x + 1)(x + 3)$:

$$(2x + 1)(x + 3) = 2x \cdot x + 2x \cdot 3 + 1 \cdot x + 1 \cdot 3$$
$$= 2x^2 + 6x + x + 3$$
$$= 2x^2 + 7x + 3$$

Since $(2x + 1)(x + 3) = 2x^2 + 7x + 3$, the factors of $2x^2 + 7x + 3$ are $2x + 1$ and $x + 3$.

Let us study these factors more closely.

$$\mathbf{F} = 2 \cdot 1 = 2 \qquad \mathbf{O + I} = (2 \cdot 3) + (1 \cdot 1) = 7 \qquad \mathbf{L} = 1 \cdot 3 = 3$$

Notice that the product of the coefficient of the first terms in the multiplication of the binomials equals 2, the coefficient of the squared term. The sum of the products of the coefficients of the outer and inner terms equals 7, the coefficient of the x term. The product of the last terms equals 3, the constant.

A procedure to factor expressions of the form $ax^2 + bx + c$, $a \neq 1$ follows.

> ### To Factor Trinomial Expressions of the Form $ax^2 + bx + c$, $a \neq 1$
> **1.** Write all pairs of factors of the coefficient of the squared term, a.
> **2.** Write all pairs of factors of the constant, c.
> **3.** Try various combinations of these factors until the sum of the products of the outer and inner terms is bx.

▶ **Example 5**

Factor $3x^2 + 17x + 10$.

Solution: The only factors of 3 are 1 and 3. Therefore we write

$$3x^2 + 17x + 10 = (3x \qquad)(x \qquad).$$

The number 10 has both positive and negative factors. However, since both the constant, 10, and the sum of the products of the outer and inner terms, 17, are positive, the two factors must be positive. Why? The positive factors of 10 are 1(10) and 2(5). Below is a listing of the possible factors:

Posssible Factors	Sum of Products of Outer and Inner Terms
$(3x + 1)(x + 10)$	$31x$
$(3x + 10)(x + 1)$	$13x$
$(3x + 2)(x + 5)$	$17x$ ← Correct middle term
$(3x + 5)(x + 2)$	$11x$

Thus $3x^2 + 17x + 10 = (3x + 2)(x + 5)$.

Note that factoring problems of this type may be checked by using the FOIL method of multiplication. We will check the results to Example 5:

$$(3x + 2)(x + 5) = 3x \cdot x + 3x \cdot 5 + 2 \cdot x + 2 \cdot 5$$
$$= 3x^2 + 15x + 2x + 10$$
$$= 3x^2 + 17x + 10$$

Since we obtained the expression we started with, our factoring is correct.

▶ **Example 6**

Factor $6x^2 - 11x - 10$.

Solution: The factors of 6 will be either $6 \cdot 1$ or $2 \cdot 3$. Therefore the factors may be of the form $(6x \quad)(x \quad)$ or $(2x \quad)(3x \quad)$. When there is more than one set of factors for the first term, we generally try the medium-sized factors first. If this does not work, we try the other factors. Thus we write

$$6x^2 - 11x - 10 = (2x \qquad)(3x \qquad).$$

The factors of -10 are $(-1)(10)$, $(1)(-10)$, $(-2)(5)$, and $(2)(-5)$. There will be eight different pairs of possible factors of the trinomial $6x^2 - 11x - 10$. Can you list them?

The correct factoring is $6x^2 - 11x - 10 = (2x - 5)(3x + 2)$.

Note that in Example 6 we first tried factors of the form $(2x \quad)(3x \quad)$. If we had not found the correct factors using these, we would have tried $(6x \quad)(x \quad)$.

Solving Quadratic Equations by Factoring

To solve a quadratic equation by factoring set one side of the equation equal to 0, then use the *zero-factor* property.

Zero-Factor Property

$$\text{If } a \cdot b = 0, \text{ then } a = 0 \text{ or } b = 0.$$

The zero-factor property indicates that if the product of two factors is 0, then one (or both) of the factors must have a value of 0.

▶ Example 7

Solve the equation $(x - 2)(x + 4) = 0$.

Solution: Using the zero-factor property, either $x - 2$ or $x + 4$ must equal 0 for the product to equal 0. Thus we set each individual factor equal to 0 and solve each resulting equation for x:

$$(x - 2)(x + 4) = 0$$
$$x - 2 = 0 \quad \text{or} \quad x + 4 = 0$$
$$x = 2 \qquad\qquad x = -4$$

Thus the solutions are 2 and -4.

Check:

$x = 2$	$x = -4$
$(x - 2)(x + 4) = 0$	$(x - 2)(x + 4) = 0$
$(2 - 2)(2 + 4) = 0$	$(-4 - 2)(-4 + 4) = 0$
$0\,(6) = 0$	$(-6)(0) = 0$
$0 = 0$ True	$0 = 0$ True

To Solve a Quadratic Equation by Factoring

1. Use the addition or subtraction property to make one side of the equation equal to 0.

2. Factor the side of the equation not equal to 0.

3. Use the zero-factor property to solve the equation.

Examples 8 and 9 illustrate this procedure.

▶ **Example 8**

Solve the equation $x^2 - 8x = -15$.

Solution: First add 15 to both sides of the equation to make the right side of the equation equal to 0:

$$x^2 - 8x = -15$$
$$x^2 - 8x + 15 = -15 + 15$$
$$x^2 - 8x + 15 = 0$$

Factor the left side of the equation. The object is to find two numbers whose product is 15 and whose sum is -8. Since the product of the numbers is positive and the sum of the numbers is negative, the two numbers must both be negative. The numbers are -3 and -5. Note that $(-3)(-5) = 15$ and $-3 + (-5) = -8$.

$$x^2 - 8x + 15 = 0$$
$$(x - 3)(x - 5) = 0$$

Now use the zero-factor property to find the solution:

$$x - 3 = 0 \quad \text{or} \quad x - 5 = 0$$
$$x = 3 \qquad\qquad x = 5$$

The solutions are 3 and 5.

▶ **Example 9**

Solve the equation $2x^2 - 11x + 12 = 0$.

Solution: $2x^2 - 11x + 12$ factors into $(2x - 3)(x - 4)$. Thus we write

$$2x^2 - 11x + 12 = 0$$
$$(2x - 3)(x - 4) = 0$$
$$2x - 3 = 0 \quad \text{or} \quad x - 4 = 0$$
$$2x = 3 \qquad\qquad x = 4$$
$$x = \frac{3}{2}$$

The solutions are $\frac{3}{2}$ and 4.

Solving Quadratic Equations Using the Quadratic Formula

Not all quadratic equations can be solved by factoring. When a quadratic equation cannot be easily solved by factoring, we can solve the equation with the **quadratic formula**. The quadratic formula can be used to solve any quadratic equation.

Quadratic Formula

For a quadratic equation in standard form, $ax^2 + bx + c = 0$, $\quad a \neq 0$, the quadratic formula is

$$x = \frac{-b \pm \sqrt{b^2 - 4ac}}{2a}$$

To use the quadratic formula, first write the quadratic equation in standard form. Then determine the values for a (the coefficient of the squared term), b (the coefficient of the x term), and c (the constant). Finally substitute the values of a, b, and c into the quadratic formula and evaluate the expression.

▶ **Example 10**

Solve the equation $x^2 + 6x - 16 = 0$ using the quadratic formula.

Solution: In this equation, $a = 1$, $b = 6$, and $c = -16$. Substituting these values into the quadratic formula gives

$$x = \frac{-b \pm \sqrt{b^2 - 4ac}}{2a} = \frac{-6 \pm \sqrt{6^2 - 4(1)(-16)}}{2(1)}$$

$$= \frac{-6 \pm \sqrt{36 + 64}}{2}$$

$$= \frac{-6 \pm \sqrt{100}}{2}$$

$$= \frac{-6 \pm 10}{2}$$

$$\frac{-6 + 10}{2} = \frac{4}{2} = 2 \qquad \frac{-6 - 10}{2} = \frac{-16}{2} = -8$$

The solutions are 2 and -8.

Note that Example 10 can also be solved by factoring. We suggest that you solve Example 10 by factoring now.

▶ **Example 11**

Solve $2x^2 - 4x = 5$ using the quadratic formula.

Solution: Begin by writing the equation in standard form by subtracting 5 from both sides of the equation:

$$2x^2 - 4x - 5 = 0$$

$$a = 2, \qquad b = -4, \qquad c = -5$$

$$x = \frac{-b \pm \sqrt{b^2 - 4ac}}{2a} = \frac{-(-4) \pm \sqrt{(-4)^2 - 4(2)(-5)}}{2(2)}$$

$$x = \frac{4 \pm \sqrt{16 + 40}}{4}$$

$$= \frac{4 \pm \sqrt{56}}{4}$$

Since $\sqrt{56} = \sqrt{4} \cdot \sqrt{14} = 2\sqrt{14}$ (see Section 5.4) we write

$$\frac{4 + \sqrt{56}}{4} = \frac{4 + 2\sqrt{14}}{4} = \frac{\overset{1}{\cancel{2}}(2 + \sqrt{14})}{\underset{2}{\cancel{4}}} = \frac{2 + \sqrt{14}}{2}$$

$$\frac{4 - \sqrt{56}}{4} = \frac{4 - 2\sqrt{14}}{4} = \frac{\overset{1}{\cancel{2}}(2 - \sqrt{14})}{\underset{2}{\cancel{4}}} = \frac{2 - \sqrt{14}}{2}$$

The solutions are $\dfrac{2 + \sqrt{14}}{2}$ and $\dfrac{2 - \sqrt{14}}{2}$.

Note that the solutions to Example 11 are irrational numbers. It is also possible for a quadratic equation to have no real solution. In solving an equation by the quadratic formula, if the radicand (the expression inside the square root) is a negative number, then the quadratic equation has **no real solution**.

Section 6.8 Exercises

1. In your own words, state the zero-factor property.
2. Have you memorized the quadratic formula? If not you need to do so. Without looking at the book write the quadratic formula.

Factor the trinomial in each exercise. If the trinomial cannot be factored, so state.

3. $x^2 + 8x + 12$
4. $x^2 + 5x + 6$
5. $x^2 - x - 12$
6. $x^2 + x - 12$
7. $x^2 + 2x - 24$
8. $x^2 - 6x + 8$
9. $x^2 + 3x - 10$
10. $x^2 - 2x + 1$
11. $x^2 - 10x + 21$
12. $x^2 - 25$
13. $x^2 - 7x - 8$
14. $x^2 - x - 30$
15. $x^2 + 3x - 28$
16. $x^2 + 4x - 32$
17. $x^2 + 16x + 63$
18. $x^2 - 14x + 40$

Factor the trinomial. If the trinomial cannot be factored, so state.

19. $2x^2 + 3x - 2$
20. $2x^2 - 7x - 15$
21. $3x^2 + 2x - 1$
22. $4x^2 + 4x + 1$
23. $5x^2 + 12x + 4$
24. $2x^2 - 9x + 10$
25. $5x^2 - 7x - 6$
26. $4x^2 + 16x + 15$

27. $5x^2 - 14x + 8$
28. $2x^2 - 17x + 30$
29. $3x^2 - 14x - 24$
30. $6x^2 + 5x + 1$

Solve each equation using the zero-factor property.

31. $(x - 6)(x + 3) = 0$
32. $(2x - 3)(3x + 1) = 0$
33. $(3x + 5)(4x - 3) = 0$
34. $(x - 6)(5x - 4) = 0$

Solve each equation by factoring.

35. $x^2 + 3x + 2 = 0$
36. $x^2 - 3x + 2 = 0$
37. $x^2 - 8x + 15 = 0$
38. $x^2 + 2x - 8 = 0$
39. $x^2 - 15 = 2x$
40. $x^2 - 7x = -6$
41. $x^2 = 4x - 3$
42. $x^2 - 13x + 40 = 0$
43. $x^2 - 18 = 7x$
44. $x^2 = 11x - 28$
45. $x^2 + 5x - 36 = 0$
46. $x^2 + 12x + 20 = 0$
47. $2x^2 + x - 3 = 0$
48. $3x^2 + 13x = -4$
49. $5x^2 + 11x = -2$
50. $2x^2 = -5x + 3$
51. $3x^2 - 4x = -1$
52. $5x^2 + 16x + 12 = 0$
53. $4x^2 - 9x + 2 = 0$
54. $6x^2 + x - 2 = 0$

Solve each equation using the quadratic formula. If the equation has no real solution, so state.

55. $x^2 - 9x + 20 = 0$
56. $x^2 - 4x - 21 = 0$

57. $x^2 - 4x + 3 = 0$
58. $x^2 + 8x + 15 = 0$
59. $x^2 - 8x = 9$
60. $x^2 = -8x + 15$
61. $x^2 - 2x + 3 = 0$
62. $2x^2 - x - 3 = 0$
63. $x^2 - 5x = 4$
64. $2x^2 - 7x + 2 = 0$
65. $3x^2 - 2x = -5$
66. $2x^2 + x = 5$
67. $4x^2 - x - 1 = 0$
68. $4x^2 - 5x - 3 = 0$
69. $2x^2 + 7x + 5 = 0$
70. $3x^2 = 9x - 5$
71. $3x^2 - 10x + 7 = 0$
72. $4x^2 + 7x - 1 = 0$

73. $2x^2 + 6x - 3 = 0$
74. $3x^2 - 2x - 6 = 0$

Problem Solving

75. The radicand in the quadratic formula, $b^2 - 4ac$, is called the **discriminant**. How many solutions will the quadratic equation have if the discriminant is (a) greater than 0, (b) equal to 0, or (c) less than zero. Explain your answer.

6.9 Functions and Their Graphs

The concepts of relations and functions are extremely important in mathematics. Functions are a common thread found in many mathematics courses. A **relation** is any set of ordered pairs. Therefore every graph will be a relation. A function is a special type of relation. Suppose you are purchasing oranges at a supermarket where each orange costs $.20. Then one orange would cost $.20, two oranges $2 \times $.20 = $.40$, three oranges $.60, and so on. We can indicate this using a table of values.

Number of Oranges	Cost
0	0.00
1	0.20
2	0.40
3	0.60
⋮	⋮
10	2.00
⋮	⋮

In general we see that the cost for purchasing n oranges will be 20 cents times the number of oranges, or $0.20n$. We can represent the cost, c, of n items by the equation $c = 0.20n$. Since the value of c depends on the value of n we refer to n as the **independent variable** and c as the **dependent variable**. *Notice for each value of the independent variable,* n, *there is one and only one value of the dependent variable,* c. When this happens the equation is called a **function**. In the equation $c = 0.20n$, since the value of c depends on the value of n, we say that "c is a function of n."

> A **function** is a special type of relation where each value of the independent variable corresponds to a unique value of the dependent variable.

The set of values that can be used for the independent variable is called the **domain** of the function and the resulting set of values obtained for the de-

Figure 6.27

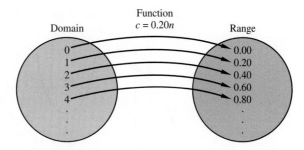

pendent variable is called the **range**. The domain and range for the function $c = 0.20n$ are illustrated in Fig. 6.27.

When we graphed equations of the form $ax + by = c$ in Section 6.6 we found they were straight lines. For example, the graph of $y = 2x - 1$ is illustrated in Fig. 6.28.

Is the equation $y = 2x - 1$ a function? To answer this question we must ask, "Does each value of x correspond to a unique value of y?" The answer is yes; therefore this equation is a function.

For the equation $y = 2x - 1$, we say that "y is a function of x" and write $y = f(x)$. The notation $f(x)$ is read "f of x." When we are given an equation that is a function, we may replace the y in the equation with $f(x)$, since $f(x)$ represents y. Thus $y = 2x - 1$ may be written $f(x) = 2x - 1$.

To evaluate a function for a specific value of x, replace each x in the function with the given value, then evaluate. For example, to evaluate $f(x) = 2x - 1$ when $x = 8$, we do the following:

$$f(x) = 2x - 1$$
$$f(8) = 2(8) - 1 = 16 - 1 = 15$$

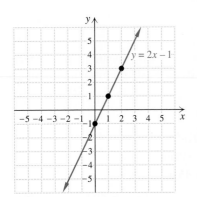

Figure 6.28

Thus $f(8) = 15$. Since $f(x) = y$, when $x = 8$, $y = 15$. What is the domain and range of $f(x) = 2x - 1$? Since x can be any real number the domain is the set of real numbers, symbolized $\mathbb{R}$. The range is also $\mathbb{R}$.

We can determine if a graph is a function by using the **vertical line test**: If a vertical line can be drawn so that it intersects the graph at more than one point then each x does not have a unique y, and the graph is not a function. If a vertical line cannot be made to intersect the graph in at least two different places, then the graph is a function.

▶ **Example 1**

Use the vertical line test to determine which of the following graphs are functions.

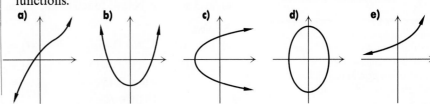

Solution: a), b), and e) are functions but c) and d) are not, as illustrated below.

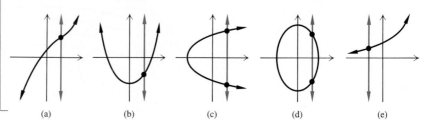

(a)   (b)   (c)   (d)   (e)

There are many real-life applications of functions—in fact, all of the applications illustrated in Sections 6.2 and 6.3 were functions. Let us consider two more applications.

▶ **Example 2**

The average stopping distance, d, in meters, for a car traveling v kilometers per hour is given by the function $d = f(v) = 0.16v + 0.01v^2$. Find the stopping distance for a car traveling at 60 km/hr.

Solution: Substitute 60 for v in the function.

$$f(v) = 0.16v + 0.01v^2$$
$$f(60) = 0.16(60) + 0.01(60)^2$$
$$= 9.6 + 36 = 45.6$$

Thus the average stopping distance of a car traveling at 60 km/hr is 45.6 meters.

▶ **Example 3**

On July 20, 1969, Neil Armstrong became the first person to walk on the moon. The velocity, v, of his spacecraft, the *Eagle*, in meters per second, was a function of time before touchdown, t, given by

$$v = f(t) = 3.2t + 0.45$$

The height of the spacecraft, h, above the moon's surface, in meters, was also a function of time before touchdown, given by

$$h = g(t) = 1.6t^2 + 0.45t$$

What was the velocity of the spacecraft and its distance from the surface of the moon

a) at 3 seconds before touchdown,   **b)** at touchdown (0 sec).

Solution:

a) $v = f(t) = 3.2t + 0.45,$ $h = g(t) = 1.6t^2 + 0.45t$

$f(3) = 3.2(3) + 0.45$ $g(3) = 1.6(3)^2 + 0.45(3)$

$= 9.6 + 0.45$ $= 1.6(9) + 1.35$

$= 10.05$ m/sec $= 14.4 + 1.35$

$= 15.75$ m

Apollo II *touched down at Mare Tranqillitatis, the Sea of Tranquillity. The rock samples taken here placed the age of the rock at 3.5 billion years old—as old as the oldest known earth rocks.*

The velocity 3 seconds before touchdown was 10.05 m/sec and the height 3 seconds before touchdown was 15.75 meters.

b) $v = f(t) = 3.2t + 0.45,$ $h = g(t) = 1.6t^2 + 0.45t$

$$f(0) = 3.2(0) + 0.45 \qquad g(0) = 1.6(0)^2 + 0.45(0)$$
$$= 0 + 0.45 \qquad\qquad = 0 + 0$$
$$= 0.45 \text{ m/sec} \qquad\qquad = 0 \text{ m}$$

The touchdown velocity was 0.45 m/sec. At touchdown the *Eagle* is on the moon, and therefore the distance from the moon is 0.

In Section 6.3 we discussed exponential equations. All exponential equations are also functions.

▶ **Example 4**

The power supply of a satellite is a radioisotope. The power output, p, in watts (w), remaining in the power supply is a function of time the satellite is in space. If there are originally 100 grams of the isotope, the power remaining after t days is $p = 100e^{-0.001t}$. What will be the remaining power after 1 year in space?

Solution: Substitute 365 days for t in the function, then evaluate using a calculator as described in Section 6.3.

$$p = 100e^{-0.001t}$$
$$= 100e^{-0.001(365)}$$
$$= 100e^{-0.365}$$
$$\approx 100(0.694)$$
$$\approx 69.4 \text{ watts}$$

Graphs of Quadratic Functions

Equations of the form $y = ax + b$ are linear functions. The graphs of linear functions are straight lines that will pass the vertical line test.

Equations of the form $y = ax^2 + bx + c$, $a \neq 0$, are **quadratic functions**. Examples of quadratic functions are $y = 2x^2 + 6x - 3$ and $y = -\frac{1}{2}x^2 + 6$.

The graph of every quadratic function is a **parabola**. Two parabolas are illustrated in Fig. 6.29. A parabola opens upward when the coefficient of the

Figure 6.29

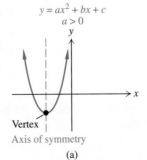

$y = ax^2 + bx + c$
$a > 0$

Vertex

Axis of symmetry

(a)

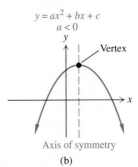

$y = ax^2 + bx + c$
$a < 0$

Vertex

Axis of symmetry

(b)

squared term is greater than 0, Fig. 6.29(a), and it opens downward when the coefficient of the squared term is less than 0, Fig. 6.29(b).

The **vertex** of a parabola is the lowest point on a parabola that opens upward and the highest point on a parabola that opens downward. Every parabola is **symmetric** with respect to a vertical line through its vertex. This line is called the **axis of symmetry** of the parabola. The x-coordinate of the vertex and the equation of the axis of symmetry can be found by using the following equation.

Axis of Symmetry of a Parabola

$$x = \frac{-b}{2a}$$

Once the x-coordinate of the vertex has been determined, the y-coordinate can be found by substituting the value found for the x-coordinate into the quadratic equation and evaluating the equation. This procedure is illustrated in Example 5.

▶ **Example**

Consider the equation $y = 2x^2 - 4x - 6$.
a) Determine whether the graph will be a parabola that opens upward or downward.
b) Find the equation of the axis of symmetry of the parabola.
c) Find the vertex of the parabola.

Solution:
a) Since $a = 2$, which is greater than 0, the parabola opens upward.
b) To find the axis of symmetry, we use the equation $x = -b/2a$. In the equation $y = 2x^2 - 4x - 6$, $a = 2$, $b = -4$, and $c = -6$:

$$x = \frac{-b}{2a} = \frac{-(-4)}{2(2)} = \frac{4}{4} = 1$$

The equation of the axis of symmetry is $x = 1$.
c) The x-coordinate of the vertex is 1 (from part b). To find the y-coordinate, we substitute 1 for x in the equation, then evaluate:

$$y = 2x^2 - 4x - 6$$
$$y = 2(1)^2 - 4(1) - 6$$
$$y = 2(1) - 4 - 6$$
$$y = 2 - 4 - 6$$
$$y = -8$$

Therefore the vertex of the parabola is located at the point $(1, -8)$ on the graph.

General Procedure to Sketch the Graph of a Quadratic Equation

1. Determine whether the parabola opens upward or downward.
2. Determine the axis of symmetry.
3. Determine the vertex of the parabola.
4. Determine the y-intercept by substituting $x = 0$ into the equation.
5. Determine the x-intercepts (if they exist) by substituting $y = 0$ into the equation and solving for x.
6. Draw the graph making use of the information gained in steps 1–5. Remember the parabola will be symmetric with respect to the axis of symmetry.

In step 5, to determine the x-intercepts you may use either factoring or the quadratic formula.

▶ **Example 6**

Sketch the graph of the equation $y = x^2 - 6x + 8$.

Solution: We will follow the steps outlined in the general procedure.

1. We first notice that since $a = 1$, which is greater than 0, the parabola opens upward.

2. $x = \dfrac{-b}{2a} = \dfrac{-(-6)}{2(1)} = \dfrac{6}{2} = 3$

   Thus the axis of symmetry is $x = 3$.

3. $y = x^2 - 6x + 8$
   $y = (3)^2 - 6(3) + 8 = 9 - 18 + 8 = -1$
   Thus the vertex is at $(3, -1)$.

4. $y = x^2 - 6x + 8$
   $y = 0^2 - 6(0) + 8 = 8$
   Thus the y-intercept is at $(0, 8)$.

5. $0 = x^2 - 6x + 8$
   or $\quad x^2 - 6x + 8 = 0$
   We can solve this equation by factoring.
   $(x - 4)(x - 2) = 0$
   $x - 4 = 0 \quad$ or $\quad x - 2 = 0$
   $\qquad x = 4 \qquad\qquad x = 2$
   Thus the x-intercepts are at $(4, 0)$ and $(2, 0)$.

6. The graph is sketched in Fig. 6.30.

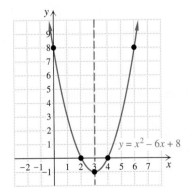

Figure 6.30

Notice that the domain of the graph in Example 6, the possible x values, is the set of all real numbers, $\mathbb{R}$. The range, the possible y values, is the set of all real numbers greater than or equal to -1. When graphing parabolas, if you feel you need additional points to graph you can always substitute values for x and find the corresponding values of y and plot those points. For example, if you substituted 1 for x, the corresponding value of y is 3. Thus you could plot the point $(1, 3)$.

▶ **Example 7**

a) Sketch the graph of the equation $y = -2x^2 + 3x + 4$.
b) Determine the domain and range of the function.

Solution:

a) Follow the steps outlined in the general procedure.

  1. Since $a = -2$, which is less than 0, the parabola opens downward.

  2. axis of symmetry: $\quad x = \dfrac{-b}{2a} = \dfrac{-(3)}{2(-2)} = \dfrac{-3}{-4} = \dfrac{3}{4}$

  3. y-coordinate of vertex: $\quad y = -2x^2 + 3x + 4$

  $$= -2\left(\frac{3}{4}\right)^2 + 3\left(\frac{3}{4}\right) + 4$$

  $$= -2\left(\frac{9}{16}\right) + \frac{9}{4} + 4$$

  $$= -\frac{9}{8} + \frac{9}{4} + 4$$

  $$= -\frac{9}{8} + \frac{18}{8} + \frac{32}{8} = \frac{41}{8} \quad \text{or} \quad 5\frac{1}{8}$$

  Thus the vertex is at $\left(\dfrac{3}{4}, 5\dfrac{1}{8}\right)$.

  4. y-intercept: $\quad y = -2x^2 + 3x + 4$
  $$= -2(0)^2 + 3(0) + 4 = 4$$
  The y-intercept is $(0, 4)$.

  5. x-intercepts: $\quad y = -2x^2 + 3x + 4$
  $$0 = -2x^2 + 3x + 4$$
  $$\text{or} \quad -2x^2 + 3x + 4 = 0$$
  Since this equation cannot be factored, we will use the quadratic formula to solve it.
  $$a = -2, \qquad b = 3, \qquad c = 4$$
  $$x = \frac{-b \pm \sqrt{b^2 - 4ac}}{2a}$$
  $$= \frac{-3 \pm \sqrt{3^2 - 4(-2)(4)}}{2(-2)}$$
  $$= \frac{-3 \pm \sqrt{9 + 32}}{-4}$$
  $$= \frac{-3 \pm \sqrt{41}}{-4}$$

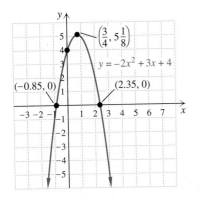

Figure 6.31

Since $\sqrt{41} \approx 6.4$,

$$x \approx \frac{-3 + 6.4}{-4} \approx \frac{3.4}{-4} \approx -0.85 \quad \text{or} \quad x \approx \frac{-3 - 6.4}{-4} \approx \frac{-9.4}{-4} \approx 2.35$$

6. Plot the vertex $(\frac{3}{4}, 5\frac{1}{8})$, the y-intercept $(0, 4)$, and the x-intercepts $(-0.85, 0)$ and $(2.35, 0)$. Then sketch the graph (Fig. 6.31).

b) The domain, the values that can be used for x, is the set of all real numbers, $\mathbb{R}$. The range, the values of y, is $y \le 5\frac{1}{8}$.

When using the quadratic formula to find the x-intercepts of a graph, if the radicand, $b^2 - 4ac$, is a negative number, then the graph has no x-intercepts and the graph will be totally above or below the x-axis.

Exponential Functions

What does the graph of an exponential function of the form $y = a^x$, $a > 0$, $a \ne 1$, look like? Examples 8 and 9 illustrate graphs of exponential functions.

▶ **Example 8**
a) Graph $y = 2^x$.   **b)** Determine the domain and range of the function.

Solution:
a) Substitute values for x and find the corresponding values of y. The graph is shown in Fig. 6.32.

$$y = 2^x$$

$$x = -3, \quad y = 2^{-3} = \frac{1}{2^3} = \frac{1}{8}$$

$$x = -2, \quad y = 2^{-2} = \frac{1}{2^2} = \frac{1}{4}$$

$$x = -1, \quad y = 2^{-1} = \frac{1}{2^1} = \frac{1}{2}$$

$$x = 0, \quad y = 2^0 = 1$$

$$x = 1, \quad y = 2^1 = 2$$

$$x = 2, \quad y = 2^2 = 4$$

$$x = 3, \quad y = 2^3 = 8$$

x	y
-3	$\frac{1}{8}$
-2	$\frac{1}{4}$
-1	$\frac{1}{2}$
0	1
1	2
2	4
3	8

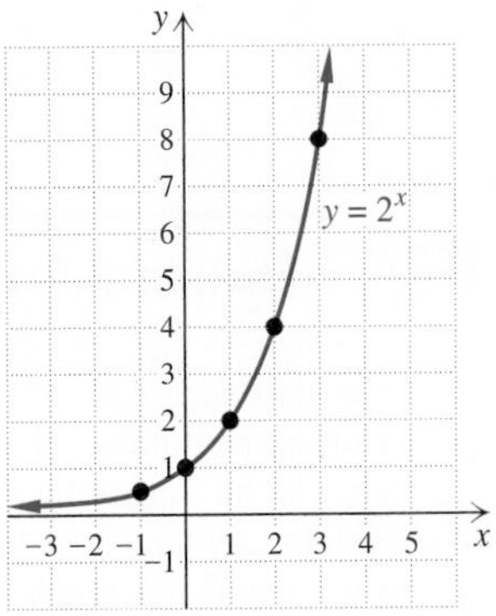

Figure 6.32

Population growth during certain time periods can be described by an exponential function. Whether it is a population of bacteria, fish, flowers, or people, the same general trend emerges: a period of rapid (exponential) growth, which is then followed by a leveling-off period.

b) The domain is all real numbers, $\mathbb{R}$. The range is $y > 0$. Notice that y can never have a value equal to 0.

All exponential functions of the form $y = a^x$, $a > 1$ will have the general shape of the graph illustrated in Example 8 (Fig. 6.32).

▶ **Example 9**
a) Graph $y = (\frac{1}{2})^x$.   **b)** Determine the domain and range of the function.

Solution: **a)** The graph is illustrated in Fig. 6.33.

$$y = \left(\frac{1}{2}\right)^x.$$

$$x = -3, \qquad y = \left(\frac{1}{2}\right)^{-3} = 2^3 = 8$$

$$x = -2, \qquad y = \left(\frac{1}{2}\right)^{-2} = 2^2 = 4$$

$$x = -1, \qquad y = \left(\frac{1}{2}\right)^{-1} = 2^1 = 2$$

$$x = 0, \qquad y = \left(\frac{1}{2}\right)^{0} = 1$$

$$x = 2, \qquad y = \left(\frac{1}{2}\right)^{2} = \frac{1}{4}$$

$$x = 3, \qquad y = \left(\frac{1}{2}\right)^{3} = \frac{1}{8}$$

x	y
−3	8
−2	4
−1	2
0	1
2	$\frac{1}{4}$
3	$\frac{1}{8}$

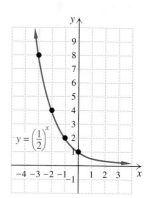

Figure 6.33

b) The domain is the set of all real numbers, $\mathbb{R}$. The range is $y > 0$.

All exponential functions of the form $y = a^x$, $0 < a < 1$, will have the general shape of the graph illustrated in Example 9 (Fig. 6.33). Can you now predict the shape of the graph of $y = e^x$? Remember: e has a value of about 2.7183.

Section 6.9 Exercises

1. What is a relation? **2.** What is a function?
3. What is the domain of a function?
4. What is the range of a function?
5. Explain how and why the vertical line test can be used to determine if a graph is a function.
6. Give three examples where one quantity is a function of another quantity.

Determine which of the following graphs are functions. For each function give the domain and range.

7.

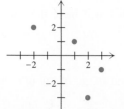

8.

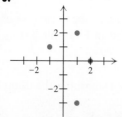

9.

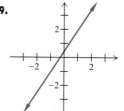

10.

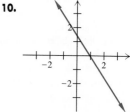

11.

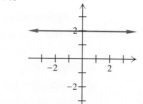

12.

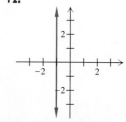

13.

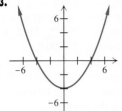

14.

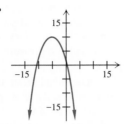

23.

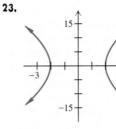

24.

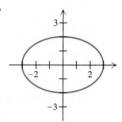

15.

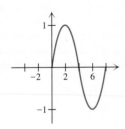

16.

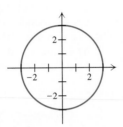

Evaluate each function for the given value of x.

25. $f(x) = x + 2, \quad x = 7$
26. $f(x) = -4x - 6, \quad x = -4$
27. $f(x) = -5x + 8, \quad x = -2$
28. $f(x) = 4x - 6, \quad x = 0$
29. $f(x) = -3x - 2, \quad x = 4$
30. $f(x) = 7x + 5, \quad x = -3$
31. $f(x) = x^2 + 2x + 4, \quad x = 6$
32. $f(x) = x^2 - 12, \quad x = 6$
33. $f(x) = 2x^2 - 2x - 8, \quad x = -2$
34. $f(x) = -x^2 + 3x + 7, \quad x = 2$
35. $f(x) = -x^2 + 5x - 12, \quad x = 4$
36. $f(x) = 4x^2 + 3x - 9, \quad x = 2$
37. $f(x) = -6x^2 - 6x - 12, \quad x = -3$
38. $f(x) = -3x^2 + 5x - 9, \quad x = -2$

17.

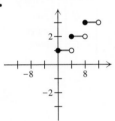

18.

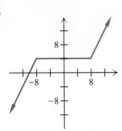

In Ex. 39–54:
a) Determine whether the parabola will open upward or downward.
b) Find the equation of the axis of symmetry.
c) Find the vertex.
d) Find the y-intercept.
e) Find the x-intercepts if they exist.
f) Sketch the graph.
g) Find the domain and range of the function.

39. $y = x^2 + 1$ **40.** $y = x^2 - 4$
41. $y = -x^2 - 9$ **42.** $y = -x^2 + 25$
43. $y = -x^2 + 2$ **44.** $y = -2x^2 + 8$
45. $y = 2x^2 - 3$ **46.** $y = -3x^2 - 6$
47. $y = x^2 - 2x - 6$ **48.** $y = x^2 + 2x - 12$
49. $y = x^2 + 5x + 6$ **50.** $y = x^2 - 7x - 8$
51. $y = -x^2 + 4x - 6$ **52.** $y = -x^2 + 8x - 8$
53. $y = 2x^2 - 2x - 12$ **54.** $y = -3x^2 - 5x + 2$

19.

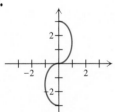

20.

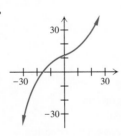

For each function draw the graph and state the domain and range.

55. $y = 3^x$ **56.** $y = 4^x$ **57.** $y = \left(\frac{1}{3}\right)^x$
58. $y = \left(\frac{1}{4}\right)^x$ **59.** $y = 2^x + 1$ **60.** $y = 2^x - 1$
61. $y = 2^{x+1}$ **62.** $y = 3^{x-1}$ **63.** $y = 3^x - 1$
64. $y = 2^{2x}$ **65.** $y = 2^{2x} - 1$ **66.** $y = \left(\frac{1}{2}\right)^x + 1$

21.

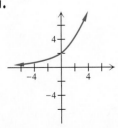

22.

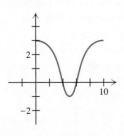

67. A. G. Edwards, a roofing contractor, determines the price, p, in dollars, for roofing a private house by the function $p = 112.6u$, where u is the number of units of tile required (1 unit is about 100 ft^2). Find the price of roofing a house that requires 20 units of roofing tiles.

68. The annual growth, g, in inches, of a certain type of maple tree is a function of its age. The younger a tree is, the greater its growth. The annual growth may be approximated by the function $g = 5.2 + 0.01t - 0.008t^2$, where t is years and $1 \le t \le 20$. Estimate the growth of the maple tree when the tree is (a) 4 years old and (b) 20 years old.

69. When an object is tossed up with a velocity of 64 ft/sec from the top of a 180-ft tall building, its distance from the ground, d, in feet, can be found by the function $d = -16t^2 + 64t + 180$ where t is time in seconds. Find the distance the object is from the ground (a) after 2 sec and (b) after 5 sec.

70. A checkerboard contains 64 squares. If two pennies are placed on the first square, 4 pennies on the second square, 8 pennies on the third square, 16 pennies on the fourth square, and so on, the number of pennies on the nth square can be found by the function $A = 2^n$.
   a) Find the number of pennies placed on the 30th square.
   b) What is this amount in dollars?

71. The atmospheric pressure, p, in pounds per square inch (psi), at an elevation of x feet above sea level can be found by the formula $p = 14.7e^{-0.00004x}$. Find the atmospheric pressure at an elevation of 4000 ft.

72. The spacing of the frets on the neck of a classical guitar is determined from the equation $d = (21.9)(2)^{(20-x)/12}$, where $x =$ the fret number and $d =$ the distance in centimeters of the xth fret from the bridge.
   a) Determine how far the 19th fret should be from the bridge (rounded off to one decimal place).
   b) Determine how far the fourth fret should be from the bridge (rounded off to one decimal place).
   c) The distance of the nut from the bridge can be found by letting $x = 0$ in the given exponential equation. Find that distance (rounded off to one decimal place).

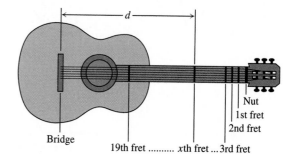

CHAPTER 6 SUMMARY

Key Terms

6.1
algebra
algebraic expression
base
check
constant
equation
evaluate
exponent
order of operations
solution to an equation
solve an equation
variable

6.2
algorithm
identity
inconsistent equation
like terms
linear (or first degree) equation
numerical coefficient
proportion
ratio
simplify
term
unlike terms

6.3
exponential formula
formula
subscript

6.5
compound inequality
inequality
order (or sense) of inequality

6.6
x- and y-axes
Cartesian (or rectangular) coordinate system
collinear
graph
x- and y-intercepts
ordered pair
origin
quadrants
slope of a line

6.7
half-plane

6.8
binomial
factoring
FOIL method
parabola
quadratic equation
trinomial

6.9
domain
exponential function
function
linear function
quadratic function
range
relation
vertical line test

Important Facts

Properties used to solve equations

If $a = b$ then $a + c = b + c$	Addition property
If $a = b$ then $a - c = b - c$	Subtraction property
If $a = b$ then $ac = bc$	Multiplication property
If $a = b$ then $a/c = b/c$, $c \neq 0$	Division property

Inequality symbols

$a < b$, means a is less than b
$a \leq b$, means a is less than or equal to b
$a > b$, means a is greater than b.
$a \geq b$, means a is greater than or equal to b.

Intercepts

To find the x-intercept, set $y = 0$ and solve the resulting equation for x.
To find the y-intercept, set $x = 0$ and solve the resulting equation for y.

Slope

Slope (m): $m = \dfrac{y_2 - y_1}{x_2 - x_1}$

Slope-intercept form of a line: $y = mx + b$

Equations and Formulas

Linear equation in two variables:
$$ax + by = c, \qquad a \neq 0, \quad b \neq 0$$
Quadratic equation in one variable:
$$ax^2 + bx + c = 0, \quad a \neq 0$$
Quadratic equation (or function) in two variables:
$$y = ax^2 + bx + c, \quad a \neq 0$$
Exponential equation (or function):
$$y = a^x, \qquad a \neq 1, \quad a > 0$$
Exponential growth or decay formula:
$$P = P_0 a^{kt}, \qquad a \neq 1, \quad a > 0$$
Quadratic formula:
$$x = \frac{-b \pm \sqrt{b^2 - 4ac}}{2a}$$

Axis of symmetry of a parabola:
$$x = \frac{-b}{2a}$$

Zero-factor property

If $a \cdot b = 0$ then $a = 0$ or $b = 0$.

CHAPTER 6 REVIEW EXERCISES

6.1

Evaluate each expression for the given value(s) of the variable.

1. $x^2 + 6$, $x = 5$
2. $-x^2 - 5$, $x = -4$
3. $5x^2 - 3x + 1$, $x = 4$
4. $-2x^2 + 9x - 2$, $x = -3$
5. $5x^3 + 11x - 2$, $x = \frac{1}{2}$
6. $4x^2 - 2xy + 3y^2$, $x = 2$, $y = -1$

6.2

Combine like terms.

7. $3x + 4 - 6 - x$
8. $x + 3(x - 2) - 5x$
9. $2(x - 4) + \frac{1}{2}(2x + 3)$

Solve the following equations for the given variable.

10. $3s + 7 = 22$
11. $2r + 3 = 5r - 15$

12. $\dfrac{x + 6}{4} = \dfrac{x + 3}{2}$

13. $4(x - 2) = 3 + 5(x + 4)$

14. $\dfrac{x}{3} + \dfrac{2}{5} = 4$

15. $3x + 0.03 = 0.41$

16. A mason lays 120 blocks in 1 hr 40 min. How long will it take her to lay 450 blocks?

6.3

Use the formula to find the value of the indicated variable for the values given.

17. $A = \frac{1}{2}bh$
    Find A when $b = 6$ and $h = 12$. (geometry)

18. $F_R = \sqrt{F_x^2 + F_y^2}$
    Find F_R when $F_x = 12$ and $F_y = 5$. (physics)

19. $Z = \dfrac{\bar{x} - \mu}{\dfrac{\sigma}{\sqrt{n}}}$
    Find $\bar{x}$ when $Z = 2$, $\mu = 100$, $\sigma = 3$, and $n = 16$. (statistics)

20. $V = \frac{1}{3}\pi r^2 h$
    Find h when $r = 3$, $\pi = 3.14$, and $V = 56.52$ (geometry)

Solve for y.

21. $2x - y = 6$
22. $3x + 4y = 8$
23. $2x - 3y + 52 = 30$
24. $-x - y + 2z = 0$

Solve for the variable indicated.

25. $A = lw$ for l
26. $P = 2l + 2w$ for w

27. $A = \dfrac{x_1 + x_2 + x_3}{3}$ for x_2

28. $a_n = a_1 + (n - 1)d$ for d

6.4

Write the following in mathematical terms.

29. 8 decreased by 7 times x
30. 4 more than the sum of 9 and x
31. 7 more than 6 times r
32. 11 less than 8 divided by q

Write an equation that can be used to solve the problem. Solve the equation and find the desired value(s).

33. Twelve decreased by 3 times a number is 21.
34. Twice a number decreased by 11 is the same as 4 more than the number.
35. Nine times the sum of a number and 4 is 72.
36. Fourteen more than 10 times a number is the same as 8 times the sum of the number and 12.

Write the equation and then find the solution.

37. Gina wishes to build a bookcase whose height is 3 ft more than the width. The bookcase is to have two shelves, not including the top and bottom pieces. If she has 24 ft of lumber with which to build the bookcase, what will be its height and width?
38. Sixty hours of overtime must be split among three workers. One worker must be assigned twice the number of hours assigned to the other two. Find the number of hours of overtime for each worker.
39. In a sales contest the first prize is $300 more than two times the second prize. The total amount of prize money is $1200. Find the amount allocated for first and second prizes.
40. A tiller can be rented for $30 an hour and purchased for $480. How many hours would the tiller have to be rented for the rental cost to equal the cost of purchasing a tiller?

6.5

Graph the solution set for the set of real numbers.

41. $4 + 5x > -16$ **42.** $3x + 7 \geq 5x + 9$

43. $3(x + 9) \leq 4x + 11$ **44.** $-2 \leq x < 4$

Graph the solution set for the set of integers.

45. $2 + 7x > -12$ **46.** $11 + 3x \leq 32$

47. $-5 \leq x \leq 2$ **48.** $-1 < x \leq 7$

6.6

Graph the following ordered pairs.

49. $(3, 5)$ **50.** $(-3, 2)$

51. $(-3, -4)$ **52.** $(6, -7)$

Points A, B, and C are vertices of a rectangle. Plot the points. Find the coordinates of the fourth point, D, to complete the rectangle. Find the area of the rectangle.

53. $A(-2, 2)$, $B(3, 2)$, $C(-2, -3)$

54. $A(-3, 1)$, $B(-3, -2)$, $C(4, -2)$

Graph the equation by plotting points.

55. $x + y = 3$ **56.** $2x + 3y = 12$

57. $x = y$ **58.** $x = 3$

Graph each equation using the x- and y-intercepts.

59. $x + 2y = 4$ **60.** $3x - 2y = 6$

61. $3x - y = 6$ **62.** $2x + 3y = 9$

Find the slope of the line through the given points.

63. $(3, 5)$, $(-1, 6)$ **64.** $(4, 2)$, $(6, -3)$

65. $(-1, -4)$, $(5, 3)$ **66.** $(6, 2)$, $(-6, -2)$

Graph each equation by plotting the y-intercept and then plotting a second point by making use of the slope.

67. $y = 3x - 2$ **68.** $y = 2x + 3$

69. $2y + x = 8$ **70.** $y = -x - 1$

Determine the equation of the graph.

71.

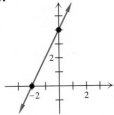

72.

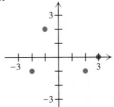

73. The monthly disability income, I, that Nadja receives is $I = 460 - 0.5e$, where e is her monthly earnings for her part-time job for the previous month.

a) Draw a graph of disability income versus earnings for earnings up to $920.

b) If Nadja earns $600 in January, how much disability income will she receive in February?

c) If she received $380 disability income in November, how much did she earn in October?

74. The monthly rental cost, c, in dollars, for space in the Galleria Mall can be approximated by the equation $C = 1.70A + 3000$, where A is the area, in square ft, of space rented.

a) Draw a graph of monthly rental cost versus square feet for up to 12,000 ft^2.

b) Determine the monthly rental cost if 2000 ft^2 are rented.

c) If the rental cost is $10,000 per month, how many square feet are rented?

6.7

Graph each of the following inequalities.

75. $5x + 7y < 35$ **76.** $3x + 2y \geq 12$

77. $2x - 3y > 12$ **78.** $-3x - 5y \leq 15$

6.8

Factor the trinomial. If the trinomial cannot be factored, so state.

79. $x^2 + 9x + 20$ **80.** $x^2 + x - 6$

81. $x^2 - 10x + 24$ **82.** $x^2 - 11x + 24$

83. $2x^2 - 7x - 15$ **84.** $3x^2 + 5x - 2$

Solve each equation by factoring.

85. $x^2 + 3x + 2 = 0$ **86.** $x^2 - 6x = -5$

87. $2x^2 - 9x - 18 = 0$ **88.** $3x^2 = -7x - 2$

Solve each equation using the quadratic formula. If the equation has no real solution, so state.

89. $x^2 - 5x + 1 = 0$ **90.** $x^2 - 3x + 2 = 0$

91. $x^2 + 2x + 6 = 0$ **92.** $x^2 - x - 3 = 0$

6.9

Determine whether each of the following is a function. If the graph is a function give the domain and range.

93.

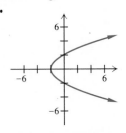

94.

95.

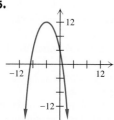

96.

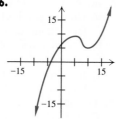

Evaluate $f(x)$ for the given value of x.

97. $f(x) = 3x - 2$, $x = 2$

98. $f(x) = -4x + 5$, $x = -3$

99. $f(x) = 2x^2 - 3x + 4$, $x = 5$

100. $f(x) = -3x^2 - 6x + 8$, $x = 2$

For each function:

a) Determine whether the parabola will open upward or downward.

b) Find the equation of the axis of symmetry.

c) Find the vertex.

d) Find the y-intercept.

e) Find the x-intercepts if they exist.

f) Sketch the graph.

g) Determine the domain and range.

101. $y = -x^2 + 7$ **102.** $y = 3x^2 - 24x - 30$

Draw the graph of each function and state the domain and range.

103. $y = 2^x$ **104.** $y = (\frac{1}{2})^x$

6.2, 6.3, 6.9

105. The gas mileage, m, of a specific car can be estimated by the equation (or function)

$$m = 30 - 0.002n^2, \quad 20 \le n \le 80,$$

where n is the speed of the car in miles per hour. Estimate the gas mileage when the car travels at 60 mph.

106. The approximate number of accidents in one month, n, involving drivers between 16 and 30 years of age can be approximated by the equation (or function)

$$n = 2a^2 - 80a + 5000, \quad 16 \le a \le 30,$$

where a is the age of the driver. Approximate the number of accidents in one month that involved

a) 18-year-olds   **b)** 25-year-olds.

107. The percent of light filtering through Swan Lake, p, can be approximated by the equation (or function) $P = 100(0.92)^x$, where x is the depth in feet. Find the percent of light filtering through at 4.5 ft.

CHAPTER TEST

1. Evaluate $-5x^2 + 2x - 7$ when $x = -3$.

Solve the equation.

2. $-2x - 4 = 2(4x + 8)$

3. $-(x - 2) + 4x = 3x - 2(x + 5)$

In Ex. 4 and 5, write an equation to represent the problem, then solve the equation.

4. Four times a number decreased by 5 is 19. Find the number.

5. A salesperson must select between two payment plans. Plan 1 is a $100-a-week salary plus a 6% commission on sales. Plan 2 is a $300-a-week salary plus 4% commission on sales. What must the weekly sales be if the two plans are to result in the same salary?

6. Evaluate $A = L^2(W - B)$ when $L = 9$, $W = 11$, and $B = 3$.

7. Solve $3x - 2y = 6$ for y.

8. Graph the solution set of $4x - 6 < 2x + 3$ on the real number line.

9. Find the slope of the line through the points $(6, 2)$ and $(-3, 5)$.

In. Ex. 10 and 11, graph the equation.

10. $y = 4x - 1$

11. $4x - 3y = 12$

12. Graph the inequality $2y \geq -2x + 4$.

13. Solve the equation $x^2 - 6x = -8$ by factoring.

14. Solve the equation $2x^2 + 3x - 4 = 0$ using the quadratic formula.

15. Determine whether the following graph is a function. Explain your answer.

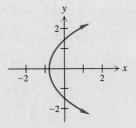

16. Evaluate $f(x) = -5x^2 + 3x - 8$ when $x = -3$.

17. For the equation $y = x^2 - 2x + 4$,
   a) determine whether the parabola will open upward or downward.
   b) find the equation of the axis of symmetry.
   c) find the vertex.
   d) find the y-intercept.
   e) find the x-intercepts if they exist.
   f) sketch the graph.
   g) determine the domain and range.

SYSTEMS OF LINEAR EQUATIONS & INEQUALITIES

I magine that you and a friend have hit upon a great idea for a new T-shirt. First you make a few to give away to friends, using your own money. Then other students see the shirt, and pretty soon everyone on campus wants one. Suddenly, you have entered the T-shirt business. To make a profit in your business, you need to keep track of the cost of your materials, the quantity of shirts sold, and the price at which you sell them — a relatively straightforward calculation.

But now suppose that you've got three other ideas for other designs, and you want to put them on sweatshirts as well as T-shirts, and you want to offer a variety of colors: black, white, blue, and maroon. The equation for finding the profitability of your venture becomes more complicated because there are more variables. To track your profits, you may need to develop and solve systems of equations.

In fact, student entrepreneurs often face these problems. Consider the case of David Mays and Jon Shecter, who met during their first week at Harvard in 1986. They started writing a free newsletter, *The Source*, to promote their rap music radio program. By their senior year, it had

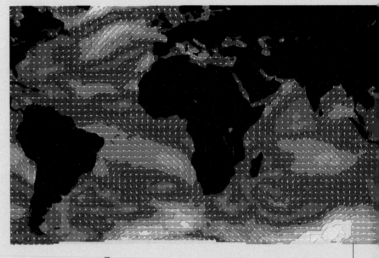

turned into a magazine with a circulation of 10,000. Following their graduation in 1990, they moved to New York, borrowed money, and now are trying to increase subscription rates and advertising income. The growth of their business requires juggling many financial details, sophisticated accounting, and projecting profitability. In order for the magazine to survive, one expert predicts that they must have half a million subscribers within the next two or three years. The magazine seems to be well on its way: Mays and Shecter hope that it will be the *Rolling Stone* of the next generation.

Mathematics can provide the means to handle and systematize a large body of numerical data. Here a matrix is used to display "vectors" that show wind speed and direction for the world's oceans. NASA used radar to collect 350,000 wind measurements over a three-day period.

For most businesses, numerous factors must be considered to determine not only whether the business is profitable, but also how much they should charge their customers, what production method is most efficient, what return they can expect by placing advertisements, etc. Many small business owners routinely make these calculations using their own experience, mathematics, and sometimes computer programs. Larger companies often employ inventory analysts, quality control engineers, and efficiency experts to aid them, along with computers to keep track of the vast quantity of data.

Linear programming provides businesses and governments with a mathematical form of decision making that makes the most efficient use of time and resources. Telecommunications companies use it to route calls through regional switching centers so few of their customers will reach a "no circuits available" message.

7.1 Systems of Linear Equations

In Chapter 6 we discussed linear equations in two variables. In algebra it is often necessary to find the common solution to two or more such equations. We refer to the equations in this type of problem as a **system of linear equations** or as **simultaneous linear equations**. A **solution to a system of equations** is the ordered pair or pairs that satisfy all equations in the system. A system of linear equations may have exactly one solution, no solution, or infinitely many solutions.

The solution to a system of linear equations may be found by a number of different techniques.

This section illustrates how a system of equations may be solved by graphing. Section 7.2 illustrates two algebraic methods, substitution and the addition method, for solving a system of linear equations.

▶ **Example 1**

Determine which of the ordered pairs is a solution to the following system of equations.

$$2x - 5y = 10$$
$$3x + y = 15$$

a) $(0, -2)$ **b)** $(5, 0)$ **c)** $(4, 3)$

Solution: For the ordered pair to be a solution to the system it must satisfy each equation in the system.

a)
$$2x - 5y = 10 \qquad\qquad 3x + y = 15$$
$$2(0) - 5(-2) = 10 \qquad\qquad 3(0) + (-2) = 15$$
$$10 = 10 \quad \text{True} \qquad\qquad -2 = 15 \quad \text{False}$$

Since $(0, -2)$ does not satisfy both equations, it is not a solution to the system.

b)
$$2x - 5y = 10 \qquad\qquad 3x + y = 15$$
$$2(5) - 5(0) = 10 \qquad\qquad 3(5) + 0 = 15$$
$$10 = 10 \quad \text{True} \qquad\qquad 15 = 15 \quad \text{True}$$

Since $(5, 0)$ satisfies both equations, it is a solution to the system.

c)
$$2x - 5y = 10 \qquad\qquad 3x + y = 15$$
$$2(4) - 5(3) = 10 \qquad\qquad 3(4) + 3 = 15$$
$$-7 = 10 \quad \text{False} \qquad\qquad 15 = 15 \quad \text{True}$$

Since $(4, 3)$ does not satisfy both equations in the system, it is not a solution to the system.

To find the solution to a system of linear equations graphically, we graph both of the equations on the same set of axes. The point or points of intersection of the graphs are the solution to the system of equations. When two equations are graphed, three situations are possible. The two lines may intersect at one point, as in Example 2; or the two lines may be parallel and not intersect, as in Example 3; or the two equations may represent the same line, as in Example 4.

Procedure for Solving a System of Equations by Graphing

1. Determine three ordered pairs that satisfy each equation.
2. Graph both of the equations on the same set of axes.
3. The point or points of intersection of the graphs are the solution to the system of equations.

Since the solution to a system of equations may not be integer values you may not be able to obtain the exact solution by graphing.

▶ **Example 2**

Find the solution to the following system of linear equations graphically.

$$x + 2y = 4$$
$$2x - 3y = 1$$

Solution: To find the solution, graph both $x + 2y = 4$ and $2x - 3y = 1$ on the same set of axes (Fig. 7.1). Three points that satisfy each equation are illustrated in the table.

$x + 2y = 4$

x	y
0	2
2	1
4	0

$2x - 3y = 1$

x	y
0	$-\frac{1}{3}$
$\frac{1}{2}$	0
2	1

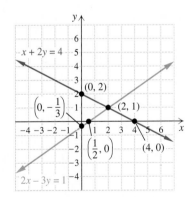

Figure 7.1

The graphs intersect at $(2, 1)$, which is the solution. This is the only point that satisfies *both* equations.

Check:

$x + 2y = 4$	$2x - 3y = 1$
$2 + 2(1) = 4$	$2(2) - 3(1) = 1$
$2 + 2 = 4$	$4 - 3 = 1$
$4 = 4$ True	$1 = 1$ True

The system of equations in Example 2 is an example of a **consistent system of equations**. A consistent system of equations is one that has a solution.

▶ **Example 3**

Find the solution to the following system of equations graphically.

$$2x + y = 3$$
$$2x + y + 5 = 0$$

Solution: Three ordered pairs that satisfy the equation $2x + y = 3$ are $(0, 3)$, $(\frac{3}{2}, 0)$ and $(-1, 5)$. Three ordered pairs that satisfy the equation $2x + y + 5 = 0$ are $(0, -5)$, $(-\frac{5}{2}, 0)$ and $(1, -7)$. The graphs of both equations are given in Fig. 7.2. Since the two lines are parallel they do not intersect; therefore the system has *no solution*.

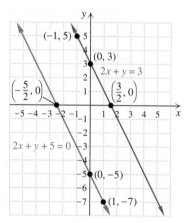

Figure 7.2

The system of equations in Example 3 has no solution. A system of equations that has no solution is called an **inconsistent system**.

▶ **Example 4**

Find the solution to the following system of equations graphically:

$$x - \frac{1}{2}y = 2$$

$$y = 2x - 4$$

Solution: Three ordered pairs that satisfy the equation $x - \frac{1}{2}y = 2$ are $(1, -2)$, $(2, 0)$, and $(-1, -6)$. Three ordered pairs that satisfy the equation $y = 2x - 4$ are $(0, -4)$, $(-2, -8)$, and $(3, 2)$. Graph the equations on the same set of axes (Fig. 7.3). We see that all six points are on the same line. The two equations represent the same line. Therefore every ordered pair that is a solution for one equation is also a solution for the other equation. Every point on the line satisfies both equations; thus this system has an *infinite number of solutions*.

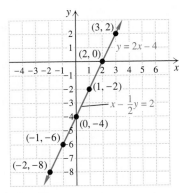

Figure 7.3

When a system of equations has an infinite number of solutions, as in Example 4, it is called a **dependent system**. Note that a dependent system is also a consistent system, since it has a solution.

Figure 7.4 summarizes the three possibilities for a system of linear equations.

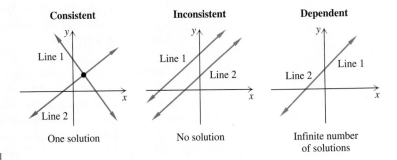

Figure 7.4

▶ **Example 5**

A college is considering purchasing one of two types of computer systems. System 1 is a minicomputer that costs $12,000 with terminals that cost $2000 each. System 2 is a networking system in which the networking device costs $4000 and the terminals cost $3000 each.

a) Write a system of equations to represent the cost of the two types of computer systems each with n terminals.

b) Graph both equations on the same set of axes and determine the number of terminals needed for both systems to have the same cost.

c) If ten terminals are to be used, which system is the least expensive?

Solution: Let n = the number of terminals used. The total cost of each computer system is the initial cost plus the cost of the terminals.

a) minicomputer system: $c = 12{,}000 + 2000n$
      network system: $c = 4000 + 3000n$

b)

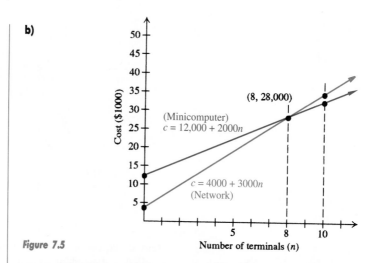

Figure 7.5

We have graphed the cost, c, versus the number of terminals, n, for up to 12 terminals (Fig. 7.5). On the graph the lines intersect at the point (8, 28,000). Thus for eight terminals, both systems would have the same cost, $28,000.

c) From the graph we can see that for more than eight terminals the mini-computer is the least expensive system. Thus for 10 terminals the minicom-puter is the least expensive.

Manufacturers use a technique called **break-even analysis** to determine how many units of an item must be sold for the business to "break even"—that is, for its total revenue to equal its total cost. Suppose we let the horizontal axis represent the number of units manufactured and sold and the vertical axis represent dollars. Then linear equations for cost, C, and revenue, R, can both be sketched (Fig. 7.6) on the set of axes since both are expressed in dollars and both are a function of the number of units.

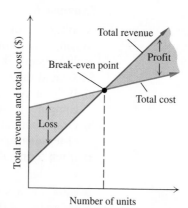

Figure 7.6

Initially the cost curve is higher than the revenue curve because of fixed overhead costs. During low levels of production the manufacturer suffers a loss

(the cost curve is greater). During higher levels of production the manufacturer realizes a profit (the profit curve is greater). The point at which the two curves intersect is called the **break-even point**. It is at that number of units sold that revenue is equal to cost and the manufacturer breaks even.

▶ **Example 6**

Charlie's Little Box company can sell a particular box for $2 per unit. The costs for making the box are a fixed overhead of $3 and a production cost of $1 per unit.

a) How many boxes must Charlie sell to break even?
b) Determine whether Charlie makes a profit or loss if he sells five boxes. What is the profit or loss?
c) How many units must Charlie sell to make a profit of $9?

Solution: Let x denote the number of boxes made and sold. The revenue is given by the equation

$$R = 2x \quad \text{($2 times the number of units)}$$

and the cost is given by the equation

$$C = 3 + x \quad \text{($3 plus $1 times the number of units)}$$

a) The break-even point is the point where the revenue and cost graphs intersect. In Fig. 7.7 the graphs intersect at the point $(3, 6)$, which is the break-even point. Thus for Charlie to break even he must manufacture three boxes. When three boxes are made and sold the cost and revenue are both $6.

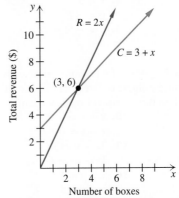

Figure 7.7

b) Examining the graph we can see that if Charlie sells five boxes he will have a profit, P, which is the revenue minus the cost. The profit formula is:

$$\begin{aligned}
P &= R - C \\
&= 2x - (3 + x) \\
&= 2x - 3 - x \\
&= x - 3
\end{aligned}$$

The profit formula is $P = x - 3$. For five boxes

$$P = x - 3$$
$$= 5 - 3 = 2$$

Thus Charlie has a profit of $2 if he sells five boxes.

c) We can determine the number of boxes Charlie must sell to have a profit of $9 by using the profit formula. Substituting 9 for P, we obtain

$$P = x - 3$$
$$9 = x - 3$$
$$12 = x$$

Thus Charlie must sell 12 boxes to make a profit of $9.

Section 7.1 Exercises

1. Define a consistent system of equations.
2. Define an inconsistent system of equations.
3. Define a dependent system of equations.
4. Define a break-even point.

In Ex. 5–8, solve the following systems of equations graphically.

5. $x = 2$
  $y = 6$
6. $x = 6$
  $y = -2$
7. $x = -4$
  $y = 5$
8. $x = -5$
  $y = -3$

In Ex. 9–24, solve each system of equations graphically. If the system does not have a single ordered pair as a solution, state whether the system is inconsistent or dependent.

9. $y = x + 3$
  $y = -x - 5$
10. $x + y = -2$
  $x - y = 0$
11. $2x - y = 6$
  $x + 2y = -2$
12. $3x - y = 3$
  $3y - 4x = 6$
13. $y - x = 3$
  $x - y = 4$
14. $x + y = 1$
  $x - y = 5$
15. $y = 2x - 4$
  $2x + y = 0$
16. $2x + y = 3$
  $2y = 6 - 4x$
17. $2x - y = -3$
  $2x + y = -9$
18. $y = x + 3$
  $y = -1$
19. $x = 1$
  $x + y + 3 = 0$
20. $3x + 2y = 6$
  $6x + 4y = 12$
21. $2x - 3y = 6$
  $3y - 2x = 3$
22. $x = \frac{1}{2}y - 3$
  $y - 2x = 3$
23. $y = \frac{1}{3}x - 2$
  $x - 3y = 6$
24. $2(x - 1) + 2y = 0$
  $3x + 2(y + 2) = 0$

25. a) If the two lines in a system of equations have different slopes, how many solutions will the system have? Explain your answer.
  b) If the two lines in a system of equations have the same slope but different y-intercepts, how many solutions will the system have? Explain.
  c) If the two lines in a system of equations have the same slope and the same y-intercept, how many solutions will the system have? Explain.

After answering question 25, determine without graphing whether each system of equations has exactly one solution, no solution, or an infinite number of solutions.

26. $2x + y = 6$
  $-2x + y = 4$
27. $3x + 2y = 4$
  $4y = -6x + 6$
28. $x - y - 7 = 0$
  $4x + 4y = 5$
29. $4x - y = 5$
  $4y - x = 5$
30. $3x - y = 7$
  $y = -2x + 3$
31. $2x - 3y = 6$
  $x - \frac{3}{2}y = 3$
32. $x - 4y = 12$
  $x = 4y + 3$
33. $3x = 6y + 5$
  $y = \frac{1}{2}x - 3$
34. $3y = 6x + 4$
  $-2x + y = \frac{4}{3}$
35. $x - 2y = 6$
  $-x + 2y = -6$
36. $12x - 5y = 4$
  $3x + 4y = 6$

Two lines are **perpendicular** when they meet at a right angle (90° angle). Two lines are perpendicular to each other when their slopes are **negative reciprocals** of each other. The negative reciprocal of 2 is $-\frac{1}{2}$, the negative reciprocal of $\frac{3}{5}$ is $-1/(\frac{3}{5})$ or $-\frac{5}{3}$, and so on. If a represents any real number, except 0, its negative reciprocal is $-1/a$. Note that the product of a number and its negative reciprocal is -1. In Ex. 37–40, determine, by finding the slope of each line, whether the lines will be perpendicular to each other when graphed.

37. $3x - 2y = 5$
$2x - 3y = 5$

38. $2x + 3y = 8$
$2x - 3y = 6$

39. $4x + y = 6$
$4y = x + 12$

40. $6x - 5y = 2$
$-10x = 2 + 12y$

41. In Example 5, if the minicomputer costs $10,000 with terminals that cost $3000, and the networking device costs $5000 with terminals that cost $4000:
  a) Write the system of equations to represent the cost of the two types of computer systems.
  b) Graph both equations (for up to and including eight terminals) on the same set of axes.
  c) Determine the number of terminals that must be used for both systems to have the same cost.

42. The total cost of printing a book consists of a setup charge and an additional fee for material for each book printed. The ABC Printing Company charges a $1600 setup fee plus $6 per book it prints. The XYZ Printing Company has a setup fee of $1200, plus $8 per book it prints.
  a) Write a system of equations to represent the cost of printing the books with each company.
  b) Graph both equations (for up to and including 300 books) on the same set of axes.
  c) Determine the number of books that need to be printed for both companies to have the same cost.
  d) If 100 books are to be printed, which is the less expensive printer?

43. Ms. Ly, a salesperson, has the option of selecting the method by which her weekly salary will be determined. Option 1 is a straight 8% of her total dollar sales, and option 2 is a salary of $200 plus 4% of her total dollar sales.

  a) Write a system of equations to represent her salary by each option.
  b) Graph both equations (for up to and including $10,000 in sales) on the same set of axes.
  c) Find the total dollar sales needed for her salary from option 1 to equal her salary from option 2.
  d) If she expects to average $8000 per week in sales, which option should she choose?

44. Talking Ed's Little Horse Harness company can sell a simple halter for $20 per unit. The costs for making the halter are a fixed overhead of $300 and a production cost of $10 per unit (see Example 6).
  a) Write the cost and revenue equations.
  b) Graph both equations, for up to and including 50 units, on the same set of axes.
  c) How many halters must Talking Ed's company sell to break even?
  d) Determine whether the company makes a profit or loss if they sell 10 halters. What is the profit or loss?
  e) How many halters must Ed's company sell to realize a profit of $1000?

45. A manufacturer can sell a certain speaker for $165 per unit. Manufacturing costs consist of a fixed overhead of $8400 and a production cost of $95 per unit.
  a) Write the cost and revenue equations.
  b) Graph both equations, for up to and including 150 units, on the same set of axes.
  c) How many units must the manufacturer sell to break even?
  d) What is the manufacturer's profit or loss if 100 units are sold?
  e) How many units must the manufacturer sell to make a profit of $1250?

46. Explain how you can determine if a system of two equations will be consistent, dependent, or inconsistent without graphing the equations.

7.2 Solving Systems of Equations by the Substitution and Addition Methods

Having solved systems of equations by graphing in Section 7.1, we are ready for two other methods used to solve systems of linear equations: the substitution method and the addition method.

Substitution Method

Procedure for Solving a System of Equations Using the Substitution Method

1. Solve one of the equations for one of the variables. If possible solve for a variable with a numerical coefficient of 1. By doing this you may avoid working with fractions.
2. Substitute the expression found in step 1 into the other equation. This yields an equation in terms of a single variable.
3. Solve the equation found in step 2 for the variable.
4. Substitute the value found in step 3 into the equation you rewrote in step 1, and solve for the remaining variable.

Examples 1, 2, and 3 illustrate the substitution method. These systems of equations are the same as in Examples 2, 3, and 4 in Section 7.1.

▶ **Example 1**

Solve the following system of equations by substitution.

$$x + 2y = 4$$
$$2x - 3y = 1$$

Solution: The numerical coefficient of the x term in the equation $x + 2y = 4$ is 1. Therefore solve this equation for x.

Step 1.
$$x + 2y = 4$$
$$x + 2y - 2y = 4 - 2y$$
$$x = 4 - 2y$$

Step 2. Substitute $4 - 2y$ for x in the other equation and solve for y.

$$2x - 3y = 1$$
$$2(4 - 2y) - 3y = 1$$

Step 3. Now solve the equation for y.

$$8 - 4y - 3y = 1$$
$$8 - 7y = 1$$
$$8 - 8 - 7y = 1 - 8$$
$$-7y = -7$$
$$\frac{-7y}{-7} = \frac{-7}{-7}$$
$$y = 1$$

Step 4. Substitute $y = 1$ in the equation solved for x, and solve for the variable x:

$$x = 4 - 2y$$
$$x = 4 - 2(1)$$
$$x = 2$$

Thus the solution is the ordered pair $(2, 1)$. This answer checks with the solution obtained graphically in Section 7.1, Example 2.

▶ **Example 2**

Solve the following system of equations by substitution:

$$2x + y = 3$$
$$2x + y + 5 = 0$$

Solution: Solve for y in the first equation:

$$2x + y = 3$$
$$2x - 2x + y = 3 - 2x$$
$$y = 3 - 2x$$

Now substitute $3 - 2x$ in place of y in the second equation:

$$2x + y + 5 = 0$$
$$2x + (3 - 2x) + 5 = 0$$
$$2x + 3 - 2x + 5 = 0$$
$$8 = 0 \quad \text{False}$$

Since 8 cannot be equal to 0, there is no solution to the system of equations. Thus the system of equations is inconsistent. This answer checks with the solution obtained graphically in Section 7.1, Example 3.

▶ **Example 3**

Solve the following system of equations by substitution:

$$x - \frac{1}{2}y = 2$$
$$y = 2x - 4$$

Solution: The second equation is already solved for y, so we will substitute $2x - 4$ for y in the first equation:

$$x - \frac{1}{2}y = 2$$

$$x - \frac{1}{2}(2x - 4) = 2$$

$$x - x + 2 = 2$$

$$2 = 2 \quad \text{True}$$

Since 2 is equal to 2, the system has infinitely many solutions. Thus the system of equations is dependent. This answer checks with the solution obtained in Section 7.1, Example 4.

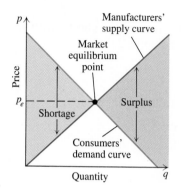

Figure 7.8

An application in economics involving systems of equations is the **law of supply and demand**. The market price p of a commodity is a determining factor in the number of units of the commodity that manufacturers are willing to supply and consumers are willing to buy. Generally, as the market price increases, manufacturers' supply increases and consumers' demand decreases. An example of a pair of linear supply and demand curves is sketched in Fig. 7.8. The vertical axis represents price, p. The horizontal axis represents quantity, q. Quantity depends on price, making price the independent variable. However, for this type of graph economists display the dependent variable quantity on the horizontal axis. This is just the opposite of what we have learned, but it best serves their purpose.

The point of intersection of the supply and demand curve is called the **market equilibrium point**. The price coordinate of this point, p_e, called the **equilibrium price**, is the market price at which supply equals demand. When supply equals demand, the market price is such that there will not be a surplus or a shortage of the commodity.

Although we will not graph supply and demand curves we can still find the market equilibrium price using the substitution method to solve a system of equations, as is illustrated in the next example.

▶ **Example 4**

The supply, S, and demand, D, equations for a certain commodity are given by $S = 2p - 5$ and $D = 20 - 3p$ respectively, where p represents the price of the commodity in dollars. Find the value of p when the supply equals the demand (the equilibrium price).

Solution: When supply equals demand, $S = D$, and we can substitute $2p - 5$ for D in the equation $D = 20 - 3p$, this gives

$$2p - 5 = 20 - 3p$$
$$2p + 3p - 5 = 20 - 3p + 3p$$
$$5p - 5 = 20$$
$$5p - 5 + 5 = 20 + 5$$
$$5p = 25$$
$$p = 5$$

Thus supply equals demand when the price of the commodity is $5.

If neither of the equations in a system of linear equations has a variable with a coefficient of 1, it is generally easier to solve the system by using the **addition (or elimination) method**.

Economics, a science dependent on mathematics, dates back to just before the Industrial Revolution of the eighteenth century. Technologies were being invented and applied to the manufacture of cloth, iron, transportation, and agriculture. These new technologies led to the development of mathematically based economic models. French economist Jules Dupuit (1804–1866) suggested a method to use to calculate the value of railroad bridges; the Irish economist Dionysis Larder (1793–1859) showed railroad companies how to structure their rates in order to increase their profits.

To solve a system of linear equations by the addition method it is necessary to obtain two equations whose sum will be a single equation containing only one variable. To achieve this goal, rewrite the system of equations as two equations where the coefficients of one of the variables are opposites (or additive inverses) of each other. For example, if one equation has a term of $2x$, we might rewrite the other equation so that its x term will be $-2x$. To obtain the desired equations, it might be necessary to multiply one or both equations in the original system by a number. When an equation is to be multiplied by a number, we will place brackets around the equation and place the number that is to multiply the equation before the brackets. For example, if we write $4[2x + 3y = 6]$ it means that each term in the equation $2x + 3y = 6$ is to be multiplied by 4:

$$4[2x + 3y = 6] \quad \text{gives} \quad 8x + 12y = 24$$

This notation will make our explanations much more efficient and easier for you to follow.

Procedure for Solving a System of Equations by the Addition Method

1. If necessary rewrite the equations so that the variables appear on one side of the equals sign and the constants on the other side of the equals sign.

2. If necessary multiply one or both equations by a constant(s), so that when you add them, the result will be an equation containing only one variable.

3. Add the equations to obtain a single equation in one variable.

4. Solve the equation in step 3 for the variable.

5. Substitute the value found in step 4 into either of the original equations and solve for the other variable.

▶ **Example 5**

Solve the following system of equations by the addition method:

$$x + 2y = 4$$
$$-x + y = 2$$

Solution: Since the coefficients of the x terms, 1 and -1, are additive inverses, the sum of the x terms will be zero. Thus the sum of the two equations will contain only one variable, y. Add the two equations to obtain one equation in one variable. Then solve for the remaining variable:

$$
\begin{array}{r}
x + 2y = 4 \\
-x + \ y = 2 \\
\hline
3y = 6
\end{array}
$$

$$\frac{3y}{3} = \frac{6}{3}$$

$$y = 2$$

Now substitute 2 for y in either of the original equations to find the value of x:

$$x + 2y = 4$$
$$x + 2(2) = 4$$
$$x + 4 = 4$$
$$x = 0$$

The solution to the system is $(0, 2)$

▶ **Example 6**

Solve the following system of equations by the addition method:

$$2x + y = 8$$
$$3x + y = 5$$

Solution: We want the sum of the two equations to have only one variable. We can eliminate the variable y by multiplying either equation by -1 and then adding. We will multiply the first equation by -1:

$$-1[2x + y = 8] \quad \text{gives} \quad -2x - y = -8$$
$$3x + y = 5 \qquad\qquad\qquad 3x + y = 5$$

$$\begin{array}{r} -2x - y = -8 \\ 3x + y = 5 \\ \hline x = -3 \end{array}$$

Now solve for y by substituting -3 for x in either of the original equations:

$$2x + y = 8$$
$$2(-3) + y = 8$$
$$-6 + y = 8$$
$$y = 14$$

The solution is $(-3, 14)$.

▶ **Example 7**

Solve the following system of equations by the addition method:

$$2x + y = 6$$
$$3x + 3y = 9$$

Solution: We can multiply the top equation by -3 and then add to eliminate the variable y:

$$-3[2x + y = 6] \quad \text{gives} \quad -6x - 3y = -18$$
$$3x + 3y = 9 \qquad\qquad\qquad 3x + 3y = 9$$

$$
\begin{array}{r}
-6x - 3y = -18 \\
3x + 3y = 9 \\
\hline
-3x = -9 \\
x = 3
\end{array}
$$

Now find y:

$$
\begin{aligned}
2x + y &= 6 \\
2(3) + y &= 6 \\
6 + y &= 6 \\
y &= 0
\end{aligned}
$$

The solution is $(3, 0)$.

Note that in Example 7 we could have eliminated the variable x by multiplying the top equation by 3 and the bottom equation by -2, then adding. Try this now.

▶ **Example 8**

Solve the following system of equations by the addition method:

$$
\begin{aligned}
3x - 4y &= 8 \\
2x + 3y &= 9
\end{aligned}
$$

Solution: In this system we cannot eliminate a variable by multiplying only one equation by an integer value and then adding. To eliminate a variable, we will need to multiply each equation by a different number. To eliminate the variable x, we can multiply the top equation by 2 and the bottom by -3 (or the top by -2 and the bottom by 3) and then add the two equations. If we wish, we can instead eliminate the variable y by multiplying the top equation by 3 and the bottom by 4 and then adding the two equations. We will eliminate the variable x:

$$
\begin{array}{ll}
2[3x - 4y = 8] & \text{gives} \quad 6x - 8y = 16 \\
-3[2x + 3y = 9] & \phantom{\text{gives}} \quad -6x - 9y = -27
\end{array}
$$

$$
\begin{array}{r}
6x - 8y = 16 \\
-6x - 9y = -27 \\
\hline
-17y = -11
\end{array}
$$

$$
y = \frac{11}{17}
$$

We could now find x by substituting $\frac{11}{17}$ for y in either of the original equations.

Although it can be done, it gets pretty messy. Instead, we will solve for x by eliminating the variable y from the two original equations. To accomplish this, we can multiply the first equation by 3 and the second equation by 4, as illustrated below.

$$3[3x - 4y = 8] \quad \text{gives} \quad 9x - 12y = 24$$
$$4[2x + 3y = 9] \qquad\qquad 8x + 12y = 36$$

$$\begin{aligned} 9x - 12y &= 24 \\ 8x + 12y &= 36 \\ \hline 17x \qquad &= 60 \end{aligned}$$

$$x = \frac{60}{17}$$

The solution to the system is

$$\left(\frac{60}{17}, \frac{11}{17}\right).$$

When solving a system of linear equations by the addition method, if you obtain the equation $0 = 0$, it indicates that the system is dependent (both equations represent the same line) and there are an infinite number of solutions. When solving, if you obtain an equation such as $0 = 6$, or any other equation that is false, it means that the system is inconsistent (the two equations represent parallel lines), and there is no solution.

▶ **Example 9**

Hertz automobile rental agency charges $26 a day plus 15 cents a mile for a specific car. For the same car, Avis charges $18 a day plus 20 cents a mile. How far would you have to drive in one day for the total cost of Hertz to equal the total cost of Avis?

Solution: We are asked to find the number of miles that must be driven for each company to have the same total cost, c. First write a system of equations to represent the total cost for each of the companies in terms of the daily fee and the mileage charge.

Let x = number of miles driven

$$\text{Total cost} = \text{daily fee} + \text{mileage charge}$$
$$\text{Hertz:} \quad c = 26 + 0.15x$$
$$\text{Avis:} \quad c = 18 + 0.20x$$

Since we wish to determine when the cost will be the same, we will set the two costs equal to each other (substitution method) and solve the resulting equation:

$$26 + 0.15x = 18 + 0.20x$$
$$26 - 18 + 0.15x = 18 - 18 + 0.20x$$
$$8 + 0.15x = 0.20x$$
$$8 + 0.15x - 0.15x = 0.20x - 0.15x$$
$$8 = 0.05x$$
$$\frac{8}{0.05} = \frac{0.05x}{0.05}$$
$$160 = x$$

Thus if you travel 160 mi, the two companies' costs will be the same. If we construct a graph, Fig. 7.9, of the two cost equations, the point of intersection is (160, 50). This means that if you were to drive 160 mi in either car the cost would be $50.

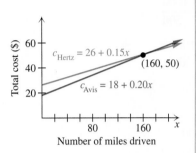

Figure 7.9

▶ Example 10

A druggist needs 100 milliliters (ml) of a 10% phenobarbital solution. She has only a 5% solution and a 25% solution available. How many milliliters of each solution should she mix to obtain the desired solution?

Solution: First we will set up a system of equations. The unknown quantities are the amount of the 5% solution and the amount of the 25% solution that must be used.

Let
$$x = \text{number of ml of 5\% solution}$$
$$y = \text{number of ml of 25\% solution}$$

We know that 100 ml of solution are needed. Thus

$$x + y = 100.$$

The total amount of pure phenobarbital in a solution is determined by multiplying the percent of phenobarbital by the number of milliliters of solution. The second equation comes from the fact that

$$\begin{pmatrix} \text{total amount of} \\ \text{phenobarbital in} \\ \text{5\% solution} \end{pmatrix} + \begin{pmatrix} \text{total amount of} \\ \text{phenobarbital in} \\ \text{25\% solution} \end{pmatrix} = \begin{pmatrix} \text{total amount of} \\ \text{phenobarbital} \\ \text{in 10\% mixture} \end{pmatrix}$$

$$0.05x \qquad + \qquad 0.25y \qquad = \qquad 0.10(100)$$

$$\text{or} \quad 0.05x + 0.25y = 10$$

The system of equations is

$$x + y = 100$$
$$0.05x + 0.25y = 10$$

We will solve this system of equations using the addition method:

$$-5[x + y = 100] \quad \text{gives} \quad -5x - 5y = -500$$
$$100[0.05x + 0.25y = 10] \qquad\qquad 5x + 25y = 1000$$

$$
\begin{aligned}
-5x - 5y &= -500 \\
5x + 25y &= 1000 \\
\hline
20y &= 500
\end{aligned}
$$

$$\frac{20y}{20} = \frac{500}{20}$$

$$y = 25$$

Now find x:

$$x + y = 100$$
$$x + 25 = 100$$
$$x = 75$$

Therefore 75 ml of a 5% phenobarbital solution must be mixed with 25 ml of a 25% phenobarbital solution to obtain 100 ml of a 10% phenobarbital solution.

Example 10 can also be done by using substitution. Try solving Example 10 by substitution now.

Section 7.2 Exercises

1. How will you know, when solving a system of equations by either the substitution or the addition method, if the system is dependent?
2. How will you know, when solving a system of equations by either the substitution or the addition method, if the system is inconsistent?

In Ex. 3–20, solve each system of equations by the substitution method. If the system does not have a single ordered pair as a solution, state whether the system is inconsistent or dependent.

3. $y = -x + 6$
   $y = x + 2$
5. $x - 2y = 6$
   $2x + y = -3$
7. $y - x = 4$
   $x - y = 3$
9. $x = 4y + 12$
   $2y + x = 0$

4. $y = 2x + 3$
   $y = x - 2$
6. $3x + y = 3$
   $4y + 3x = 6$
8. $x + y = 3$
   $y - x = 5$
10. $2y + x = 3$
    $2x = 6 - 4y$

11. $y - 2x = 3$
    $2y = 4x + 6$
13. $y = 2$
    $y + x + 3 = 0$
15. $y = 4x - 3$
    $y = 2x + 8$
17. $x = 3y + 1$
    $y = 2x - 1$
19. $6x - y = 5$
    $y = 6x - 3$

12. $x = y + 3$
    $x = -3$
14. $x + 2y = 6$
    $y = 2x + 3$
16. $x - 2y - 6 = 0$
    $3x - y = 5$
18. $x + 3y = 8$
    $2x - y - 4 = 0$
20. $x + 3y = 6$
    $y = -\frac{1}{3}x + 2$

In Ex. 21–36, solve each system of equations by the addition method. If the system does not have a single ordered pair as a solution, state whether the system is inconsistent or dependent.

21. $4x - y = 8$
    $2x + y = 4$
23. $3x + 2y = 6$
    $-3x + y = 0$

22. $3x - y = 4$
    $x + y = 4$
24. $x - 3y = 4$
    $-x + 2y = -5$

25. $2x - y = -4$
$-3x - y = 6$

26. $x + y = 6$
$-2x + y = -3$

27. $2x + y = 6$
$3x + y = 5$

28. $2x + y = 11$
$x + 3y = 18$

29. $2x - 3y = 4$
$2x + y + 4 = 0$

30. $2x - y = 7$
$3x + 2y = 0$

31. $5x - 2y = -4$
$4y = -3x + 34$

32. $4x - 2y = 6$
$4y = 8x - 12$

33. $4x + y = 6$
$-8x - 2y = 13$

34. $2x + 3y = 6$
$5x - 4y = -8$

35. $5x + 4y = 3$
$-3x - 5y = -7$

36. $3x + 4y = 15$
$5x + 2y = 11$

In Ex. 37–45, write a system of equations that can be used to solve the problem. Then solve the system and determine the solution.

37. The car rental agency, Rent-A-Wreck, charges $14 a day plus 15 cents per mile. The other agency in town, No Frills, charges $20 a day plus 5 cents per mile.
   a) How many miles would one have to drive in a day for the cost of both companies to be the same?
   b) For a 375-mi trip which rental agency will be the least expensive?

38. A plane can travel 280 mph with the wind and 240 mph against the wind. Find the speed of the plane in still air and the speed of the wind.

39. Jill can make a weekly salary of $200 plus 5% commission on sales, or a weekly salary consisting of a straight 15% commission on sales. Determine the amount of sales necessary for the 15% straight commission salary to equal the $200 plus 5% commision salary.

40. Ramon wishes to mix 30 lb of coffee to sell for a total cost of $200. To obtain the mixture, he will mix coffee that sells for $6 per pound with coffee that sells for $8 per pound. How many pounds of each type of coffee should he use?

41. The Guidas own a dairy. They have milk that is 5% butterfat and skim milk without butterfat. How much of the 5% milk and how much of the skim milk should they mix to make 100 gal of milk that is 3.5% butterfat?

42. In chemistry class, Mark has an 80% acid solution and a 50% acid solution. How much of each solution should he mix to get 100 liters of a 75% acid solution?

43. Animals in an experiment are to be kept on a strict diet. Each animal is to receive, among other things, 20 g of protein and 6 g of carbohydrates. The scientist has only two food mixes available of the following compositions:

	Protein (%)	Carbohydrates (%)
Mix A	10	6
Mix B	20	2

How many grams of each mix should be used to obtain the right diet for a single animal?

44. Membership in country club A costs $2500 per year and entitles a member to play a round of golf for a greens fee of $5.00. At country club B, membership costs $2200 per year and the greens fee is $8.75.
   a) How many rounds must a golfer play in a year in order for the costs in the two clubs to be the same?
   b) If Ms. Passaro planned to play 30 rounds of golf in a year, which club would be the least expensive?

45. The charge for maintaining a checking account at bank A is $6 per month plus 10 cents for each check that is written. The charge at bank B is $2 per month and 20 cents per check.
   a) How many checks would a customer have to write in a month for the total charges to be the same at both banks?
   b) If Mr. Scrooge planned to write 14 checks per month, which bank would be the least expensive?

In Ex. 46 and 47, supply and demand equations are given. Find the equilibrium price. See Example 4.

46. $S = 1000 + 5x$
$D = 200 + 20x$

47. $S = 1200 - 2.8x$
$D = 800 + 3.2x$

7.3 Matrices

We have discussed solving systems of equations by graphing, using substitution, and using the addition method. In Section 7.4 we will discuss solving systems of linear equations by using matrices. So that you will become familiar with matrices, in this section we explain how to add, subtract, and multiply matrices.

We will also explain how a matrix may be multiplied by a real number. Matrix techniques are easily adapted to computers.

A **matrix** is a rectangular array of elements. An array is a systematic arrangement of numbers or symbols in rows and columns. In this text brackets are used to indicate a matrix. Matrices (the plural of matrix) may be used to display information and to solve systems of linear equations.

The following matrix displays information about a survey of 500 registered voters.

	Democrats	Republicans	Conservatives	Liberals	Others
Men	100	93	20	35	4
Women	80	92	21	47	8

You are already familiar with matrices, although you may not be aware of it. The matrix is a good way of displaying numerical data as illustrated on this trail sign in Yosemite National Park.

The numbers inside the brackets are called **elements** of the matrix. The matrix above contains 10 elements. Because it has two rows and five columns, it is referred to as a 2 by 5 (2 × 5) matrix. A matrix that contains the same number of rows and columns is called a **square matrix**. Following are examples of 2 × 2 and 3 × 3 square matrices.

$$\begin{bmatrix} 2 & 3 \\ 5 & 2 \end{bmatrix} \qquad \begin{bmatrix} 4 & 6 & -1 \\ 2 & 3 & 0 \\ 5 & 2 & 1 \end{bmatrix}$$

Two matrices are equal if and only if they have the same elements in the same relative positions.

▶ **Example 1**

Given $A = B$, find x and y:

$$A = \begin{bmatrix} 1 & 4 \\ 6 & 3 \end{bmatrix}, \qquad B = \begin{bmatrix} x & 4 \\ 6 & y \end{bmatrix}$$

Solution: Since the corresponding elements must be the same, $x = 1$ and $y = 3$.

Addition of Matrices

Two matrices can be added only if they have the same dimensions (same number of rows and same number of columns). To obtain the sum of two matrices with the same dimensions, add the corresponding elements of the two matrices.

▶ **Example 2**

$$A = \begin{bmatrix} 1 & 4 \\ -2 & 6 \end{bmatrix}, \qquad B = \begin{bmatrix} 3 & 8 \\ 6 & 0 \end{bmatrix}. \quad \text{Find } A + B.$$

Solution:
$$A + B = \begin{bmatrix} 1 & 4 \\ -2 & 6 \end{bmatrix} + \begin{bmatrix} 3 & 8 \\ 6 & 0 \end{bmatrix}$$
$$= \begin{bmatrix} 1 + 3 & 4 + 8 \\ -2 + 6 & 6 + 0 \end{bmatrix} = \begin{bmatrix} 4 & 12 \\ 4 & 6 \end{bmatrix}$$

The following example illustrates an application of addition of matrices.

▶ **Example 3**

The Ski Swap Corporation owns and operates two stores. The number of pairs of downhill skis and cross-country skis sold in each store for the months of January through June and July through December are illustrated in the matrices below:

$$
\begin{array}{cc}
& \text{Store } X \\
& \text{CC} \quad \text{DH}
\end{array}
\qquad
\begin{array}{cc}
& \text{Store } Y \\
& \text{CC} \quad \text{DH}
\end{array}
$$

$$
\begin{array}{c}
\text{Jan.–June} \\
\text{July–Dec.}
\end{array}
\begin{bmatrix} 200 & 230 \\ 452 & 500 \end{bmatrix} = A
\qquad
\begin{bmatrix} 230 & 190 \\ 377 & 502 \end{bmatrix} = B
$$

Find the total number of each type of ski sold by the corporation during each time period.

Solution: To solve the problem, we add the matrices A and B:

$$
\begin{array}{ccc}
\text{CC} & \text{DH} & \qquad \text{CC} \quad \text{DH}
\end{array}
$$

$$
\begin{array}{c}
\text{Jan.–June} \\
\text{July–Dec.}
\end{array}
\begin{bmatrix} 200 + 230 & 230 + 190 \\ 452 + 377 & 500 + 502 \end{bmatrix} = \begin{bmatrix} 430 & 420 \\ 829 & 1002 \end{bmatrix}
$$

We can see from the sum matrix that during the period from January through June, 430 pairs of cross-country skis and 420 pairs of downhill skis were sold. During the period from July through December a total of 829 pairs of cross-country skis and 1002 pairs of downhill skis were sold.

The matrix

$$
I = \begin{bmatrix} 0 & 0 \\ 0 & 0 \end{bmatrix}
$$

is the **additive identity matrix** for 2×2 matrices. We will denote this matrix with the letter I. Note that for any 2×2 matrix A, $A + I = I + A = A$.

Subtraction of Matrices

Only matrices with the same dimensions may be subtracted. To subtract matrices with the same dimensions, subtract each entry in one matrix from the corresponding entry in the other matrix.

▶ **Example 4**

Find $A - B$ if

$$
A = \begin{bmatrix} 4 & 6 \\ 5 & -1 \end{bmatrix}, \qquad B = \begin{bmatrix} 3 & -4 \\ 7 & -3 \end{bmatrix}.
$$

Solution: $A - B = \begin{bmatrix} 4 & 6 \\ 5 & -1 \end{bmatrix} - \begin{bmatrix} 3 & -4 \\ 7 & -3 \end{bmatrix}$

$= \begin{bmatrix} 4-3 & 6-(-4) \\ 5-7 & -1-(-3) \end{bmatrix} = \begin{bmatrix} 1 & 10 \\ -2 & 2 \end{bmatrix}$

Multiplying a Matrix by a Real Number

A matrix may be multiplied by a real number by multiplying each entry in the matrix by the real number. Sometimes when we multiply a matrix by a real number, we call that real number a **scalar**.

▶ **Example 5**

For matrices A and B, find (a) $3A$ and (b) $3A - 2B$.

$$A = \begin{bmatrix} 4 & 6 \\ -3 & 5 \end{bmatrix}, \qquad B = \begin{bmatrix} -1 & 5 \\ 2 & 6 \end{bmatrix}$$

Solution:

a) $3A = 3 \begin{bmatrix} 4 & 6 \\ -3 & 5 \end{bmatrix} = \begin{bmatrix} 3(4) & 3(6) \\ 3(-3) & 3(5) \end{bmatrix} = \begin{bmatrix} 12 & 18 \\ -9 & 15 \end{bmatrix}$

b) We found $3A$ in part (a). Now we will find $2B$:

$$2B = 2 \begin{bmatrix} -1 & 5 \\ 2 & 6 \end{bmatrix} = \begin{bmatrix} 2(-1) & 2(5) \\ 2(2) & 2(6) \end{bmatrix} = \begin{bmatrix} -2 & 10 \\ 4 & 12 \end{bmatrix}$$

$$3A - 2B = \begin{bmatrix} 12 & 18 \\ -9 & 15 \end{bmatrix} - \begin{bmatrix} -2 & 10 \\ 4 & 12 \end{bmatrix}$$

$$= \begin{bmatrix} 12-(-2) & 18-10 \\ -9-4 & 15-12 \end{bmatrix} = \begin{bmatrix} 14 & 8 \\ -13 & 3 \end{bmatrix}$$

Multiplication of Matrices

Multiplication of matrices is slightly more difficult than addition of matrices. Multiplication of matrices is possible only when the number of *columns* of the first matrix, A, is the same as the number of *rows* of the second matrix, B. We will use the notation

$$\begin{array}{c} A \\ 3 \times 4 \end{array}$$

to indicate that matrix A has three rows and four columns. Suppose matrix A is a 3×4 matrix and matrix B is a 4×5 matrix, then

$$\begin{array}{cc} A & B \\ 3 \times 4 & 4 \times 5 \end{array}$$

$$\boxed{\text{Same}}$$

Product matrix 3×5

This indicates that matrix A has four columns and matrix B has four rows. Therefore we can multiply these two matrices. The product matrix will have the same number of rows as matrix A and the same number of columns as matrix B. Thus the dimensions of the product matrix are 3×5.

▶ **Example 6**

Determine which of the following pairs of matrices can be multiplied.

a) $A = \begin{bmatrix} 3 & 2 \\ 5 & 7 \end{bmatrix}$, $B = \begin{bmatrix} 0 & 6 \\ 4 & 1 \end{bmatrix}$

b) $A = \begin{bmatrix} 2 & 3 \\ 5 & 6 \end{bmatrix}$, $B = \begin{bmatrix} 2 & 4 & -1 \\ 6 & 8 & 0 \end{bmatrix}$

c) $A = \begin{bmatrix} 2 & 1 & 4 \\ 3 & 2 & 8 \end{bmatrix}$, $B = \begin{bmatrix} 2 & 1 & 3 \\ 1 & 0 & -2 \end{bmatrix}$

Solution:

a)
$$\begin{array}{cc} A & B \\ 2 \times 2 & 2 \times 2 \end{array}$$
Same

Since matrix A has two columns and matrix B has two rows, the two matrices can be multiplied. The product is a 2×2 matrix.

b)
$$\begin{array}{cc} A & B \\ 2 \times 2 & 2 \times 3 \end{array}$$
Same

Since matrix A has two columns and matrix B has two rows, the two matrices can be multiplied. The product is a 2×3 matrix.

c)
$$\begin{array}{cc} A & B \\ 2 \times 3 & 2 \times 3 \end{array}$$
Not same

Since matrix A has three columns and matrix B has two rows, the two matrices cannot be multiplied.

To explain matrix multiplication, we will use matrices A and B given below:

$$A = \begin{bmatrix} 3 & 2 \\ 5 & 7 \end{bmatrix}, \qquad B = \begin{bmatrix} 0 & 6 \\ 4 & 1 \end{bmatrix}$$

Since A contains two rows and B contains two columns, the product matrix will contain two rows and two columns. To multiply two matrices, we use a row-column scheme of multiplying. The numbers in the first row of matrix A are multiplied by the numbers in the first column of matrix B.

$$A \times B = \begin{bmatrix} 3 & 2 \\ 5 & 7 \end{bmatrix} \begin{bmatrix} 0 & 6 \\ 4 & 1 \end{bmatrix}$$

First row   First column

$$\begin{bmatrix} 3 & 2 \\ 5 & 7 \end{bmatrix} \qquad \begin{bmatrix} 0 & 6 \\ 4 & 1 \end{bmatrix}$$

$$(3 \times 0) + (2 \times 4) = 0 + 8$$
$$= 8$$

The 8 is placed in the first-row, first-column position of the product matrix. The other numbers in the product matrix are obtained in a similar way, as illustrated in the matrix that follows.

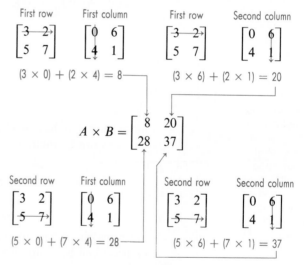

First row   First column   First row   Second column

$$\begin{bmatrix} 3 & 2 \\ 5 & 7 \end{bmatrix} \quad \begin{bmatrix} 0 & 6 \\ 4 & 1 \end{bmatrix} \qquad \begin{bmatrix} 3 & 2 \\ 5 & 7 \end{bmatrix} \quad \begin{bmatrix} 0 & 6 \\ 4 & 1 \end{bmatrix}$$

$$(3 \times 0) + (2 \times 4) = 8 \qquad (3 \times 6) + (2 \times 1) = 20$$

$$A \times B = \begin{bmatrix} 8 & 20 \\ 28 & 37 \end{bmatrix}$$

Second row   First column   Second row   Second column

$$\begin{bmatrix} 3 & 2 \\ 5 & 7 \end{bmatrix} \quad \begin{bmatrix} 0 & 6 \\ 4 & 1 \end{bmatrix} \qquad \begin{bmatrix} 3 & 2 \\ 5 & 7 \end{bmatrix} \quad \begin{bmatrix} 0 & 6 \\ 4 & 1 \end{bmatrix}$$

$$(5 \times 0) + (7 \times 4) = 28 \qquad (5 \times 6) + (7 \times 1) = 37$$

We can shorten the procedure as follows:

$$A \times B = \begin{bmatrix} 3 & 2 \\ 5 & 7 \end{bmatrix} \begin{bmatrix} 0 & 6 \\ 4 & 1 \end{bmatrix}$$

$$= \begin{bmatrix} 3(0) + 2(4) & 3(6) + 2(1) \\ 5(0) + 7(4) & 5(6) + 7(1) \end{bmatrix}$$

$$= \begin{bmatrix} 8 & 20 \\ 28 & 37 \end{bmatrix}$$

In general, if

$$A = \begin{bmatrix} a & b \\ c & d \end{bmatrix} \quad \text{and} \quad B = \begin{bmatrix} e & f \\ g & h \end{bmatrix}$$

Three mathematicians played important roles in the development of matrix theory. They are **James Sylvester** (1814–1897), **William Rowan Hamilton** (1805–1865), and **Arthur Cayley** (1821–1895). Sylvester and Cayley were good friends, and although they did not work together, they cooperated in developing matrix theory. Sylvester was the first to use the term "matrix." Hamilton, a noted physicist, astronomer, and mathematician, also used what was essentially the algebra of matrices under the name of "linear and vector functions." The mathematical concept of "vector space" grew out of Hamilton's work on the algebra of vectors.

then

$$A \times B = \begin{bmatrix} a & b \\ c & d \end{bmatrix} \begin{bmatrix} e & f \\ g & h \end{bmatrix} = \begin{bmatrix} ae + bg & af + bh \\ ce + dg & cf + dh \end{bmatrix}.$$

Let's do one more multiplication.

▶ **Example 7**

Find $A \times B$, given

$$A = \begin{bmatrix} 2 & 3 \\ 5 & 6 \end{bmatrix} \quad \text{and} \quad B = \begin{bmatrix} 2 & 4 & -1 \\ 6 & 8 & 0 \end{bmatrix}.$$

Solution: Matrix A contains two columns and matrix B contains two rows. Thus the matrices can be multiplied. Since A contains two rows and B contains three columns, the product matrix will contain two rows and three columns.

$$A \times B = \begin{bmatrix} 2 & 3 \\ 5 & 6 \end{bmatrix} \begin{bmatrix} 2 & 4 & -1 \\ 6 & 8 & 0 \end{bmatrix}$$

$$= \begin{bmatrix} 2(2) + 3(6) & 2(4) + 3(8) & 2(-1) + 3(0) \\ 5(2) + 6(6) & 5(4) + 6(8) & 5(-1) + 6(0) \end{bmatrix}$$

$$= \begin{bmatrix} 22 & 32 & -2 \\ 46 & 68 & -5 \end{bmatrix}$$

It should be noted that multiplication of matrices *is not* commutative, that is, $A \times B \neq B \times A$, except in special instances.

We previously discussed the additive identity matrix. **Square matrices** also have a **multiplicative identity matrix**. The multiplicative identity matrices for a 2×2 and a 3×3 matrix, denoted by I, follow. Note that in any multiplicative identity matrix 1s go diagonally from top left to bottom right, and all other elements in the matrix are 0s.

$$I = \begin{bmatrix} 1 & 0 \\ 0 & 1 \end{bmatrix}, \quad I = \begin{bmatrix} 1 & 0 & 0 \\ 0 & 1 & 0 \\ 0 & 0 & 1 \end{bmatrix}$$

Note that for any square matrix, A, $A \times I = I \times A = A$.

▶ **Example 8**

Using the multiplicative identity matrix for a 2×2 matrix and matrix A, show that $A \times I = A$.

$$A = \begin{bmatrix} 4 & 3 \\ 2 & 1 \end{bmatrix}$$

Solution: The identity matrix is $I = \begin{bmatrix} 1 & 0 \\ 0 & 1 \end{bmatrix}$.

$$A \times I = \begin{bmatrix} 4 & 3 \\ 2 & 1 \end{bmatrix} \begin{bmatrix} 1 & 0 \\ 0 & 1 \end{bmatrix}$$

$$= \begin{bmatrix} 4(1) + 3(0) & 4(0) + 3(1) \\ 2(1) + 1(0) & 2(0) + 1(1) \end{bmatrix}$$

$$= \begin{bmatrix} 4 & 3 \\ 2 & 1 \end{bmatrix} = A$$

The following example illustrates an application of multiplication of matrices.

▶ **Example 9**

A manufacturer of bathing suits produces three types of bathing suits. On a given day the firm produces 20 of type X, 30 of type Y, and 50 of type Z. Each suit of type X requires four units of material and 1 hour of work to produce; each of type Y requires three units of material and 2 hours of work to produce; each of type Z requires five units of material and 3 hours to produce. Use multiplication of matrices to determine the total number of units of material and the total number of hours needed for a day's production.

Solution: Let matrix A represent the number of each type of bathing suit produced:

$$\begin{array}{cccc} \text{Type} & X & Y & Z \\ A & = & [20 & 30 & 50]. \end{array}$$

The units of material and time requirements for each type are indicated in matrix B:

$$\begin{array}{ccc} & \text{Material} & \text{Hours} \\ B = & \begin{bmatrix} 4 & 1 \\ 3 & 2 \\ 5 & 3 \end{bmatrix} & \begin{array}{l} \text{Type } X \\ \text{Type } Y \\ \text{Type } Z \end{array} \end{array}$$

The product of A and B, $A \times B$, will give the total number of units of material and the total number of hours of work needed for the day's production.

$$A \times B = [20 \quad 30 \quad 50] \begin{bmatrix} 4 & 1 \\ 3 & 2 \\ 5 & 3 \end{bmatrix}$$

$$= [20(4) + 30(3) + 50(5) \quad 20(1) + 30(2) + 50(3)]$$

$$= [420 \quad 230]$$

Thus a total of 420 units of material and a total of 230 hours are required per day.

Section 7.3 Exercises

1. Bring to class an article that shows information illustrated in matrix form.
2. **a)** In your own words, explain the procedure used to add matrices.
   b) Using the procedure given in part (a),

   $$\text{add } \begin{bmatrix} 5 & 3 & -1 \\ 0 & 2 & 4 \end{bmatrix} \text{ and } \begin{bmatrix} 4 & 5 & 6 \\ -1 & 3 & 2 \end{bmatrix}.$$

3. **a)** In your own words, explain the procedure used to subtract matrices.
   b) Using the procedure given in part (a),

   $$\text{subtract } \begin{bmatrix} 5 & 3 & -1 \\ 0 & 2 & 4 \end{bmatrix} \text{ from } \begin{bmatrix} 4 & 5 & 6 \\ -1 & 3 & 2 \end{bmatrix}.$$

4. In order to multiply two matrices, what must be true about the dimensions of the matrices to be multiplied?
5. **a)** In your own words, explain the procedure used to multiply matrices.
   b) Using the procedure given in part (a),

   $$\text{multiply } \begin{bmatrix} 6 & -1 \\ 5 & 0 \end{bmatrix} \text{ by } \begin{bmatrix} 2 & -3 \\ 1 & -4 \end{bmatrix}.$$

6. Records are kept each day for a month at a local cinema that houses three movie theaters A, B, and C. The daily average Monday through Sunday receipts for the three theaters are as follows.

   A: \$654, \$785, \$458, \$345, \$1478, \$2109, \$543;
   B: \$764, \$778, \$568, \$451, \$1024, \$1689, \$853;
   C: \$567, \$764, \$873, \$407, \$2034, \$2432, \$567.

   Express this information in the form of a 3×7 matrix.

In Exercises 7–10, find $A + B$.

7. $A = \begin{bmatrix} 7 & 3 \\ 6 & 2 \end{bmatrix}$ $B = \begin{bmatrix} -3 & -2 \\ 0 & 3 \end{bmatrix}$

8. $A = \begin{bmatrix} 1 & 0 & 2 \\ 0 & 2 & 2 \end{bmatrix}$ $B = \begin{bmatrix} 5 & 7 & 3 \\ 3 & -2 & 4 \end{bmatrix}$

9. $A = \begin{bmatrix} 4 & 1 \\ -6 & 4 \\ 2 & 7 \end{bmatrix}$ $B = \begin{bmatrix} -2 & 3 \\ 0 & 5 \\ 2 & 1 \end{bmatrix}$

10. $A = \begin{bmatrix} 4 & 0 & 2 \\ 5 & 1 & -3 \\ -1 & 4 & -6 \end{bmatrix}$ $B = \begin{bmatrix} -4 & 2 & -1 \\ 5 & -3 & 6 \\ -2 & 0 & 6 \end{bmatrix}$

In Ex. 11–14, find $A - B$.

11. $A = \begin{bmatrix} 7 & 4 \\ 3 & -1 \end{bmatrix}$ $B = \begin{bmatrix} 4 & -2 \\ -3 & 5 \end{bmatrix}$

12. $A = \begin{bmatrix} 5 & -3 & 6 \\ 2 & 4 & 2 \end{bmatrix}$ $B = \begin{bmatrix} 0 & -4 & -5 \\ 1 & 3 & 8 \end{bmatrix}$

13. $A = \begin{bmatrix} 5 & 3 & -1 \\ 7 & 4 & 2 \\ 6 & -1 & -5 \end{bmatrix}$ $B = \begin{bmatrix} 4 & 3 & 6 \\ -2 & -4 & 9 \\ 0 & -2 & 4 \end{bmatrix}$

14. $A = \begin{bmatrix} -4 & 3 \\ 6 & 2 \\ 1 & -5 \end{bmatrix}$ $B = \begin{bmatrix} -6 & -8 \\ -10 & -11 \\ 3 & -7 \end{bmatrix}$

In Ex. 15–20, let

$$A = \begin{bmatrix} 1 & 2 \\ 0 & 5 \end{bmatrix} \quad B = \begin{bmatrix} 4 & 7 \\ 0 & 2 \end{bmatrix} \quad C = \begin{bmatrix} -2 & 3 \\ 4 & 0 \end{bmatrix}.$$

Find the following.

15. $2B$
16. $-3B$
17. $2B + 3C$
18. $2B + 3A$
19. $3B - 2C$
20. $4C - 2A$

In Ex. 21–26, find $A \times B$.

21. $A = \begin{bmatrix} 5 & 3 \\ 6 & 2 \end{bmatrix}$ $B = \begin{bmatrix} 4 & 2 \\ 0 & 3 \end{bmatrix}$

22. $A = \begin{bmatrix} -1 & 1 \\ 0 & 3 \end{bmatrix}$ $B = \begin{bmatrix} 4 & 1 \\ 0 & -1 \end{bmatrix}$

23. $A = \begin{bmatrix} 2 & 3 & -1 \\ 0 & 4 & 6 \end{bmatrix}$ $B = \begin{bmatrix} 2 \\ 4 \\ 1 \end{bmatrix}$

24. $A = \begin{bmatrix} 2 & 1 \\ 1 & 1 \end{bmatrix}$ $B = \begin{bmatrix} 1 & -1 \\ -1 & 2 \end{bmatrix}$

25. $A = \begin{bmatrix} 2 & 5 \\ 1 & 3 \end{bmatrix}$ $B = \begin{bmatrix} 3 & -5 \\ -1 & 2 \end{bmatrix}$

26. $A = \begin{bmatrix} 2 & 3 & 1 \\ -2 & -1 & 0 \\ 4 & 5 & 6 \end{bmatrix}$ $B = \begin{bmatrix} 1 & 0 & 0 \\ 0 & 1 & 0 \\ 0 & 0 & 1 \end{bmatrix}$

In Ex. 27–32, if possible, find $A + B$ and $A \times B$. If an operation cannot be performed, explain why.

27. $A = \begin{bmatrix} 0 & 3 \\ 2 & 4 \end{bmatrix}$ $B = \begin{bmatrix} 1 & 2 & 3 \\ 2 & -2 & 3 \end{bmatrix}$

28. $A = \begin{bmatrix} 2 & 3 & 4 \\ 4 & 6 & -2 \end{bmatrix}$ $B = \begin{bmatrix} 2 & -3 & 4 \\ 0 & 4 & 3 \end{bmatrix}$

29. $A = \begin{bmatrix} 4 & 5 & 3 \\ 6 & 2 & 1 \end{bmatrix}$ $B = \begin{bmatrix} 3 & 2 \\ 4 & 6 \\ -2 & 0 \end{bmatrix}$

30. $A = \begin{bmatrix} 1 & 2 \\ 3 & 4 \\ 5 & 6 \end{bmatrix}$ $B = \begin{bmatrix} 1 & 2 \\ 3 & 4 \\ 5 & 6 \end{bmatrix}$

31. $A = \begin{bmatrix} 1 & 2 \\ 3 & 4 \end{bmatrix}$ $B = \begin{bmatrix} 5 \\ 6 \end{bmatrix}$

32. $A = \begin{bmatrix} 5 \\ 6 \end{bmatrix}$ $B = \begin{bmatrix} 1 & 2 \\ 3 & 4 \end{bmatrix}$

In Ex. 33–35, show that the *commutative property of addition*, $A + B = B + A$, holds for matrices A and B.

33. $A = \begin{bmatrix} 6 & 3 \\ -1 & -2 \end{bmatrix}$ $B = \begin{bmatrix} 8 & 5 \\ 6 & 1 \end{bmatrix}$

34. $A = \begin{bmatrix} -4 & 3 \\ 5 & 7 \end{bmatrix}$ $B = \begin{bmatrix} -5 & -8 \\ 0 & -7 \end{bmatrix}$

35. $A = \begin{bmatrix} 0 & -1 \\ 3 & -4 \end{bmatrix}$ $B = \begin{bmatrix} 8 & 1 \\ 3 & -4 \end{bmatrix}$

36. Make up two matrices with the same dimensions, A and B, and show that $A + B = B + A$.

In Ex. 37–39, show that the *associative property of addition*, $(A + B) + C = A + (B + C)$ holds for the matrices given.

37. $A = \begin{bmatrix} 4 & -3 \\ -1 & 5 \end{bmatrix}$ $B = \begin{bmatrix} 5 & 7 \\ 3 & 4 \end{bmatrix}$ $C = \begin{bmatrix} 0 & -4 \\ 5 & 1 \end{bmatrix}$

38. $A = \begin{bmatrix} -9 & -8 \\ -7 & -6 \end{bmatrix}$ $B = \begin{bmatrix} -5 & -4 \\ -3 & -2 \end{bmatrix}$ $C = \begin{bmatrix} -1 & 0 \\ 1 & 2 \end{bmatrix}$

39. $A = \begin{bmatrix} 7 & 4 \\ 9 & -36 \end{bmatrix}$ $B = \begin{bmatrix} 5 & 6 \\ -1 & -4 \end{bmatrix}$ $C = \begin{bmatrix} -7 & -5 \\ -1 & 3 \end{bmatrix}$

40. Make up three matrices with the same dimensions, A, B, and C, and show that $(A + B) + C = A + (B + C)$.

In Ex. 41–45, determine whether the *commutative property of multiplication*, $A \times B = B \times A$, holds for the matrices given.

41. $A = \begin{bmatrix} 1 & 2 \\ 3 & -1 \end{bmatrix}$ $B = \begin{bmatrix} 2 & 0 \\ 1 & 3 \end{bmatrix}$

42. $A = \begin{bmatrix} 1 & -1 \\ 3 & 5 \end{bmatrix}$ $B = \begin{bmatrix} 4 & 2 \\ -1 & 3 \end{bmatrix}$

43. $A = \begin{bmatrix} 4 & 2 \\ 1 & -3 \end{bmatrix}$ $B = \begin{bmatrix} -4 & -2 \\ -1 & 3 \end{bmatrix}$

44. $A = \begin{bmatrix} -3 & 2 \\ 6 & -5 \end{bmatrix}$ $B = \begin{bmatrix} -\frac{5}{3} & -\frac{2}{3} \\ -2 & -1 \end{bmatrix}$

45. $A = \begin{bmatrix} 3 & 2 & 1 \\ 4 & 2 & 0 \\ 0 & -2 & 5 \end{bmatrix}$ $B = \begin{bmatrix} 1 & 0 & 0 \\ 0 & 1 & 0 \\ 0 & 0 & 1 \end{bmatrix}$

46. Make up two matrices with the same dimensions, A and B, and determine whether $A \times B = B \times A$.

In Ex. 47–51, show that the *associative property of multiplication*, $(A \times B) \times C = A \times (B \times C)$, holds for the matrices given.

47. $A = \begin{bmatrix} 1 & 2 \\ 4 & 0 \end{bmatrix}$ $B = \begin{bmatrix} 2 & 3 \\ 1 & 0 \end{bmatrix}$ $C = \begin{bmatrix} 4 & 2 \\ 3 & 1 \end{bmatrix}$

48. $A = \begin{bmatrix} -2 & 3 \\ 0 & 4 \end{bmatrix}$ $B = \begin{bmatrix} 4 & 5 \\ 7 & 2 \end{bmatrix}$ $C = \begin{bmatrix} 3 & 4 \\ -2 & 5 \end{bmatrix}$

49. $A = \begin{bmatrix} 4 & 3 \\ -6 & 2 \end{bmatrix}$ $B = \begin{bmatrix} 1 & 2 \\ 0 & 1 \end{bmatrix}$ $C = \begin{bmatrix} 4 & 3 \\ 0 & -2 \end{bmatrix}$

50. $A = \begin{bmatrix} -1 & -2 \\ -3 & -4 \end{bmatrix}$ $B = \begin{bmatrix} 1 & 0 \\ 0 & 1 \end{bmatrix}$ $C = \begin{bmatrix} 0 & 0 \\ 0 & 0 \end{bmatrix}$

51. $A = \begin{bmatrix} 3 & 4 \\ -1 & -2 \end{bmatrix}$ $B = \begin{bmatrix} 0 & 1 \\ 1 & 0 \end{bmatrix}$ $C = \begin{bmatrix} 2 & 0 \\ 3 & 0 \end{bmatrix}$

52. Make up three matrices with the same dimensions, A, B, and C, and show that $(A \times B) \times C = A \times (B \times C)$.

53. The Original Cookie Factory bakes and sells four types of cookies: chocolate chip, sugar, molasses, and peanut butter. Matrix A shows the number of units of various ingredients used in baking a dozen of each type of cookie:

$$A = \begin{array}{c} \\ \\ \\ \\ \end{array} \begin{bmatrix} 2 & 2 & \frac{1}{2} & 1 \\ 3 & 2 & 1 & 2 \\ 0 & 1 & 0 & 3 \\ \frac{1}{2} & 1 & 0 & 0 \end{bmatrix} \begin{array}{l} \text{Chocolate chip} \\ \text{Sugar} \\ \text{Molasses} \\ \text{Peanut butter} \end{array}$$

with columns labeled Sugar, Flour, Milk, Eggs

The cost, in cents per cup or per egg, for each ingredient when purchased in large quantities and in small quantities is given in matrix B:

$$B = \begin{array}{c} \quad \begin{array}{cc} \text{Large} & \text{Small} \\ \text{quantities} & \text{quantities} \end{array} \\ \begin{bmatrix} 10 & 12 \\ 5 & 8 \\ 8 & 8 \\ 4 & 6 \end{bmatrix} \begin{array}{l} \text{Sugar} \\ \text{Flour} \\ \text{Milk} \\ \text{Eggs} \end{array} \end{array}$$

Use matrix multiplication to find a matrix representing the comparative cost per item for large and small quantities purchased.

In Ex. 54 and 55, use the information given in Ex. 53. Suppose that a typical day's order consists of 30 dozen chocolate chip cookies, 20 dozen sugar cookies, 8 dozen molasses cookies, and 15 dozen peanut butter cookies.

54. Express these orders as a 1×4 matrix, and use matrix multiplication to determine the amount of each ingredient needed to fill the day's order.

55. Use matrix multiplication to determine the cost under the two purchase options (large and small quantities) to fill the day's order.

56. Food orders from the Geology Club and the Rugby Club are summarized in matrix A.

$$A = \begin{array}{c} \quad \begin{array}{ccc} \text{Burger} & \text{Fries} & \text{Cola} \end{array} \\ \begin{array}{c} \text{Geology Club} \\ \text{Rugby Club} \end{array} \begin{bmatrix} 28 & 32 & 25 \\ 33 & 26 & 31 \end{bmatrix} \end{array}$$

The prices (in dollars) of a burger, fries, and a cola at three fast food restaurants are summarized in matrix B.

$$B = \begin{array}{c} \quad \begin{array}{ccc} & \text{Burger} & \\ \text{McDougal's} & \text{Prince} & \text{Mendy's} \end{array} \\ \begin{array}{c} \text{Burger} \\ \text{Fries} \\ \text{Cola} \end{array} \begin{bmatrix} 2.45 & 2.95 & 3.15 \\ 1.35 & 0.99 & 1.00 \\ 1.40 & 0.92 & 1.20 \end{bmatrix} \end{array}$$

a) Multiply the two matrices to form a 2×3 matrix

that shows the amount each club would be charged by the fast food restaurant.

b) Determine which fast food restaurant offers each club the best total price.

Two matrices whose sum is the additive identity matrix are said to be **additive inverses**. That is, if $A + B = B + A = I$, where I is the additive identity matrix, then A and B are additive inverses. Determine whether A and B are additive inverses.

57. $A = \begin{bmatrix} 6 & 3 \\ 4 & -2 \end{bmatrix} \quad B = \begin{bmatrix} -6 & -3 \\ -2 & 4 \end{bmatrix}$

58. $A = \begin{bmatrix} 4 & 6 & 3 \\ 2 & 3 & -1 \\ -1 & 0 & 6 \end{bmatrix} \quad B = \begin{bmatrix} -4 & -6 & -3 \\ -2 & -3 & 1 \\ 1 & 0 & -6 \end{bmatrix}$

Two matrices whose product is the multiplicative identity matrix are said to be **multiplicative inverses**. That is, if $A \times B = B \times A = I$, where I is the multiplicative identity matrix, then A and B are multiplicative inverses. Determine whether A and B are multiplicative inverses.

59. $A = \begin{bmatrix} 5 & -2 \\ -2 & 1 \end{bmatrix} \quad B = \begin{bmatrix} 1 & 2 \\ 2 & 5 \end{bmatrix}$

60. $A = \begin{bmatrix} 7 & 3 \\ 2 & 1 \end{bmatrix} \quad B = \begin{bmatrix} 1 & -3 \\ -2 & 7 \end{bmatrix}$

Problem Solving

Determine whether the following statements are true or false. Give an example to support your answer.

61. $A - B = B - A$

62. For scalar a and matrices B and C, $a(B + C) = aB + aC$.

Research Activity

63. Read Chapter 9, then determine whether the set of all 2×2 matrices under the operation of addition form a commutative group.

7.4 Solving Systems of Equations Using Matrices

Section 7.3 introduced matrices. Now we will discuss the procedure to solve a system of linear equations using matrices.

We will illustrate how to solve a system of two equations and two un-

knowns. Systems of equations containing three equations and three unknowns (called third-order systems) and higher-order systems can also be solved by using matrices.

The first step in solving a system of equations using matrices is to write the **augmented matrix**. An augmented matrix is one made up of two smaller matrices—one for the coefficients of the variables in the equations and one for the constants in the equations. To determine the augmented matrix, first write each equation in standard form, $ax + by = c$. For the system of equations below on the left, its augmented matrix is shown to its right.

System of Equations	Augmented Matrix
$a_1x + b_1y = c_1$	$\begin{bmatrix} a_1 & b_1 & c_1 \\ a_2 & b_2 & c_2 \end{bmatrix}$
$a_2x + b_2y = c_2$	

Following is another example.

System of Equations	Augmented Matrix
$x + 2y = 8$	$\begin{bmatrix} 1 & 2 & 8 \\ 3 & -1 & 7 \end{bmatrix}$
$3x - y = 7$	

Note that the bar in the augmented matrix separates the numerical coefficients from the constants. Since the matrix is just a shortened way of writing the system of equations, we can solve the system of equations using matrices in a manner very similar to solving a system of equations using the addition method.

To solve a system of equations using matrices, we use *row transformations* to obtain new matrices that have the same solution as the original system. We will discuss three row transformation procedures.

Procedures for Row Transformations

1. Any two rows of a matrix may be interchanged (this is the same as interchanging any two equations in the system of equations).

2. All the numbers in any row may be multiplied by any nonzero real number. (This is the same as multiplying both sides of an equation by any nonzero real number.)

3. All the numbers in any row may be multiplied by any nonzero real number, and these products may be added to the corresponding numbers in any other row of numbers.

To solve a system of equations using matrices, we use row transformations to obtain an augmented matrix whose numbers to the left of the vertical bar are the same as in the *multiplicative identity matrix*. From this type of augmented matrix we can determine the solution to the system of equations. For example, if we get

$$\begin{bmatrix} 1 & 0 & 3 \\ 0 & 1 & -2 \end{bmatrix}$$

it tells us that $1x = 3$ or $x = 3$, and $1y = -2$ or $y = -2$. Thus the solution to the system of equations that yielded this augmented matrix is $(3, -2)$.

Now let's work an example.

▶ **Example 1**

Solve the following system of equations using matrices:

$$x + 2y = 5$$
$$3x - y = 8$$

Solution: First write the augmented matrix:

$$\left[\begin{array}{cc|c} 1 & 2 & 5 \\ 3 & -1 & 8 \end{array}\right]$$

Our goal is to obtain a matrix of the form

$$\left[\begin{array}{cc|c} 1 & 0 & c_1 \\ 0 & 1 & c_2 \end{array}\right]$$

where c_1 and c_2 may represent any real numbers. It is generally easier to work by columns. Therefore we will try to get the first column of the augmented matrix to be $\begin{smallmatrix}1\\0\end{smallmatrix}$ and the second column to be $\begin{smallmatrix}0\\1\end{smallmatrix}$. Since the element in the top left position is already a 1, we must work to change the 3 into a 0. We will use row transformation procedure 3 to change the 3 into a 0. If we multiply the top row of numbers by -3 and add these products to the second row of numbers, the element in the second row, first column will become a 0:

$$\left[\begin{array}{cc|c} 1 & 2 & 5 \\ 3 & -1 & 8 \end{array}\right]$$

The top row of numbers multiplied by -3 gives

$$1(-3) \qquad 2(-3) \qquad 5(-3)$$

Now add these products to their respective numbers in row 2.

$$\left[\begin{array}{cc|c} 1 & 2 & 5 \\ 3 + 1(-3) & -1 + 2(-3) & 8 + 5(-3) \end{array}\right] = \left[\begin{array}{cc|c} 1 & 2 & 5 \\ 0 & -7 & -7 \end{array}\right]$$

The next step is to obtain a 1 in the second row, second column. At present, -7 is in this position. To change the -7 to a 1, we will use row transformation procedure 2. If we multiply -7 by $-\frac{1}{7}$, the product will be 1. Therefore we will multiply all the numbers in the second row by $-\frac{1}{7}$. This gives

$$\left[\begin{array}{cc|c} 1 & 2 & 5 \\ 0(-\frac{1}{7}) & -7(-\frac{1}{7}) & -7(-\frac{1}{7}) \end{array}\right] = \left[\begin{array}{cc|c} 1 & 2 & 5 \\ 0 & 1 & 1 \end{array}\right].$$

The next step is to obtain a 0 in the first row, second column. At present a 2 is in this position. Multiplying the numbers in the second row by -2 and adding the products to the corresponding numbers in the first row gives a 0 in the desired position:

$$\begin{bmatrix} 1 + 0(-2) & 2 + 1(-2) & | & 5 + 1(-2) \\ 0 & 1 & | & 1 \end{bmatrix} = \begin{bmatrix} 1 & 0 & | & 3 \\ 0 & 1 & | & 1 \end{bmatrix}$$

We now have the desired augmented matrix:

$$\begin{bmatrix} 1 & 0 & | & 3 \\ 0 & 1 & | & 1 \end{bmatrix}$$

With this matrix we see that $1x = 3$ or $x = 3$ and $1y = 1$ or $y = 1$. The solution to the system is (3, 1).

Check:

$$\begin{array}{ll} x + 2y = 5 & 3x - y = 8 \\ 3 + 2(1) = 5 & 3(3) - 1 = 8 \\ 5 = 5 \quad \text{True} & 8 = 8 \quad \text{True} \end{array}$$

To Change an Augmented Matrix to the Form $\begin{bmatrix} 1 & 0 & | & c_1 \\ 0 & 1 & | & c_2 \end{bmatrix}$

1. Change the element in the first column, first row to a 1.
2. Change the element in the first column, second row to a 0.
3. Change the element in the second column, second row to a 1.
4. Change the element in the second column, first row to a 0.

Generally, when changing an element in the augmented matrix to a 1 we use step 2 in the row transformation box, and when changing an element to a 0 we use step 3 in the row transformation box.

▶ **Example 2**

Solve the following system of equations using matrices:

$$\begin{array}{l} 2x - 3y = 10 \\ 2x + 2y = 5 \end{array}$$

Solution: First write the augmented matrix:

$$\begin{bmatrix} 2 & -3 & | & 10 \\ 2 & 2 & | & 5 \end{bmatrix}$$

To obtain a 1 in the first row, first column, multiply the numbers in the first row by $\frac{1}{2}$.

$$\begin{bmatrix} 1 & -\frac{3}{2} & | & 5 \\ 2 & 2 & | & 5 \end{bmatrix}$$

To get a 0 in the second row, first column, multiply the numbers in the first row by -2, and add the products to the corresponding numbers in the second row:

$$\begin{bmatrix} 1 & -\frac{3}{2} & | & 5 \\ 2 + 1(-2) & 2 + (-\frac{3}{2})(-2) & | & 5 + 5(-2) \end{bmatrix} = \begin{bmatrix} 1 & -\frac{3}{2} & | & 5 \\ 0 & 5 & | & -5 \end{bmatrix}$$

To obtain a 1 in the second row, second column, multiply the numbers in the second row by $\frac{1}{5}$:

$$\begin{bmatrix} 1 & -\frac{3}{2} & \bigm| & 5 \\ 0(\frac{1}{5}) & 5(\frac{1}{5}) & \bigm| & -5(\frac{1}{5}) \end{bmatrix} = \begin{bmatrix} 1 & -\frac{3}{2} & \bigm| & 5 \\ 0 & 1 & \bigm| & -1 \end{bmatrix}$$

To obtain a 0 in the first row, second column, multiply the numbers in the second row by $\frac{3}{2}$, and add the products to the corresponding numbers in the first row:

$$\begin{bmatrix} 1 + \frac{3}{2}(0) & -\frac{3}{2} + \frac{3}{2}(1) & \bigm| & 5 + (-1)(\frac{3}{2}) \\ 0 & 1 & \bigm| & -1 \end{bmatrix} = \begin{bmatrix} 1 & 0 & \bigm| & \frac{7}{2} \\ 0 & 1 & \bigm| & -1 \end{bmatrix}$$

The solution to the system of equations is $\left(\dfrac{7}{2}, -1 \right)$.

Section 7.4 Exercises

In Ex. 1–12, use matrices to solve each system of equations.

1. $x - y = -1$
   $2x - y = 2$
2. $x - y = -2$
   $x + y = 2$
3. $x + 2y = -4$
   $2x - y = -3$
4. $x - y = 3$
   $2x + 5y = -1$
5. $x + y = 6$
   $2x - y = 3$
6. $x + 2y = 8$
   $2x - 3y = 2$
7. $x + 3y = 1$
   $-2x + y = 5$
8. $2x + 4y = 6$
   $4x - 2y = -8$
9. $2x + 4y = 6$
   $3x - y = 2$
10. $3x + 2y = 4$
    $5x + 5y = 9$
11. $5x + y = 3$
    $6x - y = 8$
12. $2x - 5y = 10$
    $3x + y = 15$

In Ex. 13–16, use matrices to solve each problem.

13. If Max buys 2 lb of chocolate-covered cherries and 3 lb of chocolate-covered mints, his total cost is $23. If he buys 1 lb of chocolate-covered cherries and 2 lb of chocolate-covered mints, his total cost is $14. Find the cost of 1 lb of chocolate-covered cherries and 1 lb of chocolate-covered mints.

14. The length of a rectangle is 3 ft greater than its width. If its perimeter is 14 ft, find the length and width.

15. The Chemistry Club and the German Club had their meetings at the local pizza parlor. The Chemistry Club had three large pizzas and four pitchers of beer for a cost of $61.00. The German Club had two large pizzas and three pitchers of beer for a cost of $42.00. Find the cost of a pizza, and the cost of a pitcher of beer.

16. Peoplepower, Inc., a daily employment agency, charges $10 per hour for a truck driver and $8.00 per hour for a laborer. On a certain job the laborer worked two more hours than the truck driver, and together they cost $144. How many hours did each work?

Systems of Linear Inequalities

In earlier sections we showed how to find the solution to a system of linear equations in two variables. Now we are going to explore the techniques of finding the solution set to a system of linear inequalities in two variables.

The solution set of a system of linear inequalities is the set of points that satisfy all inequalities in the system. The solution set of a system of linear inequalities may consist of infinitely many ordered pairs. To determine the solution set to a system of linear inequalities, graph each inequality on the same set of axes. The ordered pairs common to all of the inequalities are the solution set to the system.

Procedure for Solving a System of Linear Inequalities

1. Select one of the inequalities. Replace the inequality symbol with an equality symbol, and draw the graph of the equation. Draw the graph with a dashed line if the inequality is $<$ or $>$ and with a solid line if the inequality is $\leq$ or $\geq$.
2. Select a test point on one side of the line and determine whether the point is a solution to the inequality. If so, shade the area on the side of the line containing the point. If the point is not a solution shade the area on the other side of the line.
3. Repeat steps 1 and 2 for the other inequality.
4. The intersection of the two shaded areas and any solid line common to both inequalities form the solution set to the system of inequalities.

▶ **Example 1**

Graph the following system of inequalities and indicate the solution set:

$$x + y < 1$$
$$x - y < 5$$

Solution: Our method of attack will be to graph both inequalities on the same set of axes. First draw the graph of $x + y < 1$. When drawing the graph, remember to use a dashed line, since the inequality is "less than" (see Fig. 7.10a). If you have forgotten how to graph inequalities, review Section 6.7.

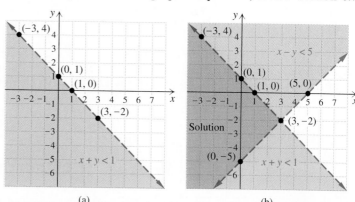

Figure 7.10

      (a)   (b)

Now, on the same set of axes, we find the half-plane determined by the inequality $x - y < 5$ (see Fig. 7.10b). The solution set consists of all the points common to the two shaded half-planes. These are the points in the region on the graph containing both color shadings. As we can see in Fig. 7.10(b) the two lines intersect at $(3, -2)$. This ordered pair can also be found by any of the algebraic methods discussed in Sections 7.2 and 7.3.

▶ **Example 2**

Graph the following system of inequalities and indicate the solution set:

$$2x + 3y \geq 4$$
$$2x - y > -6$$

Solution: Graph the inequality $2x + 3y \geq 4$. Remember to use a solid line, since the inequality is "greater than or equal to" (see Fig. 7.11a).

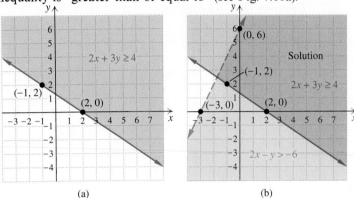

Figure 7.11   (a)   (b)

On the same set of axes, draw the graph of $2x - y > -6$. Use a dashed line, since the inequality is "greater than" (see Fig. 7.11b). The solution is the region of the graph that contains both color shadings and the part of the solid line that satisfies the inequality $2x - y > -6$. Note that the point of intersection of the two lines is not a part of the solution set.

▶ **Example 3**

Graph the following system of inequalities and indicate the solution set:

$$x \leq 5$$
$$y > -1$$

Solution: Graph the inequality $x \leq 5$ (Fig. 7.12a). On the same set of axes,

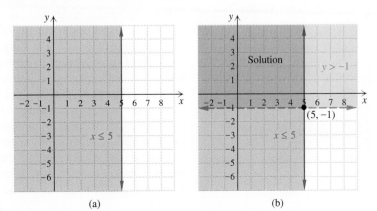

Figure 7.12   (a)   (b)

graph $y > -1$ (Fig. 7.12b). The solution set is that region of the graph that is shaded in both colors and the part of the solid line that satisfies the inequality $y > -1$. The point of intersection of the two lines, $(5, -1)$, is not part of the solution, since it does not satisfy the inequality $y > -1$.

Section 7.5 Exercises

In Ex. 1–16, graph each system of linear inequalities and indicate the solution set.

1. $y > 3x$
$\quad y > x + 2$

2. $y \le 2x + 4$
$\quad y > -2x + 4$

3. $x + y < 3$
$\quad x - y < 7$

4. $3x + 2y > 6$
$\quad x + 2y < 4$

5. $x + y \le 3$
$\quad 2x + 3y < 6$

6. $2x - y \le 3$
$\quad x - y < 3$

7. $x - 3y \le 3$
$\quad x + 2y \ge 4$

8. $x + 2y \ge 4$
$\quad 3x - y \ge -6$

9. $y \le 2x$
$\quad x \ge 2y$

10. $y \ge 3$
$\quad x + y < 1$

11. $x \ge 1$
$\quad y \le 1$

12. $x \le 0$
$\quad y \le 0$

13. $x \ge 0$
$\quad y \ge 0$

14. $x \le 1$
$\quad x + 2y < -1$

15. $3x \ge 2y + 6$
$\quad x \le y + 7$

16. $2x - 5y < 7$
$\quad x > 2y + 1$

17. Is it possible for a system of linear inequalities to have no solution? Explain your answer, giving an example to support it.

18. Is it possible for a system of linear inequalities to have as its solution all the points on the coordinate plane? Explain your answer, giving an example to support it.

7.6 Linear Programming

Government, business, and industry often require decision makers to find cost effective solutions to a variety of problems. Linear programming often serves as a method of expressing the relationships in many of these problems and utilizes a system of linear inequalities.

The typical linear programming problem has many variables and is generally so lengthy that it is solved on a computer by a technique called the **simplex algorithm**. The simplex method was developed in the 1940s by George B. Dantzig. Linear programming is used to solve problems in the social sciences, health care, land development, nutrition, and so on.

We will not discuss the simplex method in this textbook. We will merely give a brief introduction to how linear programming works. The student can find a detailed explanation in books on finite mathematics.

In a linear programming problem, there are restrictions called **constraints**. Each constraint is represented as a linear inequality. The list of constraints forms a system of linear inequalities. When the system of inequalities is graphed we often obtain a region bounded on all sides by line segments (Fig. 7.13). Since the boundaries of the region form a polygon, the region is called a **polygonal region**. The points where two or more boundaries intersect are called the **vertices** of the polygon. The points on the boundary of the region and the points inside

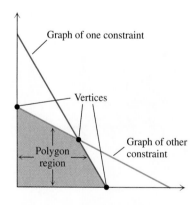

Figure 7.13

the polygonal region are the solution set for the system of inequalities. From all the vertices there is at least one point that, when substituted in an equation of the form $K = Ax + By$, will give us the maximum value for K. A different point from among the vertices will give us a minimum value of K. (K is a quantity we wish to maximize or minimize.) The linear programming problem is to determine which ordered pair will yield this maximum (or minimum) value. The fundamental principle of linear programming provides a rule for finding the maximum and minimum values.

Fundamental Principle of Linear Programming

If a linear equation of the form $K = Ax + By$ is evaluated at each point in a closed polygonal region, the maximum and minimum values of the equation occur at vertices of the region.

Did You

Know...

THE LOGISTICS OF D-DAY

Linear programming was first used to deal with the age-old military problem of logistics: obtaining, maintaining, and transporting military equipment and personnel. George Dantzig developed the simplex method for the Allies of World War II to do just that. Consider the logistics of the Allied invasion of Normandy. Meteorologic experts had settled upon three possible dates in June 1944. It had to be a day when low tide and first light would coincide; the winds should not exceed 8–13 mph; visibility had to be not less than 3 miles. A force of 170,000 assault troops were to be geared up and moved to 22 airfields in England where 1200 air transports and 700 gliders would then take them to the coast of France to converge with 5000 ships of the D-Day armada. The code name for the invasion was Operation Overlord, but it is known to most as D-Day.

Winston Churchill called the invasion of Normandy "the most difficult and complicated operation that has ever taken place."

The following example illustrates how the fundamental principle is used to solve a linear programming problem.

▶ **Example 1**

A company makes two types of rocking chairs, a plain chair and a fancy chair. Each rocking chair must be assembled and then finished. The plain chair takes 4 hr to assemble and 4 hr to finish. The fancy chair takes 8 hr to assemble and 12 hr to finish. The company can provide at most 160 worker-hours of assembling and 180 worker-hours of finishing a day. If the profit on the plain chair is $10.00 and the profit on the fancy chair is $18.00, how many rocking chairs of each type should the company make per day to maximize profits? What is the maximum profit?

Solution: From the information given above, we can establish the following facts.

	Assembly Time (hr)	Finishing Time (hr)	Profit ($)
Plain chair	4	4	10.00
Fancy chair	8	12	18.00

Let

$$x = \text{the number of plain chairs}$$
$$y = \text{the number of fancy chairs}$$
$$10x = \text{profit on the plain chairs}$$
$$18y = \text{profit on the fancy chairs}$$
$$P = \text{the total profit}$$

The total profit is the sum of the profit on the plain chairs and the profit on the fancy chairs. Since $10x$ is the profit on the plain chairs and $18y$ is the profit on the fancy chairs, the profit equation is $P = 10x + 18y$.

The maximum profit, P, is dependent on several conditions, called constraints. The number of chairs manufactured each day cannot be a negative amount. This gives us the constraints $x \geq 0$ and $y \geq 0$. Another constraint is determined by the total number of hours allocated for assembling. Since it takes 4 hr to assemble the plain chair, the total number of hours per day to assemble x plain chairs is $4x$. It takes 8 hr to assemble a fancy chair; therefore the total number of hours needed to assemble y fancy chairs is $8y$. The maximum number of hours allocated for assembling is 160. Thus the third constraint is $4x + 8y \leq 160$. The final constraint is determined by the number of hours allotted for finishing. It takes 4 hr to finish a plain chair, or $4x$ hr to finish x plain chairs. It takes 12 hr to finish a fancy chair, or $12y$ hr to finish y fancy chairs. The total number of hours allotted for finishing is

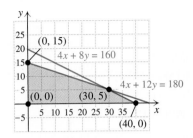

Figure 7.14

180. Therefore the fourth constraint is $4x + 12y \leq 180$. Thus the four constraints are

$$x \geq 0$$
$$y \geq 0$$
$$4x + 8y \leq 160$$
$$4x + 12y \leq 180$$

The list of constraints is a system of linear inequalities in two variables. The solution to the system of inequalities is the set of ordered pairs that satisfies all the constraints. These points are illustrated in Fig. 7.14. Note that the solution to the system consists of the colored region and the solid boundaries. The points (0, 0), (0, 15), (30, 5), and (40, 0) are the points where the boundaries intersect. These points can also be found by the substitution method described in Section 7.2.

The object in this example is to maximize the profit. The total profit is given by the formula $P = 10x + 18y$. According to the fundamental principle, the maximum profit will be found at one of the vertices of the polygonal region.

Calculate P for each one of the vertices:

$$P = 10x + 18y$$
$$\text{At } (0, 0), \quad P = 10(0) + 18(0) = 0$$
$$\text{At } (40, 0), \quad P = 10(40) + 18(0) = 400$$
$$\text{At } (30, 5), \quad P = 10(30) + 18(5) = 390$$
$$\text{At } (0, 15), \quad P = 10(0) + 18(15) = 270$$

The maximum profit is at (40, 0), which means that the company should manufacture 40 plain rocking chairs and no fancy rocking chairs. The maximum profit would be $400. The minimum profit would be at (0, 0), when no rocking chairs of either style were manufactured.

A variation of the original problem could be that the company knows that it cannot sell more than 15 plain rocking chairs per day. With this additional constraint, we now have the following set of constraints:

$$x \geq 0$$
$$x \leq 15$$
$$y \geq 0$$
$$4x + 8y \leq 160$$
$$4x + 12y \leq 180$$

The graph of these constraints is shown in Fig. 7.15.

We see that the vertices of the polygonal region are (0, 0,), (0, 15), (15, 10), and (15, 0). To determine the maximum profit, we calculate P for each of these

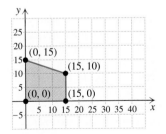

Figure 7.15

vertices:

$$P = 10x + 18y$$

$$\text{At } (0, 0), \quad P = 10(0) + 18(0) = 0$$
$$\text{At } (0, 15), \quad P = 10(0) + 18(15) = 270$$
$$\text{At } (15, 10), \quad P = 10(15) + 18(10) = 330$$
$$\text{At } (15, 0), \quad P = 10(15) + 18(0) = 150$$

With this set of constraints we see that the maximum profit of $330.00 occurs when the company manufactures 15 plain rocking chairs and 10 fancy rocking chairs.

Section 7.6 Exercises

In Ex. 1–6, a set of constraints and a profit formula are given.
a) Draw the graph of the constraints, and find the points of intersection of the boundaries.
b) Use these points of intersection to determine the maximum and minimum profit.

1. $x + y \le 5$
$x + 3y \le 9$
$x \ge 0$
$y \ge 0$
$P = 2x + 4y$

2. $x + y \le 7$
$x + 2y \le 10$
$x \ge 0$
$y \ge 0$
$P = 4x + 5y$

3. $x + y \le 4$
$x + 3y \le 6$
$x \ge 0$
$y \ge 0$
$P = 6x + 7y$

4. $x + y \le 50$
$x + 3y \le 90$
$x \ge 0$
$y \ge 0$
$P = 20x + 40y$

5. $3x + 2y \ge 6$
$3x + 4y \le 24$
$x \ge 0$
$y \ge 0$
$y \le 3$
$P = 1.5x + 3.5y$

6. $x + 2y \le 12$
$4x + 5y \ge 20$
$x \ge 0$
$y \ge 0$
$x \le 8$
$P = 5x + 7y$

7. The P³ Company manufactures two types of power lawn mowers. Type *A* is self-propelled, and type *B* must be pushed by the operator. The company can produce a maximum of 18 mowers per week. It can make a profit of $20 on mower *A* and a profit of $25 on mower *B*. The company's planners want to make at least two mowers of type *A* but not more than five. To keep certain customers happy, they must make at least two mowers of type *B*.
a) List the constraints.
b) List the profit formula.
c) Graph the set of constraints.
d) Find the vertices of the polygonal region.
e) How many of each type should be made to maximize the profit?
f) Find the maximum profit.

8. A craftsman makes wooden toy cars and trucks. He wants to make at least two cars and two trucks but no more than five trucks. He can make a total of 15 toys in his spare time in a month. At craft fairs he makes a profit of $8.00 per car and $10.00 per truck sold. How many of each type should he make each month to maximize his profit?

Problem Solving

9. To make one package of all-beef hot dogs, a manufacturer uses 1 lb of beef; to make one package of regular hot dogs, the manufacturer uses $\frac{1}{2}$ lb each of beef and pork. The profit on regular hot dogs is 30 cents per pack and for the all-beef hot dogs is 40 cents per pack. If there are 200 lb of beef and 150 lb of pork available, how many packs of all-beef and regular hot dogs should the manufacturer make to maximize the profit? What is the profit?

CHAPTER 7 SUMMARY

Key Terms

7.1
break-even analysis
break-even point
consistent system
dependent system
inconsistent system
simultaneous linear equations
system of linear equations

7.2
addition (or elimination) method
equilibrium price
law of supply and demand
market equilibrium
substitution method

7.3
additive identity matrix
matrix
multiplicative identity matrix

scalar
square matrices

7.4
augmented matrix

7.6
constraints
polygonal region
vertices

Important Facts

Systems of equations

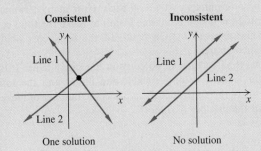

Consistent

Line 1

Line 2

One solution

Inconsistent

Line 1

Line 2

No solution

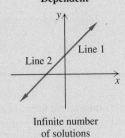

Dependent

Line 1

Line 2

Infinite number
of solutions

Methods of solving systems of equations
1. Graphing
2. Substitution
3. Addition (or elimination) method

Additive identity matrix

$$\begin{bmatrix} 0 & 0 \\ 0 & 0 \end{bmatrix}$$

Multiplicative identity matrix

$$\begin{bmatrix} 1 & 0 \\ 0 & 1 \end{bmatrix}$$

Fundamental principle of linear programming
If a linear equation of the form $K = Ax + By$ is evaluated at each point in a closed polygonal region, the maximum and minimum values of the equation occur at vertices of the region.

CHAPTER 7 REVIEW EXERCISES

7.1

Solve each system of equations graphically. If the system does not have a single ordered pair as a solution, state whether the system is inconsistent or dependent.

1. $x = 5$
$y = 6$

2. $2x - y = -1$
$3x + 2y = 9$

3. $x = 3$
$x + y = 5$

4. $x + 2y = 5$
$2x + 4y = 4$

Determine without graphing whether the system of equations has exactly one solution, no solution, or an infinite number of solutions.

5. $x + 3y = 4$
$2x - 3y = -15$

6. $2x - 3y = 6$
$-2x + 3y = -6$

7. $y = 4x - 6$
$4x - y = 4$

8. $2x - 4y = 8$
$-2x + y = 6$

7.2

Solve each system of equations by the substitution method. If the system does not have a single ordered pair as a solution, state whether the system is inconsistent or dependent.

9. $x + 3y = 5$
$2x - y = 3$

10. $x - 2y = 4$
$y = 3x - 2$

11. $2x - y = 4$
$3x - y = 2$

12. $3x + y = 1$
$3y = -9x - 4$

Solve each system of equations by the addition method. If the system does not have a single ordered pair as a solution, state whether the system is inconsistent or dependent.

13. $x + y = 12$
$-x + 2y = -3$

14. $2x - y = 2$
$3x - y = 5$

15. $2x - 3y = 1$
$x + 2y = -3$

16. $3x + 4y = 6$
$2x - 3y = 4$

17. $3x - 5y = 15$
$2x - 4y = 0$

18. $3x + y = 6$
$-6x - 2y = -12$

7.3

Given $A = \begin{bmatrix} 2 & 3 \\ 4 & 5 \end{bmatrix}$, $B = \begin{bmatrix} 1 & -6 \\ -7 & 8 \end{bmatrix}$, find the following.

19. $A + B$

20. $A - B$

21. $3A$

22. $2A - 3B$

23. $A \times B$

24. $B \times A$

7.4

Use matrices to solve each system of equations.

25. $x + 2y = 4$
$x + y = 2$

26. $-x + y = 4$
$x + 2y = 2$

27. $2x + y = 4$
$4x - y = 2$

28. $2x + 3y = 2$
$4x - 9y = 4$

7.1–7.4

29. A taxi charges a fixed fee plus a mileage fee. If a 10-mi trip costs $22 and a 5-mi trip costs $14.50, find the fixed fee and the mileage fee.

30. A chemist has a 30% acid solution and an 80% acid solution. How much of each should he mix together to obtain five gallons of a 50% solution?

31. A hospital is conducting a study to determine the most economically efficient way of replacing an outdated mainframe computer. The study committee has two options. Option 1 is a $40,000 minicomputer whose terminals cost $600 each. Option 2 is a $10,000 network system whose terminals cost $1200 each. How many terminals would the hospital have to install to make the total cost of the network system equal to the total cost of the minicomputer?

32. Cheryl has a total of 60 coins in her piggy bank. If the coins consist of only nickels and dimes and the total value of the coins is $5.75, what is the number of nickels and the number of dimes?

7.5

Graph each system of linear inequalities and indicate the solution set.

33. $2x + y < 8$
$y \geq 2x - 1$

34. $y < -2x + 3$
$y \geq 4x - 2$

35. $x + 3y \leq 6$
$2x - 7y \geq 14$

36. $x - y > 5$
$6x + 5y \leq 30$

7.6

37. A set of constraints and a profit formula are given below. Graph the constraints and find the points of intersection of the boundaries. Use these points of intersection to determine the maximum profit.

$$x + y \leq 10$$
$$2x + 1.8y \leq 18$$
$$x \geq 0$$
$$y \geq 0$$
$$P = 6x + 5y$$

Chapter 7

CHAPTER TEST

1. Solve the system of equations graphically:

$$y = 2x - 5$$
$$2x - 3y = 3$$

2. Determine without graphing whether the system of equations has exactly one solution, no solution, or an infinite number of solutions:

$$4x - 5y = -6$$
$$3x + 5y = 13$$

Solve the system of equations by the method indicated.

3. $3x - 5y = -3$
$x + y = 7$
(substitution)

4. $y = 4x - 6$
$y = -2x + 18$
(substitution)

5. $x + y = 4$
$2x + y = 3$
(addition)

6. $4x + 3y = 5$
$2x + 4y = 10$
(addition)

7. $2x - 3y = 5$
$6y - 4x = 7$
(addition)

8. $x - 3y = -4$
$-5x + 7y = 4$
(matrices)

Given $A = \begin{bmatrix} -6 & 2 \\ 3 & -5 \end{bmatrix}$ $B = \begin{bmatrix} -1 & 5 \\ 4 & -2 \end{bmatrix}$, find the following.

9. $A + B$

10. $3A - B$

11. $A \times B$

12. Graph the system of linear inequalities and indicate the solution set.

$$y \geq -2x + 4$$
$$2x - y < 8$$

13. A grocer plans to mix candy that sells for $3.40 per pound with candy that sells for $4.60 per pound to get 10 lb of a mixture to sell for $4.00 per pound. How many pounds of each type of candy should she mix to obtain the desired mixture?

14. A health club has an initiation fee plus a monthly membership fee. The total paid by Inez for 6 months of membership including the initiation fee is $390. The total paid by Raymond for 9 months of membership including the initiation fee is $435. What is the initiation fee and what is the monthly charge?

GEOMETRY

The geometry of fractals now provides mathematicians with a means of describing objects in nature in which a pattern endlessly repeats itself in smaller and smaller versions. This mountain scene, which looks eerily real, is actually a computer-generated fractal.

As a child, you probably learned to draw a house as a rectangle with a triangle on top of it, and to draw the sun as a circle with rays radiating from it. The simple geometry of this child's-eye view powerfully illustrates the fact that humans universally recognize the world to be composed of planes, angles, straight lines, and smooth curves — that is, based on Euclidean geometry. We perceive most objects to be three-dimensional in nature, that is, as objects with length, width, and depth.

Indeed, most everyday objects can be described in terms of Euclidian geometry: the desk you work on, the frisbee you toss to your dog, the soccer ball you kick across a rectangular field. Geometric patterns have been used since the beginning of human history to decorate everyday objects such as quilts, floor-tiling, and decorative containers. Patterns have remained because they can be varied nearly endlessly to provide a rhythm that pleases the eye and comforts the mind.

Islamic artists perfected the art of ornamentation and mosaic using geometric forms during the Middle Ages because Islamic law strictly prohibited the use of the human form in art. Mosaic patterns at the Alhambra, a palace and fortress built in Grenada, Spain, illustrate such mathematical concepts as symmetries, tessellations, reflections, and rotations.

In the twentieth century, however, those who study mathematics, physics, and other disciplines have discovered that there are other geometries that shape the universe. Einstein's theory of relativity, for example, is based upon a non-Euclidean geometry, and adds a fourth dimension — time — to the three dimensions of space. The work of Benoit Mandelbrot opened up new fields of mathematics with the discovery of fractal geometry, which has been used to describe the shapes of items such as snowflakes, coastlines, and tree branches.

Wherever humans go, geometry is sure to follow. The patterns and variations of geometry act almost as a rhythm to human life, comparable to music or poetry. Sonya Kovalevskaya, a mathematician whose many contributions included studies of geometric sequences, differential equations, and a prize-winning paper titled "On the rotation of a solid body about a fixed point," remarked on this connection between mathematics and poetry. Also gifted in literature, she explained, "… it is impossible to be a mathematician without being a poet in soul…. It seems to me that the poet has only to perceive that which others do not perceive, to look deeper than others look. And the mathematician must do the same thing."

Nature displays geometric forms at even the most microscopic level. A recently discovered molecule called the buckministerfullerene, is formed from 12 pentagons and 20 hexagons. Because of the molecule's resiliency, scientists anticipate that it will have applications in conductors and insulators. The molecule was named after Buckminister Fuller, the famous inventor of geodesic domes such as Disney World's Epcot Center.

Architects make use of a variety of polygons in designing both the basic structure and the ornamentation of a building. Frank Lloyd Wright, the American architect known for his command of space, pioneered twentieth century residential design by creating homes where rooms flowed together in uninterrupted space that was, in effect, continued outside.

8.1 Points, Lines, Planes, and Angles

Human beings recognized shapes, sizes, and physical forms long before geometry was developed. Geometry as a science is said to have begun in the Nile Valley of ancient Egypt. The Egyptians used geometry to measure land and to build pyramids and other structures.

The word "geometry" is derived from two Greek words, *ge*, meaning earth, and *metron*, meaning measure. Thus geometry means "earth measure" or "measurement of the earth."

Unlike the Egyptians, the Greeks were interested in more than just the applied aspects of geometry. The Greeks attempted to apply their knowledge of logic to geometry. Thales of Miletus around 600 B.C. was the first to be credited with using deductive methods to develop geometric concepts. Another outstanding Greek geometer was Pythagoras, who continued the systematic development of geometry Thales began.

Around 300 B.C., Euclid collected and summarized most of the Greek mathematics of his time. In a set of 13 books called *Elements*, Euclid laid the foundation for plane geometry, which is also called **Euclidean geometry**.

Euclid is credited with being the first mathematician to use the **axiomatic method** in developing a branch of mathematics. First, Euclid introduced **undefined terms**, such as point, line, plane, and angle. He related these to physical space by such statements as "A line is length without breadth," so that we may intuitively understand them. Because such statements play no further role in his system, they constitute primitive or undefined terms.

Second, Euclid introduced certain **definitions**. The definitions are introduced when needed and are often based on the undefined terms. Some terms that Euclid introduced and defined include triangle, right angle, and hypotenuse.

Third, Euclid asserted certain primitive propositions called **postulates** (now called **axioms***) about the undefined terms and definitions. The reader is asked to accept these statements as true on the basis of their "obviousness" and their relationship with the physical world. For example, the Greeks accepted all right angles as being equal. This is Euclid's fourth postulate.

Fourth, Euclid proved, using deductive reasoning (Section 1.1), other propositions called **theorems**. One theorem that Euclid proved is known as the Pythagorean theorem: "The sum of the areas of the squares constructed on the arms of a right triangle is equal to the area of the square constructed on the hypotenuse." He also proved that the sum of the angles of a triangle is 180°.

Using only 10 axioms, Euclid deduced 465 propositions (or theorems) in plane and solid geometry, number theory, and Greek geometrical algebra.

*The man who made mathematics into an orderly system of definitions, axioms, theorems, and proofs was the Greek mathematician **Euclid** (320–225 B.C.). It is often said that, next to the Bible, Euclid's Elements may be the most translated, published, and studied of all the books produced in the Western world.*

---

* The concept of the axiom has changed significantly since Euclid's time. Now any statement may be designated as an axiom, whether it is self-evident or not. All axioms are *accepted* as true. A set of axioms forms the foundation for a mathematical system.

"It is the glory of geometry that from so few principles, fetched from without, . . . it is able to accomplish so much."

I. Newton

Point and Line

Three basic terms in geometry are **point**, **line**, and **plane**. These three terms are not given a formal definition, but we have a common idea of what each is.

Let's consider some properties of a line. Assume that a line means a straight line unless otherwise stated.

1. A line is a set of points. Each point is on the line and the line passes through each point.
2. Any two distinct points determine a unique line (see Fig. 8.1a). The arrows at both ends of the line indicate that the line continues in each direction. The line in Fig. 8.1(a) is symbolized by $\overleftrightarrow{AB}$ or $\overleftrightarrow{BA}$.
3. Any point on a line separates the line into three parts—the point itself and two **half lines**. For example, in Fig. 8.1(a), point B separates the line into the point B, and two half lines. Half line AB is symbolized as $\overset{\circ\!\!\longrightarrow}{AB}$. The open circle above the A indicates that point A is not included in the half line.

Look at the half line $\overset{\circ\!\!\longrightarrow}{AB}$ in Fig. 8.1(b). If the **end point**, A, is included with the set of points on the half line, the result is called a **ray**. Ray AB, symbolized by $\overrightarrow{AB}$, is illustrated in Fig. 8.1(c). Ray BA is illustrated in Fig. 8.1(d).

A **line segment** is that part of a line between two points, including the end points. Line segment AB, symbolized $\overline{AB}$, is illustrated in Fig. 8.1(e).

An open line segment is the set of points on a line between two points, excluding the end points. Open line segment AB, symbolized $\overset{\circ\!-\!\circ}{AB}$, is illustrated in Fig. 8.1(f).

Figure 8.1(g) illustrates a half open line segment, AB, symbolized $\overset{\circ\!-}{AB}$.

In Chapter 2 we discussed intersection of sets. Recall that the intersection ($\cap$) of two sets is the set of elements (points in this case) common to both sets.

Description	Diagram	Symbol
(a) Line AB		$\overleftrightarrow{AB}$
(b) Half line AB		$\overset{\circ\!\!\longrightarrow}{AB}$
(c) Ray AB		$\overrightarrow{AB}$
(d) Ray BA		$\overrightarrow{BA}$
(e) Line segement AB		$\overline{AB}$
(f) Open line segement AB		$\overset{\circ\!-\!\circ}{AB}$
(g) Half open line segement AB		$\overset{-\!\circ}{AB}$ $\overset{\circ\!-}{AB}$

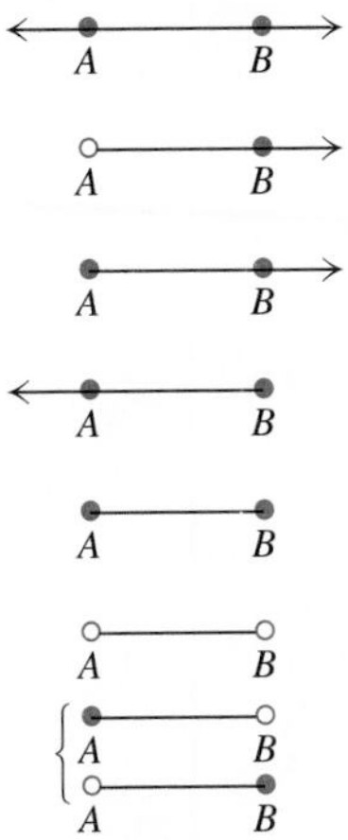

Figure 8.1

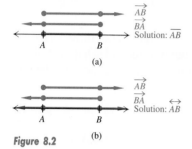

(a)

(b)

Figure 8.2

Consider the lines in Fig. 8.2(a). The intersection of $\overrightarrow{AB}$ and $\overleftarrow{BA}$ is line segment $\overline{AB}$. Thus $\overrightarrow{AB} \cap \overleftarrow{BA} = \overline{AB}$.

The union of two sets was also discussed in Chapter 2. The union ($\cup$) of two sets is the set of elements (points in this case) that belong to either of the sets or both sets. The union of $\overrightarrow{AB}$ and $\overleftarrow{BA}$ is $\overleftrightarrow{AB}$ (Fig. 8.2b). Thus $\overrightarrow{AB} \cup \overleftarrow{BA} = \overleftrightarrow{AB}$.

▶ **Example 1**

Use the line illustrated to find the following.

a) $\overrightarrow{AB} \cup \overleftarrow{DC}$ **b)** $\overline{BC} \cap \overrightarrow{CD}$ **c)** $\overline{AB} \cap \overrightarrow{CD}$ **d)** $\overline{AD} \cup \overset{\circ}{C}\overrightarrow{A}$

Solution:

a) $\overrightarrow{AB} \cup \overleftarrow{DC}$

Ray AB and ray DC are shown here. The union of these two rays is the entire line AD.

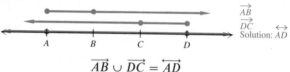

$$\overrightarrow{AB} \cup \overleftarrow{DC} = \overleftrightarrow{AD}$$

b) $\overline{BC} \cap \overrightarrow{CD}$

The intersection of line segment BC and ray CD is point C.

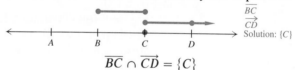

$$\overline{BC} \cap \overrightarrow{CD} = \{C\}$$

c) $\overline{AB} \cap \overrightarrow{CD}$

Since line segment AB and ray CD have no points in common, their intersection is empty.

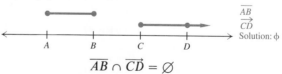

$$\overline{AB} \cap \overrightarrow{CD} = \varnothing$$

d) $\overline{AD} \cup \overset{\circ}{C}\overrightarrow{A}$

The union of line segment AD and half line CA is ray DA (or $\overrightarrow{DB}$ or $\overrightarrow{DC}$).

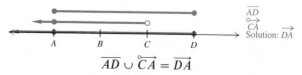

$$\overline{AD} \cup \overset{\circ}{C}\overrightarrow{A} = \overrightarrow{DA}$$

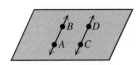

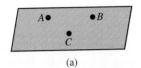

Figure 8.3

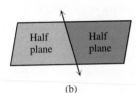

(a)

Half plane | Half plane

(b)

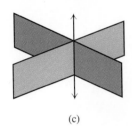

(c)

Figure 8.4

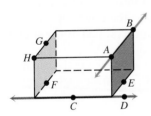

Figure 8.5

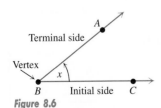

Figure 8.6

Plane

Even though the term "plane" is not formally defined, we can think of it as a two-dimensional surface that extends infinitely in both directions, like an infinitely large blackboard. Euclidean geometry is called **plane geometry** because it is the study of two-dimensional figures in a plane.

Two lines in the same plane that do not intersect are called **parallel lines**. Figure 8.3 illustrates two parallel lines in a plane ($\overleftrightarrow{AB}$ is parallel to $\overleftrightarrow{CD}$).

Some of the properties of planes are as follows:

1. Any three points that are not on the same line (noncollinear points) determine a unique plane (Fig. 8.4a).
2. A line in a plane divides the plane into three parts—the line and two half planes (Fig. 8.4b).
3. Any line and a point not on the line determines a unique plane.
4. The intersection of two planes is a line (Fig. 8.4c).

Two planes that do not intersect are said to be **parallel planes**. For example, in Fig. 8.5, plane ABE is parallel to plane GHF.

Two lines that do not lie in the same plane and do not intersect are called **skewed lines**. Figure 8.5 illustrates skewed lines ($\overleftrightarrow{AB}$ and $\overleftrightarrow{CD}$).

Angles

An **angle**, denoted by $\angle$, is the union of two rays with a common end point (Fig. 8.6).

$$\overrightarrow{BA} \cup \overrightarrow{BC} = \angle ABC \text{ (or } \angle CBA)$$

An angle can be formed by the rotation of a ray about a point. An angle has an initial side and a terminal side: The initial side indicates the position of the ray prior to rotation; the terminal side indicates the position of the ray after rotation. The point common to both rays is called the **vertex** of the angle. The letter designating the vertex is always the middle one of the three letters designating an angle. The rays that make up the angle are called its **sides**. Note that an angle is not the space between the two sides.

There are a number of ways to name an angle. The angle in Fig. 8.6 can be denoted as follows:

$$\angle ABC, \qquad \angle CBA, \qquad \angle B.$$

An angle divides a plane into three distinct parts: the angle itself, its interior, and its exterior.

▶ **Example 2**

Refer to Fig. 8.7 on page 364. Find the following.

a) $\overrightarrow{BF} \cup \overrightarrow{BD}$ **b)** $\angle ABF \cap \angle FBE$ **c)** $\overleftrightarrow{AC} \cap \overleftrightarrow{DE}$ **d)** $\overrightarrow{BA} \cup \overrightarrow{BC}$

Solution:

a) $\overrightarrow{BF} \cup \overrightarrow{BD} = \angle FBD$

b) $\angle ABF \cap \angle FBE = \overrightarrow{BF}$

c) $\overleftrightarrow{AC} \cap \overleftrightarrow{DE} = \{B\}$

d) $\overrightarrow{BA} \cup \overrightarrow{BC} = \overleftrightarrow{AC}$

Figure 8.7

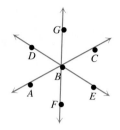

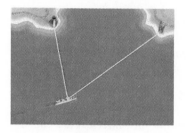

The **measure of an angle**, symbolized by the letter m, is the amount of rotation from its initial side to its terminal side. In Fig. 8.6 on page 363 the letter x represents the measure of $\angle ABC$, therefore we may write $m \angle ABC = x$.

Angles can be measured in **degrees**, radians, or gradients. In this text we will discuss only the degree unit of measurement. The symbol for degrees is the same as the symbol for temperature degrees. An angle of 45 degrees is written $45°$. A protractor is used to measure angles. The angle shown being measured by the protractor in Fig. 8.8 is $50°$.

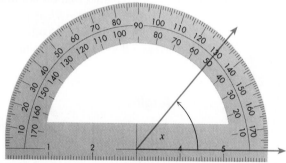

Figure 8.8

Consider a circle whose circumference is divided into 360 equal parts. If a line is drawn from each mark on the circumference to the center of the circle, we get 360 wedge-shaped pieces. The measure of an angle formed by the straight sides of each wedge-shaped piece is defined to be 1 degree.

Angles are classified by their degree measurement, as shown in the summary box. A **right angle** is $90°$; an **acute angle** is less than $90°$; an **obtuse angle** is greater than $90°$ but less than $180°$; a **straight angle** is $180°$.

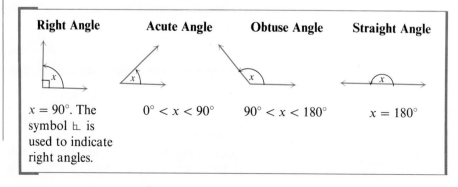

Right Angle	Acute Angle	Obtuse Angle	Straight Angle
$x = 90°$. The symbol ⌐ is used to indicate right angles.	$0° < x < 90°$	$90° < x < 180°$	$x = 180°$

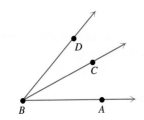

Figure 8.9

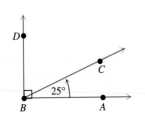

Figure 8.10

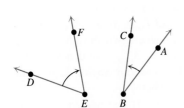

Figure 8.11

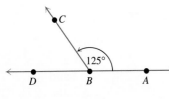

Figure 8.12

Two angles in the same plane are **adjacent angles** when they have a common vertex and a common side but no common interior points. $\angle DBC$ and $\angle CBA$ in Fig. 8.9 are adjacent angles. However, $\angle DBA$ and $\angle CBA$ are not adjacent angles.

Two angles the sum of whose measures is 90° are called **complementary angles**. Each angle is called a complement of the other.

▶ **Example 3**

If $m\angle ABC = 25°$, and $\angle ABC$ and $\angle CBD$ are complementary angles, determine $m\angle CBD$ (Fig. 8.10).

Solution: The sum of the two angles must be 90°:

$$m\angle CBD = 90° - m\angle ABC$$
$$= 90° - 25°$$
$$= 65°$$

▶ **Example 4**

If $\angle ABC$ and $\angle DEF$ are complementary angles and $\angle DEF$ is twice as large as $\angle ABC$, determine the measure of each angle (Fig. 8.11).

Solution: If $m\angle ABC = x$, then $m\angle DEF = 2 \cdot x$ or $2x$, since it is twice as large. The sum of the measures of these angles must be 90°:

$$m\angle ABC + m\angle DEF = 90°$$
$$x + 2x = 90°$$
$$3x = 90°$$
$$x = 30°$$

Therefore $m\angle ABC = 30°$ and $m\angle DEF = 2 \cdot 30°$, or 60°.

Two angles the sum of whose measures is 180° are said to be **supplementary angles**. Each angle is called a supplement of the other.

▶ **Example 5**

If $\angle ABC$ and $\angle CBD$ are supplementary angles and $m\angle ABC = 125°$, find $m\angle CBD$ (Fig. 8.12).

Solution: The sum of the two angles must be 180°:

$$m\angle CBD = 180° - m\angle ABC$$
$$= 180° - 125°$$
$$= 55°$$

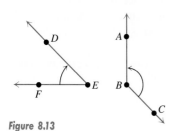

Figure 8.13

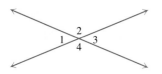

Figure 8.14

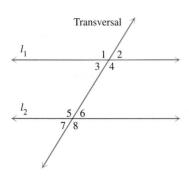

Figure 8.15

▶ **Example 6**

If $\angle ABC$ and $\angle DEF$ are supplementary angles and $\angle ABC$ is three times as large as $\angle DEF$, determine $m\angle ABC$ and $m\angle DEF$ (Fig. 8.13).

Solution: If $m\angle DEF = x$, then $m\angle ABC = 3x$.

$$m\angle ABC + m\angle DEF = 180°$$
$$3x + x = 180°$$
$$4x = 180°$$
$$x = 45°$$

Thus $m\angle DEF = 45°$, and $m\angle ABC = (3)(45°) = 135°$.

When two straight lines intersect, the nonadjacent angles formed are called **vertical angles**. In Fig. 8.14, $\angle 1$ and $\angle 3$ are vertical angles, and $\angle 2$ and $\angle 4$ are vertical angles. We can show that vertical angles have the same measure, that is, they are equal. For example, in Fig. 8.14 we see that

$$m\angle 1 + m\angle 2 = 180° \qquad \text{Why?}$$
$$m\angle 2 + m\angle 3 = 180° \qquad \text{Why?}$$

Since $\angle 2$ has the same measure in both cases, $m\angle 1$ must equal $m\angle 3$.

A line that intersects two different lines, l_1 and l_2, at two different points is called a **transversal**. Figure 8.15 illustrates that when two parallel lines are cut by a transversal, eight angles are formed. Angles 3, 4, 5, and 6 are called **interior angles** and angles 1, 2, 7, and 8 are called **exterior angles**. Eight pairs of supplementary angles are formed. Can you list them?

Special names are given to the angles formed by a transversal crossing two parallel lines.

Name	Description	Illustration	Pairs of Angles Meeting Criteria
Alternate interior angles	Interior angles on opposite sides of the transversal		$\angle 3$ and $\angle 6$ $\angle 4$ and $\angle 5$
Alternate exterior angles	Exterior angles on opposite sides of the transversal		$\angle 1$ and $\angle 8$ $\angle 2$ and $\angle 7$
Corresponding angles	One interior and one exterior angle on the same side of the transversal		$\angle 1$ and $\angle 5$ $\angle 2$ and $\angle 6$ $\angle 3$ and $\angle 7$ $\angle 4$ and $\angle 8$

When two parallel lines are cut by a transversal, alternate interior angles have the same measure, alternate exterior angles have the same measure, and corresponding angles have the same measure.

▶ **Example 7**

Figure 8.16 shows two parallel lines cut by a transversal. Determine the measure of ∡1 through ∡7.

Solution:

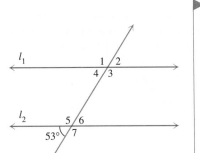

Figure 8.16

$m\angle 6 = 53°$	∡8 and ∡6 are vertical angles.
$m\angle 5 = 127°$	∡8 and ∡5 are supplementary angles.
$m\angle 7 = 127°$	∡5 and ∡7 are vertical angles.
$m\angle 1 = 127°$	∡1 and ∡7 are alternate exterior angles.
$m\angle 4 = 53°$	∡4 and ∡6 are alternate interior angles.
$m\angle 2 = 53°$	∡6 and ∡2 are corresponding angles.
$m\angle 3 = 127°$	∡3 and ∡1 are vertical angles.

Section 8.1 Exercises

In Ex. 1–6, identify each as a line, half line, ray, line segment, open line segment, or half open line segment. Denote each by its appropriate symbol.

1.

2.

3.

4.

5.

6.

In Ex. 7–18, use the figure to find each of the following.

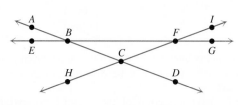

7. $\overleftrightarrow{AB} \cap \overrightarrow{HC}$ **8.** $\overleftrightarrow{BA} \cap \overrightarrow{BE}$

9. $\overrightarrow{FB} \cup \overrightarrow{FC}$ **10.** $\overset{\circ}{FE} \cup \overrightarrow{FG}$

11. $\overrightarrow{BC} \cup \overline{CD}$ **12.** $\overleftrightarrow{AD} \cup \overline{BC}$

13. $\overleftrightarrow{AD} \cap \overline{BC}$ **14.** ∡$HCD \cap$ ∡ACF

15. $\overset{\circ}{BD} \cup \overset{\circ}{CB}$ **16.** $\overset{\circ}{BD} \cap \overset{\circ}{CB}$

17. $\{C\} \cap \overset{\circ}{CH}$ **18.** $\overline{BC} \cup \overline{CF} \cup \overline{FB}$

In Ex. 19–30, use the figure to find each of the following.

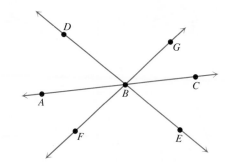

19. ∡$ABE \cup \overset{\circ}{AB}$ **20.** $\overrightarrow{BF} \cup \overrightarrow{BE}$

21. $\overrightarrow{BF} \cap \overrightarrow{BE}$ **22.** $\overrightarrow{BD} \cup \overset{\circ}{BE}$

23. ∡$GBC \cap$ ∡CBE **24.** $\overrightarrow{DE} \cup \overrightarrow{BE}$

25. ∡$CBE \cap$ ∡EBC **26.** $\{B\} \cap \overrightarrow{BA}$

27. $\overset{\circ}{AC} \cap \overline{AC}$ **28.** $\overset{\circ}{AC} \cap \overrightarrow{BE}$

29. $\overset{\circ}{EB} \cap \overrightarrow{BE}$ **30.** $\overrightarrow{GF} \cap \overleftrightarrow{AB}$

In Ex. 31–36, classify the angle as acute, right, straight, or obtuse.

31. **32.** **33.**

34. **35.** **36.**

Find the complementary angle of each of the following.

37. 30° **38.** 45° **39.** $67\frac{1}{2}$° **40.** 34.4°
41. 87° **42.** 1° **43.** 5° **44.** $3\frac{1}{2}$°

Find the supplementary angle of each of the following.

45. 91° **46.** 4° **47.** 105° **48.** 179°
49. $13\frac{3}{4}$° **50.** $89\frac{1}{4}$° **51.** 114.7° **52.** 137.9°

In Ex. 53–58, match the name of the angles with angles 1 and 2 shown in (a)–(f).

a) **b)**

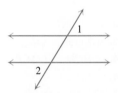

c) **d)**

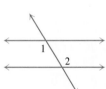

e) **f)**

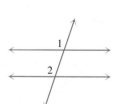

53. Vertical angles
54. Supplementary angles
55. Complementary angles
56. Alternate interior angles
57. Alternate exterior angles
58. Corresponding angles

59. If ∡1 and ∡2 are complementary angles and ∡1 is four times as large as ∡2, find the measures of ∡1 and ∡2.
60. If ∡1 and ∡2 are complementary angles and ∡1 is five times as large as ∡2, find the measures of ∡1 and ∡2.
61. If ∡1 and ∡2 are supplementary angles and ∡1 is four times as large as ∡2, find the measures of ∡1 and ∡2.
62. Given the set of parallel lines cut by a transversal shown here, determine the measures of ∡1 through ∡7.

63. What are parallel lines?
64. What are skewed lines?
65. How many planes can be drawn through a given line.

In Ex. 66–69, the angles are supplementary angles. Find the measures of ∡1 and ∡2.

66. **67.**

68. **69.**

In Ex. 70–73, the angles are complementary angles. Find the measures of ∡1 and ∡2.

70. **71.**

72. **73.**

74. a) How many lines can be drawn through a given point?
b) How many planes can be drawn through a given point?
75. What is the intersection of two distinct non-parallel planes?

76. What is the intersection of a plane and a line (on a non-parallel plane).

77. a) Will three noncollinear points *A*, *B*, and *C* always determine a plane?

    b) Is it possible to determine more than one plane with three noncollinear points?

    c) How many planes can be constructed through three collinear points?

78. Mark three noncollinear points *A*, *B*, and *C*. Draw the ray $\overrightarrow{AB}$ and the ray $\overrightarrow{AC}$. What geometric figure is produced?

This figure suggests a number of lines and planes. The lines may be described by pairs of points, and the planes may be described by naming three points. In Ex. 79–88 use the figure to name:

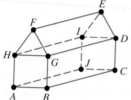

79. A pair of parallel planes
80. A pair of planes whose intersection is a line
81. Three planes whose intersection is a single point
82. Three planes whose intersection is a line
83. A line and a plane whose intersection is a point

84. A line and plane whose intersection is a line
85. A pair of parallel lines
86. A pair of skew lines
87. Three lines that intersect at a single point
88. Four planes whose intersection is a single point

89. What geometrical advantage is there in building chairs with three legs rather than four legs?

90. Why are Euclid's *Elements* important?

91. a) List the four parts of an axiomatic system as used by Euclid in developing Euclidean geometry.

    b) Discuss each of the four parts.

Problem Solving

92. Suppose that you have three distinct lines, all lying in the same plane. Find all the possible ways in which the three lines can be related. Sketch each case (four cases).

93. Suppose that you have three distinct planes in space. Find all the possible ways these planes can be related. Sketch each case (five cases).

94. Suppose that you have three distinct lines in space (not necessarily in the same plane). Find all the possible ways in which these three lines can be related. Sketch each case (nine cases).

Research Activity

95. Write a report on the contributions of Euclid in the field of mathematics.

8.2 Polygons

A **polygon** is a closed figure in a plane determined by three or more straight line segments. Examples of polygons are given in Fig. 8.17.

    The straight line segments that form the polygon are called its **sides** and a point where two sides meet is called a **vertex** (plural **vertices**). The union of the sides of a polygon and its interior is called a **polygonal region**. A **regular polygon** is one whose sides are all the same length and whose interior angles all have the same measure. Figures 8.17(b) and (d) are regular polygons.

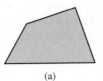
(a)

(b)

(c)

(d)

Figure 8.17

Table 8.1

Number of Sides	Name
3	triangle
4	quadrilateral
5	pentagon
6	hexagon
7	heptagon
8	octagon
9	nonagon
10	decagon
12	dodecagon
20	icosagon

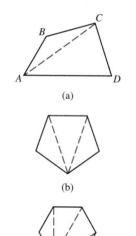

(a)

(b)

(c)

Figure 8.19

Table 8.2

Sides	Triangles	Sum of the Measures of the Interior Angles
3	1	$1(180°) = 180°$
4	2	$2(180°) = 360°$
5	3	$3(180°) = 540°$
6	4	$4(180°) = 720°$

Polygons are named according to their number of sides. The names of some polygons are given in Table 8.1.

One of the most important polygons is the triangle. The sum of the measures of the interior angles of a triangle is 180°. To illustrate this, consider triangle ABC given in Fig. 8.18. The triangle is formed by drawing two transversals through two parallel lines l_1 and l_2 with the two transversals intersecting at a point on l_1.

Suppose the transversals make angles of 130° and 110° with l_2 as illustrated in Fig. 8.18.

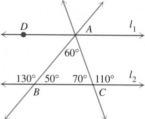

Figure 8.18

Then $\angle ABC$ must measure 50° and $\angle ACB$ must measure 70°. By alternate interior angles $\angle DAC$ must measure 110° and $\angle DAB$ must measure 50°. If we subtract 50° from 110° we get 60° for the measure of $\angle BAC$. Now by adding the measures of $\angle ABC$, $\angle BCA$, and $\angle CAB$, we see that the sum of the measures of the interior angles of triangle ABC is 180°.

Consider the quadrilateral $ABCD$ (Fig. 8.19a). Drawing a straight line segment between any two vertices forms two triangles. Since the sum of the measures of the angles of a triangle is 180°, the sum of the measures of the interior angles of a quadrilateral is $2 \cdot 180°$, or 360°.

Now let's examine a pentagon (Fig. 8.19b). As can be seen from the figure, we can draw two straight line segments to form three triangles. Thus the sum of the measures of the interior angles of a five-sided figure is $3 \cdot 180°$, or 540°. Figure 8.19(c) illustrates that four triangles can be drawn in a six-sided figure. Table 8.2 summarizes this information.

If we continue this procedure, we can see that for an n-sided polygon the sum of the measures of the interior angles is $(n - 2)180°$.

> The **sum** of the measures of the interior angles of an n-sided polygon is $(n - 2)180°$.

▶ Example 1

A gazebo is in the shape of a regular octagon (Fig. 8.20). Determine (a) the measure of an interior angle and (b) the measure of exterior angle 1.

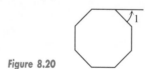

Figure 8.20

Solution:

a) Using the formula $(n - 2)180°$, we can determine the sum of the measures of the interior angles of an octagon.

The sum of the measures of the interior angles

$$= (8 - 2)180°$$
$$= 6(180°)$$
$$= 1080°$$

The measure of an interior angle of a regular polygon can be determined by dividing the sum of the interior angles by the number of angles.

The interior angle of a regular octagon $= \dfrac{1080°}{8} = 135°$

b) Since $\angle 1$ is the supplement of an interior angle,

$$m\angle 1 = 180° - 135° = 45°$$

In order to discuss area in the next section we must be able to identify various types of triangles and quadrilaterals. The summary box illustrates certain types of triangles and their characteristics.

Acute Triangle	**Obtuse Triangle**	**Right Triangle**
All angles are acute angles.	One angle is obtuse.	hypotenuse One angle is a right angle.
Isosceles Triangle	**Equilateral Triangle**	**Scalene Triangle**
Two equal sides Two equal angles	Three equal sides Three equal angles (60° each)	No two sides are equal in length.

Similar Figures

In everyday living it is often necessary to deal with geometric figures that have the "same shape" but are of different sizes. For example, an architect will make a small-scale drawing of a floor plan, or a photographer will make an enlargement of a photograph. Figures that have the same shape but may be of different sizes are called **similar figures**. A set of similar figures is illustrated in Fig. 8.21.

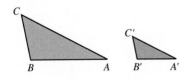

Figure 8.21

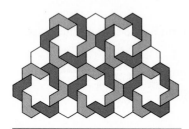

An interesting application of polygons is in the design of tiling patterns—patterns that must fit together to completely cover a flat surface. The pattern here can be found in the Taj Mahal.

Similar figures have **corresponding angles** and **corresponding sides**. In Fig. 8.21 triangle ABC has angles A, B, and C. Their respective corresponding angles in triangle $A'B'C'$ are angles A', B', and C'. Sides AB, BC, and AC in triangle ABC have corresponding sides $A'B'$, $B'C'$, and $A'C'$ respectively in triangle $A'B'C'$.

> Two polygons are **similar** if their corresponding angles have the same measure and their corresponding sides are in proportion.

In Fig. 8.21 $\angle A$ and $\angle A'$ have the same measure, as do $\angle B$ and $\angle B'$, and $\angle C$ and $\angle C'$. Also the corresponding sides of similar triangles are in proportion:

$$\frac{\overline{AB}}{\overline{A'B'}} = \frac{\overline{CB}}{\overline{C'B'}} = \frac{\overline{AC}}{\overline{A'C'}}$$

▶ **Example 2**

Consider the similar figures in Fig. 8.22.

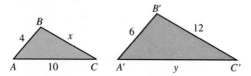

Figure 8.22

Determine the following:
a) The length of side $\overline{BC}$　　**b)** The length of side $\overline{A'C'}$

Solution:

a) We will represent side $\overline{BC}$ with the letter x. Since the corresponding sides must be in proportion, we can write a proportion (as explained in Section 6.2) to find side $\overline{BC}$. Since corresponding sides $\overline{AB}$ and $\overline{A'B'}$ are known, we will use them as one ratio in the proportion:

$$\frac{\overline{AB}}{\overline{A'B'}} = \frac{\overline{BC}}{\overline{B'C'}}$$

$$\frac{4}{6} = \frac{x}{12}$$

Now solve for x.

$$4 \cdot 12 = 6 \cdot x$$
$$48 = 6x$$
$$8 = x$$

Thus the length of $\overline{BC} = 8$.

b) Represent $\overline{A'C'}$ with the letter y:

$$\frac{\overline{AB}}{\overline{A'B'}} = \frac{\overline{AC}}{\overline{A'C'}}$$

$$\frac{4}{6} = \frac{10}{y}$$

$$4 \cdot y = 6 \cdot 10$$

$$4y = 60$$

$$y = 15$$

Thus the length of $\overline{A'C'} = 15$.

▶ Example 3

Consider the similar triangles ABC and $AB'C'$ (Fig. 8.23). Determine

a) the length of side $\overline{BC}$ **b)** the length of side $\overline{AB'}$

c) $m \angle AB'C'$ **d)** $m \angle ACB$

e) $m \angle ABC$

Solution:

a) We will set up a proportion to find the length of side $\overline{BC}$. Since corresponding sides $\overline{AC}$ and $\overline{AC'}$ are known, use them as one ratio in the proportion:

$$\frac{\overline{AC}}{\overline{AC'}} = \frac{\overline{BC}}{\overline{B'C'}}$$

$$\frac{8}{4} = \frac{x}{3}$$

$$24 = 4x$$

$$6 = x$$

Therefore the length of $\overline{BC} = 6$.

b)

$$\frac{\overline{AC}}{\overline{AC'}} = \frac{\overline{AB}}{\overline{AB'}}$$

$$\frac{8}{4} = \frac{6}{y}$$

$$8y = 24$$

$$y = 3$$

Therefore the length of $\overline{AB'} = 3$.

c) Since the sum of the angles in a triangle must measure $180°$, $m \angle AB'C'$ must be $180° - (46° + 46°)$, or $88°$.

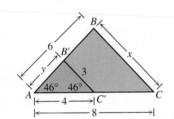

Figure 8.23

d) Since triangles ABC and $AB'C'$ are similar, their corresponding angles must have the same measure. Therefore $\angle ACB$ must have the same measure as its corresponding angle $\angle AC'B'$. Thus $m\angle ACB = 46°$.

e) Since $\angle ABC$ and $\angle AB'C'$ are corresponding angles of similar triangles, $m\angle ABC = m\angle AB'C'$. Using the results from part (c), the measure of $\angle ABC$ is $88°$.

Congruent Figures

If the corresponding sides of two similar triangles are the same length, the figures are called **congruent figures**. Corresponding angles of congruent figures have the same measure, and the corresponding sides are equal in length. Two congruent figures coincide when placed one upon the other.

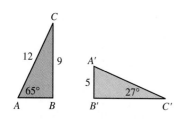

Figure 8.24

▶ **Example 4**

Triangles ABC and $A'B'C'$ in Fig. 8.24 are congruent. Find
a) the length of side $\overline{A'C'}$ **b)** the length of side $\overline{AB}$
c) $m\angle C'A'B'$ **d)** $m\angle ACB$
e) $m\angle ABC$

Solution: We are given that $\triangle ABC$ is congruent to $\triangle A'B'C'$. Using this fact we know that the corresponding sides and angles are equal.
a) $\overline{A'C'} = \overline{AC} = 12$
b) $\overline{AB} = \overline{A'B'} = 5$
c) $m\angle C'A'B' = m\angle CAB = 65°$
d) $m\angle ACB = m\angle A'C'B' = 27°$
e) The sum of the angles of a triangle is $180°$. Since $m\angle BAC = 65°$ and $m\angle ACB = 27°$, $m\angle ABC = 180° - 65° - 27° = 88°$.

Quadrilaterals

Quadrilaterals are four-sided polygons. Quadrilaterals may be classified according to their characteristics, as illustrated in the summary box on page 375. The sum of the angles in any quadrilateral measures $360°$.

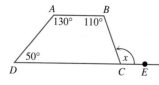

Figure 8.25

▶ **Example 5**

Find the measure of angle x in the trapezoid in Fig. 8.25.

Solution: Since sides AB and CD are parallel, BC may be considered a transversal. Then by alternate interior angles, $x = 110°$.

A second method to solve Example 5 is to find the measure of the unknown interior angle of the trapezoid by subtracting the sum of the measures of the three given angles from $360°$. $m\angle BCD = 360° - (50° + 130° + 110°) = 360° - 290° = 70°$. Since $\angle DCE$ is a straight angle, $m\angle x$ must be $180° - 70° = 110°$.

Trapezoid	Parallelogram	Rhombus
Two sides are parallel.	Both pairs of opposite sides are parallel.	Both pairs of opposite sides are parallel. The four sides are equal in length.

Rectangle	Square
Both pairs of opposite sides are parallel. The angles are right angles.	Both pairs of opposite sides are parallel. The four sides are equal in length. The angles are right angles.

 ## Section 8.2 Exercises

1. What are similar figures?
2. What are congruent figures?

In Ex. 3–7, name each of the polygons.

3.

4.

5.

6.

7.

In Ex. 8–12, identify each of the triangles as acute, obtuse, or right.

8.

9.

10.

11.

12.

In Ex. 13–17, identify each of the triangles as scalene, isosceles, or equilateral.

13.

14.

15.

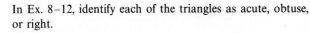

16.　**17.**

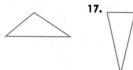

In Ex. 18–22, identify each of the quadrilaterals.

18.　　　　**19.**　　　　**20.**

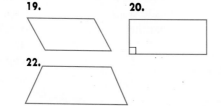

21.　　　　**22.**

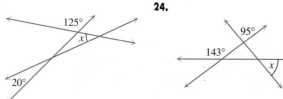

In Ex. 23–25, find the measure of each angle indicated by an *x*.

23.　　　　　　　**24.**

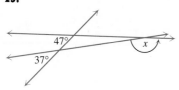

25.

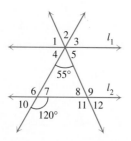

26. Given that l_1 and l_2 are parallel in the figure, determine the measures of $\angle 1$ through $\angle 12$.

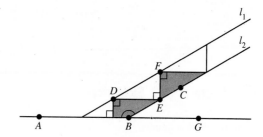

27. Sketch an octagon.
   a) Show that the figure can be divided into six triangles.
   b) Use the figure to determine the sum of the interior angles.
   c) Use the formula to verify the results obtained in part (b).

28. Sketch a decagon. Answer similar questions for this figure as asked in Ex. 27.

Find the sum of the interior angles of each polygon.

29. Nonagon　　　　　　　**30.** Octagon
31. Hexagon　　　　　　　**32.** Pentagon
33. Decagon　　　　　　　**34.** Icosagon

Find the measure of an interior angle of each regular polygon. If a side of the polygon is extended, find the supplement of the interior angle. See Example 1.

35. Triangle　　　**36.** Quadrilateral　　**37.** Heptagon
38. Decagon　　　**39.** Pentagon　　　　**40.** Octagon

In Ex. 41–43, determine the measure of the angle. In the figure $\angle ABC$ makes an angle of $125°$ with the floor and l_1 and l_2 are parallel.

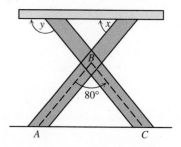

41. $\angle GBC$　　　　**42.** $\angle EDF$　　　　**43.** $\angle DFE$

44. The legs of a picnic table form an isosceles triangle as indicated in the figure. If $\angle ABC = 80°$, find $\angle x$ and $\angle y$ so that the top of the table will be parallel to the ground.

A shopping cart is illustrated here. Assuming that *wz* is parallel to *xy*, find

45. $m \angle w$ **46.** $m \angle x$ **47.** $m \angle y$ **48.** $m \angle z$

In Ex. 49–51, the figures are similar. Find the length of all sides colored red.

49.

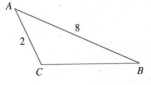

50.

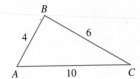

51.

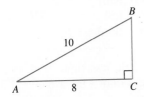

Triangles *ABC* and *A'B'C'* are similar figures. Find the length of the following.

52. Side $\overline{A'C}$ **53.** Side $\overline{BC}$ **54.** Side $\overline{BB'}$

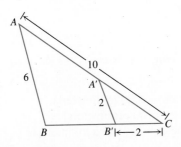

Quadrilaterals *ABCD* and *A'B'C'D'* are similar figures. Find the length of the following.

55. Side $\overline{AB}$ **56.** Side $\overline{BC}$ **57.** Side $\overline{D'C'}$

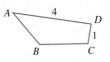

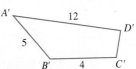

In Ex. 58–62, find the measures of the angles given that pentagons *ABCDE* and *A'B'C'D'E'* are similar figures.

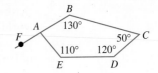

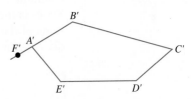

58. $\angle BAE$ **59.** $\angle B'A'E'$
60. $\angle B'C'D'$ **61.** $\angle FAE$
62. $\angle F'A'E'$

In Ex. 63–68, find the length of the sides and the measures of the angles given that triangles *ABC* and *A'B'C'* are congruent.

63. the length of side $\overline{A'B'}$
64. the length of side $\overline{A'C'}$
65. the length of side $\overline{BC}$
66. $\angle B'A'C'$
67. $\angle ACB$
68. $\angle ABC$

In Ex. 69–74, find the length of the sides and the measures

of the angles given that quadrilaterals $ABCD$ and $A'B'C'D'$ are congruent.

69. the length of side $\overline{A'B'}$
70. the length of side $\overline{AD}$
71. the length of side $\overline{B'C'}$
72. $\angle BCD$
73. $\angle A'D'C'$
74. $\angle DAB$

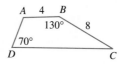
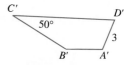

8.3 | Perimeter and Area

Units of Measurement

Before we discuss perimeter and area, let us briefly discuss measurement. In the world today there are basically two measurement systems: the U.S. system and the metric system. The metric system (officially called the International System of Units, or SI for its name in French: *Système international d' unités*) is the most common measurement system used in the world today. In fact, the United States is the only Western country that does not use it. The summary box lists some of the more commonly used units of length in both systems. (See Appendix A for more details on the metric system.)

Commonly Used Units of Length	
U.S. System	**Metric System**
inch (in.)	millimeter (mm)
foot (ft)	centimeter (cm)
yard (yd)	meter (m)
mile (mi)	kilometer (km)

Inches
Centimeters
(10 mm = 1 cm)

For comparisons, a millimeter (1 mm) is a little less than $\frac{1}{16}$ in. (see the figure to the left); a centimeter (1 cm) is a little less than $\frac{1}{2}$ in.; a meter (1 m) is a little larger than a yard; and a kilometer (1 km) is about $\frac{6}{10}$ of a mile.

The list that follows indicates common conversion factors used in the metric system.

Selected Metric Conversion Factors
milli means $\frac{1}{1000}$ of base unit
centi means $\frac{1}{100}$ of base unit
kilo means 1000 times base unit

For example, in the metric system, *the meter is the base unit of length,* therefore

1 millimeter $= \frac{1}{1000}$ meter   (or 1 meter $=$ 1000 millimeters)
1 centimeter $= \frac{1}{100}$ meter   (or 1 meter $=$ 100 centimeters)
1 kilometer $=$ 1000 meters   (or 1 meter $= \frac{1}{1000}$ of a kilometer)

In this chapter we will use both U.S. and metric measurements.

Perimeter and Area

Geometric shapes abound in the natural world and the world made by human beings. For example, a basketball court is a rectangle, a basketball is a sphere, a can of food is a cylinder, and a ream of paper is a rectangular solid.

The **perimeter**, P, of a two-dimensional figure, is the sum of the lengths of the sides of the figure. In Fig. 8.26 the sum of the red lines is the perimeter. Perimeters are measured in the same units as the sides. For example, if the sides of a figure are measured in feet, the perimeter will be measured in feet.

The **area**, A, is the region within the boundaries of the figure. The blue color in Fig. 8.26 indicates the area of the figure. Area is measured in square units. For example, if the sides of a figure are measured in centimeters, then the area of the figure will be measured in square centimeters (cm^2). See the summary box for common units of area in the U.S. and metric systems.

Figure 8.26

Common Units of Measurement of Area	
U.S. System	**Metric System**
square inch (in.^{2})	
square foot (ft^2)	square centimeter (cm^2)
square yard (yd^2)	square meter (m^2)
square mile (mi^2)	are (a) (100 m^2)
acre (43,560 ft^2)	hectare (ha) (100 a)

Consider the rectangle in Fig. 8.27. Two sides of the rectangle have length, l, and two sides of the rectangle have width w. Thus, if we add the length of the four sides to get the perimeter, we find $P = l + w + l + w = 2l + 2w$.

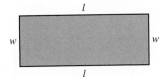

Figure 8.27

Perimeter of a Rectangle

$$P = 2l + 2w$$

Consider a rectangle of length 5 units and width 3 units (Fig. 8.28). Counting the number of 1-unit by 1-unit squares within the figure we see that the area of the rectangle is 15 square units. The area can also be obtained by multiplying the number of units of length by the number of units of width, 5 units × 3 units = 15 square units. We can find the area by the formula Area = length × width.

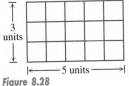

Figure 8.28

Area of a Rectangle

$$A = l \times w$$

Figure 8.29

Using the formula for the area of a rectangle, we can determine the formulas for the area of other figures.

A square (Fig. 8.29) is a rectangle that contains four equal sides. Therefore, the length equals the width. Using the formula for the area of a rectangle we can determine the area of a square. Call both the length and the width of the square s, then

$$A = l \times w$$
$$A = s \times s = s^2$$

Area of a Square

$$A = s^2$$

A parallelogram with height h and base b is shown in Fig. 8.30(a).

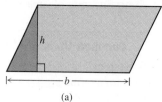

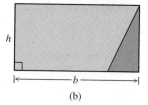

(a)   (b)

Figure 8.30

If we were to cut off the red portion of the parallelogram on the left, Fig. 8.30(a), and attach it to the right side of the figure, the resulting figure would be a rectangle shown in Fig. 8.30(b). Since the area of the rectangle is $b \times h$, the area of the parallelogram is also $b \times h$.

Area of a Parallelogram

$$A = b \times h$$

Now consider the trapezoid in Fig. 8.31(a). We can convert a trapezoid to a rectangle as shown in Fig. 8.31(a–c). The trapezoid has been changed to a rectangle with base $\frac{1}{2}(b_1 + b_2)$ and height h. The area of the rectangle is $A = h\left[\frac{1}{2}(b_1 + b_2)\right]$ or $A = \frac{1}{2}h(b_1 + b_2)$. Since the trapezoid has the same area as the rectangle we can reason that the area of a trapezoid is $A = \frac{1}{2}h(b_1 + b_2)$.

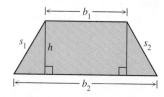

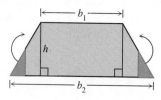

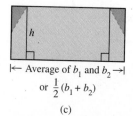

Figure 8.31   (a)   (b)   (c)

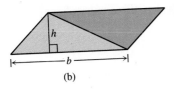

(a)

(b)

Figure 8.32

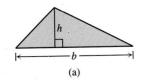

Italy

New Mexico

Which geographic area, Italy or New Mexico, do you think has the greatest land area? As a matter of fact, the land areas of Italy and New Mexico are very similar: Italy's is 116,304 sq mi and New Mexico's is 121,593 sq mi. However, Italy's perimeter, 5812 mi, is almost five times greater than New Mexico's 1200 mi.

Area of a Trapezoid

$$A = \frac{1}{2}h(b_1 + b_2)$$

Consider the triangle with height, h, and base, b, in Fig. 8.32. Using the triangle in Fig. 8.32(a) and a second identical triangle we can construct a parallelogram, Fig. 8.32(b). The area of the parallelogram is bh. The area of the triangle is one-half that of the parallelogram. Therefore, the area of the triangle is $\frac{1}{2}$(base)(height).

Area of a Triangle

$$A = \frac{1}{2}bh$$

Following is a summary box on the areas and perimeters of selected figures.

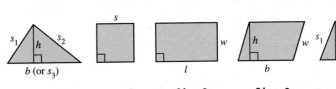

Perimeters and Areas

Triangle	Square	Rectangle	Parallelogram	Trapezoid
$p = s_1 + s_2 + s_3$ $(s_3 = b)$	$p = 4s$	$p = 2l + 2w$	$p = 2b + 2w$	$p = s_1 + s_2$ $+ b_1 + b_2$
$A = \frac{1}{2}bh$	$A = s^2$	$A = lw$	$A = bh$	$A = \frac{1}{2}h(b_1 + b_2)$

▶ **Example 1**
You are coating your driveway with blacktop sealer. One can of sealer costs $15 and covers 200 ft². If your driveway is 50 ft long and 10 ft wide, **(a)** find the area of your driveway, **(b)** determine how many cans of blacktop sealer are needed, and **(c)** determine the cost of sealing the driveway.

Solution:
a) The area of your driveway in square feet is

$$A = l \cdot w = 50 \cdot 10 = 500 \text{ ft}^2.$$

The area of the driveway is in square feet because both the length and width are measured in feet.

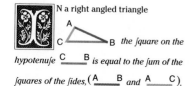

N a right angled triangle the ſquare on the hypotenuſe is equal to the ſum of the ſquares of the ſides, (A——B and A——C).

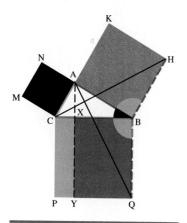

The Pythagorean theorem has probably been proven in more ways than any other theorem in mathematics. The book Pythagorean Propositions *contains 370 different proofs of the Pythagorean theorem. Another book, the nineteenth-century English-language edition of Euclid's elements by Oliver Byrne illustrated here, described by one critic as "one of the oddest and most beautiful books of the whole [nineteenth] century," uses color and shapes to illustrate a proof of the Pythagorean theorem.*

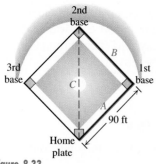

Figure 8.33

b) To determine the number of cans needed, divide the area of the driveway by the area covered by one can of sealer:

$$\frac{500}{200} = 2.5$$

Since you cannot purchase a half can of sealer, three cans of sealer are needed.

c) The cost of the sealer is 3 × $15, or $45.

Pythagorean Theorem

The Pythagorean theorem was introduced in Chapter 5. The Pythagorean theorem is an important tool for finding the perimeter and area of triangles. For this reason we restate it at this time.

Pythagorean Theorem

The sum of the squares of the lengths of the legs of a right triangle is equal to the square of the length of the hypotenuse.

Symbolically, if a and b represent the lengths of the legs and c represents the length of the hypotenuse (the side opposite the right angle), then $a^2 + b^2 = c^2$.

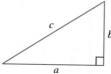

▶ **Example 2**

The bases on a baseball field are 90 ft apart (Fig. 8.33). Find the distance from home plate to second base.

Solution: The line from home plate to second base is the hypotenuse of a right triangle with sides of 90 ft. By the Pythagorean theorem,

$$c^2 = a^2 + b^2$$
$$c^2 = (90)^2 + (90)^2$$
$$c^2 = 8100 + 8100 = 16{,}200$$
$$c = \sqrt{16{,}200} \approx 127.3 \text{ ft.}$$

Therefore the distance from home plate to second base is about 127 ft.

Circles

A commonly used plane figure that is not a polygon is a **circle**. A circle is a set of points equidistant from a fixed point called the center. A **radius** , r, of a circle

Circumference
$C = 2\pi r$

Radius

Area
$A = \pi r^2$

Diameter

Figure 8.34

is a line segment from the center of the circle to any point on the circle (Fig. 8.34). A **diameter**, d, of a circle is a line segment through the center of a circle with both end points on the circle. Note that the diameter of the circle is twice its radius. The **circumference** is the length of the simple closed curve that forms the circle. The formulas for the area and circumference of a circle are given in Fig. 8.34. We introduced the symbol pi, π, in Chapter 5. Recall that π is approximately 3.14.

▶ Example 3

At the local pizza shop the price for a medium plain pizza is $7.50, and the price for a large plain pizza is $14.00. The diameters of the large and medium pizzas are 18 inches and 12 inches, respectively, and both pizzas are the same thickness. To get the most for your money should you buy two medium or one large pizza?

Solution: We begin by finding the surface area of the two pizzas. The radius of the larger pizza is 9 in. and the radius of the smaller pizza is 6 in.

Area of a large pizza	Area of a medium pizza
$A = \pi r^2$	$A = \pi r^2$
$= 3.14(9)^2$	$= 3.14(6)^2$
$= 3.14(81) = 254.34 \text{ in.}^2$	$= 3.14(36) = 113.04 \text{ in.}^2$

The two medium pizzas have an area of $2(113.04) = 226.08 \text{ in.}^2$

	One large pizza	Two medium pizzas
Area	254.34 in.^2	226.08 in.^2
Cost	$14	$15

One large pizza has a greater area and a lower cost. Therefore one large pizza should be purchased.

▶ Example 4

You plan to fertilize your lawn. The shapes and dimensions of your lot, house, pool, and shed are shown in Fig. 8.35. One bag of fertilizer costs $15.95 and covers 5000 ft². Determine how many bags of fertilizer are needed to fertilize your lawn and the total cost of the fertilizer.

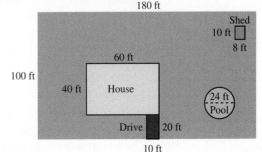

Figure 8.35

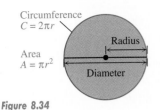

Solution: The total area of the lot is 100 · 180, or 18,000 ft². From this subtract the area of the house, driveway, pool, and shed.

$$\text{Area of house} = 60 \cdot 40 = 2400 \text{ ft}^2$$
$$\text{Area of driveway} = 20 \cdot 10 = 200 \text{ ft}^2$$
$$\text{Area of shed} = 8 \cdot 10 = 80 \text{ ft}^2$$

Since the diameter of the pool is 24 ft, its radius is 12 ft.

$$\text{Area of pool} = \pi r^2 = 3.14(12)^2 = 3.14(144) = 452.16 \text{ ft}^2$$

The total area of the house, driveway, shed, and pool is approximately 2400 + 200 + 80 + 452, or 3132 ft². The area to be fertilized is 18,000 − 3132, or 14,868 ft². The number of bags of fertilizer is found by dividing the total area to be fertilized by the number of square feet covered per bag.

The number of bags of fertilizer is 14,868/5000, or 2.97 bags. Therefore three bags are needed. At $15.95 per bag, the total cost is $47.85.

Sometimes it may be necessary to convert an area from one square unit of measure to a different square unit of measure. For example, you may have to change square feet of carpet to square yards of carpet before placing an order with the dealer. Example 5 shows how to do this.

▶ Example 5

a) Convert one square foot to square inches.
b) Convert 7.5 square feet to square inches.

Solution:

a) 1 ft = 12 in.
   Therefore 1 ft² = 12 in. × 12 in. = 144 in.².
b) Since 1 ft² = 144 in.², 2 ft² = 2(144 in.²) = 288 in.², and 7.5 ft² is

$$7.5 \text{ ft}^2 = 7.5 \times 144 = 1080 \text{ in.}^2$$

▶ Example 6

The area of a four-wall racquetball court is 115,200 in.². How large is this in square feet?

Solution: Since 1 ft² = 144 in.², divide 115,200 in.² by 144 in.² to obtain the number of square feet:

$$115,200 \text{ in.}^2 = \frac{115,200}{144} = 800 \text{ ft}^2$$

▶ Example 7

One meter equals 100 centimeters.
a) Convert one square meter to square centimeters.
b) Convert 4.2 m² to square centimeters.

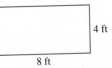

Figure 8.36

Solution:

a) One square meter is a square whose sides are each 1 m in length. Since 1 m equals 100 cm, replace 1 m with 100 cm on each side of the square (Fig. 8.36). The area $1 \text{ m}^2 = 1 \text{ m} \cdot 1 \text{ m} = 100 \text{ cm} \times 100 \text{ cm} = 10,000 \text{ cm}^2$. Thus the area of one square meter is 10,000 times the area of one square centimeter.

b) Since $1 \text{ m}^2 = 10,000 \text{ cm}^2$,

$$4.2 \text{ m}^2 = 4.2 \times 10,000 \text{ cm}^2$$
$$= 42,000 \text{ cm}^2$$

When multiplying units of length, be very careful to make sure that the units are the same. You can multiply feet by feet to get square feet or yards by yards to get square yards. However you cannot get a valid answer if you multiply numbers expressed in feet by numbers expressed in yards.

Section 8.3 Exercises

Find the area of each of the triangles.

1.

2.

3.

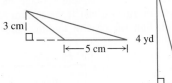

9.

10.

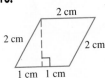

Find the area and perimeter of each of the quadrilaterals.

4.

5.

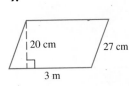

6.

7.

8.

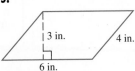

Find the area and circumference of each of the circles.

11.

12.

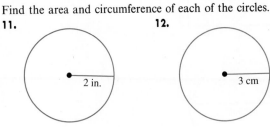

13.

14.

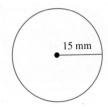

For each right triangle, use the Pythagorean theorem to determine the third side.

15.

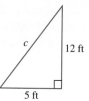

12 ft

c

5 ft

16.

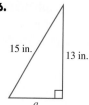

15 in.

13 in.

a

17.

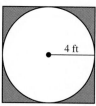

7 m

b

5 m

18.

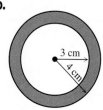

2 cm

8 cm

c

In Ex. 19–26, find the shaded area.

19.

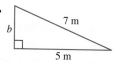

4 ft

20.

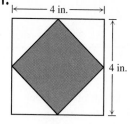

3 cm

4 cm

21.

|← 4 in. →|

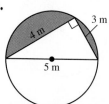

4 in.

22.

|←2 ft→|

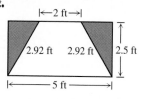

2.92 ft   2.92 ft   2.5 ft

|← 5 ft →|

23.

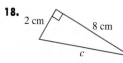

3 m

4 m

5 m

24.

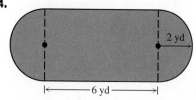

2 yd

|← 6 yd →|

25.

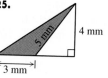

4 mm

5 mm

|← 3 mm →|

26.

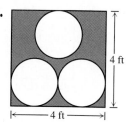

4 ft

|← 4 ft →|

One square yard equals 9 square feet. Use this information to convert the following.

27. 6.9 ft² to square yards **28.** 302 ft² to square yards
29. 18.3 yd² to square feet **30.** 23.6 yd² to square feet

One square meter equals 10,000 square centimeters. Use this information to convert the following.

31. 12.6 m² to cm² **32.** 21.6 m² to cm²
33. 608 cm² to m² **34.** 1075 cm² to m²

35. One method of estimating the cost of building a new house is to use square footage of living space (using exterior dimensions). If the current rate is $60 per square foot, find the cost of building the house in this plan:

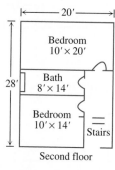

36. Determine the cost of putting a hardwood floor in the living/dining area of the house shown in Ex. 35. The cost of the hardwood flooring selected is $1.99 per square foot. Add 5% to the cost for waste.

37. Determine the cost of covering the three bedroom floors in the house shown with carpet. The price of the carpet selected is $14.95 per square yard. (Add 5% for waste.)

38. A building 40 m by 64 m is surrounded by a walk 1.5 m wide. Find the dimensions of the region covered by the building and the walk. Find the area of the walk.

39. A gallon of wood sealer covers 300 ft². Ms. Zenna plans to seal the floor of a gymnasium that is 100 ft by 160 ft. How many gallons of sealer are needed?

40. Ms. Thong Seng has selected carpet that costs $19.99 per square yard to use in the living room and family room. The living room measures 15 ft by 21 ft, and the family room measures 18 ft by 16 ft. How many square yards of carpeting will she need? What will it cost?

41. The Pella's lot is illustrated here.

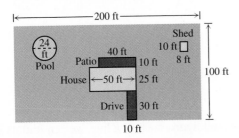

The Pella's son Tony is taking a college mathematics course and so Mr. Pella asks him the following questions. Can you answer them?

a) How many square feet of lawn do they have?

b) If a bag of fertilizer covers 5000 ft², how many bags of fertilizer will they need to cover the lawn?

c) If a bag of fertilizer costs $19.95, what will be the total cost of the fertilizer?

In Ex. 42–49, explain your answer.

42. What is the effect on the area of a square if a side is doubled in length?

43. What is the effect on the area of a square if a side is tripled?

44. What is the effect on the area of a square if a side is cut in half?

45. What is the effect on the area of a rectangle if its length is doubled and its width is left unchanged?

46. What is the effect on the area of a rectangle if its length and width are both doubled?

47. What is the effect on the area of a rectangle if its length is doubled and its width is cut in half?

48. What is the effect on the area of a triangle if the base is doubled and its altitude is left unchanged?

49. What is the effect on the area of a triangle if the base is doubled and its altitude is cut in half?

50. An isosceles right triangle has an area of 50 cm². What are the lengths of the sides?

51. Herman's garden is 13.4 m by 16.5 m. How large is his garden in hectares?

52. The shape and measurements of the base of a free-standing fireplace are shown. The base will be covered with a special stone. Determine the number of square yards of stone the mason must order.

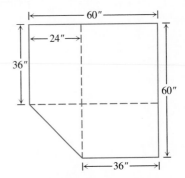

53. Which hamburger has the larger surface area: a square hamburger 3 inches on a side from Wendy's or a $3\frac{1}{2}$-in.-diameter round hamburger from Burger King? Explain your answer and give the difference in their surface areas.

54. Is it possible for two figures to have the same perimeters and areas and not be congruent figures? Explain, or show how you determined your answer.

Problem Solving

55. The Kents are painting their house this summer. The dimensions of the house are illustrated here.

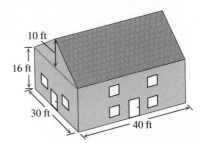

a) Find the total number of square feet to be painted. (Assume that there are 12 windows, each 2 ft 3 in. ×

3 ft 4 in., and two doors, each 80 in. × 36 in. The roof will not be painted.)

b) Determine the number of gallons needed if two coats of paint are required. For each coat assume one gallon of paint covers 500 ft^2.

c) Determine the cost of the paint if paint sells for $18.95 per gallon.

56. Find the surface area of the right circular cylinder in the figure.

57. Find the surface area of the rectangular solid in the figure.

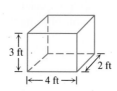

In Ex. 58–60, write an expression that represents the shaded area.

58.

59.

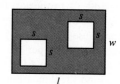

60.

8.4 Volume

When discussing a one-dimensional figure, such as a line, we can find its length. When discussing a two-dimensional figure, such as a rectangle, we can find its area. When discussing a three-dimensional figure such as a sphere, we can find its volume. **Volume** is a measure of the capacity of a figure. The measure of volume may be confusing because we use different units to measure different types of volumes. For example water and other liquids may be measured in ounces, quarts, or gallons. A volume of topsoil may be measured in cubic yards. In the metric system, liquid may be measured in liters or milliliters, and topsoil may be measured in cubic meters.

Solid geometry is the study of three-dimensional solid figures (also called space figures). Volumes of three-dimensional figures are measured in cubic units, such as cubic feet or cubic meters. Some common measurements of volume in both the U.S. system and the metric system are shown here.

Common Measurements of Volume	
U.S. System	**Metric System**
cubic inches (in.^{3})	cubic centimeters (cm^3 or cc)
cubic feet (ft^3)	cubic meters (m^3)
cubic yards (yd^3)	

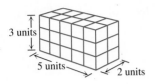

Figure 8.37

We will begin our discussion with the **rectangular solid**. If the length of the solid is 5 units, the width is 2 units and the height is 3 units, the total number of cubes is 30 (Fig. 8.37). Thus the volume is 30 cubic units. The volume of a rectangular solid can also be found by multiplying its length times width times height; 5 units × 2 units × 3 units = 30 cubic units. In general, the volume of any rectangular solid, as shown in part (a) of the summary box on page 390, is $V = l \times w \times h$.

Volume of a Rectangular Solid

$$V = l \times w \times h$$

A **cube** is a rectangular solid where the length, width, and height all measure the same (part b of the box). If we call the side of a cube s, and use the formula for a rectangular solid, substituting s for l, w, and h, we obtain $V = s \cdot s \cdot s = s^3$.

Volume of a Cube

$$V = s^3$$

Now consider the right circular cylinder (part c of the box). The base is a circle with area πr^2. When we add height, h, the figure becomes a cylinder. For the given circular base, the greater the height, the greater the volume. The volume of the right circular cylinder is found by multiplying the area of the base, πr^2, by the height h.

Volume of a Cylinder

$$V = \pi r^2 h$$

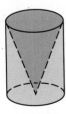

Figure 8.38

A cone is illustrated in part (d) of the box. Imagine a cone inside a cylinder, sharing the same circle as base. The volume of the cone is less than the volume of the cylinder that has the same base and height (Fig. 8.38). In fact, the volume of the cone is one-third the volume of the cylinder.

Volume of a Cone

$$V = \frac{1}{3}\pi r^2 h$$

The last shape we will discuss in this section is the sphere (part e of the box). Basketballs, golf balls, and so on have the shape of a sphere. The formula for the volume of a sphere is as follows.

Volume of a Sphere

$$V = \frac{4}{3}\pi r^3$$

The following box summarizes the volumes of selected three-dimensional figures.

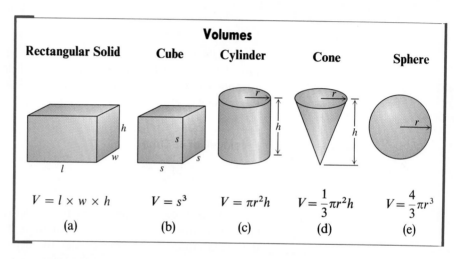

Volumes

Rectangular Solid	Cube	Cylinder	Cone	Sphere
$V = l \times w \times h$	$V = s^3$	$V = \pi r^2 h$	$V = \frac{1}{3}\pi r^2 h$	$V = \frac{4}{3}\pi r^3$
(a)	(b)	(c)	(d)	(e)

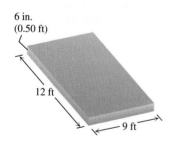

6 in.
(0.50 ft)

12 ft

9 ft

Figure 8.39

▶ **Example 1**

The base of a patio is to be a slab of concrete 12 ft long by 9 ft wide by 6 in. thick (Fig. 8.39).
a) How many cubic yards of concrete are needed?
b) If one cubic yard of concrete costs $46.50, how much will the concrete for the patio cost?

Solution:
a) When the calculations are performed, all the measurements must be in the same units. We will therefore convert all units to yards. There are 36 inches in a yard. Therefore 6 in. is $\frac{6}{36}$ or $\frac{1}{6}$ yd. The length of 12 ft is equal to 4 yd, and the width of 9 ft is equal to 3 yd.

$$V = l \cdot w \cdot h$$
$$= 4 \cdot 3 \cdot \tfrac{1}{6}$$
$$= 2 \text{ yd}^3$$

Note that since all three measurements were in terms of yards, the answer is in cubic yards.
b) Since one cubic yard of concrete costs $46.50, two cubic yards will cost 2 × $46.50, or $93.00

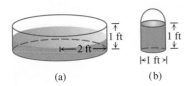

(a)　　　　(b)

Figure 8.40

▶ **Example 2**

Myra and Belinda and their friends are at a picnic at the town park. They have brought a children's wading pool in the shape of a right circular cylinder with a radius of 2 ft and a height of 1 ft into which they will put cold water to keep the soda cold (Fig. 8.40a). They carry the water from the faucet to the pool in a bucket that is also in the shape of a right circular cylinder, with a diameter of 1 ft and a height of 1 ft (Fig. 8.40b).

a) How many buckets of water are needed to fill the pool to a height of $\frac{1}{2}$ ft?
b) If one cubic foot of water weighs 62.5 lb, what is the weight of the water in the pool?
c) If there are 7.5 gal of water per cubic foot, how many gallons of water are in the pool?

Solution:

a) Volume of Water in Pool　　　Volume of Water in Bucket

$$V = \pi r^2 h \qquad\qquad\qquad V = \pi r^2 h$$
$$= 3.14(2)^2(0.5) \qquad\qquad = 3.14(0.5)^2(1)$$
$$= 3.14(4)(0.5) = 6.28 \text{ ft}^3 \qquad = 3.14(0.25)(1) = 0.785 \text{ ft}^3$$

To find the number of buckets needed, divide

$$\frac{6.28}{0.785} = 8.0.$$

Thus eight bucketfuls are needed.

b) Since 1 ft³ of water weighs 62.5 lb, 6.28 ft³ of water weighs 6.28 × 62.5, or 392.5 lb.

c) Since 1 ft³ of water contains 7.5 gal, 6.28 ft³ of water contains 6.28 × 7.5, or 47.1 gal of water. Thus there are 47.1 gal of water in the pool.

A **polyhedron** is a closed surface formed by the union of polygonal regions. Each polygonal region is called a **face** of the polyhedron. Figure 8.41 illustrates some polyhedrons.

The line segment formed by the intersection of two faces is called an **edge**.

Polyhedrons

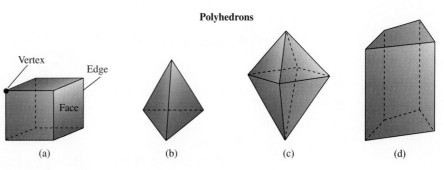

(a)　　　(b)　　　(c)　　　(d)

Figure 8.41

Gems, like all crystals, are composed of atoms arranged in regular three-dimensional geometric patterns. Many will split naturally into flat, well-defined surfaces, like this rough-cut piece of amethyst. Gems are prized for their beauty, durability, and rarity. They are traditionally cut into certain polyhedral shapes to bring out their beauty.

The point at which two or more edges intersect is called a **vertex**. In Fig. 8.41(a) there are 6 faces, 12 edges, and 8 vertices, and we see that

$$\text{number of vertices} - \text{number of edges} + \text{number of faces} = 2$$
$$8 \quad - \quad 12 \quad + \quad 6 \quad = 2$$

This formula, credited to Leonard Euler, is true for any polyhedron.

> ### Euler's Formula for Polyhedron
>
> Number of vertices − number of edges + number of faces = 2

We suggest that you verify that this formula holds for Fig. 8.41(b), (c), and (d).

A **regular polyhedron** is one whose faces are all regular polygons of the same size and shape. Figure 8.41(a) and (b) are regular polyhedrons.

A **prism** is a special type of polyhedron whose bases are congruent polygons and whose sides are parallelograms. These parallelogram regions are called the **lateral faces** of the prism. If all the lateral faces are rectangles, the prism is said to be a **right prism**. Some right prisms are illustrated in Fig. 8.42.

Prisms

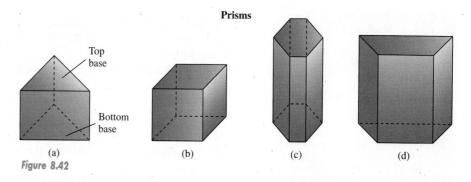

(a) (b) (c) (d)

Figure 8.42

The volume of any prism can be found by multiplying the area of the base B by the height h of the prism.

> ### Volume of a Prism
>
> $$V = Bh$$
>
> where B is the area of a base and h is the height.

Figure 8.43

▶ **Example 3**

Find the volume of the triangular prism shown in Fig. 8.43.

Solution: First find the area of the triangular base:

$$\text{Area of base} = \tfrac{1}{2}bh$$
$$= \tfrac{1}{2}(10)(4) = 20 \text{ square units}$$

Now find the volume by multiplying the area of the triangular base by the height of the prism:

$$V = B \cdot h$$
$$= 20 \cdot 8 = 160 \text{ cubic units}$$

▶ **Example 4**

Find the volume of the remaining solid after the cylinder, triangular prism, and square prism have been cut from the solid (Fig. 8.44).

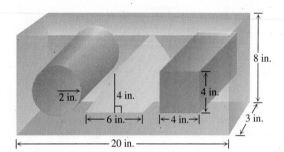

Figure 8.44

Solution: To find the volume of the remaining solid, find the volume of the rectangular solid. Then subtract the volume of the two prisms and the cylinder that were cut out.

$$\text{Volume of rectangular solid} = l \cdot w \cdot h$$
$$= 20 \cdot 3 \cdot 8 = 480 \text{ in.}^3$$
$$\text{Volume of circular cylinder} = \pi r^2 h$$
$$= (3.14)(2^2)(3)$$
$$= (3.14)(4)(3) = 37.68 \text{ in.}^3$$
$$\text{Volume of triangular prism} = \text{area of the base} \cdot \text{height}$$
$$= \tfrac{1}{2}(6)(4)(3) = 36 \text{ in.}^3$$
$$\text{Volumme of square prism} = s^2 \cdot h$$
$$= 4^2 \cdot 3 = 48 \text{ in.}^3$$
$$\text{Volume of solid} = 480 - 37.68 - 36 - 48$$
$$= 358.32 \text{ in.}^3$$

Another special category of polyhedrons is **pyramids**. Unlike prisms, pyramids have only one base. Some pyramids are illustrated in Fig. 8.45 on page 394. Note that all but one face of a pyramid intersect a common vertex.

Pyramids

(a)   (b)   (c)   (d)

Figure 8.45

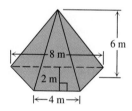

Figure 8.46

If a pyramid is drawn inside a prism, as in Fig. 8.46, we can see that the volume of the pyramid is less than that of the prism. In fact, the volume of the pyramid is one-third the volume of the prism.

Volume of a Pyramid

$$V = \frac{1}{3}Bh$$

where B is the area of the base, and h is the height.

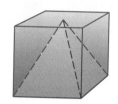

Figure 8.47

▶ **Example 5**

Find the volume of the pyramid shown in Fig. 8.47.

Solution: First find the area of the base of the pyramid. Since the base of the pyramid is a trapezoid.

$$\begin{aligned}
\text{Area of base} &= \tfrac{1}{2}h(b_1 + b_2) \\
&= \tfrac{1}{2}(2)(4 + 8) \\
&= 12 \text{ m}^2
\end{aligned}$$

Now find the volume:

$$\begin{aligned}
V &= \tfrac{1}{3}B \cdot h \\
&= \tfrac{1}{3}(12)(6) \\
&= 24 \text{ m}^3
\end{aligned}$$

It might be necessary in certain situations to convert volume from one cubic unit to a different cubic unit. For example, when purchasing topsoil you might have to change the amount of topsoil from cubic feet to cubic yards prior to placing your order.

▶ **Example 6**

a) Convert 1 yd^3 to cubic feet.
b) Convert 8.3 yd^3 to cubic feet.

Solution:

a) 1 yd = 3 ft

$1 \text{ yd}^3 = 3 \text{ ft} \cdot 3 \text{ ft} \cdot 3 \text{ ft} = 27 \text{ ft}^3$

b) $1 \text{ yd}^3 = 27 \text{ ft}^3$

Thus $8.3 \text{ yd}^3 = 8.3 \times 27 = 224.1 \text{ ft}^3$.

▶ **Example 7**

Sivle has built a flower box with the dimensions shown in Fig. 8.48. How many cubic yards of topsoil are needed to fill the box?

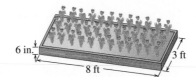

Figure 8.48

Solution: To find the volume, all the measurements must be in the same units. Since 6 in. = 0.5 ft,

$$V = lwh$$
$$= 8(3)(0.5) = 12 \text{ ft}^3$$

Now we must convert 12 ft³ to cubic yards. How shall we do this? We know from Example 6 that $1 \text{ yd}^3 = 27 \text{ ft}^3$. Thus 12 ft³ must be less than 1 yd³. If we divide the total number of cubic feet, 12, by the number of cubic feet in one cubic yard, 27, we will get a number less than 1.

$$\frac{12}{27} \approx 0.44$$

Thus approximately 0.44 yd³ of topsoil are needed.

Section 8.4 Exercises

Find the volume of each of the solids.

1.

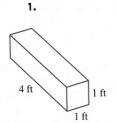

4 ft   1 ft   1 ft

2.

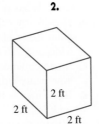

2 ft   2 ft   2 ft

3.

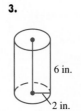

6 in.   2 in.

4.

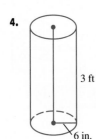

3 ft   6 in.

5.

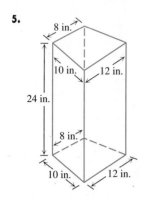

6.

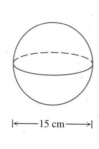

13.

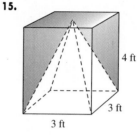

14.

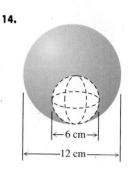

Find the volume of each of the pyramids. Assume the bases in the artwork in Ex. 7–9 contain a right angle.

7.

8.

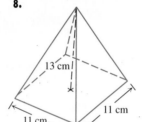

15.

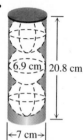

16.

9.

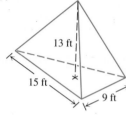

10.

One cubic yard (yd³) equals 27 cubic feet (ft³). Use this information to make the following conversions.

17. 4 yd³ to ft³   **18.** 3.2 yd³ to ft³

19. 212 ft³ to yd³   **20.** 84 ft³ to yd³

One cubic meter (m³) equals 1,000,000 cubic centimeters (cm³). Use this information to make the following conversions.

21. 1.5 m³ to cm³   **22.** 6.4 m³ to cm³

23. 4,000,000 cm³ to m³   **24.** 6,800,000 cm³ to m³

25. The dimensions of the interior of an upright freezer are height 46 in., width 25 in., depth 25 in. Find its volume **a)** in cubic inches   **b)** in cubic feet.

26. The dimensions of a square Wendy's hamburger are length and width 4 in, thickness $\frac{3}{16}$ in. The dimensions of a Ground Round circular hamburger are diameter $4\frac{1}{2}$ in., thickness $\frac{1}{4}$ in. Which hamburger has the greatest volume? What is the difference in their volumes?

27. a) How many cubic centimeters of water will a rectangular fish tank hold if the tank is 60 cm long, 45 cm wide, and 20 cm high?

b) If one cubic centimeter holds one milliliter of liquid, how many milliliters will the tank hold?

c) If one liter = 1000 milliliters, how many liters will the tank hold?

28. a) What is the volume of water in a swimming pool that

In Ex. 11–16, find the volume of the shaded area.

11.

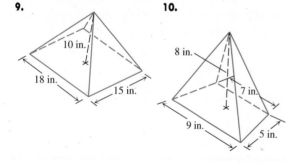

12.

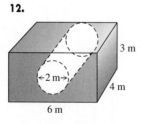

is 18 m long and 10 m wide and has an average depth of 2.5 m? Answer in cubic meters.

b) If 1 m³ = 1 kiloliter, how many kiloliters of water will it take to fill the pool?

29. a) How many cubic feet of dirt are needed to fill a ditch that is 6 ft wide by 8 ft long by 27 in. deep?

b) How many cubic yards of dirt are needed?

30. Mr. Lockhart's driveway is 60 ft long by 12.5 ft wide. He plans to lay 4 in. of blacktop on the driveway.

a) How many cubic feet of blacktop are needed?

b) How many cubic yards?

31. The Pyramid of Cheops in Egypt has a square base measuring 720 ft on a side. Its height is 480 ft. What is its volume?

32. A gallon is a measure of liquid volume. It is equivalent to 231 in.³. Measure the base and the height *of the liquid* in a half-gallon paper carton of milk and compute its volume in cubic inches. Do your results check with the value given?

33. The ball at the end of a ballpoint pen has a diameter of 1.5 mm. Find the volume of the ball.

34. What is the total piston displacement of a two-cylinder engine if each piston has a diameter of $2\frac{1}{4}$ in. and a $3\frac{1}{2}$-in. stroke?

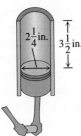

35. In baking a cake you have to choose between a round pan with a 9-in. diameter and a 7-in. × 9-in. rectangular pan.

a) Determine the area of the base of each pan.

b) If both pans are 2 in. deep, determine the volume of each pan.

36. A flower box is 4 ft long, and its ends are in the shape of a trapezoid. The upper and lower bases of the trapezoid measure 12 in. and 8 in., respectively, and the height is 9 in. Find the volume of the flower box

a) in cubic inches **b)** in cubic feet

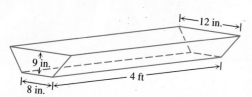

37. What is the smallest number of cuts that would divide a cube of wood whose edge measures 3 in. into cubes, each of whose edges measures 1 in.?

In Ex. 38–43, find the missing value indicated by the question mark. Use the following formula.

$$\left(\begin{array}{c}\text{number of}\\\text{vertices}\end{array}\right) - \left(\begin{array}{c}\text{number of}\\\text{edges}\end{array}\right) + \left(\begin{array}{c}\text{number}\\\text{of faces}\end{array}\right) = 2$$

	Number of Vertices	Number of Edges	Number of Faces
38.	10	12	?
39.	6	?	2
40.	?	8	4
41.	9	16	?
42.	12	?	6
43.	?	10	4

44. What effect does doubling the length of each edge of a cube have on its volume? Explain.

45. What effect does doubling the radius of the base of a cylinder have on the volume? Explain.

46. What effect does doubling the radius of a sphere have on its volume? Explain.

47. A half-inch cube is a cube measuring $\frac{1}{2}$ in. on a side. A half cubic inch is half the volume of a cube measuring 1 in. on a side. Is the volume of a half-inch cube the same as the volume of a half cubic inch? Explain.

48. A king-size waterbed mattress is 7 ft long, 7 ft wide, and 9 in. high.

a) How many cubic feet of water are needed to fill the waterbed?

b) If 1 ft³ of water weighs 62.5 lb, what is the weight of the water in the mattress?

c) If 1 gal of water weighs 8.3 lb, how many gallons of water does the mattress hold?

49. A hot-water heater is in the shape of a circular cylinder. It has a height of 5 ft 6 in. and a diameter of 15 in.

a) How many cubic feet of water will the tank hold?

b) How many gallons of water does the tank hold? (1 ft³ ≈ 7.5 gal.)

50. The general procedure for determining the number of rolls of wallpaper needed to wallpaper a given rectangular room is as follows:

a) Determine the room's overall perimeter, in feet, by measuring the distance around the room where the wall meets the floor. Ignore windows and doors at this time.

b) Multiply the perimeter by the height in feet of the room. This gives the number of square feet of wall space.

c) Divide this number by 30, the approximate number of usable square feet in a roll of wallpaper.

d) From this number, deduct one roll for every two averaged-sized openings (windows, doors, fireplace, and so on).

Mrs. Latvis plans to wallpaper her kitchen. It is 12 ft long, 10 ft wide, and 8 ft high, and it contains two doorways and two windows. How many rolls of wallpaper will she need for her kitchen?

51. One U.S. gallon of liquid has a volume of 231 in.^{3} If a cylindrical jar has a base whose radius is 2 in., how tall must the jar be in order to hold 1 gal?

Problem Solving

52. a) Explain how to show, using the adjacent cube, that

$$(a + b)^3 = a^3 + 3a^2b + 3ab^2 + b^3.$$

b) What is the volume in terms of a and b of each numbered piece in the figure? An eighth piece is not illustrated. What is its volume?

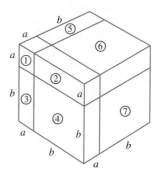

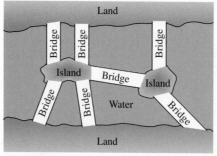

8.5 Graph Theory

In the eighteenth-century Prussian town of Königsberg (now the city of Kaliningrad) there were seven bridges along the Prigel River (Fig. 8.49a). Individuals in the area tried to determine if it was possible to walk a path that would cross each of the seven bridges exactly once. They found that they ended up either not crossing one of the bridges, or crossing one of the bridges more than once. The problem was brought to the attention of the Swiss mathematician Leonhard Euler (pronounced 'öiler,' 1707–1783). His study of this problem, now known as the Königsberg bridge problem, laid the groundwork for a modern branch of mathematics called **topology**, the study of the features of shapes and objects. In solving the problem, Euler drew figures called **graphs** or **networks**, like that shown in red in Fig. 8.49(b). In the figure, each dot represented a plot of land and each line a bridge or path. Can you find a path that crosses each bridge exactly once?

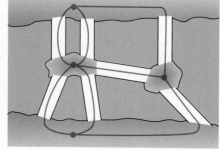

Figure 8.49   (a)   (b)

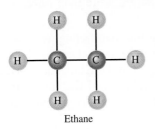
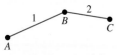

Odd vertices: *A* and *C*
Even vertex: *B*

(a)

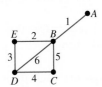

Odd vertices: *A* and *D*
Even vertices: *B*, *C*, and *E*

(b)

Figure 8.51

Before we determine whether there is a path that will cross each bridge exactly once, let us consider the following examples.

▶ **Example 1**

In Fig. 8.50 start at any point and try to trace each figure without retracing a line and without removing your pencil from the paper. If you succeed, indicate your starting point and ending point.

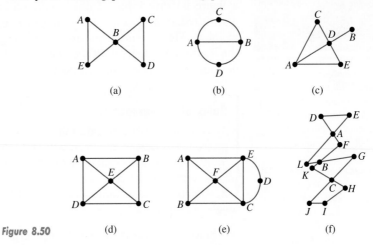

Figure 8.50   (d)   (e)   (f)

Solution:

a) The figure can be traced if you start at any point. You will end at the point at which you started.

b) The figure can be traced, but only if you start at point *A* or point *B*. If you start at point *A*, you will end at point *B*, and vice versa.

c) The figure can be traced, but only if you start at point *A* or point *B*. If you start at point *A*, you will end at point *B*, and vice versa.

d) It is not possible to trace the figure without retracing a line.

e) The figure can be traced, but only if you start at point *A* or point *B*. If you start at point *A*, you will end at point *B*, and vice versa.

f) The figure can be traced if you start at any point. You will end at the point at which you started.

In order for you to be able to answer the Königsberg bridge problem and understand why we were able to trace only some of the figures in Example 1 without retracing a line, we must introduce a number of new terms and concepts.

A **vertex** is any designated point. An **arc** (or an **edge**) is any line, either straight or curved, that begins and ends at a vertex. Figure 8.51(a) has three designated vertices, *A*, *B*, and *C*, and two arcs, 1 and 2. Figure 8.51(b) has five designated vertices, *A*, *B*, *C*, *D*, and *E*, and six arcs. A vertex with an odd number of attached arcs is called an **odd vertex**. A vertex with an even number of attached arcs is called an **even vertex**. Figure 8.51(a) has two odd vertices and one even vertex. Figure 8.51(b) has two odd vertices and three even vertices.

Today, graph (network) theory is a flourishing field, useful for dealing with a variety of everyday problems. It is used by the airline industry to arrange the complex network of flights connecting cities, and by sports officials planning tournaments and playoff schedules. City planners use it to work out new roads and traffic patterns. Tests are underway in traffic-clogged cities like Los Angeles to link cars, via an on-board navigational computer, to the city's central traffic-control computer. In this way motorists can be rerouted to less congested roadways.

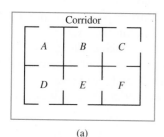

(a)

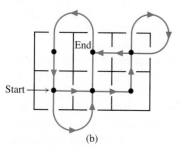

(b)

Figure 8.52

A **network** is any continuous (not broken) system of arcs and vertices.

> A network is said to be **traversable** if it can be traced without removing the pencil from the paper and without tracing an arc more than once.

After completing Example 1, do you have an intuitive feeling as to when a figure is traversable? Think about odd and even vertices. Leonhard Euler discovered an important scientific principle concealed in the Königsberg bridge problem. He presented his simple and ingenious solution of that problem to the Russian Academy at St. Petersburg in 1735. The rules of traversability Euler developed in solving the problem follow.

> ### Rules of Traversability
> 1. A network with no odd (all even) vertices is traversable; you may start from any vertex, and you will end where you began.
> 2. A network with exactly two odd vertices is traversable; you must start at either of the odd vertices and finish at the other.
> 3. A network with more than two odd vertices is not traversable.

Since the network in the Königsberg bridge problem (Fig. 8.49b) has four odd vertices, it cannot be traversed. It is therefore impossible to cross each bridge only once. Note that it is impossible for a network to contain an odd number of odd vertices (If you don't believe this, try to construct such a network.)

Now go back to Example 1 and determine which figures are traversable, using these rules. Notice that Figs. 8.50(a) and 8.50(f) are traversable from any point since they contain only even vertices. Figures 8.50(b), (c), and (e) have exactly two odd vertices and can be traversed but only by starting at one of the odd vertices, either point *A* or point *B*. Figure 8.50(d) contains more than two odd vertices and therefore cannot be traversed.

▶ Example 2

The floor plan of a six-gallery art museum is shown in Fig. 8.52(a). The openings represent doors, and the letters represent galleries.

a) Determine the galleries that contain an odd number of doors; an even number of doors.

b) Each gallery can be represented as an odd or even vertex. Use this information to determine whether it is possible to walk through each gallery using each door only once.

c) Determine a path to walk through each gallery using each door only once.

Solution:

a) Galleries *B* and *D* contain three doors each. Galleries *A* and *F* contain two doors each. Galleries *C* and *E* have four doors.

b) Since there are only two odd vertices, *B* and *D*, the figure is traversable, and you can walk through the museum using each door only once.

c) You must start in either Gallery *B* or *D* (see Fig. 8.52b). If you start in *B*, you will end in *D*, and vice versa. When you leave Gallery *E*, you can leave by either door.

Section 8.5 Exercises

In Ex. 1–4, determine the number of vertices and the number of arcs.

1.

2.

3.

4.

5. Explain why the following two figures represent the same graph.

6. In your own words, explain the rules for determining if a graph is traversable.

In Ex. 7–14, determine whether or not the networks are traversable. If a network is traversable, state the points from which you may start and end.

7.

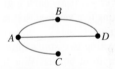

8.

9.

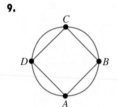

10.

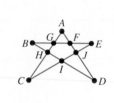

11.

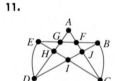

12.

13.

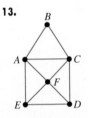

14.

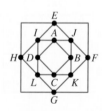

In Ex. 15–22, the floor plans of various buildings are shown. For each plan, do the following.

a) Determine the rooms that contain an odd number of doors, an even number of doors.

b) Use the rules of traversability to determine whether it is possible to walk through the building using each door only once.

c) Determine a path through the building using each door exactly once.

15.

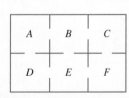

16.

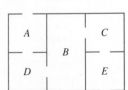

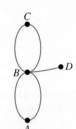

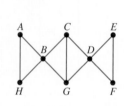

17.

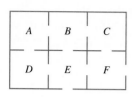

18.

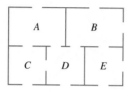

19.

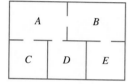

20.

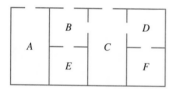

21.

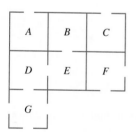

22.

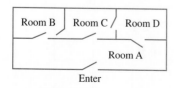

23. The floor plan for a suite of rooms is shown in the figure. Add an exit door in one of the rooms so that the security guard can enter through the door marked enter, pass through each door only once locking it behind him, and then exit by the door you added. Explain why there is only one possible room in which the door may be placed.

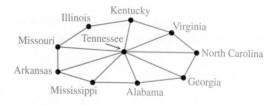

In Ex. 24 and 25, is it possible to cross each bridge exactly once? Explain each answer, and show the path if it is possible.

24.

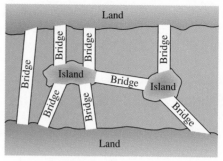

25.

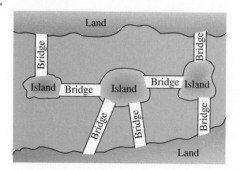

26. Draw a graph that contains four vertices and is traversable from exactly two points.

27. Can you draw a graph that contains an odd number of odd vertices? If you answer yes draw such a graph.

28. Draw a graph that contains five vertices that is traversable from exactly two points.

29. In the figure, a line connecting two states indicates that the two states share a common border. Which of these states share a common border with (a) Tennessee, (b) North Carolina?

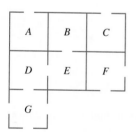

30. The remaining games in a soccer league are illustrated in the following graph.

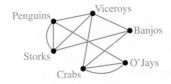

a) How many games do the Viceroys have remaining?
b) How many games do the Penguins have remaining?
c) How many games still need to be played?

31. Dawn, Jessica, Pam, Bill, Ed, and Scott go to a dance. Dawn dances with Bill and Scott. Jessica dances with all the boys and Pam dances only with Scott. Draw a graph that displays this information.

32. France has common borders with Belgium, Germany, Switzerland, Italy, and Spain. Belgium has common borders with the Netherlands, France, and Germany. Germany has common borders with Poland, Czechoslovakia, Austria, Switzerland, and France. Switzerland has common borders with France, Italy, Austria, and Germany. Draw a graph that displays this information.

Problem Solving

33. Networks (a) and (b) illustrate relationships of arcs, regions, and vertices. Network (a) has three vertices (A, B, and C), four arcs (1, 2, 3, and 4), and three regions (x, y, and z). Network (b) has five vertices (A, B, C, D, and E), seven arcs (1, 2, 3, 4, 5, 6, and 7), and four regions (w, x, y, and z). Using these networks and others that you can make up yourself, develop a formula expressing the number of arcs in terms of the number of vertices and regions. This formula is known as **Euler's formula for networks.**

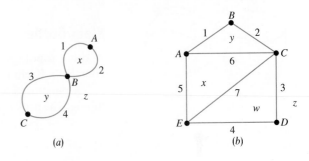

(a)   (b)

The Möbius Strip, Klein Bottle, and Maps

One of the first pioneers of topology was the German astronomer and mathematician August Ferdinand Möbius (1790–1866). A student of Gauss, Möbius was the director of the University of Leipzig's observatory. He spent a great deal of time studying geometry and played an essential part in the systematic development of projective geometry. He is most known for his studies of the properties of one-sided surfaces, including the one called the Möbius strip.

Möbius Strip

If you place a pencil on one surface of a sheet of paper and do not remove it from the sheet, you must cross the edge to get to the other surface. Thus a sheet of paper has one edge and two surfaces. The sheet retains these properties even when crumpled into a ball. The **Möbius strip** is a one-sided, one-edged surface. You can construct one, as shown in Fig. 8.53, by (a) taking a strip of paper, (b) giving one end a half twist, and (c) taping the ends together.

The Möbius strip has some very interesting properties. In order to better understand these properties, perform the following experiments.

Experiment 1 Make a Möbius strip using a strip of paper and tape as illustrated in Fig. 8.53. Place the point of a felt tip pen on the edge of the strip (Fig. 8.54). Pull the strip slowly so that the pen marks the edge—do not remove the pen from the edge. Continue pulling the strip and observe what happens.

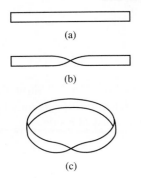

(a)

(b)

(c)

Figure 8.53

Figure 8.54

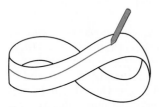

Figure 8.55

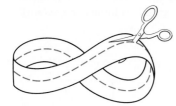

Figure 8.56

Figure 8.57

Experiment 2 Make a Möbius strip. Place the tip of a felt tip pen on the surface of the strip (Fig. 8.55). Pull the strip slowly so that the pen marks the surface. Continue and observe what happens.

Experiment 3 Make a Möbius strip. With a scissors make a small slit in the middle of the strip. Starting at the slit, cut along the strip, keeping the scissors in the middle of the strip (Fig. 8.56). Continue cutting and observe what happens.

Experiment 4 Make a Möbius strip. Make a small slit at a point about one-third of the width of the strip. Cut along the strip, keeping the scissors the same distance from the edge (Fig. 8.57). Continue cutting and observe what happens.

Klein Bottle

Another topological object is the punctured **Klein bottle**. This object, named after Felix Klein (1849–1925), resembles a bottle but only has one side.

A punctured Klein bottle can be made by stretching a hollow piece of glass tubing. The neck is then passed through a hole and joined to the base (Fig. 8.58).

Figure 8.58

A close look at the model of the Klein bottle in Fig. 8.58 will show that the bottle could not hold any water. The punctured Klein bottle has only one edge (the hole) and no outside or inside because it has just one side.

Maps

Maps have fascinated topologists for years because of the many challenging problems they present. Mapmakers have known for a long time that regardless of the complexity of the map, whether drawn on a flat surface or a sphere, only four colors are needed to differentiate each country (or state) from its immediate neighbors. This means that it is possible to draw every map using only four colors so that no two countries with a common border will have the same color. Regions that meet at only one point (such as the states of Arizona, Colorado, Utah, and New Mexico) are not considered to have a common border.

In Fig. 8.59(a) no two states with a common border are marked with the same color.

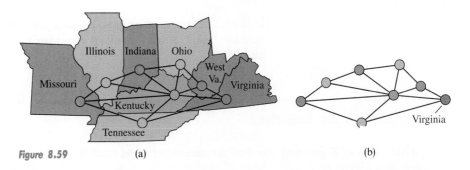

Figure 8.59   (a)   (b)

In 1976 Kenneth Appel and Wolfgang Haken of the University of Illinois, using their ingenuity, logic, and 1200 hours of computer time, succeeded in proving that only four colors are needed to draw a map. They solved the four-color map problem by reducing any map to a graph as Euler did with the Königsberg bridge problem. They replaced each country with a point. Two countries with a common border were connected with a straight line (Fig. 8.59b). They then showed that the points of any graph in the plane could be colored using only four colors in such a way that no two points connected by the same line were the same color.

There are certain cases in which more than four colors may be needed to draw a map. For example, it has been proved that a map drawn on a Möbius strip requires a maximum of six colors and a map drawn on a torus (the shape of a doughnut) requires a maximum of seven colors:

Jordan Curves

A **Jordan curve** is a topological object that can be thought of as a circle twisted out of shape (Fig. 8.60a–c). Like a circle, it has an inside and an outside. To get from one side to the other at least one line must be crossed. Consider the Jordan curve in part (d) of Fig. 8.60. Are points A and B inside or outside the curve?

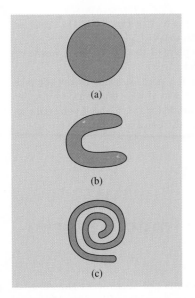

(a)

(b)

(c)

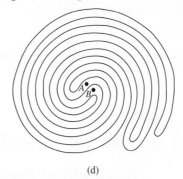

Figure 8.60   (d)

A quick way to tell whether the two dots are inside or outside the curve is to draw a straight line from each dot to a point that is clearly outside the curve. If the straight line crosses the curve an even number of times, the dot is outside. If the straight line crosses the curve an odd number of times, the dot is inside the curve. Can you explain why this procedure works? Determine whether point *A* and point *B* are inside or outside the curve (see Ex. 14 and 15).

Section 8.6 Exercises

1. Use the result of Experiment 1 to find the number of edges on a Möbius strip.
2. Use the result of Experiment 2 to find the number of surfaces on a Möbius strip.
3. How many separate strips are obtained in Experiment 3?
4. How many separate strips are obtained in Experiment 4?
5. **a)** Take a strip of paper, give it one full twist, and connect the ends. Is this a Möbius strip with only one side? Explain.
   b) Determine the number of edges, as in Experiment 1.
   c) Determine the number of surfaces, as in Experiment 2.
   d) Cut the strip down the middle. What is the result?
6. Make a Möbius strip. Cut it one-third of the way from the edge, as in Experiment 4. You should get two loops, one going through the other. Determine whether either (or both) of these loops is itself a Möbius strip.
7. Take a strip of paper, make one whole twist and another half twist, and then tape the ends together. Test by a method of your choice to see whether this has the same properties as a Möbius strip.
8. Can you see any advantage in a Möbius conveyor belt? Explain.

In Ex. 9–13:
a) Color the map using a maximum of four colors, so that no two regions with a common border have the same color.

b) Show how each map can be reduced to a graph.

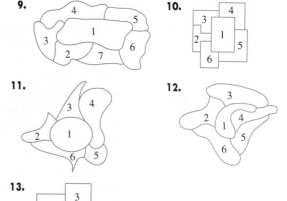

14. Determine whether point *A* in Fig. 8.60(d) is inside or outside of the Jordan curve.
15. Determine whether point *B* in Fig. 8.60(d) is inside or outside of the Jordan curve.
16. When testing to see if a point is inside or outside of a Jordan curve, explain why if you count an odd number of lines the point is inside the curve, and if you count an even number of lines the point is outside the curve.

8.7 Non-Euclidean Geometry and Fractal Geometry

Non-Euclidean Geometry

In Section 8.1 we stated that postulates or axioms are statements that are to be accepted as true. In his book *The Elements*, Euclid's fifth postulate was "If a straight line falling on two straight lines makes the interior angles on the same side less than two right angles, the two straight lines, if produced

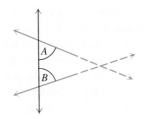

Figure 8.61

indefinitely, meet on that side on which the angles are less than the two right angles."

Euclid's fifth axiom may be better understood by observing Fig. 8.61. The sum of angles A and B is less than the sum of two right angles ($180°$). Therefore the two lines will meet if extended.

John Playfair (1748–1819), a Scottish physicist and mathematician, wrote a geometry book that was published in 1795. In his book, Playfair gave a logically equivalent interpretation of Euclid's fifth postulate. This version is often referred to as Playfair's postulate or the Euclidean parallel postulate.

The Euclidean Parallel Postulate

Given a line and a point not on the line, one and only one line can be drawn through the given point parallel to the given line (Fig. 8.62).

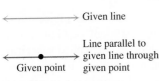

Given line

Line parallel to given line through given point

Given point

Figure 8.62

The Euclidean parallel postulate may be better understood by looking at Fig. 8.62.

Many mathematicians after Euclid believed that this postulate was not as self-evident as the other nine. Others believed that this postulate could be proved from the other nine postulates and therefore was not needed at all. Of the many attempts to prove that the fifth postulate was not needed, the most noteworthy one was presented by Girolamo Saccheri (1667–1733), a Jesuit in Italy. In the course of his elaborate chain of deductions, Saccheri proved many of the theorems of what is now called hyperbolic geomtery. However, Saccheri did not realize what he had done. He believed that Euclid's geometry was the only "true" geometry and concluded that his own work was in error. Thus Saccheri narrowly missed receiving credit for a great achievement—the founding of non-Euclidean geomtery.

In 1763 the German mathematician Georg Simon Klügel listed nearly 30 attempts to prove axiom 5 from the remaining axioms, and he rightly concluded that all the alleged proofs were unsound. Fifty years later a new generation of geometers were becoming more and more frustrated at their inability to prove Euclid's fifth postulate. One of them, a Hungarian named Farkos Bolyai, wrote a letter to his son, Janos Bolyai. "I entreat you leave the science of parallels alone. . . . I have traveled past all reefs of this infernal dead sea and have always come back with a broken mast and torn sail." The son, refusing to heed his father's advice, continued to think about parallels until, in 1823, he saw the whole truth and enthusiastically declared, "I have created a new universe from nothing." He saw that geometry branches out in two directions, depending on whether Euclid's fifth postulate is applied or not. He recognized two different geometries and published his discovery as a 24-page appendix to a textbook written by his father. The famous mathematician George Bruce Halsted called it "the most extraordinary two dozen pages in the whole history of thought." Farkos Bolyai proudly presented a copy of his son's work to his friend Carl Friedrich Gauss, then Germany's greatest mathematician, whose reply to the father had a devastating effect on the son. Gauss wrote, "I am

Ivanovich Lobachevsky

G. F. Bernhard Riemann

unable to praise this work.... To praise it would be to praise myself. Indeed, the whole content of the work, the path taken by your son, the results to which he is led, coincides almost entirely with my meditations which occupied my mind partly for the last thirty or thirty-five years." We now know from his earlier correspondence that Gauss had indeed been familiar with **hyperbolic geometry** even before Janos was born. In his letter, Gauss also indicated that it was his intention not to let his theory be published during his lifetime, but to record it so that the theory would not perish with him. It is believed that the reason Gauss did not publish his work was that he feared being ridiculed by other prominent mathematicians of his time.

At about the same time as Bolyai's publication, the Russian Nikolay Ivanovich Lobachevsky published a paper that was remarkably like Bolyai's, although it was quite independent of it. Lobachevsky made a deeper investigation and wrote several books. In marked contrast to Bolyai, who received no recognition during his lifetime, Lobachevsky received great praise and became a professor at the University of Kazan.

After the initial discovery, little attention was paid to the subject until 1854, when G. F. Bernhard Riemann (1826–1866), a student of Gauss, suggested a second type of non-Euclidean geometry, which is now called **spherical, elliptical,** or **Riemannian geometry**. The hyperbolic geometry of his predecessors was synthetic; that is, it was not based on or related to any concrete model when it was developed. Riemann's geometry was closely related to the theory of surfaces. A **model** may be considered a physical interpretation of the undefined terms that satisfies the axioms. A model may be a picture or an actual physical object.

The two types of non-Euclidean geometries that we have mentioned are elliptical geometry and hyperbolic geometry. The major difference among the three geometries lies in the fifth axiom. The fifth axiom of the three geometries is summarized here.

The Fifth Axiom of Geometry

Euclidean	**Elliptical**	**Hyperbolic**
Given a line and a point not on the line, one and only one line can be drawn parallel to the given line through the given point.	Given a line and a point not on the line, no line can be drawn through the given point parallel to the given line.	Given a line and a point not on the line, two or more lines can be drawn through the given point parallel to the given line.

To understand the fifth axiom of the two non-Euclidean geometries, remember that the term "line" is undefined. Thus a line can be interpreted differently in different geometries. A model for Euclidean geometry is a plane, such as a blackboard (Fig. 8.63a). A model for elliptical geometry is a sphere (Fig. 8.63b). A model for hyperbolic geometry is a pseudosphere (Fig. 8.63c). A

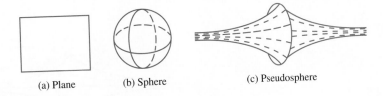

(a) Plane   (b) Sphere   (c) Pseudosphere

pseudosphere is similar to two trumpets placed bell to bell. It should be obvious that a line on a plane cannot be the same as a line on any of the other two figures.

Elliptical Geometry

A circle on the surface of a sphere is called a great circle if it divides the sphere into two equal parts. If we were to cut through a sphere along a great circle, we would have two identical pieces. If we interpret a line to be a great circle, we can see that the fifth axiom of elliptical geometry is true. Two great circles on a sphere must intersect, hence there can be no parallel lines (Fig. 8.64a).

    If we were to construct a triangle on a sphere, the sum of its angles would be greater than 180° (Fig. 8.64b). The theorem "The sum of the measures of the angles of a triangle is greater than 180°" has been proved by means of the axioms of elliptical geometry. The sum of the measures of the angles varies with the area of the triangle and gets closer to 180° as the area decreases.

Hyperbolic Geometry

The lines in hyperbolic geometry are represented by geodesics on the surface of the pseudosphere. A geodesic is the shortest and least-curved arc between two points on a surface. Figure 8.65 illustrates two lines on the surface of a pseudosphere.

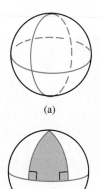

(a)

(b)

Figure 8.64

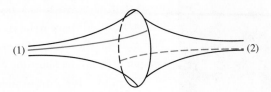

(1)   (2)

Figure 8.65

    Figure 8.66(a) illustrates the fifth axiom of hyperbolic geometry.* Note that through the given point, two lines are drawn parallel to the given line. If we

---

* A formal discussion of hyperbolic geometry is beyond the scope of this text.

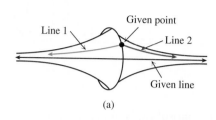

Line 1

Given point

Line 2

Given line

(a)

Figure 8.66

(b)

were to construct a triangle on a pseudosphere, the sum of the measures of the angles would be less than 180° (Fig. 8.66b). The theorem "The sum of the measures of the angles of a triangle is less than 180°" has been proved by means of the axioms of hyperbolic geometry.

Did You Know...

IT'S ALL RELATIVE

Einstein's General Theory of Relativity, published in 1916, approached space and time differently from our everyday understanding of them. Einstein's theory unites the three dimensions of space with one of time in a four-dimensional space-time continuum. His theory dealt with the path that light and objects take while moving through space under the force of gravity. Einstein conjectured that mass (such as stars and planets) caused space to be curved.

Einstein's theory was confirmed by the solar eclipses of 1919 and 1922.

The greater the mass, the greater the curvature. Also, in the region nearer to the mass, the curvature of the space is greater.

To prove his conjecture, Einstein exposed himself to Riemann's non-Euclidean geometry. Einstein felt that the trajectory of a particle in space represents not a straight line but the straightest curve possible, a geodesic.

Space-time is now thought to be a combination of three different types of curvature: spherical (described by Riemannian geometry), flat (described by Euclidean geometry), or saddle-shaped (described by hyperbolic geometry).

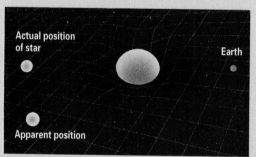

Actual position of star

Earth

Apparent position

You can visualize Einstein's theory by thinking of space as a rubber sheet pulled taut on which a mass is placed, causing the rubber sheet to bend.

We have stated that the sum of the measures of the angles of a triangle is 180°, is greater than 180°, and is less than 180°. Which statement is correct? Each statement is correct *in its own system*. Many theorems hold true for all three geometries—vertical angles still have the same measure, one can uniquely bisect a line segment with a straightedge and compass alone, and so on.

The many theorems based on the fifth postulate may differ in each geometry. It is important for you to realize that each theorem proved is true *in its own system* because each is logically deduced from the given set of axioms of the system. No one system is the "best" system. Euclidean geometry may appear to be the one to use in the classroom, where the blackboard is flat. However, in discussions involving the earth as a whole, elliptical geometry may be the most useful, since the earth is a sphere. If the object under consideration has the shape of a saddle or pseudosphere, hyperbolic geometry may be the most useful.

"The Great Architect of the universe now appears to be a great mathematician."
British physicist Sir James Jeans

Fractal Geometry

We are familiar with one-, two-, and three-dimensional figures. However, there are many objects that are difficult to categorize as one-, two-, or three-dimensional. For example, how would you classify the irregular shapes we see in nature such as a coast line, or the bark on a tree, or a mountain, or a path followed by lightning? For a long time mathematicians assumed that making realistic geometric models of natural shapes and figures was almost impossible. However, the development of **fractal geometry** now makes it possible. The color photos on this page were each made using fractal geometry. The discovery and study of fractal geometry has been one of the hottest mathematical topics of the 1980s and 1990s.

The word **fractal** (from the Latin word *fractus*, "broken up, fragmented") was first used in the mid-1970s by mathematician Benoit Mandelbrot to describe shapes that had several common characteristics, including some form of "self-similarity," as will be seen shortly in the Koch snowflake.

Typical fractals are extremely irregular curves or surfaces that wiggle enough so that they are not considered one-dimensional. Fractals do not have integer dimensions; their dimensions are between 1 and 2. For example, a fractal may have a dimension of 1.26. Fractals are developed using the same rule over and over again with the end point of each simple step becoming the starting point for the next step, a process called **recursion**.

Using the recursive process we will develop a famous fractal called the **Koch snowflake** for Helga von Koch, a Swedish mathematician who first discovered its remarkable characteristics. The Koch snowflake illustrates a property of all fractals called "self-similarity"—that is, each smaller piece of the curve resembles the whole curve.

To develop the Koch snowflake:

Fractal Images

1. Start with an equilateral triangle.
2. Whenever you see an edge —— replace it with ⌐⌐. The Koch snowflake is developed in Fig. 8.67 on page 412.

Step 1   Step 2

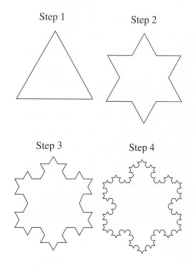

Step 3   Step 4

Figure 8.67

What is the perimeter of the snowflake, and what is the area of the snowflake? A portion of the boundary of the Koch snowflake known as the Koch curve or the snowflake curve is represented in Fig. 8.68.

Figure 8.68

Because the Koch curve consists of an infinite number of pieces of the form ⌐⌐, the perimeter is also infinite. The area of the Koch snowflake is 1.6 times the area of the starting equilateral triangle. Thus the area of the snowflake is finite. The Koch snowflake has a finite area enclosed by an infinite boundary! This fact may seem difficult to accept, but it is true. However, the Koch snowflake, like other fractals, is not an everyday run-of-the-mill geometric shape.

Section 8.7 Exercises

In Ex. 1–5, list the accomplishments of each of the following.
1. Janos Bolyai
2. Carl Friedrich Gauss
3. Nikolay Ivanovich Lobachevsky
4. Girolamo Saccheri
5. G. F. Bernhard Riemann

6. State the fifth axiom of each of the following.
   a) Euclidean geometry
   b) Hyperbolic geometry
   c) Elliptical geometry
7. State the theorem concerning the sum of the measures of the angles of a triangle in each of the following.
   a) Euclidean geometry
   b) Hyperbolic geometry
   c) Elliptical geometry
8. What model is often used in describing and explaining Euclidean geometry?
9. What model is often used in describing and explaining elliptical geometry?
10. What model is often used in describing and explaining hyperbolic geometry?
11. What do we mean when we say that there is no one axiomatic system of geometry that is "best"?

12. Explain why Einstein used non-Euclidean geometry to explain his General Theory of Relativity.
13. List the three types of curvature of space and the types of geometry that correspond to them.
14. In our discussion of the Koch snowflake we indicated that it had a finite area enclosed in an infinite perimeter. Explain how this can be true.
15. a) Develop a fractal by beginning with a square and replacing each side, ——, with a ⌐⌐. Repeat this process twice.
    b) If this process is continued will the fractal's perimeter be finite or infinite? Explain.
    c) Will the fractal's area be finite or infinite? Explain.

Research Activities

16. Write a paper on the life and achievements of Euclid. (Use encyclopedias and books on the history of mathematics as sources.)
17. There are other types of geometries; one example is projective geometry. Determine what projective geometry is, and investigate some of its common uses. (An art teacher or encyclopedia may offer information and other references.)

8.8 Logo

The computer programming language **Logo** is very useful in teaching concepts of geometry, especially to children. Using Logo in the classroom is also an enjoyable way to introduce young children to the computer. Using Logo commands, the programmer has the ability to move a small triangle called the **turtle** around the computer screen. The turtle leaves a trail behind it as it moves, creating geometric shapes and other patterns. Through this process children also learn to understand the Cartesian coordinate system.

Logo was developed by Seymour Papert and the Logo Group at the Massachusetts Institute of Technology. The word Logo is derived from the Greek word for "thought." The goal of Papert and the MIT Logo Group was to place Piaget's theories about natural learning in a powerful artificial intelligence environment. Logo has no age restrictions. It can be used by preschool children as well as by computer scientists.

The two basic modes of Logo are the edit and draw modes. In the **edit mode**, one can write programs and store the programs for future use. In the **draw mode** it is possible to draw geometric shapes and patterns.

Figure 8.69

To get started, the user may insert a Logo Language diskette into the disk drive of the computer. The message "Welcome to Logo" will appear on the screen followed by a question mark and the cursor. The question mark indicates that the computer is waiting for a Logo command. The cursor indicates where the typing will appear on the screen. Type the command SHOW TURTLE, or the abbreviation ST, press return, and the computer will be in the draw mode. At this point the turtle appears in the middle of the screen. This position of the turtle will be referred to as **home** (Fig. 8.69).

The turtle is directed with a set of word commands. Word commands, called **primitives**, are listed in the summary box titled Logo Commands on page 414. Use the abbreviations, not the words. These commands are for Apple Logo, which runs on Apple and Apple Macintosh computers. A version of Logo is available for IBM and IBM-compatible computers.

The draw mode is also called the **splitscreen mode**. In the splitscreen mode the prompt (a question mark) and the cursor are located at the bottom of the screen. The bottom four lines of the screen are for typing commands. The top part of the screen is for graphics.

The summary box on page 414 shows that the Logo command FD means "forward." The command FD 40 sends the turtle forward 40 turtle steps (40 units). To move the turtle backward 15 units, we type the command BK 15. The command RT 45 changes the direction of the turtle by turning it 45 degrees to the right. The command LT 80 turns the turtle 80 degrees to the left. The turtle is limited to about 240 steps from the top to the bottom of the screen, and about 280 steps from the left to the right of the screen. Example 1 demonstrates how to draw with the turtle using basic Logo commands.

Using the simple computer programming language Logo together with the building toy LEGO, young children can actually program the computer to operate model cars, elevators, toasters, and other mechanical devices that they have built themselves. Researchers Stephen Ocko and Mitchel Resnick of MIT developed an extended version of Logo to provide an outlet for students' creativity and to give them the chance to be an inventor.

Logo Commands

Graphics Command	Abbreviation	Effect
FORWARD	FD	Sends the turtle forward a given number of turtle steps
BACK	BK	Sends the turtle backward a given number of turtle steps
RIGHT	RT	Turns the turtle to the right a given number of degrees
LEFT	LT	Turns the turtle to the left a given number of degrees
HIDE TURTLE	HT	The turtle does not appear on screen; the turtle draws faster
SHOW TURTLE	ST	Puts the computer in draw mode and shows the turtle
CLEAR SCREEN	CS	Clears the screen, sends the turtle home, and sets it facing upward
PRINT	PR	Prints the words or symbols to the right of double quotation marks on the screen
HEADING		In the draw mode, tells the turtle's heading in degrees
PENUP	PU	In the draw mode, allows the turtle to move without leaving a trail
PENDOWN	PD	In the draw mode, causes the turtle to leave a trail
PRINT POSITION	PRPOS	Prints coordinates of the turtle
SET POSITION	SETPOS	Places the turtle at indicated coordinates
HOME		Returns the turtle to the center of the screen heading north

▶ **Example 1**

Illustrate the movement of the turtle for each of the following commands.

a) CS

b) FD 50

c) LT 90

d) FD 60

e) RT 45

f) FD 40

g) RT 45

h) BK 60

i) HOME

Solution: The results of these commands are shown in Fig. 8.70.

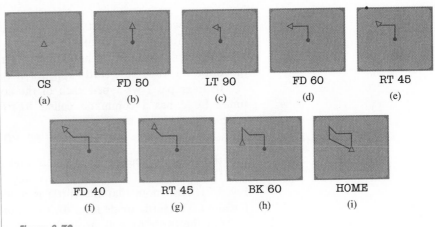

CS	FD 50	LT 90	FD 60	RT 45
(a)	(b)	(c)	(d)	(e)

FD 40	RT 45	BK 60	HOME
(f)	(g)	(h)	(i)

Figure 8.70

▶ **Example 2**

List the sequence of Logo commands that will have the turtle draw a square whose sides have a length of 80 units.

Solution: The list of commands below will result in drawing the square in Fig. 8.71.

Figure 8.71

```
CS
FD 80
RT 90
FD 80
RT 90
FD 80
RT 90
FD 80
RT 90
```

Note that it is necessary to press the return key after each command when you wish to list the command vertically.

You can type a whole list of commands on a single line. After you have completed the list and pressed the return key, the turtle will carry out the list of commands in their proper order. For example, typing the following list of commands,

FD 40 RT 45 FD 50 RT 90 FD 50 LT 45 BK 50

and then pressing the return key will cause the pattern in Fig. 8.72(a) to be displayed. Figure 8.72(b) is a detailed drawing to show exactly what happens.

We indicated earlier that the turtle is limited to about 240 steps from the top to the bottom of the screen and about 280 steps from the left to the right of the screen.

What happens when the turtle is in standard position (at home, heading

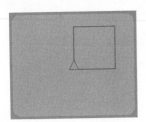

(a)

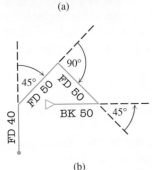

(b)

Figure 8.72

north) and we enter FD 300? The turtle disappears off the top of the screen and reappears at the bottom of the screen. This is called **wraparound**. If you move too far to the right, the turtle appears on the left of the screen. If you move too far down, the turtle appears at the top of the screen.

In Example 2 we typed each of the commands FD 80 and RT 90 four times. Logo has a command called **REPEAT** that reduces the work. The instruction

REPEAT 4[FD 80 RT 90]

tells the turtle to repeat the movement four times, FORWARD 80 (steps) then RIGHT 90 (degrees). Press the return key, and the square is displayed, as in Fig. 8.71. Also note that the turtle is at home and heading north (original position). The turtle made four 90° turns, or a total of 360°.

For the turtle to walk around any closed path, regardless of the number of sides, and return to its starting point, it must turn a total of 360°. Therefore when programming in Logo we consider the sum of the interior angles of *any polygon* to contain 360°.

The symbols for arithmetic calculations in Logo are as follows:

Operation	Logo Symbol	Logo Example	Result
Addition	+	2 + 3	5
Subtraction	−	5 − 4	1
Multiplication	*	9 * 3	27
Division	/	15/3	5
Exponentiation	∧	5 ∧ 2	25
Square root	SQRT	SQRT(16)	4

The use of these symbols is explained in the following examples.

▶ **Example 3**

Write a set of commands that instruct the turtle to draw a regular pentagon. A regular pentagon is a figure with five equal sides and five equal angles. The length of each side is to be 40 turtle steps.

Solution: Remember, in Logo the sum of the interior angles of any polygon is 360°. Therefore the turtle must make five turns whose sum of the angles total 360°. Dividing 360° by 5 gives 72°. Thus the turtle must make five turns of 72 degrees. The arithmetic for calculating the angle can be included as part of the instruction. A single instruction that will result in a pentagon is

REPEAT 5[FD 40 RT 360/5]

Pressing the return key causes the pentagon given in Fig. 8.73 to be drawn.

Figure 8.73

Note that in Example 3, part of the command is RT 360/5. The computer divides 360 by 5 to obtain 72 before the first right turn is executed.

The words listed in the earlier summary as Logo Commands, or primitives,

Computer-aided design is rapidly taking over the traditional field of drafting to create blueprints for cars, bicycles, machine parts, or buildings. Once these designs took hours or days spent with T-squares, triangles, compasses, and protractors; with computers they can be generated in minutes. The designer sketches with a light pen directly on the computer screen and inputs numerical specifications for the object. The computer converts the data into accurate geometric forms and can rotate or magnify the object, allowing the designer to view it from many angles or in cross section.

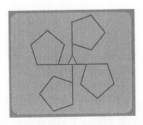

Figure 8.74

are placed in the memory of the computer by the software. Logo users can develop and store in the computer's memory a new set of instructions and assign a name to the set. The computer can recall these instructions from its memory when the name is typed. Each of these new sets of instructions is called a **procedure**.

To create a procedure, the user must leave the draw mode and enter the edit mode. To enter the edit mode, type **EDIT** followed by a space, a single set of double quotation marks, and the name of the procedure, and then press the return key.

To illustrate the steps for writing a procedure, we will define a procedure called PENTAGON.

1. Type EDIT, double quotation marks, and the procedure name.

```
EDIT "PENTAGON
```

2. Type the body of the procedure.

```
REPEAT 5[FD 40  RT 360/5]
END
```

3. Type END (on a separate line) to indicate that the procedure is complete.

4. Press CTRL-C to send the procedure to memory. (Hold down the control key while pressing the C key.) The computer will exit the edit mode.

When the computer leaves the edit mode, the procedure is entered into the computer's memory. Logo then reenters the edit mode and responds with the message: PENTAGON DEFINED.

An alternative way to enter the edit mode to define a procedure is to type EDIT (ED) and press the return key. The edit mode is entered, and TO PENTAGON can be typed, followed by the lines in steps 2–4.

To have the computer execute the instructions of a procedure in memory (when it is in draw mode), we do the following: (1) Type the name of the procedure and (2) press the return key. For example, for the procedure PENTAGON, type PENTAGON and press the return key, and the figure will be drawn as in Fig. 8.73.

The procedure called PENTAGON can now be used in the same manner as the primitives FORWARD, BACK, RIGHT, and LEFT in drawing pictures. For example, the command

```
REPEAT 4[FD 40 PENTAGON BK 40 RT 360/4]
```

will generate the design in Fig. 8.74.

▶ **Example 4**

a) Write a procedure called SEGMENT that will generate a line segment whose length is 30 turtle steps and return the turtle to the original position.

b) Write a procedure called SQUARE that will generate a square with sides of length 30 turtle steps.

c) Write a procedure called TRIANGLE that will generate an equilateral triangle with sides of length 30 turtle steps.

Solution:

a) EDIT "SEGMENT
    FD 30   BK 30
    END

b) EDIT "SQUARE
    REPEAT 4[FD 30   RT 360/4]
    END

c) EDIT "TRIANGLE
    REPEAT 3[FD 30   RT 360/3]
    END

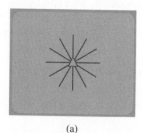

(a)

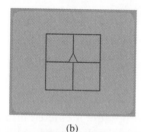

(b)

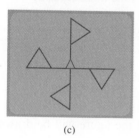

(c)

Figure 8.75

There is no limit to the patterns and figures you can construct with the procedures PENTAGON, TRIANGLE, SEGMENT, and SQUARE and the primitives.

▶ **Example 5**

Predict what happens when each of the following commands is executed. Assume the procedures SEGMENT, SQUARE, and TRIANGLE have been defined as above.

a) REPEAT 12[SEGMENT   RT 360/12]
b) REPEAT 4[SQUARE   RT 360/4]
c) REPEAT 4[FD 20   TRIANGLE   BK 20   RT 360/4]

Solution: Figure 8.75 (a)–(c) illustrates the solution.

It is possible to draw a figure that looks like a circle with Logo. It will not be a true circle, since each point on the circumference will not be exactly the same distance from the center. To construct the circle, we instruct the turtle to draw a regular polygon with 360 sides. Since the computer must draw the turtle for each turn, it will take a long time to draw the circle. The time can be shortened by adding the command HIDE TURTLE to our procedure. Typing HT, the abbreviation for this command, into the computer causes the turtle to disappear from the screen, and subsequent movements are made more quickly. A procedure for drawing a circle is given below, and the result is illustrated in Fig. 8.76.

                ED "CIRCLE
                HT
                REPEAT 360[FD 1   RT 1]
                END

Plotting Points with Logo

Logo is a means of learning to plot points on a coordinate plane. Each point on the screen is associated with an ordered pair of numbers (a, b). The number a is the x-coordinate. A positive value of a indicates a point to the right of the

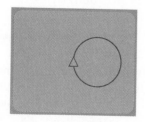

Figure 8.76

vertical line through home, and a negative value of *a* indicates a point to the left of the vertical line. The number *b* is the *y*-coordinate. A positive value of *b* indicates a point above the horizontal line through home, and a negative value of *b* indicates a point below the horizontal line. The origin, the home position of the turtle, has coordinates (0, 0).

To move the turtle in a horizontal direction to a particular position requires inputting an *x*-coordinate. The command to use is SETX. For example, if the turtle is at home, typing SETX 30 moves the turtle to the right horizontally to the point where the *x*-coordinate is 30. Likewise, typing SETX −40 moves the turtle horizontally to the left to the point on the screen where the *x*-coordinate is −40.

The command SETY works in a similar manner upon the *y*-coordinate. For example, if the turtle is at home, typing SETY 35, moves the turtle vertically (up) to the point where the *y*-coordinate is 35. Typing SETY −40 moves the turtle down to the point where the *y*-coordinate is −40. With SETX and SETY the turtle's heading is not changed.

▶ **Example 6**

Use SETX and SETY to graph a rectangle with vertices at the points (−30, 40) and (0, 0). Start with the turtle in home position.

Solution: Type

SETX −30   SETY 40   SETX 0   SETY 0

The path of the turtle is given in Fig. 8.77.

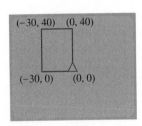

Figure 8.77

The coordinates of the turtle can be determined by typing the primitive POS. **POS** (or POSITION) is a primitive that can provide a list of two items. The first is the *x*-coordinate, and the second is the *y*-coordinate of the turtle's position. Typing **POS** for the final position in Example 6 yields the output (0, 0).

The primitive SETPOS can place the turtle at any set of coordinates in one step. SETPOS requires two numbers as inputs; the first number is the *x*-coordinate, and the second number is the *y*-coordinate. For example, typing SETPOS [−30, 40] moves the turtle from its present position to the point with coordinates (−30, 40) without changing the turtle's heading (see Fig. 8.78).

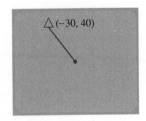

Figure 8.78

▶ **Example 7**

Write a set of commands that will generate a triangle with coordinates (30, 0), (20, 40), (60, 0).

Solution: The following commands will result in the triangle in Fig. 8.79.

```
PENUP
SETPOS [30, 0]
PENDOWN
SETPOS [20, 40]
SETPOS [60, 0]
SETPOS [30, 0]
```

Figure 8.79

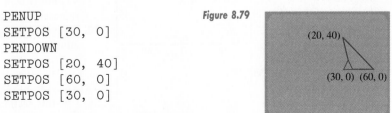

With the primitives and the procedures developed in this section you are now ready to try Logo. If you have access to a Logo disk, check your work with the computer.

Section 8.8 Exercises

In Ex. 1–10, we will assume that each sketch starts with the turtle at its home position. Sketch the turtle's trail as a result of the commands. Use the scale shown here for turtle distances. If possible, check your answers with a computer.

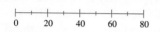

1. FD 40   LT 90   FD 30   RT 90   FD 20
   RT 90   FD 30
2. FD 40   RT 90   BK 30   LT 90   FD 30
   RT 90   HOME
3. BK 30   LT 45   FD 20   RT 45   FD 20
   RT 90   FD 10
4. FD 30   RT 120   FD 40   RT 30   FD 10
   RT 90   FD 20
5. BK 30   RT 3*60   FD 30   LT 45   FD 20
   RT 90   BK 10
6. RT 60   FD 22 + 18   RT 60   FD 80/2
   RT 60   FD 40   RT 60   FD 20*2   RT 60
   FD 120/3   RT 60   FD 40
7. REPEAT 3[FD 30   RT 60   FD 20   RT 90]
8. REPEAT 5[FD 25   BK 25   RT 72]
9. REPEAT 2[FD 40   RT 60   FD 40
   RT 60   FD 40   RT 60]
10. REPEAT 12[FD 30   RT 360/12]

In Ex. 11–20, write a list of commands that will generate the indicated figures. The numbers on the figures indicate the number of turtle steps. Note there is more than one correct answer.

11.

12.

13.

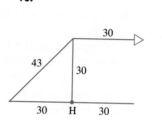

14.

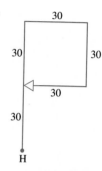

15.

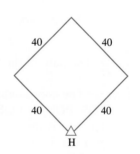

16.

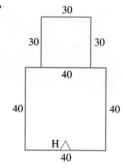

17.

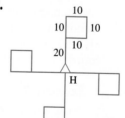

18.

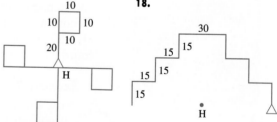

19. A rectangle with a width of 30 units and a length of 40 units (where home is in the center of the line segment representing the length)
20. A regular hexagon (six equal sides) with sides of length 30 units

In Ex. 21–30, assume that the procedures PENTAGON, TRIANGLE, SEGMENT, SQUARE, and CIRCLE as defined in this section are in the memory of the computer.

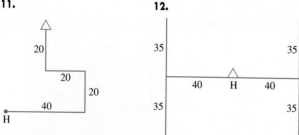

Predict the design that will result from each exercise. If possible check your results on a computer.

21. SEGMENT FD 30 TRIANGLE
22. SQUARE FD 30 RT 30 TRIANGLE
23. RT 90 TRIANGLE FD 20 RT 180
   CIRCLE
24. REPEAT 6[FD 40 TRIANGLE BK 40
   RT 360/6]
25. REPEAT 4[SQUARE RT 360/4]
26. REPEAT 20[SEGMENT RT 360/20]
27. REPEAT 5[FD 30 PENTAGON BK 30
   RT 360/5]
28. REPEAT 4[TRIANGLE FD 30 LT 90]
29. REPEAT 6[SQUARE RT 60]
30. REPEAT 360[FD 2 RT 1]

In Ex. 31–34, (a) write a set of commands using SETX and SETY that will instruct the turtle to move from Home to each of the points in the order given, and (b) illustrate the path of the turtle.

31. (50, 30), (−40, −10)
32. (−30, −20), (30, −20), (30, 0), (0, 0)
33. (40, 40), (0, 40), (0, 10), (20, 10)
34. (0, 50), (40, 50), (40, 30), (0, 30)

In Ex. 35–38, (a) write a set of commands using SETPOS to generate the figure (only lines that are part of the figure should show), and (b) show the figure that is generated by the commands.

35. Triangle with coordinates (−10, 0), (−40, 0),
   (−40, 50)
36. Parallelogram with coordinates (0, 0), (20, 30),
   (60, 30), (40, 0)
37. Trapezoid with coordinates (0, 0), (20, 40), (70, 40),
   (90, 0)
38. Square with coordinates (0, −40), (−40, 0), (0, 40),
   (40, 0)

In Ex. 39–44, use the procedures developed in this section and any primitives to write instructions to generate each figure. In the figures Home is indicated with the letter H. Note there is more than one correct answer.

39.

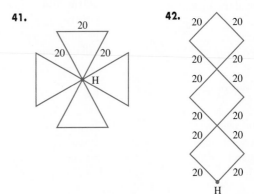

40.

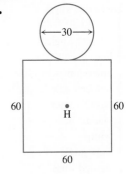

41.

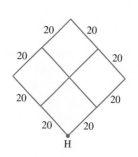

42.

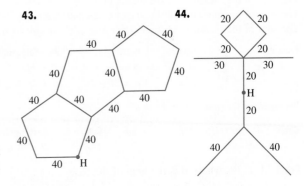

43.

44.

CHAPTER 8 SUMMARY

Key Terms

8.1
acute angle
adjacent angles
alternate exterior angles

alternate interior angles
angle
axiomatic method
complementary angles

corresponding angles
definitions
degrees
end point

Euclidean (or plane) geometry
exterior angle
half line
interior angle
line
line segment
obtuse angle
parallel lines
plane
point
right angle
skewed lines
supplementary angles
theorem
transversal
undefined terms
vertex
vertical angles

8.2
congruent figures
polygon
quadrilateral

regular polygon
similar figures

8.3
area
circumference
diameter
metric (or SI) system
perimeter
radius

8.4
polyhedron
prism
pyramid
solid geometry
volume

8.5
arc
even vertex
graph
network

odd vertex
topology
traversable

8.6
Jordan curve
Klein bottle
map
Möbius strip

8.7
fractal
hyperbolic geometry
model
non-Euclidean geometry
parallel postulate
recursion
Riemannian (or spherical) geometry

8.8
draw mode
edit mode
primitives
procedure

Important Facts

The sum of the measures of the angles of a triangle is 180°.
The sum of the measures of the angles of a quadrilateral is 360°.
The sum of the measures of the interior angles of an n-sided polygon is $(n-2)180°$.

Triangle	Square	Rectangle	Parallelogram	Trapezoid
$A = \frac{1}{2}bh$	$A = s^2$	$A = lw$	$A = bh$	$A = \frac{1}{2}h(b_1 + b_2)$
$p = s_1 + s_2 + s_3$	$p = 4s$	$p = 2l + 2w$	$p = 2b + 2w$	$p = s_1 + s_2 + b_1 + b_2$

Pythagorean theorem $a^2 + b^2 = c^2$

Circle $A = \pi r^2$, $C = 2\pi r$ or $C = \pi d$ **Cube** $V = s^3$ **Rectangular solid** $V = lwh$ **Cylinder** $V = \pi r^2 h$ **Cone** $V = \frac{1}{3}\pi r^2 h$ **Sphere** $v = \frac{4}{3}\pi r^3$

Prism $V = Bh$, where B is the area of the base   **Pyramid** $V = \frac{1}{3}Bh$, where B is the area of the base

Rules of traversability

1. A network with no odd (all even) vertices is traversable; you may start from any vertex, and you will end where you began.

2. A network with exactly two odd vertices is traversable; you must start at either of the odd vertices and finish at the other.

3. A network with more than two odd vertices is not traversable.

Fifth postulate in Euclidean geometry   Given a line and a point not on the line, only one line can be drawn through the given point parallel to the given line.

Fifth postulate in elliptical geometry Given a line and a point not on the line, no line can be drawn through the given point parallel to the given line.

Fifth postulate in hyperbolic geometry Given a line and a point not on the line, two or more lines can be drawn through the given point parallel to the given line.

CHAPTER 8 REVIEW EXERCISES

8.1

In Ex. 1–6, use the figure shown to find the following.

1. $\overleftrightarrow{HB} \cap \overline{BF}$
2. $\triangle ABH \cap \triangle HBC$
3. $\overleftrightarrow{EG} \cap \overleftrightarrow{BD}$
4. $\overline{BF} \cup \overline{FC} \cup \overline{BC}$
5. $\overleftrightarrow{AB} \cup \overleftrightarrow{DC}$
6. $\overrightarrow{FI} \cap \overrightarrow{FB}$

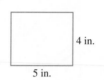

8.2

In Ex. 7–10, use the similar triangles ABC and $A'B'C$ shown to find the following.

7. The length of $\overline{A'B'}$
8. The length of $\overline{BC}$
9. The measure of $\triangle ABC$
10. The measure of $\triangle BAC$

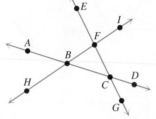

11. In the following figure, l_1 and l_2 are parallel lines. Find $m \triangle 1$ through $m \triangle 6$ as indicated.

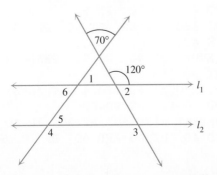

8.3

Find the area of each of the figures.

12.

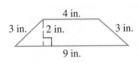

13.

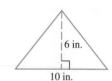

14.

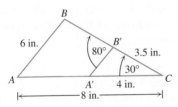

15.

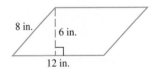

16.

8.4

Find the volume of each of the figures.

17.

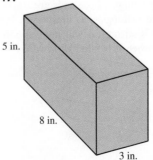

18.

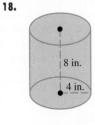

19. Find the area of the isosceles triangle.

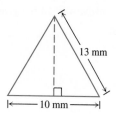

13 mm

10 mm

Find the volume of each solid.

20.

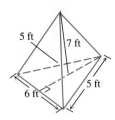

5 ft

7 ft

5 ft

6 ft

21.

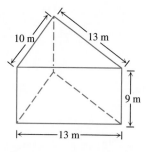

10 m

13 m

9 m

13 m

22.

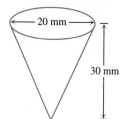

20 mm

30 mm

23.

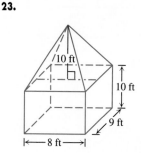

10 ft

10 ft

9 ft

8 ft

24. Determine the total cost of covering a 15 ft by 20 ft kitchen floor with tile. The cost of the tile selected is $18.50 per square yard. (Add 5% of cost for waste.)

25. Farmer Jones has a water trough whose ends are trapezoids and whose sides are rectangles, as illustrated. He is afraid that the base it is sitting on will not support the weight of the trough when it is filled with water.

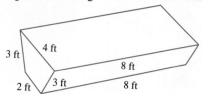

3 ft

4 ft

8 ft

2 ft

3 ft

8 ft

a) Find the number of cubic feet of water contained in the through.

b) Find the total weight, assuming that the trough weighs 375 lb and the water weighs 62.5 lb per cubic foot.

c) If 1 gal of water weighs 8.3 lb, how many gallons of water will the trough hold?

8.5

Determine whether the networks in Ex. 26–30 are traversable. If they are, indicate from which points you may start.

26.

A

B

C

H

G

F

D

E

27.

A

B

E

D

C

28.

A

B

C

E

D

29.

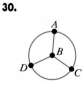

A

B

C

E

D

30.

A

B

D

C

8.7

31. Summarize the history of non-Euclidean geometry.

32. State the fifth axiom of Euclidean, elliptical, and hyperbolic geometry.

8.8

Assume that the Logo turtle starts at the Home position. Sketch the turtle's trail as a result of the commands. Use the following scale for turtle distances.

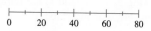

0   20   40   60   80

33. FD 40   RT 90   FD 50   RT 90   FD 40

34. RT 180   FD 80   LT 90   FD 50   RT 90
    BK 40

35. PENUP   FD 40   RT 90   PENDOWN   FD 50
    RT 90   FD 30

36. RT 270/3   FD 160/4   LT 30*3
    FD 90+10   RT 90-60

37. REPEAT 2[FD 50   LT 90   FD 40   RT 90
    FD 40   RT 90   FD 40   BK 10   RT 90]

CHAPTER TEST

In Ex. 1–4, use the figure below to describe the following sets of points:

1. $\overset{\circ}{AF} \cap \overline{AE}$

2. $\overset{\circ}{BC} \cup \overline{BC}$

3. $\angle ABF \cap \angle CBD$

4. $\overrightarrow{AE} \cup \overrightarrow{EA}$

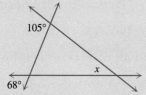

5. $m \angle A = 37.4°$. Find the measure of the complement of $\angle A$.

6. $m \angle B = 93.7°$. Find the measure of the supplement of $\angle B$.

7. In the following figure, find the measure of $\angle x$.

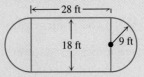

8. Find the sum of the interior angles of a pentagon.

9. Triangles ABC and $A'B'C'$ are similar figures. Find the length of side $\overline{B'C'}$.

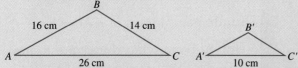

10. Right triangle ABC has one leg of length 12 in. and a hypotenuse of length 20 in. (See art on top right.)
a) Find the length of the other leg.
b) Find the perimeter of the triangle.
c) Find the area of the triangle.

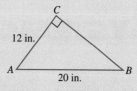

11. Find the volume of a sphere of diameter 12 in.

12. The following sketch shows the dimensions of an ice skating rink with semicircular ends. How many cubic feet of water are needed to fill the rink to a depth of 4 in.?

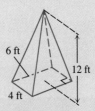

13. Find the volume of the pyramid.

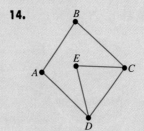

Determine whether each network below is traversable. If the network is traversable, state at which points you may start.

14.

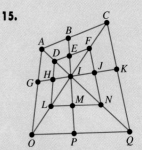

15.

MATHEMATICAL SYSTEMS

We have, throughout human history, devised different ways of making sense out of our world. For primitive peoples, explanations of why things happen the way they do took the form of mythology. With the philosophers of ancient Greece we see a more reasoned, scientific approach used to discover the essence of things. Aristotle described this essence in terms of earth, air, fire, and water. Today, education is highly specialized and fragmented into different disciplines, including physics, mathematics, chemistry, biology, psychology, and sociology. Yet the driving force behind all of these disciplines is still the same, to understand the most basic laws and patterns of the subject.

In the last 200 years, much of the focus of scientific study of the fundamental laws of nature has shifted from what things are to how they change in space and time. The mathematics used for this purpose belongs to a branch of mathematics

Physicists use group theory to categorize the different elementary particles that spin off when subatomic particles such as protons and neutrons are smashed together. This photo shows a computer simulation of the tracks made by such a collision.

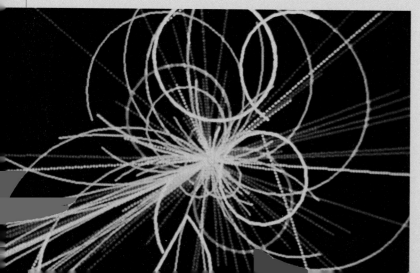

known as group theory. A group is a collection of fundamentally basic elements: It could be numbers, it could be the elementary particles of physics, it could be a pattern of repeating geometric designs. The way in which the elements change or remain the same when acted upon by some operation or transformation defines membership in the group. There exists a certain underlying pattern and symmetry in this.

To appreciate why group theory is so useful to scientists, consider the physicists who are trying to piece together the history of the universe. They have a clear sense of what the universe is like today, but what about 20 billion years ago when scientists believe the universe exploded into existence in what is called the Big Bang? Group theory may enable us to derive what the initial conditions of the universe may have been from knowledge of what matter is like today. Group theory can also be used to describe number systems as well as the basic building blocks of crystalline solids. It can even be used to define the symmetries that appear in the artifacts of a given culture. No wonder the physicist Sir Arthur Stanley Eddington called group theory the "supermathematics."

The concepts of group theory can be found in music. Consider the musical form known as the fugue. A single theme is repeated over time, in several voices, but changed in key and pitch to create a densely layered piece of music that eventually returns to the original theme.

Despite the seemingly endless variety of patterns the human imagination can devise, group theory can be used to catalog and define patterns by the way in which the design elements are transformed and positioned.

9.1 Groups

Many different types of mathematical systems have been developed. Some systems are used in solving everyday problems, such as planning work schedules. Others are more abstract and are used primarily in research, for example, in the study of chemistry, physical structure, matter, and the nature of genes.

When you learned how to add integers, you were introduced to a mathematical system. When you learned how to multiply integers, you became familiar with a second mathematical system. The set of integers with the operation of subtraction and the set of integers with the operation of division are two other examples of mathematical systems. Addition, subtraction, multiplication, and division are called binary operations. A **binary operation** is an operation, or rule, that can be performed on two and only two elements of a set. The result is a single element. When we add *two* integers, the sum is *one* integer. When we multiply *two* integers, the product is *one* integer. Is finding the reciprocal of a number a binary operation? No, it is an operation on a single element of a set.

The mixing of two paints is a binary operation. When two colors are mixed—for example, yellow and blue—a single color results, in this case green.

> A **mathematical system** consists of a set of elements and at least one binary operation.

Commutative and Associative Properties

Once a mathematical system is defined, its structure may display certain properties. Consider the set of integers:

$$I = \{\ldots, -3, -2, -1, 0, 1, 2, 3, \ldots\}$$

Recall that the ellipsis, the three dots, at each end of the set indicates that the set continues in the same manner.

The set of integers can be studied with the operations of addition, subtraction, multiplication, or division as separate mathematical systems. For example, when we study the set of integers under the operations of addition or multiplication, we see that the commutative and associative properties hold. The general forms of the properties are shown here.

For any elements a, b, and c	Addition	Multiplication
Commutative property	$a + b = b + a$	$a \cdot b = b \cdot a$
Associative property	$(a + b) + c = a + (b + c)$	$(a \cdot b) \cdot c = a \cdot (b \cdot c)$

The integers *are commutative* under the operations of *addition and multiplication.* For example,

$$2 + 4 = 4 + 2 \quad \text{and} \quad 2 \cdot 4 = 4 \cdot 2$$
$$6 = 6 \qquad\qquad\qquad 8 = 8$$

However, the integers *are not commutative* under the operations of *subtraction and division.* For example,

$$4 - 2 \neq 2 - 4 \quad \text{and} \quad 4 \div 2 \neq 2 \div 4$$
$$2 \neq -2 \qquad\qquad\qquad 2 \neq \tfrac{1}{2}$$

The integers *are associative* under the operations of *addition and multiplication.* For example,

$$(1 + 2) + 3 = 1 + (2 + 3) \quad \text{and} \quad (1 \cdot 2) \cdot 3 = 1 \cdot (2 \cdot 3)$$
$$3 + 3 = 1 + 5 \qquad\qquad\qquad 2 \cdot 3 = 1 \cdot 6$$
$$6 = 6 \qquad\qquad\qquad\qquad 6 = 6$$

However, the integers *are not associative* under the operations of *subtraction and division.* See Ex. 5 and 6 at the end of this section.

To say that a set of elements is commutative under a given operation means that the commutative property holds for *any* elements a and b in the set. Similarly, to say that a set of elements is associative under a given operation means that the associative property holds for *any* elements a, b, and c in the set.

Consider the mathematical system consisting of the set of integers under the operation of addition. Because the set of integers is infinite, this mathematical system is an example of an **infinite mathematical system**. We will study certain properties of this mathematical system. The first property that we will examine is *closure.*

Closure

The sum of any two integers is an integer. Therefore the set of integers is said to be **closed**, or to satisfy the **closure property**, under the operation of addition.

> If a binary operation is performed on any two elements of a set and the result is an element of the set, then that set is **closed** (or has **closure**) under the given binary operation.

Is the set of integers closed under the operation of multiplication? The answer is yes. When any two integers are multiplied, the product will be an integer.

Is the set of integers closed under the operation of subtraction? Again the answer is yes. The difference of any two integers is an integer.

Is the set of integers closed under the operation of division? The answer is no because it is possible to find two integers whose quotient is not an integer.

For example, if we select the integers 2 and 3, the quotient of 2 divided by 3 is $\frac{2}{3}$, which is not an integer. Thus the integers are not closed under the operation of division.

We showed that the set of integers was not closed under the operation of division by finding two integers whose quotient was not an integer. A specific example illustrating that a specific property is not true is called a **counterexample**. Mathematicians and scientists often try to find a counterexample to confirm that a specific property is not always true.

Identity Element

Now we will discuss the identity element for the set of integers under the operation of addition. Is there an element in the set that, when added to any given integer, results in a sum that is the given integer? The answer is yes. The sum of 0 and any integer is the given integer. For example, $1 + 0 = 0 + 1 = 1$, $2 + 0 = 0 + 2 = 2$, and so on. For this reason we call 0 the **additive identity element** for the set of integers. Note that for any integer a, $a + 0 = 0 + a = a$.

> An **identity element** is an element in a set such that when a binary operation is performed on it and any given element in the set, the result is the given element.

Is there an identity element for the set of integers under the operation of multiplication? The answer is yes; it is the number 1. Note that $2 \cdot 1 = 1 \cdot 2 = 2$, $3 \cdot 1 = 1 \cdot 3 = 3$, and so on. For any integer a, $a \cdot 1 = 1 \cdot a = a$. For this reason, 1 is called the **multiplicative identity element** for the set of integers.

Inverses

What integer when added to 4 gives a sum of 0; that is, $4 + \boxed{} = 0$? The shaded area is to be filled in with the integer -4: $4 + (-4) = 0$. We say that -4 is the additive inverse of 4, and 4 is the additive inverse of -4. Note that the sum of the element and its additive inverse gives the additive identity element 0. What is the additive inverse of 12? Since $12 + (-12) = 0$, -12 is the additive inverse of 12.

Here are some other examples of integers and their additive inverses:

Element	+	Additive Inverse	=	Identity Element
0	+	0	=	0
2	+	(−2)	=	0
−5	+	5	=	0

Note that for the operation of addition, every integer a has a unique inverse, $-a$, such that $a + (-a) = -a + a = 0$.

When a binary operation is performed on two elements in a set and the result is the identity element for the binary operation, then the two elements are said to be **inverses** of each other.

Does every integer have an inverse under the operation of multiplication? For multiplication the product of an integer and its inverse must yield the multiplicative identity element, 1. What is the multiplicative inverse of 2? That is, 2 times what number gives 1?

$$2 \cdot ? = 1 \qquad 2 \cdot \tfrac{1}{2} = 1$$

However, since $\tfrac{1}{2}$ is not an integer, 2 does not have a multiplicative inverse in the set of integers.

Group

Let us review what we have learned about the mathematical system consisting of the set of integers under the operation of addition.

1. The set of integers is *closed* under the operation of addition.
2. The set of integers has an *identity element* under the operation of addition.
3. Each element in the set of integers has an *inverse* under the operation of addition.
4. The *associative property* holds for the set of integers under the operation of addition.

The set of integers under the operation of addition is an example of a group. The properties of a group can be summarized as shown in the box.

Properties of a Group

Any mathematical system that meets the following four requirements is called a **group**.
1. The set of elements is *closed* under the given operation.
2. There exists an *identity element* for the set.
3. Every element in the set has an *inverse*.
4. The set of elements is *associative* under the given operation.

It is often very time consuming to show that the associative property holds for all cases. In many examples that follow we will state that the associative property holds for the given set of elements under the given operation.

Commutative Group

The commutative property does not need to hold for a mathematical system to be a group. However, if a mathematical system meets the four requirements of a group and is also commutative under the given operation, then the mathematical system is a **commutative** or **abelian group**.

Niels Abel *(1802–1829)*

Évartiste Galois *(1811–1832)*

*Untimely ends: Important contributions to the development of group theory were made by two young men who would not live to see their work gain acceptance: **Niels Abel**, a Norwegian, and **Évariste Galois**, a Frenchman. While their work showed brilliance, neither was noticed in his lifetime by the mathematics community. Abel lived in poverty and died of malnutrition and tuberculosis at the age of 26, just two days before a letter arrived with the offer of a teaching position. Galois died of complications from a gunshot wound he received in a duel over a love affair. The 20-year-old man had spent the night before summarizing his theories in the hope that eventually someone would find it "profitable to decipher this mess." After their deaths, both men had important elements of group theory named after them.*

> A group that satisfies the commutative property is called a **commutative group** or **abelian group**.

Since the commutative property holds for the set of integers under the operation of addition, the set of integers under the operation of addition is not only a group, but it is a commutative group.

Properties of a Commutative Group

A mathematical system is a commutative group if all five conditions hold.
1. The set of elements is *closed* under the given operation.
2. There exists an *identity element* for the set.
3. Every element in the set has an *inverse*.
4. The set of elements is *associative* under the given operation.
5. The set of elements is *commutative* under the given operation.

To determine whether a given mathematical system is a group under a given operation, check, in the following order, to see whether (a) the system is closed under the given operation, (b) there is an identity element in the set for the given operation, (c) every element in the set of elements has an inverse under the given operation, and (d) the associative property holds under the given operation. If any of these four requirements is *not* met, stop and state that the mathematical system is not a group. If asked to determine whether the mathematical system is a commutative group, then check to see if the commutative property holds for the given operation.

▶ **Example 1**

Determine whether the set of rational numbers under the operation of multiplication forms a group.

Solution: Recall from Chapter 5 that the rational numbers are the set of numbers of the form p/q where p and q are integers, $q \neq 0$. All fractions and integers are rational numbers.
1. *Closure* The product of two rational numbers is a rational number. Therefore the rational numbers are closed under the operation of multiplication.
2. *Identity Element* The multiplicative identity element for the set of rational numbers is 1. Note, for example, $3 \cdot 1 = 1 \cdot 3 = 3$, and $\frac{3}{8} \cdot 1 = 1 \cdot \frac{3}{8} = \frac{3}{8}$. For any rational number a, $a \cdot 1 = 1 \cdot a = a$.
3. *Inverse Elements* For the mathematical system to be a group under the operation of multiplication, *each and every* rational number must have a multiplicative inverse in the set of rational numbers. Remember that for the operation of multiplication, the product of a number and its inverse must give the multiplicative identity element, 1. Let's check a few rational

numbers:

Rational Number	·	Inverse	=	Identity Element
3	·	$\dfrac{1}{3}$	=	1
$\dfrac{2}{3}$	·	$\dfrac{3}{2}$	=	1
$-\dfrac{1}{5}$	·	-5	=	1

Looking at these examples one might deduce that each rational number does have an inverse. However, there is one rational number, 0, that does not have an inverse.

$$0 \cdot ? = 1$$

Because there is no rational number that, when multiplied by 0, gives 1, 0 does not have a multiplicative inverse. Since every rational number does not have an inverse, this mathematical system is not a group.

There is no need at this point to check the associative property, since we have already shown that the mathematical system of rational numbers under the operation of multiplication is not a group.

Section 9.1 Exercises

1. What is a binary operation?
2. What are the parts of a mathematical system?
3. Give an example to show that the commutative property does not hold for the set of integers under the operation of subtraction.
4. Give an example to show that the commutative property does not hold for the set of integers under the operation of division.
5. Give an example to show that the associative property does not hold for the set of integers under the operation of subtraction.
6. Give an example to show that the associative property does not hold for the set of integers under the operation of division.
7. What properties are required for a mathematical system to be a group?
8. What properties are required for a mathematical system to be a commutative group?

In Ex. 9–17, explain your answer.

9. Is the set of positive integers a commutative group under the operation of addition?
10. Is the set of integers a group under the operation of multiplication?

11. Is the set of rational numbers a commutative group under the operation of addition?
12. Is the set of positive integers a group under the operation of subtraction?
13. Is the set of integers a group under the operation of subtraction?
*14. Is the set of irrational numbers a group under the operation of addition?
*15. Is the set of irrational numbers a group under the operation of multiplication?
*16. Is the set of real numbers a group under the operation of addition?
*17. Is the set of real numbers a group under the operation of multiplication?

Research Activity

18. There are other classifications of mathematical systems besides groups. For example, there are *rings* and *fields*. Do research to determine the requirements that must be met for a mathematical system to be (a) a ring and (b) a field. (c) Is the set of real numbers, under the operations of addition and multiplication, a field?

9.2 Finite Mathematical Systems

In the previous section we presented infinite mathematical systems. In this section we present some finite mathematical systems. A **finite mathematical system** is one whose set contains a finite number of elements.

Clock Arithmetic

Figure 9.1

Let us develop a finite mathematical system called clock arithmetic. The set of elements in this system will be the hours on a clock: {1, 2, 3, 4, 5, 6, 7, 8, 9, 10, 11, 12}. The binary operation that we will use is addition, which we define as movement of the hour hand in a clockwise direction. Assume that it is 4 o'clock: What time will it be in nine hours? (See Fig 9.1.) If we add 9 hours to 4 o'clock, the clock will read 1 o'clock. Thus $4 + 9 = 1$ in clock arithmetic. Would $9 + 4$ be the same as $4 + 9$? Yes, $4 + 9 = 9 + 4 = 1$.

Table 9.1 is the addition table for clock arithmetic. Its elements were determined using the definition of addition as previously illustrated. For example, we found the sum of 9 and 4 was 1, so a 1 was put in the table where the row to the right of the 9 intersects the column below the 4. Likewise, since the sum of 5 and 3 is 8, an 8 was put in the table where the row to the right of the 5 intersects the column below the 3.

Table 9.1

+	1	2	3	4	5	6	7	8	9	10	11	12
1	2	3	4	5	6	7	8	9	10	11	12	1
2	3	4	5	6	7	8	9	10	11	12	1	2
3	4	5	6	7	8	9	10	11	12	1	2	3
4	5	6	7	8	9	10	11	12	1	2	3	4
5	6	7	8	9	10	11	12	1	2	3	4	5
6	7	8	9	10	11	12	1	2	3	4	5	6
7	8	9	10	11	12	1	2	3	4	5	6	7
8	9	10	11	12	1	2	3	4	5	6	7	8
9	10	11	12	1	2	3	4	5	6	7	8	9
10	11	12	1	②	3	4	5	6	7	8	9	10
11	12	1	2	3	4	5	6	7	8	9	10	11
12	1	2	3	4	5	6	7	8	9	10	11	12

The binary operation of this system is defined by the table. It is denoted by the symbol $+$. To determine the value of $a + b$, where a and b are any two numbers in the set, find a in the left-hand column and find b along the top row. Assume that there is a horizontal line through a and a vertical line through b;

the point of intersection of these two lines is where you find the value of $a + b$. For example, $10 + 4 = 2$ has been circled in Table 9.1. Note that $4 + 10$ also equals 2, but this will not necessarily be so for all examples in this chapter.

▶ **Example 1**

Determine whether the clock arithmetic system under the operation of addition is a commutative group.

Solution: Check the five requirements that must be satisfied for a commutative group.

1. *Closure* Is the set of elements in clock arithmetic closed under the operation of addition? Yes, since Table 9.1 contains only the elements in the set $\{1, 2, 3, 4, 5, 6, 7, 8, 9, 10, 11, 12\}$. If Table 9.1 had contained an element other than the numbers 1 through 12, the set would not have been closed under addition.

2. *Identity Element* Is there an identity element for clock arithmetic? If it is currently 4 o'clock, how many hours have to pass before it is 4 o'clock again? Twelve hours: $4 + 12 = 12 + 4 = 4$. In fact, given any hour, in 12 hours the clock will return to the starting point. Therefore 12 is the additive identity element in clock arithmetic.

   In examining Table 9.1 we see that the row of numbers next to the 12 in the left-hand column is identical to the row of numbers along the top. We also see that the column of numbers under the 12 in the top row is identical to the column of numbers on the left. The search for such a column and row is one technique for determining whether there is an identity element for a system defined by a table.

3. *Inverses* Is there an inverse for the number 4 in clock arithmetic for the operation of addition? Recall that the identity element in clock arithmetic is 12. What number when added to 4 gives 12; that is, $4 + \boxed{} = 12$? By examining Table 9.1 we see that $4 + 8 = 12$ and also $8 + 4 = 12$. Thus 8 is the additive inverse of 4, and 4 is the additive inverse of 8.

   To find the additive inverse of 7, find 7 in the left-hand column. Look to the right of the 7 until you come to the identity element 12. Determine the number at the top of this column. The number is 5. Since $7 + 5 = 5 + 7 = 12$, 5 is the inverse of 7, and 7 is the inverse of 5. The other inverses can be found in the same way, as shown in Table 9.2. Note that each element in the set has an *inverse*.

4. *Associative Property* Now consider the associative property. Does $(a + b) + c = a + (b + c)$ for all values a, b, and c of the set? Remember to always evaluate the values within the parentheses first. Let us select some values for a, b, and c. Let $a = 2$, $b = 6$, and $c = 8$. Then

$$(2 + 6) + 8 = 2 + (6 + 8)$$
$$8 + 8 = 2 + 2$$
$$4 = 4 \quad \text{True}$$

Table 9.2

Element	+	Inverse	=	Identity Element
1	+	11	=	12
2	+	10	=	12
3	+	9	=	12
4	+	8	=	12
5	+	7	=	12
6	+	6	=	12
7	+	5	=	12
8	+	4	=	12
9	+	3	=	12
10	+	2	=	12
11	+	1	=	12
12	+	12	=	12

Let $a = 5$, $b = 12$, and $c = 9$. Then

$$(5 + 12) + 9 = 5 + (12 + 9)$$
$$5 + 9 = 5 + 9$$
$$2 = 2 \quad \text{True}$$

Randomly selecting *any* elements a, b, and c of the set reveals that $(a + b) + c = a + (b + c)$. Thus the system of clock arithmetic is associative under the operation of addition. Note that if there is just one set of values a, b, and c such that $(a + b) + c \neq a + (b + c)$, then the system is not associative. Normally you will not be asked to check every case to determine if the associative property holds.

5. *Commutative Property* Does the commutative property hold under the given operation? Does $a + b = b + a$ for all elements a and b of the set? Let us randomly select some values for a and b to see whether the commutative property appears to hold. Let $a = 5$ and $b = 8$; then Table 9.1 shows that

$$5 + 8 = 8 + 5$$
$$1 = 1 \quad \text{True}$$

Let $a = 9$ and $b = 6$; then

$$9 + 6 = 6 + 9$$
$$3 = 3 \quad \text{True}$$

The commutative property holds for these two specific cases. In fact, if we were to randomly select *any* values for a and b, we would find that $a + b = b + a$. Thus the commutative property of addition is true in clock arithmetic. Note that if there is just one set of values a and b such that $a + b \neq b + a$, then the system is not commutative.

Since this system satisfies the five properties required for a mathematical system to be a commutative group, clock arithmetic under the operation of addition is a commutative group.

Table 9.3

+	0	1	2	3	4
0	0	1	2	3	4
1	1	2	3	4	0
2	2	3	4	0	1
3	3	4	0	1	2
4	4	0	1	2	3

One method that can be used to determine whether a system defined by a table is commutative under the given operation is to determine whether the elements in the table are symmetric about the main diagonal. The main diagonal is the diagonal from the upper left-hand corner to the lower right-hand corner of the table. In Table 9.3 the main diagonal is shaded in color.

If the elements are symmetric about the main diagonal, then the system is commutative. If the elements are not symmetric about the main diagonal, then the system is not commutative. If you examine the system in Table 9.3 you will see that its elements are symmetric about the main diagonal. Therefore this mathematical system is commutative.

It is possible to have groups that are not commutative. Such groups are called **noncommutative** or **nonabelian groups**. However, a *noncommutative group defined by a table must be at least a six-element by six-element table*. Nonabelian groups are illustrated in Ex. 51 and 54 at the end of this section. We suggest that you do at least Ex. 51.

Now we will look at another finite mathematical system.

▶ **Example 2**

Consider the mathematical system defined by Table 9.4. Assume that the associative property holds for the given operation.

a) List the elements in the set of this mathematical system.
b) Identify the binary operation.
c) Determine whether this mathematical system is a commutative group.

Solution:

a) The set of elements for this mathematical system consists of the elements found on the top (or left-hand side) of the table. The set of elements is $\{1, 3, 5, 7\}$.
b) The binary operation is $\odot$.
c) We must determine whether the five requirements for a commutative group are satisfied.

1. *Closure* Since all the elements in the table are in the original set of elements, $\{1, 3, 5, 7\}$, the system is closed.
2. *Identity Element* The identity element is 5. Note that the column of elements under the 5 is identical to the left-hand column *and* the row of elements to the right of the 5 is identical to the top row.
3. *Inverse Elements* When an element operates on its inverse element, the result is the identity element. For this example the identity element is 5. To determine the inverse of 1, find the element to replace the question mark:

$$1 \odot ? = 5$$

Since $1 \odot 1 = 5$, 1 is the inverse of 1. Thus 1 is its own inverse.

To find the inverse of 3, find the element to replace the question mark:

$$3 \odot ? = 5$$

Since $3 \odot 7 = 7 \odot 3 = 5$, 7 is the inverse of 3 (and 3 is the inverse of 7). The elements and their inverses are shown in Table 9.5. Every element has a unique inverse.
4. *Associative Property* It is given that the associative property holds for the given operation. One example of the associative property is

$$(7 \odot 3) \odot 1 = 7 \odot (3 \odot 1)$$
$$5 \odot 1 = 7 \odot 7$$
$$1 = 1 \quad \text{True}$$

5. *Commutative Property* Since the elements in the table are symmetric about the main diagonal, the commutative property holds for the operation

Table 9.4

$\odot$	1	3	5	7
1	5	7	1	3
3	7	1	3	5
5	1	3	5	7
7	3	5	7	1

Table 9.5

Element	$\odot$	Inverse	=	Identity Element
1	$\odot$	1	=	5
3	$\odot$	7	=	5
5	$\odot$	5	=	5
7	$\odot$	3	=	5

of $\odot$. One example of the commutative property is

$$3 \odot 5 = 5 \odot 3$$
$$3 = 3 \quad \text{True}$$

Since the five necessary properties hold, this mathematical system is a commutative group.

Mathematical Systems Without Numbers

Thus far all the systems we have discussed have been based on sets of numbers. The example below illustrates a mathematical system using symbols other than numbers as its elements.

▶ Example 3

Given the mathematical system defined by Table 9.6, determine:
a) The set of elements
b) The binary operation
c) Closure or nonclosure of the system
d) The identity element
e) The inverse of #
f) $\# \otimes L$ and $L \otimes \#$
g) $(\# \otimes \triangle) \otimes L$ and $\# \otimes (\triangle \otimes L)$

Table 9.6

$\otimes$	#	$\triangle$	L
#	$\triangle$	L	#
$\triangle$	L	#	$\triangle$
L	#	$\triangle$	L

Solution:
a) The set of elements of this mathematical system is $\{\#, \triangle, L\}$.
b) The binary operation is $\otimes$.
c) Since the table does not contain any symbols except #, $\triangle$, and L, the system is closed under $\otimes$.
d) The identity element is L, since the row next to L in the left-hand column is the same as the top row, and the column under L is identical to the left-hand column. Note that

$$\# \otimes L = L \otimes \# = \#$$
$$\triangle \otimes L = L \otimes \triangle = \triangle$$
$$L \otimes L = L$$

e) Element $\otimes$ inverse element = identity element
$$\# \otimes ? = L$$

To find the inverse of #, we must determine the element to replace the question mark. Since $\# \otimes \triangle = L$ and $\triangle \otimes \# = L$, $\triangle$ is the inverse of #.
f) $\# \otimes L = \# \qquad L \otimes \# = \#$
g) $(\# \otimes \triangle) \otimes L = L \otimes L \qquad \# \otimes (\triangle \otimes L) = \# \otimes \triangle$
$$\qquad\qquad\qquad = L \qquad\qquad\qquad\qquad\qquad = L$$

Group theory can be fun as well as educational, at least so thought Erno Rubik, a Hungarian teacher of architecture and design. In 1976 he presented the world with a puzzle in the form of a cube with six faces, each of a different color. Each face is divided into nine squares; each row and column of each face can rotate so that in no time at all six faces are transformed into a mix of colors. Your job, should you choose to accept it, is to get the cube back to its original condition.

Table 9.7

*	A	B	C	D
A	D	A	B	C
B	A	B	C	D
C	B	C	D	A
D	C	D	A	B

Table 9.8

Element		Inverse		Identity
A	*	C	=	B
B	*	B	=	B
C	*	A	=	B
D	*	D	=	B

Table 9.9

$\odot$	x	y	z
x	x	z	y
y	z	y	x
z	y	x	z

▶ **Example 4**

Determine whether the mathematical system in Table 9.7 in which * is an associative operation is a commutative group.

Solution:

1. *Closure* The system is closed.
2. *Identity Element* The identity element is *B*.
3. *Inverse* Each element has an inverse as illustrated in Table 9.8.
4. *Associative Property* It is given that the associative property holds. An example illustrating the associative property is

$$(D * A) * C = D * (A * C)$$
$$C * C = D * B$$
$$D = D \quad \text{True}$$

5. *Commutative Property* By examining the table we can see that it is symmetric about the main diagonal. Thus the system is commutative under the given operation. One example of the commutative property is

$$D * C = C * D$$
$$A = A \quad \text{True}$$

Since all five properties are satisfied, the system is a commutative group.

▶ **Example 5**

Determine whether the mathematical system in Table 9.9 is a commutative group under the operation of $\odot$.

Solution:

1. The system is closed.
2. Since there is no row identical to the top row, there is no identity element. Therefore this mathematical system is *not a group*. There is no need to go any further, but for practice, let's look at a few more items.
3. Since there is no identity element, there can be no inverses.
4. The associative property does not hold. The following example illustrates that the associative property is not true for each and every case:

$$(x \odot y) \odot z \neq x \odot (y \odot z)$$
$$z \odot z \neq x \odot x$$
$$z \neq x$$

5. The table is symmetrical about the main diagonal. Therefore the commutative property does hold for the operation of $\sim$.

Note that the associative property does not hold even though the commutative property does hold. This can happen when there is no identity element and every element does not have an inverse, as in this example.

Table 9.10

*	a	b	c
a	a	b	c
b	b	b	a
c	c	a	c

▶ **Example 6**

Determine if the mathematical system in Table 9.10 is a commutative group under the operation of ∗.

Solution:

1. The system is closed.
2. There is an identity element, *a*.
3. Each element has an inverse; *a* is the inverse of *a*, *b* is the inverse of *c* and *c* is the inverse of *b*.
4. Since every element in the set does not appear in every row and every column of the table, we need to check the associative property carefully. There are many specific cases where the associative property does hold.

Did You

Know...

CREATING PATTERNS
BY DESIGN

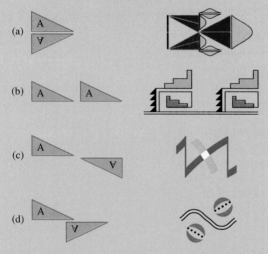

(a)

(b)

(c)

(d)

Patterns adapted from *Symmetries of Culture* by Dorothy K. Washburn and Donald W. Crowe (University of Washington Press, 1988)

There are just four geometric motions that generate all two-dimensional patterns: (a) reflection, (b) translation, (c) rotation, and (d) glide reflection. How these motions are applied is the basis of pattern analysis.

What makes group theory such a powerful tool is that it can be used to reveal the underlying structure of just about any physical phenomenon that involves symmetry and patterning, such as wallpaper or quilt patterns. Interest in the formal study of symmetry in design came out of the Industrial Revolution in the late nineteenth century. The new machines of the Industrial Revolution could vary any given pattern almost indefinitely. Designers needed a way to systematically describe and manipulate patterns. At the same time, explorers were discovering artifacts of other cultures, which stimulated interest in categorizing patterns. Shown here are the four geometric motions that generate all two-dimensional patterns: (a) reflection, (b) translation, (c) rotation, and (d) glide reflection. How these motions are applied, or not applied, is the basis of the pattern analysis.

However the following counterexample illustrates that the associative property does not hold for every case.

$$(b * b) * c \neq b * (b * c)$$
$$b * c \neq b * a$$
$$a \neq b$$

5. The commutative property holds since there is symmetry about the main diagonal.

Since we have shown that the associative property does not hold under the operation of *, this system is not a commutative group.

Section 9.2 Exercises

In Ex. 1–12, use Table 9.1 to determine the following sums in clock arithmetic.

1. $6 + 5$
2. $8 + 6$
3. $9 + 11$
4. $12 + 10$
5. $9 + 10$
6. $11 + 11$
7. $4 + (5 + 8)$
8. $(7 + 8) + 6$
9. $(3 + 4) + 7$
10. $(11 + 9) + 12$
11. $(5 + 7) + (8 + 6)$
12. $(8 + 9) + (12 + 11)$

In Ex. 13–18, determine the value of the following in clock arithmetic by starting at the first number and counting counterclockwise on the clock the number of units given by the second number.

13. $9 - 6$
14. $12 - 7$
15. $6 - 9$
16. $4 - 7$
17. $3 - 8$
18. $2 - 11$

19. Can you find another way to obtain the answers for Ex. 13–18? Explain.
20. Develop an addition table for the five-hour clock in Fig. 9.2.

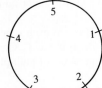

Figure 9.2

In Ex. 21–32, find the following sums and differences in five-hour clock arithmetic.

21. $2 + 4$
22. $5 + 3$
23. $4 + 2$
24. $3 + 3$
25. $1 + 4$
26. $1 + 5$

27. $5 - 3$
28. $1 - 3$
29. $2 - 4$
30. $1 - 5$
31. $(3 - 4) - 2$
32. $3 + (2 - 3)$

33. Determine whether five-hour clock arithmetic under the operation of addition is a commutative group. Explain.
34. A mathematical system is defined by a 3-element by 3-element table where every element in the set appears in each row and each column. Must the mathematical system be a commutative group? Explain.
35. Consider the mathematical system indicated by the following table in which the operation * is associative.

*	0	1	2	3
0	0	1	2	3
1	1	2	3	0
2	2	3	0	1
3	3	0	1	2

a) What are the elements of the set in this mathematical system?
b) What is the binary operation?
c) Is the system closed? Explain.
d) Is there an identity element for the system under the given operation? If so, what is it?
e) Does every element in the system have an inverse? If so, give each element and its corresponding inverse.
f) Give an example to illustrate the associative property.
g) Is the system commutative? Give an example to verify your answer.
h) Is the mathematical system a commutative group? Explain.

In Ex. 36–38, repeat questions (a)–(h) of Ex. 35 for the mathematical system given by each of the tables. Assume that the operations are associative.

36.

∅	3	5	7	9
3	7	9	3	5
5	9	3	5	7
7	3	5	7	9
9	5	7	9	3

37.

%	19	23	27
19	19	23	27
23	23	27	19
27	27	19	23

38.

~	2	6	8	4
2	6	8	4	2
6	8	4	2	6
8	4	2	6	8
4	2	6	8	4

39. a) Is the following mathematical system a group? Explain your answer.
b) Find an example showing that the operation is not associative for the given set of elements.

⊙	1	2	3	4
1	1	3	4	2
2	3	4	2	1
3	4	2	1	3
4	2	1	3	4

40. Given the mathematical system

!	z	p	o	n
z	z	p	o	n
p	p	o	n	z
o	o	n	z	p
n	n	z	p	o

determine the following.
a) The elements in the set
b) The binary operation
c) Closure or nonclosure of the system
d) $(z\,!\,o)\,!\,p$
e) $p\,!\,(n\,!\,z)$
f) The identity element
g) The inverse of z
h) The inverse of p

In Ex. 41–46, for each of the following mathematical systems determine which of the five properties of a commutative group do not hold.

41.

3	○	△	□
○	○	△	□
△	△	○	□
□	□	△	○

42.

∧	w	x	y
w	w	y	x
x	y	x	w
y	x	a	y

43.

φ	~	*	?	L	P
~	L	P	~	*	*
*	P	~	*	~	L
?	~	*	?	L	P
L	*	~	L	P	?
P	*	L	P	?	~

44.

☺	a	b	□	○	△
a	a	□	○	b	△
b	□	b	△	a	○
□	b	△	□	○	a
○	○	a	b	△	□
△	△	○	a	□	b

45.

⊙	a	b	c	d	e
a	c	d	e	a	b
b	d	e	a	b	c
c	e	a	b	c	d
d	a	b	c	e	d
e	b	c	d	e	a

46.

#	0	1	2	3	4	5
0	0	0	0	0	0	0
1	0	1	2	3	4	5
2	0	2	4	0	2	4
3	0	3	0	3	0	3
4	0	4	2	0	4	2
5	0	5	4	3	2	1

47. a) Consider the set consisting of two elements $\{E, O\}$, where E stands for an even number and O stands for an odd number. For the operation of addition, complete the following table.

+	E	O
E		
O		

b) Determine whether this mathematical system forms a commutative group under addition. Explain your answer.

48. a) Let E and O represent even numbers and odd numbers, respectively, as in Ex. 47. Complete the following table for the operation of multiplication.

×	E	O
E		
O		

b) Determine whether this mathematical system forms a

commutative group under the operation of multiplication. Explain your answer.

Make up your own mathematical systems that are groups. List the identity element and the inverses of each element. Do this with sets containing:

49. Three elements
50. Four elements

51. The following table is an example of a noncommutative or nonabelian group:

∞	1	2	3	4	5	6
1	5	3	4	2	6	1
2	4	6	5	1	3	2
3	2	1	6	5	4	3
4	3	5	1	6	2	4
5	6	4	2	3	1	5
6	1	2	3	4	5	6

a) Show that this is a group. (It would be very time consuming to prove that the associative property holds, but you can give some examples to show that it appears to hold.)

b) Find a counterexample to show that the commutative property does not hold.

Problem Solving

52. If a mathematical system is defined by a 3-element by 3-element table, how many specific cases must be illustrated to prove the set of elements is associative under the given operation?

53. Repeat Ex. 52 for a 4-element by 4-element table.

54. Suppose that three books numbered 1, 2, and 3 are placed next to one another on a shelf. If we remove volume 3 and place it before volume 1, the new order of books is 3, 1, 2. Let us call this replacement R. We can write

$$R = \begin{pmatrix} 1 & 2 & 3 \\ 3 & 1 & 2 \end{pmatrix},$$

which indicates the books were switched in order from 1, 2, 3 to 3, 1, 2. Other possible replacements are S, T, U, V, and I, as indicated.

$$S = \begin{pmatrix} 1 & 2 & 3 \\ 2 & 1 & 3 \end{pmatrix}, \quad T = \begin{pmatrix} 1 & 2 & 3 \\ 3 & 2 & 1 \end{pmatrix}, \quad U = \begin{pmatrix} 1 & 2 & 3 \\ 1 & 3 & 2 \end{pmatrix},$$

$$V = \begin{pmatrix} 1 & 2 & 3 \\ 2 & 3 & 1 \end{pmatrix}, \quad I = \begin{pmatrix} 1 & 2 & 3 \\ 1 & 2 & 3 \end{pmatrix}$$

Replacement set I indicates that the books were removed from the shelves and placed back in their original order. Consider the mathematical system with the set of elements R, S, T, U, V, I, with the operation $*$.

To evaluate $R * S$, write

$$R * S = \begin{pmatrix} 1 & 2 & 3 \\ 3 & 1 & 2 \end{pmatrix} * \begin{pmatrix} 1 & 2 & 3 \\ 2 & 1 & 3 \end{pmatrix}$$

R replaces 1 with 3, and S replaces 3 with 3 (no change), so $R * S$ replaces 1 with 3 (Fig. 9.3). R replaces 2 with 1, and S replaces 1 with 2, so $R * S$ replaces 2 with 2 (no change), R replaces 3 with 2, and S replaces 2 with 1, so $R * S$ replaces 3 with 1. $R * S$ replaces 1 with 3, 2 with 2, and 3 with 1 as shown in the figure.

Figure 9.3

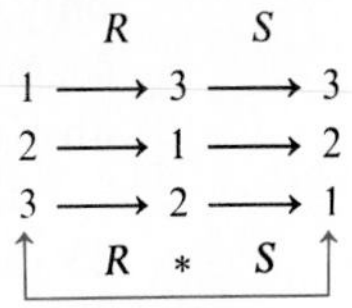

$$R * S = \begin{pmatrix} 1 & 2 & 3 \\ 3 & 2 & 1 \end{pmatrix} = T$$

Since this is the same as replacement set T, we write $R * S = T$.

a) Complete the following table for the operation using the procedure outlined.

*	R	S	T	U	V	I
R		T				
S						
T						
U						
V						
I						

b) Is this mathematical system a group? Explain.

c) Is this mathematical system a commutative group? Explain.

Research Activity

55. In Section 7.3 we introduced matrices. Show that 2×2 matrices under the operation of addition form a commutative group.

56. Show that 2×2 matrices under the operation of multiplication do not form a commutative group.

9.3 Modular Arithmetic

Figure 9.4

Figure 9.5

Table 9.11

+	0	1	2	3	4	5	6
0	0	1	2	3	4	5	6
1	1	2	3	4	5	6	0
2	2	3	4	5	6	0	1
3	3	4	5	6	0	1	2
4	4	5	6	0	1	2	③
5	5	6	0	1	2	3	4
6	6	0	1	2	3	4	5

Figure 9.6

The clock arithmetic we discussed in the previous section is similar to modular arithmetic. The set of elements $\{0, 1, 2, 3, 4, 5, 6, 7, 8, 9, 10, 11\}$ together with the operation of addition is called a modulo 12 or mod 12 system. There is one difference in notation between clock arithmetic and modulo 12 arithmetic. In the modulo 12 system the symbol 12 is replaced with the symbol 0.

A **modulo m system** consists of m elements, 0 through $m - 1$, and a binary operation. In this section we will discuss modular arithmetic systems and their properties.

If today is Sunday, what day of the week will it be in 23 days? The answer, Tuesday, is arrived at by dividing 23 by 7 and observing the remainder of 2. Twenty-three days represent three weeks plus two days. Since we are interested only in the day of the week that the twenty-third day will fall on, the three-week segment is unimportant to the answer. The remainder of 2 indicates that the answer will be two days later than Sunday, which is Tuesday.

If we place the days of the week on a clock face as shown in Fig. 9.4, in 23 days the hand would have made three complete revolutions and end on Tuesday. If we replace the days of the week with numbers, a modulo 7 arithmetic system will result. See Fig. 9.5: Sunday = 0, Monday = 1, Tuesday = 2, and so on. If we start at 0 and move the hand 23 places, we will end at 2. Table 9.11 shows a modulo 7 addition table.

If we start at 4 and add 6, we will end at 3 on the clock in Fig. 9.6. This number is circled in Table 9.11. The other numbers can be obtained in the same way.

A second method of determining the sum of 4 + 6 in modulo 7 arithmetic is to divide the sum, 10, by 7 and observe the remainder:

$$10 \div 7 = 1, \quad \text{remainder } 3$$

The remainder, 3, is the sum of 4 + 6 in a modulo 7 arithmetic system.

The concept of congruence is important in modular arithmetic.

> a **is congruent to** b modulo m, written, $a \equiv b \pmod{m}$, if a and b have the same remainder when divided by m.

We can show, for example, that $10 \equiv 3 \pmod 7$ by dividing both 10 and 3 by 7 and observing that we obtain the same remainder in each case.

$$10 \div 7 = 1, \quad \text{remainder } 3 \quad \text{and} \quad 3 \div 7 = 0, \quad \text{remainder } 3.$$

Since the remainders are the same, 3 in each case, 10 is congruent to 3 modulo 7, and we may write $10 \equiv 3 \pmod 7$.

Table 9.12 Modulo 7 Classes

0	1	2	3	4	5	6
0	1	2	3	4	5	6
7	8	9	10	11	12	13
14	15	16	17	18	19	20
21	22	23	24	25	26	27
28	29	30	31	32	33	34
⋮	⋮	⋮	⋮	⋮	⋮	⋮

Now consider $37 \equiv 5 \pmod 8$. If we divide both 37 and 5 by 8, each will have the same remainder, 5.

In any modulo system we can develop a set of **modulo classes** by placing all numbers with the same remainder in the appropriate modulo class. In a modulo 7 system, every number must have a remainder of either 0, 1, 2, 3, 4, 5, or 6. Thus there are seven modulo classes in a modulo 7 system. The seven classes are given in Table 9.12.

Every number is congruent to a number from 0 to 6 in mod 7. For example, $24 \equiv 3 \pmod 7$ because 24 is in the same modulo class as 3.

The solution to a problem in modular arithmetic, if it exists, will always be a number from 0 through $m - 1$, where m is the **modulus** of the system. For example, in a modulo 7 system, since 7 is the modulus, the solution will be a number from 0 through 6.

▶ **Example 1**

Determine which number, from 0 through 6, the following numbers are congruent to in modulo 7.

a) 84   **b)** 65   **c)** 47

Solution: We could determine the answer by listing more entries in Table 9.12. Another method of finding the answer is to divide the given number by 7 and observe the remainder.

a) $84 \equiv ? \pmod 7$

To determine the value that 84 is congruent to in mod 7, divide 84 by 7.

$$84 \div 7 = 12, \quad \text{remainder } 0.$$

Thus $84 \equiv 0 \pmod 7$.

b) $65 \equiv ? \pmod 7$

$$65 \div 7 = 9, \quad \text{remainder } 2.$$

Thus $65 \equiv 2 \pmod 7$.

c) $47 \equiv ? \pmod 7$

$$47 \div 7 = 6, \quad \text{remainder } 5.$$

Thus $47 \equiv 5 \pmod 7$.

Your car has an odometer that shows your mileage to be a little over 28,000 miles, but this is not the first time you've seen this reading. In fact, this is the third time around. The car's odometer uses a modulo 100,000 system.

▶ **Example 2**

Evaluate each of the following in mod 5.

a) $4 + 3$ **b)** $4 - 3$ **c)** $2 \cdot 4$

Solution:

a) $4 + 3 \equiv ? \pmod 5$

$7 \equiv ? \pmod 5$

Since $7 \div 5 = 1$, remainder 2, $4 + 3 \equiv 2 \pmod 5$

b) $4 - 3 \equiv ? \pmod 5$

$ 1 \equiv ? \pmod 5$

$ 1 \equiv 1 \pmod 5$

Remember that we wish to replace the question mark by a number between 0 and 4, inclusive. Thus $4 - 3 \equiv 1 \pmod 5$

c) $2 \cdot 4 \equiv ? \pmod 5$

$ 8 \equiv ? \pmod 5$

Since $8 \div 5 = 1$, remainder 3, $\quad 8 \equiv 3 \pmod 5$. Thus $2 \cdot 4 \equiv 3 \pmod 5$.

Notice that in Table 9.12 every number in the same modulo class differs by a multiple of the modulo, in this case a multiple of 7. Adding (or subtracting) a multiple of the modulo number to (or from) a given number does not change the modulo class or congruence of the given number. For example, 3, 3 + 1(7), 3 + 2(7), 3 + 3(7), ..., 3 + n(7) are all in the same modulo class, namely 3. We use this fact in the solution to Example 3.

▶ **Example 3**

Find the replacement for the question mark that makes each of the following true.

a) $3 - 15 \equiv ? \pmod 7$ **b)** $2 - 4 \equiv ? \pmod 5$ **c)** $4 - ? \equiv 6 \pmod 8$

Solution:

a) In mod 7, adding 7 or a multiple of 7 to a number results in a sum that is in the same modulo class. Thus if we add 7, 14, 21, ... to 3, the result will be a number in the same modulo class. We wish to replace 3 with an equivalent mod 7 number that is greater than 15. Adding 14 to 3 yields a sum of 17, which is greater than 15:

$$3 - 15 \equiv ? \pmod 7$$
$$(3 + 14) - 15 \equiv ? \pmod 7$$
$$17 - 15 \equiv ? \pmod 7$$
$$2 \equiv 2 \pmod 7$$

Therefore ? = 2.

b) $2 - 4 \equiv ? \pmod 5$

In mod 5 adding 5 (or a multiple of 5) to a number will result in a sum in the same modulo class. Thus we can add 5 to 2 so that $2 - 4 \equiv ? \pmod 5$ becomes $7 - 4 \equiv ? \pmod 5$:

$$7 - 4 \equiv ? \pmod 5$$
$$3 \equiv ? \pmod 5$$
$$3 \equiv 3 \pmod 5$$

Thus $2 - 4 \equiv 3 \pmod 5$. This could be demonstrated with a modulo 5 clock. The minus sign indicates counting in a counterclockwise direction.

Amalie Emmy Noether
(1882–1935) was described by Albert Einstein as the "most significant creative mathematical genius thus far produced since the higher education of women began." Because she was a woman, she was never to be granted a regular professorship in her native Germany. In 1933, with the rise of the Nazis to power, she and many other mathematicians and scientists emigrated to the United States. This brought her to Bryn Mawr College, at the age of 51, to her first full-time faculty appointment. She was to die just two years later.

Her special interest lay in abstract algebra, which includes such topics as groups, rings, and fields. She developed a number of theories that helped draw together many important mathematical concepts. Her innovations in higher algebra gained her recognition as one of the most creative abstract algebraists of modern times.

c) $4 - ? \equiv 6 \pmod{8}$

In mod 8, adding 8 (or a multiple of 8) to a number results in a sum in the same modulo class. Thus we can add 8 to 4 so that the statement becomes

$$(8 + 4) - ? \equiv 6 \pmod 8$$
$$12 - ? \equiv 6 \pmod 8$$

We can see that $12 - 6 = 6$. Therefore $? = 6$.

▶ **Example 4**

Find all replacements for the question mark that make the statements true.

a) $4 \cdot ? \equiv 3 \pmod 5$ **b)** $3 \cdot ? \equiv 0 \pmod 6$ **c)** $3 \cdot ? \equiv 2 \pmod 6$

Solution:

a) One method of determining the solution is to replace the question mark with the numbers 0–4 and then find the equivalent modulo class of the product. We use the numbers 0–4 because we are working in modulo 5.

$$4 \cdot ? \equiv 3 \pmod 5$$
$$4 \cdot 0 \equiv 0 \pmod 5$$
$$4 \cdot 1 \equiv 4 \pmod 5$$
$$\boxed{4 \cdot 2 \equiv 3 \pmod 5}$$
$$4 \cdot 3 \equiv 2 \pmod 5$$
$$4 \cdot 4 \equiv 1 \pmod 5$$

Therefore $? = 2$ since $4 \cdot 2 \equiv 3 \pmod 5$.

b) Replace the question mark with the numbers 0–5 and follow the procedure used in part (a).

$$3 \cdot ? \equiv 0 \pmod 6$$
$$\boxed{3 \cdot 0 \equiv 0 \pmod 6}$$
$$3 \cdot 1 \equiv 3 \pmod 6$$
$$\boxed{3 \cdot 2 \equiv 0 \pmod 6}$$
$$3 \cdot 3 \equiv 3 \pmod 6$$
$$\boxed{3 \cdot 4 \equiv 0 \pmod 6}$$
$$3 \cdot 5 \equiv 3 \pmod 6$$

Therefore replacing the question mark with 0, 2, or 4 results in true statements. The answers are 0, 2, and 4.

c) $3 \cdot ? \equiv 2 \pmod 6$

Examining the products in part (b) shows that there are no values that satisfy the statement. The answer is "no solution."

Modular arithmetic systems under the operation of addition are commutative groups as illustrated in Example 5.

▶ **Example 5**

Construct a mod 5 addition table and show that the mathematical system is a commutative group. Assume that the associative property holds for the given operation.

Solution: The set of elements in modulo 5 arithmetic is $\{0, 1, 2, 3, 4\}$; the binary operation is $+$:

+	0	1	2	3	4
0	0	1	2	3	4
1	1	2	3	4	0
2	2	3	4	0	1
3	3	4	0	1	2
4	4	0	1	2	3

To determine if this is a commutative group, we must see whether the five properties of a commutative group are satisfied.

1. *Closure* Since every entry in the table is a member of the set $\{0, 1, 2, 3, 4\}$, the system is closed under addition.

2. *Identity Element* An easy way to determine whether there is an identity element is to see whether there is a row in the table that is identical to the elements at the top of the table. Note that the row next to 0 is identical to the top of the table. This indicates that 0 *might be* the identity element. Now look at the column under the 0 at the top of the table. If this column is identical to the left-hand column, then 0 is the identity element. Since the column under 0 is the same as the left-hand column, 0 is the additive identity element in modulo 5 arithmetic.

Element	+	Identity	=	Element
0	+	0	=	0
1	+	0	=	1
2	+	0	=	2
3	+	0	=	3
4	+	0	=	4

3. *Inverses* Does every element have an inverse? Recall that an element plus its inverse must equal the identity element. In this example the identity element is 0. Therefore for each of the given elements 0, 1, 2, 3, and 4, we must find the element that when added to it results in a sum of zero. These elements will be the inverses.

Element	+	Inverse	=	Identity	
0	+	?	=	0	Since $0 + 0 = 0$, 0 is its own inverse
1	+	?	=	0	Since $1 + 4 = 0$, 4 is the inverse of 1.
2	+	?	=	0	Since $2 + 3 = 0$, 3 is the inverse of 2.
3	+	?	=	0	Since $3 + 2 = 0$, 2 is the inverse of 3.
4	+	?	=	0	Since $4 + 1 = 0$, 1 is the inverse of 4.

Note that each element has an inverse.

4. *Associative Property* It is given that the associative property holds. One example that illustrates the associative property is

$$(2 + 3) + 4 = 2 + (3 + 4)$$
$$0 + 4 = 2 + 2$$
$$4 = 4$$

5. *Commutative Property* Is $a + b = b + a$ for *all* elements a and b of the given set? From the table we can see that the system is commutative, since the elements in the table are symmetric about the main diagonal. We will give one example to illustrate the commutative property.

$$4 + 2 = 2 + 4 = 1$$

Since each of the five properties is satisfied, modulo 5 arithmetic under the operation of addition forms a commutative group.

Section 9.3 Exercises

In Ex. 1–6, assume today is Wednesday (day 3). Determine the day of the week it will be at the end of each of the following periods. (Assume no leap years.)

1. 18 days   **2.** 134 days   **3.** 365 days
4. 3 years   **5.** 2 years 38 days   **6.** 376 days

In Ex. 7–12, consider the 12 months as a modulo 12 system. Determine the month it will be in the specified number of months. Use the current month as your reference point.

7. 13 months   **8.** 33 months
9. 2 years 7 months   **10.** 3 years 17 months
11. 7 years   **12.** 73 months

In Ex. 13–24, determine what number each of the following is congruent to in mod 5.

13. $8 + 5$ **14.** $7 + 10$
15. $5 + 7 + 9$ **16.** $8 - 5$
17. $3 - 8$ **18.** $5 \cdot 4$
19. $4 \cdot 3$ **20.** $10 \cdot 16$
21. $6 - 9$ **22.** $5 - 7$
23. $(13 \cdot 3) - 6$ **24.** $(5 - 7) \cdot 6$

In Ex. 25–36, find the modulo class to which each of the following numbers belongs for the indicated modulo system.

25. 13, mod 6   **26.** 25, mod 8

27. 96, mod 14
28. 14, mod 7
29. 35, mod 8
30. 41, mod 9
31. 36, mod 7
32. 55, mod 4
33. −6, mod 7
34. −5, mod 4
35. −13, mod 11
36. −11, mod 13

In Ex. 37–50, find all replacements (less than the modulus) for the question mark that make each of the following true.

37. $4 + 3 \equiv ? \pmod 5$
38. $? + 9 \equiv 6 \pmod 8$
39. $3 - ? \equiv 4 \pmod 5$
40. $4 \cdot ? \equiv 2 \pmod 6$
41. $4 - ? \equiv 5 \pmod 6$
42. $3 \cdot 5 \equiv ? \pmod 7$
43. $2 \cdot ? \equiv 7 \pmod 9$
44. $3 \cdot ? \equiv 5 \pmod 7$
45. $2 \cdot ? \equiv 3 \pmod 4$
46. $3 \cdot ? \equiv 3 \pmod{12}$
47. $3 \cdot ? \equiv 2 \pmod 8$
48. $4 - 6 \equiv ? \pmod 8$
49. $5 - 7 \equiv ? \pmod 9$
50. $6 - ? \equiv 8 \pmod 9$

51. Leap years are always presidential election years. Some leap years are . . . 1984, 1988, 1992, . . .
  a) List the next five presidential election years.
  b) What will be the first election year after the year 3000?
  c) List the election years between the years 2500 and 2525.

52. A pilot is scheduled to fly for five consecutive days and rest for three consecutive days. If today is the second day of her rest shift, determine whether she will be flying
  a) 60 days from today.
  b) 90 days from today.
  c) 240 days from today.
  d) Was she flying 6 days ago?
  e) Was she flying 20 days ago?

53. A nurse's work pattern at XYZ Hospital consists of working the 7 A.M.–3 P.M. shift for three weeks and then the 3 P.M.–11 P.M. shift for two weeks.
  a) If this is the third week of the pattern, what shift will the nurse be working six weeks from now?
  b) If this is the fourth week of the pattern, what shift will the nurse be working seven weeks from now?
  c) If this is the first week of the pattern, what shift will the nurse be working eleven weeks from now?

54. A truck driver's routine is as follows: drive three days from New York to Chicago; rest one day in Chicago; drive three days from Chicago to Los Angeles; rest two days in Los Angeles; drive five days to return to New York; rest three days in New York. Then the cycle begins again.
  If the truck driver is starting his trip to Chicago today, what will he be doing
  a) 20 days from today?
  b) 60 days from today?
  c) 1 year from today?

55. a) Construct a modulo 4 addition table.
  b) Is the system closed? Explain.
  c) Is there an identity element for the system? If so, what is it?
  d) Does every element in the system have an inverse? If so, list the elements and their inverses.
  e) The associative property holds for the system. Give an example.
  f) Does the commutative property hold for the system? Give an example.
  g) Is the system a commutative group?
  h) Will every modulo system under the operation of addition be a commutative group? Explain.

56. Construct a modulo 8 addition table. Answer questions (b)–(h) in Ex. 55.

57. a) Construct a modulo 4 multiplication table.
  b) Is the system closed under the operation of multiplication?
  c) Is there an identity element in the system? If so, what is it?
  d) Does every element in the system have an inverse? Make a list showing the elements that have a multiplicative inverse, and list the inverses.
  e) The associative property holds for the system. Give an example.
  f) Does the commutative property hold for the system? Give an example.
  g) Is this mathematical system a commutative group? Explain.

58. Construct a modulo 7 multiplication table. Answer questions (b)–(g) in Ex. 57.

Problem Solving

59. A perpetual calendar is to be made by separately listing all of the individual calendars needed to cover all possible situations that may occur. How many individual calendars are needed to develop the perpetual calendar?

We have not discussed division in modular arithmetic. Can you replace the question marks with the number or numbers that make the statement true?

60. $5 \div 7 \equiv ? \pmod 9$
63. $1 \div 2 \equiv ? \pmod 5$
61. $? \div 5 \equiv 5 \pmod 9$
64. $2 \div ? \equiv 4 \pmod 6$
62. $? \div ? \equiv 1 \pmod 4$

Solve for x where k is any counting number.

65. $5k \equiv x \pmod 5$
66. $5k + 3 \equiv x \pmod 5$
67. $4k - 3 \equiv x \pmod 4$

68. One important use of modular arithmetic is coding. One type of coding circle is given in Fig. 9.7. To use this code, the person you are sending the message to must know the code key to decipher the code. The code key to this message is *j*. Can you decipher this code? (*Hint:* Subtract the code key from the code numbers.)

$$19 \quad 10 \quad 22 \quad 19 \quad 21 \quad 15 \quad 10 \quad 26 \quad 19 \quad 9 \quad 9 \quad 11$$

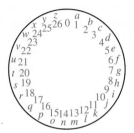

Figure 9.7

69. The procedure of *casting out nines* can be used to check arithmetic problems. The procedure is based on the fact that any whole number is congruent to the sum of its digits modulo 9. For example, the number 5783 and the sum of its digits, $5 + 7 + 8 + 3$, or 23, are both congruent to $5 \pmod 9$: $5783 \equiv 5 \pmod 9$ and $23 \equiv 5 \pmod 9$.

a) Explain why the procedure works.

b) Perform the indicated operations and check your solutions by casting out nines.

$$\begin{array}{r} 4236 \\ + 3784 \\ \hline \end{array}$$

70. Find a number divisible by 5 to which 2 is congruent to in modulo 6.

CHAPTER 9 SUMMARY

Key Terms

9.1
abelian group
associative property
binary operation
closure property
commutative group
commutative property

identity element
inverse element
mathematical system

9.2
clock arithmetic
finite mathematical system

9.3
congruent
modulo *m* system
modulus

Important facts
If the elements in a table are symmetric about the main diagonal, then the system is commutative.

	Addition	**Multiplication**
Commutative property	$a + b = b + a$	$a \cdot b = b \cdot a$
Associative property	$(a + b) + c$ $= a + (b + c)$	$(a \cdot b) \cdot c = a \cdot (b \cdot c)$

Properties of a group and a commutative group
A mathematical system is a group if the first four conditions hold and a commutative group if all five conditions hold.

1. The set of elements is *closed* under the given operation.
2. There exists an *identity element* for the set.
3. Every element in the set has an *inverse*.
4. The set of elements is *associative* under the given operation.
5. The set of elements is *commutative* under the given operation.

CHAPTER 9 REVIEW EXERCISES

9.1–9.2

1. List the parts of a mathematical system.

Determine the value of the following in clock arithmetic.

2. $8 + 9$ **3.** $4 + 9$ **4.** $5 - 9$

5. $5 + 7 + 9$ **6.** $7 - 10$ **7.** $8 - 3 + 5$

8. $3 - 4 - 5$

9. List the properties of a group, and explain what each property means.

10. What is an abelian group?

In Ex. 11–14 explain your answer.

11. Determine whether the set of integers under the operation of addition forms a group.

12. Determine whether the set of integers under the operation of multiplication forms a group.

13. Determine whether the set of rational numbers under the operation of addition forms a group.

14. Determine whether the set of rational numbers under the operation of multiplication forms a group.

In Ex. 15–17, for each of the mathematical systems, determine which of the five properties of a commutative group do not hold.

15.

:	$*$	?	$\triangle$	p
$*$	?	$\triangle$	$*$	p
?	$\triangle$	p	?	$*$
$\triangle$	$*$	?	$\triangle$	p
p	p	$*$	p	$*$

16.

$*$	a	c	b
a	a	b	c
c	b	c	a
b	c	a	b

17.

?	4	#	L	P
4	#	4	P	L
#	4	P	L	#
L	P	L	#	4
P	L	#	4	P

18. Consider the mathematical system indicated by the following table, in which the operation is associative.

$\otimes$	1	$\bigcirc$	?	$\triangle$
1	1	$\bigcirc$	?	$\triangle$
$\bigcirc$	$\bigcirc$	?	$\triangle$	1
?	?	$\triangle$	1	$\bigcirc$
$\triangle$	$\triangle$	1	$\bigcirc$	?

a) What are the elements of the set in this mathematical system?

b) What is the binary operation?

c) Is the system closed? Explain.

d) Is there an identity element for the system under the given operation?

e) Does every element in the system have an inverse? If so, give each element and its corresponding inverse.

f) Give an example to illustrate the associative property.

g) Is the system commutative? Give an example.

h) Is this mathematical system a commutative group? Explain.

9.3

In Ex. 19–27, find the modulo class to which each of the following numbers belongs for the indicated modulo system.

19. 20, mod 3   **20.** 24, mod 5

21. 19, mod 8   **22.** 59, mod 6

23. 74, mod 12   **24.** 85, mod 4

25. 37, mod 5   **26.** 83, mod 14

27. 97, mod 13

Find all replacements (less than the modulus) for the question mark that make each of the following true.

28. $5 + 8 \equiv ? \pmod 9$ **29.** $? - 2 \equiv 0 \pmod 4$

30. $4 \cdot ? \equiv 3 \pmod 6$ **31.** $4 - ? \equiv 5 \pmod 7$

32. $? \cdot 4 \equiv 0 \pmod 8$ **33.** $41 \equiv ? \pmod{12}$

34. $7 - 9 \equiv ? \pmod 8$ **35.** $? \cdot 5 \equiv 3 \pmod 6$

36. $4 \cdot ? \equiv 6 \pmod 7$ **37.** $7 \cdot ? \equiv 1 \pmod 8$

38. Construct a modulo 6 addition table. Then determine whether the modulo 6 system forms a commutative group under the operation of addition.

39. Construct a modulo 4 multiplication table. Then determine whether the modulo 4 system forms a commutative group under the operation of multiplication.

40. Toni's work pattern at the fast food restaurant is as follows: She works three evenings, then has two evenings off, then she works two evenings, and then she has three evenings off; then the pattern repeats. If today is the first day of the work pattern,

a) will Toni be working 18 days from today?

b) will Toni have the evening off for a party that is being held in 38 days?

CHAPTER TEST

1. What is a mathematical system?

2. List the requirements needed for a mathematical system to be a commutative group.

3. Is the set of whole numbers a commutative group under the operation of addition? Explain your answer completely.

4. Develop a three-hour clock arithmetic addition table.

5. Is three-hour clock arithmetic under the operation of addition a commutative group? Assume that the associative property holds. Explain your answer completely.

Determine whether the following mathematical systems are commutative groups. Explain your answer completely.

6.

*	a	b	c
a	a	b	c
b	b	b	a
c	c	a	d

7.

?	1	2	3
1	1	3	2
2	3	2	1
3	2	1	3

Determine whether the mathematical system is a commutative group. Assume that the associative property holds.

8.

W	□	△	L	Z
□	Z	□	△	L
△	□	△	L	Z
L	△	L	Z	□
Z	L	Z	□	△

Determine the modulo class to which each of the following numbers belongs for the indicated modulo system.

9. 73, mod 8

10. 47, mod 11

Find all replacements for the question mark, less than the modulus, that make each of the following true.

11. $5 + 8 \equiv ? \pmod 7$

12. $? - 8 \equiv 3 \pmod 5$

13. $2 - ? \equiv 6 \pmod 9$

14. $3 \cdot 5 \equiv ? \pmod 6$

15. $2 \cdot ? \equiv 4 \pmod 6$

16. $79 \equiv ? \pmod 7$

17. a) Construct a modulo 3 multiplication table.
    b) Is this mathematical system a commutative group? Explain your answer completely.

CONSUMER
MATHEMATICS

Managing the way you spend money, however much or little of it you have, takes a great deal of thought and planning. First, you need to know what your needs are on a daily basis. How much will you spend on transportation to get to school? How much on food? Do you have enough to go see a movie tonight? On a monthly and yearly basis, the questions are broader: What kind of house or apartment can you afford? Will you be able to save enough to buy a car, or go on a vacation? How much money do you need to keep on hand for unexpected emergencies? And then there are questions about your long-term goals. What would you like to accomplish financially 3, 5, and 10 years in the future? Will you be able to pay your student loans or other debts? Do you want to own a home? Will you need to save money for raising children?

The financial decisions you make on a daily basis all contribute to your longer range plans: As the saying goes,

Due to the use of telecommunications and computers, trading in international stock markets now continues 'round the clock. When the trading day ends in London, it begins again in Tokyo; when it ends there, it begins again in Chicago.

"Take care of the pennies and the dollars will take care of themselves." For example, say you ride your bicycle every day instead of taking the bus. Over the course of a year, if the bus fare is 75 cents, you may save as much as $180, which you could use to buy something immediately, or you could invest your savings and earn even more money.

Many consumers shop carefully, using coupons and watching for sales. Sometimes, however, what is on "sale" isn't necessarily a bargain. For example, suppose that your local department store is having a sale on winter coats. You don't really need a new coat, but you see a big rack marked 25% off. Curious, you try one on. Its original price reads $68, but the sale brings the price of the coat down to $51. If you buy the coat, it may seem that you are saving $17. But since you were not planning to buy a coat, you really are not saving $17, you're spending $51!

The national economy reflects the sum of all of the consumer decisions made. One variable studied by economists to determine the health of the national economy is a measure called the Index of Consumer Confidence. When consumers are confident in their financial prospects, they purchase houses, cars, and retail goods, leading economists to predict economic growth for the nation.

In order to make the best use of your money, you need to be able to make intelligent judgments. A solid understanding of consumer mathematics — including loans, interest rates, and mortgages — can help you plan your financial goals and how to achieve them.

Every year, an enormous number of magazines devoted to consumerism and financial planning are published. Most of these resources are available free of charge at your local library.

10.1 Percent

The study of mathematics is crucial to understanding how to make better financial decisions. A basic topic necessary for understanding the material in this chapter is percent. This section will give you a better understanding of the meaning of percent and its use in real-life situations.

The word "percent" comes from the Latin *per centum*, meaning "per hundred." A **percent** is simply a ratio of some number to 100. Thus $\frac{15}{100} = 15\%$ and $x/100 = x\%$.

Percents are useful in making comparisons. Consider Maureen and Howard, who both invested in the stock market and want to determine which one made the better investment. One way to compare their investments is for each person to write a ratio of the amount of money they made, called the **return on investment**, to the amount they invested. Then they can convert these ratios to percents and compare the percents. These percents are called the **percent of return on the investment**. Suppose that Maureen invested $500 and made $300 and Howard invested $400 and made $200. Who made the better investment? The fact that Maureen made more money does not necessarily mean that Maureen's was the better investment, since they invested different amounts. We will now find the percent of return for each investment by (a) writing for each person a ratio of the return on investment to amount invested, (b) writing these ratios with a denominator of 100, and (c) expressing the ratios as percents.

Maureen

$$\frac{\text{Return on investment}}{\text{Investment}} = \frac{300}{500} = \frac{300 \div 5}{500 \div 5} = \frac{60}{100} = 60\%$$

Howard

$$\frac{\text{Return on investment}}{\text{Investment}} = \frac{200}{400} = \frac{200 \div 4}{400 \div 4} = \frac{50}{100} = 50\%$$

By changing the results of both investments to percent we have a common standard for comparison. Since Maureen made 60% or $60 per $100 dollars invested and Howard made 50% or $50 per $100 invested, Maureen made the better investment.

Another procedure to change a fraction to a percent follows.

> **Procedure to Change a Fraction to a Percent**
> **1.** Divide the numerator by the denominator.
> **2.** Multiply the quotient by 100 (which has the effect of moving the decimal point two places to the right).
> **3.** Add a percent sign.

IVORY SOAP [ICON] 99⁴⁴⁄₁₀₀% PURE
IT FLOATS

99 and 44/100% Pure: *It is a slogan that has been around since 1881, ever since Procter & Gamble decided to market its laundry soap, Ivory Soap, as a cosmetic soap. "Pure what?" you may ask. Soap, as opposed to detergent, is a mixture of fatty acids and alkali, so the makeup of Ivory Soap is 99.44% soap and only 0.56% impurities. A semanticist, who studies language and its meaning, might say that's pure baloney. That which is described as pure, by definition, cannot be a mixture at all, and it is either 100% pure or not pure at all.*

▶ **Example 1**

Change $\frac{3}{4}$ to percent.

Solution: Follow the steps listed in the procedure box.
1. $3 \div 4 = 0.75$ **2.** $0.75 \times 100 = 75$ **3.** 75%
Thus $\frac{3}{4} = 75\%$.

Procedure to Change a Decimal Number to a Percent
Multiply the decimal by 100 and add a percent sign.

The procedure for changing a decimal number to a percent is equivalent to moving the decimal point two places to the right and adding a percent sign.

▶ **Example 2**

Change 0.235 to percent.

Solution: $0.235 = (0.235 \times 100)\% = 23.5\%$

Procedure to Change a Percent to a Decimal Number
1. Divide the number by 100.
2. Remove the percent sign.

The procedure for changing a percent to a decimal number is equivalent to moving the decimal point two places to the left and removing the percent sign.

▶ **Example 3**

a) Change 35% to a decimal number.   **b)** Change $\frac{1}{2}\%$ to a decimal number.

Solution:

a) $35\% = \frac{35}{100} = 0.35$
Thus $35\% = 0.35$.

b) First change $\frac{1}{2}$ to 0.5:

$$\frac{1}{2}\% = 0.5\% = \frac{0.5}{100} = 0.005$$

Thus $\frac{1}{2}\% = 0.005$.

▶ **Example 4**

Of 673 people surveyed, 418 said they favored the school budget. Find the percent in favor of the school budget.

Solution: The percent in favor is the ratio of the number in favor to the total number surveyed. To find the percent in favor, divide the number in favor by the total. Move the decimal point two places to the right and add a percent sign:

$$\text{Percent in favor} = \frac{418}{673} = 0.6211$$

$$= 62.11\% \text{ or } 62.1\% \quad \text{(to the nearest tenth of a percent).}$$

All answers in this section will be given to the nearest tenth of a percent. When rounding answers off to the nearest tenth of a percent, we carry the division out to four places after the decimal point and then round to the nearest thousandths position.

▶ **Example 5**

At Metropolitan Community College the grades of students completing Mathematics 100 were as follows: 185 received A, 348 received B, 785 received C, 216 received D, and 96 received F. Find the percent that received each grade.

Solution: The total completing the course is

$$185 + 348 + 785 + 216 + 96 = 1630.$$

$$\text{Percent receiving A} = \frac{185}{1630} = 0.1134 = 11.3\%$$

$$\text{Percent receiving B} = \frac{348}{1630} = 0.2134 = 21.3\%$$

$$\text{Percent receiving C} = \frac{785}{1630} = 0.4815 = 48.2\%$$

$$\text{Percent receiving D} = \frac{216}{1630} = 0.1325 = 13.3\%$$

$$\text{Percent receiving F} = \frac{96}{1630} = 0.0588 = \frac{5.9\%}{100.0\%}$$

Note that every possible outcome was considered, so the sum of the percents is 100%. In some problems there may be a slight round-off error.

The percent increase or decrease, or percent change, over a period of time is found by the following formula:

$$\textbf{Percent Change} = \frac{\left(\begin{array}{c}\text{Amount in}\\\text{latest period}\end{array}\right) - \left(\begin{array}{c}\text{Amount in}\\\text{previous period}\end{array}\right)}{\text{Amount in previous period}} \times 100$$

If the latest amount is greater than the previous amount, the answer will be positive and will indicate a percent increase. If the latest amount is smaller than the previous amount, the answer will be negative and will indicate a percent decrease.

▶ **Example 6**

In 1992 Cook County Ford sold 937 new cars. In 1991 it sold 812 new cars. Find the percent increase or decrease in new car sales from 1991 to 1992.

Solution: The previous period is 1991, and the latest period is 1992.

$$\text{Percent change} = \frac{937 - 812}{812} \times 100$$

$$= \frac{125}{812} \times 100$$

$$= 0.1539 \times 100$$

$$= 15.4\% \text{ increase}$$

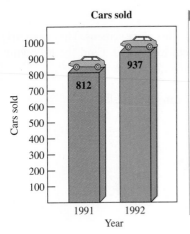

Cars sold

▶ **Example 7**

In 1991 Ms. Gadziala received a salary of $20,000 and a commission of $8000. In 1992 she received a salary of $22,000 and a commission of $15,000. Find the following:

a) The change in Ms. Gadziala's income from 1991 to 1992.
b) The percent change in her income.

Solution:

a) Ms. Gadziala's total income in 1991 was her salary plus commission, $20,000 + $8,000 = $28,000. Her total income in 1992 was $22,000 + $15,000 = $37,000. The change in her total income from 1991 to 1992 was $37,000 − $28,000 = $9,000.

b) $\text{Percent change} = \dfrac{\$37,000 - \$28,000}{\$28,000} \times 100$

$$= \frac{\$9,000}{\$28,000} \times 100$$

$$= 0.321 \times 100$$

$$= 32.1\%$$

Ms. Gadziala's income rose 32.1%.

A similar formula is used to calculate percent markup or markdown on cost. A positive answer indicates a markup and a negative answer indicates a markdown.

$$\textbf{Percent Markup on Cost} = \frac{\text{Selling price} - \text{Dealer's cost}}{\text{Dealer's cost}} \times 100$$

▶ **Example 8**

Modern Hardware stores pay $48.76 for glass fireplace screens. They regularly sell them for $79.88. At a sale they sell them for $69.99. Find the following.
a) The percent markup on the regular price
b) The percent markup on the sale price
c) The percent decrease of the sale price from the regular price

Solution:
a) Percent markup on the regular price

$$= \frac{\$79.88 - \$48.76}{\$48.76} \times 100$$

$$= 0.6382 \times 100$$

$$= 63.8\%$$

b) Percent markup on the sale price

$$= \frac{\$69.99 - \$48.76}{\$48.76} \times 100$$

$$= 0.4353 \times 100$$

$$= 43.5\%$$

c) Percent decrease of the sale price from the regular price

$$= \frac{\$69.99 - \$79.88}{\$79.88} \times 100$$

$$= -0.1238 \times 100$$

$$= -12.4\%$$

The sale price is 12.4% lower than the regular price.

In daily life we may need to know how to solve any one of the following three types of problems involving percent.

1. What is a 15% tip on a restaurant bill of $24.66? The problem can be stated as

15% of $24.66 is what number?

2. If Claryce made a sale of $500 and received a commission of $25, what percent of the sale is the commission? The problem can be stated as

What percent of $500 is $25?

3. If the price of a jacket was reduced by 25% or $12.50, what was the original price of the jacket? The problem can be stated as

25% of what number is $12.50?

To answer these questions we will write each problem as an equation. The word "is" means "is equal to," $=$. In each problem we will represent the unknown quantity with the letter x. Therefore the preceding problems can be represented as

1. 15% of $24.66 = x$ **2.** $x\%$ of $500 = \$25$ **3.** 25% of $x = \$12.50$

The word "of" in such problems indicates multiplication. To solve each problem change the percent to a decimal number, and express the problem as an equation; then solve the equation for the variable x. The solutions follow.

1. 15% of $24.66 = x$

$0.15(24.66) = x$ (15% is written as 0.15 in decimal form.)

$3.699 = x$

Since 15% of 24.66 is 3.699, the tip would be $3.70.

2. $x\%$ of $500 = 25$

$(0.01x)500 = 25$ ($x\%$ is written as $0.01x$ in decimal form.)

$5x = 25$

$$\frac{5x}{5} = \frac{25}{5}$$

$x = 5$

Since 5% of $500 is $25, the commission is 5% of the sale.

3. 25% of $x = 12.50$

$0.25(x) = 12.50$

$$\frac{0.25x}{0.25} = \frac{12.50}{0.25}$$

$x = 50$

Since 25% of 50 is 12.50, the original price of the jacket was $50.00.

▶ **Example 9**

a) A sofa was purchased on an installment plan for $875. The down payment was 20% of the purchase price. How much was the down payment?

b) Nineteen, or 95%, of the students in a first-grade class have a pet. How many students are in the class?

c) Lalo scored 48 points on a 60-point test. What percent did he get correct?

Solution:

a) We want to find the amount of the down payment. Let x = the down payment.

$$\text{Down payment} = 20\% \text{ of purchase price}$$
$$x = 20\% \text{ of } 875$$
$$x = 0.20(875)$$
$$x = 175$$

The down payment is $175.

b) We want to find the number of students in the first-grade class. We know that 19 students represent 95% of the class. Let x = the number of students in the first-grade class.

$$95\% \text{ of } x = 19$$
$$0.95x = 19$$
$$\frac{0.95x}{0.95} = \frac{19}{0.95}$$
$$x = 20$$

The number of students in the class is 20.

c) We want to determine the percent of answers that Lalo got correct. We need to determine what percent of 60 is 48. Let x = the percent that Lalo got correct.

$$x\% \text{ of } 60 = 48$$
$$0.01x(60) = 48$$
$$0.60x = 48$$
$$x = \frac{48}{0.60} \quad \text{or} \quad 80$$

Lalo scored 80% on the test.

Section 10.1 Exercises

Change the following numbers to percent. Express your answer to the nearest tenth of a percent.

1. $\dfrac{1}{4}$

2. $\dfrac{3}{8}$

3. $\dfrac{5}{12}$

4. $\dfrac{5}{8}$

5. 0.4568

6. 0.006345

7. 13.678

8. 2.98

9. 342.67

Change the following percents to decimal numbers.

10. 23%

11. 25.4%

12. 6.75%

13. 0.004%

14. 125.7%

15. 7.3468%

16. $\dfrac{1}{5}\%$

17. $\dfrac{3}{4}\%$

18. $\dfrac{5}{8}\%$

19. A family estimates that $7000 of its total income of $42,700 is used to purchase food. What percent of the family's income is spent on food?

20. Of approximately 53.03 million employed American women, 5.9% are holding more than one job. Find the number of employed American women that are holding more than one job.

21. In 1990 the farmers in Nebraska produced approximately 1,007.5 million bushels of grain. Of this total approximately 0.7 million bushels were barley, 851 million bushels were corn, and 8.6 million bushels were oats. What percent of grain produced was
a) barley?　　**b)** corn?　　**c)** oats?

The circle graph below illustrates sales of athletic and sport footwear in 1989.

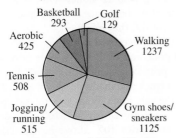

Athletic and sport footwear sales–1989 (in millions of dollars)

Basketball 293, Golf 129, Aerobic 425, Walking 1237, Tennis 508, Jogging/running 515, Gym shoes/sneakers 1125

Total sales = $4232 million

Find the percent of athletic and sport footwear sales for
22. walking　　　　**23.** aerobics

The circle graph below illustrates the sales of personal copiers in the United States in 1991.

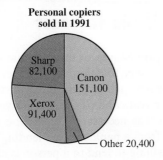

Personal copiers sold in 1991

Sharp 82,100, Canon 151,100, Xerox 91,400, Other 20,400

Find the percent of sales of personal copiers for
24. Xerox　　　　**25.** Canon

26. The population of the United States rose from approximately 226.5 million in 1980 to approximately 248.7 million in 1990. Find the percent increase from 1980 to 1990.

27. The consumer price index (CPI) in 1982–1984 was 100, in 1987 it was 113.6, and in 1990 it was 130.7. Find the following:
a) The percent increase in the CPI from 1982–1984 to 1987
b) The percent increase from 1987 to 1990

28. The graph below compares the median family income in the United States in 1950 and in 1990. Find
a) the increase in personal income from 1950 to 1990
b) the percent increase in personal income for the same period

Median family income in 1990 dollars

$35,353, $16,557

Major supermarkets are leaving the inner cities in record numbers, as illustrated by the following graph.

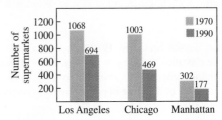

Number of supermarkets: Los Angeles 1068/694, Chicago 1003/469, Manhattan 302/177. (1970/1990)

Find (a) the decrease in the number of supermarkets and (b) the percent decrease in the number of supermarkets from 1970 to 1990 for these cities:
29. Manhattan　　**30.** Chicago　　**31.** Los Angeles

32. The maximum monthly social security benefits for 1985 and 1990 are shown in the following graph. Find (a) the dollar increase and (b) the percent increase in the maximum social security benefit from 1985 to 1990.

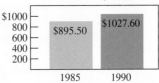

Maximum monthly Social Security benefits

$895.50 (1985), $1027.60 (1990)

Write each statement as an equation and solve.

33. What percent of 70 is 14?

34. Twenty-four is 15% of what number?

35. What is 17% of 400?

36. What is 3% of 9?

37. Seven is 20% of what number?

38. What percent of 292 is 73?

In Ex. 39–44, write the problem as an equation, then solve.

39. What is the sales tax on a dinner that cost $32.50 if the sales tax rate is 5%?

40. If the number of employees at the Royal Thai restaurant was increased by 20%, or six, what was the original number of employees?

41. The Fastlock Company is hiring 57 new employees, which increases its staff by 30%. What was the original number of employees?

42. Eighteen students received a grade of A on the third test, which is 150% of the students who received a grade of A on the second test. How many students received a grade of A on the second test?

43. In a survey of 400 people, 17% prefer blue cheese dressing on their salad. How many people in the group prefer blue cheese dressing?

44. Mr. Brown's present salary is $32,000. He is getting an increase of 8% in his salary next year. What will his new salary be?

45. In the United States in 1980 there were 29.3 million people classified as living in poverty. In 1989 there were 31.5 million classified as living in poverty. Find the percent increase or decrease of people living in poverty from 1980 to 1989.

46. A Kirby vacuum cleaner dealership sold 430 units in 1991 and 407 units in 1992. Find the percent increase or decrease in units sold.

47. If the Kirby dealership in Ex. 46 pays $320 for each unit and sells it for $699, what is the percent markup?

48. The regular price of a Zenith color TV is $539.62. During a sale, Hill TV is selling it for $439. Find the percent decrease in the price of this TV.

49. The Tadoseans had eight grandchildren in 1987. In 1992 they had 13 grandchildren. Find the percent increase in the number of grandchildren from 1987 to 1992.

50. The number of deaths in the United States due to AIDS was 19,559 in 1988, 25,591 in 1989, and 26,619 in 1990.
  a) Find the percent increase from 1988 to 1989.
  b) Find the percent increase from 1988 to 1990.

51. The cost of a fish dinner to the owner of the Golden Wharf restaurant is $6.95. The fish dinner is sold for $9.75. Find the percent markup.

52. A football team played 12 games. They lost one game.
  a) What percent of the games did they lose?
  b) What percent did they win?

53. A furniture store advertised a table at a 15% discount. The original price was $115, and the sale price was $100. Was the sale price consistent with the ad? Explain.

54. Charlie sold a truck and made a profit of $675. His profit was 18% of the sale price. What was the sale price?

55. Quincy purchased a used car for $1000. He decided to sell the car at a price 10% above his purchase price. Quincy could not sell the car so he reduced his asking price by 10%. If he sells the car at the reduced price will he have a profit, a loss, or break even? Explain how you arrived at your answer.

10.2 Promissory Notes and Simple Interest

Often consumers want to buy clothing, appliances, or furniture but do not have the cash to do so. It then becomes necessary to determine whether the cost of borrowing the money is worth the convenience or pleasure of having the item today versus waiting several months or years and paying cash. If you are faced with this choice, you may decide that it is necessary to have the item today, and may choose to borrow the money from a bank or other lending institution. The amount of money that a bank is willing to lend to a person is called **credit**, or the **principal of the loan**.

The amount of credit and the interest rate that you may obtain depend on the assurance you can give the lender that you will be able to repay the loan. Your credit is determined by your business reputation for honesty, by your earning power, and by what you can pledge as security to cover the loan. **Security** is anything of value pledged by the borrower that the lender may sell or keep if the borrower does not repay the loan. Acceptable security may be a business, a mortgage on a property, the title to an automobile, savings accounts, or stocks or bonds. The more marketable the security, the easier it is to obtain the loan, and in some cases marketability may help in getting a lower interest rate.

Bankers sometimes grant loans without security, but they require the signature of one or more other persons, called **comakers** (or **cosigners**), who guarantee that the loan will be repaid. For either of the two types of loans, the secured loan or the comaker loan, the borrower (and comaker if there is one) must sign an agreement called a **promissory note**. This document states the terms and conditions of the loan.

The most common way for individuals to borrow money is through the installment loan or the credit card. (Installment loans and credit cards will be discussed in Section 10.4.)

The concept of simple interest is essential in the understanding of notes and installment buying. **Interest** is the money the borrower pays for the use of the lender's money. One type of interest is called simple interest. **Simple interest** is based on the entire amount of the loan for the total period of the loan. The formula used to find simple interest follows.

Simple Interest Formula

$$\text{Interest} = \text{principal} \times \text{rate} \times \text{time}$$
$$i = prt$$

In the simple interest formula the **principal**, p, is the amount of money lent, the **rate**, r, is the rate of interest expressed as a percent, and the **time**, t, is the number of days, months, or years for which the money will be lent. Time is expressed in the same period as the rate. For example, if the rate is 2% per month, then the time must be expressed in months. Typically rate means the annual rate unless otherwise stated. Principal and interest are expressed in dollars in the United States.

Ordinary Interest

The most common type of simple interest is called **ordinary interest**. When computing ordinary interest, one considers each month to have 30 days and a year to have 12 months or 360 days. On the due date of a **simple interest note** the borrower must repay the principal plus the interest. (*Note*: Simple interest will mean ordinary interest unless stated otherwise.)

▶ **Example 1**

Find the simple interest for a three-month loan of $640 at 6% per year.

Solution: Use the formula $i = prt$. We know that $p = \$640$, $r = 6\%$ (or converted to a decimal, 0.06), and t in years $= \frac{3}{12} = \frac{1}{4} = 0.25$. Substitute the appropriate values in the formula.

$$i = p \times r \times t$$
$$i = \$640 \times 0.06 \times 0.25$$
$$= \$9.60$$

The simple interest on $640 at 6% for three months is $9.60.

▶ **Example 2**

Bonnie borrowed $3000 from a bank for two months. The rate of interest charged by the bank is $0.033\overline{3}\%$ per day.

a) Find the simple interest on the loan.

b) Find the amount (principal + interest) that Bonnie will pay the bank at the end of the two months.

Solution:

a) To find the interest on the loan, we use the formula $i = prt$. Since the rate is given in terms of days the time must be expressed in terms of days. Two months = 60 days. Substituting yields

$$i = p \times r \times t$$
$$= \$3000 \times 0.000333\overline{3} \times 60$$
$$= \$60.00$$

b) The amount to be repaid is equal to the principal plus interest:

$$A = p + i$$
$$= \$3000 + \$60.00$$
$$= \$3060.00$$

▶ **Example 3**

Jean lent her friend $300. Six months later her friend repaid the original $300 plus $30.00 interest. What is the annual rate of interest Jean received?

Solution: Using the formula $i = prt$, we get

$$\$30 = \$300 \times r \times 0.50$$
$$30 = 150r$$
$$\frac{30}{150} = r$$
$$0.2 = r$$

The annual rate of interest paid is 20%.

In Examples 1, 2, and 3 we illustrated a simple interest note, where the interest and principal are paid on the due date of the note. There is another type of loan, the **discount note**, where the interest is paid at the time you receive the loan. The interest charged in advance is called the **bank discount**. Example 4 illustrates a discount note.

▶ **Example 4**

Heather borrowed $500 on a 10% discount note for a period of three months. Find the following:
a) The interest she must pay to the bank on the date she receives the loan
b) The net amount of money she receives from the bank
c) The actual rate of interest for the loan

Solution:
a) To find the interest, use the simple interest formula.

$$i = prt$$

$$= \$500(0.10)\left(\frac{3}{12}\right)$$

$$= \$12.50$$

b) Since Heather must pay $12.50 interest when she first receives the loan, the net amount she receives is $500 − $12.50 or $487.50.
c) We can calculate the actual rate of interest charged by substituting the net amount she received and the actual interest paid into the simple interest formula.

$$i = prt$$

$$\$12.50 = \$487.50 \times r \times \left(\frac{3}{12}\right)$$

$$12.50 = 121.875 \times r$$

$$\frac{12.50}{121.875} = r$$

$$0.1026 \approx r$$

Thus the actual rate of interest is about 10.3% rather than the quoted 10%.

The United States Rule

A note has a date of maturity, at which time the principal and interest are due. It is possible to make payments on a note before the date of maturity. A Supreme Court decision enacted the method by which these payments are credited. The procedure is called the **United States rule**.

The United States rule states that if a partial payment is made on the loan, interest is computed on the principal from the first day of the loan until the date of the partial payment. The partial payment is used to pay the interest first; then the rest of the payment is used to reduce the principal. The next time a partial payment is made, interest is calculated on the unpaid principal from

the date of the previous date of payment. Again the payment goes first to pay the interest, with the rest of the payment used to reduce the principal. An individual can make as many partial payments as he or she wishes; the procedure is repeated for each payment. The balance due on the date of maturity is found by computing interest due since the last partial payment and adding this interest to the unpaid principal.

The **Banker's rule** is used to calculate simple interest when applying the United States rule. In the Banker's rule a year is considered to be 360 days, and any fractional part of a year is the exact number of days of the loan.

To determine the exact number of days in a period, we can use Table 10.1. Example 5 illustrates how to use the table.

▶ **Example 5**

Using Table 10.1, find (a) the due date of a loan made on March 15 for 180 days and (b) the number of days from April 18 to July 31.

Solution:

a) To determine the due date of the note, do the following: In Table 10.1 find 15 in the left column (headed Day of Month), then move three columns to the right (heading at the top of the column is March) and find the number 74 (circled in black). This says that March 15 is the 74th day of the year. Add 74 and 180, since the note will be due 180 days after March 15:

$$74 + 180 = 254$$

Thus the due date of the note is the 254th day of the year. Find 254 (circled in blue) in Table 10.1 in the column headed September. The number in the same row as 254 in the left-hand column (headed Day of Month) is 11. Thus the due date of the note is September 11.

b) To determine the number of days from April 18 to July 31, use Table 10.1 to find that April 18 is the 108th day of the year and July 31 is the 212th day of the year. Then find the difference: $212 - 108 = 104$. Thus the number of days from April 18 to July 31 is 104 days.

▶ **Example 6**

Find the simple interest on $300 at 12% for the period March 3 to May 3 using the Banker's rule.

Solution: The exact number of days from March 3 to May 3 is 61. Thus the period of time in years is 61/360. Substituting in the simple interest formula,

$$i = prt$$

$$= \$300 \times 0.12 \times \frac{61}{360}$$

$$= \$6.10$$

The interest due is $6.10.

Table 10.1

Day of Month	Days in Each Month											
	31	28	31	30	31	30	31	31	30	31	30	31
	Jan	Feb	Mar	Apr	May	Jun	Jul	Aug	Sep	Oct	Nov	Dec
Day 1	1	32	60	91	121	152	182	213	244	274	305	335
Day 2	2	33	61	92	122	153	183	214	245	275	306	336
Day 3	3	34	62	93	123	154	184	215	246	276	307	337
Day 4	4	35	63	94	124	155	185	216	247	277	308	338
Day 5	5	36	64	95	125	156	186	217	248	278	309	339
Day 6	6	37	65	96	126	157	187	218	249	279	310	340
Day 7	7	38	66	97	127	158	188	219	250	280	311	341
Day 8	8	39	67	98	128	159	189	220	251	281	312	342
Day 9	9	40	68	99	129	160	190	221	252	282	313	343
Day 10	10	41	69	100	130	161	191	222	253	283	314	344
Day 11	11	42	70	101	131	162	192	223	(254)	284	315	345
Day 12	12	43	71	102	132	163	193	224	255	285	316	346
Day 13	13	44	72	103	133	164	194	225	256	286	317	347
Day 14	14	45	73	104	134	165	195	226	257	287	318	348
Day 15	15	46	(74)	105	135	166	196	227	258	288	319	349
Day 16	16	47	75	106	136	167	197	228	259	289	320	350
Day 17	17	48	76	107	137	168	198	229	260	290	321	351
Day 18	18	49	77	108	138	169	199	230	261	291	322	352
Day 19	19	50	78	109	139	170	200	231	262	292	323	353
Day 20	20	51	79	110	140	171	201	232	263	293	324	354
Day 21	21	52	80	111	141	172	202	233	264	294	325	355
Day 22	22	53	81	112	142	173	203	234	265	295	326	356
Day 23	23	54	82	113	143	174	204	235	266	296	327	357
Day 24	24	55	83	114	144	175	205	236	267	297	328	358
Day 25	25	56	84	115	145	176	206	237	268	298	329	359
Day 26	26	57	85	116	146	177	207	238	269	299	330	360
Day 27	27	58	86	117	147	178	208	239	270	300	331	361
Day 28	28	59	87	118	148	179	209	240	271	301	332	362
Day 29	29		88	119	149	180	210	241	272	302	333	363
Day 30	30		89	120	150	181	211	242	273	303	334	364
Day 31	31		90		151		212	243		304		365

Add 1 day for leap year if February 29 falls between the two dates.

The next example illustrates how a partial payment is credited using the United States rule. Making partial payments reduces the amount of interest paid and the cost of the loan.

▶ **Example 7**

On March 3, Benny wrote a 90-day note for $475.00 at an interest rate of 15.5%. On April 1 he made a partial payment of $200.00, and on May 10 he made a partial payment of $150.00.

a) Find the interest and the amount credited to the principal on April 1.

b) Find the interest and the amount credited to the principal on May 10.

c) Find the amount that Benny must pay on the due date.

Solution:

a) The number of days from March 3 to April 1, the date of the first partial payment, is 29. Use the Banker's rule to compute the simple interest on $475 at 15.5% for 29 days. Substituting in the simple interest formula gives

$$i = \$475 \times 0.155 \times \frac{29}{360}$$

$$= \$5.93$$

The interest of $5.93 that is due on April 1 is deducted from the payment of $200.00. The remaining amount of the payment is then applied to the principal:

$200.00	Partial payment
− 5.93	Interest
$194.07	Amount to be applied to principal

$475.00	Original principal
− 194.07	Amount to be applied to principal
$280.93	New principal after April 1 payment

b) Using the Banker's rule, calculate the interest on the unpaid principal for the period from April 1 to May 10. The number of days from April 1 to May 10 is 39.

$$i = \$280.93 \times 0.155 \times \frac{39}{360}$$

$$= \$4.72$$

Now subtract the interest of $4.72 from the partial payment of $150.00 to obtain $145.28. Thus $145.28 is applied against the unpaid principal.

$280.93	Unpaid principal
−$145.28	Amount to be applied to principal
$135.65	Balance due on principal on date of maturity of loan

c) The date of maturity of the loan is June 1. The number of days from May 10 to June 1 is 22. The interest is computed on the remaining balance of $135.65 using the simple interest formula:

$$i = \$135.65 \times 0.155 \times \frac{22}{360}$$

$$= \$1.28$$

Therefore the balance due on the maturity date of the loan is the sum of the principal and the interest, $135.65 plus $1.28, or $136.93. *Note:* The sum of the days in the three calculations, $29 + 39 + 22$, equals the total number of days of the note, 90.

Section 10.2 Exercises

Find the simple interest for the following problems. (The rate is an annual rate unless otherwise noted.)
1. $p = \$30$, $r = 8\%$, $t = 1$ year
2. $p = \$375$, $r = 4.5\%$, $t = 3$ years
3. $p = \$650$, $r = 14\%$, $t = 60$ days
4. $p = \$225.75$, $r = 12\frac{1}{2}\%$, $t = 6$ months
5. $p = \$587$, $r = .045\%$ daily rate, $t = 2$ months
6. $p = \$6742.75$, $r = 6.05\%$, $t = 90$ days
7. $p = \$4,372.80$, $r = 5.25\%$, $t = 60$ days
8. $p = \$9,752$, $r = 1\frac{1}{4}\%$ per month, $t = 4$ months
9. $p = \$10,750$, $r = 1\frac{1}{2}\%$ per month, $t = 9$ months
10. $p = \$17,852$, $r = 1\frac{3}{4}\%$ per month, $t = 1$ year

Use the simple interest formula to find the missing value.
11. $p = ?$, $r = 6\%$, $t = 60$ days, $i = \$6.00$
12. $p = \$600$, $r = ?$, $t = 3$ months, $i = \$12.00$
13. $p = \$600$, $r = 6\%$, $t = ?$, $i = \$54.00$
14. $p = ?$, $r = 6.5\%$, $t = 2$ years, $i = \$124.50$
15. $p = \$1650.00$, $r = ?$, $t = 6.5$ years, $i = \$345.57$
16. $p = ?$, $r = 10\%$, $t = 4$ months, $i = \$9.50$
17. $p = \$675$, $r = ?$, $t = 4$ months, $i = \$29.25$
18. $p = \$980$, $r = ?$, $t = 9$ months, $i = \$84.53$

19. Ward borrowed $1500 from his bank for 60 days at an interest rate of 9%. He gave the bank stock certificates as security.
   a) How much did he pay for the use of the money?
   b) What is the amount he paid to the bank on the date of maturity?
20. Nita borrowed $1500 from the bank for six months. Her friend Ms. Harris was comaker of Nita's promissory note. The bank collected $12\frac{1}{2}\%$ interest on the date of maturity.
   a) How much did she pay for the use of the money?

   b) Find the amount she repaid to the bank on the due date of the note.
21. Kwame borrowed $2500 for five months from his bank, using U.S. government bonds as security. The bank discounted the loan at 13%.
   a) How much interest did Kwame pay the bank for the use of its money?
   b) How much did he receive from the bank?
   c) What was the actual rate of interest he paid?
22. Carol borrowed $3650 from the bank for eight months. The bank discounted the loan at 10%.
   a) How much interest did Carol pay the bank for the use of its money?
   b) How much did she receive from the bank?
   c) What was the actual rate of interest she paid?
23. Enrico wants to borrow $350 for six months from his bank, using his savings account as security. The bank's policy is that the maximum amount that he can borrow is 80% of the amount in his savings account. The interest rate is 2% higher than the interest rate being paid on the savings account. The current rate is $5\frac{1}{4}\%$ on the savings account.
   a) How much money must he have in his account in order to borrow $350?
   b) What is the rate of interest the bank will charge for the loan?
   c) Find the amount Enrico must repay in six months.

Find the exact time from the first date to the second date.
24. April 4 to September 23
25. May 19 to October 4
26. June 19 to November 5
27. March 29 to August 12

28. February 15 to June 15

29. January 17 to May 30 (leap year)

Determine the due date, using the exact time.

30. 90 days after May 15

31. 120 days after June 8

32. 120 days after April 15

33. 60 days after September 1

34. 180 days after August 15

In Ex. 35–41, a partial payment is made on the date indicated. Use the United States rule to determine the balance due on the note at the date of maturity.

	Principal	Rate	Effective Date	Maturity Date	Partial Payment(s) Amount	Date
35.	$1500	14%	July 1	Oct. 15	$300	Aug. 1
36.	$2400	13%	April 15	July 1	$500	May 1
37.	$3000	14.5%	May 1	Nov. 1	$1,000	June 1
38.	$7500	12%	April 15	Oct. 1	$1,000	Aug. 1
39.	$1000	12.5%	Jan. 1	Feb. 15	$300	Jan. 15
40.	$1800	15%	Aug. 1	Nov. 1	$500	Sept. 1
					$500	Oct. 1
41.	$2000	16%	Nov. 15	Jan. 1	$800	Dec. 1
					$800	Dec. 15

42. The Zwick Balloon Company signed a $2500 note with interest at $10\frac{1}{2}$% for 180 days on March 1. The company made payments of $750 on May 1 and $835 on July 1. How much will the company owe on the date of maturity?

43. The Sweet Tooth Restaurant borrowed $3000 on a note dated May 15 with interest of 11%. The maturity date of the loan is September 1. They made partial payments of $875 on June 15 and $940 on August 1. Find the amount due on the maturity date of the loan.

44. The U.S. government borrows money by selling Treasury bills. A Treasury bill is a discounted note that has a time period of less than a year. Clint purchases a 26-week, $10,000 U.S. Treasury bill at an 8% discount.

a) How much does Clint pay for the T-bill?

b) How much will the bill be worth after 26 weeks?

c) How much interest does Clint earn on the investment?

45. Mr. Harrison borrowed $600 for three months. The banker said he must repay the loan at the rate of $200 per month plus interest. The bank was charging a rate of 2% above the "prime interest rate." The **prime interest rate** is the rate charged to preferred customers of the bank. The first month the prime rate was 11%, the second month 11.5%, and the third month 12%.

a) Find the amount he paid the bank at the end of the first month; at the end of the second month; at the end of the third month.

b) What was the total amount of interest he paid the bank?

Research Activities

46. There are three types of simple interest that may be calculated with the simple interest formula. They are ordinary, Banker's rule, and exact time.

a) Visit a local bank to determine how time is used in the simple interest formula when using exact time.

b) Compare the results for each method on a loan for $500 at 8% for the period January 2, 1992, to April 2, 1992.

47. Visit your local bank and investigate the procedure for obtaining a loan. Determine (a) what type of security the bank requires, (b) what rate of interest the bank will charge for the loan, and (c) when the loan must be repaid.

10.3 Compound Interest

Albert Einstein was once asked to name the greatest discovery of man. His reply was, "Compound interest."

In this section we discuss some ways to put your money to work for your benefit.

An **investment** is the use of money or capital for income or profit. We can divide investments into two classes: fixed investments and variable investments. In a **fixed investment** the amount invested as principal is guaranteed, and the interest is computed at a fixed rate. "Guaranteed" means that the exact amount invested will be paid back together with any accumulated interest. Examples of a fixed investment are savings accounts and certificates of deposit. Another fixed investment is a government savings bond. In a **variable investment** neither the principal nor the interest is guaranteed. Examples of variable investments are stocks and commercial bonds.

Simple interest, introduced earlier in the chapter, is calculated once for the period of a loan using the formula $i = prt$. The interest paid on savings accounts at most banks is compound interest. This means that the bank computes the interest periodically (for example, daily or quarterly) and adds this interest to the original principal. The interest for the following period is computed by using the new principal (original principal plus interest). In effect the bank is computing interest on interest, which is called compound interest.

Interest that is computed on the principal and any accumulated interest is called **compound interest**.

▶ Example 1

Find the amount, A, to which \$1000 will grow at the end of one year if the interest is 12% compounded quarterly.

Solution: Compute the interest for the first quarter using the simple interest formula. Add this interest to the principal to find the amount at the end of the first quarter:

$$i = prt$$
$$i = \$1000 \times 0.12 \times 0.25 = \$30.00$$
$$A = \$1000 + \$30 = \$1030$$

Find the amount at the end of the second quarter:

$$i = \$1030 \times 0.12 \times 0.25 = \$30.90$$
$$A = \$1030 + \$30.90 = \$1060.90$$

Find the amount at the end of the third quarter:

$$i = \$1060.90 \times 0.12 \times 0.25 = \$31.83$$
$$A = \$1060.90 + \$31.83 = \$1092.73$$

Find the amount at the end of the fourth quarter:

$$i = \$1092.73 \times 0.12 \times 0.25 = \$32.78$$
$$A = \$1092.73 + \$32.78 = \$1125.51$$

Hence the \$1000 grows to a final value of \$1125.51.

This example shows the effect of earning interest on interest, or compounding interest. In one year the amount of \$1000 has grown to \$1125.51, compared to \$1120 that would have been obtained with a simple interest rate of

12%. Thus in one year alone there is a gain of $5.51 more with compound interest than with simple interest.

A simpler and less time-consuming way to calculate compound interest is to use the compound interest formula and a calculator.

In 1626 Peter Du Minuit, representing the Dutch West India Company, traded beads and blankets to the native American inhabitants of Manhattan Island for the island. The trade was valued at 60 Dutch guilders, about $24. If at that time the $24 had been invested at 6% interest compounded annually, the investment would be worth $41,385,867,540 in 1991. As a comparison, the simple interest on $24 for 365 years at 6% is only $525.60. If the $24 had been compounded continuously, as is done in many savings accounts today, the investment would be worth approximately $115,240,000,000,000. These results illustrate the effect of compounding interest over time.

Compound Interest Formula

$$A = p\left(1 + \frac{r}{n}\right)^{nt}$$

In this formula, A is the amount, p is the principal, r is the annual rate of interest, t is the time in years, and n is the number of periods per year.

▶ **Example 2**

Calculate the amount, A, to which $1 will grow at 3% compounded annually for six years.

Solution: Since the interest is compounded annually, there is one period per year. Thus $n = 1, r = 0.03$, and $t = 6$. Substituting into the formula, we find the amount A:

$$A = p\left(1 + \frac{r}{n}\right)^{nt}$$
$$= 1\left(1 + \frac{0.03}{1}\right)^{(1)(6)}$$
$$= 1(1 + 0.03)^6$$
$$= 1(1.03)^6$$
$$= 1(1.1940523)$$
$$= \$1.19$$

In Example 2, to find the value of $(1.03)^6$ with a calculator you can use one of two methods, depending on whether your calculator is a four-function calculator or a scientific calculator with a $\boxed{y^x}$ key. On a four-function calculator multiply 1.03 by itself six times. The key strokes are

$$1.03 \boxed{\times} 1.03 \boxed{\times} 1.03 \boxed{\times} 1.03 \boxed{\times} 1.03 \boxed{\times} 1.03 \boxed{=} 1.1940523$$

On a scientific calculator the key strokes are

$$1.03 \boxed{y^x} 6 \boxed{=} 1.1940523$$

Note: Some scientific calculators have an $\boxed{x^y}$ or a $\boxed{\wedge}$ key in place of a $\boxed{y^x}$ key. In either case the procedure works the same: Enter the base then the key ($\boxed{y^x}$, $\boxed{x^y}$, or $\boxed{\wedge}$), the exponent, and then the $\boxed{=}$ key.

▶ **Example 3**

Calculate the interest on $650 at 8% compounded semiannually for three years, using the compound interest formula.

Solution: Since interest is compounded semiannually, there are two periods per year. Thus $n = 2$, $r = 0.08$, and $t = 3$. Substituting into the formula, we find the amount, A:

$$A = p\left(1 + \frac{r}{n}\right)^{nt}$$

$$= 650\left(1 + \frac{0.08}{2}\right)^{(2)(3)}$$

$$= 650(1.04)^6$$

$$= 650(1.2653190)$$

$$= \$822.46$$

Since the total amount is $822.46 and the original principal is $650, the interest must be $822.46 − $650, or $172.46.

In Example 3, the interest rate is stated as an annual rate of 8%, but the number of compounding periods per year is two. In applying the compound interest formula, the rate for one period is r/n, which was 8%/2 or 4% per period.

Now we will calculate the amount, and interest, on $1 invested at 8% compounded semiannually for 1 year. The result is

FIRST NATIONAL BANK

5.25%
RATE
5.39%
EFFECTIVE ANNUAL YIELD

$$A = 1\left(1 + \frac{0.08}{2}\right)^{(2)(1)} = 1.0816$$

$$\text{Interest} = \text{Amount} - \text{Principal}$$

$$i = 1.0816 - 1$$

$$= 0.0816$$

The interest for one year is 0.0816. This amount written as a percent, 8.16%, is called the **effective annual yield** or **effective yield**. If $1 was invested at a simple interest rate of 8.16% and $1 was invested at 8% compounded semiannually (equivalent to an effective yield of 8.16%), the interest from both investments would be the same.

Many banks compound interest daily. When computing the effective annual yield they use 360 for the number of periods in a year. In the advertisement the annual rate is 5.25% and the effective annual yield is 5.39%. Can you show that an interest rate of 5.25% compounded daily has an effective yield of 5.39%?

To find the effective annual yield for a given interest rate, calculate the amount using the compound interest formula where p is $1. Then subtract $1 from that amount. The difference-written as a percent, is the effective annual yield. This is illustrated in Example 4.

▶ **Example 4**

Determine the effective annual yield for $1 invested for one year at

a) 8% compounded daily

b) 6% compounded quarterly

Solution:

a) With daily compounding $n = 360$.

$$A = p\left(1 + \frac{r}{n}\right)^{nt}$$

$$= 1\left(1 + \frac{0.08}{360}\right)^{(1)(360)}$$

$$= 1.0832774 \approx 1.0833$$

$$i = A - 1$$

$$= 1.0833 - 1$$

$$= 0.0833$$

Thus when compounding daily, the effective annual yield is 8.33%.

b) With quarterly compounding, $n = 4$.

$$A = p\left(1 + \frac{r}{n}\right)^{nt}$$

$$= 1\left(1 + \frac{0.06}{4}\right)^{(1)(4)}$$

$$= 1.06136355 \approx 1.0614$$

$$i = 1.0614 - 1$$

$$= 0.0614$$

Thus when compounding quarterly, the effective annual yield is 6.14%.

There are many different types of savings accounts. Many savings institutions compound interest daily. Some pay interest from the day of deposit to the day of withdrawal, and others pay interest from the first of the month on all deposits made before the tenth of the month. In each of these accounts in which interest is compounded daily, interest is entered into the depositor's account only once each quarter. Some savings banks will not pay any interest on a day-to-day account if the balance falls below a set amount.

Present Value

You may often wonder what amount of money you must deposit in an account today in order to have a set amount of dollars in the future. For example, how much must you deposit in an account today at a given rate of interest so that it will accumulate to $60,000 to pay your child's college costs in 15 years? The principal, p, which would have to be invested now is called the **present value**.

Present Value Formula

$$p = \frac{A}{\left(1 + \dfrac{r}{n}\right)^{nt}}$$

In the formula A represents the amount to be accumulated in n years.

▶ **Example 5**

Marshall is planning to spend \$15,000 on a car in five years. How much principal, p, must be invested today at $8\frac{1}{2}\%$ compounded quarterly, to accumulate to \$15,000 in five years?

Solution: To determine the principal we will use the present value formula. Since the interest is compounded quarterly there are 4 periods per year. Thus $n = 4$, $r = 0.085$, $t = 5$, and $A = \$15,000$.

$$p = \frac{A}{\left(1 + \dfrac{r}{n}\right)^{nt}}$$

$$= \frac{15,000}{\left(1 + \dfrac{0.085}{4}\right)^{(4)(5)}}$$

$$= \frac{15,000}{(1.02125)^{20}}$$

$$= \frac{15,000}{1.5227948}$$

$$= 9850.309$$

Marshall needs to invest \$9850.31 today in order to accumulate \$15,000 in five years.

Section 10.3 Exercises

In Ex. 1–16, use the formula $A = P\left(1 + \dfrac{r}{n}\right)^{nt}$ to compute (a) the amount and (b) the interest earned for each investment.

1. \$2000 for 2 years at 5% compounded annually
2. \$1500 for 3 years at 6% compounded semiannually
3. \$2400 for 2 years at 8% compounded quarterly
4. \$800 for 1 year at 16% compounded quarterly
5. \$2500 for 3 years at 7% compounded annually
6. \$4500 for 4 years at 12% compounded quarterly
7. \$5500 for 5 years at 9% compounded semiannually
8. \$3500 for 3 years at 10% compounded monthly
9. \$2000 for 2 years at 6% compounded annually
10. \$2000 for 2 years at 6% compounded semiannually

11. $2000 for 2 years at 6% compounded quarterly

12. $2000 for 2 years at 6% compounded monthly

13. $2000 for 2 years at 12% compounded quarterly

14. $2000 for 2 years at 12% compounded monthly

15. $1500 for 3 years at 8.5% compounded daily. (Use 1 year = 360 days.)

16. $4500 for 3 years at 12% compounded daily. (Use 1 year = 360 days.)

17. When Steven was born, his father deposited $1000 in his name in a savings account. The account was paying 8% interest compounded annually.
  a) Assuming the rate did not change, what was the value of the account after 15 years?
  b) If the money had been invested at 8% compounded semiannually, what would the value of the account have been after 15 years?

18. When Joan was born, her father deposited $2000 in her name at a savings and loan association. At the time that bank was paying 8% interest compounded annually on savings. After 10 years the association changed to an interest rate of 8% compounded semiannually. How much had the $2000 amounted to after 18 years when the money was withdrawn for Joan to use to help pay her college expenses?

19. Michael borrowed $3000 from his brother Dave. He agreed to repay the money at the end of two years, giving Dave the same amount of interest that he would have received if the money had been invested at 8% compounded quarterly. How much money did Michael repay his brother?

20. Kathy borrowed $6000 from her daughter Lynette. She repaid the money at the end of two years. If Lynette had left the money in a bank account that paid 12% compounded monthly, how much interest would she have accumulated?

21. Joe invested $6000 at 12% compounded quarterly. Three years later he withdrew the full amount and used it for the down payment on a house. How much money did he put down on the house?

22. Compute the amount and the interest on each of the following principals at 12% compounded monthly for two years.
  a) $100   **b)** $200   **c)** $400
  d) Is there a predictable outcome when the principal is doubled?

23. Determine the amount and the interest on $1000 with interest compounded semiannually for ten years at the following rates.

a) 2%   **b)** 4%   **c)** 8%
d) Is there a predictable outcome when the rate is doubled?

24. Compute the amount and the interest on $1000 at 6% compounded semiannually for the following time periods.
  a) 6 years   **b)** 12 years   **c)** 24 years
  d) Is there a predictable outcome when the time is doubled?

25. Show that the effective annual yield of a 6.77% interest rate compounded daily is 7.00%.

26. Determine the effective annual yield for $1 invested for 1 year at 5.5% compounded daily.

27. Determine the effective annual yield for $1 invested for 1 year at 6.5% compounded quarterly.

28. Determine the effective annual yield for $1 invested for 1 year at 5.5% compounded quarterly.

29. How many dollars should parents invest at the birth of their child in order to provide their child with $30,000 at age 18? Assume the money earns 8% compounded quarterly.

30. The Pearsons are planning on retiring in 20 years and feel that they will need $100,000 in addition to their retirement plan. How much must they invest today at 8.5% compounded quarterly to accomplish their goal?

31. How many dollars must you invest today to have $18,000 in 15 years? Assume the money earns 7% compounded quarterly.

32. How many dollars must you invest today to have $18,000 in 15 years? Assume the money earns 7% compounded monthly.

Problem Solving

33. Given that the amount is $3633.39, the principal is $2000, and time period is 5 years, find r if interest is compounded monthly.

34. If the cost of a loaf of bread was $1.25 in 1988 and the annual average inflation rate is $3\frac{1}{2}$%, what will be the cost of a loaf of bread in 1998?

35. Dick borrowed $2000 from Diane. The terms of the loan are as follows: The period of the loan is 3 years, and the rate of interest is 8% compounded semiannually. What rate of simple interest would be equivalent to the rate Diane charged Dick?

36. The question may be asked: If I deposit a fixed sum of money monthly, quarterly, semiannually, or annually at a fixed rate of interest, how many dollars will I have accumulated in x years? This is called an **annuity**. The formula for determining the value of the annuity is

$$S = \dfrac{R\left[\left(1 + \dfrac{r}{n}\right)^{nt} - 1\right]}{\dfrac{r}{n}}$$

where S is the value of the annuity in t years, R is the

amount invested each period, r is the annual rate of interest, n is the number of payments per year, and t is the number of years. After Harriet's birth her parents invested $500 semiannually (every six months) at an annual rate of 5.5% compounded semiannually. What is the value of the annuity after 17 years?

Installment Buying

In Section 10.2 we discussed promissory notes and discounted notes. When borrowing money by either of these methods, the borrower normally repays the loan as a single payment at the end of the scheduled time period. There may be circumstances under which it is more convenient for the borrower to repay the loan on a weekly or monthly basis or to use some other convenient time period. One method of doing this is to borrow money on the **installment plan**.

There are two types of installment loans, open end and fixed payment. An **open-end installment loan** is a loan on which you can make variable payments each month. Credit cards such as Master Card, Visa, and Discover are actually open-end installment loans, used to purchase items such as clothing, textbooks, and meals. A **fixed installment loan** is one on which you pay a fixed amount of money for a set number of payments. Examples of items purchased with fixed-payment installment loans are college tuition loans, cars, boats, appliances, and furniture. These loans are generally repaid in 24, 36, 48, or 60 equal monthly payments.

Lenders give any individual wishing to borrow money or purchase goods or services on the installment plan a credit rating to determine if the borrower is likely to repay the loan. The lending institution determines whether you are a good "credit risk" by examining your income, assets, and liabilities, and by looking at your history of repaying debts.

The advantage of installment buying is that the buyer has the use of an article while still paying for it. If the article is essential, installment buying may serve a real need. A disadvantage is that some people buy more on the installment plan than they can afford. Another disadvantage is the large amount of interest the purchaser pays for the use of someone else's money. The method of determining the interest charged on an installment plan may vary with different lenders.

To provide the borrower with a means of comparing interest charged, Congress passed the Truth in Lending Act in 1969. The law requires that the lending institution tell the borrower two things, the annual percentage rate and the finance charge. The **annual percentage rate (APR)** is the true rate of interest charged for the loan. The APR is calculated by using a complex formula, so we will use the tables provided by the government to determine the APR. The technique of using the tables to find the APR will be illustrated in Example 1. The total **finance charge** is the total amount of money the borrower must pay for the use of money. The finance charge includes the interest plus any additional fees charged. The additional fees may include service charges, credit investigation fees, mandatory insurance premiums, and so on.

The finance charge a consumer pays when purchasing goods or services on the installment plan is the difference between the total installment price and the cash price. The **total installment price** is the sum of all the monthly payments and the down payment, if any.

Fixed Installment Loan

The next example shows how to determine the annual percentage rate on a loan that has a schedule for repaying a fixed number of dollars for each payment.

▶ **Example 1**

Jane is purchasing a new computer and accessories for $3600, including taxes. She is making a $600 down payment and will finance the balance, $3000, through a bank. The bank says that the payment will be $101 per month for 36 months.

a) Determine the finance charge.   **b)** Determine the APR.

Solution:

a) The total installment price = down payment + total monthly installment payments.

$$= 600 + (36 \times 101)$$
$$= 600 + 3636 = \$4236$$

The finance charge = total installment price − cash price.

$$= 4236 - 3600 = \$636$$

b) To determine the annual percentage rate, use Table 10.2 as follows: First divide the finance charge by the amount financed and multiply the quotient by 100. This gives the finance charge per $100 of the amount financed:

$$\frac{\text{Finance charge}}{\text{Amount financed}} \times 100 = \frac{636}{3000} \times 100 = 0.212 \times 100 = 21.2$$

This means that the borrower pays $21.20 interest for each $100 being financed. To use the table, look for 36 in the left-hand column under the heading Number of Payments. Then move across to the right until you find the value closest to $21.20. The value that is closest to $21.20 is $21.30 (circled in blue). At the top of this column is the value 13.00%. Therefore the annual percentage rate is approximately 13.00%.

Table 10.2 Annual Percentage Rate Table for Monthly Payment Plans

Number of Payments	Annual Percentage Rate												
	10.75%	11.00%	11.25%	11.50%	11.75%	12.00%	12.25%	12.50%	12.75%	13.00%	13.25%	13.50%	13.75%
	(Finance Charge per $100 of Amount Financed)												
6	3.16	3.23	3.31	3.38	3.45	3.53	3.60	3.68	3.75	3.83	3.90	3.97	4.05
12	5.92	6.06	6.20	6.34	6.48	6.62	6.76	6.90	7.04	7.18	7.32	7.46	7.60
18	8.73	8.93	9.14	9.35	9.56	9.77	9.98	10.19	10.40	10.61	10.82	11.03	11.24
24	11.58	11.86	12.14	12.42	12.70	12.98	13.26	13.54	13.82	14.10	14.38	14.66	14.95
30	14.48	14.83	15.19	15.54	15.89	16.24	16.60	16.95	17.31	17.66	18.02	18.38	18.74
36	17.43	17.86	18.29	18.71	19.14	19.57	20.00	20.43	20.87	(21.30)	21.73	22.17	22.60
48	23.48	24.06	24.64	25.23	25.81	26.40	26.99	27.58	28.18	28.77	29.37	29.97	30.57
60	29.71	30.45	31.20	31.96	32.71	33.47	34.23	34.99	35.75	36.52	37.29	38.06	38.33

Note: Much more complete APR tables similar to Table 10.2 are available at your local bank.

▶ Example 2

Gary borrowed $9800 in 1992 to purchase a car. He does not recall the APR of the loan but remembers that there are 48 payments of $255.75. If he did not make a down payment on the car, determine the APR.

Solution: First determine the finance charge by subtracting the cash price from the total amount paid:

$$\text{Finance charge} = (255.75 \times 48) - 9800$$
$$= 12276 - 9800$$
$$= \$2476$$

Next divide the finance charge by the amount of the loan and multiply this quotient by 100:

$$\frac{2476}{9800} \times 100 = 25.27$$

Now find 48 payments in the left-hand column of Table 10.2. Move to the right until you find the value that is closest to 25.27. The closest value to 25.27 is 25.23. Looking at the top of the column, we find the APR to be 11.50%.

In Example 2 Gary agreed to pay a finance charge of $2476. If he decides to repay the loan after making 30 payments, must he pay the total finance charge? The answer is no. Two methods are used to determine your finance charge when you repay an installment loan early: the **actuarial method** and the **rule of 78s**. The actuarial method uses the APR tables whereas the more commonly used method, the rule of 78s, does not.

Producers of all types of goods, from soup to nuts to cars, engage in a practice called "odd pricing." The belief is that people respond more favorably to prices that end in "odd" numbers, not nice round ones. So a consumer is supposed to feel better buying a product for $4.99 (just four dollars and change) than for $5.00. Odd pricing was originally used to force clerks to make change and thus serve as a means of cash control. A similar practice is used in the pricing of big-ticket items.

Actuarial Method

$$u = \frac{n \cdot P \cdot V}{100 + V}$$

u = unearned interest
n = number of remaining monthly
   payments (excluding
   current payment)
P = monthly payment
V = the value from the APR table
   that corresponds to the annual
   percentage rate for the number
   of remaining payments
   (excluding current payment)

Rule of 78s

$$u = \frac{f \cdot k(k + 1)}{n(n + 1)}$$

u = unearned interest
f = original finance charge
k = number of remaining
   monthly payments (excluding
   current payment)
n = original number of payments

The two formulas in the box determine the unearned interest, u, which is the interest saved by paying off the loan early.

Example 3 illustrates the actuarial method and Example 4 the rule of 78s.

▶ **Example 3**

In Example 2 we determined the APR of Gary's loan to be 11.50%. Instead of making his 30th payment Gary wishes to pay his remaining balance and terminate the loan.

a) Use the actuarial method to determine how much interest Gary will save (the unearned interest, u) by repaying the loan early.

b) What is the total amount due on that day?

Solution:

a) After 30 payments have been made, 18 payments remain to be made. Thus $n = 18$ and $P = \$255.75$. To determine V use the APR table (Table 10.2). In the Number of Payments column find the number of remaining payments, 18, and then look to the right until you are under the column headed by 11.50%, the APR. This row and column intersect at 9.35. Thus $V = 9.35$.

$$u = \frac{n \cdot P \cdot V}{100 + V}$$

$$= \frac{18 \cdot 255.75 \cdot 9.35}{100 + 9.35}$$

$$= 393.62$$

Gary will save $393.62 in interest by the actuarial method.

b) Since the remaining payments total $18 \cdot \$255.75 = \$4603.50$, Gary's remaining balance is as follows:

$4603.50	Total of remaining payments (which includes interest)
−$ 393.62	Interest saved (unearned interest)
$4209.88	Balance due

A payment of $4209.88 plus the 30th monthly payment of $255.75 will terminate Gary's installment loan. The total amount due is $4465.63.

▶ Example 4

In Example 2 we determined the APR of Gary's loan to be 11.50%. Instead of making his 30th payment Gary wishes to pay his remaining balance and terminate the loan.

a) Use the rule of 78s to determine how much interest Gary will save by repaying the loan early.

b) What is the total amount due on that day?

Solution:

a) After the 30 payments have been made, there are 18 remaining payments. Thus $k = 18$, $n = 48$, and $f = \$2476$.

$$u = \frac{f \cdot k(k + 1)}{n(n + 1)}$$

$$= \frac{2476 \cdot 18(18 + 1)}{48(48 + 1)}$$

$$= \$360.03$$

Gary will save $360.03 in interest by the rule of 78s.

b) Gary's balance is computed in a manner similar to the method in Example 3.

$4603.50	Total of remaining payments
−$ 360.03	Interest saved
$4243.47	Balance due

A payment of $4243.47 plus a regular monthly payment of $255.75 will terminate Gary's installment loan. The total amount due is $4499.22.

Most states use the rule of 78s. However, there is some discussion that the rule of 78s is outdated and is not fair to the borrower.

Open-End Installment Loan

A credit card is a popular way of making purchases or borrowing money. This is an example of an open-end installment loan. A typical charge account with a bank may have the terms in Table 10.3 on page 484.

Table 10.3 Credit Card Terms

| Type of Charge | Periodic Rates* | | Annual Percentage Rate* |
	Monthly	Daily	
Purchases	1.30%		15.60%
Cash advances		0.04273%	15.60%

* These rates vary with different charge accounts and localities.

Typically, credit card monthly statements contain the following information: balance at the beginning of the period, balance at the end of the period (or new balance), the transactions for the period, statement closing date (or billing date), payment due date and the minimum payment due. For *purchases* there is no finance or interest charge if you pay the entire new balance by the payment due date. However, if you borrow *money (or cash advances)* through this account, there will be a finance charge from the date you borrowed the money until the date the money is repaid. When you make purchases or borrow money, the minimum monthly payment is sometimes determined by dividing the balance due by 36 and rounding the answer up to the nearest whole dollar, thus ensuring repayment in 36 months. However, if the balance due for any month is less than $360, the minimum monthly payment is typically $10. These general guidelines may vary with different banks and with different stores.

▶ **Example 5**

While on vacation Sonya charged the following items to her MasterCard account:

Airplane ticket	$285
Hotel room	120
Meals	115
	$520

Her balance due on August 1 is $520 and she makes a payment of $100. The bank requires repayment within 36 months and charges an interest rate of 1.3% per month.

a) Determine the minimum payment due on August 1.

b) Find the balance due on September 1 assuming that there are no additional charges or cash advances.

Solution:

a) To determine the minimum payment we divide the balance due on August 1 by 36.

$$\frac{520}{36} = \$14.44$$

Rounding up to the nearest dollar, we determine that the minimum payment due on August 1 is $15.

b) With no additional purchases or cash advances and paying $100 on August 1 the balance on September 1 is $520 − $100 or $420. In addition she must pay interest on the outstanding balance of $420 for the month of August. The interest is calculated as follows:

$$i = prt$$
$$= 420(0.013)(1)$$
$$= \$5.46$$

Therefore the balance due on September 1 is $420 + $5.46 or $425.46.

In Example 5 there were no additional transactions in the account for the period. When additional charges are made during the period, the finance charges on open-end installment loans, or credit cards, are generally calculated in one of two ways: the *unpaid balance method* or the *average daily balance method*.

With the **unpaid balance method**, you are charged interest or a finance charge on the unpaid balance from the previous charge period. Example 6 illustrates the unpaid balance method.

▶ **Example 6**

On November 1, the billing date, Ms. Storm had a balance due of $275 on her credit card. During the month of November she charged articles totaling $320 and made a payment of $145.

a) Find the finance charge due on December 1 using the unpaid balance method. Assume the interest rate is 1.3% per month.

b) Find the new account balance on December 1.

Solution:

a) The finance charge is based on the $275 balance due on November 1. To find the finance charge due on December 1 we use the simple interest formula with a time of one month.

$$i = \$275 \times .013 \times 1 = \$3.58$$

The finance charge on December 1 is $3.58.

b) The balance due on December 1 is

$$\$275 + \$320 + \$3.58 - \$145 = \$453.58$$

The balance due on December 1 is $453.58. The finance charge on January 1 will be based on $453.58.

Many lending institutions are using the **average daily balance method** of calculating the finance charge since they believe it is a fairer method for the customer.

With the average daily balance method, a balance is found on each day of the billing period for which there is a transaction in the account. The average daily balance method is illustrated in Example 7.

▶ Example 7

The balance on the Margolius's credit card account on July 1, their billing date, was $375.80. They had the following transactions during the month of July.

July 5	Payment	$150.00
July 10	Toy store	74.35
July 18	Garage	123.50
July 28	Restaurant	42.50

a) Find the average daily balance for the billing period.
b) Find the finance charge that is due on August 1. Assume the interest rate is 1.3% per month.
c) Find the balance due on August 1.

Solution:

a) To determine the average daily balance we do the following: (i) Find the balance due for each transaction date.

July 1	$375.80
July 5	$375.80 − 150 = 225.80
July 10	$225.80 + 74.35 = $300.15
July 18	$300.15 + $123.50 = $423.65
July 28	$423.65 + $42.50 = $466.15

(ii) Find the number of days that the balance did not change between each transaction. Count the first day in the period but not the last day. (iii) Multiply the balance due by the number of days the balance did not change. (iv) Find the sum of the products.

Date	(i) Balance Due	(ii) Number of Days Balance Did Not Change	(iii) (Balance)(Days)
July 1	$375.80	4	($375.80)(4) = $ 1503.20
July 5	$225.80	5	($225.80)(5) = $ 1129.00
July 10	$300.15	8	($300.15)(8) = $ 2401.20
July 18	$423.65	10	($423.65)(10) = $ 4236.50
July 28	$466.15	4	($466.15)(4) = $ 1864.60
		31	(iv) Sum = $11,134.50

Note that from July 28 through July 31 is 4 days. The next billing cycle starts on August 1. (v) Divide this sum by the number of days in the billing cycle (the month). The number of days may be found by summing the days in column ii.

$$\frac{11134.50}{31} = \$359.18$$

Thus the average daily balance is $359.18.

b) The finance charge for the month is found using the simple interest formula with the average daily balance as the principal.

$$i = \$359.18 \times 0.013 \times 1 = \$4.67$$

c) Since the finance charge for the month is $4.67, the balance owed on August 1 is $466.15 + $4.67 or $470.82.

Example 8 illustrates how a credit card is used to borrow money.

▶ **Example 8**

Don needed $250 to pay some bills. He borrowed the money from his charge account on September 1, 1992, and repaid it on September 21, 1992. How much interest did Don pay for the loan? Assume the interest rate is 0.04273% per day.

Solution: The agreement says that the interest is calculated for the exact number of days of the loan, starting with the day the money is obtained. The time of the loan in this case is 20 days. Using the simple interest formula and remembering that the rate and time are in terms of days, we find that the interest is $2.14:

$$
\begin{aligned}
i &= prt \\
&= \$250 \times 0.0004273 \times 20 \\
&= \$2.14
\end{aligned}
$$

The amount due on September 21 was $252.14 ($250 + $2.14).

When purchasing a car or other costly items one should consider a number of different sources for a loan. Example 9 illustrates one method of making a comparison.

▶ **Example 9**

Fran needed an additional $500 to buy a sailboat. She found that the Easy Loan Company charged 8.0% simple interest on the amount borrowed for the duration of the loan. It also required that the loan, including interest, be repaid in six equal monthly payments. The Friendly Loan Company offered loans of $500 to be repaid in 12 monthly payments of $44.50.

a) How much interest would be charged for the loan from the Easy Loan Company?

b) How much interest would be charged for the loan from the Friendly Loan Company?

c) What was the APR on the Easy Loan Company loan?

d) What was the APR on the Friendly Loan Company loan?

Solution:

a) To find the interest charged by the Easy Loan Company, we use the simple interest formula:

$$i = prt$$

$$= \$500 \times 0.08 \times \frac{6}{12}$$

$$= \$20.00$$

Thus the interest charged is $20.00 for the loan.

b) To find the interest charged by the Friendly Loan Company, we must subtract the amount borrowed from the total amount repaid:

$$\text{Interest} = \text{total amount repaid} - \text{amount borrowed}$$
$$= (\$44.50 \times 12) - \$500$$
$$= \$534 - \$500$$
$$= \$34.00$$

c) To find the APR on the loan from the Easy Loan Company, divide the interest by the amount financed, multiply by 100, and determine the APR from Table 10.2.

$$\frac{20.00}{500} \times 100 = 4.00$$

The APR found in the table is 13.50%.

d) To find the APR on the loan from the Friendly Loan Company, divide the interest by the amount financed, multiply by 100, and determine the APR from Table 10.2.

$$\frac{34}{500} \times 100 = 6.8$$

The APR found in the table is 12.25%.

Notice that Fran pays fewer dollars in interest with the loan from the Easy Loan Company ($20.00 versus $34). However, she is paying a higher rate of interest with the Easy Loan Company (13.50% versus 12.25%). The reason the actual interest is less even though the APR is greater is that the time period for repayment with the Easy Loan Company is shorter than that of the Friendly Loan Company.

Section 10.4 Exercises

1. Ruth bought a microwave oven on a monthly purchase plan. The oven sold for $460. She paid $100 down and $16.95 a month for 24 months.

a) What finance charge did Ruth pay over the period of the loan?

b) What was the APR?

2. The Dechs can buy a refrigerator for a cash price of $710. The installment terms are a down payment of $170 and 18 monthly payments of $33.15 each.

a) What finance charge will the Dechs pay?

b) What is the APR?

3. Mr. and Mrs. Gerling wish to buy furniture that has a cash price of $450. On the installment plan they must pay one third of the cash price as a down payment and make six monthly payments of $51.69.
   a) What finance charge will the Gerlings pay?
   b) What is the APR?

4. A diamond ring sells for $750 cash. On the installment plan the ring may be purchased with a 25% down payment and the balance paid in 18 equal monthly installments. Compute the monthly payment if the total finance charge is $75.

5. Sara wants to purchase a new television. The purchase price is $890. If she purchases the set today and pays cash, she must take money out of her savings account. Another option is to charge the TV on her credit card, take the set home today, and pay next month. Next month she will have cash and can pay her credit card balance without paying any interest. The interest rate on her savings account is $5\frac{1}{4}\%$ compounded annually. How much is she saving by using the credit card instead of taking the money out of her savings account?

6. A certain automobile finance agency requires that the buyer make a down payment of one-third of the cash price. The agency subtracts the down payment from the cash price and then adds the cost of disability and life insurance.* The resulting sum is the unpaid balance. The total interest on the loan is computed on the unpaid balance at 6.9% simple interest for the duration of the loan. The Stearns family bought a second-hand car on these installment terms. The cash price of the car was $3600, the insurance was $30, and the balance was to be paid in 18 monthly payments.
   a) What was the down payment?
   b) What was the interest charge?
   c) What was the APR?
   d) What was the monthly payment including interest and insurance?

7. Penny Miller needs $250. She finds that the ABC Loan Company charges 7.5% simple interest on the amount borrowed for the duration of the loan and requires the loan to be repaid in 6 equal monthly payments. The XYZ Loan Company offers loans of $250 to be repaid in 12 monthly payments of $22.39.

---

* Disability insurance will cover the monthly payments if the insured individual becomes disabled and cannot work. The life insurance will pay off the balance due on the car if the insured individual dies.

a) How much interest is charged by the ABC Loan Company?
b) How much interest is charged by the XYZ Company?
c) What is the APR on the ABC Loan Company loan?
d) What is the APR on the XYZ Loan Company loan?

8. John Jones borrowed $375 against his charge account on September 12 and repaid the loan on October 14 (32 days later). Assume the interest rate is 0.04273% per day.
   a) How much interest did John pay on the loan?
   b) What amount did he pay the bank when he repaid the loan?

9. Amina bought a new Dodge van for $19,450, including sales tax. She made a down payment of $3,450 and financed the balance through her bank. She repaid the loan by paying $346.67 per month for 60 months. What was the APR?

10. Tina has a 48-month installment loan, with a fixed monthly payment of $193.75. The amount borrowed was $7500. Instead of making her 18th payment, Tina is paying the remaining balance on the loan.
   a) Determine the APR of the installment loan.
   b) How much interest will Tina save, computed by the actuarial method?
   c) What is the total amount due on that day?

11. Melissa paid $4320 for furniture for her apartment. She made a down payment of $1762 and financed the balance on a 24-month fixed payment installment loan. The monthly payments are $119.47. Instead of making her 12th payment, Melissa decides to pay the remaining balance.
   a) Determine the APR of the installment loan.
   b) How much interest will Melissa save, computed by the actuarial method?
   c) What is the total amount due on that day?

12. Tony Joseph is buying a $3000 sound system by making a down payment of $500 and 18 monthly payments of $151.39. Instead of making his 12th payment, Tony decides to pay the remaining balance.
   a) How much interest will Tony save, computed by the rule of 78s?
   b) What is the total amount due on that day?

13. Roger purchased woodworking tools for $2375. He made a down payment of $850 and financed the balance with a 12-month fixed payment installment loan. Instead of making his sixth monthly payment of $134.71 he decides to pay the remaining balance.
   a) How much interest will Roger save, computed by the rule of 78s?
   b) What is the total amount due on that day?

14. On September 1, the billing date, Verna had a balance due of $385.75 on her credit card. The transactions during the month of September were:

September 4	Payment	$275.00
September 8	Charge: Airline ticket	330.00
September 21	Charge: Hotel bill	190.80
September 28	Charge: Clothing	84.75

a) Find the finance charge on October 1 using the unpaid balance method. Assume the interest rate is 1.40% per month.

b) Find the new balance on October 1.

15. On April 5, the billing date, Lori had a balance due of $57.88 on her credit card. She is redecorating her apartment and made the following transactions.

April 7	Charge: Paint	$ 64.75
April 10	Payment	45.00
April 15	Charge: Curtains	72.85
April 21	Charge: Chair	135.50

a) Find the finance charge on May 5 using the unpaid balance method. Assume the interest rate is 1.35% per month.

b) Find the new balance on May 5.

16. The balance on the Simpsons' credit card on May 12, their billing date, was $378.50. For the period ending June 12 they made the following transactions.

May 13	Charge: Toys	$129.79
May 15	Payment	50.00
May 18	Charge: Clothing	135.85
May 29	Charge: Housewares	37.63

a) Find the average daily balance for the billing period.

b) Find the finance charge that is due on June 12. Assume an interest rate of 1.3% per month.

c) Find the balance that is due on June 12.

17. The Calusines' credit card statement shows a balance due of $1578.25 on March 23, the billing date. For the period ending April 23 they made the following transactions.

March 26	Charge: Party supplies	$ 79.98
March 30	Charge: Restaurant meal	52.76
April 3	Payment	250.00
April 15	Charge: Clothing	190.52
April 22	Charge: Car repairs	190.85

a) Find the average daily balance for the billing period.

b) Find the finance charge that is due April 23. Assume an interest rate of 1.3% per month.

c) Find the balance due on April 23.

18. The cash price of a motorboat and trailer is $7500. The package may be purchased for a 10% down payment

and 36 monthly payments. For convenience the dealer computes the finance charge at 6% simple interest on the amount financed.

a) What is the finance charge on the boat and trailer?

b) How much is the monthly payment?

c) Determine the total cost of the package on the installment plan.

d) What annual percentage rate must the dealer disclose to the buyer?

19. In Ex. 17, if the Calusines had made the payment on their account on March 24 rather than on April 3, how would it have affected

a) the average daily balance?

b) the finance charge?

Problem Solving

20. Ken bought a new car, but he cannot now remember the original purchase price. His payments are $379.50 per month for 36 months. He remembers that the salesperson said that the simple interest rate for the period of the loan was 6%. He also recalls that he was allowed $2500 on his old car. Find the original purchase price.

21. Suppose that the Gerlings in Ex. 3 use a credit card rather than the installment plan. Assume that they make the same down payment and they pay $50 a month plus the finance charge for that month. Also assume that they paid off the previous balance and there is no finance charge the first month. Finally, assume that they make no other charges on their credit card until the furniture is paid off. Assume the interest rate is 1.3% per month.

a) How many months will it take them to repay the loan?

b) How much interest will they pay on the loan?

c) Which method of borrowing will cost the Gerlings the least amount of interest, the installment loan in Ex. 3 or the credit card?

Research Activities

22. Leasing an automobile might be cheaper than buying. Determine whether this statement is true or false for your driving needs. To help you arrive at a conclusion, do the following.

a) Decide on the automobile you would like to drive, including the accessories.

b) Determine the cost of the automobile.

c) Determine the period of time you wish to keep the automobile.

d) Determine how much of the cost you would finance.

e) Determine the amount of interest you would lose from

the money withdrawn from your savings account to purchase the car.

f) Determine the cost of insurance if you purchase the car.

g) Estimate the resale value of the car when sold.

h) Determine the cost of leasing the same car for the same period of time.

i) Determine who pays for the insurance and the repairs on the leased vehicle.

j) Finally, determine which is less expensive, leasing or buying the car.

23. Assume that you are married and have a child. You do not own a washer and dryer and have no money to buy the appliances. Would it be cheaper to borrow money on an installment loan and buy the appliances or to continue going to the local coin-operated laundry for five years until you have saved enough to pay cash for a washer and dryer?

With the aid of parents or friends, establish how many loads of laundry you would be doing each week. Then determine the cost of doing that number of loads at a coin-operated laundry. (Do not forget the cost of transportation to the laundry.) Shop around for a washer and dryer, and determine the total cost on an installment plan. Do not forget to include the cost of gas, electricity, and water. This information can be obtained from a local gas and electric company. With this information you should be able to make a clear decision about whether to buy now or wait for five years.

10.5 Buying a House with a Mortgage

Buying a house is the largest purchase of a lifetime for most people. The purchaser will normally be committed to 10, 15, 20, 25, or 30 years of mortgage payments. Before selecting the "Dream House" the buyer should consider the following questions: "Can I afford it?" and "Does it suit my needs?"

The question "Can I afford the house?" must be answered carefully and accurately. If a family buys a house beyond its means, it will have a difficult time living within its income. When deciding whether to purchase a particular house, the purchaser must also consider more questions that are critical: "Do I have enough cash for the down payment?" "Can I afford the monthly payments with my current income?" These two items, down payment and mortgage payments over time, constitute the buyer's total cost of a house.

Buyers usually seek a *mortgage* from a bank or other lending institution. Before approving a mortgage, which is a long-term loan, the bank will require the buyer to have a specified minimum amount for the down payment. The **down payment** is the amount of cash the buyer must pay to the seller before the lending institution will grant the buyer a mortgage. If the buyer has the down payment and meets the other criteria for the mortgage, the lending institution prepares a written agreement called the mortgage, stating the terms of the loan. The loan specifies the repayment schedule, the duration of the loan, whether the loan can be assumed by another party, and the penalty if payments are late. The party borrowing the money accepts the terms of this agreement and gives the lending institution the title or deed to the property as security.

> **Homeowner's Mortgage**
> A long-term loan in which the property is pledged as security for payment of the difference between the down payment and the sale price.

Courtesy of Sears, Roebuck Co.

Sears and Roebuck started its mail-order business in the late 1880s. Their catalog, or "Consumer's Guide," was one of the most eagerly awaited publications of the time. You could order just about anything from the 1908 "Wish Book" including a line of "modern homes." The Langston, a modest two-story home, was offered for $1630. The houses were shipped in pieces by rail, assembly required.

The two most popular types of mortgages or loans available today are the **variable-rate loan** and the **conventional loan**. The major difference between the two is that the interest rate for a conventional loan is fixed for the duration of the loan, whereas the interest rate for the variable-rate loan may change every period, as specified in the loan. We will first discuss the requirements that are the same for both types of loans.

The size of the down payment required depends on who is lending the money, how old the property is, and whether or not it is easy to borrow money at that particular time. The down payment required by the lending institution can vary from 5% to 50% of the purchase price of the property. A larger down payment is required when money is "tight"—that is, when it is difficult to borrow money. Furthermore, most lending institutions tend to require larger down payments on older homes and smaller down payments on newer homes.

Most lending institutions may require the buyer to pay one or more points for their loan at the time of the **closing** (the final step in the sale process). **One point** amounts to 1% of the mortgage money (the amount being borrowed). By charging points, the bank reduces the rate of interest on the mortgage, thus reducing the size of the monthly payments and enabling more people to purchase houses. However, because they charge points, the rate of interest banks state is not the APR (annual percentage rate) for the loan. The APR would be determined by adding the amount paid for points to the total interest paid and then using an APR table.

Conventional Loans

Example 1 discusses purchasing a home with a conventional loan.

▶ **Example 1**

The Cohens wish to purchase a house selling for $95,000. Their bank requires a 10% down payment and a payment of two points at the time of closing.
a) What is the required down payment?
b) With a 10% down payment, what is the mortgage on the property?
c) What is the cost of two points on the mortgage determined in part (b)?

Solution:
a) The required down payment is 10% of $95,000:

$$0.10 \times \$95,000 = \$9,500$$

b) To find the mortgage on the property, subtract the down payment from the selling price of the house:

$$\$95,000 - \$9,500 = \$85,500$$

c) To find the cost of two points on a mortgage of $85,500, find 2% of $85,500:

$$0.02 \times \$85,500 = \$1,710$$

When the Cohens sign the mortgage, they must pay the bank $1710. The $9500 down payment is paid to the seller.

Banks use a formula to determine the maximum monthly payment that they feel is within the purchaser's ability to pay. The banker will first determine the buyer's **adjusted monthly income** by subtracting from the gross monthly income (total income before any deductions) any fixed monthly payments with more than six payments remaining (such as for a student loan, a car, furniture, or a television). The banker then multiplies the adjusted monthly income by 25%. (This percent may vary in different locations.) In general, this product is the maximum monthly payment the lending institution believes the purchaser can afford to pay for a house. This estimated payment must cover principal, interest, and taxes, and sometimes it must cover fire insurance as well. Note that the taxes and insurance are not necessarily paid to the bank; they may be paid directly to the tax collector and the insurance company. Example 2 shows how a bank uses the formula to determine whether the prospective buyer qualifies for a mortgage.

▶ Example 2

In Example 1, suppose the Cohens' gross monthly income is $4100, and that they have 14 remaining payments of $225 per month on their car, 48 payments of $55 per month on a college loan, and 9 payments of $65 per month on new furniture. The taxes on the house they would like to purchase are $150 per month, and the fire insurance is $42 per month.

a) What maximum monthly payment does the bank's lending officer think the Cohens can afford?

b) The Cohens would like to get a 20-year $85,500 mortgage. Do they qualify for this mortgage if the current rate of interest is 8.5%?

Solution:

a) To find the maximum monthly payment that the bank believes the Cohens can afford, first find their adjusted monthly income:

$$
\begin{array}{ll}
\$4100 & \text{Gross income} \\
-\quad 345 & \text{Monthly payments (car, college loan, furniture)} \\
\hline
\$3755 & \text{Adjusted monthly income}
\end{array}
$$

Next find 25% of the adjusted monthly income:

$$0.25 \times \$3755 = \$938.75$$

The bank determines that the Cohens can afford a maximum monthly payment of $938.75 with their current income.

b) To determine whether the Cohens qualify for a 20-year conventional mortgage with their current income, calculate the monthly mortgage payment, which includes principal and interest. All lending institutions and most lawyers have a set of tables that contain a monthly mortgage payment per thousand dollars for a specific number of years at a specific interest rate. Table 10.4 on page 494 shows selected monthly mortgage payments. With the current rate of interest of 8.5%, a 20-year loan would have a monthly payment of $8.68 per thousand dollars of mortgage (circled in blue in the table).

Table 10.4 Monthly Payment per $1000 of Mortgage, Including Principal and Interest

	Number of Years				
Rate %	20	25	30	35	40
8	$8.36	$7.72	$7.34	$7.10	$6.95
8.5	8.68	8.05	7.69	7.47	7.33
9	9.00	8.40	8.05	7.84	7.72
9.5	9.33	8.74	8.41	8.22	8.11
10	9.66	9.09	8.70	8.60	8.50
10.5	9.98	9.44	9.15	8.98	8.89
11	10.32	9.80	9.52	9.37	9.28
11.5	10.66	10.16	9.90	9.76	9.68
12	11.01	10.53	10.29	10.16	10.08
12.5	11.36	10.90	10.67	10.55	10.49
13	11.72	11.28	11.06	10.95	10.90
13.5	12.07	11.66	11.45	11.35	11.30
14	12.44	12.04	11.85	11.76	11.71
14.5	12.80	12.42	12.25	12.16	12.12
15	13.17	12.81	12.64	12.57	12.53
15.5	13.54	13.20	13.05	12.98	12.94
16	13.91	13.59	13.45	13.38	13.36
16.5	14.29	13.98	13.85	13.79	13.77
17	14.67	14.38	14.26	14.21	14.18

To determine the Cohens' payment, divide the mortgage by $1000:

$$\frac{\$85,500}{\$1000} = 85.5$$

The monthly mortgage payment is found by multiplying the number of thousands of dollars of mortgage, 85.5, by the value found in Table 10.4:

$$\$8.68 \times 85.5 = \$742.14$$

To the $742.14 add the monthly cost of fire insurance, $42, and real estate taxes of $150, for a total cost of $934.14. Since $934.14 is less than the maximum monthly payment the bank has determined that the Cohens can afford, $938.75, they will most likely be granted the loan.

What is the effect on the monthly payments when only the period of time has been changed? The total monthly payments on the mortgage in Example 2 for different periods of time are $688.28 for 25 years, $657.50 for 30 years, $638.69 for 35 years, and $626.72 for 40 years. Increasing the length of time decreases the monthly payment but results in an increase in the total interest because the borrower is paying for a longer period of time. The longer the term of a mortgage the more expensive the total cost of the house.

▶ **Example 3**

The Cohen family of Examples 1 and 2 obtains a 20-year $85,500 conventional mortgage at 8.5% on a house selling for $95,000. Their monthly mortgage payment, including principal and interest, is $742.14. They paid 2 points at the time of closing.

a) Determine the amount the Cohens will pay for their house.
b) How much of the cost will be interest?
c) How much of the first payment on the mortgage is applied to the principal?

Solution:
a) To find the total amount the Cohens will pay for the house, compute as follows:

$	742.14	Mortgage payment for one month
×	12	
$	8905.68	Mortgage payments for one year
×	20	Number of years
	$178,113.60	Mortgage payments for 20 years
	9,500.00	Down payment
+	1,710.00	Points
	$189,323.60	Total cost of the house

Note: The result might not be the exact cost of the house, since the final payment on the mortgage might be slightly more or less than the regular monthly payment.

b) To determine the amount of interest paid over 20 years, subtract the purchase price of the house and the cost of the points from the total cost:

	$189,323.60	Total cost
−	95,000.00	Purchase price
	$ 94,323.60	Total interest including points
−	1,710.00	Points
	$ 92,613.60	Total interest on mortgage payments

c) We can find the amount of the first payment that is applied to the principal by subtracting the amount of interest on the first payment from the monthly mortgage payment. We can find the interest on the first payment by using the simple interest formula:

$$i = prt$$

$$= \$85,500 \times 0.085 \times \frac{1}{12}$$

$$= \$605.63$$

Now subtract the interest for the first month from the monthly mortgage payment. The difference will be the amount paid on the principal for the first month:

$742.14	Monthly mortgage payment
− 605.63	Interest paid for the first month
$136.51	Principal paid for the first month

We see that the first payment of $742.14 consists of $605.63 in interest and $136.51 in principal. The $136.51 is applied to reduce the loan. Thus the balance due after the first payment is $85,500 − $136.51, or $85,363.49.

By repeatedly using the simple interest formula month-to-month on the unpaid balance you can calculate the principal and the interest for all the payments. This would be a tedious task, however. With a computer it is easy to prepare a list containing the payment number, payment on the interest, payment on the principal, and balance of the loan. Such a list is called a loan **amortization schedule**. Part of an amortization schedule for the Cohens' loan in Example 3 is given in Table 10.5.

Table 10.5 Amortization Schedule

Annual % rate: 8.5   Monthly payment: $742.14
Loan: $85,500   Term: Years 20, Months 0
Periods: 240

Payment Number	Interest	Principal	Balance of Loan
1	$605.63	$136.51	$85,363.49
2	604.66	137.48	85,226.01
3	603.68	138.46	85,087.55
4	602.70	139.44	84,948.11
23	582.69	159.45	82,102.93
24	581.56	160.58	81,942.35
119	428.16	313.98	60,132.43
120	425.94	316.20	59,816.23
179	262.60	479.54	36,593.57
180	259.20	482.94	36,110.63
239	9.74	732.40	642.64
240	4.55	642.64	

Adjustable-Rate Loans

Now let us consider **adjustable-rate mortgages**, ARM's, which are also called **variable-rate mortgages**. The rules for ARM's vary from state to state, and it is possible that the material we present on ARM's may not be true in your state. Generally, with adjustable-rate mortgages, the monthly mortgage payment remains the same for a one-, two-, or five-year period even though the interest rate of the mortgage may change every three months, six months, or some other predetermined period. The interest rate in an adjustable-rate mortgage may be based on an index that is determined by the Federal Home Loan Bank Association or it may be based on the interest rate of a three-month, six-month, or one-year Treasury bill. The interest rate of a three-month Treasury bill may change every three months, a six-month Treasury bill may change every six months, and so on. When the base is a Treasury bill, the actual interest rate charged for the mortgage is often determined by adding 3% to $3\frac{1}{2}$% (called the **add on rate** or **margin**) to the rate of the Treasury bill. Thus if the rate of the Treasury bill is 6% and the add on rate is 3%, then the interest rate charged is 9%.

▶ **Example 4**

The Jacksons purchased a house for $115,000 with a down payment of $23,100. They obtained a 30-year adjustable-rate mortgage with the following terms. The interest rate is based on a six-month Treasury bill, the effective interest rate is 3% above the rate of the Treasury bill on the date of adjustment (3% is the add on rate), the interest rate is adjusted every six months, the interest rate will not change more than 1% (up or down) when the interest rate is adjusted, the maximum interest rate that can be charged for the duration of the loan is 16%, there is no lower limit on the interest rate, the initial mortgage interest rate is 8%, and the monthly payment (including principal and interest) is adjusted every five years.

a) Determine the initial monthly payment.
b) Determine the adjusted interest rate in six months if the interest rate on the Treasury bill at that time is 4.5%.

Solution:

a) To determine the initial monthly payment of interest and principal, divide the amount of the loan, $91,900 ($115,000 − $23,100), by $1000. The result is 91.9. Now multiply the number of thousands of dollars of mortgage, 91.9, by the value found in Table 10.4 with $r = 8$% for 30 years:

$$\$7.34 \times 91.9 = \$674.55$$

Thus the initial monthly payment for principal and interest is $674.55. This amount will not change for the first five years of the mortgage.

b) The adjusted interest rate in six months will be the Treasury bill rate plus the add on rate.

$$4.5\% + 3\% = 7.5\%$$

In Example 4(b), note that the rate after six months is lower than the initial rate of 8%. Since the monthly payment remains the same, the additional money paid the bank is applied to reduce the principal. The monthly interest and principal payment of $674.55 would pay off the loan in 30 years if the interest remained constant at 8%. What happens if the interest rate drops and stays lower than the initial 8% for the length of the loan? In this case, at the end of each five-year period the bank reduces the monthly payment in order to pay off the loan in 30 years. What happens if the interest rates increase above the initial 8% rate? In this case part or, if necessary, all of the principal part of the monthly payment would be used to meet the interest obligation. At the end of the five-year period the bank will increase the monthly payment so that the loan can be repaid in the 30-year time period. There is generally one restriction, that the bank may not increase the monthly payment by more than 25% of the initial monthly payment. If the monthly payment cannot be increased sufficiently to repay the loan in 30 years, then the bank will increase the time period by the number of years required.

Other Types of Mortgages

Conventional mortgages and variable-rate mortgages are not the only methods of financing the purchase of a house. In the following paragraphs are brief descriptions of other methods. Two of these, FHA and VA mortgages, are backed by the government.

FHA Mortgage A house can be purchased with a smaller down payment than with a conventional mortgage if the individual qualifies for a Federal Housing Administration (FHA) loan. He or she applies for the loan through a local bank. The bank provides the money, but the FHA insures the loan. The down payment for an FHA loan is as low as 2.5% percent of the purchase price, rather than the standard 5 to 50%. Another advantage is that FHA loans can be assumed at the original rate of interest on the loan. For example, if you purchase a home today that already has an 8% FHA mortgage, you, the new buyer, will be able to assume that 8% mortgage regardless of the current interest rates. However, to be able to assume a mortgage, the purchaser must be able to make a down payment equal to the difference between the purchase price and the balance due on the original mortgage.

One drawback to FHA loans is that the borrower must pay an FHA insurance premium as part of the monthly mortgage payment. The insurance premium is calculated at a rate of one-half percent (0.5%) of the unpaid balance of the loan on the anniversary date of the loan. Thus even though the insurance premium decreases each year, it adds to the monthly payments.

VA Mortgage The Veterans Administration does not require a down payment for VA-backed loans, however the lending institution may require a down payment. Each veteran applying for a loan can receive government backing of the loan up to $36,000. The amount depends on the veteran's benefit package, which is determined from his or her tour of duty in the service. The amount the government backs can be used in lieu of part or all of the down payment.

For both VA and FHA mortgages the government sets maximum interest rates the lender can charge. A VA loan is always assumable; the seller may be asked to pay points, and there is no monthly insurance premium.

Graduated Payment Mortgage (GPM) A GPM mortgage is designed so that for the first five to ten years the size of the mortgage payment is smaller than the payments for the remaining time of the mortgage. After the 5- to 10-year period the mortgage payments remain constant for the duration of the mortgage. Depending on the size of the mortgage, the lender might find that the monthly payments made for the first few years may actually be less than the interest owed on the loan for those few years. The interest not paid during the first few years of the mortgage is then added to the original loan. This type of loan is strictly for the person who is confident that his or her annual income will be increasing at a rate commensurate with the higher mortgage payments.

Balloon-Payment Mortgage (BPM) This type of loan could be for the person who needs time to find a permanent loan. It may work this way: The individual pays the interest for three to five years, the period of the loan. At the end of the three- to five-year period, the buyer must repay the entire principal unless the lender agrees that the loan may be extended for another period of time.

For further information on any of these mortgages, consult your local banker.

Section 10.5 Exercises

1. Eric is buying a house selling for $130,000. The bank is requiring a minimum down payment of 25%. The current mortgage rate is 9%.
   a) Determine the amount of the required down payment.
   b) Determine the monthly mortgage payment for a 35-year loan with a minimum down payment.
2. Ann and Ted Thomas are buying a house selling for $150,000. The bank is requiring a minimum down payment of 10%. The current mortgage rate is 9.5%.
   a) Determine the amount of the required down payment.
   b) Determine the monthly mortgage payment for a 40-year loan with a minimum down payment.
3. Barbara is buying a condominium selling for $98,000. The bank is requiring a minimum down payment of 10%. To obtain a 30-year mortgage at 9%, she must pay two points at the time of closing.
   a) What is the required down payment?
   b) With the 10% down payment, what is the amount of the mortgage on the property?
   c) What is the cost of two points on the mortgage determined in part (b)?
4. The Imans are buying a house that sells for $120,000. The bank is requiring a minimum down payment of 15%. To obtain a 30-year mortgage at 9.5%, they must pay three points at the time of closing.
   a) What is the required down payment?
   b) With the 15% down payment, what is the amount of the mortgage on the property?

c) What is the cost of three points on the mortgage determined in part (b)?
5. Marlon's gross monthly income is $2500. He has 12 remaining payments of $95 on furniture and appliances. The taxes and fire insurance on the house are $105 per month.
   a) What maximum monthly payment does the bank feel that Marlon can afford?
   b) Marlon would like to get a 30-year, $52,200 mortgage. Does he qualify for this mortgage with a 9% interest rate?
6. The Dechs' gross monthly income is $6200. They have 15 remaining payments of $290 on a new car. The taxes on the house are $150 per month and the fire insurance is $45 per month.
   a) What maximum monthly payment does the bank feel that the Dechs can afford?
   b) The Dechs would like to get a 30-year, $140,000 mortgage. Do they qualify for this mortgage with a 9% interest rate?
7. Andrea obtains a 30-year, $56,250 conventional mortgage at 8.5% on a house selling for $75,000. Her monthly payment, including principal and interest, is $432.56.
   a) Determine the total amount Andrea will pay for her house.
   b) How much of the cost will be interest?
   c) How much of the first payment on the mortgage is applied to the principal?

8. Mr. and Mrs. Georgi obtain a 25-year, $102,000 conventional mortgage at 11.5% on a house selling for $160,000. Their monthly mortgage payment, including principal and interest, is $1036.32.

a) Determine the amount the Georgis will pay for their house.

b) How much of the cost will be interest?

c) How much of the first payment on the mortgage is applied to the principal?

9. The Schollers purchased a home selling for $135,000 with a 20% down payment and were required to pay two points at the time of closing. The period of the conventional mortgage is 35 years, and the rate of interest is 9.5%.

a) Determine the amount of the down payment.

b) Determine the cost of the two points.

c) Determine the monthly mortgage payment.

d) Determine the total cost of the house.

e) Determine the total interest paid.

f) Determine how much of the first payment on the loan is applied to the principal.

10. The Washingtons found a house that is selling for $113,500. The taxes on the house are $1200 per year, and the fire insurance is $320 per year. They are requesting a conventional loan from the local bank. The bank is currently requiring a 25% down payment, three points, and the interest rate is 10%. The Washingtons' monthly income is $4750. They have more than 6 monthly payments remaining on a car, a boat, and furniture. The total monthly payments for these items is $420.

a) Determine the required down payment.

b) Determine the cost of the three points.

c) Determine their adjusted monthly income.

d) Determine the maximum monthly payment the bank feels that they can afford.

e) Determine the monthly payments of principal and interest for a 20-year loan.

f) Determine their total monthly payment, including insurance and taxes.

g) Determine whether the Washingtons qualify for the 20-year loan.

h) Determine how much of the first payment on the loan is applied to the principal.

11. Kathy would like to buy a condominium selling for $85,000. The taxes on the property are $900 per year, and the fire insurance is $336 per year. Kathy's gross monthly income is $3500. She has 15 monthly payments of $135 remaining on her van. She has $20,000 in her savings account. The bank is requiring 20% down and is charging 9.5% interest.

a) Determine the required down payment.

b) Determine the maximum monthly payment that the bank believes Kathy can afford.

c) Determine the monthly payment of principal and interest for a 35-year loan.

d) Determine her total monthly payments, including insurance and taxes.

e) Does Kathy qualify for the loan?

f) Determine how much of the first payment on the mortgage is applied to the principal.

g) Determine the total amount she pays for the condominium with a 35-year conventional loan. (Do not include taxes or insurance.)

h) Determine the total interest paid for the 35-year loan.

12. The Christophers are negotiating with two banks for a mortgage to buy a house selling for $105,000. The terms at bank A are a 10% down payment, an interest rate of 10%, a 30-year conventional mortgage, and three points to be paid at the time of closing. The terms at bank B are a 20% down payment, an interest rate of 11.5%, a 25-year conventional mortgage and no points. Which loan should the Christophers select in order for the total cost of the house to be less?

13. The Kellys purchased a house for $95,000 with a down payment of $13,000. They obtained a 30-year adjustable rate mortgage. The terms of the mortgage are as follows: The interest rate is based on a three-month Treasury bill, the effective interest rate is 3.25% above the rate of the Treasury bill on the date of adjustment, the interest rate is adjusted every three months, the interest rate will not change more than 1% (up or down) when the interest rate is adjusted, the maximum interest rate that can be charged for the duration of the loan is 16%, there is no lower limit on the interest rate, the initial mortgage interest rate is 8.5%, and the monthly payment of interest and principal is adjusted annually.

a) Determine the initial monthly payment for interest and principal.

b) Determine the interest rate in three months if the interest rate on the Treasury bill at the time is 5.65%.

c) Determine the interest rate in six months if the interest rate on the Treasury bill at the time is 4.85%.

Problem Solving

14. The Wongs can afford to pay $950 a month in mortgage payments. If the bank will give them a 25-year conventional mortgage at 9% and requires a 25% down payment, what is (a) the maximum mortgage the bank will grant the Wongs? (b) the highest-price house they can afford?

15. The Kalecks purchased a house for $95,000 with a down payment of $6000. They obtained a 30-year adjustable-rate mortgage. The terms of the mortgage are as follows: The interest rate is based on a three-month Treasury bill, the effective interest rate is 3.25% above the rate of the Treasury bill on the date of adjustment, the interest rate is adjusted every three months, the interest rate will not change more than 1% (up or down) when the interest rate is adjusted, the maximum interest rate that can be charged for the duration of the loan is 16%, there is no lower limit on the interest rate, the initial mortgage interest rate is 9%, and the monthly payment of interest and principal is adjusted semiannually.

a) Determine the initial monthly payment for interest and principal.

b) Determine an amortization schedule for months 1–3.

c) Determine the interest rate for months 4–6 if the interest rate on the Treasury bill at the time is 6.13%.

d) Determine an amortization schedule for months 4–6.

e) Determine the interest rate for months 7–9 if the interest rate on the Treasury bill at the time is 6.21%.

Research Activities

16. An important part of buying a house is the closing. The exact procedures for the closing differ with individual cases and in different parts of the country. In any closing, however, there are expenses to both parties, the buyer and the seller. To determine what is involved in the closing of a property in your community, contact a lawyer, a real estate agent, or a banker. Explain that you are a student and that your objective is to understand the procedure for closing a real estate purchase and the costs to both the buyer and the seller. Select a specific piece of property that is for sale. List the asking price, and then determine the total closing costs to both the buyer and the seller. Here is a partial list of the most common costs. Consider these in your research.

a) Fee for title search and title insurance

b) Credit report on buyer

c) Fees to the lender for services in granting the loan

d) Fee for property survey

e) Fee for recording of the deed

f) Appraisal fee

g) Lawyer's fee

h) Escrow accounts (taxes, insurance)

i) Mortgage assumption fee

17. Visit local banks and investigate the different types of mortgages that are available. List the advantages and disadvantages of each type. Determine which type you would select and explain your decision.

CHAPTER 10 SUMMARY

Key Terms

10.1
percent
percent change
percent markup on cost

10.2
bank discount
Banker's rule
comaker
cosigner
credit
discount note
interest

note
ordinary interest
principal
promissory note
rate
simple interest
time
United States rule

10.3
compound interest
effective annual yield

10.4
annual percentage rate (APR)
average daily balance method
fixed installment loan
open-end installment loan
unpaid balance method

10.5
adjustable-rate mortgage
amortization schedule
conventional loan
down payment
homeowner's mortgage

Important Facts

Ordinary interest When computing ordinary interest, each month is considered to have 30 days and a year is considered to have 360 days.

Banker's rule When computing interest with the banker's rule a year is considered to have 360 days and any fractional part of a year is the exact number of days.

$$\text{Percent change} = \frac{\left(\begin{array}{c}\text{amount in}\\\text{latest period}\end{array}\right) - \left(\begin{array}{c}\text{amount in}\\\text{previous period}\end{array}\right)}{\text{amount in previous period}} \times 100$$

$$\text{Percent markup on cost} = \frac{\text{selling price} - \text{dealer's cost}}{\text{dealer's cost}} \times 100$$

Simple interest formula

Interest = principal × rate × time
or $i = prt$

Compound interest formula

$$A = p\left(1 + \frac{r}{n}\right)^{nt}$$

Present value formula

$$p = \frac{A}{\left(1 + \frac{r}{n}\right)^{nt}}$$

Actuarial method   **Rule of 78s**

$$u = \frac{n \cdot P \cdot V}{100 + V} \qquad u = \frac{f \cdot k(k + 1)}{n(n + 1)}$$

CHAPTER 10 REVIEW EXERCISES

10.1

Change the following to percents. Express your answer to the nearest tenth of a percent.

1. $\frac{1}{8}$ **2.** $\frac{3}{5}$ **3.** $\frac{3}{6}$

4. 0.045   **5.** 0.0075   **6.** 1.25

Change the following percents to decimal numbers.

7. 26%   **8.** 11.5%   **9.** 0.056%

10. $\frac{2}{5}\%$ **11.** $\frac{5}{6}\%$ **12.** 0.00045%

13. In 1985 Corinne purchased an automobile for $15,000. The 1992 model of the same automobile cost $22,500. What is the percent increase in the price of the automobile?

14. Hugo the happy salesman earned a commission of $47,800 in 1989 and $42,870 in 1992. What is the percent change in his commission from 1989 to 1992?

In Ex. 15–19, write each statement as an equation and solve.
15. What percent of 80 is 20?
16. Forty-four is 20% of what number?
17. What is 18% of 540?

18. If your restaurant bill is $48.93 and the suggested tip is 15%, what is the tip on the meal?
19. If the number of people in your card club increased by 20% or 7, what was the original number of people in the club?

20. The football team had 85 players and increased the number of players to 110. What is the percent increase in the number of players?

10.2
Find the missing quantity using the simple interest formula.
21. $p = \$1500$, $r = 6\%$, $t = 30$ days, $i = ?$
22. $p = \$1275$, $r = ?$, $t = 100$ days, $i = \$35.42$
23. $p = ?$, $r = 8\frac{1}{2}\%$, $t = 3$ years, $i = \$114.75$
24. $p = \$5500$, $r = 11\frac{1}{2}\%$, $t = ?$, $i = \$316.25$

25. Bobby borrowed $4500 from the bank for 12 months. His mother cosigned the promissory note. The rate of interest on the note was $11\frac{1}{2}\%$. How much did Bobby pay the bank on the date of maturity?
26. Sue borrowed $2300 from her bank for 180 days at a simple interest rate of 12%.
   a) How much interest did she pay for the use of the money?

b) How much did she pay the bank on the date of maturity?

27. The Manbecks borrowed $4000 for 18 months from the bank, using stock as security. The bank discounted the loan at $11\frac{1}{2}\%$.

  a) How much interest did the Manbecks pay the bank for the use of the money?

  b) How much did they receive from the bank?

  c) What was the actual rate of interest?

28. Charlotte borrowed $800 for six months from her bank, using her savings account as security. A bank rule limits the amount that can be borrowed in this manner to 85% of the amount in the savings account. The rate of interest is 2 percent higher than the interest rate being paid on the savings account. The current rate on the savings account is $5\frac{1}{2}\%$.

  a) What rate of interest will the bank charge for the loan?

  b) Find the amount that Charlotte must repay in six months.

  c) How much money must she have in her account in order to borrow $800?

29. How many dollars must you invest today to have $28,000 in 20 years? Assume the money earns 6% interest compounded quarterly.

10.3

30. Determine the amount and the interest when $500 is invested for three years at 6%

  a) compounded annually.

  b) compounded semiannually.

  c) compounded quarterly.

31. Donald deposited $5000 in a savings account that pays $5\frac{1}{4}\%$ interest compounded quarterly. What will be the total amount of money in the account 8 years from the day of deposit.

32. Determine the effective annual yield of an investment if the interest is compounded daily at an annual percentage rate of 5.6%.

10.4

33. Wanda has a 48-month installment loan, with a fixed monthly payment of $193.75. The amount borrowed was $7500. Instead of making her 24th payment, Wanda is paying the remaining balance on the note.

  a) Determine the APR of the installment loan.

  b) How much interest will Wanda save, computed by the actuarial method?

  c) What is the total amount due on that day?

34. Reggie is buying a book collection that costs $4000. He

is making a down payment of $500 and 24 monthly payments of $163.33. Instead of making his 12th payment, Reggie decides to pay the total remaining balance and terminate the loan.

  a) How much interest will Reggie save, computed by the rule of 78s?

  b) What is the total amount due on that day?

35. Jessica paid $3420 for a new wardrobe. She made a down payment of $860 and financed the balance on a 24-month fixed payment installment loan. The monthly payments are $119.47. Instead of making her 12th payment, Jessica decides to pay the total remaining balance and terminate the loan.

  a) Determine the APR of the installment loan.

  b) How much interest will Jessica save, computed by the actuarial method?

  c) What is the total amount due on the day?

36. On November 1, the billing date, Lydia had a balance due of $485.75 on her credit card. The transactions during the month of November are:

November 4	Payment	$375.00
November 8	Charge: Airline ticket	370.00
November 21	Charge: Hotel bill	175.80
November 28	Charge: Clothing	184.75

  a) Find the finance charge on December 1 using the unpaid balance method. Assume the interest rate is 1.3% per month.

  b) Find the new account balance on December 1.

37. On March 5, the billing date, Janine had a balance due of $75.92 on her credit card. She is redecorating her apartment and made the following transactions.

March 8	Charge: Clothing	$ 85.75
March 10	Payment	65.00
March 15	Charge: Draperies	72.85
March 21	Charge: Chair	275.00

  a) Find the finance charge on April 5 using the average balance method. Assume the interest rate is 1.4% per month.

  b) Find the new account balance on April 5.

38. Patrick bought a new Dodge Spirit for $12,500 . He was required to make a 25% down payment. He financed the balance with the dealer on a 48-month payment plan. The salesperson told him that the interest rate was 7.5% simple interest on the principal for the duration of the loan.

  a) Find the amount of the down payment.

  b) Find the amount to be financed.

c) Find the total interest paid.

d) Find the APR.

39. Joyce can buy a cross-country skiing outfit for $135. The store is offering the following terms: $35 down and 12 monthly payments of $8.95.

a) Find the interest paid.

b) Find the APR.

10.5

40. The Clars have decided to build a new house. The contractor has quoted them a price of $135,700. The taxes on the house will be $2350 per year, and fire insurance will be $350 per year. They have applied for a conventional loan from a local bank. The bank is requiring a 35% down payment, and the interest rate on the loan is 9.5%. The Clars' annual income is $54,000. They have more than 6 monthly payments remaining on each of the following: $218 on a car, $120 on new furniture, and $190 on a camper.

a) Determine the required down payment.

b) Determine their adjusted monthly income.

c) Determine the maximum monthly payment the bank feels that they can afford.

d) Determine the monthly payment of principal and interest for a 40-year loan.

e) Determine their total monthly payment, including insurance and taxes.

f) Do the Clars qualify for the mortgage?

41. The Bakunins purchased a home selling for $89,900 with a 25% down payment. The period of the mortgage is 30 years and the interest rate is 13.5%. Determine the following.

a) The amount of the down payment

b) The monthly mortgage payment

c) The amount of the first payment that is applied to the principal

d) The total cost of the house

e) The total interest paid.

42. The Browns purchased a house for $105,000 with a down payment of $26,250. They obtained a 30-year adjustable-rate mortgage. The terms of the mortgage are as follows: The interest rate is based on the 6-month Treasury bill, the effective interest rate is 3.50% above the rate of the Treasury bill on the date of adjustment, the interest rate is adjusted every six months, the interest rate will not change more than 1% (up or down) when the interest rate is adjusted, the maximum interest rate that can be charged for the duration of the loan is 16%, there is no lower limit on the interest rate, the initial mortgage interest rate is 9.5%, the monthly payment of interest and principal is adjusted annually.

a) Determine the initial monthly payment for interest and principal.

b) Determine the interest rate in 6 months if the interest rate on the Treasury bill at the time is 5.65%.

c) Determine the interest rate in 6 months if the interest rate on the Treasury bill at the time is 5.85%.

CHAPTER TEST

Find the missing quantity using the simple interest formula.

1. $i = ?, \quad p = \$500, \quad r = 8\%, \quad t = 7$ months

2. $i = \$280, \quad p = \$500, \quad r = 8\%, \quad t = ?$

Greg borrowed $3000 from a bank for 11 months. The rate of interest charged is 11.4%.

3. How much interest did he pay for the use of the money?

4. What is the amount he repaid to the bank on the due date of the loan?

A new computer sells for $4350. To finance the computer through a bank, the bank will require a down payment of 15% and monthly payments of $172.50 for 24 months.

5. How much money will the purchaser borrow from the bank?

6. What finance charge will the individual pay the bank?

7. What is the APR?

8. Jacques purchased a fishing boat for $3375. He made a down payment of $875 and financed the balance with a 12-month fixed-payment installment note. Instead of making the sixth monthly payment of $220.34 he decides to pay the remaining balance.

    a) How much interest will Jacques save, computed by the rule of 78s?

    b) What is the total amount due on that day?

9. The balance on the Bishops' credit card on May 8, their billing date, was $478.50. For the period ending June 8 they made the following transactions.

May 13	Charge: Gifts	$276.79
May 15	Payment	$150.00
May 18	Charge: Books	$144.85
May 29	Charge: Restaurant bill	$ 35.63

    a) Find the finance charge on June 8 by the average daily balance method. Assume the interest rate is 1.3% per month.

    b) Find the new balance on June 8.

10. The O'Grady's credit card statement shows a balance due of $578.25 on March 23, the billing date. For the period ending April 23 they made the following transactions.

March 26	Charge: Auto supplies	$ 75.89
March 30	Charge: Restaurant bill	$ 48.76
April 3	Payment	$250.00
April 15	Charge: Clothing	$ 90.52
April 22	Charge: Lumber	$250.85

    a) Find the finance charge due on March 23 using the unpaid balance method. Assume the interest rate is 1.4% per month.

    b) Find the new account balance on April 23.

Brian signed a $4400 note with interest at 12.5% for 90 days on August 1. Brian made a payment of $2000 on September 15.

11. How much did he owe the bank on the date of maturity?

12. What was the total amount of interest paid on the loan?

Compute the amount and the compound interest for each of the following:

	Principal	Time	Rate	Compounded
13.	$3600	4 years	12%	Quarterly
14.	$4800	2 years	16%	Semiannually

The Klines decided to build a new house. The contractor quoted them a price of $154,500 including the lot. The taxes on the house would be $2875 per year, and fire insurance would cost $450 per year. They have applied for a conventional loan from a bank. The bank is requiring a 25% down payment, and the interest rate is $10\frac{1}{2}\%$. The Klines' annual income is $76,500. They have more than six monthly payments remaining on each of the following: $220 for a car, $175 for new furniture, and $210 on a college education loan.

15. What is the required down payment?

16. Determine their adjusted monthly income.

17. What is the maximum monthly payment the bank feels the Klines can afford?

18. Determine the monthly payments of principal and interest for a 20-year loan.

19. Determine their total monthly payments including insurance and taxes.

20. Does the bank feel the Klines meet the requirements for the mortgage?

21. Find the total cost of the house after 20 years of payments (do not include taxes and insurance).

22. How much of the total cost of the house is interest?

The laws of probability have applications in the science of genetics. The larger and more diverse the population, the greater the probability that the breed will have the characteristics necessary to adapt to changes in the environment. The cheetah, the world's fastest-running mammal, faces extinction because it lacks the genetic diversity necessary to survive disease. Once found worldwide, the breed now lives wild in a few areas of Africa.

PROBABILITY

"With his elbows propped, his head between his hands, he seemed to lose himself in the study of an abstruse problem in mathematics." So sat the character George Massy in the Joseph Conrad story *The End of the Tether* (1902). Massy had won the second-greatest prize in the Manila lottery. So obsessed was he with winning again that he pored over lists of winning numbers. "There was a hint there of a definite rule! He was afraid of missing some recondite principle in the overwhelming wealth of his material. What could it be?" It could be, and indeed is, a principle belonging to an area of mathematics known as probability.

The fascination that the fictional character George Massy displayed in a game of chance is shared in some degree by the millions of Americans each year who "gamble" on state lotteries or play bingo. In 1988 some $211 billion was legally wagered in the

It is now legal to place some kind of bet in all but 2 of the 50 United States (Utah and Hawaii are the exceptions).

United States. If you play the lottery, your hope is
to beat the odds and be the person with the winning numbers. Mathematicians of the sixteenth,
seventeenth, and eighteenth centuries, not satisfied
with leaving things to chance, invented the study of
probability to use mathematics to determine the
likelihood of an event (winning numbers, for example) occurring.

Although the rules of probability were first
applied to gaming, there are many other applications. The quality of the food you eat, the pedigree
of your cat or dog, the cost of your car insurance
— all these have to do with probability. In agriculture, animals and plants are bred to ensure that
certain desirable traits will be passed on from generation
to generation. In the business of insurance
underwriting, the likelihood of an event
like an automobile accident, given the
age, sex, and location of the driver,
are facts used in determining the
cost of the insurance.

Insurers use the laws of
probability to set rates and
define risk. One of the best-
known insurers is Lloyd's of
London. Lloyd's has the reputation of insuring just about
anything, from the dia-
monds of actress
Elizabeth Taylor to the
riches of the Tut-Ankhamen
exhibition.

Governments and private industry
apply the counting principle of
probability when they set the patterns of
numbers and letters used to identify individuals.
This includes social security numbers, telephone
numbers, and license plates.

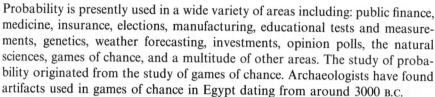

11.1 The Nature of Probability

History

Probability is presently used in a wide variety of areas including: public finance, medicine, insurance, elections, manufacturing, educational tests and measurements, genetics, weather forecasting, investments, opinion polls, the natural sciences, games of chance, and a multitude of other areas. The study of probability originated from the study of games of chance. Archaeologists have found artifacts used in games of chance in Egypt dating from around 3000 B.C.

Mathematical problems relating to games of chance were studied by a number of mathematicians of the Renaissance. Italy's Girolamo Cardano (1501–1576) in his *Liber de Ludo Aleae* (Book on the games of chance) presents one of the first systematic computations of probabilities. Although it is basically a gambler's manual, many consider it the first book ever written on probability. A short time later, two French mathematicians, Blaise Pascal (1623–1662) and Pierre de Fermat (1601–1665), worked together studying "the geometry of the die." In 1657 the Dutch mathematician Christian Huygens (1629–1695) published *De Ratiociniis in Luno Aleae* (On ratiocination in dice games), which contained the first documented reference to the concept of mathematical expectation (see Section 11.4). The Swiss mathematician Jacob Bernoulli (1654–1705), whom many consider the founder of probability theory, is said to have fused pure mathematics with the empirical methods used in statistical experiments. The works of Pierre-Simon de Laplace (1749–1827) dominated probability throughout the nineteenth century.

Jacob Bernoulli (1654–1705), the Swiss mathematician considered to be a pioneer of probability theory, was part of a famous family of mathematicians and scientists. In Ars Conjectand, published posthumously in 1713, he proposed that an increased degree of accuracy can be obtained by increasing the number of trials of an experiment. This theorem, called Bernoulli's theorem (of probability), is also known as the law of large numbers. Bernoulli's theorem of fluid dynamics, used in aircraft wing design, was developed by *Daniel Bernoulli* (1700–1782), Jacob's second son.

The Nature of Probability

Before we discuss the meaning of the word "probability" and learn how to calculate probabilities, we must introduce a few definitions.

> An **experiment** is a controlled operation that yields a set of results.

The process in which medical researchers administer experimental drugs to patients to determine their reaction is one type of experiment.

> The possible results of an experiment are called its **outcomes**.

For example, when administering an experimental drug the possible outcomes may be a favorable reaction, no reaction, or an adverse reaction.

> An **event** is a subcollection of the outcomes of an experiment.

"The laws of probability, so true in general, so fallacious in particular."
Edward Gibbon, 1796

For example, when a die is rolled, the event of rolling a number greater than 2 can be satisfied by any one of four outcomes—3, 4, 5, or 6. The event of rolling a 5 can be satisfied by only one outcome, the 5 itself. The event of rolling an even number can be satisfied by any of three outcomes—2, 4, or 6.

Probability is classified as either **empirical** (experimental) or **theoretical** (mathematical). **Empirical probability** is the relative frequency of occurrence of the event and is determined by actual observations of an experiment. **Theoretical probability** is determined through a study of the possible *outcomes* that can occur for the given experiment. We will indicate the empirical probability of an event E as $F(E)$. We will use the letter F for empirical probability since it is computed using the frequency of an event occurring. The theoretical probability of event E is symbolized by $P(E)$.

In this section we will briefly discuss empirical probability. The emphasis in the remaining sections is on theoretical probability. Following is the formula for computing empirical probability, or relative frequency.

Empirical Probability (Relative Frequency)

$$F(E) = \frac{\text{number of times event } E \text{ has occurred}}{\text{total number of times the experiment has been performed}}$$

The probability of an event, whether empirical or theoretical, will always be a number between 0 and 1, inclusive, and may be expressed as a decimal, a fraction, or a percent. An empirical probability of 0 indicates that the event has never occurred. An empirical probability of 1 indicates that the event has always occurred.

▶ **Example 1**

In 100 tosses of a fair coin, 58 landed heads up. Find the empirical probability of the coin landing heads up.

Solution: Let E be the event that the coin lands heads up. Then

$$F(E) = \frac{58}{100} = 0.58.$$

▶ **Example 2**

A pharmaceutical company is testing a new drug that is supposed to reduce cholesterol. The drug is given to 500 individuals with the following outcome:

Cholesterol Reduced	Cholesterol Unchanged	Cholesterol Increased
307	120	73

If this drug is given to an individual, find the empirical probability that the cholesterol level is (a) reduced, (b) not affected, (c) increased.

Solution:

a) Let E be the event that the cholesterol level is reduced.

$$F(E) = \frac{307}{500} = 0.614$$

b) Let E be the event that the cholesterol level is not affected.

$$F(E) = \frac{120}{500} = 0.24$$

c) Let E be the event that the cholesterol level is increased.

$$F(E) = \frac{73}{500} = 0.146$$

Empirical probability is used in situations where probabilities cannot be theoretically calculated. For example, life insurance companies use empirical probabilities to determine the chance of an individual in a certain profession, with certain risk factors, living to age 65.

Empirical Probability in Genetics

Using empirical probability, Gregor Mendel (1822–1884) developed the laws of heredity by crossbreeding different types of "pure" pea plants and observing the relative frequencies of the resulting offspring. These laws became the foundation for the study of genetics. For example, when he crossbred a pure yellow pea plant and a pure green pea plant, the resulting offspring (the first generation) were always yellow (see Fig. 11.1a). When he crossbred a pure round-seeded pea plant and a pure wrinkled-seeded pea plant, the resulting offspring (the first generation) were always round (Fig. 11.1b).

Richard von Mises (1883–1953) and **Hilda Geiringer von Mises** (1893–1973) are known for their contributions in applying probability theory to biology and genetics. They began working together at the University of Berlin in 1921, when Geiringer became von Mises's assistant. Von Mises proposed that all probability is grounded in real phenomena—that only after a large number of observations can one make a statement about the probability of an event. Von Mises and Geiringer fled Germany in 1933 and settled in the United States, where he taught at Harvard and she chaired the math department at Wheaton College. They were married in 1943. After Richard von Mises's death, his wife took up his early work, reworking and expanding his controversial views.

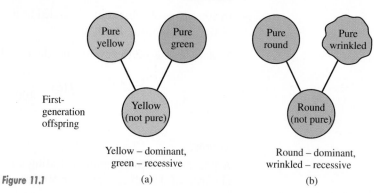

Figure 11.1

Yellow – dominant, green – recessive

(a)

Round – dominant, wrinkled – recessive

(b)

Traits such as yellow color and round seeds Mendel called **dominant** because they overcame or "dominated" the other trait. The green and wrinkled traits he labeled **recessive**.

Mendel then crossbred the offspring of the first generation. The resulting second generation had both the dominant and the recessive traits of their

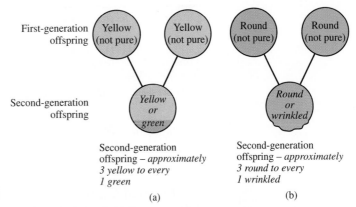

First-generation offspring — Yellow (not pure) · Yellow (not pure) · Round (not pure) · Round (not pure)

Second-generation offspring — *Yellow or green* · *Round or wrinkled*

Second-generation offspring – *approximately* 3 yellow to every 1 green

Second-generation offspring – *approximately* 3 round to every 1 wrinkled

Figure 11.2

(a)

(b)

grandparents (see Fig. 11.2a and b). What's more, these traits always appeared in approximately a 3 to 1 ratio of dominant to recessive.

Did You **Know...**

THE ROYAL DISEASE

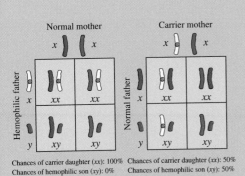

Normal mother

x) (x

Carrier mother

x) (x

Hemophilic father

x — xx · xx

y — xy · xy

Normal father

x — xx · xx

y — xy · xy

Chances of carrier daughter (xx): 100%
Chances of hemophilic son (xy): 0%

Chances of carrier daughter (xx): 50%
Chances of hemophilic son (xy): 50%

In humans genes are located on 23 pairs of *chromosomes.* Each parent contributes one member of each pair to a child. The gene that affects blood clotting is carried on the X chromosome. Females have two X chromosomes; males have one X chromosome and one Y chromosome.

The effect of genetic inheritance is dramatically demonstrated by the occurrence of hemophilia in some of the royal families of Europe. Great Britain's Queen Victoria (1819–1901) was the initial carrier. The disease was subsequently introduced into the royal lines of Prussia, Russia, and Spain through the marriages of her children.

Hemophilia is a disease that keeps blood from clotting. As a result, even a minor bruise or cut can be dangerous. It is also a recessive sex-linked disease. Females have a second gene that enables the blood to clot, which blocks the effects of the recessive carrier gene; males do not. So even though both males and females are carriers, the disease afflicts only the males.

Queen Victoria had 9 children, 26 grandchildren, and 34 great grandchildren. Among them 1 son and 9 grandsons were hemophiliacs, and 2 daughters and 4 granddaughters were carriers of the gene for hemophilia. The genetic line of the present-day royal family is free of the disease.

Table 11.1 Second Generation Offspring					
Dominant Trait	Number with Dominant Trait	Recessive Trait	Number with Recessive Trait	Ratio of Dominant to Recessive	F (dominant trait)
Yellow seeds	6022	Green seeds	2001	3.01 to 1	$\dfrac{6022}{8023} = 0.75$
Round seeds	5474	Wrinkled seeds	1850	2.96 to 1	$\dfrac{5474}{7324} = 0.75$

Table 11.1 lists some of the results of Mendel's experiments with pea plants. Note that the ratio of dominant trait to recessive trait in the second generation offspring is about 3 to 1 for each of the experiments. The empirical probability of the dominant trait has also been calculated. How would you find the empirical probability of the recessive trait?

From his work Mendel concluded that the sex cells (now called gametes) of the pure yellow (dominant) pea plant carried some factor that caused the offspring to be yellow, and the gametes of the green variety had a variant factor that "induced the development of green plants." In 1909 the Danish geneticist W. Johannsen called these factors "genes." Mendel's work led to the understanding that each pea plant contains two genes for color, one that comes from the mother and the other from the father. If the two genes are alike, for instance both for yellow plants or both for green plants, the plant will be that color. If the genes for color are different the plant will grow the color of the dominant gene. Thus if one parent contributes a gene for the plant to be yellow (dominant) and the other parent contributes a gene for the plant to be green (recessive), the plant will be yellow.

The Law of Large Numbers

Most of us accept the fact that if a "fair coin" is tossed many, many times, it will land heads up approximately half of the time. Intuitively, one can guess that the probability that a fair coin will land heads up is $\frac{1}{2}$. Does this mean that if a coin is tossed twice, it will land heads up exactly once? If a fair coin is tossed 10 times, will there necessarily be five heads? The answer is clearly no. What then does it mean when we state that the probability that a fair coin will land heads up is $\frac{1}{2}$? To answer this question, let us examine Table 11.2, which shows what may occur when a fair coin is tossed a given number of times.

The last column of Table 11.2, the relative frequency of heads, is a ratio of the number of heads observed to the total number of tosses of the coin. The relative frequency is the empirical probability, as defined earlier. Note that as the number of tosses increases, the relative frequency of heads gets closer and closer to $\frac{1}{2}$, or 0.5, which is what we expect.

Table 11.2			
Number of Tosses	Expected Number of Heads	Actual Number of Heads Observed	Relative Frequency of Heads
10	5	4	$\dfrac{4}{10} = 0.4$
100	50	43	$\dfrac{43}{100} = 0.43$
1,000	500	540	$\dfrac{540}{1,000} = 0.54$
10,000	5,000	4,852	$\dfrac{4,852}{10,000} = 0.4852$
100,000	50,000	49,770	$\dfrac{49,770}{100,000} = 0.49770$

If Ryne Sandberg gets three hits in his first three at bats of the season, he is batting a thousand (1.000). But over the course of the 162 games of the season (with three or four at bats per game), his batting average will fall closer to .306 (his 1990 average). In 1990 out of 615 at bats, Sandberg had 188 hits, an average above most players' but much less than 1.000. His batting average is a relative frequency (or empirical probability) of hits to at bats. It is only the long-term average that we take seriously, based on the law of large numbers.

The nature of probability is summarized by the law of large numbers.

> The **law of large numbers** states that probability statements apply in practice to a large number of trials—not to a single trial. It is the relative frequency over the long run that is accurately predictable, not the individual events or precise totals.

What does it mean to say that the probability of rolling a 2 on a die is $\frac{1}{6}$? It means that over the long run, on the average, one out of every six rolls will result in a 2.

Section 11.1 Exercises

1. A probability statement is often interpreted differently by different people. Consider the statement "The National Weather Service predicts that the probability of rain today in the Denver, Colorado, area is 70%." Below are five possible interpretations of that statement. Which, if any, do you think is its actual meaning?
   a) In 70% of the Denver area there will be rain.
   b) There is a 70% chance that at least any one point in the Denver area will receive rain.
   c) It has rained on 70% of the days with similar weather conditions in the Denver area.
   d) There is a 70% chance that a specific location in the Denver area will receive rain.

   e) There is a 70% chance that the entire Denver area will receive rain.
2. Which mathematician is considered by many to be the founder of probability theory?
3. In order to determine premiums, life insurance companies must compute the probable date of death. On the basis of a great deal of research it is determined that Mr. Bennett, age 35 will live another 42.34 years. Does this mean that Mr. Bennett will live until he is 77.34 years old? If not, what does it mean?
4. The probability of rolling a 4 on a die is $\frac{1}{6}$. Does this mean that if a die is rolled six times, one 4 will appear? If not, what does it mean?

5. If you roll a die many times, what would you expect to be the relative frequency of rolling an even number?

6. Briefly explain the meaning of empirical probability.

7. Explain how you would find the empirical probability of rolling a 4 on a die.

8. Find the empirical probability of rolling a 4 by rolling a die 50 times.

9. Explain, in your own words, the law of large numbers. Give an example to illustrate your explanation.

10. Roll a die 50 times and record the results. Find the empirical probability of rolling (a) a 1 and (b) a 6. (c) Does it appear that the probability of rolling a 1 is the same as the probability of rolling a 6?

11. Roll a pair of dice 60 times and record the sums. Compute the empirical probability of rolling a sum of (a) 2 and (b) 7. (c) Does it appear that the probability of rolling a sum of 2 is the same as the probability of rolling a sum of 7?

12. Toss two coins 50 times and record the number of times exactly one head was obtained. Compute the empirical probability of tossing exactly one head.

13. The last 20 birds that fed at the Zwick's bird feeder were 10 finches, 7 cardinals, and 3 blue jays. Use this information to determine the empirical probability that the next bird to feed from the feeder is a
   a) finch
   b) cardinal
   c) blue jay

14. In a sample of 10,000 births, 65 were found to have syndrome A. Find the empirical probability that Mrs. Gardina's first child will be born with this syndrome.

15. Of 100 brand X carrot seeds planted, 92 germinated and produced carrots.
   a) Find the empirical probability of germination of brand X carrot seeds.
   b) How many of these seeds should you plant if you want to grow 368 carrots?

16. A certain community recorded 3000 births last year, of which 1450 were females. What is the empirical probability that the next child born in the community will be
   a) female
   b) male

17. There is a faulty candy machine that accepts dollar bills next to the classroom. A student obtains the information to form the following chart by observing fellow students' experience with the machine.
   a) If the machine is tried a 41st time, what is the empirical probability that it will work properly?
   b) What is the empirical probability of obtaining candy from the machine?

Outcome	Frequency of Occurrence
Gives the candy and returns the proper change	25
Gives the candy but does not return the proper change	8
Does not give the candy and keeps the money	7
	40

18. In a survey, a sample of 1600 people were asked whether the president's performance was good, fair, or poor. The results were as follows.

Number of People	Rating
380	good
602	fair
490	poor
128	no opinion

If this sample is representative of the general population, find the empirical probability that the next person surveyed
   a) rates the president good
   b) rates the president poor
   c) has no opinion

19. An experimental serum was injected into 500 guinea pigs. Initially, 150 of the guinea pigs had circular cells, 250 had elliptical cells, and 100 had irregularly shaped cells. After the serum was injected, it was observed that none of the guinea pigs with circular cells were affected, 50 with elliptical cells were affected, and all of those with irregular cells were affected. Find the empirical probability that a guinea pig with (a) circular cells, (b) elliptical cells, and (c) irregular cells will be affected by injection of the serum.

20. Jim finds an irregularly shaped five-sided rock. He labels each side and tosses the rock 100 times. The result of his tosses is illustrated below. Find the empirical probability that the rock will land on side 4 if tossed again.

Side	1	2	3	4	5
Frequency	32	18	15	13	22

21. In one of Mendel's experiments, he crossbred nonpure purple flower pea plants. These purple pea plants had

two traits for flowers, purple (dominant) and white (recessive). The result of this crossbreeding was 705 second generation plants with purple flowers and 224 second generation plants with white flowers. Find the empirical probability of a second generation plant having (a) white flowers, (b) purple flowers.

22. In another experiment Mendel crossbred nonpure tall pea plants. The results of the second generation off-springs were 787 tall plants and 277 short plants. Find the empirical probability of a second generation plant being (a) tall, (b) short.

23. In the United States there are generally more male infants born than female. In 1988 there were 2,002,000 males born and 1,907,000 females born (to the neasest thousand). Find the empirical probability of an individual being born (a) male, (b) female.

Research Activities

24. Write a report on how insurance companies use empirical probabilities in determining insurance premiums. An insurance agent may be able to direct you to a source of information to answer this question.

25. Write a report on how Gregor Mendel's use of empirical probability led to the development of the science of genetics.

11.2 Theoretical Probability

To be able to do the problems in this section and the remainder of the chapter, you must have a thorough understanding of fractions. If you have forgotten how to use fractions, we strongly suggest that you review Section 5.3 before beginning this section.

Should you spend the 29 cents for a stamp to return the *Reader's Digest* Sweepstakes ticket? What are your chances of winning a lottery? If you go to a carnival or bazaar, which games provide the greatest chance of winning? These and similar questions can be answered once you have an understanding of theoretical probability. *In the remainder of this chapter, the word "probability" will refer to theoretical probability.*

Recall from Section 11.1 that the results of an experiment are called outcomes. When you roll a die and observe the number of points that face up, the possible outcomes are 1, 2, 3, 4, 5, and 6. It is equally likely that you will roll any one of the possible numbers.

> If each outcome of an experiment has the same chance of occurring as any other outcome, they are said to be **equally likely outcomes**.

Can you think of a second set of equally likely outcomes when a die is rolled? An odd number is as likely to be rolled as an even number. Therefore odd and even are another set of equally likely outcomes.

If an event has **equally likely outcomes**, the probability of event E, symbolized by $P(E)$, may be calculated with the following formula.

"It is remarkable that a science which began with the consideration of games of chance should have become the most important object of human knowledge."
Marquis de Laplacé (1749–1827)

> **Probability**
>
> $$P(E) = \frac{\text{number of outcomes favorable to } E}{\text{total number of possible outcomes}}$$

Let us do a problem illustrating how this formula is used.

▶ **Example 1**

A die is rolled. Find the probability of rolling

a) a 5   **b)** an odd number   **c)** a number greater than 4

d) a 7   **e)** a number less than 7

Solution:

a) There are six possible equally likely outcomes: 1, 2, 3, 4, 5 and 6. The event of rolling a 5 can occur in only one way.

$$P(5) = \frac{\text{the number of outcomes that will result in a 5}}{\text{total number of possible outcomes}} = \frac{1}{6}$$

b) The event of rolling an odd number can occur in three ways (1, 3, 5):

$$P(\text{odd number}) = \frac{\text{number of outcomes that result in an odd number}}{\text{total number of possible outcomes}}$$

$$= \frac{3}{6} = \frac{1}{2}$$

c) There are two numbers greater than 4, namely, 5 and 6:

$$P(\text{number greater than 4}) = \frac{2}{6} = \frac{1}{3}$$

d) There are no outcomes that will result in a 7. Thus the event cannot occur and the probability is 0.

$$P(7) = \frac{0}{6} = 0$$

e) All of the outcomes 1 through 6 are less than 7. Thus the event must occur and the probability is 1.

$$P(\text{number less than 7}) = \frac{6}{6} = 1$$

Four important facts on probability are boxed below.

Some theorists have proposed that the best way to make money in the stock market is to randomly select a large, diversified portfolio of stocks. This hypothesis, known as "a random walk down Wall Street," states that the probability of a stock price going up or down is as random as flipping a coin, and that commonly available information cannot help you predict the way it will go.

---

1. The probability of an event that cannot occur is 0.

2. The probability of an event that must occur is 1.

3. Every probability will be a number between 0 and 1 inclusive, that is, $0 \leq P(E) \leq 1$.

4. The sum of the probabilities of all possible outcomes of an experiment is 1.

Table 11.3 Astronauts
Edwin Aldrin, Jr.
Neil Armstrong
Frank Borman
Scott Carpenter
Michael Collins
Charles Conrad, Jr.
L. Gordon Cooper
John Engle
James Erwin
John Glenn
Richard Gordon, Jr.
Vergil Grissom
Jim Lovell
James McDivitt
Sally Ride
Walter Schirra
Alan Shepard
Thomas Stafford
Edward White
John Young

Let us consider a second example.

▶ **Example 2**

The names of 20 astronauts are listed in Table 11.3. Each name is written on a slip of paper, and the 20 slips are deposited in a bag. One name is to be selected at random from the bag. Find the probability that

a) the last name starts with C;
b) the name of the first person to step on the moon is selected;
c) the name of one of the three astronauts on the first moon-walk mission is selected.

Solution:

a) Four of the 20 astronauts have last names that start with C:

$$P(\text{last name starts with C}) = \frac{4}{20} = \frac{1}{5}$$

b) Neil Armstrong was the first person to step on the moon on July 20, 1969:

$$P(\text{the name of the first person to step on the moon}) = \frac{1}{20}$$

c) Neil Armstrong, Edwin Aldrin, Jr., and Michael Collins were on the first moon-walk mission:

$$P(\text{one of the three astronauts on the first moon-walk mission}) = \frac{3}{20}$$

In any given experiment an event must either occur or not occur. *The sum of the probability that an event will occur and the probability that it will not occur is 1.* Thus for any event A we conclude that

$$P(A) + P(\text{not } A) = 1$$
$$\text{or} \quad P(\text{not } A) = 1 - P(A).$$

For example, if the probability that event A will occur is $\frac{5}{12}$, the probability that event A will not occur is $1 - \frac{5}{12}$, or $\frac{7}{12}$. Similarly, if the probability that event A will not occur is 0.3, the probability that event A will occur is $1 - 0.3 = 0.7$ or $\frac{7}{10}$. We make use of this concept in the following example.

▶ **Example 3**

A deck of 52 playing cards is illustrated on page 518. The deck consists of four suits: spades, diamonds, hearts, and clubs. Each suit has 13 cards. Hearts and diamonds are red; clubs and spades are black. There are 12 picture cards,

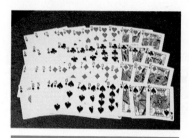

A standard deck of cards consists of 52 cards. There are four suits: hearts, diamonds, clubs, and spades. For each suit, there are 13 cards, including numbered cards from ace (1) through 10 and three picture (or face) cards—the jack, the queen, and the king.

consisting of 4 jacks, 4 queens, and 4 kings. One card is to be selected at random from the deck of cards. Find the probability that the card selected is

a) a 3   **b)** not a 3

c) a spade   **d)** a jack *or* queen *or* king (a picture card)

e) a heart *and* a club   **f)** a card greater than 6 *and* less than 9

Solution:

a) There are four 3s in a deck of 52 cards:

$$P(3) = \frac{4}{52} = \frac{1}{13}$$

b) $P(\text{not a } 3) = 1 - P(3)$

$$= 1 - \frac{1}{13} = \frac{12}{13}$$

This could also have been found by noting that there are 48 cards that are not 3s in a deck of 52 cards:

$$P(\text{not a } 3) = \frac{48}{52} = \frac{12}{13}$$

c) There are 13 spades in the deck:

$$P(\text{spade}) = \frac{13}{52} = \frac{1}{4}$$

d) There are 4 jacks, 4 queens, and 4 kings, or a total of 12 picture cards.

$$P(\text{jack } or \text{ queen } or \text{ king}) = \frac{12}{52} = \frac{3}{13}$$

e) The word "and" means that "both" events must occur. Since it is not possible to select one card that is both a heart and a club, the probability is 0.

$$P(\text{heart and club}) = \frac{0}{52} = 0$$

f) The cards that are both greater than 6 and less than 9 are 7s and 8s. There are four 7s and four 8s, or a total of eight cards:

$$P(\text{greater than 6 } and \text{ less than 9}) = \frac{8}{52} = \frac{2}{13}$$

 Section 11.2 Exercises

A multiple choice test has five possible answers for each question.

1. If you guess at an answer, what is the probability you select the correct answer for one particular question?

2. In Ex. 1, if you eliminate one of the five possible answers and guess from the remaining possibilities, what is the probability you select the correct answer to that question?

One card is selected at random from a deck of cards. Find the probability that the card selected is
3. a 9
4. a 9 or a 10
5. not a 9
6. the ace of spades
7. a red card
8. a red card or a black card
9. a red card and a black card
10. a card greater than 6 and less than 10
11. a king and a diamond

In Ex. 12–16, assume the spinner cannot land on a line. Find the probability the spinner lands on (a) red, (b) blue, (c) green.

12. **13.** **14.**

15. **16.**

In Ex. 17–20, a package of flower seeds contains 100 seeds for white flowers, 50 for red flowers, 25 for pink flowers, and 25 for blue flowers. One seed is selected at random. Find the probability that it will produce
17. a red flower
18. a red or white flower
19. a white or red or pink or blue flower
20. a green flower

In Ex. 21–24, a pond is stocked with 300 fish: 50 bass, 100 trout, 80 pike, and 70 perch. Janet Jones plans to fish until she catches one fish. Find the probability that it will be
21. a bass
22. a trout
23. a pike or perch
24. a shark

In Ex. 25–28, fifteen record albums, including six Metallica, five Mariah Carey, three Hammer, and one New York Philharmonic are wrapped in plain brown paper. Ms Gilligan selects one album at random. Find the probability that the album selected is
25. a Metallica album

26. a New York Philharmonic album
27. either a Hammer or a Mariah Carey album
28. not a Metallica album

In Ex. 29–32, a traffic light is red for 30 sec, yellow for 5 sec, and green for 40 sec. What is the probability that when you reach the light,
29. the light is red
30. the light is yellow
31. the light is not red
32. the light is not red or yellow

In Ex. 33–38, each individual letter of the word "pfeffernuesse" is placed on a piece of paper, and all 13 pieces of paper are placed in a hat. If one letter is selected at random from the hat find the probability that
33. the letter f is selected
34. the letter f is not selected
35. a vowel is selected
36. the letter f or e is selected
37. the letter e is not selected
38. the letter g is selected

In Ex. 39–42, there are 12 pencils in a cup. Five are sharpened and have erasers, four are sharpened but do not have erasers, and three are not sharpened but have erasers. If one pencil is selected at random from the cup, find the probability that the pencil
39. is sharpened
40. is not sharpened
41. has an eraser
42. does not have an eraser

In Ex. 43–47, the minimum age required to obtain a regular driver's license in 1991 (without restrictions) in the 50 states and the District of Columbia is summarized as follows:

Minimum Age for License*	Number of Locations
15	2
16	20
$16\frac{1}{2}$	1
17	3
18	24
19	1

* In some states, licenses may be obtained at a younger age if applicant completes a driver education course.

If one of these locations is selected at random, find the probability that the minimum age needed for a regular driver's license is

43. 15
44. less than 17
45. 16 or $16\frac{1}{2}$
46. greater than 16
47. greater than or equal to 16

In Ex. 48–50, a dart is thrown randomly and sticks on the circular dart board illustrated below. Assuming the dart cannot land on the black area or on a border between colors, find the probability that the dart lands on

48. the area marked 25
49. a green area
50. an area marked with a number that is greater than or equal to 24.

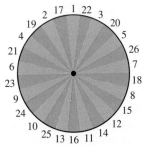

In Ex. 51–56, refer to the chart below, which contains information about a psychology class.

	Freshmen	Sophomores	Total
Females	8	6	14
Males	12	9	21
Total	20	15	35

If one individual is selected at random from this class, find the probability that the individual selected is a

51. female
52. male
53. freshman
54. sophomore
55. a female sophomore
56. a male freshman

In Ex. 57–61, refer to the chart below, which summarizes the travel of clients at the World Travel Agency.

	Domestic Travel	Foreign Travel	Total
Cost under $2000	20	12	32
Cost $2000 or more	8	10	18
Total	28	22	50

If one client is selected at random, find the probability that the

57. client traveled within the United States
58. client traveled outside the United States
59. client's cost was under $2000
60. client's cost was under $2000 on foreign travel
61. client's cost was $2000 or more on domestic travel

In Ex. 62–66, a bean bag is randomly thrown onto the square table and does not touch a line. Find the probability that the bean bag lands on

62. a red area
63. a green area
64. a blue area
65. a red or green area
66. a green or blue area

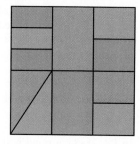

11.3 Odds

The odds against winning a lottery are 89,000 to 1; the odds against being audited by the IRS this year are 50 to 1. We see the word "odds" daily in newspapers and magazines and often use it ourselves. Yet there is a widespread misunderstanding of its meaning. This section will explain the meaning of odds.

The odds given at horse races, at craps, and at all gambling games in Las Vegas and other casinos throughout the world are always odds against unless they are otherwise specified. The odds *against* an event is a ratio of the probability that the event will fail to occur (failure) to the probability that the event will occur (success). Thus **in order to find odds you must first know or determine the probability of success and the probability of failure.**

$$\text{Odds against event} = \frac{P(\text{event fails to occur})}{P(\text{event occurs})} = \frac{P(\text{failure})}{P(\text{success})}$$

▶ **Example 1**

Find the odds against rolling a 3 on one roll of a die.

Solution: Before we can determine the odds, we must first determine the probability of rolling a 3 (success) and the probability of not rolling a 3 (failure). When a die is rolled there are six possible outcomes: 1, 2, 3, 4, 5, and 6.

$$P(\text{rolls a 3}) = \frac{1}{6} \qquad P(\text{fails to roll a 3}) = \frac{5}{6}$$

Now that we know the probabilities of success and failure, we can determine the odds against rolling a 3.

$$\text{Odds against rolling a 3} = \frac{P(\text{fails to roll a 3})}{P(\text{rolls a 3})}$$

$$= \frac{5/6}{1/6} = \frac{5}{6} \cdot \frac{6}{1} = \frac{5}{1}$$

The ratio 5/1 is commonly written as 5 : 1 and is read "5 to 1." Thus the odds against rolling a 3 are 5 to 1.

Note: The denominators of the probabilities in an odds problem will always divide out.

In Example 1 consider the possible outcomes of the die—1, 2, 3, 4, 5, 6. Over the long run, one out of every six rolls will result in a 3, and five out of every six rolls will result in a number other than 3. Therefore for each dollar bet in favor of the rolling of a 3, five dollars should be bet against the rolling of a 3 if it is to be a fair game. The person betting in favor of the rolling of a 3 will either lose one dollar (if a number other than a 3 is rolled) or win five dollars (if a 3 is rolled). The person betting against the rolling of a 3 will either win one dollar (if a number other than a 3 is a rolled) or lose five dollars (if a 3 is rolled). If this game is played for a long enough period, each player will theoretically break even.

"We figured the odds as best we could, and then we rolled the dice."
Jimmy Carter, on his decision to run for President

Archaelogists have found evidence of gambling in all cultures, from the Stone Age Australian aborigines to the ancient Egyptians, and across cultures touched by the Roman Empire. There is an equally long history of moral and legal opposition to gambling.

▶ **Example 2**

It is estimated that 3 out of every 10 new businesses in a specific location go bankrupt within their first year of operation. If you open a new business, what are the odds against its going bankrupt?

Solution: The probability that your business will go bankrupt is $\frac{3}{10}$. Therefore the probability that your business will not go bankrupt is $1 - \frac{3}{10}$, or $\frac{7}{10}$.

$$\begin{array}{l}\text{Odds against your}\\ \text{business going bankrupt}\end{array} = \frac{P(\text{the business fails to go bankrupt})}{P(\text{the business goes bankrupt})}$$

$$= \frac{7/10}{3/10} = \frac{7}{\cancel{10}} \cdot \frac{\cancel{10}}{3} = \frac{7}{3} \text{ or } 7:3$$

Although odds are generally given against an event, there are times when they may be given in favor of an event. The odds *in favor of* an event are expressed as a ratio of the probability that the event will occur to the probability that the event will fail to occur.

$$\text{Odds in favor of event} = \frac{P(\text{event occurs})}{P(\text{event fails to occur})} = \frac{P(\text{success})}{P(\text{failure})}$$

If the odds *against* an event are $a:b$, then the odds *in favor of* the event will be $b:a$.

▶ **Example 3**

Two percent of the U.S. population is born gifted (IQ of 130 or above). Find
a) the odds against an individual's being born gifted;
b) the odds in favor of an individual's being born gifted.

Solution:
a) The probability of being born gifted is 0.02 or 2/100. The probability of not being born gifted is therefore $1 - (2/100) = 98/100$.

$$\text{Odds against being born gifted} = \frac{P(\text{not born gifted})}{P(\text{born gifted})}$$

$$= \frac{98/100}{2/100}$$

$$= \frac{\overset{49}{\cancel{98}}}{\cancel{100}} \cdot \frac{\overset{1}{\cancel{100}}}{\underset{1}{\cancel{2}}} = \frac{49}{1} \text{ or } 49:1$$

b) The odds in favor of being born gifted are 1:49.

Finding Probabilities from Odds

When odds are given, either in favor or against a particular event, it is possible to determine the probabilities of that event. The denominators of the probabilities are found by adding the numbers in the odds statement. The numerators of the probabilities are the numbers given in the odds statements.

▶ **Example 4**

The odds against Jennifer being admitted to the college of her choice are $9 : 2$. Find the probability that (a) Jennifer is admitted and (b) Jennifer is not admitted.

Solution:

a) We have been given odds against and have been asked to find probabilities.

$$\text{Odds against being admitted} = \frac{P(\text{fails to be admitted})}{P(\text{is admitted})}$$

Since the odds statement is 9:2, the denominators of both the probability of success and failure must be $9 + 2$ or 11. To get the odds ratio of 9:2 the probabilities must be $\frac{9}{11}$ and $\frac{2}{11}$. Since odds against is a ratio of failure to success, the $\frac{9}{11}$ and $\frac{2}{11}$ represent the probabilities of failure and success respectively. Thus, the probability that Jennifer is admitted (success) is $\frac{2}{11}$.

b) The probability that Jennifer is not admitted (failure) is $\frac{9}{11}$.

Odds and probability statements are sometimes stated incorrectly. For example, consider the statement "The odds of being selected to represent the district are 1 in 5." Odds are given using the word "to," not "in." Thus there is a mistake in this statement. The correct statement might be "The odds of being selected to represent the district are 1 to 5" or "The probability of being selected to represent the district is 1 in 5." Without additional information it is not possible to tell which is the correct interpretation.

Before you buy your next lottery ticket, consider the odds. Compare the odds of being struck by lightning (approximately 700,000 to 1) with your chances of winning a lottery. In Colorado, for example, where you must select 6 of 42 numbers, the odds against winning the lottery are about 5.2 million to 1. Thus you are about 7.5 times more likely to be struck by lightning than to win the lottery. Given those odds, consider the case of a Colorado man, Don Whittman, 29, who won half of a $4 million jackpot on December 23, 1989, and won a $2.2 million jackpot again on October 22, 1991!

Section 11.3 Exercises

1. The odds in favor of winning the door prize are 1 to 25. Find the odds against winning the door prize.

2. The odds against Hot Dog winning the greyhound race are 3:2. Find the odds in favor of Hot Dog winning.

3. Mikhail's closet contains 15 shirts, of which six are white. Mikhail selects one shirt at random. Find
   a) the probability it is white
   b) the probability it is not white
   c) the odds against it being white
   d) the odds in favor of it being white

4. There are three golden retrievers, five poodles, and two huskies in the animal shelter. If Rasheed selects one dog at random find
   a) the probability it is a poodle
   b) the probability it is not a poodle
   c) the odds in favor of it being a poodle
   d) the odds against it being a poodle

In Ex. 5–8, a die is tossed. Find the odds against rolling
5. a 4

6. an even number

7. a number less than 3

8. a number greater than 4

In Ex. 9–12, a card is picked from a deck of cards. Find the odds against and the odds in favor of selecting

9. a 7

10. a picture card

11. a heart

12. a red card

In Ex. 13–16, assume the spinner cannot land on a line. Find the odds against the spinner landing on the color red.

13.

14.

15.

16.

17. A million tickets are sold for a lottery. If you purchase one ticket, what are your odds (a) against winning, (b) in favor of winning?

18. One person is selected at random from a class of 18 males and 11 females. Find the odds against selecting (a) a female, (b) a male.

In Ex. 19–23, the results of test scores for a class of 28 students are five As, six Bs, thirteen Cs, three Ds, and one F. If one student is selected at random, find

19. the odds in favor of the student receiving a grade of C

20. the odds in favor of the student passing the test

21. the odds against a student receiving a grade of C or higher

22. the odds against a student receiving a grade of A

23. the odds against a student receiving a grade of F

24. The odds in favor of an event are 8:3. Find the probability that the event (a) occurs, (b) does not occur.

25. The odds in favor of Becky winning a raquetball tournament are 1:8. Find the probability that Becky will (a) win the tournament, (b) not win the tournament.

26. The odds against Vic getting a teaching position next fall are 2:9. Find the probability that Vic gets a teaching position next fall.

27. The odds against Man of Peace winning the fifth race are 5:2.
   a) Find the probability that Man of Peace wins.
   b) Find the probability that Man of Peace loses.

28. Suppose the probability that you sell your car this week is 0.4. Find the odds against selling your car this week.

29. Suppose the probability that you are asked to work overtime this week is 3/8. Find the odds in favor of being asked to work overtime.

30. Suppose the probability that the mechanic fixes the car right the first time is 0.8. Find the odds against the car being repaired right on the first attempt.

31. Gout constitutes about 5 percent of all systemic arthritis, and it is uncommon in women. The male to female ratio of gout is estimated as 20 to 1.
   a) If J. Douglas has gout, what are the odds against J. Douglas's being female?
   b) If J. Douglas has gout, what is the probability that J. Douglas is a male?

32. One in 40 individuals in the $10,000–$40,000 tax range will be randomly selected to have their income tax returns audited for this year. Mr. Frank is in this income tax range. Find
   a) the probability that Mr. Frank is audited
   b) the odds against Mr. Frank's being audited

11.4 Expected Value (Expectation)

Expected value is often used to determine the expected results of an experiment or business venture *over the long run.* Expectation is used to make important decisions in many different areas. In business, for example, expectation is used to predict future profits of a new product. In the insurance industry, expectation is used to determine how much each insurance policy should cost for the company to make an overall profit. Expectation is also used to predict the expected gain or loss in games of chance such as the lottery, roulette, craps, and slot machines.

Consider the following: Tim tells Barbara he will give her $1 if she can roll an even number on a single die. If she fails to roll an even number, she must give Tim $1. Who would win money in the long run if this game were played many times? We would expect in the long run that half the time Tim would win $1 and half the time he would lose $1, therefore breaking even. Mathematically, we could find Tim's expected gain or loss by the following procedure:

$$\text{Tim's expected gain or loss} = P\left(\begin{matrix}\text{Tim}\\\text{wins}\end{matrix}\right)\left(\begin{matrix}\text{amount}\\\text{Tim wins}\end{matrix}\right) + P\left(\begin{matrix}\text{Tim}\\\text{loses}\end{matrix}\right)\left(\begin{matrix}\text{amount}\\\text{Tim loses}\end{matrix}\right)$$

$$= \tfrac{1}{2}(\$1) + \tfrac{1}{2}(-\$1) = \$0$$

Note that the loss is written as a negative number. This procedure indicates that Tim has an expected gain or loss (or expected value) of $0. The expected value of zero indicates that he would indeed break even, as we had anticipated. Thus the game is a *"fair game."* If his expected value were positive, it would indicate a gain; if negative, a loss.

The **expected value** can be used to determine the expected gain or loss of an experiment or business venture *over the long run*. The expected value, *E*, is calculated by multiplying the probability of an event occurring by the **net** amount that will be gained or lost if the event occurs. If there are a number of different events and amounts to be considered, we use the following formula.

Expected Value

$$E = P_1 A_1 + P_2 A_2 + P_3 A_3 + \cdots + P_n A_n$$

The symbol P_1 represents the probability that the first event will occur; and A_1 represents the net amount won or lost if the first event occurs. P_2 is the probability of the second event; and A_2 is the net amount won or lost if the second event occurs; and so on. The sum of these products of the probabilities and their respective amounts is the expected value. The expected value is the average (or mean) result that would be obtained if the experiment were performed a great many times.

▶ **Example 1**

Smitty's Construction Company, after considerable research, is planning to bid on a building contract. From past experience, Smitty estimates that if the company's bid is accepted, there is a 50% chance of making a $500,000 profit, a 10% chance of breaking even, and a 40% chance of losing $200,000, depending on weather conditions, inflation, and a possible strike. How much can Smitty "expect" to make or lose on this contract if his bid is accepted?

Solution: There are three amounts to be considered: a gain of $500,000, breaking even at $0, and a loss of $200,000. The probability of gaining

The concept of expected value can be used to help evaluate the consequences of many decisions. You use this concept when you consider whether to double park your car for a few minutes. The probability of being caught may be low, but the penalty, a parking ticket, may be high. You weigh these factors when you decide whether to spend time looking for a legal parking spot.

$500,000 is 0.5, the probability of breaking even is 0.1, and the probability of losing $200,000 is 0.4.

$$\text{Smitty's expectation} = \overset{\text{gain}}{\overbrace{P_1A_1}} + \overset{\substack{\text{Break}\\\text{even}}}{\overbrace{P_2A_2}} + \overset{\text{loss}}{\overbrace{P_3A_3}}$$
$$= (0.5)(\$500,000) + (0.1)(\$0) + (0.4)(-\$200,000)$$
$$= \$250,000 + \$0 - \$80,000$$
$$= \$170,000$$

In the long run, Smitty would have an average gain of $170,000 on each bid of this type. However, there is still a 40% chance that he would lose $200,000 on this *particular* bid.

▶ **Example 2**

Maria is taking a multiple choice exam where there are five possible answers for each question. The instructions indicate that you will be awarded 2 points for each correct response, lose $\frac{1}{2}$ point for each incorrect response, and no points will be added or subtracted for answers left blank.

a) If Maria does not know the correct answer to a question is it to her advantage or disadvantage to guess at an answer?

b) If she can eliminate one of the possible choices, is it to her advantage or disadvantage to guess at the answer?

Solution:

a) Let us determine the expected value if Maria guesses at an answer. There is only one of five possible answers that is correct.

$$P(\text{guesses correctly}) = \frac{1}{5} \qquad P(\text{guesses incorrectly}) = \frac{4}{5}$$

$$\text{Maria's expectation} = \overset{\substack{\text{Guesses}\\\text{correctly}}}{\overbrace{P_1A_1}} + \overset{\substack{\text{Guesses}\\\text{incorrectly}}}{\overbrace{P_2A_2}}$$
$$= \frac{1}{5}(2) + \frac{4}{5}\left(-\frac{1}{2}\right)$$
$$= \frac{2}{5} - \frac{2}{5} = 0$$

Thus Maria's expectation is zero when she guesses. This means that over the long run she will neither gain nor lose points by guessing.

b) If Maria can eliminate one possible choice then there will be one of four answers that will be correct.

$$P(\text{guesses correctly}) = \frac{1}{4}$$

$$P(\text{guesses incorrectly}) = \frac{3}{4}$$

$$\text{Maria's expectation} = \overbrace{P_1A_1}^{\substack{\text{Guesses}\\\text{correctly}}} + \overbrace{P_2A_2}^{\substack{\text{Guesses}\\\text{incorrectly}}}$$

$$= \frac{1}{4}(2) + \frac{3}{4}\left(-\frac{1}{2}\right)$$

$$= \frac{2}{4} - \frac{3}{8} = \frac{4}{8} - \frac{3}{8} = \frac{1}{8}$$

Since the expectation is positive $\frac{1}{8}$ Maria will, on the average, gain $\frac{1}{8}$ point each time she guesses when she can eliminate one possible choice.

The amounts in the expectation formula are **net** amounts. The net amounts are the actual amounts won or lost. Consider the following example, which illustrates the use of the net amounts.

▶ **Example 3**

One hundred lottery tickets are sold for $2 each. One prize of $50 will be awarded. Nick purchases one ticket. Find his expectation.

Solution: There are two amounts to be considered: the net amount Nick may win and the cost of the ticket. If Nick wins, his net or actual winnings are $48 (the $50 prize minus his $2 cost of the ticket). If he does not win, he loses the $2 paid for the ticket. Since only one prize will be awarded, Nick's probability of winning is $\frac{1}{100}$. His probability of losing is therefore $\frac{99}{100}$.

$$E = P(\text{Nick wins})(\text{amount wins}) + P(\text{Nick loses})(\text{amount loses})$$

$$E = \frac{1}{100}(48) + \frac{99}{100}(-2)$$

$$= \frac{48}{100} - \frac{198}{100} = \frac{-150}{100} = -\$1.50$$

Nick's expectation is $-\$1.50$ per ticket.

The **fair price** of a game of chance is the amount that should be charged for the game to be fair and result in an expectation of 0. If the fair price and cost to play a game are known, the expectation may be found by the formula

Expectation = fair price − cost to play

If any two of the three items in the formula are known, the third may be found. For instance, in Example 3, Nick's expectation is $-\$1.50$ and the cost of a ticket is $2.00; thus the fair price may be calculated as follows:

$$\text{Expectation} = \text{fair price} - \text{cost to play}$$

$$-1.50 = f - 2.00$$

$$2.00 - 1.50 = f - 2.00 + 2.00$$

$$0.50 = f$$

Thus the fair price of a ticket is $0.50, or 50¢. This makes sense, since if the price of a ticket were reduced by $1.50 to 50¢ Nick's expectation would be $0.00.

Expectation problems where money must be paid in advance as in Example 3, can be done by an alternative procedure. First determine the fair price to play using the formula

$$\text{Fair price} = P_1G_1 + P_2G_2 + \cdots + P_nG_n$$

where each P represents the probability of winning and each corresponding G represents the **gross amount** won. The gross amounts do not include the cost to play the game. After determining the fair price, use the formula

$$\text{Expectation} = \text{fair price} - \text{cost to play}$$

The expectation of the lottery ticket in Example 3 can be found as follows:

$$\text{Fair price} = P_1G_1 = \frac{1}{100}(\$50) = \frac{50}{100} = \$0.50$$

$$\text{Expectation} = \text{fair price} - \text{cost to play}$$
$$= \$0.50 - \$2.00 = -\$1.50$$

Using either procedure we see that Nick's expectation is $-\$1.50$.

When finding the fair price using gross amounts where more than one amount is awarded, find the sum of the products of the probabilities and their respective amounts. If, for example, there are two amounts awarded, we would use the formula:

$$\text{Fair price} = P(\text{winning amount 1})(\text{gross amount 1})$$
$$+ P(\text{winning amount 2})(\text{gross amount 2})$$

▶ **Example 4**

One thousand lottery tickets are sold for $1 each. One grand prize of $500 and two consolation prizes of $100 will be awarded. Using *net amounts*, find

a) Carol's expectation if she purchases one ticket;

b) Carol's expectation if she purchases five tickets.

c) What is the fair price of a ticket?

Solution:

a) There are three amounts to be considered: the net gain in winning the grand prize, the net gain in winning the consolation prize, and the loss of the cost of the ticket. If Carol wins the grand prize, her net gain is $499 ($500 minus $1 spent for the ticket). If Carol wins the consolation prize, her net gain is $99 ($100 minus $1). We will assume that the winners' names are replaced in the pool after being selected. The probability that Carol wins the grand prize is $\frac{1}{1000}$. Since two consolation prizes will be awarded, the probability that she wins a consolation prize is $\frac{2}{1000}$. The probability that she does not win either prize is $1 - \frac{3}{1000} = \frac{997}{1000}$.

$$E = P_1A_1 + P_2A_2 + P_3A_3$$
$$E = \frac{1}{1000}(\$499) + \frac{2}{1000}(\$99) + \frac{997}{1000}(-\$1)$$
$$= \frac{499}{1000} + \frac{198}{1000} - \frac{997}{1000} = \frac{-300}{1000} = -\$0.30 \text{ or } -30¢$$

b) On the average Carol loses 30¢ on each ticket purchased. On five tickets her expectation is $(-\$0.30)(5)$, or $-\$1.50$.

c) We learned in part (a) that Carol's expectation on one ticket is $-\$0.30$.

$$\text{Expectation} = \text{fair price} - \text{cost}$$
$$-0.30 = F - 1.00$$
$$0.70 = F$$

The fair price of a ticket is 70¢.

The fair price and expectation of the ticket in Example 4 may also be found using *gross amounts* as follows:

$$\text{Fair price} = \frac{1}{1000}(\$500) + \frac{2}{1000}(\$100) = \frac{500}{1000} + \frac{200}{1000} = \frac{700}{1000} = \$0.70$$

$$\text{Expectation} = \text{fair price} - \text{cost to play}$$
$$= \$0.70 - \$1.00 = -\$0.30$$

▶ **Example 5**

A highway crew repairs 30 potholes a day in dry weather and 12 potholes a day in wet weather. If the weather in this region is wet 40% of the time, find the expected (average) number of potholes that can be repaired per day.

Solution: The amounts in this problem are the number of potholes repaired. Since the weather is wet 40% of the time, it will be dry 60% (100% − 40%) of the time:

$$E = P(\text{dry})(\text{amount repaired}) + P(\text{wet})(\text{amount repaired})$$
$$= 0.60(30) + 0.40(12) = 18.0 + 4.8 = 22.8$$

Thus the average or expected number of potholes repaired per day is 22.8.

Section 11.4 Exercises

1. Explain the meaning of expected value.
2. If on a $1 bet John's expected value is $-\$0.30$,
   a) what is John's expected value on a $5 bet?
   b) How much can John expect to win or lose if he places a $5 bet? Explain.
3. An investment club is considering purchasing a given stock option. After considerable research the club members determine that there is a 40% chance of making $5000, a 10% chance of breaking even, and a 50% chance of losing $3500. Find the expectation of this purchase.
4. An investment counselor is advising her client on a particular investment. She estimates that if the tax law does not change the client will make $60,000, but if the tax law changes the client will lose $15,000. Find the client's expected value if there is a 60% chance that the tax law will change.

5. In July in Seattle, the grass grows $\frac{1}{2}$ in. a day on a sunny day and $\frac{1}{4}$ in. a day on a cloudy day. In Seattle in July 75% of the days are sunny and 25% are cloudy.
   a) Find the expected amount of grass growth on a typical day in July in Seattle.
   b) Find the expected total grass growth in the month of July in Seattle.
6. Bob and Larry play the following game: Larry picks a card from a deck of cards. If he selects a club, Bob gives him $10. If not, he gives Bob $4.
   a) Find Larry's expectation.
   b) Find Bob's expectation.
7. A multiple-choice exam has four possible answers for each question. For each correct answer you are awarded 5 points. For each incorrect answer, 2 points are subtracted from your score. For answers left blank, no points

are added or subtracted.

a) If you do not know the correct answer to a particular question, is it to your advantage to guess?

b) If you do not know the correct answer but can eliminate one possible choice, is it to your advantage to guess?

8. A cookie jar contains six $1 bills, three $5 bills, and one $10 bill. Corine will select and keep one bill from the jar.

a) Find Corine's expected gain.

b) What is a fair price for Corine to pay to play this game?

9. One thousand raffle tickets are sold for $1 each. One prize of $500 is to be awarded.

a) Olga purchases one ticket. Find her expected value.

b) Find the fair price of a ticket.

c) If the vendor sells all 1000 tickets, how much profit will he make?

10. Ten thousand raffle tickets are sold for $5 each. Four prizes will be awarded—one for $10,000, one for $5,000, and two for $1,000. Sidhardt purchases one of these tickets.

a) Find his expected value.

b) Find the fair price of a ticket.

In Ex. 11–14, assume you spin the pointer and you are awarded the amount indicated by the pointer.

a) Find the fair price to play the game.

b) If it costs $2 to play the game, find the expectation of a person who plays the game.

11.

12.

13.

14.

15. According to a mortality table, the probability that a 20-year-old woman will survive one year is 0.994, and the probability that she will die within one year is 0.006. If she buys a $10,000 one-year term policy for $100, what is the company's expected gain or loss?

16. It will cost an oil drilling company $30,000 to sink a test well. If they hit oil the company will make a net profit of $500,000. If they hit natural gas, the net profit will be $100,000. If they hit nothing they will lose their $30,000. If the probability of hitting oil is 0.08 and the proba-

bility of hitting gas is 0.20, what is the expectation of the oil drilling company? Should they sink the test well? Explain.

17. An instrumentation contractor is considering making a bid on a water pollution control project. She determines that if her bid is accepted and she completes the project on schedule her net profit would be $420,000. If she completes the project between zero and three months after the contract deadline date her net profit drops to $200,000. If she completes the project more than three months late her net loss is $600,000. The probability she completes the job on schedule is 0.5, the probability she completes the job between zero and three months late is 0.3, and the probability she completes the job more than three months late is 0.2. Find her expected gain or loss for this bid.

18. A die is rolled many times, and the points facing up are recorded. Find the expected (average) number of points facing up over the long run.

19. American Airlines is planning its staffing needs for next year. On January 1 the Civil Aeronautics Board will inform American Airlines whether they will be granted the new routes they have requested. If the new routes are approved, they will hire 850 new employees. If the new routes are not granted, they will hire only 150 new employees. If the probability that the Board will grant American Airlines' request is 0.25, what is the expected number of new employees to be hired by American Airlines?

20. Manufacturer XYZ is negotiating a contract with its employees. The probability that the union will go on strike is 5%. If the union goes on strike, the company estimates that it will lose $345,000 for the year. If the union does not go on strike, the company estimates that it will make $1.2 million. Find the company's expected gain or loss for the year.

21. On a clear day in Reading the AAA makes an average of 125 service calls for motorist assistance; on a rainy day it makes an average of 190 service calls; and on a snowy day it makes an average of 280 service calls. If the weather in Reading is clear 200 days of the year, rainy 100 days of the year, and snowy 65 days of the year, find the expected number of service calls made by the AAA in a given day.

22. The expenses for Dale, a realtor, to list, advertise, and attempt to sell a house are $1000. If Dale succeeds in selling the house, he will receive 6% of the sales price. If a realtor with a different company sells the house, Dale will still receive 3% of the sales price. If the house is unsold after three months, Dale loses the listing and receives nothing. Suppose the probability Dale sells a

$100,000 house is 0.2, the probability another realtor sells the house is 0.5, and the probability the house is unsold after three months is 0.3. Find Dale's expectation if he accepts this house for listing. Should Dale list the house? Explain.

23. An insurance company has written life insurance policies on people of age 25. There are 200 policies of $10,000, 400 policies of $5000, and 1000 policies of $1000. Experience shows that the probability of an individual dying at age 25 is 0.002. Determine the amount the insurance company can expect to pay out on these policies this year.

24. During those years when financial responsibilities are greatest, such as when children are in school or there is a mortgage on the house, a person may choose to buy a term life insurance policy. The insurance company will pay the face value of the policy if the insured person dies during the term of the policy. For how much should an insurance company sell a 10-year term policy with a face value of $40,000 to a 30-year-old male in order for the company to make a profit? The probability of a 30-year-old man living to age 40 is 0.97. Explain your answer.

Problem Solving

A roulette wheel typically contains slots with numbers from 1 through 36, and slots marked 0 and 00. A ball is spun on

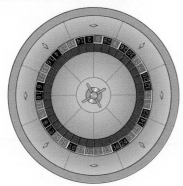

the wheel and comes to rest in one of the 38 slots. Eighteen numbers are colored red and 18 numbers are colored black. The 0 and 00 are colored green. If you bet on one particular number and that number is obtained the house pays off odds of 35 to 1. If you bet on a red number or black number to win, and win, the house pays 1 to 1 (even money).

25. Find the expected value of betting on a particular number.

26. Find the expected value of betting on red.

27. One of the more popular sweepstakes is the *Reader's Digest* sweepstakes. The prizes and the approximate probability of winning the prize listed in the sweepstakes package are:

$10—one in 824
$100—one in 21,446
$500—one in 160,852
$5,000—one in 1,836,000
$10,000—one in 8,042,600
$25,000—one in 16,085,000
$250,000—one in 23,423,000

Find your expectation if you must use a 29¢ stamp to enter the sweepstakes.

28. The dealer shuffles five black cards and five red cards and spreads them out on the table face down. You choose two at random. If both cards are red or both cards are black, you win a dollar. Otherwise, you lose a dollar. Determine whether the game favors you, is fair, or favors the dealer. Explain your answer.

29. A dealer has three cards. One is red on both sides, another is black on both sides, and the third is red on one side and black on the other side. The cards are shuffled in a hat, and one is drawn at random and placed flat on the table. The side showing is red. If the other side is red, you win a dollar. If the other side is black, you lose a dollar. Determine whether the game favors the player, is fair, or favors the dealer.

11.5 Tree Diagrams

We stated earlier that the possible results of an experiment are called its outcomes. In order to solve more difficult probability problems we must first be able to determine all the possible outcomes of the experiment. The counting principle can be used to determine the number of outcomes of an experiment, and will also be helpful in constructing tree diagrams to be discussed shortly.

Fashion experts often suggest that you select your wardrobe from pieces of clothing that can be mixed and matched. If, for example, you have 7 shirts, 3 sweaters, and 4 pairs of pants, you actually have 84 possible outfits. If you wear only a sweater or a shirt, but not both, the possibilities go down to 40.

Counting Principle

If a first experiment can be performed in M distinct ways and a second experiment can be performed in N distinct ways, then the two experiments in that specific order can be performed in $M \cdot N$ distinct ways.

If we wished to find the number of possible outcomes when a coin is tossed and a die is rolled we can reason like this: The coin has two possible outcomes—heads or tails. The die has six possible outcomes—1, 2, 3, 4, 5, and 6. Thus the two experiments together have $2 \cdot 6$ or 12 possible outcomes.

A listing of all the possible outcomes of an experiment is called a **sample space**. Each individual outcome in the sample space is called a **sample point**. Tree diagrams are helpful in determining sample spaces.

A tree diagram illustrating all the possible outcomes when a coin is tossed and a die is rolled (see Fig. 11.3) will have two initial branches, one for each of the possible outcomes of the coin. Each of these branches will have six branches emerging from them, one for each of the possible outcomes of the die. This will give a total of 12 branches, the same number of possible outcomes found using the counting principle. We can obtain the sample space by listing all the possible combinations of branches. Notice this sample space consists of 12 sample points.

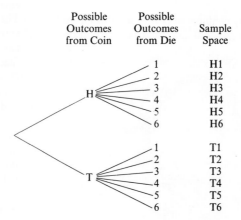

Figure 11.3

Example 1 uses the phrase "without replacement." This tells us that once an item is selected it cannot be selected again, making it impossible to select the same exact item twice.

▶ Example 1

Two balls are to be selected *without replacement* from a bag that contains one red, one blue, one green, and one orange ball (see Fig. 11.4).

a) Use the counting principle to determine the number of points in the sample space.

b) Construct a tree diagram and list the sample space.

Figure 11.4

c) Find the probability that one red ball is selected.

d) Find the probability that a green ball followed by a red ball is selected.

Solution:

a) The first selection may be any one of the four balls. Once the first ball is selected, only three balls remain for the second selection. Thus there are $4 \cdot 3$, or 12, sample points in the sample space.

b) The first ball selected can be red, blue, green, or orange. Since this experiment is done without replacement, the same colored ball cannot be selected twice. For example, if the first ball selected is red, the second ball selected must be either blue, green, or orange. The tree diagram and sample space are given in Fig. 11.5. The sample space contains 12 points. That result checks with the answer obtained by the counting principle.

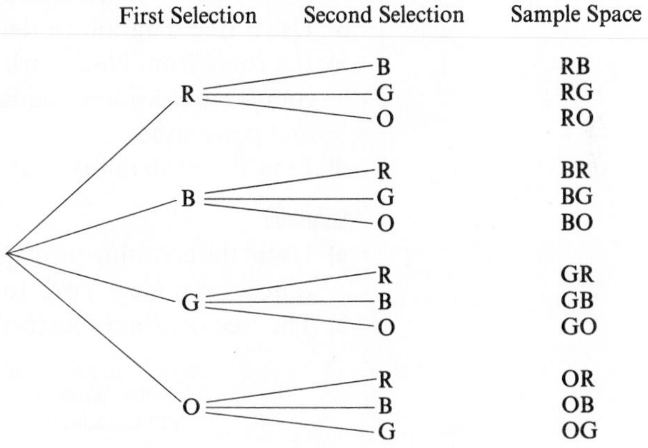

First Selection	Second Selection	Sample Space
R	B	RB
	G	RG
	O	RO
B	R	BR
	G	BG
	O	BO
G	R	GR
	B	GB
	O	GO
O	R	OR
	B	OB
	G	OG

Figure 11.5

c) If we know the sample space we can compute probabilities using the formula

$$P(E) = \frac{\text{number of outcomes favorable to } E}{\text{total number of outcomes}}.$$

The total number of outcomes will be the number of points in the sample space. From Fig. 11.5 we determine that there are 12 possible outcomes. The number of outcomes that have one red ball is six (RB, RG, RO, BR, GR, and OR).

$$P(\text{one red ball is selected}) = \frac{6}{12} = \frac{1}{2}$$

d) There is one possible outcome that meets the criteria of the problem (GR).

$$P(\text{green followed by red}) = \frac{1}{12}$$

The counting principle can be extended to any number of experiments, as illustrated in Example 2.

▶ **Example 2**

There are two highways from New York to Cleveland, three highways from Cleveland to Chicago, and two highways from Chicago to San Francisco as illustrated in Fig. 11.6.

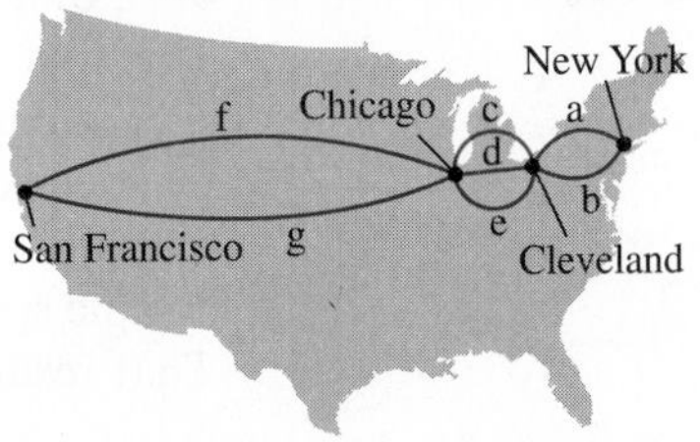

Figure 11.6

a) Use the counting principle to determine the number of different routes from New York to San Francisco.
b) Use a tree diagram to determine the routes.
c) If a route from New York to San Francisco is selected at random, and all routes are considered equally likely, find the probability that both routes a and g are used.
d) Find the probability that neither routes d nor f are used.

Solution:

a) Using the counting principle we can determine that there are $2 \cdot 3 \cdot 2$, or 12, routes from New York to San Francisco.
b) The tree diagram illustrating the 12 possibilities is given in Fig. 11.7.

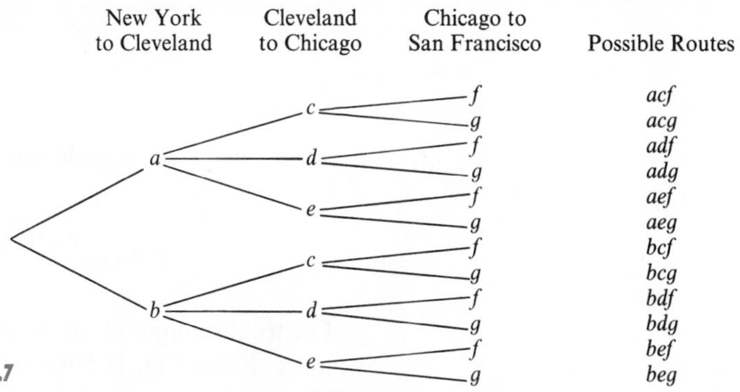

Figure 11.7

c) Of the 12 possible routes three use both a and g (acg, adg, aeg).

$$P(\text{routes } a \text{ and } g \text{ are both used}) = \frac{3}{12} = \frac{1}{4}$$

d) Of the 12 possible routes, four use neither d nor f (acg, aeg, bcg, beg).

$$P(\text{neither } d \text{ nor } f \text{ is used}) = \frac{4}{12} = \frac{1}{3}$$

Section 11.5 Exercises

In Ex. 1–14, use the counting principle to determine the number of points in the sample space before constructing the tree diagram. Assume each event is equally likely to occur.

1. Two coins are tossed:
  a) Construct a tree diagram and list the sample space.
  Find the probability that
  b) no heads are tossed
  c) exactly one head is tossed
  d) two heads are tossed

2. A hat contains three chips: one red, one blue, and one green. Two chips are to be selected at random with replacement.
  a) Construct a tree diagram and determine the sample space.
  Find the probability that
  b) two red chips are selected
  c) a red chip and then a blue chip is selected
  d) at least one green chip is selected.

3. Repeat Ex. 2 without replacement.

4. A hat contains four marbles, one yellow, one red, one blue, and one green. Two marbles are to be selected at random without replacement from the hat.
  a) Construct a tree diagram and list the sample space.
  Find the probability of selecting
  b) exactly one red marble
  c) at least one marble that is not red
  d) no green marbles

5. A couple plan to have exactly two children.
  a) Construct a tree diagram and list the sample space of the possible arrangements of boys and girls.
  Find the probability that the family has
  b) two girls
  c) at least one girl

6. Cecelia owns three cats including a Siamese, Angora, and Persian, and four dogs including a collie, poodle, beagle, and husky. She selects one cat and one dog at random to bring with her to the park.
  a) Construct a tree diagram and determine the sample space.
  Find the probability that she selects
  b) the Persian cat
  c) not the Persian cat
  d) the Siamese cat and either the collie or poodle
  e) the poodle or the husky

7. Suzette has three dolls including a Raggedy Ann, a Raggedy Andy, and a Barbie doll, and five stuffed animals including a white teddy bear, a brown teddy bear, a gorilla, Miss Piggy, and Kermit the frog. Her dad tells her that she can bring two toys with her to the store. Suzette wants to bring one of her dolls and one stuffed animal.
  a) Construct a tree diagram and determine the sample space.
  Find the probability that Suzette selects the
  b) Raggedy Ann or Raggedy Andy doll
  c) the Barbie doll or the gorilla
  d) the Barbie doll and the gorilla
  e) the Barbie doll and one of the teddy bears

8. Gustave Ruckert is starting his own publishing company. He is considering six different printers including: Abe's, Bonnie's, Charlie's, Do It Right, Economy, and First Class; and three different binders including: Goldman's, Hunt's, and Ivanhoe's. He needs to select a printer and a binder.
  a) Construct a tree diagram of the possible combinations of printers and binders and list the sample space.
  Find the probability that
  b) Goldman's is selected
  c) Abe's or Economy is selected
  d) Charlie's and Ivanhoe's are not both selected

9. Two dice are rolled.
  a) Construct a tree diagram and list the sample space.
  Find the probability that
  b) a double is rolled
  c) a sum of 7 is rolled
  d) a sum of 2 is rolled
  e) Are you as likely to roll a sum of 2 as you are of rolling a sum of 7? Explain your answer.

10. Three coins are tossed.
  a) Construct a tree diagram and list the sample space.
  Find the probability that
  b) exactly three heads are tossed
  c) exactly one head is tossed
  d) at least one head is tossed

11. A couple plans to have exactly three children.
  a) Construct a tree diagram and list the sample space.
  Find the probability that the family has
  b) no boys
  c) at least one girl
  d) either exactly two boys or two girls

12. Two coins are tossed and a die is rolled
  a) Construct a tree diagram and list the sample space.
  Find the probability of
  b) tossing two heads and rolling the number 3
  c) tossing exactly one head

d) tossing at least one head and rolling an even number

13. A bag contains three chips: one red, one blue, and one green. Three chips are to be selected at random with replacement.
a) Construct a tree diagram and list the sample space. Find the probability of selecting
b) three red chips
c) a red chip, then a blue chip, then a green chip
d) at least one chip which is not blue

14. Repeat Ex. 13 without replacement.

15. An individual can be classified as male or female with red, brown, black, or blonde hair and with brown, blue, or green eyes.
a) How many different classifications are possible?
b) Construct a tree diagram to determine the sample space (for example, red-headed, blue-eyed, and male).
c) If each outcome is equally likely, find the probability

that the individual will be a male with black hair and blue eyes.
d) Find the probability that the individual will be a female with blonde hair.

16. A pea plant must have exactly one of each of the following pairs of traits: short (*s*) or tall (*t*); round (*r*) or wrinkled (*w*) seeds; yellow (*y*) or green (*g*) peas; and white (*wh*) or purple (*p*) flowers (for example, short, wrinkled, green pea with white flowers).
a) How many different classifications of pea plants are possible?
b) Use a tree diagram to determine all the classifications possible.
c) If each characteristic is equally likely, find the probability that the pea plant will have round peas.
d) Find the probability that the pea plant will be short, be wrinkled, have yellow seeds, and have purple flowers.

11.6 "Or" and "And" Problems

Section 11.5 showed how to work probability problems by constructing sample spaces. Often it is inconvenient or too time-consuming to solve a problem by first constructing a sample space. For example, if an experiment consists of selecting two cards with replacement from a deck of 52 cards, there would be $52 \cdot 52$ or 2704 points in the sample space. Trying to list all these sample points could take hours. In this section we will learn how to solve **compound probability problems** that contain the words "and" or "or" without constructing a sample space.

"Or" Problems

The "or" type of problem requires obtaining a "successful" outcome for *at least one* of the given events. For example suppose we roll one die and we are interested in finding the probability of rolling an even number *or* a number greater than 4. For this situation rolling either a 2, 4, or 6 (an even number) or a 5 or 6 (a number greater than 4) would be considered successful. Note that the number 6 satisfies both criteria. A formula for finding the probability of event *A* or event *B*, symbolized $P(A \text{ or } B)$, is

$$P(A \text{ or } B) = P(A) + P(B) - P(A \text{ and } B).$$

Since we add (and subtract) probabilities to find $P(A \text{ or } B)$ this formula is sometimes referred to as the *addition formula*. We will explain the use of the "or" formula in Example 1.

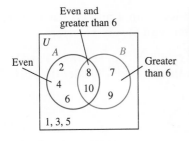

Figure 11.8

▶ Example 1

Each of the numbers 1, 2, 3, 4, 5, 6, 7, 8, 9, and 10 is written on a separate piece of paper. The 10 pieces of paper are then placed in a hat, and one piece is randomly selected. Find the probability that the piece of paper selected contains an even number or a number greater than 6.

Solution: We are asked to find the probability that the number selected *is even* or *greater than 6.* Let us use set A to represent the statement "the number is even," and set B to represent the statement "the number is greater than 6." Figure 11.8 is a diagram with sets A (Even) and B (Greater than 6). There are a total of 10 numbers of which five are even (2, 4, 6, 8, and 10). Thus the probability of selecting an even number is $\frac{5}{10}$. There are four numbers that are greater than 6: the 7, 8, 9, and 10. Thus the probability of selecting a number greater than 6 is $\frac{4}{10}$. There are two numbers that are both even and greater than 6: the 8 and 10. Thus the probability of selecting a number that is both even and greater than 6 is $\frac{2}{10}$.

If we substitute the appropriate statements for A and B in the formula we obtain

$$P(A \text{ or } B) = P(A) + P(B) - P(A \text{ and } B)$$

$$P\left(\begin{array}{c}\text{even or}\\\text{greater than 6}\end{array}\right) = P(\text{even}) + P\left(\begin{array}{c}\text{greater}\\\text{than 6}\end{array}\right) - P\left(\begin{array}{c}\text{even and}\\\text{greater than 6}\end{array}\right)$$

$$= \frac{5}{10} + \frac{4}{10} - \frac{2}{10}$$

$$= \frac{7}{10}$$

Example 1 illustrates that when finding the probability of A or B, we find the sum of the probabilities of events A and B and then subtract the probability of both events occurring simultaneously.

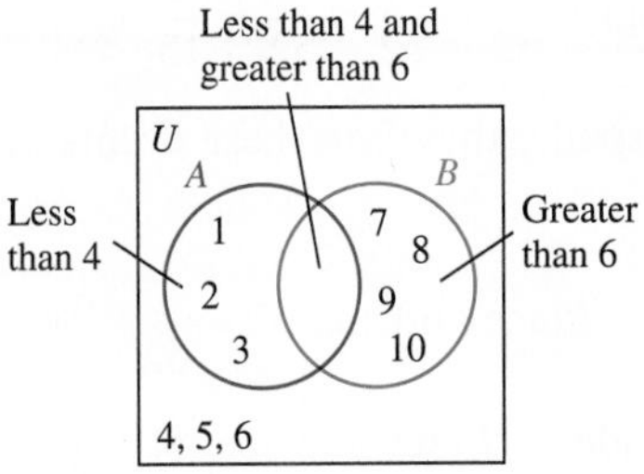

▶ Example 2

Consider the same sample space, the numbers 1 through 10, as in the preceding example. If one piece of paper is selected, find the probability that it contains a number less than 4 or a number greater than 6.

Solution: Let A represent the statement "the number is less than 4" and let B represent the statement "the number is greater than 6." A diagram illustrating these statements is given in Fig. 11.9.

$$P(\text{selecting a number less than 4}) = \frac{3}{10}$$

$$P(\text{selecting a number greater than 6}) = \frac{4}{10}$$

Figure 11.9

Since there are no numbers that are both less than 4 and greater than 6, P(selecting a number less than 4 and greater than 6) $= 0$. Therefore

$$P\binom{\text{less than 4 or}}{\text{greater than 6}} = P(\text{less than 4}) + P\binom{\text{greater}}{\text{than 6}} - P\binom{\text{less than 4 and}}{\text{greater than 6}}$$

$$= \frac{3}{10} + \frac{4}{10} - 0 = \frac{7}{10}$$

In Example 2 it is impossible to select a number that is both less than 4 and greater than 6 when only one number is to be selected. Events such as these are said to be mutually exclusive.

> Two events A and B are **mutually exclusive** if it is impossible for both events to occur simultaneously.

If events A and B are mutually exclusive, then $P(A \text{ and } B) = 0$, and the addition formula simplifies to $P(A \text{ or } B) = P(A) + P(B)$.

▶ Example 3

One card is selected from a deck of cards. Determine whether the following pairs of events are mutually exclusive, and find $P(A \text{ or } B)$.

a) A = an ace, B = a king
b) A = an ace, B = a spade
c) A = a red card, B = a black card
d) A = a picture card, B = a red card

Solution:

a) It is impossible to select both an ace and a king when only one card is selected. Therefore these events are mutually exclusive.

$$P(\text{ace or king}) = P(\text{ace}) + P(\text{king}) = \frac{4}{52} + \frac{4}{52} = \frac{8}{52} = \frac{2}{13}$$

b) The ace of spades is both an ace and a spade; therefore these events are not mutually exclusive.

$$P(\text{ace}) = \frac{4}{52} \qquad P(\text{spade}) = \frac{13}{52} \qquad P(\text{ace and spade}) = \frac{1}{52}$$

$$P(\text{ace or spade}) = P(\text{ace}) + P(\text{spade}) - P(\text{ace and spade})$$

$$= \frac{4}{52} + \frac{13}{52} - \frac{1}{52}$$

$$= \frac{16}{52} = \frac{4}{13}$$

c) It is impossible to select one card that is both a red card and a black card. Therefore the events are mutually exclusive.

$$P(\text{red or black}) = P(\text{red}) + P(\text{black})$$

$$= \frac{26}{52} + \frac{26}{52} = \frac{52}{52} = 1$$

Therefore a red card or a black card must be selected.

d) There are six picture cards that are red: jack, queen, and king of hearts and jack, queen, and king of diamonds. Thus these events are not mutually exclusive.

$$P\left(\begin{array}{c}\text{picture card}\\\text{or red card}\end{array}\right) = P\left(\begin{array}{c}\text{picture}\\\text{card}\end{array}\right) + P\left(\begin{array}{c}\text{red}\\\text{card}\end{array}\right) - P\left(\begin{array}{c}\text{picture card}\\\text{and red card}\end{array}\right)$$

$$= \frac{12}{52} + \frac{26}{52} - \frac{6}{52}$$

$$= \frac{32}{52} = \frac{8}{13}$$

"And" Problems

A second type of problem is the "and" type of problem, which requires obtaining a favorable outcome in *each* of the given events. For example suppose *two* cards are to be selected from a deck of cards and we are interested in the probability of selecting two aces (one ace *and* then a second ace). Only if *both* cards selected are aces would this experiment be considered successful. A formula for finding the probability of events A and B, symbolized $P(A \text{ and } B)$, follows.

$$P(A \text{ and } B) = P(A) \cdot P(B)$$

Since this type of problem requires obtaining a favorable outcome in *both* of the given events, *we must assume that event A has occurred before calculating the probability of event B*. Since we multiply to find $P(A \text{ and } B)$ this formula is sometimes referred to as the *multiplication formula*.

▶ **Example 4**

Two cards are to be selected with replacement from a deck of cards. Find the probability that two aces will be selected.

Solution: Since the deck of 52 cards contains four aces, the probability of selecting an ace on the first draw is $\frac{4}{52}$. The card selected is then returned to the deck. Therefore the probability of selecting an ace on the second draw remains $\frac{4}{52}$.

If we let A represent the selection of the first ace and B represent the selection of the second ace then the formula may be written

$$P(A \text{ and } B) = P(A) \cdot P(B)$$
$$P(2 \text{ aces}) = P(\text{ace } and \text{ ace}) = P(\text{ace } 1) \cdot P(\text{ace } 2)$$
$$= \frac{4}{52} \cdot \frac{4}{52}$$
$$= \frac{1}{13} \cdot \frac{1}{13} = \frac{1}{169}$$

▶ **Example 5**

Repeat Example 4 without replacement.

Solution: The probability of selecting an ace on the first draw is $\frac{4}{52}$. When calculating the probability of selecting the second ace, we must assume that the first ace has been selected. Once this first ace has been selected only 51 cards, including 3 aces, remain in the deck. The probability of selecting an ace on the second draw becomes $\frac{3}{51}$. The probability of selecting two aces without replacement is

$$P(2 \text{ aces}) = P(\text{ace } 1) \cdot P(\text{ace } 2)$$
$$= \frac{4}{52} \cdot \frac{3}{51}$$
$$= \frac{1}{13} \cdot \frac{1}{17} = \frac{1}{221}.$$

> Event A and event B are **independent events** if the occurrence of either event in no way affects the probability of occurrence of the other.

Rolling dice and tossing coins are examples of independent events. In Example 4 the events are independent since the first card was returned to the deck. The probability of selecting an ace on the second draw was not affected by the first selection. The events in Example 5 are not independent since the probability of the selection of the second ace was affected by removing the first ace selected from the deck. Such events are called **dependent events**. Experiments done with replacement will result in independent events while those done without replacement will result in dependent events.

▶ **Example 6**

A bowl of jelly beans contains 20 green and 15 orange jelly beans. Samantha, a mathematician, is going to pick and eat three jelly beans at random one

The odds against a family having eight girls in a row is 255 to 1. This means that for every 256 families with eight children, on the average, one family will have all girls and 255 will not. However, should another child be born to that one family, the probability of the ninth child being a girl is still $\frac{1}{2}$!

after the other. She is interested in determining the probability that she picks three orange jelly beans. Are the events of picking and eating the jelly beans independent or dependent events in this experiment? Explain your answer.

Solution: The events are dependent events since each time she picks one orange jelly bean it changes the probability that the next jelly bean selected will be an orange jelly bean.

▶ **Example 7**

A package of 25 zinnia seeds contains 8 seeds for red flowers, 12 seeds for white flowers, and 5 seeds for yellow flowers. Three seeds are randomly selected and planted. Find the probability of each of the following.

a) All three seeds will produce red flowers.

b) The first seed selected will produce a red flower, the second seed will produce a white flower, and the third seed will produce a red flower.

c) None of the seeds will produce red flowers.

Solution: Each time a seed is selected and planted, the number of seeds remaining decreases by one.

a) The probability that the first seed selected produces a red flower is $\frac{8}{25}$. If the first seed selected is red, only 7 red seeds in 24 are left. The probability of selecting a second red seed is $\frac{7}{24}$. If the second seed selected is red, only 6 red seeds in 23 are left. The probability of selecting a third red seed is $\frac{6}{23}$.

$$P(3 \text{ red seeds}) = P(\text{red seed 1}) \cdot P(\text{red seed 2}) \cdot P(\text{red seed 3})$$

$$= \frac{8}{25} \cdot \frac{7}{24} \cdot \frac{6}{23} = \frac{14}{575}$$

b) The probability that the first seed selected produces a red flower is $\frac{8}{25}$. Once a seed for a red flower is selected, only 24 seeds are left. Twelve of the remaining 24 will produce white flowers. Thus the probability that the second seed selected will produce a white flower is $\frac{12}{24}$. After the second seed had been selected, there are 23 seeds left, 7 of which will produce red flowers. The probability that the third seed produces a red flower is therefore $\frac{7}{23}$.

$$P\begin{pmatrix} \text{first red and second} \\ \text{white and third red} \end{pmatrix} = P(\text{first red}) \cdot P(\text{second white}) \cdot P(\text{third red})$$

$$= \frac{8}{25} \cdot \frac{12}{24} \cdot \frac{7}{23} = \frac{28}{575}$$

c) If no flowers are to be red, they must either be white or yellow. There are 17 seeds that will not produce red flowers (12 for white and 5 for yellow). The probability that the first seed does not produce a red flower is $\frac{17}{25}$. After the first seed has been selected, 16 of the remaining 24 seeds will not produce red flowers. After the second seed has been selected, 15 of the remaining 23 seeds will not produce red flowers.

$$P(\text{none red}) = P(\text{first not red}) \cdot P(\text{second not red}) \cdot P(\text{third not red})$$

$$= \frac{17}{25} \cdot \frac{16}{24} \cdot \frac{15}{23} = \frac{34}{115}$$

Which Formula to Use

It is sometimes difficult to determine when to use the "or" formula and when to use the "and" formula. The following information may be helpful in deciding which formula to use.

OR Formula

OR problems will almost always contain the word "or" in the statement of the problem. For example, find the probability of selecting a heart *or* a 6. OR problems in this book generally involve only *one* selection. For example, "one card is selected" or "one die is rolled."

AND Formula

AND problems will often *not* use the word "and" in the statement of the problem. For example, "find the probability that both cards selected are red," or "find the probability that none of those selected is a banana" are both AND type problems. AND problems in this section will generally involve *more than 1* selection. For example, the problem may read "two cards are selected" or "three coins are flipped."

Section 11.6 Exercises

1. An individual is selected at random. Let event A be that the individual is happy. Let event B be that the individual is healthy.
   a) Are events A and B mutually exclusive events? Explain.
   b) Are they independent events? Explain.
2. A family is selected at random. Let event A be that the father smokes. Let event B be that the mother smokes.
   a) Are events A and B mutually exclusive events? Explain.
   b) Are they independent events? Explain.
3. If events A and B are mutually exclusive events, explain why the formula $P(A \text{ or } B) = P(A) + P(B) - P(A \text{ and } B)$ can be simplified to $P(A \text{ or } B) = P(A) + P(B)$.

In Ex. 4–8, a single die is rolled. Find the probability of rolling each of the following.
4. A 2 or 3
5. An odd number
6. An odd number or a number greater than 2
7. A number greater than 5 or less than 3
8. A number greater than 3 or less than 5

In Ex. 9–14, one card is selected from a deck of cards. Find the probability of selecting each of the following.
9. An ace or a king
10. A king or a heart
11. A picture card or a black card
12. A club or a red card

13. A card less than 8 or a diamond (*Note:* The ace is considered a low card.)

14. A card greater than 9 or a black card

In Ex. 15–22, two cards are selected from a deck of cards. Find the probability of the following (a) with replacement, (b) without replacement.

15. They are both jacks.

16. They are both red.

17. They are both spades.

18. The first is a king and the second is a queen.

19. The first is a heart and the second is a black card.

20. They are both picture cards.

21. Neither is a picture card.

22. The first is a picture card and the second is not a picture card.

In Ex. 23–32, assume the pointer cannot land on the line and assume that each spin is independent.

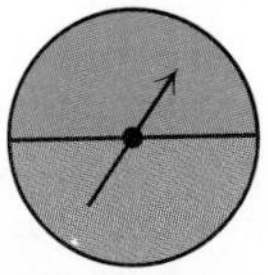

Figure 11.10

If the pointer in Fig. 11.10 is spun twice, find the probability the pointer lands on

23. red on both spins   **24.** red and then blue

If the pointer in Fig. 11.11 is spun twice, find the probability the pointer lands on

25. red and then green   **26.** green on both spins

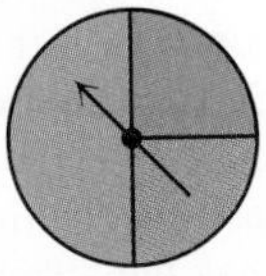

Figure 11.11

If the pointer in Fig. 11.12 is spun twice, find the probability the pointer lands on

27. blue on both spins

28. a color other than red on both spins

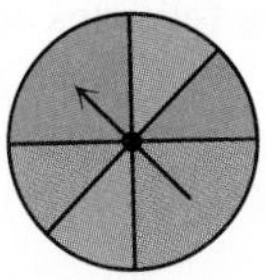

Figure 11.12

If the pointer in Fig. 11.10 is spun, and then the pointer in Fig. 11.11 is spun, find the probability of the pointers landing on

29. red on both spins

30. red on the first spin and blue on the second spin

31. a color other than blue on both spins

32. blue on the first spin and a color other than blue on the second spin

33. The Haefners plan to have five children. Find the probability that all their children will be boys. (Assume that $P(\text{boy}) = \frac{1}{2}$, and assume independence.)

34. The McDonalds presently have five boys. Mrs. McDonald is expecting another child. Find the probability that it will be a boy.

In Ex. 35–37, a family has three children. Assuming independence, find the probability of the following.

35. All three are girls.

36. All three are boys.

37. The youngest child is a boy, and the older children are girls.

In Ex. 38–41, a box contains four marbles: two red, one blue, and one green. Two marbles will be selected at random. Find the probability of selecting each of the following (a) with replacement, (b) without replacement.

38. a red marble and then a blue marble

39. two blue marbles

40. no red marbles

41. no green marbles

In Ex. 42–46, each individual letter of the word "pfeffernuesse" is placed on a piece of paper, and all 13 pieces of paper are placed in a hat. Three letters are selected at random from the hat. Find the probability of selecting each of the following (a) with replacement, (b) without replacement.

42. three fs

43. no fs

44. three vowels

45. The first letter selected is an f, the second is an e, and the third letter is an e.

46. The first letter selected is not a vowel and the second and third letters selected are vowels.

In Ex. 47–50, a sample of 150 people indicates that 50 favor candidate A, 65 favor candidate B, 20 favor candidate C, and 15 have no opinion. Three *different* people from the sample are randomly selected. Find the probability of each of the following.

47. The first favors candidate A, the second favors candidate B, and the third favors candidate C.

48. The first two have no opinion, and the third favors candidate C.

49. They all favor candidate A.

50. None favor candidate C.

In Ex. 51–54, an experimental drug was given to a sample of 100 hospital patients with an unknown sickness. Of the total, 70 patients reacted favorably, 10 reacted unfavorably, and 20 were unaffected by the drug. Assume that this sample is representative of the entire population. If this drug is given to Mr. and Mrs. Connors and their son Jimmy, what is the probability of each of the following? (Assume independence.)

51. Mrs. Connors reacts favorably.

52. Mr. and Mrs. Connors react favorably, and Jimmy is unaffected.

53. They all react favorably.

54. None reacts favorably.

In Ex. 55–59, each question of a five-question multiple-choice exam has four possible answers. Gurshawn picks an answer at random for each question. Find the probability that he selects the correct answer on

55. any one given question

56. only the first question

57. only the third and fourth questions

58. all five questions

59. none of the questions

In Ex. 60–64, consider the slot machine illustrated below.

Most people who play slot machines end up losing money since the machines are designed to favor the casino (the house). A slot machine generally has three or four reels that spin independently of each other. When you pull the handle the reels spin and the pictures on the reels come to rest in a

random order. Below is a list of the number of symbols of each type on each of the three reels (for one particular slot machine).

Pictures on Reels	Reels		
	1	2	3
Cherries	2	4	4
Oranges	5	4	5
Plums	6	3	4
Bells	2	4	3
Melons	2	2	2
Bars	2	2	1
7s	1	1	1

For this slot machine, find the probability of obtaining

60. a cherry on the first reel

61. oranges on all three reels

62. bars on all three reels

63. no 7s

64. three 7s

65. Certain birth defects and syndromes are polygenetic in nature. In the typical polygenetic affliction, the chance that an offspring will be born with one of these afflictions is small. Once an offspring is born with the affliction, the probability that future offspring of the same parents will be born with the same affliction increases. Let us assume that the probability of a child's being born with affliction A is 0.001. If a child is born with this affliction, the probability of a future child's being born with the same affliction becomes 0.04.

  a) Are the events of the births of two children in the same family with affliction A independent? Explain.

  b) A couple plans to have one child. Find the probability that the child will be born with this affliction.

  c) A couple plans to have two children. Find the probability that (i) both children will be born with the affliction; (ii) the first has the affliction and the second does not; (iii) neither has the affliction.

In Ex. 66–69, the probability that a heat-seeking torpedo will hit its target is 0.4. If the first torpedo hits its target, the probability that the second torpedo will hit the target increases to 0.9 because of the extra heat generated by the first explosion. If two heat-seeking torpedos are fired at a target, find the probability that

66. neither hits the target

67. the first hits the target and the second misses the target

68. both hit the target

69. the first misses the target and the second hits the target

In Ex. 70–74, use the fact that the Internal Revenue Service claims that 24 in every 1000 people in the $10,000–$50,000 income bracket are audited yearly. Assuming that the returns to be audited are selected at random and that each year's selections are independent of the previous year's selections, find the probability that a person in this income bracket will be audited

70. this year
71. next year
72. the next two years in succession
73. this year but not next year
74. neither this year nor next year

75. Ms. Runningdeer has a lottery ticket with a three-digit number in the range 000 to 999. Three balls are to be selected at random with replacement from a bin. There are an equal number of balls marked with the digits 0, 1, 2, . . . , 9. Find the probability that Ms. Runningdeer's number is selected.

Problem Solving

76. A bag contains five red chips, three blue chips, and two yellow chips. Two chips are selected from the bag without replacement. Find the probability that two chips of the same color are selected.

77. Ron has ten coins from Japan: three 1-yen coins, one 10-yen coin, two 20-yen coins, one 50-yen coin, and three 100-yen coins. He selects two coins at random without replacement. Assuming that each coin is equally likely to be selected, find the probability that Ron selects at least one 1-yen coin.

78. An investment advisor is considering the purchase of stock from five drug companies and ten computer companies. Knowing that each company had the same record of gain over the last year, the investment advisor decides to select three stocks at random from the fifteen stocks. What is the probability that two stocks are computer companies and one is a drug company?

79. Two playing cards are dealt to you from a well-shuffled deck of 52 cards. If either card is a diamond, or both are, you win; otherwise, you lose. Determine whether this game favors you, is fair, or favors the dealer. Explain your answer.

80. You have three cards: an ace, a king, and a queen. A friend shuffles the cards, selects two of them at random, and discards the third. You ask your friend to show you a picture card, and she turns over the king. What is the probability that she also has the queen?

Conditional Probability

In Section 11.6 we indicated that events are independent when the outcome of either event has no effect on the outcome of the other event. For example, selecting two cards from a deck of cards *with replacement* represented independent events. However, not all events are independent events. Consider this problem: Find the probability of selecting two aces from a deck of cards *without replacement*. The probability of selecting the first ace is $\frac{4}{52}$. The probability that the second ace is selected becomes $\frac{3}{51}$ since we assume an ace was removed from the deck with the first selection. Since the probability of the second ace being selected is affected by the first ace being selected, these two events are *dependent events*. Probability problems involving dependent events can be solved using conditional probability.

> ### Conditional Probability
> In general, the probability of event E_2 occurring, given that an event E_1 has happened (or will happen—the time relationship does not matter) is called a **conditional probability** and is written $P(E_2 | E_1)$.

The symbol $P(E_2|E_1)$, read "the probability of E_2 given E_1," represents the probability of E_2 occurring assuming that E_1 has already occurred (or will occur).

▶ Example 1

A single card is selected from a deck of cards. Find the probability it is a club given that it is black.

Solution: We are told that it is black. Thus only 26 cards are possible, of which 13 are clubs. Therefore

$$P(\text{Clubs}|\text{Black}) \text{ or } P(C|B) = \frac{13}{26} = \frac{1}{2}$$

▶ Example 2

Given a family with two children, and assuming that boys and girls are equally likely, find the probability that the family has
a) two girls
b) two girls if you know at least one of the children is a girl
c) two girls given that the older child is a girl

Solution:
a) The sample space of two children, BB, BG, GB, GG, can be determined by a tree diagram, see Fig. 11.13.

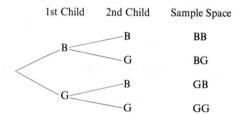

| 1st Child | 2nd Child | Sample Space |

Figure 11.13

Since there are four possibilities, of which only one has two girls,

$$P(2 \text{ girls}) = \frac{1}{4}$$

b) We are told that at least one of the children is a girl. Therefore, for this problem the sample space is BG, GB, GG. Since there are three possibilities, of which only one has two girls,

$$P(\text{both girls}|\text{at least one is a girl}) = \frac{1}{3}$$

c) If the older child is a girl the sample space reduces to GB, GG. Thus,

$$P(\text{both girls}|\text{older child is a girl}) = \frac{1}{2}$$

There are a number of formulas that can be used to find conditional probabilities. The one we will use in this book follows.

Did You
Know...

Dice-y Music

During the eighteenth century, composers were fascinated with the idea of creating compositions by rolling dice. A musical piece attributed by some to Mozart, titled "Musical Dice Game," consists of 176 numbered musical fragments of three-quarter beats each. Two charts show the performer how to order the fragments given a particular roll of two dice. Modern composers can create music using the same random technique and a computer; however, as in Mozart's time, the music created by random selection of notes or bars is still considered by musicologists to be esthetically inferior to the work of a creative composer.

Conditional Probability

For any two events E_1 and E_2,

$$P(E_2|E_1) = \frac{n(E_1 \text{ and } E_2)}{n(E_1)}$$

In the formula $n(E_1 \text{ and } E_2)$ represents the number of sample points common to both event 1 and event 2, and $n(E_1)$ is the number of sample points in event E_1, the given event. Since the intersection of E_1 and E_2, $E_1 \cap E_2$, represents the sample points common to both E_1 and E_2 the formula can also be expressed as

$$P(E_2|E_1) = \frac{n(E_1 \cap E_2)}{n(E_1)}$$

Figure 11.14 is helpful in explaining conditional probability.

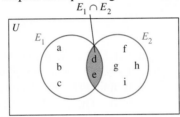

Figure 11.14

In the figure, the number of elements in E_1 is five, the number of elements in E_2 is six, and the number of elements in both E_1 and E_2, or $E_1 \cap E_2$, is two.

$$P(E_2|E_1) = \frac{n(E_1 \text{ and } E_2)}{n(E_1)} = \frac{2}{5}$$

Thus, for this given situation the probability of selecting an element from E_2 given that the element is in E_1 is $\frac{2}{5}$.

▶ **Example 3**

Each person in a sample of 200 residents in Washington County was asked whether he or she favored having a single countywide police department. The county consists of one large city and a number of small townships. The response of those sampled is given in the following chart.

Residence	Favor	Oppose	Total
Live in city	80	50	130
Live outside city	60	10	70
Total	140	60	200

If one person from the sample is selected at random, find the probability that the person

a) favors a single combined police force

b) favors a single combined police force if the person lives in the city

c) opposes a single countywide police force if the person lives outside the city

Solution:

a) The total number of respondents is 200, of which 140 favor a single police force.

$$P(\text{favors a single police force}) = \frac{140}{200} = \frac{7}{10}$$

b) Let E_1 be the given information "the person lives in the city." Let E_2 be "the person favors a single countywide police force." We are being asked to find $P(E_2|E_1)$. The number of people who live in the city, $n(E_1)$, is 130. The number of people who live in the city and favor a single countywide police force, $n(E_1 \text{ and } E_2)$, is 80. Thus by the formula

$$P(E_2|E_1) = \frac{n(E_1 \text{ and } E_2)}{n(E_1)} = \frac{80}{130} = \frac{8}{13}.$$

c) Let E_1 be the given information "the person lives outside the city." Let E_2 be "the person opposes a single countywide police force." We are asked to find $P(E_2|E_1)$. The number of people who live outside the city, $n(E_1)$, is 70. The number of people who live outside the city and oppose the countywide police force, $n(E_1 \text{ and } E_2)$, is 10.

$$P(E_2|E_1) = \frac{n(E_1 \text{ and } E_2)}{n(E_1)} = \frac{10}{70} = \frac{1}{7}$$

Section 11.7 Exercises

In Ex. 1–4, assume one die is rolled. Find the probability of rolling

1. a 6 given the number is even.

2. a 3 given the number is even.

3. an even number given the number is greater than 2.

4. a number less than 2 given the number is less than 5.

In Ex. 5–10, assume one card is selected from a deck of cards. Find the probability of selecting

5. a heart given the card is red

6. a club given the card is red

7. a jack given the card is a picture card

8. a picture card given the card is red

9. a 6 given the card is black

10. a 4 given the card is not a picture card

In Ex. 11–14, assume that a hat contains four bills: a $1, a $5, a $10, and a $20 bill. Two bills are to be selected at random with replacement. Construct a sample space as was done in Example 2 and find the following probabilities.

11. Both bills are $5 bills.

12. Both bills are $5 if the first selected is a $5 bill.

13. Both bills are $5 if at least one of the bills is a $5 bill.

14. Both bills are greater than $5 if the second bill was a $10 bill.

In Ex. 15–18, assume a couple has three children and that boys and girls are equally likely. Construct a sample space and find the probability that

15. they have three boys
16. they have at least two boys if the first born is a boy
17. they have three girls if the first and second born are girls
18. they have no girls if the first is a boy

In Ex. 19–24, two dice are rolled. Construct a sample space and find the probability the sum of the points on the dice total

19. 7
20. 7 if the first die is a 1
21. 7 if the first die is a 3
22. an even number if the second die is a 2
23. a number greater than 7 if the second die is a 5
24. a 7 or 11 if the first die is a 5

The results of a survey of the service at a local restaurant are summarized below.

Customer	Service Good	Service Poor	Total
Males	40	10	50
Females	55	20	75
Total	95	30	125

In Ex. 25–28, use the survey to find the probability that the service was rated

25. good
26. good given that the customer was a male
27. poor given that the customer was a female
28. poor given that the customer was a male

In Ex. 29–34, a quality control inspector is checking a sampling of light bulbs for defects. The chart that follows summarizes her findings.

Wattage	Good	Defective	Total
20	80	15	95
50	100	5	105
100	120	10	130
Total	300	30	330

If one of the light bulbs is selected at random, find the probability that the light bulb is

29. good
30. good given it is 50 watts
31. defective given it is 20 watts
32. good given it is 100 watts
33. good given it is 50 or 100 watts
34. defective given it is not 50 watts

In Ex. 35–39, two hundred and ten individuals are asked which evening news they watch most often. The results are summarized in the following chart.

	ABC	NBC	CBS	Other	Total
Men	30	20	40	25	115
Women	50	10	20	15	95
Total	80	30	60	40	210

If one of these individuals is selected at random find the probability they watch

35. ABC or NBC
36. ABC given the individual is a woman
37. ABC or NBC given the individual is a man
38. A station other than CBS given the individual is a woman
39. ABC, NBC, or CBS given the individual is a man

11.8 The Counting Principle and Permutations

In Section 11.5 we introduced the counting principle, which is repeated here for your convenience.

> **Counting Principle**
>
> If a first experiment can be performed in M distinct ways and a second experiment can be performed in N distinct ways, then the two experiments in that specific order can be performed in $M \cdot N$ distinct ways.

The counting principle is illustrated in Examples 1 and 2.

▶ **Example 1**

A license plate is to consist of two letters followed by three digits. Determine how many different license plates are possible if

a) repetition of letters and digits is permitted;

b) repetition of letters and digits is not permitted;

c) the first letter must be a vowel (a, e, i, o, u) and the first digit cannot be a 0, and repetition of letters and digits is not permitted.

Solution: There are 26 letters and 10 digits (0–9). We have five positions to fill as indicated below:

$$\overline{\text{L}} \quad \overline{\text{L}} \quad \overline{\text{D}} \quad \overline{\text{D}} \quad \overline{\text{D}}$$

a) Since repetition is permitted, there are 26 possible choices for both the first and second positions. There are 10 possible choices for the third, fourth, and fifth positions:

$$\frac{26}{\text{L}} \quad \frac{26}{\text{L}} \quad \frac{10}{\text{D}} \quad \frac{10}{\text{D}} \quad \frac{10}{\text{D}}$$

Since $26 \cdot 26 \cdot 10 \cdot 10 \cdot 10 = 676{,}000$, there are 676,000 different possible arrangements.

b) There are 26 possibilities for the first position. Since repetition of letters is not permitted, there are only 25 possibilities for the second position. The same reasoning is used when determining the number of digits for positions 3 through 5.

$$\frac{26}{\text{L}} \quad \frac{25}{\text{L}} \quad \frac{10}{\text{D}} \quad \frac{9}{\text{D}} \quad \frac{8}{\text{D}}$$

Since $26 \cdot 25 \cdot 10 \cdot 9 \cdot 8 = 468{,}000$, there are 468,000 different possible arrangements.

c) Since the first letter must be an a, e, i, o, or u, there are five possible choices for the first position. The second position can be filled by any of the letters except for the vowel selected for the first position. Therefore there are 25 possibilities for the second position.

Since the first digit cannot be a 0, there are nine possibilities for position 3. The fourth position can be filled by any digit except the one selected for position 3. Thus there are nine possibilities for position 4. Since the last position cannot be filled by any of the two digits previously used, there are eight possibilities for the last position:

$$\frac{5}{L} \quad \frac{25}{L} \quad \frac{9}{D} \quad \frac{9}{D} \quad \frac{8}{D}$$

Since $5 \cdot 25 \cdot 9 \cdot 9 \cdot 8 = 81{,}000$, there are 81,000 different arrangements that meet the conditions specified.

▶ Example 2

At a baseball card show, Kristen DiMarco wishes to display five different Mickey Mantle cards.
a) In how many different ways can she place the five cards in a straight row?
b) If she wishes to place Mickey Mantle's rookie card in the middle, in how many ways can she arrange the cards?

Solution:
a) There are five positions to fill using the five cards. In the first position, on the left, she can use any one of the five cards. In the second position she can use any of the four remaining cards. In the third position she can use any of the three remaining cards, and so on. The number of distinct possible arrangements is

$$\underline{5} \cdot \underline{4} \cdot \underline{3} \cdot \underline{2} \cdot \underline{1} = 120$$

b) We begin by satisfying the specific requirements stated. In this case the rookie card must be placed in the middle. Therefore there is only one possibility for the middle position:

$$\underline{} \quad \underline{} \quad \underline{1} \quad \underline{} \quad \underline{}$$

For the first position there are now four possibilities. For the second position, there will be three possibilities. For the fourth position there will be two possibilities. Finally, in the last position, there is only one possibility:

$$\underline{4} \cdot \underline{3} \cdot \underline{1} \cdot \underline{2} \cdot \underline{1} = 24$$

Thus under the condition stated there are 24 different possible arrangements.

Permutations

Now we introduce the definition of a permutation.

A **permutation** is any *ordered arrangement* of a given set of objects.

Peter, Paul, Mary and Mary, Paul, Peter represent two different permutations of the same three names. In Example 2(a) we can say there are 120 different ordered arrangements, or permutations, of the five Mickey Mantle cards. In Example 2(b) we see there are 24 different arrangements, or permutations, possible if Mickey Mantle's rookie card must be displayed in the middle.

When determining the number of permutations possible we assume that repetition of an item is not permitted. To help you understand and visualize permutations, we illustrate the various permutations possible when a triangle, rectangle, and circle are to be placed in a line.

Six Permutations

Note that for this set of three shapes there are six different arrangements or six permutations possible. We can obtain the number of permutations using the counting principle. For the first position there are three choices. There are then two choices for the second position, and only one choice left for the third position.

$$\text{Number of permutations} = 3 \cdot 2 \cdot 1 = 6$$

The product $3 \cdot 2 \cdot 1$ is referred to as 3 factorial, and written 3!. Thus

$$3! = 3 \cdot 2 \cdot 1 = 6.$$

Number of Permutations

The number of permutations of n distinct items is n factorial symbolized $n!$, where

$$n! = n(n - 1)(n - 2) \cdots (3)(2)(1)$$

It is important to note that 0! is defined to be 1. Many calculators have a factorial key, generally symbolized as $\boxed{x!}$ or $\boxed{n!}$. To find 5! on your calculator you would press:

$$5 \; \boxed{x!} \; 120$$

⎣—— displayed

▶ **Example 3**

In how many ways can eight different books be arranged on a shelf?

Solution: Since there are eight different books, the number of permutations is 8!

$$8! = 8 \cdot 7 \cdot 6 \cdot 5 \cdot 4 \cdot 3 \cdot 2 \cdot 1 = 40,320$$

Example 4 illustrates how to use the counting principle to determine the number of permutations possible when only a part of the total number of items is to be selected and arranged.

▶ **Example 4**

Consider the five letters p, q, r, s, t. In how many distinct ways can three letters be selected and arranged if repetition is not allowed?

Solution: We are asked to select and arrange only three of the five possible letters. Using the counting principle, we find that there are five possible letters for the first choice, four possible letters for the second choice, and three possible letters for the third choice:

$$5 \cdot 4 \cdot 3 = 60$$

Thus there are 60 different possible ordered arrangements, or permutations.

In Example 4 we determined the number of different ways in which we could select and arrange three of the five items. We can indicate this using the notation $_5P_3$. The notation $_5P_3$ is read "the number of permutations of five items taken three at a time." *The notation $_nP_r$ is read "the number of permutations of n items taken r at a time."*

We use the counting principle in the following problems to evaluate permutations of the form $_nP_r$.

$$_7P_4 = 7 \cdot 6 \cdot 5 \cdot \overset{\overbrace{\text{one more than } 7-4}}{4}$$

$$_9P_3 = 9 \cdot 8 \cdot \overset{\overbrace{\text{one more than } 9-3}}{7}$$

$$_{10}P_5 = 10 \cdot 9 \cdot 8 \cdot 7 \cdot \overset{\overbrace{\text{one more than } 10-5}}{6}$$

In general, the number of permutations of n items taken r at a time, $_nP_r$, may be found by the formula

$$_nP_r = n(n-1)(n-2) \cdots \overbrace{(n-r+1)}^{\text{one more than } n-r}$$

Therefore when evaluating $_{20}P_{15}$ we would find the product of consecutive decreasing integers from 20 to $(20-15+1)$ or 6, which is written as $20 \cdot 19 \cdot 18 \cdot 17 \cdot \cdots \cdot 6$.

Now let us develop an alternative formula that can be used to find the number of permutations possible when r objects are selected from n objects:

$$_nP_r = n(n-1)(n-2) \cdots (n-r+1)$$

or

$$_nP_r = n(n-1)(n-2) \cdots (n-r+1) \times \frac{(n-r)!}{(n-r)!}$$

For example,

$$_{10}P_5 = 10 \cdot 9 \cdot \cdots \cdot 6 \times \frac{5!}{5!}$$

or

$$_{10}P_5 = \frac{10 \cdot 9 \cdot \cdots \cdot 6 \times 5!}{5!}.$$

The expression for $_nP_r$ on page 553 can be rewritten as

$$_nP_r = \frac{n(n-1)(n-2) \cdots (n-r+1)\overbrace{(n-r)(n-r-1) \cdots (3)(2)(1)}^{(n-r)!}}{(n-r)!}$$

Since the numerator of the expression above is $n!$, we can see that

$$_nP_r = \frac{n!}{(n-r)!}.$$

For example, $_{10}P_5 = \dfrac{10!}{(10-5)!}.$

> The number of permutations possible when r objects are selected from n objects is found by the **permutation formula**
>
> $$_nP_r = \frac{n!}{(n-r)!}.$$

In Example 4 we found that when selecting three of five letters there were 60 permutations. We can obtain the same result using the permutation formula.

$$_5P_3 = \frac{5!}{(5-3)!} = \frac{5!}{2!} = \frac{5 \cdot 4 \cdot 3 \cdot \cancel{2 \cdot 1}}{\cancel{2 \cdot 1}} = 60$$

▶ **Example 5**

You are among eight people forming a bicycle club. Collectively you decide to put everyone's name in a hat and to randomly select a president, vice-president, and a secretary. How many different arrangements or permutations of officers are possible?

Solution: Since there are eight people, $n = 8$, of which three are to be selected; thus $r = 3$.

$$_8P_3 = \frac{8!}{(8-3)!} = \frac{8!}{5!} = \frac{8 \cdot 7 \cdot 6 \cdot \cancel{5 \cdot 4 \cdot 3 \cdot 2 \cdot 1}}{\cancel{5 \cdot 4 \cdot 3 \cdot 2 \cdot 1}} = 336$$

Thus with eight people there can be 336 different arrangements for president, vice-president, and secretary.

In Example 5, the fraction $\dfrac{8 \cdot 7 \cdot 6 \cdot \cancel{5 \cdot 4 \cdot 3 \cdot 2 \cdot 1}}{\cancel{5 \cdot 4 \cdot 3 \cdot 2 \cdot 1}}$ can also be expressed as $\dfrac{8 \cdot 7 \cdot 6 \cdot \cancel{5!}}{\cancel{5!}} = 336$. Example 5, like other permutation problems, can also be answered using the counting principle.

▶ **Example 6**

The Stereo Shop's warehouse has 10 different receivers in stock. The owner of the chain calls the warehouse requesting that a different receiver be sent to each of its 6 stores. How many ways can the distribution of the receivers to the stores be made?

Solution: There are ten receivers from which six are to be selected and distributed. Thus $n = 10$ and $r = 6$.

$$_{10}P_6 = \frac{10!}{(10 - 6)!} = \frac{10!}{4!} = \frac{10 \cdot 9 \cdot 8 \cdot 7 \cdot 6 \cdot 5 \cdot \cancel{4!}}{\cancel{4!}}$$

$$= 151{,}200$$

There are 151,200 different permutations of receivers possible.

We have worked permutation problems (selecting and arranging, without replacement, *r* items out of *n distinct* items) using the counting principle and using the permutation formula. When given a permutation problem, unless specified by your instructor, you may use either technique to determine its solution.

Permutations of Duplicate Items

So far all of the examples we have discussed in this section have involved arrangements with distinct items. Now we will consider permutation problems in which some of the items to be arranged are duplicates. For example, the word BOB contains three letters, of which the two Bs are duplicates. How many permutations of the word BOB are possible? If the two Bs were distinguishable (one red and the other blue), there would be six permutations.

BOB   BBO   OBB

BOB   BBO   OBB

However if the Bs are not distinguishable (replacing all colored Bs with black print), we see that there are only three permutations:

BOB   BBO   OBB

The number of permutations of BOB can be computed as follows:

$$\frac{3!}{2!} = \frac{3 \cdot \cancel{2 \cdot 1}}{\cancel{2 \cdot 1}} = 3$$

where 3! represents the number of permutations of three letters assuming that none are duplicates, and 2! represents the number of ways the two items that are duplicates can be arranged (BB or BB). In general, we have the following rule.

> **Permutations of Duplicate Objects**
> The number of distinct permutations of n objects where n_1 of the objects are identical, n_2 of the objects are identical, . . . , n_r of the objects are identical is found by the formula
>
> $$\frac{n!}{n_1! n_2! \cdots n_r!}.$$

▶ **Example 7**

In how many different ways can the letters of the word "statistics" be arranged?

Solution: There are 10 letters including three t's, three s's, and two i's. The number of possible arrangements is

$$\frac{10!}{3! 3! 2!} = \frac{10 \cdot 9 \cdot 8 \cdot 7 \cdot 6 \cdot 5 \cdot 4 \cdot 3 \cdot 2 \cdot 1}{3 \cdot 2 \cdot 1 \cdot 3 \cdot 2 \cdot 1 \cdot 2 \cdot 1} = 50,400.$$

There are 50,400 different possible arrangements of the letters in the word "statistics."

Section 11.8 Exercises

1. In your own words describe what is meant by a permutation.
2. In your own words explain how to find $n!$ for any whole number n.

Evaluate each of the following:

3. 6!
4. 8!
5. 7!
6. 0!
7. $_5P_2$
8. $_6P_4$
9. $_9P_0$
10. $_5P_0$
11. $_8P_7$
12. $_4P_4$
13. $_8P_8$
14. $_9P_6$

15. The daily double at most race tracks consists of selecting the winning horse in both the first and second races. If the first race has seven entries and the second race has eight entries, how many daily double tickets must you purchase to guarantee a win?

16. To use an automated teller machine, you generally must enter a four-digit code using the digits 0–9. How many four-digit codes are possible if repetition of digits is permitted?

17. Secured doors at airports and other locations are opened by pressing the correct sequence of numbers on a control panel. If the control panel contains the digits 0–9 and a three-digit code must be entered (repetition is permitted).
   a) How many different codes are possible?
   b) If one code is entered at random, find the probability the correct code is entered.

18. Some doors on cars can now be opened by pressing the correct sequence of buttons. A display of buttons by the door handle of a car follows.

| 0–1 | 2–3 | 4–5 | 6–7 | 8–9 |

The correct sequence of five buttons must be pressed to unlock the door. If the same button may be pressed consecutively
   a) How many possible ways can the five buttons be pressed (repetition is permitted)?
   b) If five buttons are pressed at random, find the probability a sequence that unlocks the door will be entered.

19. A social security number consists of nine digits. How many different social security numbers are possible if repetition of digits is permitted?

20. A college identification number consists of a letter followed by four digits. How many different identification numbers are possible if repetition is not permitted?

21. The trifecta at most race tracks consists of selecting the first-, second-, and third-place finishers in a particular race in their proper order. If there are six entries in the trifecta race, how many tickets must you purchase to guarantee a win?

22. The operator of the Audio Warehouse is planning a grand opening. He wishes to advertise that the store has many different sound systems available. They stock 10 different tape decks, 12 different receivers, and 9 different speakers. Assuming that a sound system will consist of one of each, and that all pieces are compatible, how many different sound systems can they advertise?

23. In how many different ways can five dogs be lined up to be displayed at dog shows?

24. In how many different ways can six books be arranged on a shelf?

25. If a club consists of eight members, how many different arrangements of president and vice-president are possible?

26. A $100, a $50, and a $20 prize are to be awarded to three different people. If seven people are being considered for the prizes, how many different arrangements are possible?

27. Each book registered in the Library of Congress must have an ISBN code number. For an ISBN number of the form D-DD-DDDDDD-D, where D represents a digit from 0 through 9, how many different ISBN numbers are possible if repetition of digits is allowed? (See research activity in Ex. 56.)

28. At the reception line of a wedding, the bride, the groom, the best man, the maid of honor, the four ushers, and the four bridesmaids must line up to receive the guests.

   a) If the individuals listed above can line up in any order, how many arrangements are possible?

   b) If the groom must be the last in line and the bride must be next to the groom, and the others can line up in any order, how many arrangements are possible?

   c) If the groom is to be last in line, the bride next to the groom, and males and females are to alternate, how many arrangements are possible?

In Ex. 29–32, an identification code is to consist of two letters followed by two digits. How many different codes are possible if

29. repetition is permitted?

30. repetition is not permitted?

31. the first letter must be a W, X, Y, or Z and repetition is permitted?

32. the first two entries must both be the same letter and repetition of the digits is not permitted?

In Ex. 33–36, a license plate is to consist of four digits followed by two letters. Determine the number of different license plates possible under the following conditions:

33. Repetition of numbers and letters is permitted.

34. Repetition of numbers and letters is not permitted.

35. The first and second digits must be odd, and repetition is not permitted.

36. The first digit cannot be zero, and repetition is not permitted.

37. A telephone number consists of seven digits with the restriction that the first digit cannot be 0 or 1.

   a) How many distinct telephone numbers are possible?

   b) How many distinct telephone numbers are possible using three-digit area codes preceding the 7 digit number, where the first digit of the area code is not 0 or 1?

   c) With the increasing use of cellular phones and paging systems our society is beginning to run out of usable phone numbers. A number of phone companies are developing phone numbers that use 11 digits instead of 7. How many distinct phone numbers can be made with 11 digits assuming the area code remains three digits and the first digit of the area code and the phone number cannot be 0 or 1?

38. Find the number of permutations of the letters in the word "HELP."

39. A teacher decides to give six different prizes to six of the 30 students in her class. In how many ways can she do this?

40. A bookcase contains 20 different books. Three books will be selected at random and given away as prizes to three individuals. In how many ways can this be done?

41. A night guard visits 10 different offices every hour. The pattern is varied each night so that the guard will not follow a specific routine. In how many different ways can this be done?

42. Six different horses are to be lined up at the starting gate. How many different arrangements are possible?

43. How many different ways are there to arrange the digits 0, 1, 2, 3, 4, 5, 6, and 7?

44. A man has 8 pairs of pants, 12 shirts, 15 ties, and 6 sport coats. Assuming he must wear one of each, how many different outfits can he wear?

45. **a)** To open a combination lock, you must know the lock's three-number sequence in its proper order. Repetition of numbers is permitted. Why is this more like a permutation lock than a combination lock? Why is it not a true permutation problem?

b) Assuming that a combination lock has 40 numbers, determine how many different three-number arrangements are possible if repetition of numbers is allowed?

c) Answer the question in part (b) if repetition is not allowed.

46. In how many ways can the letters in the word "HORSE" be arranged?

47. In how many ways can the letters in the word "WINNING" be arranged?

48. In how many ways can the letters in the word "KISSIMMEE" be arranged?

49. In how many ways can the letters in the word "SASSAFRAS" be arranged?

50. In how many ways can the winner and first and second runners-up be selected from seven remaining finalists in a science project contest?

51. In one question of a history test the student is asked to match ten dates with 10 events; each date can only be matched with one event. In how many different ways can this question be answered?

52. Five different colored flags will be placed on a pole, one beneath another. The arrangement of the colors indicates the message. How many messages are possible if five flags are to be selected from eight different colored flags?

53. In how many ways can the manager of a baseball team arrange his batting order of nine players if the pitcher must bat last?

54. Five Rembrandts are to be displayed in a museum.
a) In how many different ways can they be arranged if they must be next to one another?
b) In how many different ways can they be displayed if a specific one is to be in the middle?

Problem Solving

55. On a preferential ballot, a person is asked to rank three of seven candidates for recommendation for promotion, giving their first, second, and third choices (no ties). What is the minimum number of ballots that must be cast in order to guarantee that at least two ballots are the same?

Research Activity

56. When a book is published it is assigned a 10-digit code number called the International Standard Book Number (ISBN). The digits are used to code the language, the publisher, the book, and a check digit. Do research to determine the details of how this works and write a report on your findings. Ask a school librarian or check the source: *The UMAP Journal*, 6(1): 41–54, 1985 for an article entitled International Standard Code Number by P. Tuchinsky.

 Combinations

When the order of the selection of the items is important to the final outcome, the problem is a permutation problem, and when the order of the selection of the items is unimportant to the final outcome the problem is a **combination** problem.

Recall from Section 11.8 that permutations are *ordered* arrangements. Thus, for example, *a, b, c* and *b, c, a* are two different permutations because the ordering of the three letters is different. The letters *a, b, c* and *b, c, a* represent the same combination of letters because the *same letters* are used in each set. However, the letters *a, b, c* and *a, b, d* represent two different combinations of letters because the letters contained in each set are different.

> A **combination** is a distinct group (or set) of objects without regard to their arrangement.

▶ **Example 1**

Determine whether the situation represents a permutation or combination problem.

a) A group of five friends, Arline, Inez, Judy, Dan, and Eunice, are forming a club. The group will elect a president and a treasurer. In how many different ways can the president and treasurer be selected?

b) Of the five individuals listed above two will be attending a meeting together. How many different ways can this be done?

Solution:

a) Since the president's position is different from the treasurer's position, this problem is a permutation problem. Judy as president with Dan as treasurer is different from Dan as president and Judy as treasurer.

b) Since the order in which the two individuals selected to attend the meeting is not important this is a combination problem. There is no difference if Judy is selected and then Dan is selected, or if Dan is selected and then Judy is selected.

In the previous section we learned that $_nP_r$ represents the number of permutations when r items are selected from n distinct items. *Similarly $_nC_r$ represents the number of combinations when r items are selected from n distinct items.*

Consider the set of elements $\{a, b, c, d, e\}$. The number of permutations of two letters from the set is represented as $_5P_2$ and the number of combinations of two letters from the set is represented as $_5C_2$. Below we show that there are 20 permutations of two letters possible, and 10 combinations of two letters possible from these five letters. Thus $_5P_2 = 20$ and $_5C_2 = 10$.

Permutations	Combinations
$\left.\begin{array}{l} ab,\ ba,\ ac,\ ca,\ ad,\ da,\ ae,\ ea,\ bc,\ cb, \\ bd,\ db,\ be,\ eb,\ cd,\ dc,\ ce,\ ec,\ de,\ ed \end{array}\right\}\ 20$	$\left.\begin{array}{l} ab,\ ac,\ ad,\ ae,\ bc, \\ bd,\ be,\ cd,\ ce,\ de \end{array}\right\}\ 10$

When discussing both combination and permutation problems we always assume the experiment is performed without replacement. This is why duplicate letters such as *aa* or *bb* are not included above.

Notice that from one combination of two letters two permutations can be formed. For example, the combination *ab* gives the permutations *ab* and *ba*. This results in twice as many permutations as combinations. Thus for this example we may write

$$_5P_2 = 2(_5C_2).$$

Since $2 = 2!$ we may write

$$_5P_2 = 2!(_5C_2).$$

If we repeated this same process for comparing the number of permutations in $_nP_r$ with the number of combinations in $_nC_r$ we would find that

$$_nP_r = r!(_nC_r).$$

Dividing both sides of the equation by $r!$ gives

$$_nC_r = \frac{_nP_r}{r!}.$$

Since $_nP_r = n!/(n-r)!$, the combination formula may be expressed as

$$_nC_r = \frac{n!/(n-r)!}{r!} = \frac{n!}{(n-r)!r!}.$$

> The number of combinations possible when r objects are selected from n objects is found by the **combination formula**
>
> $$_nC_r = \frac{n!}{(n-r)!r!}.$$

▶ **Example 2**

Eight colleagues staying at the same hotel wish to go to a restaurant that is not within walking distance. There is only one car available that seats five. Assuming that any of the eight can drive the car, determine the number of different groups of people that can ride in the car.

Solution: This problem is a combination problem because the order in which the five people are selected is unimportant. There are a total of eight people; thus $n = 8$. Five are to be selected; thus $r = 5$.

$$_8C_5 = \frac{8!}{(8-5)!5!} = \frac{8!}{3!5!} = \frac{8 \cdot 7 \cdot 6 \cdot 5 \cdot 4 \cdot 3 \cdot 2 \cdot 1}{3 \cdot 2 \cdot 1 \cdot 5 \cdot 4 \cdot 3 \cdot 2 \cdot 1} = 56$$

Thus 56 different combinations are possible when selecting five from a group of eight.

▶ **Example 3**

An exam consists of six questions. Any four may be selected for answering. In how many ways can this be done?

Solution: This is a combination problem because the order in which the four questions are answered is immaterial.

$$_6C_4 = \frac{6!}{(6-4)!4!} = \frac{6!}{2!4!} = \frac{\overset{3}{6} \cdot 5 \cdot 4 \cdot 3 \cdot 2 \cdot 1}{2 \cdot 1 \cdot 4 \cdot 3 \cdot 2 \cdot 1} = 15$$

There are 15 different ways that four of the six questions can be selected. Can you list them? Do so now.

▶ **Example 4**

At Su Wong's Chinese Restaurant, dinner for eight consists of three items from column A, four items from column B, and three items from column C. If columns A, B, and C have five, seven, and six items, respectively, how many different dinner combinations are possible?

In popular card games there is such a variety of possible combinations of cards that a player rarely gets the same hand twice. The total number of different 5-card poker hands is 2,598,960, and for 13-card bridge hands this number increases to 635,013,559,600.

Solution: For column A, three items out of five items must be selected. This can be represented as $_5C_3$. For column B, four items out of seven must be selected. This can be represented as $_7C_4$. For column C, three items out of six must be selected, or $_6C_3$.

$$_5C_3 = 10 \qquad _7C_4 = 35 \qquad _6C_3 = 20$$

Using the counting principle, we can determine the total number of dinner combinations by multiplying the number of choices from columns A, B, and C together:

$$\text{Total number of dinner choices} = {_5C_3} \cdot {_7C_4} \cdot {_6C_3}$$
$$= 10 \cdot 35 \cdot 20 = 7000$$

Therefore 7000 different combinations are possible under these conditions.

We have studied various counting schemes including the counting principle, permutations, and combinations. You will often need to decide which method to use to solve a problem. Table 11.4 may help you in selecting the procedure to use.

Table 11.4

Counting Principle	Permutations	Combinations
Used when determining the number of different ways two or more experiments can occur. Used when there are specific placement requirements, such as the first digit must be a 0 or 1.	Used when determining number of ways of selecting r items from n items. Repetition not permitted.	
Repetition is permitted. *Counting Principle:* If a first experiment can be performed in M ways and a second experiment can be performed in N ways, then the two experiments, in that specific order, can be performed in $M \times N$ ways.	Order is important. For example, a, b, c and b, c, a are two different permutations of the same three letters. $$_nP_r = \frac{n!}{(n-r)!}$$ Problems worked using the permutation formula may also be worked using the counting principle.	Order is not important. For example, a, b, c and b, c, a are the same combination of three letters. But a, b, c and a, b, d are two different combinations of three letters. $$_nC_r = \frac{n!}{r!(n-r)!}$$

Section 11.9 Exercises

1. In your own words explain what is meant by a combination.

In Ex. 2–12, evaluate each of the following.

2. $_5C_4$ 3. $_6C_2$

4. $_7C_3$ 5. $_8C_0$

6. $_9C_5$ 7. $_{12}C_8$

8. $_{10}C_3$ 9. $_7C_7$

10. $\dfrac{_5C_3}{_5P_3}$ 11. $\dfrac{_6C_2}{_6P_2}$

12. $\dfrac{_8C_5}{_8C_2}$

13. An ice cream parlor has 15 different flavors. Cynthia orders a banana split and has to select three different flavors. How many different selections are possible?

14. Mr. Ballou, a teacher, decides to give identical prizes to six of the 30 students in his class. In how many ways can he do this?

15. A student must select and answer four of five essay questions on a test. In how many ways can this be done?

16. A bookcase contains 20 valuable and different books. Three books will be selected at random and donated to a public television auction. How many different sets of three books can be donated?

17. Three of a sample of 10 radios will be selected and tested for defects. In how many ways can this be done?

18. Three of eight finalists will be selected and awarded $5000 scholarships. In how many ways can this be done?

19. A disc jockey has 20 records that she wishes to play during the next hour, but she only has time to play 15 of them. How many different ways can she select the batch of 15?

20. At the beginning of the semester your instructor informs the class that she marks on a curve and that exactly four of the 25 students in the class will receive a final grade of A. In how many ways can this be done?

21. Find the number of heart flushes possible when there are 13 hearts in a deck of cards. (A heart flush is any set of five hearts.)

22. If there are four jacks in a deck of cards, how many different pairs of jacks are possible?

23. A five-card poker hand is to be dealt from a deck of 52 cards. How many different poker hands are possible?

24. A quinella bet consists of selecting the first- and second-place winners, in any order, in a particular event. For example, suppose you select a 2–5 quinella. If 2 wins and 5 finishes second, or if 5 wins and 2 finishes second, you win. In the game of jai alai, 8 two-man teams play against one another. How many quinella tickets are necessary to guarantee a win?

25. On an English test, Tito must write an essay for three of the five questions in Part 1 and four of the six questions in Part 2. How many different combinations of questions can he answer?

26. The singing group Def Leppard is planning to make a compact disc consisting of six fast songs and four slow songs. If they have 10 fast songs and seven slow songs to choose from, how many different possible combinations do they have?

27. How many ways are there of selecting three kings and two queens when five cards are selected at random from a deck of cards without replacement?

28. An editor has eight manuscripts for mathematics books and five manuscripts for computer science books. If he is to select five mathematics and three computer science books for publication, how many different choices does he have?

29. How many ways are there of selecting three diamonds and two clubs when five cards are selected at random from a deck of cards without replacement?

30. Michael is sent to the store to get five different bottles of regular soda and three different bottles of diet soda. If there are 10 different regular sodas and seven different diet sodas to choose from, how many different choices does Michael have?

31. How many different committees can be formed from six teachers and 50 students if the committee is to consist of two teachers and three students?

32. A teacher is constructing a mathematics test consisting of 10 questions. She has a pool of 28 questions, which are classified by level of difficulty as follows: 6 difficult questions, 10 average questions, and 12 easy questions. How many different 10-question tests can she construct from the pool of 28 questions if her test is to have 3 difficult, 4 average, and 3 easy questions?

33. General Mills is testing six oat cereals, five wheat cereals, and four rice cereals. If it plans to market three of the oat cereals, two of the wheat cereals, and two of the rice cereals, how many different combinations are possible?

34. A catering service is making up trays of hors d'oeuvres. The hors d'oeuvres are categorized as inexpensive, average, and expensive. If the client must select three of the seven inexpensive, five of the eight average, and two of the four expensive hors d'oeuvres, how many different choices are possible?

Problem Solving

35. Consider a 10-question test in which each question can be answered either right or wrong.
   a) How many different ways are there to answer the questions so that exactly eight are right and two are wrong?
   b) How many ways are there to answer the questions so that at least eight are right?

36. The notation $_nC_r$ may be written $\binom{n}{r}$.

   a) Using this notation, evaluate each of the combinations in the following array. Form a triangle of the results, similar to the one given, by placing the answer to each combination in the same relative position in the triangle.

$$\binom{0}{0}$$
$$\binom{1}{0} \quad \binom{1}{1}$$
$$\binom{2}{0} \quad \binom{2}{1} \quad \binom{2}{2}$$
$$\binom{3}{0} \quad \binom{3}{1} \quad \binom{3}{2} \quad \binom{3}{3}$$
$$\binom{4}{0} \quad \binom{4}{1} \quad \binom{4}{2} \quad \binom{4}{3} \quad \binom{4}{4}$$

   b) Using the number pattern in (a), can you find the next row of numbers of the triangle (known as **Pascal's triangle**)?

37. a) Four people at dinner make a toast. If each person is to tap glasses with each other person, how many taps will take place?
   b) Repeat part (a) with five people.
   c) How many taps will there be if there are n people at the dinner table?

(11.10) Probability Problems Using Combinations

In the previous section we discussed combination problems. Now we will use combinations in working probability problems.

Suppose we want to find the probability of selecting two picture cards (jacks, queens, or kings) when two cards are selected, without replacement, from a deck of 52 cards. Using the "and" probability formula discussed in Section 11.6 we could reason as follows:

$$P(2 \text{ picture cards}) = P(1\text{st picture card}) \cdot P(2\text{nd picture card})$$

$$= \frac{12}{52} \cdot \frac{11}{51} = \frac{132}{2652} \quad \text{or} \quad \frac{11}{221}$$

Since the order of the two picture cards that are selected is not important to the final answer, this can be considered a combination probability problem. We can find the probability of selecting two picture cards, using combinations, by finding the number of possible successful outcomes (selecting two picture cards) and dividing that answer by the total number of possible outcomes (selecting any two cards).

The number of ways in which two picture cards can be selected from the

12 picture cards in a deck is $_{12}C_2$:

$$_{12}C_2 = \frac{12!}{(12-2)!2!} = \frac{\overset{6}{\cancel{12}} \cdot 11 \cdot \cancel{10!}}{\cancel{10!} \cdot \cancel{2} \cdot 1} = \frac{66}{1} = 66$$

The number of ways in which two cards can be selected from a deck of 52 cards is $_{52}C_2$:

$$_{52}C_2 = \frac{52!}{(52-2)!2!} = \frac{\overset{26}{\cancel{52}} \cdot 51 \cdot \cancel{50!}}{\cancel{50!} \cdot \cancel{2} \cdot 1} = 1326$$

Thus

$$P(\text{selecting 2 picture cards}) = \frac{_{12}C_2}{_{52}C_2} = \frac{66}{1326} = \frac{11}{221}.$$

Notice that the same answer is obtained by using either method. To give you more practice with counting techniques, we will work the problems in this section using counting techniques.

▶ **Example 1**

A club consists of four men and five women. Three members are to be selected at random to form a committee. What is the probability that the committee will consist of three women?

Solution:

$$P\left(\begin{array}{c}\text{committee consists}\\ \text{of 3 women}\end{array}\right) = \frac{\text{number of possible committees with 3 women}}{\text{total number of possible 3-member committees}}$$

The number of possible committees with 3 women is $_5C_3 = 10$.
The total number of possible 3-member committees is $_9C_3 = 84$.

$$P(\text{committee consists of 3 women}) = \frac{10}{84} = \frac{5}{42}.$$

The probability of randomly selecting a committee with three women is $\frac{5}{42}$.

▶ **Example 2**

There are 10 seats left aboard an airplane: six aisle seats and four window seats. If three people about to board the plane are given seat numbers at random, what is the probability that they are all given aisle seats?

Solution:

$$P\left(\begin{array}{c}\text{all 3 given}\\ \text{aisle seats}\end{array}\right) = \frac{\text{number of possible combinations of aisle seats for 3 people}}{\text{total number of possible combinations of seats for 3 people}}$$

The number of possible combinations of seating the 3 people in the 6 aisle seats is $_6C_3 = 20$. The total number of possible combinations of seating the 3 people in the 10 available seats is $_{10}C_3 = 120$.

$$P(\text{all 3 given aisle seats}) = \frac{_6C_3}{_{10}C_3} = \frac{20}{120} = \frac{1}{6}$$

Thus the probability of all three passengers getting aisle seats is $\frac{1}{6}$.

▶ Example 3

You are dealt a flush in the game of poker when you are dealt five cards of the same suit. If you are dealt a five-card hand, find the probability that you will be dealt a flush in hearts.

Solution:

$$P(\text{flush in hearts}) = \frac{\text{number of possible 5-card heart flushes}}{\text{total number of 5-card hands possible}}$$

The order in which the five cards are received is immaterial. Thus the number of possible hands can be found by the combination formula. Since there are 13 hearts in a deck of cards, the number of possible five-card flush hands is $_{13}C_5 = 1287$. The total number of possible five-card hands in a deck of 52 cards is $_{52}C_5 = 2,598,960$.

$$P(\text{flush in hearts}) = \frac{_{13}C_5}{_{52}C_5} = \frac{1287}{2,598,960} = \frac{33}{66,640}$$

The probability of being dealt a heart flush is $\frac{33}{66,640}$.

▶ Example 4

The Kellogg Company is testing 12 new cereals for possible production. They are testing three oat cereals, four wheat cereals, and five rice cereals. If we assume that each of the 12 cereals has the same chance of being selected and four new cereals will be produced, find the probability that
a) no wheat cereals are selected,
b) at least one wheat cereal is selected,
c) two wheat cereals and two rice cereals are selected.

Solution:
a) If no wheat cereals are to be selected, then only oat and rice cereals must be selected. There are a total of eight cereals that are oat or rice. Thus the number of ways that four oat or rice cereals may be selected out of the eight possible oat or rice cereals is $_8C_4$. The total number of possible selections is $_{12}C_4$.

$$P(\text{no wheat cereals}) = \frac{_8C_4}{_{12}C_4} = \frac{70}{495} = \frac{14}{99}$$

b) When four cereals are selected, the choice must contain either no wheat cereal or at least one wheat cereal. Since one of these outcomes must occur, the sum of the probabilities must be 1:

$$P(\text{no wheat cereal}) + P(\text{at least 1 wheat cereal}) = 1$$

Therefore

$$P(\text{at least 1 wheat cereal}) = 1 - P(\text{no wheat cereal})$$

$$= 1 - \frac{14}{99} = \frac{99}{99} - \frac{14}{99} = \frac{85}{99}$$

Note that the probability of selecting no wheat cereals was found in part (a).

c) The number of ways of selecting two wheat cereals out of four wheat cereals is $_4C_2$, which equals 6. The number of ways of selecting two rice cereals out of five rice cereals is $_5C_2$, which equals 10. The total number of possible selections when four cereals are selected from the 12 choices is $_{12}C_4$. Since both the two wheat *and* the two rice cereals must be selected, the probability is calculated as follows:

$$P(2 \text{ wheat and } 2 \text{ rice}) = \frac{_4C_2 \cdot {}_5C_2}{_{12}C_4} = \frac{6 \cdot 10}{495} = \frac{60}{495} = \frac{4}{33}$$

Section 11.10 Exercises

In Ex. 1–41, the problems are to be done without replacement. Use combinations to determine probabilities.

1. Each of the numbers 1 through 6 is written on a piece of paper, and the six pieces of paper are placed in a hat. If two numbers are selected at random, find the probability that both numbers selected are even.

2. An urn contains five red balls and four blue balls. You plan to draw three balls at random. Find the probability of selecting three red balls.

3. A hat contains four 1-dollar bills, three 5-dollars bills, and one 10-dollar bill. If you select two bills at random from the hat, find the probability of selecting two 1-dollar bills.

4. There are four good and four defective batteries in a box. If you select three at random, find the probability that you select three good batteries.

5. Each of the digits 0, 1, 2, 3, 4, 5, 6, 7, 8, and 9 is placed on a slip of paper, and the slips are placed in a hat. If three slips of paper are selected at random, find the probability that the three numbers selected are greater than 4.

6. The 10 finalists in a scholarship contest consist of six women and four men. If two equal prizes are to be awarded, find the probability that both prizes are awarded to women. Assume that each applicant has an equal chance of being selected.

7. A three-person committee is to be selected at random from five Democrats and three Republicans. Find the probability that all three selected are Democrats.

8. A committee of four is to be randomly selected from a group of seven teachers and eight students. Find the probability that the committee will consist of four students.

9. You are dealt five cards from a deck of 52 cards. Find the probability that you are dealt five red cards.

10. You are dealt five cards from a deck of 52 cards. Find the probability that you are dealt no aces.

In Ex. 11–13, a television game show has five doors, of which the contestant must pick two. Behind two of the doors are expensive cars, and behind the other three doors are consolation prizes. The contestant gets to keep the items behind the two doors she selects. Find the probability that the contestant wins

11. both cars
12. no cars
13. at least one car

In Ex. 14–16, your friend agrees to give you a Spider Man comic book in exchange for three of your record albums he selects at random. You have a total of 10 albums including three Beach Boys, four Boston Symphony, and three Garth Brooks (you like a wide variety of music). Assuming each album has the same chance of being selected, find the probability that:

14. Three of your Beach Boys albums will be selected.
15. Two of your Boston Symphony and one Garth Brooks albums will be selected.
16. None of your Garth Brooks albums will be selected.

In Ex. 17–20, a hat contains three red, four white, and five blue chips. If three chips are selected at random, find the probability of selecting

17. two red chips and one blue chip
18. two blue chips and one white chip
19. at least one red chip
20. one red, one white, and one blue chip

In Ex. 21–24, drug A is given to five patients. Drug B is give to four patients, and drug C is given to six patients. If four of these 15 patients are selected at random, find the probability that

21. two were given drug A and two were given drug C
22. three were given drug C and one was given drug A
23. at least one patient was given drug C
24. one was given drug A, two were given drug B, and one was given drug C

In Ex. 25–28, five men and six women are going to be assigned to a specific row of seats in the theater. If the 11 tickets for the numbered seats are given out at random, find the probability that

25. five women are given the first five seats next to the center aisle.
26. at least one woman is in one of the first five seats
27. exactly one woman is in one of the first five seats
28. three women are seated in the first three seats and two men are seated in the next two seats

In Ex. 29–31, determine the probability of winning the state lottery if you must select

29. exactly six specific numbers out of a total of 40 numbers
30. exactly six specific numbers out of a total of 44 numbers
31. any six of eight specific numbers out of a total of 40 numbers

32. A full house in poker consists of getting three of one card and two of another card in a five-card hand. For example, if a hand contains three kings and two 5s, it is a full house. If five cards are dealt at random from a deck of 52 cards, without replacement, find the probability of getting three kings and two 5s.
33. A royal flush consists of the ace, king, queen, jack, and 10 all in the same suit. If seven cards are dealt at random from a deck of 52 cards, find the probability of getting a
   a) royal flush in spades
   b) royal flush in any suit

Problem Solving

In Ex. 34–37, the first reel of a three-reel slot machine consists of three oranges, four cherries, two lemons, six plums, two bells, and one bar. The second reel consists of two oranges, five cherries, three lemons, five plums, two bells, and one bar. The third reel consists of four oranges, four cherries, three lemons, three plums, two bells, and two bars. When you pull the handle of the slot machine, find the probability of obtaining

34. cherries on all three reels
35. cherry, cherry, bell
36. bars on all three reels
37. a cherry on at least one reel

38. If three men and three women are to be assigned at random to six seats in a row at a theater, find the probability that they will alternate by sex.
39. A club consists of 15 people including Ali, Kendra, Ted, Alice, Marie, Dan, Linda, and Frank. From the 15 members a president, vice-president, and treasurer will be selected at random, and an advisory committee of five other individuals will also be selected at random.
   a) Find the probability that Ali is selected president, Kendra is selected vice-president, Ted is selected treasurer, and the other five individuals named above form the advisory committee.
   b) Find the probability that three of the eight individuals named above are selected for the three officers' positions and the other five are selected for the advisory board.

40. A pair of aces and a pair of 8s is often referred to as the "dead man's hand." (See the Did You Know in this section.)

    a) Find the probability of being dealt the dead man's hand when dealt five cards, without replacement, from a deck of 52 cards.

    b) The actual cards Hickok was holding when shot were the aces of spades and clubs, the 8s of spades and clubs, and the 9 of diamonds. If you are dealt five cards without replacement, find the probability of being dealt this exact hand.

41. A number is written with a magic marker on each card of a deck of 52 cards. The numeral 1 is put on the first card, 2 on the second, and so on. The cards are then shuffled and cut. What is the probability that the top four cards will be in ascending order? (For example, the top card is 12, the second 22, the third 41, and the fourth 51.)

CHAPTER 11 SUMMARY

Key Terms

11.1
empirical probability
event
experiment
outcome
probability
relative frequency
theoretical (mathematical) probability

11.2
equally likely outcomes

11.3
odds against
odds in favor

11.4
expected value (expectation)
fair price

11.5
counting principle
sample point
sample space
tree diagram

11.6
"and" probability problems
compound probability
dependent events
independent events
mutually exclusive events
"or" probability problems

11.7
conditional probability

11.8
permutation

11.9
combination

Important Facts
Empirical probability

$$F(E) = \frac{\text{number of times event } E \text{ has occurred}}{\text{total number of times the experiment has been performed}}$$

The law of large numbers Probability statements apply in practice to a large number of trials, not to a single trial. It is the relative frequency over the long run that is accurately predictable, not the individual events or precise totals.

Theoretical probability

$$P(E) = \frac{\text{number of outcomes favorable to } E}{\text{total number of possible outcomes}}$$

The probability of an event that cannot occur is 0.
The probability of an event that must occur is 1.
Every probability must be a number between 0 and 1 inclusively, that is,

$$0 \leq P(E) \leq 1$$

The sum of the probabilities of all possible outcomes of an event is 1.

$$P(A) + P(\text{not } A) = 1$$

Odds against an event

$$\text{Odds against} = \frac{P(\text{event fails to occur})}{P(\text{event occurs})} = \frac{P(\text{failure})}{P(\text{success})}$$

Odds in favor of an event

$$\text{Odds in favor} = \frac{P(\text{event occurs})}{P(\text{event fails to occur})} = \frac{P(\text{success})}{P(\text{failure})}$$

Expected value

$$E = P_1 A_1 + P_2 A_2 + P_3 A_3 + \cdots + P_n A_n$$

Expected value = Fair price − Cost to play

Counting principle If a first experiment can be performed in M distinct ways and a second experiment can be performed in N distinct ways, then the two experiments in that specific order can be performed in $M \cdot N$ distinct ways.

OR and AND problems

$$P(A \text{ or } B) = P(A) + P(B) - P(A \text{ and } B)$$
$$P(A \text{ and } B) = P(A) \cdot P(B)$$

Conditional probability $\quad P(E_2 | E_1) = \dfrac{n(E_1 \text{ and } E_2)}{n(E_1)}$

The **number of permutations** of n items is $n!$

$$n! = n(n-1)(n-2) \cdots (3)(2)(1)$$

Permutation formula $\quad {}_n P_r = \dfrac{n!}{(n-r)!}$

The number of different permutations of n objects where $n_1, n_2, \ldots, n_r$ of the objects are identical is

$$\frac{n!}{n_1! n_2! \cdots n_r!}$$

Combination formula $\quad {}_n C_r = \dfrac{n!}{(n-r)! r!}$

CHAPTER 11 REVIEW EXERCISES

11.1–11.10

1. In your own words explain the law of large numbers.
2. Of 100 tosses of a trick coin, 80 resulted in heads. Find the empirical probability of the coin landing heads up.
3. Select a card from a deck of cards 50 times with replacement and compute the empirical probability of selecting a spade.
4. In 100 births at Collins Hospital 62 were males. Find

the empirical probability that the next baby born at the hospital will be a male.

In Ex. 5–8, each of the digits 0, 1, 2, 3, 4, 5, 6, 7, 8, 9 is written on a piece of paper, and all the pieces of paper are placed in a hat. One number is selected at random. Find the probability that the number selected is

5. odd

6. even or greater than 4

7. greater than 4 or less than 6

8. even and greater than 4

In Ex. 9–12, assume the ages of a sampling of children in a junior high school resulted in the following data:

Age	Number
11	10
12	42
13	50
14	60
15	38

If one individual is selected at random from this sample, find the probability that the individual selected is

9. 14 years old

10. less than 14 years old

11. 12 or 13 years old

12. greater than 13 years old

13. The probability that Horst wins a prize at a carnival is $\frac{1}{3}$. Find the odds (a) against and (b) in favor of Horst winning a prize.

14. A Radio Shack science fair kit has 150 projects, of which 12 are defective. If one project is selected at random, find the odds against selecting a defective project.

15. Assuming that the odds against winning the game of Monopoly are 1:2, find the probability of winning the game.

16. The probability that Dorothy and Toto get lost on their way to Auntie Em's house is 0.4. Find the odds in favor of their getting lost.

17. A thousand raffle tickets are sold at $2 each. Three prizes of $200 and two prizes of $100 will be awarded.
  a) Find the expectation of a person who purchases a ticket.
  b) Find the expectation of a person who purchases three tickets.
  c) Find the fair price to pay for a ticket.

18. If Pete selects a picture card from a deck of cards, George will give him $6. If Pete does not select a picture card, he must give George $2.
  a) Find Pete's expectation.
  b) Find George's expectation.
  c) If Pete plays this game 100 times, how much can he expect to lose or gain?

19. If it is a sunny day then 1000 people will attend the baseball game. If the day is cloudy then only 500 people will attend, and if it rains then only 100 people will attend the game. The local meteorologist predicts the probability of a sunny day is 0.4, of a cloudy day is 0.5, and of a rainy day is 0.1. Find the number of people that are expected to attend.

20. Tina, Jake, Gina, and Carla form a club. They plan to select a president and a vice-president.
  a) Construct a tree diagram showing all the possible outcomes.
  b) List the sample space.
  c) Find the probability that Gina is selected president and Jake is selected vice-president.

21. A coin is flipped and then a marble is to be selected from a bag. The bag contains four marbles, one red, one blue, one green and one purple.
  a) Construct a tree diagram showing all the possible outcomes.
  b) List the sample space.
  c) Find the probability that a head is flipped and either a red or purple marble is selected.

Each of the digits 0, 1, 2, 3, 4, 5, 6, 7, 8, 9 is written on a piece of paper, and all the pieces of paper are placed in a hat. Two pieces of paper are to be selected. Find the following probabilities (a) with replacement, (b) without replacement.

22. Both numbers are even.

23. Both are the number 2.

24. Both numbers are prime.

25. Both numbers are less than 3.

A dog pound has six dogs to give away—two collies, one German shepherd, and three poodles. A family decides to select at random two of these dogs. Find the probability of each of the following.

26. They are both collies.

27. They are both German shepherds.

28. The first selected is a poodle and the second a collie.

In Ex. 29–32, assume the spinner cannot land on a line. If spun once, find

29. the probability the spinner lands on red

30. the odds against and in favor of the spinner landing on red
31. If you are awarded $5 if the spinner lands on red, find a fair price to play the game.
32. If the spinner is spun twice, find the probability that it lands on red and then gray (assume independence).

In Ex. 33–36, assume the spinner below cannot land on a line. If spun once, find

33. the probability the spinner does not land on gray.
34. the odds in favor and against the spinner landing on gray.
35. A person wins $10 if the spinner lands on gray, wins $5 if the spinner lands on red, and loses $20 if the spinner lands on blue. Find the expectation of a person who plays this game.
36. If the spinner is spun three times, find the probability that at least one spin lands on red.

In Ex. 37–40, a sampling of 160 new cars were checked for defects. The following chart shows the results of the survey.

Car	Fewer than 5 Defects	Five or More Defects	Total
American built	84	12	96
Foreign built	50	14	64
Total	134	26	160

Find the probability that if one car is selected from this sample the car has:

37. Fewer than five defects if it is American built.
38. Fewer than five defects if it is foreign built.
39. Five or more defects if it is foreign built.
40. Five or more defects if it is American built.

In Ex. 41–44, a sampling of voters is made to determine which of three candidates, Mr. Spock, Captain Kirk, or Dr. McCoy, should run against the incumbent governor. The results are shown in the chart below.

	Mr. Spock	Captain Kirk	Dr. McCoy	Total
Male	60	40	70	170
Female	20	100	50	170
Total	80	140	120	340

If one voter is selected from this sample, find the probability the voter favors

41. Mr. Spock if the voter is male
42. Captain Kirk if the voter is female
43. a candidate other than Dr. McCoy if the voter is male
44. Mr. Spock or Captain Kirk if the voter is female

45. Five finalists remain in a lottery drawing. The prizes to be awarded among the finalists are $100, $1000, $10,000, $100,000, $1,000,000.
   a) In how many different ways can the winners be selected?
   b) What is the expectation of a finalist?
46. Six finalists remain in a lottery drawing. Three will receive $1000 each, two will receive $10,000 each, and one will receive $100,000. How many different arrangements of prizes are possible?
47. Each of Mr. Gonzalez's three children is planning to select his or her own pet rabbit from a litter of five rabbits. In how many ways can this be done?
48. Three astronauts out of nine must be selected for a mission. One will be the captain, one will be the navigator, and one will explore the surface of the moon. In how many ways can a three-person crew be selected such that each person has a different assignment?
49. Dr. Fox has three doses of serum for influenza type A. Six patients in the office require the serum. In how many ways can Dr. Fox dispense the serum?
50. a) Ten of 15 huskies are to be selected to pull a dogsled. In how many ways can this selection be made?
   b) How many different arrangements of the ten huskies on a dogsled are possible?
51. In the New York State lottery game called Lotto, you must pick 6 from a possible 54 numbers. If all 6 numbers you picked earlier are selected, you win the grand prize. Find the probability of winning the grand prize at Lotto.
52. A club consists of six males and eight females. Three males and three females are to be selected to represent

their club at a dinner. How many different combinations are possible?

53. In a psychology research laboratory, one room contains eight men, numbered 1 through 8, and another room contains five women, numbered 1 through 5. Three men and two women are to be selected at random to be given a psychological test. How many different combinations of people are possible?

54. Two cards are selected at random, without replacement, from a deck of 52 cards. Find the probability that two aces are selected (use combinations).

In Ex. 55–58, a bag contains five red chips, three white chips, and two blue chips. Three chips are to be selected at random, without replacement. Find the probability that

55. they are all red

56. the first two are red and the third is blue

57. the first is red, the second is white, and the third is blue

58. at least one is red

In Ex. 59–62, Agway Lawn and Garden Center carries five pine, six maple, and three birch trees. Gene Rodenberry plans to select three trees at random from the 14 mentioned. Assuming each tree has the same chance of being selected, find the probability that

59. three maple trees are selected

60. two pine trees and one maple tree is selected

61. no pine trees are selected

62. at least one pine tree is selected

CHAPTER TEST

1. In a certain community, of the last 180 births, 85 were male. Find the empirical probability that the next child born in this community is a male.

In Ex. 2–5, each of the numbers 1 through 8 is written on a sheet of paper, and the eight sheets of paper are placed in a hat. If *one* sheet of paper is selected at random from the hat, find the probability that the number selected is

2. greater than 2

3. even

4. even or greater than 4

5. odd and greater than 4

In Ex. 6–9, if *two* sheets of paper are selected, without replacement, from the hat discussed above, find the probability that

6. both numbers are greater than 4

7. both numbers are even

8. the first number is odd and the second number is even

9. neither of the numbers is greater than 6

10. One card is selected at random from a deck of cards. Find the probability that the card selected is a red card or a picture card.

In Ex. 11–15, one die is rolled and one letter—either *a*, *b*, or *c*—is selected at random.

11. Use the counting principle to determine the number of sample points in the sample space.

12. Construct a tree diagram illustrating all the possible outcomes, and list the sample space.

By observing the sample space, determine the probability of obtaining

13. the number 6 and the letter *a*

14. the number 6 or the letter *a*

15. an even number or the letter *c*

16. A code is to consist of three digits followed by two letters. Find the number of possible codes if the first digit cannot be a 0 or 1 and replacement is not permitted.

17. A class consists of 18 females and 10 males. If one person is selected at random from the class, find the odds against the person's being female.

18. The odds against Jane winning the raquetball tournament are 5:2. Find the probability that Jane wins the tournament.

19. You get to select one card at random from a deck of cards. If you pick a club, you win $10. If you pick a heart, you win $5. If you pick any other suit, you lose $7. Find your expectation for this game.

20. A biologist is studying the habits of squirrels at two national parks. The chart below indicates the number and type of squirrels that were studied at the two parks.

	Gray Squirrels	Other Types of Squirrel	Total
Great Smoky Mountain	82	33	115
Yosemite	60	45	105
Total	142	78	220

If one of the squirrels under study is spotted at Yosemite, find the probability it is a gray squirrel.

21. There are six people from whom three are to be selected and given small prizes. One will be given a book, one will be given a calculator, and the third will be given a $10 dollar bill. In how many different ways can these prizes be awarded?

In Ex. 22 and 23, a bin has a total of 15 batteries of which six are defective. If you select two at random without replacement, find the probability that

22. none are good   **23.** at least one is good

24. There are five green apples and seven red apples in a bucket. Five apples are to be selected at random without replacement. Find the probabiliy that three red apples and two green apples are selected.

STATISTICS

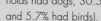

Now in its 111th edition and weighing over 2 pounds, the *Statistical Abstract of the United States* is a compilation of facts and figures taken from the U.S. Census. In it you'll find what Americans own (in 1990, 92% of households had a TV and 63% had a VCR) and owe (in 1989, 956.9 million credit cards were used to charge $430.4 billion worth of goods) and what kinds of pets Americans have (38.2% of households had dogs, 30.5% had cats, and 5.7% had birds).

Benjamin Disraeli (1804–1881), once Prime Minister of Britain, said there are three kinds of lies: "Lies, damned lies, and statistics." Do numbers lie? Numbers are, after all, the foundation of all statistical information. The "lie" comes in when, either intentionally or carelessly, a number is used in such a way as to lead us to a conclusion that is unjustified or incorrect.

The desire to record data about the human population and its activities has been with us for thousands of years. The earliest censuses were undertaken either to provide a basis for taxation or to register individuals for military service. The first survey on a large scale was commissioned in 1086 by William the Conqueror of England. Called the Domesday Book (meaning "day of judgment"), the survey was intended to establish the Crown's holdings and to list the resources of the country for taxation purposes. Completed in only one year, the survey details the property holdings of over 13,000 medieval towns and villages. The first modern census was taken in the United States in August 1790. This was

■ Wetter
□ Drier

National Center for Atmospheric Research

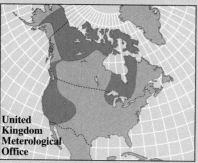

United Kingdom Meterological Office

One of the most pressing environmental concerns involves a much debated warming of the atmosphere due to the greenhouse effect. Scientists using the same statistical data sometimes make predictions that may vary considerably. The two maps shown differ considerably in predicting the effects the warming has on soil moisture.

the first time a census was taken to be used as a basis for government representation. A census has been taken in the United States every 10 years since that time, and is used for many purposes, including determining the number of seats in the House of Representatives allotted to each state.

The results of statistical surveys are closely related to the way the questions are asked. A classic example is a 1940 survey by Elmer Roper in which people were asked essentially the same question in two different ways: "Do you think the United States should forbid public speeches against democracy?" (about half the interviewees responded "yes"); "Do you think the United States should allow public speeches against democracy?" (here three-quarters of the interviewees reponded "no").

Census taking represents a very real attempt to gather statistical data household by household, a practice that is time-consuming and costly. Today there are methods by which statistics obtained from a numerically small sample are used as representative of the much larger population. In fact, a sample as small as 1500 may be used to predict the outcome of a national election. Such information gathering is conducted by a variety of people, including medical researchers, scientists, advertisers, and political pollsters. You may find many examples of statistics every day on the front page of your newspaper.

When evaluating statistical information, remember that you need to judge the sampling method as well as the numbers given. Ask yourself who conducted the study and whether they have a bias, how large the sample was and whether it was representative, where the study appeared and whether there was an opposing side. Numbers may not lie, but they can be manipulated and misinterpreted.

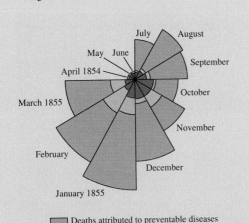

Deaths attributed to preventable diseases
Deaths attributed to war wounds
All other causes

Florence Nightingale (1820–1910), the British nurse considered to be the founder of modern medical nursing, used statistical data and graphs as evidence in her fight to improve sanitary conditions in hospitals. She used this polar-area diagram (her own invention) to show dramatically that British soldiers hospitalized in the Crimean War were far more likely to die of infection and disease than of war wounds. Using these graphs, she lobbied the British government and public to make sanitary reforms which saved many lives.

12.1 Sampling Techniques

Statistics is the art and science of gathering, analyzing, and making inferences (predictions) from numerical information obtained in an experiment. This numerical information is referred to as **data**. Statistics, originally associated with numbers gathered for governments, has grown significantly and is now used in all walks of life.

Governments use statistics to estimate the amount of unemployment and the cost of living. Thus statistics has become an indispensable tool in attempts to regulate the economy. In psychology and education the statistical theory of tests and measurements has been developed to compare achievements of individuals from diverse places and backgrounds. Another use of statistics that we are all familiar with is the public opinion poll. Newspapers and magazines carry the results of different Harris and Gallup polls on topics ranging from the president's popularity to the number of cans of beer consumed. In recent years these polls have had a high degree of accuracy. The A. C. Nielson rating is a public opinion poll that determines the country's most and least watched TV shows. Statistics is used in scores of other professions; in fact, it is difficult to find one that does not depend on some aspect of statistics.

Statistics is divided into two main branches: descriptive and inferential. **Descriptive statistics** is concerned with the collection, organization, and analysis of data. **Inferential statistics** is concerned with making generalizations or predictions from the data collected.

Probability and statistics are closely related. Someone in the field of probability is interested in computing the chance of occurrence of a particular event when all the possible outcomes are known. A statistician's interest lies in drawing conclusions about what the possible outcomes may be through observations of only a few particular events.

If a probability expert and a statistician find identical boxes, the probability expert might open the box, observe the contents, replace the cover, and proceed to compute the probability of randomly selecting a specific object from the box. The statistician, on the other hand, might select a few items from the box without looking at the contents and make a prediction as to the total contents of the box. The few items the statistician randomly selected from the box constitute what is referred to as a sample. The entire contents of the box constitute what is referred to as the **population**. The statistician often uses a subset of the population, the **sample**, to make predictions concerning the population.

When a statistician draws a conclusion from a sample, there is always the possibility that the conclusion is incorrect. For example, suppose a jar contains 90 blue marbles and 10 red marbles, Fig. 12.1. If the statistician selects a random sample of five marbles from the jar and they are all blue, he or she may wrongly conclude that the jar contains all blue marbles. If the statistician takes a larger sample, say 15 marbles, he or she is likely to select some red marbles. At that point the statistician may make a prediction about the contents of the

Figure 12.1

Did You
Know...

The Birth of Inferential Statistics

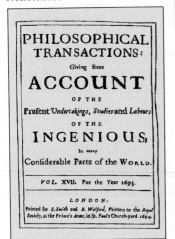

The London merchant John Gaunt is credited with first making statistical predictions, or inferences, from a set of data rather than basing them simply on the laws of chance. He studied the vital statistics (births, deaths, marriages) contained in the Bills of Mortality published during the years of the Great Plague and observed that more males were born than females, and women lived longer than men. From these observations, he made predictions about life expectancies. The keeping of mortality statistics was stimulated considerably by the growth of the insurance industry.

jar based on the sample selected. Of course, the most accurate result would occur if every object in the jar, the entire population, was sampled. However, in most statistical experiments this is not practical.

Consider the task of determining the political strength of a certain candidate running in a national election. It is not possible for pollsters to ask each of the approximately 113 million registered voters his or her preference for candidates. Thus pollsters must select and use a sample of the population to obtain their information. How large a sample do you think they use to make predictions about an upcoming national election? You might be surprised to learn that pollsters use only about 1600 registered voters in their national sample. How can a pollster using such a small percentage of the population make an accurate prediction?

The answer lies in the fact that when pollsters select a sample, they use sophisticated statistical techniques to obtain a sample that is unbiased. An **unbiased sample** is one that is a small replica of the entire population with regard to income, education, sex, race, religion, political affiliation, age, and so on. The procedures that statisticians use to obtain reliable samples are quite complex. The following sampling techniques will give you a brief idea of how statisticians obtain unbiased samples.

Random Sampling

If a sample is drawn in such a way that each time an item is selected, each item in the population has an equal chance of being drawn, the sample is said to be a **random sample**. Under these conditions, one combination of a specified number of items will have the same probability of being selected as any other combination. When all of the items in the population are similar with regard to the specific characteristic we are interested in, a random sample can be expected to produce satisfactory results. For example, consider a large container holding 300 tennis balls that are identical except for color. One-third of the balls are yellow, one-third are white, and one-third are green. If the balls can be thoroughly mixed between each draw of a tennis ball so that each ball has an equally likely chance of being selected, then randomness is not difficult to achieve. However, if the objects or items are not all the same size, shape, or texture, it might be impossible to obtain a random sample by reaching into a container and selecting an object.

The best procedure for selecting a random sample is to use a random number generator or a table of random numbers. A random number generator is a device, usually a calculator or computer program, that produces a list of random numbers. A random number table is a collection of random digits in which each digit has an equal chance of appearing. To select a random sample, first assign a number to each element in the population. Numbers are usually assigned in order. Then select the number of random numbers needed, which is determined by the sample size. Each numbered element from the population that corresponds to a selected random number becomes part of the sample.

Systematic Sampling

When a sample is obtained by drawing every *n*th item on a list or production line, the sample is a **systematic sample**. The first item should be determined by using a random number.

It is important that the list from which a systematic sample is chosen include the entire population one wants to study. See the Did You Know on the *Literary Digest*. Another problem that one must avoid when using this method of sampling is the constantly recurring characteristic. For example, on an assembly line, every tenth item could be the work of robot X. If only every tenth item is checked, the work of other robots doing the same job may not be checked and may be defective.

Cluster Sampling

The **cluster sample** is sometimes referred to as an *area sample* because it is frequently applied on a geographical basis. Essentially, the sampling consists of a random selection of groups of units. For example, on a geographical basis we might select blocks of a city or some other geographical subdivision to use as a sample unit. Another example would be to select *x* boxes of screws from a whole order, count the number of defective screws in the *x* boxes, and use this number to determine the expected number of defective screws in the whole order.

Stratified Sampling

When a population is divided into parts, called strata, for the purpose of drawing a sample, the procedure is known as **stratified sampling**. Stratified sampling involves dividing up the population by characteristics called *stratifying factors*, such as sex, race, religion, or income. When a population has varied characteristics, it is desirable to separate the population into classes with similar characteristics and then take a random sample from each stratum (or class).

The use of stratified sampling requires some knowledge of the population. For example, to obtain a cross section of voters in a city, we must know where various groups are located and the approximate numbers in each location.

 Section 12.1 Exercises

1. Explain the difference between descriptive and inferential statistics.
2. Define statistics in your own words.
3. When you hear the word "statistics," what specific words or ideas come to mind?
4. Name five areas other than those mentioned in which statistics is used.

5. Attempt to list at least two professions in which no aspect of statistics is used.
6. Explain the difference between probability and statistics.
7. **a)** What is a population?
   b) What is a sample?
8. **a)** What is a random sample?
   b) How might a random sample be selected?

9. **a)** What is a systematic sample?
   b) How might a systematic sample be selected?
10. **a)** What is a cluster sample?
   b) How might a cluster sample be selected?
11. **a)** What is a stratified sample?
   b) How might a stratified sample be selected?
12. The principal of an elementary school wishes to determine the "average" family size of the children who attend the school. To obtain a sample, the principal visits each room and selects the four students closest to each corner of the room. The principal asks each of these students how many people are in his or her family.
   a) Will this technique result in an unbiased sample? Explain your answer.
   b) If the sample is biased, will the average be greater than or less than the true family size? Explain.
13. **a)** Select a topic of interest to which a random sampling technique can be applied to obtain data.
   b) Explain how you would obtain a random sample for your topic of interest.
   c) Actually obtain the sample by the procedure stated in part (b).

The Misuses of Statistics

Statistics, when used properly, is a valuable tool to society. However, many individuals, businesses, and advertising firms misuse statistics to their own advantage. You should examine statistical statements very carefully before accepting them as fact. Two questions you should ask yourself are: Was the sample used to gather the statistical data unbiased and of sufficient size? Is the statistical statement ambiguous; that is, can it be interpreted in more than one way?

Let us examine two advertisements. "Three out of five dentists recommend sugarless gum for their patients who chew gum." In this advertisement, it is unknown how large a sample was used or how many times this experiment was performed before the desired results were found. The advertisement does not mention that possibly only 1 of 100 dentists recommended gum at all.

In a golf ball commercial, a "type A" ball is hit, and a second ball is hit in the same manner. The type A ball travels farther. From this the public is to conclude that the type A is the better ball. The advertisement makes no mention of the number of times the experiment was previously performed or the results of the earlier experiments. Possible sources of bias include (1) wind speed and direction, (2) the fact that no two swings are identical, and (3) the fact that the ball may land on a rough or smooth surface.

Vague or ambiguous words also lead to statistical misuses or misinterpretations. The word "average" is one such culprit. There are at least four different "averages," some of which are discussed in Section 12.5. Each is calculated differently, and each may have a different value for the same sample. During contract negotiations it is not uncommon for an employer to state publicly that the average salary of its employees is $25,000, while the employees' union states that the average is $20,000. Who is lying? It is possible that both are telling the truth. Each will use the average that best suits its needs to arrive at the stated figures. Advertisers also use the average that most enhances their

"There are 3 kinds of lies; lies, damned lies, and statistics."
Benjamin Disraeli

IN THE BARREL...

Price per bbl. of
light crude, leaving
Saudi Arabia
on Jan. 1

April 1
$14.55

$13.34

$12.70

$12.09

$11.51

$10.46

$10.95

$2.41

'74 '75 '76 1977 1978 **1979**

'73

Visual graphics are often used to "dress up" what might otherwise be considered boring statistics. Although visually appealing, such creative displays of numerical data can be misleading. In his book The Visual Display of Quantitative Information *(Graphics Press, 1983), Edward Tufte cites this misleading graphic taken from* Time, *April 9, 1979. The data shown are the annual increase in oil prices from 1973 to 1979, an overall increase of 454%. Yet what you "see" is a volumetric increase of 27,000% in the relative size of the barrels.*

product. Consumers often misinterpret this average as the one with which they are most familiar.

Another vague word is the word "largest." For example, ABC claims that it is the largest department store in the United States. Does this mean largest profit, largest sales, largest building, largest staff, largest acreage, or largest number of outlets?

Still another deceptive technique used in advertising is to state a claim from which the public may draw irrelevant conclusions. For example, a disinfectant manufacturer claims that its product killed 40,760 germs in a laboratory in five seconds. "To prevent colds, use disinfectant A." It may well be that the germs killed in the laboratory were not related to any type of cold germ. In another example, company C claims that its paper towels are heavier than its competition's towels. Therefore they will hold more water. Is weight a measure of absorbency? A rock is heavier than a sponge, yet a sponge is more absorbent.

An insurance advertisement claims that in Duluth, Minnesota, 212 people switched to insurance company Z. One may conclude that this company is offering something special to attract these people. What may have been omitted from the advertisement is that 415 people in Duluth, Minnesota, dropped insurance company Z during the same period.

A foreign car manufacturer claims that nine out of 10 of a popular model that it sold in the United States during the previous 10 years were still on the road. From this statement the public is to conclude that this foreign car is well manufactured and would last for many years. The commercial neglects to state that this model has been selling in the United States for only a few years. The manufacturer could just as well have stated that 9 of every 10 of these cars sold in the United States in the previous 100 years were still on the road.

Charts and graphs can also be misleading or deceptive. In Fig. 12.2 two graphs show performance of two stocks. Which stock would you purchase? Actually, the two graphs present identical information; the only difference is that the vertical scale of stock B has been exaggerated.

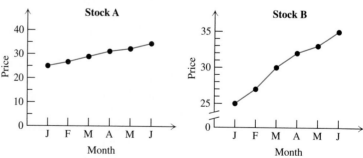

Figure 12.2

The two graphs in Fig. 12.3 show the same change. However, the graph on the left appears to show a greater increase than the graph on the right, once again because of a different scale.

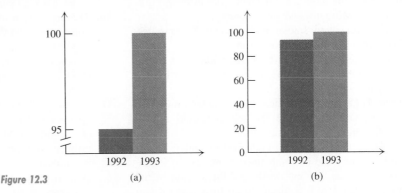

Figure 12.3   (a)   (b)

Consider a claim that if you invest $1, by next year you will have $2. This is sometimes misrepresented as in Fig. 12.4. Actually your investment has only doubled, but the area of the square on the right is four times that of the square on the left. By expressing the amounts as cubes (Fig. 12.5), you increase the volume eightfold.

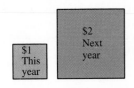

Figure 12.4

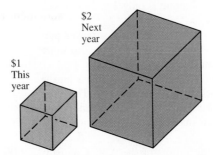

Figure 12.5

Despite the examples in this section, you should not be left with the impression that statistics is used solely for the purpose of misleading or cheating the consumer. As stated earlier, there are many important and necessary uses of statistics. Most statistical reports are accurate and useful. You should realize, however, the importance of being an aware consumer.

 ## Section 12.2 Exercises

1. Find five advertisements or commercials that may be statistically misleading. Explain why each may be misleading.

In Ex. 2–14 discuss each of the following statements, and tell what possible misuses or misinterpretations may exist.

2. Most accidents occur on Saturday night. This means that people do not drive carefully on Saturday night.

3. Morgan's is the largest department store in New York. So shop at Morgan's and save money.

4. Most doctors who smoke prefer brand X cigarettes. Therefore it is safe to smoke brand X.

5. Eighty percent of all automobile accidents occur within 10 miles of the driver's home. Therefore it is safer to take long trips.

6. At East High School, half the students are below average in mathematics. Therefore the school should receive more federal aid to raise standards.

7. Florida has the highest death rate in the country. Therefore Florida should be condemned as unsafe.

8. Arizona has the highest death rate for asthma in the country. Therefore it is unsafe to go to Arizona if you have asthma.

9. Four of every 6 dentists surveyed preferred Calgonite toothpaste. Therefore it is the best toothpaste.

10. More men than women are involved in automobile accidents. Therefore women are better drivers.

11. Brand X has 80 percent fewer calories than the best-selling brand. So eat brand X and stay healthy.

12. The average depth of the pond is only 3 ft, so it is safe to go wading.

13. A recent survey showed that 60% of those surveyed preferred Coors beer to Budweiser beer. Therefore more people buy Coors than Budweiser.

14. Females have a higher average score than males on the English part of the Scholastic Aptitude Test. Therefore on this test a particular female selected at random will outperform a particular male selected at random.

15. Below is a chart of the profits of the ABC Company from January to December of last year.

Month	Profit (%)
January	6.0
February	5.0
March	4.5
April	4.3
May	4.8
June	5.0
July	5.5
August	6.0
September	7.0
October	6.6
November	7.2
December	7.5

a) Draw a line graph that makes the profit appear stable.

b) Draw a line graph that makes the profit appear to have a high gain.

16. The population of the United States in 1980 was 226.5 million. In 1990 the population was 248.7 million

a) Draw a bar graph that appears to show a small population increase.

b) Draw a bar graph that appears to show a large population increase.

17. The week of March 16, 1992, the two most popular soap operas were "The Young and the Restless," with a rating of 7.9, and "All My Children," with a rating of 7.2. Each rating point represents 921,000 homes in which the show is watched.

a) Draw a bar graph that appears to show a small difference in the popularity of the soap operas.

b) Draw a bar graph that appears to show a large difference in the popularity of the soap operas.

18. The following chart shows the average price paid for a man's suit from 1987 to 1991.

Year	Price
1987	$254
1988	$260
1989	$268
1990	$275
1991	$285

a) Draw a line graph that makes the increase in price appear to be small.

b) Draw a line graph that makes the increase in price appear to be large.

19. The following graph indicates the time it takes for two 1992 model cars to accelerate from 0 to 60 mph.

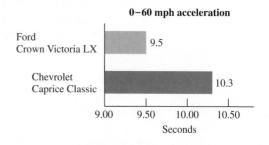

a) Draw a graph that shows the entire scale from 0 to 11 sec.

b) Does the new graph give a different impression?

Frequency Distributions

It is not uncommon for statisticians and others to have to analyze thousands of pieces of data. A **piece of data** is a single response to an experiment. When the amount of data is large, it is usually advantageous to construct a frequency distribution. A **frequency distribution** is a listing of the observed values and the corresponding frequency of occurrence of each value.

"Statistical thinking will one day be as necessary for efficient citizenship as the ability to read and write."
H. G. Wells

▶ **Example 1**

The number of children per family is recorded for 64 families surveyed. Construct a frequency distribution of the following data:

```
0   1   1   2   2   3   4   5
0   1   1   2   2   3   4   5
0   1   1   2   2   3   4   6
0   1   2   2   2   3   4   6
0   1   2   2   2   3   4   7
0   1   2   2   3   3   4   8
0   1   2   2   3   3   5   8
0   1   2   2   3   3   5   9
```

Solution: Listing the number of children (observed values) and the number of families (frequency) gives the following frequency distribution:

Number of Children (observed values)	Number of Families (frequency)
0	8
1	11
2	18
3	11
4	6
5	4
6	2
7	1
8	2
9	1
	64

There were 8 families with no children, 11 families with one child, 18 families with two children, and so on. Note that the sum of the frequencies is equal to the original number of pieces of data, 64.

Often data are grouped in "*classes*" to provide information about the distribution that would be difficult to observe if the data were ungrouped. Graphs called *histograms* and *frequency polygons* can be made of grouped data as will be explained in Section 12.4. These graphs also provide a great deal of useful information.

When data are grouped in classes, certain rules should be followed.

Rules for Data Grouped by Classes

1. The classes should be of the same "width."
2. The classes should not overlap.
3. Each piece of data should belong to one and only one class.

In addition it is often suggested that a frequency distribution should be constructed with 5 to 12 classes. If there are too few or too many classes, the distribution may become difficult to interpret.

To understand these rules, consider a set of observed values that go from a low of 0 to a high of 26. Let's assume that the first class is arbitrarily selected to go from 0 through 4. Thus any of the data with values of 0, 1, 2, 3, 4 would belong in this class. We say that this class has a "width" of 5, since there are five integral values that belong to it. Since this first class ended with 4, the second class must start with 5. If this class is to have a width of 5, at what value must it end? The answer is 9 (5, 6, 7, 8, 9). The second class is 5–9. Continuing in the same manner we obtain the following set of classes:

$$\text{Lower class limits} \left\{ \begin{array}{c} 0-4 \\ 5-9 \\ 10-14 \\ 15-19 \\ 20-24 \\ 25-29 \end{array} \right\} \text{Upper class limits}$$

Classes

We need not go beyond the 25–29 class because the largest value we are considering is 26. The classes meet our three criteria: They have the same width, there is no overlap among the classes, and each of the values from a low of 0 to a high of 26 belongs to one and only one class.

The choice of the first class, 0–4, was arbitrary. If we wanted to have more classes or fewer, we would make the class widths smaller or larger, respectively.

The numbers 0, 5, 10, 15, 20, 25 are called the **lower class limits**, and the numbers 4, 9, 14, 19, 24, 29 are called the **upper class limits**. Each class has a width of 5. Notice that the class width, 5, can be obtained by subtracting the first lower class limit from the second lower class limit: $5 - 0 = 5$. The difference between any two consecutive lower class limits and the difference between any two consecutive upper class limits is also 5.

▶ **Example 2**

The map on page 585 illustrates the number of electoral votes for president based upon the 1990 census. Construct a frequency distribution of the data letting the first class be 1–7.

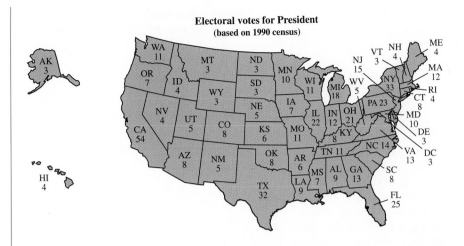

Electoral votes for President
(based on 1990 census)

Solution: There are 51 pieces of data. Rewriting the data in ascending order, that is, low to high, gives the following arrangement of the data:

3	4	5	8	10	13	32
3	4	5	8	11	14	33
3	4	6	8	11	15	54
3	4	6	8	11	18	
3	4	7	8	11	21	
3	4	7	9	12	22	
3	5	7	9	12	23	
3	5	8	10	13	25	

The first class is 1–7. The second class must therefore start at 8. If we subtract 1 from 8 we obtain a class width of 7 units. The upper class limit of the second class must therefore be 7 + 7 or 14

1–7	first class
8–14	second class

The remaining classes will be 15–21, 22–28, 29–35, 36–42, 43–49, and 50–56. Since the highest observed value is 54, there is no need to go any further. Note that each two consecutive lower class limits differ by 7, as do each two consecutive upper limits. There are 23 pieces of data in the 1–7 class (3, 3, 3, 3, 3, 3, 3, 3, 4, 4, 4, 4, 4, 4, 5, 5, 5, 5, 6, 6, 7, 7, 7). There are 19 pieces of data in the 8–14 class, three pieces in the 15–21 class, and so on. The complete frequency distribution is as shown in the chart on the left. The number of states and the District of Columbia total 51 so we have included each piece of data. There are eight classes in the frequency distribution.

Electoral Votes	Number of States and D.C.
1–7	23
8–14	19
15–21	3
22–28	3
29–35	2
36–42	0
43–49	0
50–56	1
	51

The **modal class** of a frequency distribution is the class with the greatest frequency. In Example 2 the modal class is 1–7. The **midpoint** of a class, also called the **class mark**, is found by adding the lower and upper class limits and dividing the sum by 2. The midpoint of the first class in Example 2 is

$$\frac{1+7}{2} = \frac{8}{2} = 4$$

Section 12.3 Exercises

In Ex. 1 and 2, use the given frequency distributions to determine the following.
a) The total number of observations
b) The width of each class
c) The midpoint of the second class
d) The modal class (or classes)
e) The class limits of the next class if an additional class had to be added

1.

Class	Frequency
7–13	4
14–20	2
21–27	3
28–34	0
35–41	4
42–48	6

2.

Class	Frequency
20–29	8
30–39	6
40–49	4
50–59	3
60–69	8
70–79	2

3. The results of a quiz given to a statistics class are given below. Construct a frequency distribution, letting each class have a width of 1 (as in Example 1).

```
0  1  2  3  5  6  7  8  8   9
0  1  2  3  5  6  7  8  8   9
0  1  2  4  5  6  7  8  8   9
0  1  3  4  6  6  8  8  8  10
1  1  3  4  6  6  8  8  8  10
```

4. The heights of a class of first-grade children, rounded to the nearest inch, are given below. Construct a frequency distribution, letting each class have a width of 1.

```
30  33  34  36  37
32  34  35  36  37
32  34  35  37  38
33  34  35  37  38
33  34  35  37  38
```

Note: There is no height of 31 inches. However, it is customary to include a missing value as an observed value and assign to it a frequency of 0.

In Ex. 5–8, use the following data which show the result of 50 sixth-grade I.Q. scores.

```
80  89  92  95   97  100  102  106  110  120
81  89  93  95   98  100  103  108  113  120
87  90  94  97   99  100  103  108  114  122
88  91  94  97  100  100  103  108  114  128
89  92  94  97  100  101  104  109  119  135
```

Construct a frequency distribution with a first class of
5. 80–84 **6.** 80–88 **7.** 80–90 **8.** 80–92

In Ex. 9–12, use the following data which represent the ages of the U.S. presidents at their first inauguration.

```
57  57  49  52  50  51  51  56
61  61  64  56  47  56  60  61
57  54  50  46  55  55  62  52
57  68  48  54  54  51  43  69
58  51  65  49  42  54  55  64
```

Construct a frequency distribution with a first class of
9. 40–45 **10.** 42–47 **11.** 42–46 **12.** 40–44

In Ex. 13–16, use the data in Example 2 on page 585 to construct a frequency distribution with a first class of
13. 1–8 **14.** 0–8 **15.** 0–10 **16.** 1–9

In Ex. 17–20, use the following data which represent the population of the world's 35 largest urban areas in millions (rounded to the nearest 100,000).

```
16.2  7.8  7.0  5.4  4.4  3.8  3.1
11.3  7.7  7.0  5.0  4.3  3.7  3.1
10.8  7.6  7.0  4.8  4.3  3.2  3.1
 9.3  7.4  6.0  4.7  4.1  3.2  3.0
 8.6  7.3  5.5  4.6  4.1  3.1  2.9
```

Construct a frequency distribution with a first class of
17. 2.0–3.9 **18.** 2.5–4.4 **19.** 2.9–5.8 **20.** 2.8–5.9

In Ex. 21–24, use the data in the following chart.

Density of Population by State*
(Per square mile, land area only)

State	1990	State	1990	State	1990
AL	79.6	FL	239.6	KY	92.8
AK	1.0	GA	111.9	LA	96.9
AZ	32.3	HI	172.5	ME	39.8
AR	45.1	ID	12.2	MD	489.2
CA	190.8	IL	205.6	MA	767.6
CO	31.8	IN	154.6	MI	163.6
CT	678.4	IA	49.7	MN	55.0
DE	340.8	KS	30.3	MS	54.9

State	1990	State	1990	State	1990
MO	74.3	ND	9.3	TX	64.9
MT	5.5	OH	264.9	UT	21.0
NB	20.5	OK	45.8	VT	60.8
NV	10.9	OR	29.6	VA	156.3
NH	123.7	PA	265.1	WA	73.1
NJ	1042.0	RI	960.3	WV	74.5
NM	12.5	SC	115.8	WI	90.1
NY	381.0	SD	9.2	WY	4.7
NC	136.1	TN	118.3		

Construct a frequency distribution with a first class of:
21. 0.5–100.5 **22.** 0.5–90.5 **23.** 0.5–150.5 **24.** 0.5–250.5

12.4 Statistical Graphs

Now we will study three types of graphs: the circle graph, the histogram, and the frequency polygon. Each can be used to represent statistical data.

Circle graphs (also known as pie charts) are often used to compare parts of one or more components of the whole to the whole. The circle graph in Fig. 12.6 shows where the average American's dollar goes. Since the total circle represents $1.00 (or 100%), the area marked Federal Taxes should be 28.2% of the area of the circle. Household Expenses should be 20.3% of the area of the circle, and so on.

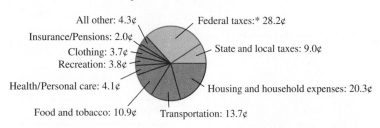

Where Your Dollar Goes
How the typical American family
will spend each dollar in 1992.

All other: 4.3¢
Insurance/Pensions: 2.0¢
Clothing: 3.7¢
Recreation: 3.8¢
Health/Personal care: 4.1¢
Food and tobacco: 10.9¢
Federal taxes:* 28.2¢
State and local taxes: 9.0¢
Housing and household expenses: 20.3¢
Transportation: 13.7¢

*Including individual and corporate income taxes, Social Security, and excise taxes. Based on a two-earner family with two dependent children earning $46,000 per year.

Figure 12.6
(Source: Tax Foundation)

In order to construct circle graphs we must be able to find central angles. A **central angle** of a circle is an angle whose vertex is at the center of the circle.

---

* Washington, D.C., has a population density of 9882.8 per square mile. The United States average is 70.3 people per square mile.

Since every circle contains 360°, we find the central angles of each class (sector) by multiplying the percent of the total of each class by 360°.

▶ **Example 1**

A survey of 2000 business executives yielded the following information:

Vacation Time (weeks)	Number of Executives
2	322
3	180
4	940
5	320
Other	238

Using this information as a sample construct a circle graph illustrating the percent in each category.

Solution: First determine the percent of the total for each of the categories. Then determine the measure of the corresponding central angle, as illustrated in Table 12.1.

Table 12.1

Weeks of Vacation	Number of Executives	Percent of Total (to nearest percent)	Measure of Central Angle
2	322	$\dfrac{322}{2000} \times 100 = 16.1\%$	$0.161 \times 360° = 58.0°$
3	180	$\dfrac{180}{2000} \times 100 = 9.0\%$	$0.09 \times 360° = 32.4°$
4	940	$\dfrac{940}{2000} \times 100 = 47.0\%$	$0.47 \times 360° = 169.2°$
5	320	$\dfrac{320}{2000} \times 100 = 16.0\%$	$0.16 \times 360° = 57.6°$
Other	238	$\dfrac{238}{2000} \times 100 = 11.9\%$	$0.119 \times 360° = 42.8°$
Total	2000	100.0%	360.0°

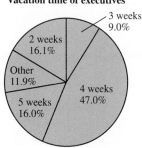

Vacation time of executives

Figure 12.7

Now use a protractor to construct the circle graph and label it properly, as illustrated in Fig. 12.7. The measure of the central angle for two-week vacations is 58.0°, the three-week sector is 32.4°, and so on.

Note that the total percent in Example 1 is 100%. Occasionally, the total percent may be slightly less than or greater than 100% due to an error caused by rounding off decimals. The total degree measure may sometimes differ slightly from 360° because of errors due to rounding off.

Histograms and frequency polygons are statistical graphs used to illustrate frequency distributions. A **histogram** is a graph with observed values on its horizontal scale and frequencies on its vertical scale. A bar is constructed above each observed value (or class when classes are used) indicating the frequency of that value. The horizontal scale need not start at zero, and the calibrations on the horizontal scale do not have to be the same as the calibrations on the vertical scale. The vertical scale must start at zero. To accommodate large frequencies on the vertical scale, it may be necessary to break the scale. Because histograms and other bar graphs are easy to interpret visually, they are used a great deal in newspapers and magazines.

▶ **Example 2**

The frequency distribution developed in Example 1 in Section 12.3 is listed below. Construct a histogram of this frequency distribution.

Number of Children (observed values)	Number of Families (frequency)
0	8
1	11
2	18
3	11
4	6
5	4
6	2
7	1
8	2
9	1

Solution: The vertical scale must extend at least to the number 18, since that is the greatest recorded frequency (see Fig. 12.8). The horizontal scale must include the numbers 0 through 9, the number of children observed. There are eight families with no children. We indicate this by constructing a bar above the number 0 on the horizontal scale extended up to 8 on the vertical scale.

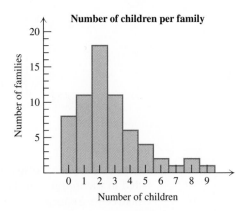

Number of children per family

Figure 12.8

Since 11 families have one child, a bar extending up to 11 is constructed above the number 1 on the horizontal scale. We continue this procedure for each observed value. The horizontal and vertical scales should both be labeled, the bars should be the same width, and the histogram should have a title. In a histogram the bars should always touch.

Frequency polygons are line graphs with scales the same as those of the histogram; that is, the horizontal scale indicates observed values and the vertical scale indicates frequency. To construct a frequency polygon, place a dot at the corresponding frequency above each of the observed values. Then connect the dots with straight line segments. In constructing frequency polygons, always put in two additional class marks, one at the lower end and one at the upper end on the horizontal scale (values for these added class marks are not needed on the frequency polygon). Since the frequency at these added class marks is 0, the endpoints of the frequency polygon will always be on the horizontal scale.

▶ Example 3

Construct a frequency polygon of the frequency distribution in Example 2.

Solution: Since there are eight families with no children, place a mark above the 0 at 8 on the vertical scale, as shown in Fig. 12.9. There are 11 families with one child. Place a mark above the 1 at the 11 on the vertical scale, and so on. Connect the dots by straight line segments, and bring the endpoints of the graph down to the horizontal scale, as shown.

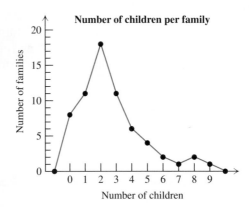

Figure 12.9

▶ Example 4

Given the frequency distribution of average gas mileage for 1992 automobiles listed in the margin on the top of page 591, construct a histogram and then construct a frequency polygon on the histogram.

Mileage (mpg)	Number of Cars
10–14	4
15–19	32
20–24	36
25–29	8
30–34	5
35–39	4
40–44	1

Solution: The histogram can be constructed with either class limits or class marks on the horizontal scale. Frequency polygons are constructed with class marks on the horizontal scale. Since we will construct a frequency polygon on the histogram, we will use class marks. Recall that class marks, or class midpoints, are found by adding the lower class limit and upper class limit and dividing the sum by 2. For the first class, the class mark is $(10 + 14)/2$, or 12. Since the class widths are 5 units, the class marks will also differ by 5 units (see Fig. 12.10).

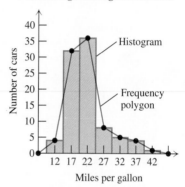

Figure 12.10

▶ Example 5

Given the histogram in Fig. 12.11, construct the frequency distribution.

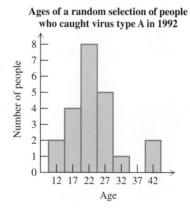

Figure 12.11

Age	Number of People
10–14	2
15–19	4
20–24	8
25–29	5
30–34	1
35–39	0
40–44	2

Solution: Since there are five units between class midpoints, each class width must also be five units. Since 12 is the midpoint of the first class, there must be two units below and two above it. The first class must be 10–14. The second class must therefore be 15–19, and so on. The frequency distribution is given in the margin.

Section 12.4 Exercises

1. The circle graph in Fig. 12.12 illustrates the percent (to the nearest tenth of 1%) of students receiving degrees in various majors at a given university over the past 10 years. If the total number of degrees awarded over the past 10 years is 3600, determine the approximate number of degrees awarded for each sector of the circle graph.

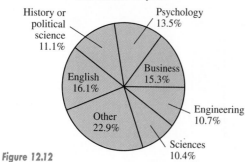

Degrees awarded at a university over the last 10 years

Figure 12.12

2. The Shortsville School District's budget of $20,962,415 for the 1991–92 school year was distributed as indicated in Fig. 12.13. Determine to the nearest dollar the amount allocated for each sector of the circle.

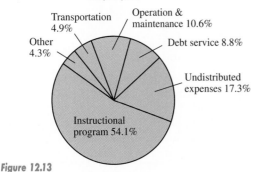

Shortsville school budget for 1991–92: $20,962,415

Figure 12.13

3. The sources of energy used in the United States in 1990 as reported by the United States Department of Energy are shown in the following chart. Construct a circle graph that illustrates this information.

Sources of Energy	Percent
Petroleum	40.3%
Natural gas	22.5
Coal	22.2
Nuclear	6.9
Other (solar, wind, hydro, etc)	8.1

4. The following chart shows a breakdown of contributions to the Commonwealth of Independent States from September 1990 to December 1991. Construct a circle graph indicating the percentage donated by each country. Include a sector of the circle for other.

Country	Donation (in billions)
Germany	$34.88
Italy	5.85
United States	4.08
European Commission	3.88
Japan	2.72
Spain	1.36
France	1.22
United Kingdom	0.07
Others	14.01

Source: National Planning Association

5. The frequency distribution below indicates the number of years the 38 employees of the ABC Company have worked for that company.

Years	Number of Employees
0	6
1	4
2	2
3	1
4	6
5	8
6	3
7	3
8	5

a) Construct a histogram of the frequency distribution.
b) Construct a frequency polygon of the frequency distribution.

6. The frequency distribution below indicates the ages of a group of 40 people attending a party.

Age	Number of People
20	6
21	3
22	0
23	4
24	6
25	3
26	8
27	10

a) Construct a histogram of the frequency distribution.
b) Construct a frequency polygon of the frequency distribution.

7. The frequency distribution below represents the annual salaries in thousands of dollars of the people in management positions at the ABC Corporation.

Salary (in $1000)	Number of People
20–25	4
26–31	6
32–37	8
38–43	9
44–49	8
50–55	5
56–61	3

a) Construct a histogram of the frequency distribution.
b) Construct a frequency polygon of the frequency distribution.

8. The frequency distribution below indicates the gross weekly salaries of families whose children receive free lunch at Villdale Elementary School.

Salary	Number of Families
133–141	8
142–150	6
151–159	8
160–168	4
169–177	3
178–186	1

a) Construct a histogram of the frequency distribution.
b) Construct a frequency polygon of the frequency distribution.

9. Using the histogram in Fig. 12.14, answer the following questions.

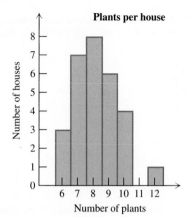

Figure 12.14

a) How many houses were surveyed?
b) In how many houses were nine plants observed?
c) What is the modal class?
d) How many plants were observed?
e) Construct a frequency distribution from this histogram.

10. Using the histogram in Fig. 12.15, answer the following questions.

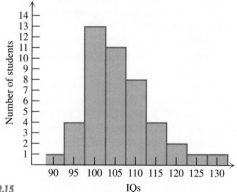

Figure 12.15

a) What are the upper and lower class limits of the first and second classes?
b) How many students had IQs in the class with a class mark of 100?
c) How many students were surveyed?
d) Construct a frequency distribution.

11. Use the frequency polygon in Fig. 12.16 to answer the following questions.

Number of visits selected families have made to the Pacific Science Center in Seattle, Washington

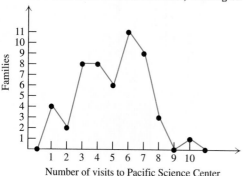

Figure 12.16

a) How many families visited the Pacific Science Center four times?

b) How many families visited the Pacific Science Center at least six times?

c) How many families were surveyed?

d) Construct a frequency distribution from the frequency polygon.

e) Construct a histogram from the frequency distribution in part (d).

12. Construct a histogram and a frequency polygon from the frequency distribution given in Exercise 1 for Section 12.3. See page 586.

13. Construct a histogram and a frequency polygon from the frequency distribution given in Exercise 2 for Section 12.3. See page 586.

14. Construct a histogram and a frequency polygon from the frequency distribution given in the solution of Example 2 in Section 12.3. See page 585.

15. Starting salaries (rounded to the nearest thousand) for chemical engineers with B.S. degrees and no experience are given below for 25 different companies.

$$25 \quad 26 \quad 27 \quad 29 \quad 31$$
$$26 \quad 26 \quad 27 \quad 29 \quad 31$$
$$26 \quad 26 \quad 28 \quad 30 \quad 31$$
$$26 \quad 27 \quad 28 \quad 30 \quad 32$$
$$26 \quad 27 \quad 28 \quad 30 \quad 32$$

a) Construct a frequency distribution.

b) Construct a histogram.

c) Construct a frequency polygon.

16. The ages of a random sample of U.S. ambassadors are

40	43	45	50	52	55	59	64
41	43	46	50	53	55	60	65
41	44	47	50	54	55	60	65
42	44	48	51	54	57	60	66
43	45	48	51	54	58	62	67

a) Construct a frequency distribution with the first class 40–44.

b) Construct a histogram.

c) Construct a frequency polygon.

17. The circulation of the 30 best-selling U.S. magazines as of December 31, 1990, follows.

a) Construct a frequency distribution with the first class 1.9–6.3 million.

b) Construct a histogram.

c) Construct a frequency polygon.

Magazine	Circulation
1. NRTA/AARP Bulletin	22,103,887
2. Reader's Digest	16,264,547
3. TV Guide	15,604,267
4. National Geographic	10,189,703
5. Better Homes & Gardens	8,007,222
6. Family Circle	5,431,779
7. Good Housekeeping	5,152,521
8. McCall's	5,020,127
9. Ladies' Home Journal	5,001,739
10. Woman's Day	4,802,842
11. Time	4,094,935
12. Redbook	3,907,221
13. National Enquirer	3,803,607
14. Playboy	3,488,006
15. Star	3,431,453
16. Sports Illustrated	3,220,016
17. Newsweek	3,211,958
18. People	3,208,668
19. American Legion	2,956,342
20. Prevention	3,022,106
21. First for Women	2,649,810
22. Cosmopolitan	2,600,971
23. Southern Living	2,341,074
24. U.S. News & World Report	2,311,534
25. Smithsonian	2,234,706
26. Glamour	2,156,157
27. Field & Stream	2,016,298
28. VFW	2,103,256
29. NEA Today	1,978,641
30. Money	1,905,053

12.5 Measures of Central Tendency

Most people have an intuitive idea of what is meant by an "average." The term is used daily in many familiar ways: "This car averages 13 miles per gallon." "The average test grade was 76." "Her batting average is .313." "On the average, one of four will be selected."

An **average** is a number that is representative of a group of data. There are at least four different averages: the mean, the median, the mode, and the midrange. Each is calculated differently and may yield different results for the same set of data. Each will result in a number near the center of the data; and for this reason, averages are commonly referred to as the **measures of central tendency**.

The **arithmetic mean**, or simply the **mean**, is symbolized either by $\bar{x}$ (read "x bar") or by the Greek letter mu, μ. The symbol $\bar{x}$ is used when the mean of a *sample* of the population is calculated. The symbol μ is used when the mean of the *entire population* is calculated. We will assume that the data studied in this book represent samples, and therefore we will always use $\bar{x}$ for the mean.

The Greek letter sigma, Σ, is used to indicate "summation." The notation Σx, read "the sum of x," is used to indicate the sum of all the data. For example, if there are five pieces of data, 4, 6, 1, 0, 5, then $\Sigma x = 4 + 6 + 1 + 0 + 5 = 16$.

Now we can discuss the procedure for finding the mean of a set of data.

> The mean, $\bar{x}$, is the sum of the data divided by the number of pieces of data. The formula for calculating the mean is
>
> $$\bar{x} = \frac{\Sigma x}{n},$$
>
> where Σx represents the sum of all the data, and n represents the number of pieces of data.

The most common use of the word average is the mean.

▶ **Example 1**

Find Tom's mean grade if he received grades of 49, 75, 40, 78, and 78 on his five statistics exams.

Solution:

$$\bar{x} = \frac{\Sigma x}{n} = \frac{49 + 75 + 40 + 78 + 78}{5} = \frac{320}{5} = 64$$

Therefore the mean, $\bar{x}$, is 64.

The mean represents "the balancing point" of a set of scores. Consider a seesaw. If a seesaw is pivoted at the mean and uniform weights are placed at points corresponding to the test scores the seesaw will balance. Figure 12.17 shows the five test scores given in Example 1 and the calculated mean.

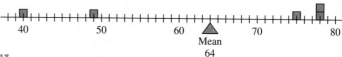

Figure 12.17

A second average is the median. To find the median of a set of data, *rank the data* from smallest to largest, or largest to smallest, and determine the value in the middle of the set of *ranked data*. This value will be the median.

> The **median** is the value in the middle of a set of *ranked data*.

▶ **Example 2**

Find the median of Tom's grades in Example 1.

Solution: Ranking the data from smallest to largest gives 40, 49, 75, 78, 78. Since 75 is the value in the middle of this set of ranked data (two pieces above and two pieces below), 75 is the median.

If there are an even number of pieces of data, the median will be halfway between the two middle pieces and will be found by adding the two middle pieces and dividing this sum by 2.

▶ **Example 3**

Find the median of the following sets of data
a) 1, 4, 6, 8, 9, 8, 3, 3
b) 2, 3, 3, 3, 4, 5

Solution:
a) Ranking the data gives 1, 3, 3, 4, 6, 8, 8, 9. There are eight pieces of data. Therefore the median will lie halfway between the two middle pieces, the 4 and the 6. The median is (4 + 6)/2, or 10/2, or 5.
b) There are six pieces of data and they are already ranked. Therefore the median will lie halfway between the two middle pieces. Both middle pieces are 3s. The median is (3 + 3)/2, or 6/2, or 3.

A third average is the mode.

> The **mode** is the piece of data that occurs most frequently.

▶ **Example 4**

Find the mode of Tom's grades in Example 1.

Solution: Tom's grades were 49, 75, 40, 78, 78. The grade 78 is the mode, since it occurs twice and the other values occur only once.

If no one piece of data occurs more frequently than every other piece of data, the set of data has no mode. For example, neither of the following two sets of data has a mode:

$$1, 2, 3, 4, 5 \qquad \text{no mode}$$
$$1, 1, 2, 3, 3, 4, 5 \qquad \text{no mode*}$$

The last average that we will discuss is the **midrange**. The midrange is the value halfway between the lowest and highest values in a set of data. It is found by adding the lowest and highest values and dividing the sum by 2. A formula for finding the midrange follows.

$$\textbf{Midrange} = \frac{\text{lowest value} + \text{highest value}}{2}$$

▶ **Example 5**

Find the midrange of Tom's grades given in Example 1: 49, 75, 40, 78, 78.

Solution: Tom's lowest grade is 40, and his highest grade is 78.

$$\text{Midrange} = \frac{\text{lowest} + \text{highest}}{2}$$

$$= \frac{40 + 78}{2} = \frac{118}{2} = 59$$

Tom's "average" grade for the exam scores of 49, 75, 40, 78, 78 can be considered any one of the following values: 64 (mean), 75 (median), 78 (mode), or 59 (midrange). Which average do you feel is most representative of his grades? We will discuss this question later in this section.

▶ **Example 6**

The salaries of eight selected chemical engineers rounded to the nearest thousand dollars are 40, 25, 28, 35, 42, 60, 60, and 73. For this set of data find the (a) mean, (b) median, (c) mode, and (d) midrange, and then (e) rank the measures of central tendency from lowest to highest.

---

* Some textbooks refer to sets of data of this type as bimodal.

Solution:

a) $\bar{x} = \dfrac{40 + 25 + 28 + 35 + 42 + 60 + 60 + 73}{8} = \dfrac{363}{8} = 45.375$

b) Listing the data from the smallest to largest gives

$$25, 28, 35, 40, 42, 60, 60, 73.$$

Since there are an even number of pieces of data, the median is halfway between 40 and 42. The median $= (40 + 42)/2 = 82/2 = 41$.

c) The mode is the piece of data that occurs most frequently. The mode is 60.

d) The midrange $= (L + H)/2 = (25 + 73)/2 = 98/2 = 49$.

e) The averages from lowest to highest are the median, mean, midrange, and mode. Their values are 41, 45.375, 49, and 60 respectively.

Did You

Know...

MEASURES OF LOCATION

When you took the Scholastic Aptitude Test (SAT) before applying to college, your score was described as a **measure of location** rather than a measure of central tendency. Measures of location are often used to make comparisons, such as comparing the scores of individuals from different populations, and are generally used when the amount of data is large.

Two measures of location are **percentiles** and **quartiles**. Percentiles divide the set of data into 100 equal parts (see Fig. a). For example, suppose you scored 490 on the math half of the SAT, and that the score of 490 was reported to be in the 78th percentile of high school students. This *does not* mean that 78% of your answers were correct; it *does* mean that you outperformed about 78% of all those taking the exam. In general, a score in the nth percentile means you outperformed about $n\%$ of the population who took the test, and that $(100 - n)\%$ of the people taking the test performed better than you did.

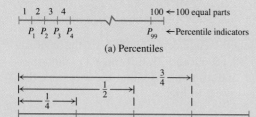

Measures of location

(a) Percentiles

(b) Quartiles

Quartiles divide data into four equal parts: The first quartile is a value higher than 1/4 or 25% of the population. It is the same as the 25th percentile. The second quartile is higher than 1/2 the population and is the same as the 50th percentile, or the median. The third quartile is higher than 3/4 of the population and is the same as the 75th percentile (see Fig. b).

At this point you should be able to calculate the four measures of central tendency. Now let's examine the circumstances in which each is used.

The mean is used when each piece of data is to be considered and "weighed" equally. It is the most commonly used average. It is the only average that can be affected by *any* change in the set of data; for this reason it is the most sensitive of all the measures of central tendency (see Exercise 14).

Occasionally, one or more pieces of data may be much greater or much smaller than the rest of the data. When this occurs, these "extreme" values have the effect of increasing or decreasing the mean significantly so that the mean will not be representative of the total set of data. Under these circumstances the median is often used in place of the mean. The median is often used in describing average family incomes, because a relatively small number of families have extremely large incomes. These few families would inflate the mean income, making it nonrepresentative of the millions of families in the population.

Consider a set of exam scores from a mathematics class: 0, 16, 19, 65, 65, 65, 68, 69, 70, 72, 73, 73, 75, 78, 80, 85, 88, 92. Which average would best represent these grades? The mean is 64.06. The median is 71. Since only three scores of the 18 fall below the mean, the mean would not be considered a good representative score, and the median of 71 would probably be the better average to use.

The mode is the piece of data, if any, that occurs most frequently. Builders planning houses are interested in the most common family size. Retailers ordering shirts are interested in the most common shirt size. An individual purchasing a thermometer might select one with the most common reading. These examples illustrate how the mode may be used.

The midrange is sometimes used as the average when the item being studied is constantly fluctuating. Average daily temperature, used to compare temperature in different areas, is calculated by adding the lowest and highest temperatures for the day and dividing the sum by 2. The midrange is actually the mean of the high value and the low value of a set of data. Occasionally, the midrange is used to estimate the mean, since it is much easier to calculate.

Sometimes an average itself is of little value. One must be very careful in interpreting its meaning. For example Jim is told that the average depth of Willow Pond is only 3 ft. He is not a good swimmer but decides that it is safe to go out a short distance in this shallow pond. After he is rescued, he exclaims, "I thought this pond was only three feet deep." Jim didn't realize that an average does not indicate extreme values or the spread of the values. The spread of data will be discussed in Section 12.6.

Section 12.5 Exercises

Find the mean, median, mode, and midrange of each of the following sets of data. Round answers off to the nearest tenth.

1. 1, 3, 3, 4, 5, 7, 8, 8, 8
2. 4, 5, 10, 12, 10, 9, 8, 372, 40, 37

3. 60, 72, 80, 84, 86, 45, 96
4. 5, 3, 6, 6, 6, 9, 11
5. 1, 3, 5, 7, 9, 11, 13, 15
6. 1, 7, 11, 27, 36, 14, 12, 9, 1

7. 40, 50, 30, 60, 90, 100, 140

8. 1, 1, 1, 1, 4, 4, 4, 4, 6, 8, 10, 12, 15, 21

9. 5, 7, 11, 12, 10, 9, 12, 14, 16

10. 1, 1, 1, 1, 9, 9, 9, 9

11. 237, 463, 812, 487, 387

12. 3, 2432, 16203, 962, 72

13. 78, 60, 72, 83, 92, 96

14. The mean is the "most sensitive" average, since it is affected by any change in the data.
   a) Determine the mean, median, mode, and midrange for 1, 2, 3, 5, 5, 7, 11.
   b) Change the 7 in the data above to a 10. Find the mean, median, mode, and midrange.
   c) Which averages were affected by changing the 7 to a 10 in part (b)?
   d) Which averages will be affected by changing the 11 to a 10 in part (a)?

15. In 1990 the National Center for Health Statistics indicated a new record "average life expectancy" of 75.4 years for the total U.S. population. The average life expectancy for men was 72.0 years and for women, 78.8 years. Which "average" do you think the National Center for Health is using? Explain your answer.

16. To get a grade of B, a student must have a mean average of 80. Jim has a mean average of 79 for 10 quizzes. He approaches his teacher and asks for a B, since he reasoned that he missed a B by only one point. What is wrong with Jim's reasoning?

17. The Webers' monthly gas and electric bills for the year are indicated below. Find the mean, median, mode, and midrange.

$103.64	$96.96	$100.60
$100.50	$101.08	$102.74
$69.41	$92.18	$77.95
$104.72	$95.95	$101.72

18. Below are the Ranieris' monthly telephone charges for a one-year period. Find the mean, median, mode, and midrange.

$75.11	$85.53	$28.94
$77.15	$67.78	$72.13
$85.29	$74.33	$84.19
$73.48	$74.68	$87.15

19. The following chart gives the annual cinema attendance and the per person cinema attendance for the 22 leading countries in each category.

	Annual Cinema Attendance (in millions)		Number of Visits per Person*
USSR	4,200.0	Singapore	16.8
India	2,920.0	USSR	15.9
United States	1,022.0	Brunei	14.7
Philippines	318.0	Taiwan	13.0
Italy	276.3	Guyana	12.9
Mexico	269.8	Hong Kong	12.6
Vietnam	253.0	Grenada	12.5
Taiwan	229.0	Iceland	11.4
Burma	222.5	Bulgaria	10.7
Spain	200.5	United Arab Emirates	9.8
Romania	193.6		
Pakistan	187.4	Macau	9.3
France	176.4	Mongolia	9.3
Brazil	164.8	Mauritius	8.7
Japan	164.0	Romania	8.7
West Germany	135.5	Malta	8.5
Indonesia	123.6	Burma	8.1
Poland	98.9	Philippines	7.5
Great Britain	98.8	Western Samoa	7.4
Bulgaria	95.9	Bolivia	6.8
Canada	93.2	Israel	6.5
Czechoslovakia	82.3	Chad	6.0
		Ireland	5.8

* There are 4.5 visits per person in the United States, 4.1 in Canada, 1.8 in Great Britain, and 1.4 in Japan.

a) Find the mean, median, mode, and midrange of annual cinema attendance for the 22 countries listed in the table above.

b) Find the mean, median, mode, and midrange of the number of visits per person for the 22 countries listed.

c) Why is it possible for Singapore to have the largest number of visits per person and not be included among the top 22 countries for annual cinema attendance?

20. In 1991 the Oakland Athletics had the highest average salary in the major baseball league. Using the information that follows, find the mean, median, mode, and midrange for the team. *Note:* The salaries include base salaries, prorated signing bonuses, and deferred payments but do not include extra incentives earned by individual players.

Oakland Athletics, 1992	
Pitchers	
Steve Chitren	$ 105,000
Ron Darling	1,966,667
Dennis Eckersley	3,000,000
Rick Honeycutt	1,350,000
Joe Klink	140,000
Mike Moore	1,566,687
Gene Nelson	1,100,000
Eric Show	800,000
Dave Stewart	3,500,000
Bob Welch	3,450,000
Curt Young	775,000
Catchers	
Jamie Quirk	$ 500,000
Terry Steinbach	1,050,000

Oakland Athletics, 1992	
Infielders	
Lance Blankenship	$ 125,000
Mike Bordick	105,000
Scott Brosius	100,000
Mike Gallego	565,000
Scott Hamond	102,000
Brook Jacoby	1,150,000
Carney Lansford	1,275,000
Mark McGwire	2,875,000
Ernest Pliss	792,000
Walt Weiss	780,000
Outfielders	
Harold Baines	$1,358,333
Jose Canseco	3,500,000
Dave Henderson	2,825,000
Rickey Henderson	3,250,000
Willie Wilson	1,000,000

21. Paula's mean average on six exams is 83. Find the sum of her scores.

22. Mary's mean average on five exams is 72. Find the sum of her scores.

23. In the 50 United States in 1990 there were approximately 2.17 million farms. The state with the fewest farms was Alaska with 600 farms and the state with the greatest number of farms was Texas with 186,000. Considering all the farms in the United States, determine whether it is possible to find
   a) the mean number of farms for the 50 states
   b) the median number of farms
   c) the mode
   d) the midrange
   Find all measures of central tendency that can be found with the information given and explain why the others cannot be found.

24. Construct a set of five pieces of data in which the mode has a lower value than the median and the median has a lower value than the mean.

25. Construct a set of five pieces of data with a mean of 70 with no two pieces of data the same.

26. Construct a set of six pieces of data with a mean, median, and midrange of 70 with no two pieces of data the same.

27. A mean average of 80 for five exams is needed for a final

grade of B. George's first four exam grades are 70, 76, 83, 80. What grade does George need on the fifth exam to get a B in the course?

28. A mean average of 60 on seven exams is needed to pass a course. On her first six exams, Sheryl received grades of 49, 72, 80, 60, 57, and 69.
   a) What grade must she receive on her last exam to pass the course?
   b) An average of 70 is needed to get a C in the course. Is it possible for Sheryl to get a C? If so, what grade must she receive on the seventh exam?
   c) If her lowest grade is to be dropped and only her six best exams are counted, what grade must she receive on her last exam to pass the course?
   d) If her lowest grade is to be dropped and only her six best exams are counted, what grade must she receive on her last exam to get a C in the course?

29. Which of the measures of central tendency *must* be an actual piece of data in the distribution? Explain.

30. Construct a set of six pieces of data such that if only one piece of data is changed, the mean, median, and mode will all change.

31. Consider the set of data: 1, 1, 1, 2, 2, 2. If one 2 is changed to a 3, which of the following will change: mean, median, mode, midrange? Explain.

32. Is it possible to construct a set of six different pieces of data such that by changing only one piece of data you cause the mean, median, mode, and midrange to change? Explain.

Refer to the Did You Know on Measures of Location on page 598, and then answer the following questions.

33. When given a set of data, what must be done to the data before percentiles can be determined?

34. Elisabeth scored in the 85th percentile on the verbal part of her College Board Test. What does this mean?

35. When a national sample of the heights of kindergarten children were taken, Kevin was told that he was in the 43rd percentile. Explain what this means.

36. A union leader is told that when all workers' salaries are considered, the first quartile is $14,750. Explain what this means.

37. The third quartile for the time it takes students to travel one way to Central Valley College is 47.6 minutes. Explain what this means.

38. Give the names of two other statistics that have the same value as the 50th percentile.

39. Jonathan took an admission test for the University of California and scored in the 85th percentile. The following year Jonathan's sister Kendra took a similar admission test for the University of California and scored in the 90th percentile.

a) Is it possible to determine which of the two answered the higher percent of questions correctly on their respective exams? Explain your answer.

b) Is it possible to determine which of the two was in a better relative position with respect to their respective populations? Explain.

40. The following statistics represent weekly salaries at the Midtown Construction Company:

Mean	$460	First quartile	$420
Median	$450	Third quartile	$485
Mode	$440	83rd percentile	$525

a) What is the most common salary?

b) What salary did half the employees surpass?

c) About what percent of employees surpassed $485?

d) About what percent of employees' salaries was below $420?

e) About what percent of employees surpassed $525?

f) If the company has 100 employees, what is the total weekly salary of all employees?

Research Activity

41. Two other measures of location that we did not mention in the Did You Know on page 598 are stanines and deciles. Use statistics books and books on educational testing and measurements to determine what stanines and deciles are and when percentiles, quartiles, stanines, and deciles are used.

Measures of Dispersion

The measures of central tendency by themselves do not always give sufficient information to analyze a situation and make necessary decisions. As an example, two manufacturers of airplane engines are under consideration for a contract. Manufacturer A's engines have an average (mean) life of 1000 hours of flying time before they must be rebuilt, while manufacturer B's engines have an average life of 950 hours of flying time before they must be rebuilt. If you assume that both cost the same, which engines should be purchased? The average engine life may not be the most important factor. The fact that manufacturer A's engines have an average life of 1000 hours could mean that half will last about 500 hours and the other half will last about 1500 hours. If in fact all of manufacturer B's engines have a life span of between 900 and 1000 hours, then B's engines are more consistent and reliable. If A's engines were purchased, they would all have to be rebuilt every 300 hours or so, because it would be impossible to determine which ones would fail first. If B's engines were purchased, they could go much longer before having to be rebuilt. This example is of course an exaggeration used to illustrate the importance of knowing something about the *spread*, or *variability*, of the data.

The **measures of dispersion** are used to give indications of the spread of the data. The range and standard deviation* are the measures of dispersion that will be discussed in this book.

---

* *Variance*, another measure of dispersion, is the square of the standard deviation.

Knowing the dispersion of temperature can be important in choosing an optimal climate. For example, the average maximum daily temperature in San Diego and Wichita is about the same. Does this mean that the weather in Wichita is the same as in San Diego? No, because the variability is very different: In Wichita the seasonal temperatures are more extreme, while in San Diego temperatures are more or less even all year.

The **range** is the difference between the highest and lowest values; it indicates the total spread of the data.

$$\textbf{Range} = \text{highest value} - \text{lowest value}$$

▶ **Example 1**

Find the range of the following weights: 140, 195, 203, 175, 85, 197, 220, 212, 415.

Solution: Range = highest value − lowest value = 415 − 85 = 330. The range of this data is 330 pounds.

The second measure of dispersion, the **standard deviation**, measures how much the data *differ from the mean*. It is symbolized either by the letter s or by the Greek letter sigma, σ.* The s is used when the standard deviation of a sample is calculated. The σ is used when the standard deviation of the entire population itself is calculated. Since we are assuming that all data presented here are for samples, we will always use s to represent the standard deviation in this book (note, however, that on the doctors' charts in Fig. 12.19, σ is used). The larger the spread of the data about the mean, the larger the standard deviation. Consider the following two sets of data:

$$5, 8, 9, 10, 12, 13 \qquad 8, 9, 9, 10, 10, 11$$

Both have a mean of 9.5. Which set of values on the whole do you feel differs less from the mean of 9.5? Figure 12.18 may make the answer more apparent. The scores in the second set of data are closer to the mean and have a smaller standard deviation. You will soon be able to verify such relationships yourself.

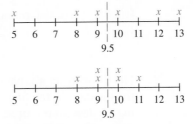

Figure 12.18

Sometimes only a very small standard deviation is desirable or acceptable. Consider a cereal box that is to contain 8 oz of cereal. If the amount of cereal put into the box varies too much—sometimes underfilling, sometimes overfilling—the manufacturer will soon be in trouble with consumer groups and government agencies.

---

* Our alphabet uses both uppercase and lowercase letters—for example, A and a. The Greek alphabet also uses both uppercase and lowercase letters. The symbol Σ is the capital Greek letter sigma, and σ is the lowercase Greek letter sigma.

At other times a larger spread of data is desirable or expected. In considering intelligence quotients (IQs) it is expected that there will be considerable spread about the mean, since we are all different.

To Find the Standard Deviation of a Set of Data:

1. Find the mean of the set of data.
2. Make a chart with three columns as follows:

> Data   Data − Mean   (Data − Mean)^2

3. List the data vertically under the column marked Data.
4. Complete the Data − Mean column for each piece of data.
5. Square the values obtained in the Data − Mean column and place these values in the (Data − Mean)^2 column.
6. Find the sum of the values in the (Data − Mean)^2 column.
7. Divide the sum obtained in step 6 by $n − 1$, where n is the number of pieces of data.*
8. Find the square root of the number obtained in step 7. This number is the standard deviation for the set of data.

Example 2 illustrates the procedure to follow to find the standard deviation of a set of data.

▶ **Example 2**

Find the standard deviation of the following set of data:

$$7, 9, 11, 15, 18$$

Solution: First, determine the mean:

$$\bar{x} = \frac{7 + 9 + 11 + 15 + 18}{5} = \frac{60}{5} = 12$$

Next, construct a table with three columns, as illustrated in Table 12.2, and list the data in the first column (it is often helpful to list the data in ascending or descending order). Complete the second column by subtracting the mean, 12 in this case, from each piece of data in the first column.

The sum of the values in column 2 should always be zero; if not, you have made an error. (If the mean is a decimal, there may be a slight round-off error.)

---

* To find the standard deviation of a sample, divide the sum of (Data − Mean)^2 column by $n − 1$. To find the standard deviation of a population, divide the sum by n. In this book we will always assume that the set of data represents a sample and divide by $n − 1$. The quotient obtained in step 7 represents a measure of dispersion called the *variance.*

Table 12.2

Data	Data — Mean	(Data — Mean)^2
7	$7 - 12 = -5$	
9	$9 - 12 = -3$	
11	$11 - 12 = -1$	
15	$15 - 12 = 3$	
18	$18 - 12 = 6$	
	$\overline{0}$	

Next, square the values in column 2 and place the squares in column 3 (Table 12.3).

Table 12.3

Data	Data — Mean	(Data — Mean)^2
7	-5	$(-5)^2 = (-5)(-5) = 25$
9	-3	$(-3)^2 = (-3)(-3) = 9$
11	-1	$(-1)^2 = (-1)(-1) = 1$
15	3	$(3)^2 = (3)(3) = 9$
18	6	$(6)^2 = (6)(6) = 36$
	$\overline{0}$	$\overline{80}$

Add the numbers in column 3. In this case the sum is 80. Divide this sum by one less than the number of pieces of data $(n - 1)$. In this case the number of pieces of data is 5. Therefore we divide by 4:

$$\frac{80}{4} = 20*$$

Finally, take the square root of this number. The square root of 20 can be found by using a calculator or the square root table in the appendix. Since $\sqrt{20} \approx 4.5$, the standard deviation, symbolized s, is 4.5 (to the nearest tenth).

Now we will develop a formula for finding the standard deviation of a set of data. If we call the individual data x and the mean $\bar{x}$, then the three column heads Data, Data — Mean, and (Data — Mean)^2 in Table 12.3 could be written:

$$x \qquad x - \bar{x} \qquad (x - \bar{x})^2$$

Let us follow the procedure we used to obtain the standard deviation in Example 2. We found the sum of the (Data — Mean)^2 column, which is the same as the sum of the $(x - \bar{x})^2$ column. We can represent the sum of the $(x - \bar{x})^2$ column

---

* 20 is the variance, symbolized s^2, of this set of data.

by using the summation notation, $\Sigma(x - \bar{x})^2$. Thus in Table 12.3 $\Sigma(x - \bar{x})^2 = 80$. We then divided this number by one less than the number of pieces of data, $n - 1$. Thus we have

$$\frac{\Sigma(x - \bar{x})^2}{n - 1}$$

Finally, we took the square root of this value to obtain the standard deviation:

Standard Deviation

$$s = \sqrt{\frac{\Sigma(x - \bar{x})^2}{n - 1}}$$

Table 12.4

x	$x - \bar{x}$	$(x - \bar{x})^2$
20	−17	289
30	−7	49
32	−5	25
35	−2	4
40	3	9
40	3	9
41	4	16
42	5	25
43	6	36
47	10	100
	0	$\Sigma(x - \bar{x})^2 = 562$

▶ **Example 3**

Find the standard deviation of the following set of data:

$$20, 30, 32, 35, 40, 40, 41, 42, 43, 47$$

Solution: The mean, $\bar{x}$, is

$$\bar{x} = \frac{\Sigma x}{n} = \frac{20 + 30 + 32 + 35 + 40 + 40 + 41 + 42 + 43 + 47}{10} = \frac{370}{10}$$

$$= 37$$

Now find $\Sigma(x - \bar{x})^2$ (see Table 12.4). There are 10 pieces of data. Thus $n - 1$ is $10 - 1$ or 9.

$$s = \sqrt{\frac{\Sigma(x - \bar{x})^2}{n - 1}} = \sqrt{\frac{562}{9}} = \sqrt{62.4} \approx 7.9$$

The standard deviation is 7.9.

Standard deviation will be used in Section 12.7 to find the percentage of data between any two values in a normal curve. Standard deviations are also often used in determining norms for a population (see Exercise 22).

Section 12.6 Exercises

1. Without actually doing the calculations, decide which of the following two sets of data will have the greater standard deviation: 5, 8, 9, 10, 12, 16 or 8, 9, 9, 10, 10, 11. Explain why.

2. Of the following two sets of data, which would you expect to have the larger standard deviation? Explain.

   $$2, 4, 6, 8, 10 \quad \text{or} \quad 102, 104, 106, 108, 110$$

3. What does it mean when the standard deviation of a set of data is zero?

4. Can you think of any situations in which a large standard deviation may be desirable?

5. Can you think of any situations in which a small standard deviation may be desirable?

Find the range and standard deviations of the following sets of data.

6. 4, 3, 0, 8, 10

7. 6, 6, 10, 12, 3, 5

8. 120, 121, 122, 123, 124, 125, 126

9. 4, 0, 3, 6, 9, 12, 2, 3, 4, 7

10. 4, 8, 9

11. 6, 6, 6, 6, 6, 6

12. 10, 9, 14, 15, 17

13. 5, 6, 5, 4, 4, 5, 6

14. 42, 40, 44, 49, 30, 33, 54, 52

15. 7, 4, 7, 4, 7, 4, 7, 8

16. 5, 7, 12, 20, 17, 8, 13, 10, 10, 8

17. 60, 72, 84, 88, 76

18. 3, 4, 5, 9, 3, 7, 4, 4, 9, 2

19. 103, 106, 109, 112, 115, 118, 121

20. a) Pick any five numbers. Compute the mean and the standard deviation of this distribution.
   b) Add 20 to each of the numbers in your original distribution and compute the mean and the standard deviation of this new distribution.
   c) Subtract 5 from each number in your original distribution and compute the mean and standard deviation of this new distribution.
   d) Can you draw any conclusions about changes in the mean and the standard deviation when the same number is added to or subtracted from each piece of data in a distribution?
   e) How will the mean and standard deviation of the numbers 6, 7, 8, 9, 10, 11, 12 differ from the mean and standard deivation of the numbers 596, 597, 598, 599, 600, 601, 602? Find the mean and standard deviation of the latter set of numbers.

21. a) Pick any five numbers. Compute the mean and standard deviation of this distribution.
   b) Multiply each number in your distribution by 4 and compute the mean and the standard deviation of this new distribution.
   c) Multiply each number in your original distribution by 9 and compute the mean and the standard deviation of this new distribution.
   d) Can you draw any conclusions about changes in the

mean and the standard deviation when each value in a distribution is multiplied by the same number?
   e) The mean and standard deviation of the distribution 1, 3, 4, 4, 5, 7 are 4 and 2, respectively. Use the conclusion drawn in part (d) to determine the mean and standard deviation of the distribution 5, 15, 20, 20, 25, 35.

22. The chart in Fig. 12.19 uses the symbol σ to represent the standard deviation. Note that 2σ represents the value that is two standard deviations above the mean; -2σ represents the value that is two standard deviations below the mean. The unshaded areas, from two standard deviations below the mean to two standard deviations above the mean, are considered the normal range. For example, the average (mean) eight-year-old boy has a height of about 50 inches. But any heights between approximately 46 inches and 55 inches are considered normal for eight-year-old boys.

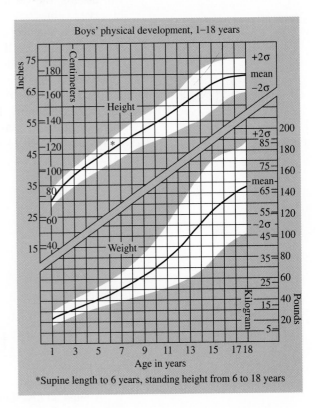

*Supine length to 6 years, standing height from 6 to 18 years

Figure 12.19

Refer to Fig. 12.19 to answer the following questions.

a) What happens to the standard deviation for weights of boys as the age of boys increases? What is the significance of this fact?

b) At age 16, what is the mean weight, in pounds, of boys?

c) What is the approximate standard deviation of boys' weights at age 16?

d) Find the mean weight and normal range for boys at age 13.

e) Find the mean height and normal range for boys at age 13.

f) Assuming that this chart was constructed so that approximately 95% of all boys are always in the normal range, determine what percentage of boys will not be in the normal range.

The Normal Curve

Certain shapes of distributions of data are more common than others. A few of the more common ones are illustrated below. In each distribution that we will discuss, the vertical scale is the frequency, and the horizontal scale is the observed values.

In a **rectangular distribution** (Fig. 12.20), all the observed values occur with the same frequency. If a die is rolled many times, one would expect the numbers 1–6 to occur with about the same frequency. The distribution representing the outcomes of the die is rectangular.

In **J-shaped distributions** the frequency is either constantly increasing (Fig. 12.21a) or constantly decreasing (Fig. 12.21b). The number of hours studied per week by students may have a distribution like Fig. 12.21(b). The bars might represent from left to right 0–5, 6–10, 11–15 hours, and so on.

Frequency

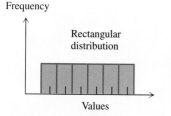

Rectangular distribution

Values

Figure 12.20

J-shaped distributions

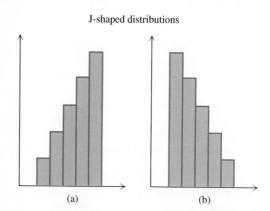

Figure 12.21

(a)   (b)

A **bimodal distribution** (Fig. 12.22) is one in which two nonadjacent values occur more frequently than any other values in a set of data. For example, if an equal number of men and women were weighed, the distribution of their weights would probably be bimodal, with one mode for the women's weights and the second for the men's weights.

The life expectancy of light bulbs has a bimodal distribution: a small peak very near 0 hours of life, resulting from the bulbs that burned out very quickly because of a manufacturing defect, and a much broader peak representing the nondefective bulbs. A bimodal frequency distribution generally means that you are dealing with two distinct populations, in this case, defective and nondefective bulbs.

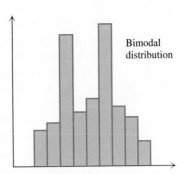

Figure 12.22

Another distribution, called a **skewed distribution**, has more of a "tail" on one side than the other. A skewed distribution with a tail on the right (Fig. 12.23a) is said to be skewed to the right. If the tail is on the left (Fig. 12.23b), the distribution is referred to as skewed to the left.

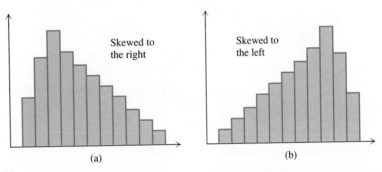

Figure 12.23

The number of children per family might be a distribution that is skewed to the right. Some families have no children, more families may have one child, the greatest percentage may have two children, fewer three children, still fewer four children, and so on.

Since few families have high incomes, distributions of family incomes might be skewed to the right.

If the histograms of the skewed distributions in Fig. 12.23 are smoothed out to form curves, we get the curves illustrated in Fig. 12.24.

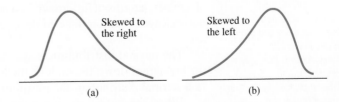

Figure 12.24

In Fig. 12.24(a) the greatest frequency appears on the left side of the curve, and the frequency decreases from left to right. Since the mode is the value with the greatest frequency, the mode would appear on the left side of the curve.

Every value in the set of data is considered in determining the mean. The values on the far right of the curve in Fig. 12.24(a) would tend to increase the value of the mean. Thus the value of the mean would be farther to the right than the mode. The median would be between the mode and the mean. The relationship between the mean, median, and mode for curves that are skewed to the right and left is given in Fig. 12.25.

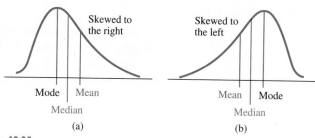

Figure 12.25

Each of these distributions is useful in describing sets of data. However, the most important distribution is the **normal** or **Gaussian distribution**, named for Carl Friedrich Gauss. The histogram of a normal distribution is illustrated in Fig. 12.26.

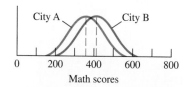

Based on the figure above, which shows the distribution of scores on the mathematics part of the SAT test for two different cities, can we say that any given person selected at random from city A has outperformed any given person selected at random from city B? Both distributions appear normal, and the mean of city A is slightly smaller than the mean of city B. But consider two randomly selected students who took this test: Sally, from city A, and Kendra, from city B. Since the graph shows that there are many students in city B who outperformed students from city A, we cannot conclude that Sally scored higher than Kendra or that Kendra scored higher than Sally.

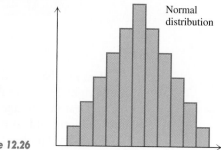

Figure 12.26

The normal distribution is important because many sets of data are normally distributed, or they closely resemble a normal distribution. Such distributions include intelligence quotients, heights and weights of males, heights and weights of females, lengths of full-grown boa constrictors, weights of watermelons, wearout mileage of automobile brakes, and life spans of refrigerators—to name a few.

The normal distribution is symmetric about the mean. If you were to fold the histogram down the middle, the left side would fit exactly on the right side. **In a normal distribution the mean, median, and mode all have the same value.** When the histogram of a normal distribution is smoothed out to form a

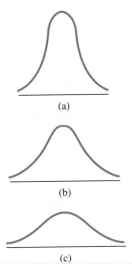

(a)

(b)

(c)

Figure 12.27

curve, the curve is bell-shaped. The bell may be high and narrow or short and wide. Each of the three curves in Fig. 12.27 represents a normal curve. Curve 12.27(a) has the smallest standard deviation (spread from the mean); curve 12.27(c) has the largest.

Since the curve is symmetric, 50% of the data always falls above (to the right of) the mean and 50% of the data below (to the left of) the mean. In addition, every normal distribution has approximately 68% of the data between the value that is one standard deviation below the mean and the value that is one standard deviation above the mean (see Fig. 12.28). Approximately 95% of the data falls between the value that is two standard deviations below the mean and the value that is two standard deviations above the mean.

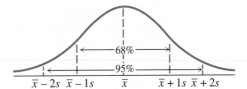

Figure 12.28

Thus if a normal distribution has a mean of 100 and a standard deviation of 10, then approximately 68% of all the data will fall between $100 - 10$ and $100 + 10$, or between 90 and 110. Approximately 95% of the data will fall between $100 - 20$ and $100 + 20$, or between 80 and 120. In fact, given any normal distribution with a known standard deviation and mean, it is possible through the use of Table 12.5 on page 613 (the z-table) to determine the percent of data between any two given values.

We use **z-scores** (or **standard scores**) to determine how far, in terms of standard deviations, a given score is from the mean of the distribution. For example, a score that has a z-value of 1.5 means that the score is 1.5 standard deviations above the mean. The standard or z-score is calculated by the formula:

$$z = \frac{\text{value of the piece of data} - \text{mean}}{\text{standard deviation}}$$

Letting x represent the value of the given piece of data, $\bar{x}$ the mean, and s the standard deviation, the formula above can be symbolized as

$$z = \frac{x - \bar{x}}{s}$$

In this book the notation z_x represents the z-score, or standard score, of the value x. For example, if a normal distribution has a mean of 86 with a standard deviation of 12, a score of 110 has a standard or z-score of

$$z_{110} = \frac{110 - 86}{12} = \frac{24}{12} = 2$$

Therefore a value of 110 in this distribution has a z-score of 2. The score of 110 is thus two standard deviations above the mean.

Data below the mean will always have negative z-scores; data above the mean will always have positive z-scores. The mean will always have a z-score of 0.

▶ **Example 1**

A normal distribution has a mean of 100 and a standard deviation of 10. Find z-scores for the following values.

a) 110   **b)** 115   **c)** 100   **d)** 88

Solution:

a)
$$z = \frac{\text{value} - \text{mean}}{\text{standard deviation}}$$

$$z_{110} = \frac{110 - 100}{10} = \frac{10}{10} = 1$$

A score of 110 is one standard deviation above the mean

b)
$$z_{115} = \frac{115 - 100}{10} = \frac{15}{10} = 1.5$$

c)
$$z_{100} = \frac{100 - 100}{10} = \frac{0}{10} = 0$$

The mean always has a z-score of 0.

d)
$$z_{88} = \frac{88 - 100}{10} = \frac{-12}{10} = -1.2$$

A score of 88 is 1.2 standard deviations below the mean.

Now we turn our attention to finding areas under the normal curve. The total area under the normal curve is 1.00. Table 12.5 will be used to determine the area under the normal curve between any two given points: The values in the table have been rounded off. **Table 12.5 gives the area under the normal curve from the mean (a z-value of 0) to a z-value to the right of the mean.**

For example, between the mean and $z = 2.00$ the table shows a value of 0.477. Thus there is 0.477 of the total area under the curve between the mean and $z = 2.00$ (Fig. 12.29). To change this area of 0.477 to a percent, simply multiply by 100%: $0.477 \times 100\%$ is 47.7%. Thus 47.7% of all scores will be between the mean and the score that is two standard deviations above the mean.

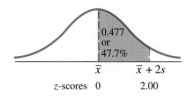

0.477
or
47.7%

$\bar{x}$ $\bar{x} + 2s$

z-scores   0   2.00

Figure 12.29

Table 12.5 Areas under the Standard Normal Curve (the z-table)

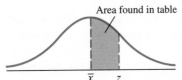

Area found in table

The column under *A* gives the area under the entire curve that is between $z = 0$ and a positive value of *z*.

$\bar{x}$ z

z	A	z	A	z	A	z	A	z	A	z	A	z	A	z	A	z	A
.00	.000	.37	.144	.74	.270	1.11	.367	1.48	.431	1.85	.468	2.22	.487	2.59	.495	2.96	.499
.01	.004	.38	.148	.75	.273	1.12	.369	1.49	.432	1.86	.469	2.23	.487	2.60	.495	2.97	.499
.02	.008	.39	.152	.76	.276	1.13	.371	1.50	.433	1.87	.469	2.24	.488	2.61	.496	2.98	.499
.03	.012	.40	.155	.77	.279	1.14	.373	1.51	.435	1.88	.470	2.25	.488	2.62	.496	2.99	.499
.04	.016	.41	.159	.78	.282	1.15	.375	1.52	.436	1.89	.471	2.26	.488	2.63	.496	3.00	.499
.05	.020	.42	.163	.79	.285	1.16	.377	1.53	.437	1.90	.471	2.27	.488	2.64	.496	3.01	.499
.06	.024	.43	.166	.80	.288	1.17	.379	1.54	.438	1.91	.472	2.28	.489	2.65	.496	3.02	.499
.07	.028	.44	.170	.81	.291	1.18	.381	1.55	.439	1.92	.473	2.29	.489	2.66	.496	3.03	.499
.08	.032	.45	.174	.82	.294	1.19	.383	1.56	.441	1.93	.473	2.30	.489	2.67	.496	3.04	.499
.09	.036	.46	.177	.83	.297	1.20	.385	1.57	.442	1.94	.474	2.31	.490	2.68	.496	3.05	.499
.10	.040	.47	.181	.84	.300	1.21	.387	1.58	.443	1.95	.474	2.32	.490	2.69	.496	3.06	.499
.11	.044	.48	.184	.85	.302	1.22	.389	1.59	.444	1.96	.475	2.33	.490	2.70	.497	3.07	.499
.12	.048	.49	.188	.86	.305	1.23	.391	1.60	.445	1.97	.476	2.34	.490	2.71	.497	3.08	.499
.13	.052	.50	.192	.87	.308	1.24	.393	1.61	.446	1.98	.476	2.35	.491	2.72	.497	3.09	.499
.14	.056	.51	.195	.88	.311	1.25	.394	1.62	.447	1.99	.477	2.36	.491	2.73	.497	3.10	.499
.15	.060	.52	.199	.89	.313	1.26	.396	1.63	.449	2.00	.477	2.37	.491	2.74	.497	3.11	.499
.16	.064	.53	.202	.90	.316	1.27	.398	1.64	.450	2.01	.478	2.38	.491	2.75	.497	3.12	.499
.17	.068	.54	.205	.91	.319	1.28	.400	1.65	.451	2.02	.478	2.39	.492	2.76	.497	3.13	.499
.18	.071	.55	.209	.92	.321	1.29	.402	1.66	.452	2.03	.479	2.40	.492	2.77	.497	3.14	.499
.19	.075	.56	.212	.93	.324	1.30	.403	1.67	.453	2.04	.479	2.41	.492	2.78	.497	3.15	.499
.20	.079	.57	.216	.94	.326	1.31	.405	1.68	.454	2.05	.480	2.42	.492	2.79	.497	3.16	.499
.21	.083	.58	.219	.95	.329	1.32	.407	1.69	.455	2.06	.480	2.43	.493	2.80	.497	3.17	.499
.22	.087	.59	.222	.96	.332	1.33	.408	1.70	.455	2.07	.481	2.44	.493	2.81	.498	3.18	.499
.23	.091	.60	.226	.97	.334	1.34	.410	1.71	.456	2.08	.481	2.45	.493	2.82	.498	3.19	.499
.24	.095	.61	.229	.98	.337	1.35	.412	1.72	.457	2.09	.482	2.46	.493	2.83	.498	3.20	.499
.25	.099	.62	.232	.99	.339	1.36	.413	1.73	.458	2.10	.482	2.47	.493	2.84	.498	3.21	.499
.26	.103	.63	.236	1.00	.341	1.37	.415	1.74	.459	2.11	.483	2.48	.493	2.85	.498	3.22	.499
.27	.106	.64	.239	1.01	.344	1.38	.416	1.75	.460	2.12	.483	2.49	.494	2.86	.498	3.23	.499
.28	.110	.65	.242	1.02	.346	1.39	.418	1.76	.461	2.13	.483	2.50	.494	2.87	.498	3.24	.499
.29	.114	.66	.245	1.03	.349	1.40	.419	1.77	.462	2.14	.484	2.51	.494	2.88	.498	3.25	.499
.30	.118	.67	.249	1.04	.351	1.41	.421	1.78	.463	2.15	.484	2.52	.494	2.89	.498	3.26	.499
.31	.122	.68	.252	1.05	.353	1.42	.422	1.79	.463	2.16	.485	2.53	.494	2.90	.498	3.27	.500
.32	.126	.69	.255	1.06	.355	1.43	.424	1.80	.464	2.17	.485	2.54	.495	2.91	.498	3.28	.500
.33	.129	.70	.258	1.07	.358	1.44	.425	1.81	.465	2.18	.485	2.55	.495	2.92	.498	3.29	.500
.34	.133	.71	.261	1.08	.360	1.45	.427	1.82	.466	2.19	.486	2.56	.495	2.93	.498	3.30	.500
.35	.137	.72	.264	1.09	.362	1.46	.428	1.83	.466	2.20	.486	2.57	.495	2.94	.498	3.31	.500
.36	.141	.73	.267	1.10	.364	1.47	.429	1.84	.467	2.21	.487	2.58	.495	2.95	.498	3.32	.500

When you are finding the area under the normal curve, it is often helpful to draw a picture such as the one in Fig. 12.29, indicating the area or percent to be found.

The normal curve is symmetric about the mean. Thus the same percent of data is between the mean and a positive z-score as between the mean and the corresponding negative z-score. For example, there is the same area under the normal curve between a z of 1.60 and the mean as between a z of -1.60 and the mean. Both have an area of 0.445 (Fig. 12.30).

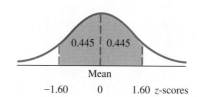

Figure 12.30

At this time we have the necessary knowledge to find the percent of data between any two values in a normal distribution.

To Find the Percent of Data Between any Two Values:

1. Use the formula $z = (x - \bar{x})/s$ to convert the given values to z-scores.
2. Look up the percent that corresponds to each z-score in Table 12.5.
3. a) When finding the percent of data between two z-scores on the opposite side of the mean (when one z-score is positive and the other is negative) you find the sum of the individual percents.
   b) When finding the percent of data between two z-scores on the same side of the mean (when both z-scores are positive or both are negative) subtract the smaller percent from the larger percent).

▶ **Example 2**

Intelligence quotients are normally distributed with a mean of 100 and a standard deviation of 15. Find the percent of individuals with IQs in the following ranges.

a) Between 100 and 115   **b)** Between 70 and 100
c) Between 70 and 115   **d)** Between 115 and 130
e) Below 130   **f)** Above 122.5

Solution:

a) We want to find the area under the normal curve between the values of 100 and 115 as illustrated in Fig. 12.31(a).

Converting 100 to a z-score yields a z-score of 0:

$$z_{100} = \frac{100 - 100}{15} = \frac{0}{15} = 0$$

Converting 115 to a z-score yields a z-score of 1.00:

$$z_{115} = \frac{115 - 100}{15} = \frac{15}{15} = 1.00$$

The percent of individuals with IQs between 100 and 115 is the same as the percent of data between z-scores of 0 and 1 (see Fig. 12.31b).

From Table 12.5 we determine that 0.341 of the area, or 34.1% of all the data, is between z-scores of 0 and 1.00. Therefore 34.1% of individuals have IQs between 100 and 115.

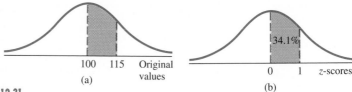

Figure 12.31

(a) 100 115 Original values

(b) 34.1% 0 1 z-scores

b)

$$z_{70} = \frac{70 - 100}{15} = \frac{-30}{15} = -2.00$$

$z_{100} = 0$ (from part a)

The percent of data between scores of 70 and 100 is the same as the percent between $z = -2$ and $z = 0$ (Fig. 12.32). The percent of data between the mean and two standard deviations below the mean is the same as between the mean and two standard deviations above the mean. 47.7% of the data is between $z = 0$ and $z = 2$. Thus 47.7% of the data is also between $z = -2$ and $z = 0$. Therefore 47.7% of all individuals have have IQs between 70 and 100.

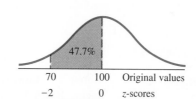

Figure 12.32

47.7% 70 100 Original values
−2 0 z-scores

c) In parts (a) and (b) we determined that $z_{115} = 1.00$ and $z_{70} = -2.00$. Since the values are on opposite sides of the mean, the percent of data between the two values is found by adding the individual percents: 34.1% + 47.7% = 81.8% (Fig. 12.33). Thus 81.8% of the IQs are between 70 and 115.

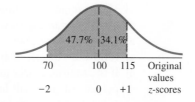

Figure 12.33

47.7% 34.1%
70 100 115 Original values
−2 0 +1 z-scores

d)
$$z_{130} = \frac{130 - 100}{15} = \frac{30}{15} = 2.00$$

$$z_{115} = 1.00 \text{ (from above)}$$

Since both values are on the same side of the mean (Fig. 12.34), the smaller percent must be subtracted from the larger percent to obtain the percent of data in the shaded area: 47.7% − 34.1% is 13.6%. Thus 13.6% of all the individuals have IQs between 115 and 130.

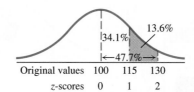

Figure 12.34

e) The percent of IQs below 130 is the same as the percent of data below a z-score of 2. Between $z = 2$ and the mean is 47.7% of the data (Fig. 12.35). To this 47.7% we add the 50% of the data below the mean to give 97.7%. Thus 50% + 47.7%, or 97.7%, of all IQs are below 130.

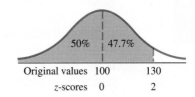

Figure 12.35

f)
$$z_{122.5} = \frac{122.5 - 100}{15} = \frac{22.5}{15} = 1.50$$

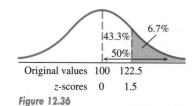

Figure 12.36

The percent of IQs above 122.5 is the same as the percent of data above $z = 1.5$ (Fig. 12.36). Fifty percent of the data is to the right of the mean. Since 43.3% of the data is between the mean and $z = 1.5$, 50% − 43.3%, or 6.7%, of the data is greater than $z = 1.5$. Thus 6.7% of all IQs are greater than 122.5.

▶ **Example 3**

The wear-out mileage of a certain tire is normally distributed with a mean of 35,000 miles and standard deviation of 2500 miles.
a) Find the percent of tires that will last at least 35,000 miles.
b) Find the percent of tires that will last between 30,000 and 37,500 miles.
c) Find the percent of tires that will last at least 39,000 miles.
d) If the manufacturer guarantees the tires to last at least 30,000 miles, what percent of tires will fail to live up to the guarantee?
e) If 200,000 tires are produced, how many will last at least 39,000 miles?

Figure 12.37

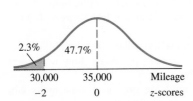

Figure 12.38

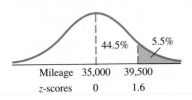

Figure 12.39

Solution:

a) In a normal distribution, half the data are always above the mean. Since 35,000 miles is the mean, half or 50% of the tires will last at least 35,000 miles.

b) Convert 30,000 miles and 37,500 miles to z-scores:

$$z_{30,000} = \frac{30,000 - 35,000}{2500} = -2.00$$

$$z_{37,500} = \frac{37,500 - 35,000}{2500} = 1.00$$

Now look up the areas in Table 12.5. The percent of tires that will last between 30,000 and 37,500 miles is 47.7% + 34.1%, or 81.8% (Fig. 12.37).

c)
$$z_{39,000} = \frac{39,000 - 35,000}{2500} = 1.60$$

44.5% of the data is between the mean and $z = 1.60$ (Fig. 12.38). Therefore the percent of data above $z = 1.60$ is 50% − 44.5% = 5.5%. Thus 5.5% of the tires will last at least 39,000 miles.

d) To solve this problem, we must find the percent of tires that last less than 30,000 miles: $z_{30,000} = -2.00$. From Table 12.5 we determine that 47.7% of the data is between $z = -2.00$ and $z = 0$ (Fig. 12.39). The percent to the left of $z = -2.00$ is found by subtracting 47.7% from 50% to obtain 2.3%. Thus 2.3% of all the tires will last less than 30,000 miles and fail to live up to the guarantee.

e) In part (c) we determined that 5.5% of all tires will last at least 39,000 miles. We now multiply 0.055 times 200,000 to determine the number of tires that will last at least 39,000 miles: $0.055 \times 200,000 = 11,000$ tires.

Section 12.7 Exercises

Try to list an example of the following types of distributions.
1. rectangular
2. J-shaped
3. skewed
4. bimodal
5. In a distribution that is skewed to the right, which has the greatest value—the mean, median, or mode? Which has the smallest value? Explain.
6. In a distribution that is skewed to the left, which has the greatest value—the mean, median, or mode? Which has the smallest value? Explain.
7. List three populations other than those noted that may be normally distributed.
8. List three populations other than those noted that may not be normally distributed.

Using Table 12.5, find the area specified in each of the following.
9. Above the mean
10. Below the mean
11. Between two standard deviations below the mean and one standard deviation above the mean
12. Between 1.20 and 1.90 standard deviations above the mean
13. To the right of $z = 1.35$
14. To the left of $z = 1.35$
15. To the left of $z = -1.74$
16. To the right of $z = -1.74$
17. To the right of $z = 2.16$

18. To the left of $z = 1.96$
19. To the left of $z = -1.21$
20. To the left of $z = -0.20$

Using Table 12.5, find the percent of data specified in each of the following.
21. Between $z = 0$ and $z = 1.60$
22. Between $z = -0.20$ and $z = -0.87$
23. Between $z = -1.40$ and $z = 2.20$
24. Less than $z = -1.90$
25. Greater than $z = -1.90$
26. Greater than $z = 2.66$
27. Less than $z = 1.96$
28. Between $z = 0.74$ and $z = 1.65$
29. Between $z = -2.65$ and $z = -0.92$
30. Between $z = -2.15$ and $z = 3.31$

Heights of kindergarten children are normally distributed with a mean of 40 in. and a standard deviation of 6 in.
31. What percent of kindergarten children are between 40 in. and 46 in. tall?
32. What percent of kindergarten children are less than 46 in. tall?
33. What percent of kindergarten children are between 34 in. and 46 in. tall?
34. If 1000 kindergarten children are selected at random, how many will be less than 43 in. tall?

The results of a statistics examination are normally distributed with a mean of 74 and a standard deviation of 8. Find the percent of individuals that scored as indicated below.
35. Between 58 and 82
36. Less than 78
37. Greater than 78
38. Between 78 and 86
39. Between 60 and 72
40. Greater than 70

The diameters of red blood cells are normally distributed with a mean diameter of 0.008 mm and a standard deviation of 0.002 mm.
41. Find the percent of red blood cells with diameters greater than 0.009 mm.
42. Find the percent of red blood cells with diameters between 0.004 mm and 0.013 mm.
43. Find the percent of red blood cells with diameters greater than 0.003 mm.
44. If 500 cells are isolated, how many will have diameters greater than 0.009 mm?

The weights of full-grown Brontosauruses, the most massive dinosaurs, were normally distributed with a mean of 33 tons and a standard deviation of 4 tons. Find the percent of Brontosauruses in each of the following categories.
45. Greater than 33 tons
46. Between 31 and 35 tons
47. Between 34 and 36 tons
48. Less than 30 tons
49. Greater than 30 tons
50. Less than 39 tons

A vending machine is designed to dispense a mean of 6.7 oz of coffee into a 7-oz cup. If the standard deviation of the amount of coffee dispensed is 0.4 oz and the amount of coffee dispensed is normally distributed, find each of the following.
51. The percent of times the machine will dispense from 6.5 to 6.8 oz
52. The percent of times it will dispense less than 6.0 oz
53. The percent of times it will dispense less than 6.8 oz
54. The percent of times the 7-oz cup will overflow

The life expectancy of nondefective brand Z light bulbs is normally distributed with a mean life of 1500 hr and a standard deviation of 100 hr.
55. Find the percent of bulbs that will last more than 1450 hr.
56. Find the percent of bulbs that will last between 1400 hr and 1550 hr.
57. Find the percent of bulbs that will last less than 1480 hr.
58. If 80,000 of these bulbs are produced, how many will last 1500 hr or more?
59. If 80,000 of these bulbs are produced, how many will last between 1400 hr and 1600 hr?

60. A weight loss clinic guarantees that its new customers will lose at least 5 lb by the end of their first month of participation or their money will be refunded. If the loss of weight of customers at the end of their first month is normally distributed with a mean of 6.7 lb and a standard deviation of 0.81 lb, find the percent of customers that will be able to claim a refund.
61. The warranty on the motor of a dishwasher is 8 years. If the breakdown times of this motor are normally distributed with a mean of 10.2 years and a standard deviation of 1.8 years, find the percent of motors that can be expected to require repair or replacement under warranty.
62. A vending machine that dispenses coffee does not appear to be working correctly. The machine rarely gives the proper amount of coffee. Some of the time the cup is underfilled, and some of the time the cup overflows. Does this indicate that the mean number of ounces dispensed has to be adjusted or that the standard deviation of the

amount of coffee dispensed by the machine is too large? Explain your answer.

63. Mr. Brittain marks his class on a normal curve. Those with z-scores above 1.8 will receive A, those between 1.8 and 1.1 will receive B, those between 1.1 and -1.2 will receive C, those between -1.2 and -1.9 will receive D, and those under -1.9 will receive F. Find the percent of grades that will be A, B, C, D, and F.

64. Professor Hart marks his classes on the normal curve. His statistics class has a mean of 72 with a standard deviation of 8. He has decided that 10% of the class will get A, 20% will get B, 40% will get C, 20% will get D, and 10% will get F. Find the following.
   a) The minimum grade needed to get an A
   b) The minimum grade needed to pass the course (D or better)
   c) The range of grades that will result in a C grade

Problem Solving

65. How can one tell if a distribution is approximately normal? A statistical theorem called **Chebyshev's theorem** states that the minimum percentage of data between plus and minus K standard deviations from the mean ($K > 1$) in *any distribution* can be found by the formula

$$\text{Minimum percent} = 1 - \frac{1}{K^2}$$

Thus, for example, between ± 2 standard deviations from the mean, there will always be a minimum of 75% of data. This minimum percent is true for any distribution:

$$\text{For } K = 2, \quad 1 - \frac{1}{2^2} = 1 - \frac{1}{4} = \frac{3}{4} \quad \text{or} \quad 75\%$$

Likewise, between ± 3 standard deviations from the mean, there will always be a minimum of 89% of the data:

$$\text{For } K = 3, \quad 1 - \frac{1}{3^2} = 1 - \frac{1}{9} = \frac{8}{9} \quad \text{or} \quad 89\%$$

The following chart lists the minimum percent of data in *any distribution* and the actual percent of data in *the normal distribution* between ± 1.1, ± 1.5, ± 2.0, and ± 2.5 standard deviations from the mean. The minimum percents of data in any distribution were calculated by using Chebyshev's theorem. The actual percents of data for the normal distribution were calculated by using the area given in the standard normal, or z, table.

	$K = 1.1$	$K = 1.5$	$K = 2$	$K = 2.5$
Minimum % (for any distribution)	17.4%	55.6%	75%	84%
Normal distribution %	72.8%	86.6%	95.4%	99.8%
Given distribution				

The third row of the chart has been left blank for you to fill in percents, as will be explained shortly.

Consider the following 30 pieces of data obtained from a quiz:

1, 1, 1, 1, 2, 2, 2, 2, 3, 3, 4, 4, 4, 5, 6, 6, 6,
7, 7, 7, 7, 8, 8, 8, 8, 9, 9, 9, 10, 10

   a) Find the mean of the set of scores.
   b) Find the standard deviation of the set of scores.
   c) Determine the values that correspond to 1.1, 1.5, 2, and 2.5 standard deviations above the mean. (For example, the value that corresponds to 1.5 standard deviations above the mean is $\bar{x} + 1.5s$.)
      Then determine the values that correspond to 1.1, 1.5, 2, and 2.5 standard deviations below the mean. (For example, the value that corresponds to 1.5 standard deviations below the mean is $\bar{x} - 1.5s$.)
   d) By observing the 30 pieces of data, determine the actual percent of quiz scores between

      ± 1.1 standard deviations from the mean
      ± 1.5 standard deviations from the mean
      ± 2 standard deviations from the mean
      ± 2.5 standard deviations from the mean

   e) Place the percents found in part (d) in the third row of the chart.
   f) Compare the percents in the third row of the chart with the minimum percents in the first row and the normal percents in the second row, then make a judgment as to whether this set of 30 scores is approximately normally distributed. Explain your answer.

Research Activities

66. Obtain a set of test scores from your teacher.
   a) Find the mean, median, mode, and midrange of the test scores.
   b) Find the range and standard deviation of the set of

scores. (You may round the mean off to the nearest tenth when finding the standard deviation.)

c) Construct a frequency distribution of the set of scores. Select your first class so that there will be between 5 and 12 classes.

d) Construct a histogram and frequency polygon of the frequency distribution in part (c).

e) Does the histogram in part (d) appear to represent a normal distribution?

f) Use the procedure explained in Ex. 65 to determine whether the set of scores represented is a normal distribution. Explain.

67. In this project you actually become the statistician.
   a) Select a project of interest to you in which data must be collected.
   b) Write a proposal to submit to your teacher for approval. In the proposal, discuss the aims of your proj-ect and how you plan to gather the data to make your sample unbiased.

c) After your proposal is approved, gather 50 pieces of data by the method you proposed.

d) Rank the data from smallest to largest.

e) Compute the mean, median, mode, and midrange.

f) Determine the range and standard deviation of the data. You may round the mean off to the nearest tenth when computing the standard deviation.

g) Construct a frequency distribution, histogram, and frequency polygon of your data. Select your first class so that there will be between 5 and 12 classes. Make sure to label your histogram and frequency polygon.

h) Does your distribution appear to be normal? Explain your answer. Does it appear to be another type of distribution discussed? Explain.

i) Determine whether your distribution is approximately normal by using the technique discussed in Ex. 65.

CHAPTER 12 SUMMARY

Key Terms

12.1
cluster sample
data
descriptive statistics
inferential statistics
population
random sample
sample
stratified sample
systematic sample
unbiased sample

12.3
classes
class mark
frequency distribution
lower class limit
midpoint of a class
modal class
piece of data
upper class limit

12.4
central angle
circle graph
frequency polygon
histogram

12.5
mean
median
midrange
mode
percentile
quartile
ranked data

12.6
measures of dispersion
standard deviation

12.7
bimodal distribution

Gaussian distribution
J-shaped distribution
normal distribution
rectangular distribution
skewed distribution
standard score
z-score

The **median** is the value in the middle of a set of ranked data.

The **mode** is the piece of data that occurs most frequently (if there is one).

The **midrange** is the value halfway between the lowest and highest values: midrange $= \dfrac{L + H}{2}$.

Important Facts

Rules for data grouped by classes

1. The classes should be the same width.
2. The classes should not overlap.
3. Each piece of data should belong to one and only one class.

Measures of dispersion

The **range** is the difference between the highest value and lowest value in a set of data.

The **standard deviation**, s, is a measure of the spread of a set of data about the mean: $s = \sqrt{\dfrac{\Sigma(x - \bar{x})^2}{n - 1}}$.

z-scores $z = \dfrac{x - \bar{x}}{s}$

Measures of central tendency

The **mean** is the sum of the data divided by the number of pieces of data: $\bar{x} = \dfrac{\Sigma x}{n}$.

Chebyshev's theorem   Minimum percentage $= 1 - \dfrac{1}{k^2}$, $k > 1$

CHAPTER 12 REVIEW EXERCISES

12.1
1. a) What is a population?
   b) What is a sample?
2. What is a random sample?

12.2
In Ex. 3 and 4, tell what possible misuses or misinterpretations may exist in the statements.
3. The Stay Healthy Candy Bar indicates on its label that it has no cholesterol. Therefore it is safe to eat as many of these candy bars as you want.
4. More copies of *Time* magazine are sold than are copies of *Money* magazine. Therefore *Time* is a more profitable magazine than *Money*.
5. The value of a 1-ounce gold coin on December 31, 1990, was $402. The value of the same coin on December 31, 1991, was $380. Draw a graph that appears to show (a) a small decrease in value, and (b) a large decrease in value.

12.3–12.4
6. Consider the following set of data:

20	21	22	23	26	28	30
20	21	23	24	26	29	30
20	21	23	24	26	30	30
20	21	23	25	27	30	30
21	21	23	26	28	30	30

   a) Construct a frequency distribution.
   b) Construct a histogram.
   c) Construct a frequency polygon.

7. Consider the following set of data:

30	40	52	58	63	72	78	84
32	44	53	60	67	72	78	86
36	46	54	60	68	74	78	88
38	48	55	60	70	76	80	92
39	50	57	62	72	78	82	98

a) Construct a frequency distribution with the first class 30–38.
b) Construct a histrogram.
c) Construct a frequency polygon.

12.5–12.6
Given the set of data 64, 66, 74, 82, 85, 91, find each of the following:

8. Mean
9. Median
10. Mode
11. Midrange
12. Range
13. Standard deviation

Given the set of data 4, 5, 12, 14, 19, 7, 12, 23, 7, 17, 15, 21, find each of the following.

14. Mean
15. Median
16. Mode
17. Midrange
18. Range
19. Standard deviation

12.7
Anthropologists have determined that a certain type of primitive animal had a mean head circumference of 40 cm with a standard deviation of 5 cm. Given that head sizes were normally distributed, determine the percent of heads that are

20. Between 35 and 45 cm
21. Between 30 and 50 cm
22. Less than 48 cm
23. Greater than 48 cm
24. Greater than 37 cm

The heights of fully grown redwood trees are normally distributed with a mean height of 270 ft and a standard deviation of 30 ft. Find the percent of redwood trees that are

25. Greater than 270 ft
26. Between 270 and 290 ft
27. Less than 275 ft
28. Between 210 and 300 ft

12.5–12.7
A study was made on the weights of adult men. The following statistics were obtained from that study:

Mean	187 lb	First quartile	173 lb
Median	180 lb	Third quartile	227 lb
Mode	175 lb	86th percentile	234 lb
Standard deviation	23 lb		

29. What is the most common weight?
30. What weight did half of those surveyed surpass?
31. About what percent of those surveyed surpassed 227 lb?
32. About what percent of those surveyed weighed below 173 lb?
33. About what percent of those surveyed weighed more than 234 lb?

34. If 100 men were surveyed, what is the total weight of all the men?
35. What weight represents 2 standard deviations above the mean?
36. What weight represents 1.8 standard deviations below the mean?

12.2–12.7
Below is a listing of U.S. presidents and the number of children in their families:

Washington	0	Arthur	3
J. Adams	5	Cleveland	5
Jefferson	6	B. Harrison	3
Madison	0	McKinley	2
Monroe	2	T. Roosevelt	6
J. Q. Adams	4	Taft	3
Jackson	0	Wilson	3
Van Buren	4	Harding	0
W. H. Harrison	10	Coolidge	2
Tyler	14	Hoover	2
Polk	0	F. D. Roosevelt	6
Taylor	6	Truman	1
Fillmore	2	Eisenhower	2
Pierce	3	Kennedy	3
Buchanan	0	L. B. Johnson	2
Lincoln	4	Nixon	2
A. Johnson	5	Ford	4
Grant	4	Carter	4
Hayes	8	Reagan	4
Garfield	7	Bush	6

Determine the following.
37. Mean number of children
38. Mode
39. Median
40. Midrange
41. Range
42. Standard deviation (round the mean off to the nearest tenth)
43. Construct a frequency distribution; let the first class be 0–1.
44. Construct a histogram.
45. Construct a frequency polygon.
46. Does this distribution appear to be normal? Explain.
47. On the basis of this sample, do you think that the number of children per family in the United States would have a normal distribution? Explain.
48. Do you feel that this sample is representative of the population? Explain.

CHAPTER TEST

Given the set of data 12, 19, 24, 24, 21 find the following.

1. Mean   **2.** Median

3. Mode   **4.** Midrange

5. Range   **6.** Standard deviation

Consider the following set of data:

6	8	15	26	29	36
6	10	16	26	29	38
6	12	20	27	30	38
6	12	24	27	32	42
7	15	26	27	34	46

7. Construct a frequency distribution. Let the first class be 5–10.

8. Construct a histogram of the frequency distribution.

9. Construct a frequency polygon of the frequency distribution.

The following represent statistics of weekly salaries at the Hoperchang Publishing Company:

Mean	$480	First quartile	$450
Median	$450	Third quartile	$485
Mode	$475	79th percentile	$500
Standard deviation	$40		

10. What is the most common salary?

11. What salary did half the employees surpass?

12. About what percent of employees' salaries surpassed $450?

13. About what percent of employees' salaries was below $500?

14. If the company has 100 employees, what is the total weekly salary of all employees?

15. What salary represents 1 standard deviation above the mean?

16. What salary represents 1.5 standard deviations below the mean?

The useful life of an electric motor is normally distributed with a mean of 7.3 years and a standard deviation of 0.5 year. Find the percent of motors with a useful life:

17. Between 7.3 years and 7.8 years

18. Less than 7.1 years

19. Between 7.5 years and 8.5 years

20. If the manufacturer warranties its motors for 7 years, what percent will need to be replaced under warranty?

APPENDIX A
THE METRIC SYSTEM AND DIMENSIONAL ANALYSIS

Two systems of weights and measures exist side by side in the United States today, the **U.S. customary system** and the **SI system** (System International), commonly called the metric system. There are many advantages to using the metric system. Some of these are summarized here.

1. The metric system is the worldwide accepted standard measurement system. All industrial nations that trade internationally, except the United States, use the metric system as the official system of measurement.
2. There is only one basic unit of measurement for each physical quantity. In the customary system many units are often used to represent the same physical quantity. For example, when discussing length, we use inches, feet, yards, miles, and so on. Converting from one of these units to the other is often a tedious task (consider changing 12 miles to inches). In the metric system it is possible to make many conversions by simply moving the decimal point.
3. The SI system is based on the number 10, and there is little need for fractions, since most quantities can be expressed as decimals.

Basic Terms and Rules

Because the official definitions of many metric terms are quite technical, we will present them in lay language.

The **meter** (m) is the base unit of **length** in the metric system. The meter is a little more than a yard. Basketball players are often more than 2 meters tall.

The **kilogram** (kg) is the base unit of **mass**. The kilogram is a little more than 2 pounds. A newborn baby might have a mass of about 3 kilograms. The gram (g), a unit of mass derived from the kilogram, is used to measure small amounts. A nickel has a mass of about 5 grams.

The **liter** (ℓ) is not a base unit of the metric system, but it is commonly used to measure **volume**. A liter is a little more than a quart. The gas tank of a compact car may hold 50 liters of gasoline. The **cubic decimeter** is the official unit of volume. It is equal to a liter, but the liter is more commonly used.

$$1 \text{ m} \approx 1 \text{ yd}$$
$$1 \text{ kg} \approx 2.2 \text{ lb}$$
$$1 \ell \approx 1 \text{ qt}$$

The term **degree Celsius** (°C) is used to measure temperature in place of the official or base unit, degree Kelvin. The freezing point of water is 0°C, and the boiling point of water is 100°C. The temperature on a warm day may be 30°C.

$$0°C = 32°F$$
$$37°C = 98.6°F$$
$$100°C = 212°F$$

Prefixes

The metric system is based on the number 10, and it is therefore a decimal system. Prefixes are used to denote a multiple or part of a base unit. For example, a **deka**meter represents 10 meters, and a **centi**meter represents $\frac{1}{100}$ of a meter. In the metric system, when used outside the United States, when writing large numbers, groups of three digits are separated by a space, not a comma. For example, the number for one thousand is 1 000 and the number for nine million is 9 000 000. Commas are not used in the SI system because many countries use the comma as we use a decimal point. In this appendix we will separate numbers greater than 999 with a space rather than a comma, as is done outside the U.S. Table T.1 summarizes the more commonly used prefixes and their meanings. These prefixes pertain to all the units of measurement in the metric system. Thus 1 kiloliter = 1 000 liters, 1 kilogram = 1 000 grams, 1 centimeter = $\frac{1}{100}$ meter, and 1 milliliter = $\frac{1}{1000}$ liter.

Table T.1

Prefix	Symbol	Meaning
kilo	k	$1\ 000 \times$ base unit
hecto	h	$100 \times$ base unit
deka	da	$10 \times$ base unit
		base unit
deci	d	$\frac{1}{10}$ of base unit
centi	c	$\frac{1}{100}$ of base unit
milli	m	$\frac{1}{1000}$ of base unit

The abbreviations or symbols for units of measure are never pluralized, but full names are. For example, 5 milliliters is symbolized as 5 mℓ not 5 mℓs.

For scientific work where very large and very small quantities are used the following prefixes are also used: mega (M) is one million times the base unit, giga (G) is one billion times the base unit, tera (T) is one trillion times the base unit, micro (m) is one millionth of the base unit, nano (n) is one billionth of the base unit, and pico (p) is one trillionth of the base unit.

Conversions within the Metric System

We will use Table T.2 to help demonstrate how to change from one metric unit to another metric unit (meters to kilometers, and so on).

Table T.2

kilometer	hectometer	dekameter	meter	decimeter	centimeter	millimeter
km	hm	dam	m	dm	cm	mm
1 000 m	100 m	10 m	1 m	0.1 m	0.01 m	0.001 m

The meters in Table T.2 can be replaced by grams, liters, or any other base unit of the metric system. Regardless of which of these units we choose, the procedure will be the same. For purposes of explanation we have used the meter.

The table tells us that 1 hectometer equals 100 meters and 1 millimeter is 0.001 (or $\frac{1}{1000}$) meter. The millimeter is the smallest unit in the table. A centimeter is 10 times as large as a millimeter, a decimeter is 10 times as large as a centimeter, a meter is 10 times as large as a decimeter, and so on. Since each unit is 10 times as large as the unit on its right, converting from one unit to another is simply a matter of multiplying or dividing by powers of 10.

Changing Units within the Metric System

1. To change from a smaller unit to a larger unit (for example from meters to kilometers), move the decimal point in the original quantity one place to the left for each larger unit of measurement until you obtain the desired unit of measurement.

2. To change from a larger unit to a smaller unit (for example from kilometers to meters), move the decimal point in the original quantity one place to the right for each smaller unit of measurement until you obtain the desired unit of measurement.

▶ **Example 1**

a) Change 403.2 grams to kilograms.
b) Change 14 grams to centigrams.
c) Change 0.18 liter to milliliters.
d) Change 240 dekaliters to kiloliters.

Solution:

a) To change units from grams to kilograms, note that there are three greater units: dag, hg, and kg. Thus the decimal point must be moved three places to the left:

$$403.2 \text{ g} = 0.403\ 2 \text{ kg}$$

b) To change grams to centigrams, move the decimal point two places to the right:

$$14 \text{ g} = 1\ 400 \text{ cg}$$

c) $0.18\ \ell = 180 \text{ m}\ell$
d) $240 \text{ da}\ell = 2.40 \text{ k}\ell$

▶ **Example 2**

Arrange in order from the smallest to largest length: 3.4 m, 3 421 mm, 104 cm.

Solution: To be compared, these lengths should all be in the same units of measure. Convert all the measures to millimeters:

$$3.4 \text{ m} = 3\ 400 \text{ mm}, \qquad 3\ 421 \text{ mm}, \qquad 104 \text{ cm} = 1\ 040 \text{ mm}$$

The lengths arranged in order from smallest to largest are 104 cm, 3.4 m, 3 421 mm.

Dimensional Analysis and Conversions to and from the Metric System

You may sometimes need to change units of measurement in the metric system to equivalent units in the customary system. To do so we use **dimensional analysis**. To perform dimensional analysis you must first understand what is meant by a unit fraction. A **unit fraction** is any fraction that has a value of 1.

Examples of Unit Fractions

$$\frac{12 \text{ in.}}{1 \text{ ft}}, \quad \frac{1 \text{ ft}}{12 \text{ in.}}, \quad \frac{16 \text{ oz}}{1 \text{ lb}}, \quad \frac{1 \text{ lb}}{16 \text{ oz}}, \quad \frac{60 \text{ min}}{1 \text{ hr}}, \quad \frac{1 \text{ hr}}{60 \text{ min}}$$

In each of these examples the numerator is equal to the denominator so the value of the fraction is 1.

To convert an expression from one unit of measurement to a different unit, multiply the given expression by the unit fraction (or fractions) that will result in the answer having the units you are seeking. When two fractions are being multiplied, and the same unit appears in the numerator of one fraction and the denominator of the other fraction, then that common unit may be divided out. For example suppose we wish to convert 30 inches to feet. We consider the following:

$$30 \text{ in.} = ? \text{ ft}$$

Since inches are given we will need to eliminate them. Thus inches will need to appear in the denominator of the unit fraction. We need to convert to feet, so feet will need to appear in the numerator of the unit fraction. If we multiply a quantity in inches by a unit fraction containing feet/inches then the units will divide out as follows.

$$(\cancel{\text{in.}})\left(\frac{\text{ft}}{\cancel{\text{in.}}}\right) = \text{ft}$$

Thus the solution to the problem is found as follows.

$$30 \text{ in.} = (30 \ \cancel{\text{in.}})\left(\frac{1 \text{ ft}}{12 \ \cancel{\text{in.}}}\right) = \frac{30}{12} \text{ ft} = 2.5 \text{ ft}$$

▶ **Example 3**

Convert 80 miles per hour to feet per second.

Solution: Let us consider the units given and where we wish to end up. We are given $\frac{\text{mi}}{\text{hr}}$ and wish to end with $\frac{\text{ft}}{\text{sec}}$. Thus we need to change miles into feet and hours into seconds. Since two units need to be changed we will need to multiply the given quantity by two unit fractions, one for each conversion. Here we show how to convert the units of measurement from miles per hour to feet per second:

$$\left(\frac{\text{mi}}{\text{hr}}\right)\left(\frac{\text{ft}}{\text{mi}}\right)\left(\frac{\text{hr}}{\text{sec}}\right) \quad \text{gives an answer in} \quad \frac{\text{ft}}{\text{sec}}$$

Now multiply the given quantity by the appropriate unit fractions to obtain the answer

$$80\,\frac{\text{mi}}{\text{hr}} = \left(80\,\frac{\text{mi}}{\text{hr}}\right)\left(\frac{5280\ \text{ft}}{1\ \text{mi}}\right)\left(\frac{1\ \text{hr}}{3600\ \text{sec}}\right) = \frac{(80)(5280)}{(1)(3600)}\,\frac{\text{ft}}{\text{sec}}$$

$$\approx 117.3\,\frac{\text{ft}}{\text{sec}}$$

Note that $\left(\dfrac{80\ \text{mi}}{\text{hr}}\right)\left(\dfrac{1\ \text{hr}}{3600\ \text{sec}}\right)\left(\dfrac{5280\ \text{ft}}{1\ \text{mi}}\right)$ will also give the same answer.

▶ **Example 4**

If \$1 = 3280 pesos, what is the amount in dollars of 1,000,000 pesos?

Solution:

$$1{,}000{,}000\ \text{pesos} = (1{,}000{,}000\ \text{pesos})\left(\frac{\$1}{3280\ \text{pesos}}\right) = \frac{\$1{,}000{,}000}{3280}$$

$$\approx \$304.88$$

Thus 1,000,000 pesos has a value of about \$305.

Now we will apply dimensional analysis to the metric system.

Table T.3 on page A-6 is used in making conversions to and from the metric system. A more exact table of conversion factors may be found in many science books at your college's library. From this table we can obtain many unit fractions.

From Table T.3 we see that 1 inch = 2.54 centimeters. From this we can write the two unit fractions $\dfrac{1\ \text{in.}}{2.54\ \text{cm}}$ or $\dfrac{2.54\ \text{cm}}{1\ \text{in.}}$.

Examples of other unit fractions from Table T.3 are

$$\frac{1\ \text{yd}}{0.9\ \text{m}}, \quad \frac{0.9\ \text{m}}{1\ \text{yd}}, \quad \frac{1\ \text{gal}}{3.8\ \ell}, \quad \frac{3.8\ \ell}{1\ \text{gal}}, \quad \frac{1\ \text{lb}}{0.45\ \text{kg}}, \quad \frac{0.45\ \text{kg}}{1\ \text{lb}}$$

To change from a metric unit to a customary unit or vice versa, multiply the given quantity by the unit fraction whose product will result in the units we are seeking. For example, to convert 5 inches to centimeters, multiply the 5 inches by a unit fraction with centimeters in the numerator and inches in the

Table T.3 Conversions Table

Length

1 inch (in.) = 2.54 centimeters (cm)
1 foot (ft) = 30 centimeters (cm)
1 yard (yd) = 0.9 meter (m)
1 mile (mi) = 1.6 kilometers km)

Area

1 square inch (in.²) = 6.5 square centimeters (cm²)
1 square foot (ft²) = 0.09 square meter (m²)
1 square yard (yd²) = 0.8 square meter (m²)
1 square mile (mi²) = 2.6 square kilometers (km²)
1 acre = 0.4 hectare (ha)

Volume

1 teaspoon (tsp) = 5 milliliters (mℓ)
1 tablespoon (Tbsp) = 15 milliliters (mℓ)
1 fluid ounce (fl oz) = 30 milliliters (mℓ)
1 cup (c) = 0.24 liter (ℓ)
1 pint (pt) = 0.47 liter (ℓ)
1 quart (qt) = 0.95 liter (ℓ)
1 gallon (gal) = 3.8 liters (ℓ)
1 cubic foot (ft³) = 0.03 cubic meter (m³)
1 cubic yard (yd³) = 0.76 cubic meter (m³)

Weight (mass)

1 ounce (oz) = 28 grams (g)
1 pound (lb) = 0.45 kilogram (kg)
1 ton (T) = 0.9 tonne (t)

Temperature

$C = \frac{5}{9}(F - 32)$
$F = \frac{9}{5}C + 32$

denominator.

$$5 \text{ in.} = (5 \cancel{\text{in.}})\left(\frac{2.54 \text{ cm}}{1 \cancel{\text{in.}}}\right)$$

$$= 5(2.54) \text{ cm}$$

$$= 12.7 \text{ cm}$$

Suppose we wish to convert 150 millimeters to inches. The table does not have a conversion factor from millimeters to inches, but it does have one for inches to centimeters. Since 1 inch = 2.54 centimeters and since 1 centimeter = 10 millimeters we can reason that 1 inch = 25.4 millimeters. With this information we can solve the problem as follows.

$$150 \text{ mm} = (150 \text{ mm})\left(\frac{1 \text{ in.}}{25.4 \text{ mm}}\right) = \frac{150}{25.4} \text{ in.}$$

$$\approx 5.91 \text{ in.}$$

If we wish we can use dimensional analysis using two unit fractions as follows to make the conversion.

$$150 \text{ mm} = (150 \text{ mm})\left(\frac{1 \text{ cm}}{10 \text{ mm}}\right)\left(\frac{1 \text{ in.}}{2.54 \text{ cm}}\right) = \frac{150}{(10)(2.54)} \text{ in.}$$

$$\approx 5.91 \text{ in.}$$

▶ **Example 5**

Convert 60 fluid ounces to liters.

Solution:

$$60 \text{ fl oz} = (60 \text{ fl oz})\left(\frac{30 \text{ ml}}{1 \text{ fl oz}}\right)\left(\frac{1 \ell}{1\ 000 \text{ ml}}\right) = \frac{(60)(30)}{1\ 000} \ell$$

$$= 1.8 \ell$$

We will use this procedure in making conversions throughout this appendix.

Now we will study the base units in the metric system in a little more depth.

Length

The basic unit of length is the meter. The meter is defined to be 1,650,763.73 wavelengths in a vacuum, of the orange-red line of the spectrum of Krypton-86. Other units of length that are commonly used are the kilometer, centimeter, and millimeter. The meter, which is a little longer than one yard, is used to measure items that we measure in yards and feet. A man whose height is about 2 meters is a tall man. A tractor trailer unit (an 18-wheeler) is about 18 meters long.

The kilometer is used to measure what we measure in miles. For example, it is about 5 120 km from New York to Seattle. One kilometer is about 0.6 mile, and one mile is about 1.6 kilometers.

Centimeters and millimeters are used to measure what we measure in inches. The centimeter is a little less than $\frac{1}{2}$ inch (see Fig. A.1), and the milli-

Figure A.1

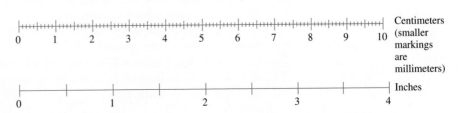

There are 27 different units of length used in the English system. How many of them can you name? Don't forget rod, mil, paris line, toise, cubit, and light year. The different units can be found in the *Handbook of Chemistry and Physics*.

meter is a little less than $\frac{1}{20}$ inch. A millimeter is about the thickness of a dime. A book may measure 20 cm by 25 cm with a thickness of about 3 cm. Millimeters are often used in scientific work and other areas in which small quantities must be measured. The length of a small insect may be measured in millimeters.

▶ **Example 6**

The distance from Boston, Massachusetts, to New Orleans, Louisiana, is about 1359 miles. Express the distance (a) in kilometers and (b) in meters.

Solution: From Table T.3 we can determine the unit fractions needed to work this problem.

a) $1\ 359 \text{ mi} = (1\ 359\ \cancel{\text{mi}})\left(\dfrac{1.6 \text{ km}}{1\ \cancel{\text{mi}}}\right) = (1\ 359)(1.6) \text{ km} = 2\ 174.4 \text{ km}$

b) Since 1 km = 1 000 m, 2 174.4 km = 2 174 400 m.

Area

The area enclosed in a square with 1 cm sides (Fig. A.2) is 1 cm × 1 cm = 1 cm². A square whose sides are 2 cm (Fig. A.3) has an area of 2 cm × 2 cm = 4 cm².

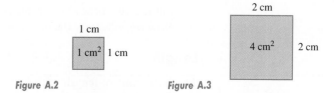

Figure A.2   Figure A.3

Areas are always expressed in square units, such as square centimeters, square kilometers, or square meters. When finding areas, be careful that all the numbers being multiplied are expressed in the same units.

In the metric system the square centimeter replaces the square inch. The square foot and square yard are replaced by the square meter. In the future you might purchase carpet or other floor covering by the square meter instead of by the square yard.

For measuring large land areas the metric system uses a square unit 100 meters on each side (a square hectometer). This unit is called a **hectare** (pronounced "hectair," symbolized ha). A hectare is about 2.5 acres. One square mile of land contains about 260 hectares.

Very large units of area are measured in square kilometers. One square kilometer is about $\frac{4}{10}$ of a square mile.

The chart on page A-9 illustrates a relationship between length, area, and volume in the metric system.

Given that 1 mile = 5280 feet can you change 1 mi³ to ft³? It would be done in the same manner as in the chart, but the numbers would not be nearly as convenient to work with.

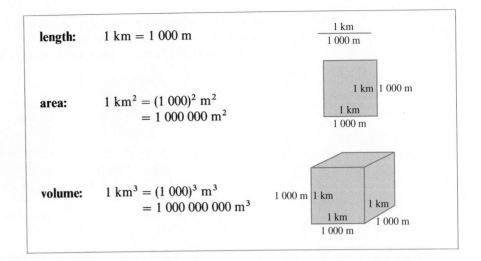

length: $1 \text{ km} = 1\,000 \text{ m}$

area: $1 \text{ km}^2 = (1\,000)^2 \text{ m}^2$
$= 1\,000\,000 \text{ m}^2$

volume: $1 \text{ km}^3 = (1\,000)^3 \text{ m}^3$
$= 1\,000\,000\,000 \text{ m}^3$

▶ **Example 7**

A farm in Ontario, Canada, has an area of 37.4 hectares. Find the area in acres.

Solution: From Table T.3 we know that 0.4 hectare = 1 acre.

$$37.4 \text{ ha} = (37.4 \text{ ha})\left(\frac{1 \text{ acre}}{0.4 \text{ ha}}\right) = \frac{37.4}{0.4} \text{ acres} = 93.5 \text{ acres}$$

Volume

When a figure has only two dimensions—length and width—we can find its area. When a figure has three dimensions—length, width, and height—we can find its volume. The volume of an item can be considered the space occupied by the item.

In the metric system, volume may be expressed in terms of liters or cubic meters, depending on what is being measured.

The volume of liquids is expressed in liters. A liter is a little larger than a quart. Liters are used in place of pints, quarts, and gallons. A liter can be divided into 1000 equal parts, each of which is called a milliliter. Figure A.4 illustrates a type of liter-container (a 1 000 m*l* graduated cylinder) that is often used in chemistry. Milliliters are used to express the volume of very small amounts of liquid. Drug dosages are often expressed in milliliters. An eight-ounce cup will hold about 240 m*l* of liquid.

The kiloliter, 1 000 liters, is used to represent the volume of large amounts of liquid. Tank trucks carrying gasoline to service stations hold about 10.5 kiloliters of gasoline.

Cubic meters are used to express the volume of large amounts of solid material. The volume of a dump truck's load of topsoil is measured in cubic

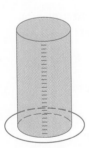

Figure A.4

meters. The volume of natural gas for heating a house may soon be measured in cubic meters instead of cubic feet.

The liquid in a liter container will fit exactly in a cubic decimeter (Fig. A.5).

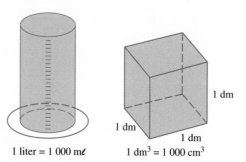

1 liter = 1 000 m𝓁   1 dm³ = 1 000 cm³

Figure A.5

Note that 1 liter = 1 000 m𝓁 and 1 cubic decimeter = 1 000 cm³. Since 1 liter = 1 cubic decimeter, *1 m𝓁 must equal 1 cm³*. Other useful facts are illustrated in Table T.4.

These conversion factors make conversions within the metric system much simpler than conversions in our own customary system. For example, how would you change cubic feet of water into gallons of water?

Nurses and doctors and all scientists should have a thorough understanding of the metric system because in medicine and the sciences all measurements are made in metric units. For example, a nurse may have to give a patient an injection of 3 cubic centimeters (3 cc) or 3 milliliters (3 m𝓁) of the drug ampicillin.

Table T.4

Volume in Cubic Units		Volume in Liters
1 cm³	=	1 m𝓁
1 dm³	=	1 𝓁
1 m³	=	1 k𝓁

▶ **Example 8**

Brian, a nurse, must administer 4 cc of codeine elixir to a patient.
a) How many milliliters of the drug will he administer?
b) How many ounces is this equivalent to?

Solution:
a) Since 1 cc = 1 m𝓁, he will administer 4 m𝓁 of the drug.
b) Since 1 fl oz = 30 m𝓁,

$$4 \text{ m}\ell = (4 \text{ m}\ell)\left(\frac{1 \text{ fl oz}}{30 \text{ m}\ell}\right) = \frac{4}{30} \text{ fl oz} \approx 0.13 \text{ fl oz}$$

Mass

Weight and mass are not the same. Mass is a measure of the amount of matter in an object. Mass is determined by the molecular structure of the object, and it will not change from place to place. Weight is a measure of the gravitational pull on an object. The gravitational pull of the earth is about six times as great as the gravitational pull of the moon. Thus a person on the moon weighs about $\frac{1}{6}$ as much as on earth, even though the person's mass remains the same. In space, where there is no gravity, a person has no weight.

Figure A.6

Figure A.7

The gravitational pull even on the earth varies from point to point. The closer you are to the center of the earth, the greater the gravitational pull. Thus a person weighs very slightly less on a mountain than in a nearby valley. Since the mass of an object does not vary with location, scientists use mass rather than weight.

Although weight and mass are not the same, on the earth they are proportional to each other (the greater the weight, the greater the mass). Therefore for our purposes we can treat weight and mass as the same.

Weight is measured on a spring scale (Fig. A.6). To measure mass, we use a balance and compare the object under consideration with a known mass (Fig. A.7).

The kilogram is the basic unit of mass in the metric system. It is a little more than 2 pounds. The official kilogram is a cylinder of platinum-iridium alloy kept by the International Bureau of Weights and Measures, located in Sevres, near Paris. Items that we measure in pounds are usually measured in kilograms. For example, an average-sized man has a mass of about 75 kilograms.

The gram (a unit that is 0.001 kilogram) is relatively small and is used in place of the ounce. A nickel has a mass of 5 grams, a cube of sugar has a mass of about 2 grams, and a large paper clip has a mass of about 1 gram.

The milligram is used extensively in the medical and scientific fields, as well as in the pharmaceutical industry. Practically all bottles of tablets are now labeled in either milligrams or grams.

The metric tonne (t) is used to express the mass of heavy items. A metric tonne is equal to 1 000 kilograms. It is a little larger than our customary ton of 2000 pounds. The mass of a large truck may be expressed in metric tonnes.

A kilogram of water has a volume of exactly 1 liter. In fact, a liter is defined to be the volume of one kilogram of water at a specified temperature and pressure. Thus mass and volume are easily interchangeable in the metric system. Converting from weight to volume is not nearly as convenient in our customary system. For example, how would you change pounds of water to cubic feet or gallons of water in our customary system?

Figure A.8 illustrates the relationship between volume of water in cubic

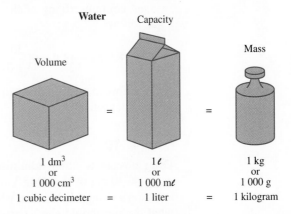

Figure A.8

decimeters, capacity in liters, and mass in kilograms. Table T.5 expands on this relationship between volume and mass of water.

Table T.5

Volume in Cubic Units		Volume in Liters		Mass of Water
1 cm^3	$=$	$1 \text{ m}\ell$	$=$	1 g
1 dm^3	$=$	1ℓ	$=$	1 kg
1 m^3	$=$	$1 \text{ k}\ell$	$=$	$1 \text{ t } (1\,000 \text{ kg})$

▶ **Example 9**

The label on a bottle of Vicks 44D Cough Syrup indicates that the active ingredient is dextromethorphan hydrobromide, and that 5 mℓ (1 teaspoon) contains 10 mg of this ingredient. If the recommended dosage for adults is 3 teaspoons determine

a) How many milliliters of cough medicine will be taken?
b) How many milligrams of the active ingredient will be taken?
c) If the bottle contains 8 fluid ounces of medicine, how many milligrams of the active ingredient are in the bottle?

Solution:

a) Since each teaspoon contains 5 mℓ and 3 teaspoons will be taken 15 mℓ of the cough medicine will be taken.

$$3 \text{ tsp} = (3 \text{ tsp})\left(\frac{5 \text{ m}\ell}{1 \text{ tsp}}\right) = 15 \text{ m}\ell$$

b) Since each teaspoon contains 10 mg of the active ingredient, 30 mg of the active ingredient will be taken.

$$3 \text{ tsp} = (3 \text{ tsp})\left(\frac{10 \text{ mg}}{1 \text{ tsp}}\right) = 30 \text{ mg}$$

c) From Table T.3 we see that each fluid ounce contains 30 mℓ. Since each 5 mℓ contains 10 mg of the active ingredient, we can work the problem as follows.

$$8 \text{ fl oz} = (8 \text{ fl oz})\left(\frac{30 \text{ m}\ell}{1 \text{ fl oz}}\right)\left(\frac{10 \text{ mg}}{5 \text{ m}\ell}\right) = \frac{8(30)(10)}{5} \text{ mg} = 480 \text{ mg}$$

Therefore there are 480 mg (or 0.48 g) of the active ingredient in the bottle of cough syrup.

▶ **Example 10**

Drug dosage is often administered according to a patient's weight. For example 30 mg of the drug Vancomicin is to be given for each kilogram of a person's weight. If Martha, who weighs 136 pounds, is to be given the drug, what dosage should she be given?

Figure A.9

Solution: First we need to convert Martha's weight into kilograms. From Table T.3 we see that 1 lb = 0.45 kilogram. We need to determine the number of milligrams of the drug for Martha's weight in kilograms. The answer may be found as follows.

$$136 \text{ lb} = (136 \text{ lb})\left(\frac{0.45 \text{ kg}}{1 \text{ lb}}\right)\left(\frac{30 \text{ mg}}{1 \text{ kg}}\right) = (136)(0.45)(30) \text{ mg} = 1\ 836 \text{ mg}$$

Thus 1 836 mg or 1.836 g of the drug should be given.

Temperature

The Celsius scale is used to measure temperatures in the metric system. Figure A.9 shows a thermometer with the Fahrenheit scale on the left and the Celsius scale on the right.

The Celsius scale was named for the Swedish astronomer Anders Celsius (1701–1744), who first devised it in 1742. On the Celsius scale, water freezes at 0°C and boils at 100°C. The Celsius thermometer in the past was called a "centigrade thermometer." Recall that *centi* means 1/100, and there are 100 degrees between the freezing point of water and the boiling point of water. Thus 1°C is 1/100 of this interval.

Table T.6 gives some common temperatures in both Celsius (°C) and Fahrenheit (°F). Table T.3 gives formulas for converting from degrees Fahrenheit to degrees Celsius, and from degrees Celsius to degrees Fahrenheit.

Table T.6		
−18°C	A very cold day	0°F
0°C	Freezing point of water	32°F
10°C	A warm winter day	50°F
20°C	A mild spring day	68°F
30°C	A warm summer day	86°F
37°C	Body temperature	98.6°F
100°C	Boiling point of water	212°F
177°C	Oven temperature for baking	350°F

▶ **Example 11**

At the doctor's office the nurse informs you that your temperature is 38.2° Celsius. What is your temperature in degrees Fahrenheit?

Solution:

$$F = \tfrac{9}{5}C + 32$$
$$F = \tfrac{9}{5}(38.2) + 32$$
$$F = 68.76 + 32 = 100.76$$

Therefore your temperature is about 101°F.

APPENDIX A EXERCISES

In Ex. 1–16, convert the quantities to the indicated units.
1. 80 lb to kg
2. 43 km to mi
3. 4.2 ft to m
4. 11 in. to cm
5. 15 yd² to m²
6. 28 mi to km
7. 150 kg to lb
8. 765 mm to in.
9. 675 ha to acres
10. 346 g to oz
11. 14.5 gal to ℓ
12. 27.4 m³ to ft³
13. 45.6 mℓ to fl oz
14. 1.6 km² to mi²
15. 120 lb to kg
16. 2.8 lb to kg

17. Without referring to any table, name as many of the metric system prefixes as you can and give their meanings. If you don't already know all the prefixes in Table T.1, memorize them now.

Fill in the blanks.
18. A meter is a little longer than a _____.
19. A kilogram is a little more than _____ pounds.
20. A nickel has a mass of about _____ grams.
21. The temperature on a warm day may be _____ °C.

In Ex. 22–27, fill in the blank with the appropriate letter, a–f.
22. deci _____
23. milli _____
24. hecto _____
25. kilo _____
26. centi _____
27. deka _____

a) 1000 times base unit
b) $\frac{1}{100}$ of base unit
c) $\frac{1}{1000}$ of base unit
d) 100 times base unit
e) $\frac{1}{10}$ of base unit
f) 10 times base unit

28. Cindy ran 100 meters, and Chuck ran 100 yards in the same length of time. Who ran faster? Explain.

Without referring to any of the tables or your notes, give the symbol and the equivalent in grams for each of the following units.
29. milligram
30. centigram
31. decigram
32. gram
33. dekagram
34. hectogram
35. kilogram

Fill in the missing values.
36. 2 m = _____ cm
37. 3 dam = _____ m
38. 15.7 hg = _____ g
39. 0.024 hℓ = _____ ℓ

40. 242.6 cm = _____ hm
41. 1.34 mℓ = _____ ℓ
42. 417 g = _____ kg
43. 14.27 hℓ = _____ ℓ
44. 1.34 hm = _____ cm
45. 0.076 mm = _____ m

Convert the given units to the units indicated.
46. 130 cm to hm
47. 8.3 m to cm
48. 1 049 mm to m
49. 94.5 kg to g
50. 3 472 mℓ to dℓ
51. 895 ℓ to mℓ
52. 14 000 mℓ to ℓ
53. 24 dm to km

Arrange each of the following in order from smallest to largest.
54. 64 dm, 67 cm, 680 mm
55. 5.6 dam, 0.47 km, 620 cm
56. 4.3 ℓ, 420 cℓ, 0.045 kℓ
57. 2.2 kg, 2 400 g, 24 300 dg
58. 0.045 kℓ, 460 dℓ, 48 000 cℓ

59. Charlie cut a piece of lumber 7.8 meters in length into six equal pieces. What was the length of each section (a) in meters? (b) in centimeters?
60. A baseball diamond is 27 meters along each side. (a) How many meters does a batter run if he hits a home run? (b) How many kilometers? (c) How many millimeters?
61. How many centimeters of picture molding should be purchased to frame two pictures? One picture is 33 centimeters by 440 millimeters, and the other is 3.4 decimeters by 44 centimeters. Molding is sold in centimeters.
62. At 70 cents a meter, what will it cost Karen to sew a lace border around a bedspread that measures 190 cm by 114 cm?
63. The filter pump on an aquarium circulates 360 milliliters of water every minute. If the aquarium holds 30 liters of water, how long will it take to circulate all the water?
64. Dale drove 1 200 kilometers and used 187 liters of gasoline. What was her average rate of gas use for the trip (a) in kilometers/liter? (b) in meters/liter?
65. A mixture of 15 grams of salt and 16 grams of baking soda is poured into 250 milliliters of water. What is the total mass of the mixture in grams?
66. The recommended dosage of the drug codeine for pediatric patients is 1 mg per each kg of a child's weight.

What dosage of codeine should be given to April, who weighs 56 pounds?

67. For each kilogram of a person's weight, 1.5 mg of the antibiotic drug gentamycin is to be administered. If Mr. Gigliotti weighs 170 pounds, how much of the drug should be given to him?

68. The recommended dosage of the drug ampicillin for pediatric patients is 200 mg per kilogram of a patient's weight. If Simpson weighs 76 pounds, how much ampicillin should he be given?

69. The recommended dosage of the drug theophylline is 5 mg per kg of weight. If Alex weighs 178 pounds, how much of the drug should he be given?

70. The recommended dosage of the generic antibiotic vancomycin is 40 mg per kg of a person's weight. If Maria weighs 125 pounds, how much of the drug should she be given?

71. For each kilogram of weight of a dog, 5 mg of the drug bretylium is to be given. If Blaster, an Irish setter, weighs 82 pounds, how much of the drug should be given?

72. The label on the bottle of Triaminic Expectorant indicates that each teaspoon (5 ml) contains 12.5 mg of the active ingredient phenylpropanolamine hydrochloride
  a) Determine the amount of the active ingredient in the recommended adult dosage of 2 teaspoons.
  b) Determine the quantity of the active ingredient in a 12 fluid ounce bottle.

73. The label on the bottle of Maximum Strength Pepto-Bismol indicates that each tablespoon contains 236 mg of the active ingredient bismuth subsalicyate.
  a) Determine the amount of the active ingredient in the recommended dosage of 2 tablespoons.
  b) If the bottle contains 8 fluid ounces, determine the quantity of the active ingredient in the bottle.

74. The label on a bottle of Robitussin Cough Syrup indicates that each teaspoon (5 ml) contains 100 mg of guaifenesin in a pleasant tasting syrup with 3.5% alcohol. If the recommended dosage is 2 teaspoons determine

  a) the amount of the active ingredient guaifenesin in the recommended dosage.
  b) the amount of alcohol, in milligrams, in the recommended dosage.
  c) the amount of guaifenesin in the 8 fl oz bottle.

75. If the temperature outside is 28°C, what is the Fahrenheit temperature?

76. If the room temperature is 68°F, what is the Celsius temperature?

77. If your outdoor thermometer shows a temperature of −6°F, what is the Celsius temperature?

78. If Ron's body temperature is 39°C, what is his Fahrenheit temperature?

Problem Solving

79. The following question was selected from a nursing exam. Can you answer it?

In caring for a patient after delivery, you are to give 0.2 mg Ergotrate Maleate. The ampule is labeled $\frac{1}{300}$ grain/mℓ. How much would you draw and give? (60 mg = 1 grain)
  a) 15 cc   **b)** 1.0 cc   **c)** 0.5 cc   **d)** 0.01 cc

80. Phil is planning a picnic and plans on purchasing 0.18 kg of ground beef for each 100 pounds of weight of guests who will be in attendance. If he expects 15 people whose average weight is 130 pounds, how many pounds of beef will he purchase?

Research Activities

81. Use an encyclopedia and other reference materials to determine how and why the different units of measurement were defined in the customary system.

82. Investigate the development of the metric system in Europe. Which groups of people had the most influence in developing the metric system?

APPENDIX B
SQUARES AND SQUARE ROOTS

No.	Sq.	Sq. Root	No.	Sq.	Sq. Root	No.	Sq.	Sq. Root	No.	Sq.	Sq. Root
1	1	1.000	31	961	5.568	61	3,721	7.810	91	8,281	9.539
2	4	1.414	32	1,024	5.657	62	3,844	7.824	92	8,464	9.592
3	9	1.732	33	1,089	5.745	63	3,969	7.937	93	8,649	9.644
4	16	2.000	34	1,156	5.831	64	4,096	8.000	94	8,836	9.695
5	25	2.236	35	1,225	5.916	65	4,225	8.062	95	9,025	9.747
6	36	2.449	36	1,296	6.000	66	4,356	8.124	96	9,216	9.798
7	49	2.646	37	1,369	6.083	67	4,489	8.185	97	9,409	9.849
8	64	2.828	38	1,444	6.164	68	4,624	8.246	98	9,604	9.899
9	81	3.000	39	1,521	6.245	69	4,761	8.307	99	9,801	9.950
10	100	3.162	40	1,600	6.325	70	4,900	8.367	100	10,000	10.000
11	121	3.317	41	1,681	6.403	71	5,041	8.426			
12	144	3.464	42	1,764	6.481	72	5,184	8.485			
13	169	3.606	43	1,849	6.557	73	5,329	8.544			
14	196	3.742	44	1,936	6.633	74	5,476	8.602			
15	225	3.873	45	2,025	6.708	75	5,625	8.660			
16	256	4.000	46	2,116	6.782	76	5,776	8.718			
17	289	4.123	47	2,209	6.856	77	5,929	8.775			
18	324	4.243	48	2,304	6.928	78	6,084	8.832			
19	361	4.359	49	2,401	7.000	79	6,241	8.888			
20	400	4.472	50	2,500	7.071	80	6,400	8.944			
21	441	4.583	51	2,601	7.141	81	6,561	9.000			
22	484	4.690	52	2,704	7.211	82	6,724	9.055			
23	529	4.796	53	2,809	7.280	83	6,889	9.110			
24	576	4.899	54	2,916	7.348	84	7,056	9.165			
25	625	5.000	55	3,025	7.416	85	7,225	9.220			
26	676	5.099	56	3,136	7.483	86	7,396	9.274			
27	729	5.196	57	3,249	7.550	87	7,569	9.327			
28	784	5.292	58	3,364	7.616	88	7,744	9.381			
29	841	5.385	59	3,481	7.681	89	7,921	9.434			
30	900	5.477	60	3,600	7.746	90	8,100	9.487			

ANSWERS

Chapter 1

Section 1.1 Exercises

1. a) 1, 2, 3, 4, 5, ... **b)** Counting numbers
3. A conjecture is a belief based on specific observations that has not been proven or disproven.
5. Deductive reasoning is the process of reasoning to a specific conclusion from a general statement.
7. $(1234 \times 9) + 5 = 11111$
9. $1 + 3 + 5 + 7 + 9 = 25$ **11.** 1 5 10 10 5 1
13. 4, 2, 0 **15.** $-1, 1, -1$ **17.** $\frac{1}{16}, \frac{1}{32}, \frac{1}{64}$ **19.** 34, 55, 89
21. 1234321 **23.** 81818181
25. Each time the number being multiplied by 9 is increased by 1, the tens digit increases by 1 and the units digit decreases by 1. (Also, the sum of the digits of each product is 9.)
27. a) 28, 36
b) To obtain the nth triangular number (after the first), add n to the previous triangular number.
c) No
29. Blue: 1, 5, 7, 10, 12 Red: 2, 4, 6, 9, 11 Yellow: 3, 8
31. a) You should obtain twice the original number.
b) You should obtain twice the original number.
c) The result is always twice the original number.
d) $n, 6n, 6n + 3, \dfrac{6n + 3}{3} = 2n + 1, 2n + 1 - 1 = 2n$
33. a) You should obtain the original number.
b) You should obtain the original number.
c) The result is always the original number.
d) $n, n + 5, \dfrac{n + 5}{5} = \dfrac{n}{5} + 1, \dfrac{n}{5} + 1 - 1 = \dfrac{n}{5}, \dfrac{n}{5} \cdot 5 = n$
35. A counterexample is 8.
37. $(3 + 5)$ is not divisible by 3.
39. a) $180°$ **b)** The sum of the measures should all be $180°$.
c) Sum of the measures of the interior angles of a triangle is $180°$.
41. b) The sum of the digits is 18.
c) The sum of the digits in the product of either a one-digit or two-digit number multiplied by 99 is 18.

Section 1.2 Exercises

(Answers in this section will vary depending on how you round your numbers. All answers are approximate.)

1. 860 **3.** 19 **5.** 110 **7.** 200 **9.** 100 **11.** $6.00
13. 20 lb **15.** $280 **17.** $480 **19.** 24 mpg **21.** $7.00
23. a) about 3.5 mi **b)** about 6 km
25. 2.4 mi **27. a)** 38% **b)** 90% **c)** 10%
29. 100 female lawyers and judges **31.** $2000 **33.** 18
35. 200 **37.** $120°$ **39.** 10% **41.** 9 **43.** 32 ft
45. 38 in.

Section 1.3 Exercises

1. 10.5 ft **3.** 12 min **5.** $11.65 **7.** $7.95 **9.** 11.5 m
11. $12.50 per week **13.** Between 5 and 6 hr **15.** 126 yd
17. $2.80 **19.** 7.5 oz **21.** 121 ft **23.** Snow ($24°$F)
25. 24 **27. a)** 4106.25 gal **b)** $17.25 **29.** 62 gal
31. $505.62
33. No, it only means that in this period there was a greater increase in pizza orders than burger orders.
35. 144 **37.** The area is 4 times as large.
39. $7 + 7 - (7 \div 7) = 13$ **41.** 36 squares
43. a) Place 5g, 1g, and 3g on one side and 9g on the other side.
b) Place 16g, 9g, and 3g on one side and 27g and 1g on the other side.
45. i) Place slice 1 and slice 2 in pan and cook for $\frac{1}{2}$ min.
ii) Turn slice 1 over and replace slice 2 with slice 3, cook $\frac{1}{2}$ min.
iii) Remove slice 1 and replace with slice 2 and turn slice 3 over and cook for $\frac{1}{2}$ min.
47.

8	6	16
18	10	2
4	14	12

49. Multiply the middle number by 3. **51.** $216 \, \text{cm}^3$ **53.** 4
55. U = 1, C = 5, A = 7, N = 2, D = 6, O = 3, T = 9, H = 8, I = 0, and S = 4. Other answers are possible.
57. Mary is the skier.

Review Exercises

1. 11, 13, 15 **2.** 25, 36, 49 **3.** 64, -128, 256
4. 25, 32, 40 **5.** 10, 4, -3 **6.** 21, 34, 55
7. a) The original number and the final number are the same.

b) The original number and the final number are the same.

c) The final number is the same as the original number.

d) n, $2n$, $2n + 10$, $\dfrac{2n + 10}{2} = n + 5$, $n + 5 - 5 = n$

8. This process will always result in an answer of 3.

9. $6^2 - 4^2 = 36 - 16 = 20$.

The answers to Ex. 10–23 will vary depending upon how you round the numbers. All answers are approximate.

10. 400,000,000 **11.** 30 **12.** 1570 **13.** 200 **15.** $8.00

16. $12.00 **17.** 3 mph **18.** $14.00 **19.** 2 mi

20. 4 units greater

21. Decreased from August 1990 to January 1991; increased from January 1991 to July 1991.

22. 12 sq units **23.** Length = 22 ft; height = 8 ft

24. $3.95 **25.** $1.70 **26.** $3.30 **27.** $1.20

28. Vendor A is cheaper by $20.00. **29.** $8.45 **30.** $3.44

31. 7.05 mg **32.** $665.00 **33.** 6 hr 45 min

34. July 26, 11:00 AM

35. a) 6.45 cm^2 **b)** 16.39 cm^3 **c)** 1 cm = 0.39 in.

36. 201 **37.** 6 **38.** 6

39. a) 6, 10, 15, 55 **b)** $\dfrac{n(n + 1)}{2}$ **c)** 210 **d)** 1275

40.

16	2	12
6	10	14
8	18	4

41.

21	7	8	18
10	16	15	13
14	12	11	17
9	19	20	6

42.

23	25	15
13	21	29
27	17	19

43.

120	140	40
20	100	180
160	60	80

44. a) 10,890 **b)** 10,890

c) The final result will be 10,890.

45. $25   Room
$ 3   Men
$ 2   Clerk
$30

46. a) 111 **b)** should be 111

c)
$$ABC$$
$$BCA$$
$$+ \; CAB$$
$$(A + B + C)100 + (B + C + A)10$$
$$+ \, (C + A + B)$$
$$= (A + B + C)(100 + 10 + 1)$$
$$= (A + B + C)(111)$$
$$\dfrac{(A + B + C)(111)}{A + B + C} = 111$$

47. 2.57 qt **48.** Higher if no change in salary.

49.

50. 140 lb **51.** 59 min 59 sec

52. Same amount of wine in red cup as water in blue cup

53. 125,250 **54.** All numbers between 1 and -1, except 0.

55. $3 + 5 + 7 + 9 = 24$ (There are other possible answers, for example $11 + 13 = 24$.)

56. 40 posts

57.

A	B	C	D	E
E	A	B	C	D
D	E	A	B	C
C	D	E	A	B
B	C	D	E	A

58. Place six coins in each pan with one coin off to the side. If it balances, then the heavier coin is the one on the side. If the pan does not balance, then take the six coins on the heavier side and split them into two groups of three. Select the three heavier coins and weigh two coins. If the pan balances, then it is the third coin. If the pan does not balance, you can identify the heavier coin.

59. If each number is written out in words, then the numbers of letters in the words form a magic square, each row, column, and diagonal having a sum of 27.

60.

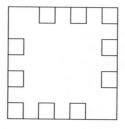

61. a) 2 **b)** 6 **c)** 24 **d)** 120

e) $n(n - 1)(n - 2) \ldots 1$, (or $n!$), (where n = the number of people in line).

Chapter 1 Test

1. 16, 20, 24 **2.** $\frac{1}{81}$, $\frac{1}{243}$, $\frac{1}{729}$

3. a) The result is the original number plus one.

b) The result is the original number plus one.

c) The result will always be the original number plus one.

d) n, $5n$, $5n + 10$, $\dfrac{5n + 10}{5} = n + 2$, $n + 2 - 1 = n + 1$

The answers for Ex. 4–6 are approximate.

4. 360 **5.** 30,000,000 **6.** 4

7. a) 42% **b)** 44% **c)** 56%

8. 113 therms **9.** $75 original cost of jacket

10. $7\frac{1}{2}$ min **11.** Approximately 70 in. by 50 in.

12. $579.38

13.

40	15	20
5	25	45
30	35	10

14. 4 times larger

15. Less time if she had driven at 45 mph.

16. $2 \cdot 6 \cdot 8 \cdot 9 \cdot 13$; 11 does not divide 11,232.

17. $\frac{161}{13} \approx 12.4$ sec

Chapter 2

Section 2.1 Exercises

1. Not well defined **3.** Well defined **5.** Well defined

7. Infinite **9.** Infinite **11.** Infinite **13.** Finite

15. {Africa, Antarctica, Asia, Australia, Europe, North America, South America}

17. $\{166, 167, 168, \ldots, 396\}$ **19.** $\{13\}$

21. {Arizona, California, New Mexico, Texas}

23. $\{3, 4, 5, 6, \ldots, 98\}$ **25.** $\{x \mid x \in N \text{ and } 1 \le x \le 6\}$

27. $\{x \mid x \in N \text{ and } x \text{ is odd}\}$ **29.** $\{x \mid x \in N \text{ and } x \text{ is even}\}$

31. $\{x \mid x \text{ is Chrysler, General Motors, or Ford}\}$

33. A is the set of natural numbers less than 8.

35. L is the set of Great Lakes in the United States.

37. S is the set of the seven dwarfs.

39. E is the set that contains no elements.

41. False **43.** True **45.** True **47.** False **49.** 4

51. 0 **53.** Both **55.** Neither **57.** Equivalent

59. a) A is the set of natural numbers greater than 2. B is the set of all numbers greater than 2.

b) Set A contains only natural numbers. Set B contains other types of numbers, including fractions and decimals.

c) $A = \{3, 4, 5, 6, \ldots\}$

d) Cannot write set B in roster form

Section 2.2 Exercises

1. False **3.** True **5.** True **7.** False **9.** True

11. True **13.** False **15.** True **17.** None

19. $A = B$, $A \subseteq B$, $B \subseteq A$ **21.** $B \subseteq A$, $B \subset A$

23. $A = B$, $A \subseteq B$, $B \subseteq A$ **25.** $\{\ \}$

27. $\{\ \}$, {pizza}, {cola}, {pizza, cola}

29. $\{\ \}$, $\{a\}$, $\{b\}$, $\{c\}$, $\{d\}$, $\{a, b\}$, $\{a, c\}$, $\{a, d\}$, $\{b, c\}$, $\{b, d\}$, $\{c, d\}$, $\{a, b, c\}$, $\{a, b, d\}$, $\{a, c, d\}$, $\{b, c, d\}$, $\{a, b, c, d\}$

31. Number of proper subsets $= 2^n - 1$ **33.** True

35. True **37.** True **39.** True **41.** True **43.** True

45. 10 **47.** $2^n = 128$

Section 2.3 Exercises

1.

3. A' is the set of all married people in the state of Texas in 1993 who have been married for 25 years or more.

5. The set of college students with a major in English and history

7. The set of college students who do not have a major in English or do have a major in biology

9. The set of college students who have a major in English, history, and biology

11. The set of cities in the state of North Carolina with a population over 50,000 or with a philharmonic orchestra

13. The set of cities in the state of North Carolina without a professional sports team and with a philharmonic orchestra

15. The set of cities in the state of North Carolina that do not have a population over 50,000 or do not have a professional sports team

17. The set of students at the University of Iowa that have a major in agriculture or who grew up on a farm

19. The set of students at the University of Iowa who do not have a major in agriculture and who did not grow up on a farm

21. The set of students at the University of Iowa who have a major in agriculture or who grew up on a farm or who are not residents of the state of Iowa

23. $\{2, 3, 4, 6\}$ **25.** $\{2, 3, 4, 5, 6, 7, 8, 9, 10\}$

27. $\{2, 3\}$ **29.** $\{7, 8\}$ **31.** {IL, NE, IN, IA, MN}

33. {IL, NE, IN, IA, MN, ND, SD, WI, KS, MI, MT, OH, OK, TX}

35. {IA, MN} **37.** {KS, MI, MT, OH, OK, TX}

39. $\{1, 2, 3, 4, 5, 6, 8\}$ **41.** $\{1, 5, 7, 8\}$ **43.** $\{7\}$

45. $\{\ \}$ **47.** $\{7\}$ **49.** $\{0, \square, \triangle, \#, \alpha, \beta\}$ **51.** $\{\alpha, \beta, \#\}$

53. $\{0, \square, \triangle\}$ **55.** U **57.** $\{0, \square, \triangle, \#\}$ **59.** $\{\alpha, \beta\}$

61. $\{\ \}$ **63.** $\{b, d, e, h, j, k\}$ **65.** $\{a, f, i\}$
67. $\{a, b, c, d, f, g, i\}$ **69.** $\{a, c, d, e, f, g, h, i, j, k\}$
71. $\{b\}$ **73. a)** $n(A \cup B) = n(A) + n(B) - n(A \cap B)$

$$8 = 4 + 6 - 2$$
$$8 = 8$$

75. A **77.** B **79.** C **81.** $\{2, 6, 10, 14, 18, \ldots\}$
83. $\{2, 6, 10, 14, 18, \ldots\}$ **85.** U **87.** A **89.** $\{\ \}$ **91.** A
93. $B \subseteq A$ **95.** A and B are disjoint sets. **97.** $A \subseteq B$
99.

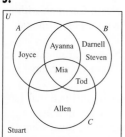

101.

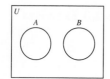

Section 2.4 Exercises

1.

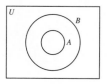

3.

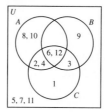

5.

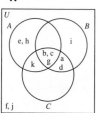

7. VI **9.** III **11.** VIII
13. IV **15.** I **17.** III
19. VIII **21.** VIII
23. III **25.** III **27.** VI
29. $\{1, 2, 3, 4, 5, 6\}$
31. $\{4, 5, 6, 7, 8, 10\}$
33. $\{3, 4, 5\}$

35. $\{1, 2, 3, 6, 9, 10, 11, 12\}$
37. $\{1, 2, 3, 4, 5, 6, 7, 8, 9, 12\}$ **39.** $\{9, 11, 12\}$
41. $\{7, 8, 9, 10, 11, 12\}$
43.

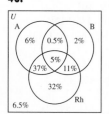

45. No **47.** No **49.** No **51.** No **53.** Yes **55.** No
57. Yes **59.** Yes **61.** No **63. a)** Yes **c)** No
65.

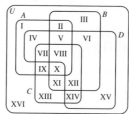

67. a)

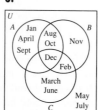

b) I $A \cap B' \cap C' \cap D'$
II $A \cap B \cap C' \cap D'$
III $A' \cap B \cap C' \cap D'$
IV $A \cap B' \cap C' \cap D$
V $A \cap B \cap C' \cap D$
VI $A' \cap B \cap C' \cap D$
VII $A \cap B' \cap C \cap D$
VIII $A \cap B \cap C \cap D$
IX $A \cap B' \cap C \cap D'$
X $A \cap B \cap C \cap D'$
XI $A' \cap B \cap C \cap D'$
XII $A' \cap B \cap C \cap D$
XIII $A' \cap B' \cap C \cap D'$
XIV $A' \cap B' \cap C \cap D$
XV $A' \cap B' \cap C' \cap D$
XVI $A' \cap B' \cap C' \cap D'$

Section 2.5 Exercises

1.

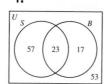

a) 57
b) 17
c) 53

3.

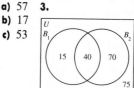

a) 15
b) 75
c) Yes

5.

a) 361
b) 176
c) 30
d) 297
e) 514

7.

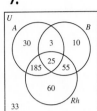

a) 185
b) 10
c) 25
d) 401

9.

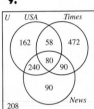

a) 472
b) 1192
c) 932
d) 58
e) 692

11.

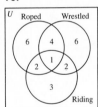

Only 99 people accounted for out of 100 interviewed

13.

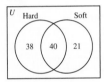

a) 6
b) 15
c) 24
d) 15

15.

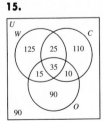

a) 410
b) 35
c) 90
d) 50

17. **a)** Color: yellow   blue   red   ivory   green
    b) Nationality: Norwegian   Afghanistani   Senegalese   Spaniard   Japanese
    c) Food: apple   cheese   banana   peach   fish
    d) Drink: vodka   tea   milk   whiskey   ale
    e) Pet: fox   horse   snail   dog   zebra
    f) The zebra's owner drinks ale.

Section 2.6 Exercises

1. $\{3, 4, 5, 6, \ldots, n + 2, \ldots\}$
$\;\;\downarrow\;\downarrow\;\downarrow\;\downarrow\qquad\;\;\downarrow$
$\{4, 5, 6, 7, \ldots, n + 3, \ldots\}$

3. $\{2, 3, 4, 5, \ldots, n + 1, \ldots\}$
$\;\;\downarrow\;\downarrow\;\downarrow\;\downarrow\qquad\;\;\downarrow$
$\{3, 4, 5, 6, \ldots, n + 2, \ldots\}$

5. $\{4,\;7, 10, 13, \ldots, 3n + 1, \ldots\}$
$\;\;\downarrow\;\;\downarrow\;\;\downarrow\;\;\downarrow\qquad\quad\downarrow$
$\{7, 10, 13, 16, \ldots, 3n + 4, \ldots\}$

7. $\{\;8\;,\;10, 12, 14, \ldots, 2n + 6, \ldots\}$
$\;\;\downarrow\quad\downarrow\;\;\downarrow\;\;\downarrow\qquad\quad\downarrow$
$\{10, 12, 14, 16, \ldots, 2n + 8, \ldots\}$

9. $\left\{1, \dfrac{1}{3}, \dfrac{1}{5}, \dfrac{1}{7}, \ldots, \dfrac{1}{2n - 1}, \ldots\right\}$
$\;\;\downarrow\;\downarrow\;\downarrow\;\downarrow\qquad\;\;\downarrow$
$\left\{\dfrac{1}{3}, \dfrac{1}{5}, \dfrac{1}{7}, \dfrac{1}{9}, \ldots, \dfrac{1}{2n + 1}, \ldots\right\}$

11. $\{1, 2, 3, 4, \ldots, n, \ldots\}$
$\;\;\downarrow\;\downarrow\;\downarrow\;\downarrow\qquad\;\downarrow$
$\{3, 6, 9, 12, \ldots, 3n, \ldots\}$

13. $\{1, 2, 3, 4, \ldots,\quad n\quad, \ldots\}$
$\;\;\downarrow\;\downarrow\;\downarrow\;\downarrow\qquad\;\;\downarrow$
$\{4, 6, 8, 10, \ldots, 2n + 2, \ldots\}$

15. $\{1, 2, 3, 4, \ldots,\quad n\quad, \ldots\}$
$\;\;\downarrow\;\downarrow\;\downarrow\;\downarrow\qquad\;\;\downarrow$
$\{2, 5, 8, 11, \ldots, 3n - 1, \ldots\}$

17. $\{1, 2, 3, 4, \ldots,\quad n\quad, \ldots\}$
$\;\;\downarrow\;\downarrow\;\downarrow\;\downarrow\qquad\;\;\downarrow$
$\{5, 8, 11, 14, \ldots, 3n + 2, \ldots\}$

19. $\{1, 2, 3, 4, \ldots,\quad n\quad, \ldots\}$
$\;\;\downarrow\;\downarrow\;\downarrow\;\downarrow\qquad\;\;\downarrow$
$\left\{\dfrac{1}{3}, \dfrac{1}{4}, \dfrac{1}{5}, \dfrac{1}{6}, \ldots, \dfrac{1}{n + 2}, \ldots\right\}$

21. $\{1, 2, 3, 4, \ldots, n, \ldots\}$
$\;\;\downarrow\;\downarrow\;\downarrow\;\downarrow\qquad\;\;\downarrow$
$\{1, 4, 9, 16, \ldots, n^2, \ldots\}$

23. $\{1, 2, 3, 4, \ldots, n, \ldots\}$
$\quad\downarrow\downarrow\downarrow\quad\downarrow\qquad\downarrow$
$\{3, 9, 27, 81, \ldots, 3^n, \ldots\}$

Review Exercises

1. True **2.** False **3.** False **4.** False **5.** False
6. True **7.** False **8.** True **9.** True **10.** True
11. True **12.** True **13.** True **14.** True
15. $\{7, 8, 9, 10, 11, 12\}$ **16.** $\{14, 15, 16, 17, \ldots\}$
17. {California, Oregon, Washington}
18. $\{357, 358, 359, \ldots, 752\}$
19. $\{x \mid x \in N \text{ and } 13 \le x \le 18\}$
20. $\{x \mid x \in N \text{ and } x > 38\}$ **21.** $\{x \mid x \in N \text{ and } 17 \le x \le 25\}$
22. $\{x \mid x \text{ is a mode of transportation}\}$
23. A is the set of letters in the English alphabet from E through M inclusive.
24. B is the set of United States coins with a value less than a dollar.
25. C is the set of the three commercial television networks in the United States with the largest number of viewers.
26. D is the set of English teachers who own a red car and do not live in Nebraska.
27. $\{4\}$ **28.** $\{1, 2, 3, 4, 5, 7\}$ **29.** $\{6, 8, 9\}$ **30.** $\{2, 4, 6, 7\}$
31. 16 **32.** 31
33.

34. $\{a, c, d, e, f, g, i, k\}$ **35.** $\{i, d\}$
36. $\{a, b, c, d, e, f, g, i, k\}$ **37.** $\{e\}$ **38.** $\{a, d, e\}$
39. $\{a, b, d, e, f, g\}$ **40.** True **41.** True **42.** \$450
43. 315
44. 10
45. 30
46. 110

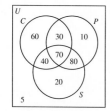

47. 13
48. 11
49. 35
50. 4
51. 1

52. $\{2, 4, 6, 8, \ldots, \quad 2n, \ldots\}$
$\quad\downarrow\downarrow\downarrow\downarrow\qquad\downarrow$
$\{4, 6, 8, 10, \ldots, 2n + 2, \ldots\}$

53. $\{3, 5, 7, 9, \ldots, 2n + 1, \ldots\}$
$\quad\downarrow\downarrow\downarrow\downarrow\qquad\downarrow$
$\{5, 7, 9, 11, \ldots, 2n + 3, \ldots\}$

54. $\{1, 2, 3, 4, \ldots, \quad n, \ldots\}$
$\quad\downarrow\downarrow\downarrow\downarrow\qquad\downarrow$
$\{5, 8, 11, 14, \ldots, 3n + 2, \ldots\}$

55. $\{1, 2, 3, 4, \ldots, \quad n, \ldots\}$
$\quad\downarrow\downarrow\downarrow\downarrow\qquad\downarrow$
$\{4, 9, 14, 19, \ldots, 5n - 1, \ldots\}$

Chapter 2 Test

1. True **2.** False **3.** True **4.** True **5.** True
6. True **7.** True **8.** True **9.** False **10.** True
11. $A = \{x \mid x \in N \text{ and } x < 6\}$
12. A is the set of the first 5 natural numbers.
13. $\{2, 4, 6, 8, 10, 12\}$ **14.** $\{6, 10\}$ **15.** $\{6\}$ **16.** 2
17. **18.** No **19.**

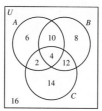

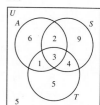

20. 9 **21.** 5 **22.** 25 **23.** 2
24. $\{3, 4, 5, 6, \ldots, n + 2, \ldots\}$
$\quad\downarrow\downarrow\downarrow\downarrow\qquad\downarrow$
$\{4, 5, 6, 7, \ldots, n + 3, \ldots\}$

25. $\{1, 2, 3, 4, \ldots, n, \ldots\}$
$\quad\downarrow\downarrow\downarrow\downarrow\qquad\downarrow$
$\{2, 4, 6, 8, \ldots, 2n, \ldots\}$

Chapter 3

Section 3.1 Exercises

1. Compound, conjunction, $\wedge$
3. Compound, biconditional, $\leftrightarrow$
5. Compound, disjunction, $\vee$ **7.** Simple
9. Compound, negation, $\sim$
11. Compound, conjunction, $\wedge$

13. Compound, conditional, →
15. Compound, conditional, →   17. No spiders are black.
19. All llamas have tails.
21. Some plants are not living things.
23. All symphony conductors use a baton.
25. Some people like squash.   27. All soda contains sugar.
29. Some nurses do not have a degree.
31. $p \land q$ 33. $p \lor \sim q$ 35. $p \land \sim q$ 37. $p \to q$
39. $\sim p \land \sim q$ 41. The horse is not five years old.
43. The horse is five years old and the horse won the race.
45. If the horse is not five years old, then the horse won the race.
47. It is false that the horse won the race or the horse is five years old.
49. The horse is not five years old or the horse has not won the race.
51. $(p \land q) \lor r$ 53. $p \to (q \lor \sim r)$ 55. $(\sim p \leftrightarrow \sim q) \lor \sim r$
57. $(r \leftrightarrow q) \land p$ 59. $(r \lor \sim q) \leftrightarrow p$
61. The sun is shining or we will go to the lake, and we will go swimming.
63. If we go to the lake then the sun is shining, or we will go swimming.
65. If we do not go swimming, then we will go to the lake and the sun is shining.
67. If we go swimming then we will go to the lake, and the sun is shining.
69. We will go to the lake if and only if the sun is shining, and we will go swimming.
71. $p \land q$ (p: Each language has a beauty of its own; q: Each language has forms of expression which are duplicated nowhere else.)
73. $(p \land q) \land r$ (p: The Constitution requires that the president be at least 35 years old; q: The Constitution requires that the president be a natural-born citizen; r: The Constitution requires that the president be a resident of the country for 14 years.)
75. $p \to q$ (p: you want to kill an idea in the world today; q: get a committee working on it.)
77. $\sim p \land q$ (p: I know the dignity of their birth; q: I know the glory of their death.)

Section 3.2 Exercises

1. F   3. T   5. T
   F   F   F
         T   T
         T   T
7. F   9. T
   F   T
   T   F
   F   T

11. T   13. T   15. T   17. $p \land q$ 19. $p \land \sim q$
T   F   T   T   F
T   F   T   F   T
T   F   F   F   F
T   F   T   F
T   F   F
F   F   T
T   F   F

21. $(p \land q) \land \sim r$ 23. $p \land (\sim q \land r)$ 25. $(\sim p \land q) \lor \sim q$
F   F   F
T   F   T
F   T   T
F   F   T
F   F
F   F
F   F
F   F

27. False   29. False   31. True   33. True   35. False
37. True   39. False   41. True   43. True   45. True

47. $\sim p \land q$ 49. $p \lor \sim q$
F   T
F   T
T   F
F   T
True in case 3 when p is false and q is true.   True in cases 1, 2, and 4, when p is true, or when p and q are both false.

51. $(p \land q) \lor r$ 53. $q \lor (p \land \sim r)$
T   T
T   T
T   F
F   T
T   T
F   T
T   F
F   F
True in cases 1, 2, 3, 5, and 7. True except when p, q, r have truth values of TFF or FTF or FFF.   True in cases 1, 2, 4, 5, and 6. True except when p, q, r have truth values TFT, FFT, or FFF.

55. Mr. Vapmek does not qualify since he makes less than $40,000. Ms. Bell qualifies. Mrs. Spatz qualifies.
57. Mr. James qualifies, the other four do not. Ms. Vela is returning on April 2. Mr. Davis is returning on a Monday. Ms. Chang is not staying over on a Saturday. Ms. Ghandi is returning on a Monday.

Section 3.3 Exercises

1. T   3. F   5. F   7. T   9. F

T	T	T	T	T
T	T	T	F	T
F	F	F	T	T

11. T   13. T   15. T   17. T   19. T

T	F	T	T	T
T	F	F	T	T
F	T	F	F	F
T	F	F	T	T
T	T	T	F	T
T	F	T	T	F
T	T	F	F	T

21. $p \rightarrow (\sim q \wedge r)$ 23. $(p \leftrightarrow q) \vee \sim r$ 25. $(p \rightarrow q) \vee \sim r$

21.	23.	25.
F	T	T
F	T	T
T	F	F
F	T	T
T	F	T
T	T	T
T	T	T
T	T	T

27. A tautology is a compound statement that is always true.
29. An implication is a conditional statement that is a tautology.
31. Tautology   33. Tautology   35. Neither
37. Implication   39. Not implication   41. Implication
43. True   45. True   47. True   49. True   51. False
53. True   55. True   57. False   59. True   61. True
63. True   65. True   67. No
69. Helen Hurry—chemist; Sam Swift—historian; Nick Nickelson—mathematician; Peter Perkins—biologist.
71. a) T F F F T F F F T F F F T T T T
    b) T F F T T T T T T T T T T T T T
(answer listed horizontally to save space)

Section 3.4 Exercises

1. Equivalent statements are statements that have exactly the same truth values in the answer columns of the truth tables.
3. The two statements are equivalent.
5. Not equivalent   7. Not equivalent   9. Equivalent
11. Not equivalent   13. Equivalent   15. Equivalent
17. Equivalent   19. Equivalent   21. Not equivalent
23. Not equivalent   27. Equivalent   29. Equivalent
31. Not equivalent   33. Not equivalent
35. Not equivalent   37. Not equivalent
39. Neil Diamond did not sing "Cherry Cherry" or AC/DC did not sing "Cherry Cherry."

41. It is not true that the new carpet is not treated with Scotch Guard and it is a good carpet.
43. It is not true that the figure is a triangle or a square.
45. If the house does not need painting, then it is not true that the siding needs repair and we will not be able to sell the house.
47. The wheat is not yellow or it is ready to be cut.
49. The frog does not have legs or it is no longer a tadpole.
51. If the mail is delivered today, then today is not Sunday.
53. A figure is a quadrilateral if and only if it has four sides.
55. If the bird flies then it has wings, and if the bird has wings then it flies.

Section 3.5 Exercises

1. a) $q \rightarrow p$ b) $\sim p \rightarrow \sim q$ c) $\sim q \rightarrow \sim p$
3. a and b   5. a, b, and c   7. b and c   9. a and c
11. If you can buy the paper bag, then you have 50 cents.
13. If it does not snow, then I will go hunting.
15. If one rents an apartment, then one buys rental insurance.
17. If one hang glides, then one buys a parachute.
19. If you get an A, then I will eat my hat.
21. If one goes to work, then one gets paid.
23. Converse: If we party, then today is Friday.
Inverse: If today is not Friday, then we will not party.
Contrapositive: If we do not party, then today is not Friday.
25. Converse: If I pass the course, then the score on the test is greater than 70.
Inverse: If the score on the test is not greater than 70, then I will not pass the course.
Contrapositive: If I do not pass the course, then the score on the test is not greater than 70.
27. Converse: If I am a monkey's uncle, then the pear tree bears fruit.
Inverse: If the pear tree does not bear fruit, then I am not a monkey's uncle.
Contrapositive: If I am not a monkey's uncle, then the pear tree doesn't bear fruit.
29. Converse: If I enjoy the food, then the restaurant is clean.
Inverse: If the restaurant is not clean, then I will not enjoy the food.
Contrapositive: If I do not enjoy the food, then the restaurant is not clean.
31. Converse: If the opposite sides are parallel and equal, then the figure is a rhombus.
Inverse: If the figure is not a rhombus, then the opposite sides are not parallel or not equal.
Contrapositive: If the opposite sides are not parallel or are not equal, then the figure is not a rhombus.

33. If the sum of the digits is divisible by 3, then the number is divisible by 3. True

35. If 2 divides the units digit of the counting number, then 2 divides the counting number. True

37. If two lines are not parallel, then the two lines intersect in at least one point. True

39. If the product of a and b is an even counting number, then a and b are both even counting numbers. False

41. Yes **43.** No

Section 3.6 Exercises

1. The conclusion necessarily follows from the premises.

3. If $(p_1 \wedge p_2) \to c$ is a tautology, then the argument is valid.

5. $p \to q$ **7.** $p \vee q$ **9.** $p \to q$
$q \to r$ $\sim p$ $\sim p$
$\therefore p \to r$ $\therefore q$ $\therefore \sim q$

11. Valid **13.** Valid **15.** Invalid **17.** Invalid
19. Valid **21.** Valid **23.** Valid **25.** Invalid
27. Valid **29.** $c \to f$ **31.** $s \to p$ **33.** $r \vee p$
c p $p \to \sim r$
$\therefore f$ $\therefore s$ $\therefore \sim p$
Valid   Invalid   Invalid

35. $\sim e$ **37.** $b \vee h$ **39.** $f \to t$
$e \to m$ $\sim h \to b$ $t \to e$
$\therefore m$ $\therefore h \vee b$ $\therefore f \to e$
Invalid   Valid   Valid

41. $\sim p \vee r$ **43.** $b \to p$ **45.** $b \wedge c$
p $\sim p$ $\sim c \vee b$
$\therefore r$ $\therefore \sim b$ $\therefore \sim b$
Valid   Valid   Invalid

47. $f \to d$ **49.** $l \to p$
$d \to \sim s$ $u \to p$
$\therefore f \to s$ $\therefore l \to u$
Invalid   Invalid

51. Therefore, you will be healthy.

53. Therefore, John fixes the car.

55. Therefore, the dog is not brown.

57. Therefore, if you overdraw your checking account, then you will have to pay a fee.

59. Therefore, if the logs are not dry, then you will have to dry the logs. (or, Therefore, if you do not have to dry the logs, then the logs are dry.)

Section 3.7 Exercises

1. The conclusion necessarily follows from the premises.

3. Valid **5.** Valid **7.** Fallacy **9.** Valid **11.** Fallacy

13. Fallacy **15.** Fallacy **17.** Valid **19.** Fallacy
21. Fallacy **23.** Valid **25.** Valid

Review Exercises

1. Some wood is not hard. **2.** Some birds are yellow.

3. All apples are red. **4.** No squirrels fly.

5. All people wear glasses. **6.** Some rabbits wear glasses.

7. I opened a can of soda and I spilled some soda.

8. I did not open a can of soda or the soda is red.

9. If I opened a can of soda, then I spilled some soda and the soda is not red.

10. I opened a can of soda if and only if I spilled some soda.

11. I opened a can of soda or I did not spill some soda, and the soda is not red.

12. I did not open a can of soda, if and only if the soda is red and I did not spill some soda.

13. $q \vee \sim r$ **14.** $p \wedge r$ **15.** $r \to (\sim p \vee q)$

16. $(q \leftrightarrow p) \wedge \sim r$ **17.** $(r \wedge p) \vee q$ **18.** $\sim (q \vee r)$

19. T **20.** F **21.** T **22.** F **23.** T **24.** F **25.** T **26.** F
   T   F   T   T   F   T   T   T
   F   F   T   T   T   F   T   T
   F   T   F   F   T   F   T   T
                        F   T   T   T
                        F   T   F   T
                        F   T   F   T
                        F   T   T   T

27. True **28.** True **29.** True **30.** True **31.** True
32. True **33.** True **34.** False **35.** False **36.** True
37. Equivalent **38.** Not equivalent **39.** Equivalent
40. Not equivalent **41.** Not equivalent **42.** Equivalent
43. a and b **44.** a and b

45. If the pencil is yellow, then the desk is orange. (or another answer is: It is false that the pencil is yellow and the desk is not orange.)

46. It is false that the Flyers do not score goals or the Flyers do not win games.

47. New Orleans is not in California and Chicago is a city.

48. The water is not blue or the fish will not bite.

49. It is not true that I went to the party or I finished my special report.

50. If you do not have to stop, then the railroad crossing light is not flashing red.

51. If the repair costs are not high, then the car is not an expensive car.

52. If I am not at work, then today is a holiday.

53. If it is not Sam's computer, then it does not have a color monitor or it does not have an extended keyboard.

54. Converse: If I get a good grade, then I studied.
Inverse: If I do not study, then I will not get a good grade.

Contrapositive: If I do not get a good grade, then I did not study.
55. Invalid **56.** Valid **57.** Invalid **58.** Invalid

Chapter 3 Test

1. $(q \vee p) \wedge r$ **2.** $(q \wedge r) \to p$ **3.** $\sim(r \leftrightarrow p)$
4. It is false that if the house is red then the owner is not a lawyer.
5. The house is red, if and only if the car is blue and the owner is a lawyer.
6. T **7.** F **8.** T

T	F	T
F	T	T
T	F	T
T	F	T
	F	F
	F	T
	F	T
	F	F

9. True **10.** False **11.** True **12.** True
13. Equivalent **14.** Equivalent **15.** Equivalent
16. $s \to f$ **17.** Valid
$f \to p$
$\therefore s \to p$
Valid
18. Some birds are not black. **19.** No people are funny.
20. Inverse: If the apple is not red, then it is not a delicious apple.
Converse: If it is a delicious apple, then the apple is red.
Contrapositive: If it is not a delicious apple, then the apple is not red.

Chapter 4

Section 4.1 Exercises

1. 313 **3.** 1512 **5.** 334,214 **7.** 99∩∩∩|||||
9. ⌄999999∩∩∩∩∩∩∩|||||||
11. ∝)))))⌄⌄9999∩∩∩∩|
13. 64 **15.** 117 **17.** 1892 **19.** 1999 **21.** 7602
23. 10,679 **25.** XXXVIII **27.** CXLVIII
29. MCMLXXVIII **31.** MMDCVI
33. MMMMCCCXXI or $\overline{\text{IV}}$CCCXXI **35.** 57 **37.** 1976
39. 2605 **41.** 四十二 **43.** 二百四十八 **45.** 九千七
47. 364 **49.** 42,505 **51.** 9007
53. γ

55. $\rho\kappa\gamma$ **57.** $\lambda'\epsilon'\upsilon\iota\beta$
63. Advantage: Ciphered system is a more compact method to write numbers.
Disadvantage: In ciphered system, a larger number of symbols must be memorized.
65. 1021, MXXI, 千二十一, $\alpha'\kappa\alpha$

67. 527, 99999∩∩|||||, DXXVII, $\phi\kappa\zeta$

Section 4.2 Exercises

1. $(4 \times 10) + (8 \times 1)$
3. $(9 \times 100) + (4 \times 10) + (2 \times 1)$
5. $(3 \times 1000) + (4 \times 100) + (5 \times 10) + (2 \times 1)$
7. $(6 \times 10,000) + (4 \times 1000) + (5 \times 100) + (2 \times 10) + (1 \times 1)$
9. $(2 \times 100,000) + (4 \times 10,000) + (5 \times 1000) + (6 \times 100) + (7 \times 10) + (2 \times 1)$
11. 32 **13.** 724 **15.** 5528 **17.** ▼◄ **19.** ▼▼ ▼
21. ▼ ▼▼▼▼ ◄◄◄◄▼▼▼ **23.** 70 **25.** 6841
27. 4000 **29.** ═ **31.** ═ **33.** ⠇⠿

35. There is no clear way of indicating the place value without a place holder.
39. 1944, ◄◄◄▼▼ ◄◄▼▼▼▼

Section 4.3 Exercises

1. 5 **3.** 7 **5.** 11 **7.** 79 **9.** 553 **11.** 1367 **13.** 193
15. 83 **17.** 4135 **19.** 1000_2 **21.** 10111_2 **23.** 626_8
25. $E93_{12}$ **27.** 1022_6 **29.** $1ET_{12}$ **31.** 1111110011_2
33. 4550_8 **35.** 1844 **37.** 27,963 **39.** 133_{16} **41.** 1566_{16}
43. 11111001000_2 **45.** 30432_5 **47.** 3710_8
49. The numeral is written incorrectly, there is no 6 in the set of numerals for base 5.
51. Written correctly **53. b)** 10213_5 **c)** 1373_8 **55.** 6

Section 4.4 Exercises

1. 122_4 **3.** 11412_5 **5.** $6T3_{12}$ **7.** 2100_3 **9.** 24001_7
11. 100_4 **13.** 1041_5 **15.** 325_{12} **17.** 11_2 **19.** 3616_7
21. 231_5 **23.** 3504_7 **25.** 20212_4 **27.** 5503_9
29. 110001_2 **31.** 42_6R4 **33.** 63_8R4 **35.** 86_{12}R1
37. 13_5R4

Section 4.5 Exercises

1. 540 **3.** 1160 **5.** 9503 **7.** 6150 **9.** 1825
11. 20,440 **13.** 900 **15.** 162,056 **17.** 112 **19.** 1316

21. 625 **23.** 18,760

Review Exercises

1. 2101 **2.** 2211 **3.** 2022 **4.** 2114 **5.** 3214 **6.** 2312
7. bbbbaaa **8.** cbbbbbbaaaaaaaa
9. cccbbbbbbbbbaaaaaaaaaa
10. dccccccccccbbbbbbbaaaaaaaa **11.** dddddddaaaa
12. dddccccbba **13.** 93 **14.** 27 **15.** 648 **16.** 273
17. 7482 **18.** 9574 **19.** dxc **20.** $ayfxg$ **21.** $dyhxi$
22. $aziygxh$ **23.** fzd **24.** bza **25.** 93 **26.** 407
27. 8298 **28.** 46,883 **29.** 70,082 **30.** 60,529 **31.** mc
32. sog **33.** vgi **34.** $BAph$ **35.** LDxqe **36.** QFvrf
37. ↊99999∩∩∩∩∩II   **38.** MCDLXII
39. 千
四
百
六
十
二

40. $\alpha'\upsilon\xi\beta$ **41.** ≪≪IIII ≪≪II **42.** ••••
                                                  ••

43. 122,025 **44.** 8254 **45.** 585 **46.** 1991 **47.** 1277
48. 1971 **49.** 34 **50.** 5 **51.** 79 **52.** 1459 **53.** 1451
54. 186 **55.** 3323_5 **56.** 2051_6 **57.** 111001111_2
58. 13033_4 **59.** 327_{12} **60.** 717_8 **61.** 137_8
62. 100011_2 **63.** 135_{12} **64.** 1023_7 **65.** 13101_5
66. 10423_8 **67.** 3611_7 **68.** 100_2 **69.** $3E4_{12}$
70. 3324_5 **71.** 1704_8 **72.** 1022_3 **73.** 202_5 **74.** 1232_4
75. 5050_{12} **76.** 12221_3 **77.** 110111_2 **78.** 13632_8
79. 1011_2 **80.** 112_4 **81.** 30_5 **82.** 433_6 **83.** 230_6R4
84. 664_8R2 **85.** 3408 **86.** 3408 **87.** 3408

Chapter 4 Test

1. A number is a quantity; a numeral is the symbol used
 to represent the quantity.
2. 1231 **3.** 1999 **4.** 1991 **5.** 9835 **6.** 99999∩∩IIIIIII
7. — **8.** ≪≪≪≪≪≪↑I ≪≪III **9.** $\omega o\epsilon$
    •
  ◯

10. *Additive system:* Symbols are added to form numbers.
 Multiplicative system: Numbers are multiplied by the
 powers of the base, where the powers of the base are
 illustrated next to the number. *Ciphered system:* A
 unique numeral is used to represent each number. The
 numerals are placed next to one another to obtain the
 number. *Place value system:* Each number is multiplied
 by a power of the base. The position of the number
 indicates the power of the base by which it is multiplied.
11. 14 **12.** 51 **13.** 605 **14.** 21 **15.** 100110_2
16. 243_5 **17.** 943_{12} **18.** 7072_8 **19.** 1110_3 **20.** 227_8
21. 2412_7 **22.** 255 **23.** 4862

Chapter 5

Section 5.1 Exercises

1. A prime number is a natural number greater than 1
 that has exactly two factors (or divisors): itself and one.
3. A conjecture is a belief that has not been proven or
 disproved.
5. a is a divisor of b if the quotient of b divided by a has
 a remainder of 0.
7. T **9.** T **11.** T **13.** F **15.** T **17.** 2, 3, 4, 6
19. 3, 5 **21.** 2, 3, 4, 5, 6, 8, 10
23. 60 (Other answers are possible.) **25.** $3 \cdot 13$ **27.** $3 \cdot 19$
29. $2^3 \cdot 3 \cdot 13$ **31.** $2 \cdot 3^3 \cdot 19$ **33.** $5 \cdot 97$ **35.** $2^3 \cdot 3 \cdot 83$
37. 15 **39.** 14 **41.** 30 **43.** 4 **45.** 8 **47.** 180
49. 168 **51.** 1800 **53.** 5088 **55.** 384 **57.** T
59. 11 and 13, 17 and 19 **65.** 29 days
67. a) All do.
 b) Every prime number greater than 3 differs by 1
  from a multiple of the number 6.
69. 5 **71.** 2 **73.** 30 **75.** Not a perfect number
77. Perfect number **79.** Yes **81.** Yes
83. No **85.** No
87. The sum of the groups that have the same three digits
 will always be divisible by 3 (i.e., $d + d + d = 3d$ and
 $3 \mid 3d$)
89. Yes

Section 5.2 Exercises

1. 2 **3.** 9 **5.** -4 **7.** -1 **9.** -7 **11.** -2
13. -12 **15.** -3 **17.** -7 **19.** -2 **21.** -35
23. 49 **25.** 60 **27.** -75 **29.** -360 **31.** 2 **33.** -4
35. -5 **37.** -3 **39.** -20 **41.** T **43.** T
45. F **47.** F **49.** F **51.** 5 **53.** -23 **55.** -9
57. -3 **59.** -6 **61.** $-4, -2, -1, 0, 3, 5$
63. $-6, -5, -4, -3, -2, -1$
65. 23 points **67.** 14,777 **69.** $+1$
71. $\dfrac{-a}{-b} = \dfrac{\cancel{-1}}{\cancel{-1}} \cdot \dfrac{a}{b} = \dfrac{a}{b}$
73. -1
75. $0 + 1 - 2 + 3 + 4 - 5 + 6 - 7 - 8 + 9 = 1$ (Other
 answers are possible.)

Section 5.3 Exercises

1. Terminating **3.** Repeating **5.** Repeating
7. Repeating **9.** Repeating **11.** 0.25 **13.** $0.\overline{714285}$
15. $0.\overline{5}$ **17.** 4.0 **19.** $5.\overline{3}$ **21.** $\frac{5}{10} = \frac{1}{2}$ **23.** $\frac{345}{1000} = \frac{69}{200}$
25. $\frac{3006}{1000} = \frac{1503}{500}$ **27.** $\frac{1246}{1000} = \frac{623}{500}$ **29.** $\frac{20,678}{10,000} = \frac{10,339}{5000}$
31. $\frac{4}{9}$ **33.** $\frac{2}{1}$ **35.** $\frac{19}{11}$ **37.** $\frac{1085}{333}$ **39.** $\frac{109}{90}$ **41.** $\frac{24}{35}$

43. $\frac{2}{5}$ **45.** $\frac{49}{81}$ **47.** $\frac{36}{35}$ **49.** $\frac{5}{14}$ **51.** $\frac{3}{16}$ **53.** $\frac{12}{23}$
55. $\frac{13}{27}$ **57.** $\frac{24}{53}$ **59.** $\frac{4}{11}$ **61.** $\frac{11}{30}$ **63.** $\frac{23}{130}$ **65.** $\frac{23}{54}$
67. $\frac{17}{144}$ **69.** $\frac{-109}{600}$ **71.** $\frac{17}{12}$ **73.** $\frac{41}{28}$ **75.** $\frac{76}{96} = \frac{19}{24}$
77. $\frac{23}{60}$ **79.** $\frac{4}{11}$ **81.** $\frac{23}{42}$

In Ex. 83–89 other answers are possible.
83. 0.475 **85.** -1.055 **87.** 3.1234505 **89.** 4.8725
91. $\frac{5}{14}$ **93.** $\frac{3}{40}$ **95.** $\frac{9}{40}$ **97.** $\frac{11}{200}$
99. Yes, between every two rational numbers there exists another rational number.
103. \$25.59 **105.** She saves $\frac{2}{15}$ of her income.
107. $1\frac{7}{8}$ yd **109.** $3\frac{1}{2}$ points **111.** $26\frac{1}{16}$ in. **113.** \$200

Section 5.4 Exercises

3. Rational **5.** Rational **7.** Irrational **9.** Rational
11. Irrational **13.** 8 **15.** 7 **17.** -13 **19.** -15
21. -9 **23.** Natural number, integer, rational number
25. Natural number, integer, rational number
27. Irrational number **29.** Rational number
31. Rational number **33.** 7 **35.** $3\sqrt{5}$ **37.** $5\sqrt{5}$
39. $4\sqrt{3}$ **41.** $3\sqrt{6}$ **43.** $5\sqrt{5}$ **45.** $-4\sqrt{5}$
47. $-13\sqrt{3}$ **49.** $4\sqrt{3}$ **51.** $23\sqrt{2}$ **53.** $\sqrt{6}$ **55.** $4\sqrt{2}$
57. $6\sqrt{2}$ **59.** $\sqrt{2}$ **61.** 3 **63.** $\dfrac{3\sqrt{2}}{2}$ **65.** $\dfrac{\sqrt{14}}{2}$ **67.** 2
69. $\dfrac{3\sqrt{2}}{2}$ **71.** $\dfrac{\sqrt{15}}{3}$ **73.** T **75.** F **77.** F
79. $\sqrt{2} + (-\sqrt{2}) = 0$ **81.** $\sqrt{3} \cdot \sqrt{3} = 3$
83. No, $\sqrt{11}$ is an irrational number and $3.316\overline{6}$ is a rational number.
85. No, 3.14 and $\frac{22}{7}$ are rational numbers, π is an irrational number.
87. ≈ 21.6 ft

Section 5.5 Exercises

1. Closed **3.** Closed **5.** Closed **7.** Not closed
9. Closed **11.** Closed **13.** Not closed **15.** Not closed
17. Closed **19.** Not closed **21.** Commutative property
23. $(-3) + (-4) = (-4) + (-3)$ **25.** No, $3 - 2 \neq 2 - 3$
27. No, $3 \div 2 \neq 2 \div 3$
29. $[(-3) + (-4)] + (-5) = (-3) + [(-4) + (-5)]$
31. No, $(1 - 2) - 3 \neq 1 - (2 - 3)$
33. No, $(5 - 3) - 2 \neq 5 - (3 - 2)$
35. No, $3 + (5 \cdot 6) \neq (3 + 5) \cdot (3 + 6)$
37. Commutative property of multiplication

39. Distributive property of multiplication over addition
41. Commutative property of addition
43. Commutative property of addition
45. Distributive property of multiplication over addition
47. Commutative property of addition
49. Commutative property of multiplication **51.** $4x + 20$
53. $2\sqrt{3} + 3\sqrt{2}$ **55.** $x\sqrt{3} + 3$
57. Distributive property of multiplication over addition
59. Distributive property of multiplication over addition
61. Commutative property of addition
63. Distributive property of multiplication over addition
65. Associative property of addition **67.** No **69.** Yes
71. No **73.** No

Section 5.6 Exercises

1. 16 **3.** 4 **5.** 1 **7.** $\frac{25}{36}$ **9.** 576 **11.** 27 **13.** $\frac{1}{8}$
15. 1 **17.** 16 **19.** $\frac{1}{4}$ **21.** 729 **23.** 25 **25.** $\frac{1}{216}$
27. 81 **29.** 3 **31.** 5.5×10^4 **33.** 9.0×10^2
35. 5.3×10^{-2} **37.** 1.9×10^4 **39.** 1.86×10^{-6}
41. 4.23×10^{-6} **43.** 1.07×10^2 **45.** 1.53×10^{-1}
47. 3100 **49.** 60,000,000 **51.** 0.0000213 **53.** 0.312
55. 9,000,000 **57.** 231 **59.** 35,000 **61.** 10,000
63. 120,000,000 **65.** 0.0153 **67.** 320 **69.** 0.0021
71. 20 **73.** 4.2×10^{12} **75.** 4.5×10^{-7} **77.** 2.0×10^3
79. 2.0×10^{-7} **81.** 3.0×10^8
83. 9.2×10^{-5}, 1.3×10^{-1}, 8.4×10^3, 6.2×10^4
85. 11.95 hr **87.** 3.2×10^7 sec **89.** 1.8×10^4
91. 1 mg $= 10^{-6}$ kg or 1 kg $= 10^6$ mg
93. a) $\approx 5.87 \times 10^{12}$ mi **b)** 500 sec or 8 min 20 sec

Section 5.7 Exercises

1. 4, 7, 10, 13, 16 **3.** -4, 1, 6, 11, 16 **5.** 6, 4, 2, 0, -2
7. 24, 16, 8, 0, -8 **9.** $\frac{1}{2}$, 1, $\frac{3}{2}$, 2, $\frac{5}{2}$ **11.** $a_5 = 36$
13. $a_8 = 52$ **15.** $a_{10} = -138$ **17.** $a_{15} = \frac{139}{2}$
19. $a_n = 4n$ **21.** $a_n = 10n - 2$ **23.** $a_n = n - \frac{7}{2}$
25. $a_n = \frac{5}{2}n - \frac{15}{2}$ **27.** $a_n = -\frac{1}{2}n - \frac{5}{2}$ **29.** 78 **31.** 110
33. -36 **35.** -85.5 **37.** -35 **39.** 3, 6, 12, 24, 48
41. 5, -10, 20, -40, 80 **43.** -2, 2, -2, 2, -2
45. -2, -1, $-\frac{1}{2}$, $-\frac{1}{4}$, $-\frac{1}{8}$ **47.** 5, 3, $\frac{9}{5}$, $\frac{27}{25}$, $\frac{81}{125}$
49. $a_6 = 486$ **51.** $a_5 = 48$ **53.** $a_4 = 216$ **55.** $a_3 = \frac{3}{4}$
57. $a_5 = 8$ **59.** $a_n = 2(2)^{n-1}$ or $a_n = 2^n$
61. $a_n = 1(3)^{n-1}$ or $a_n = 3^{n-1}$ **63.** $a_n = (-8)(1/2)^{n-1}$
65. $a_n = (-6)(-3)^{n-1}$
67. $a_n = (-4)(2/3)^{n-1}$ **69.** $s_4 = 30$
71. $s_5 = 2343$ **73.** $s_5 = 484$ **75.** $s_5 = -33$
77. $s_{50} = 2550$ **79.** $s_{20} = 630$
81. a) $a_{12} = 63$ in. **b)** $s_{12} = 954$ in. **83.** $s_{31} = \$496$
85. $a_{15} = \$45,218.00$ (or \$45,218.08, depending on the capacity of the calculator)

87. $a_5 = 12.288$ ft **89.** $s_6 = 161.4375$ **91.** $r = 3, a_1 = 8$

Section 5.8 Exercises

3. 0.618 **7. a)** 1.618 **b)** 0.618 **c)** 1.00
9. Alternate increasing then decreasing.
15. a) 3 in. × 5 in. and 5 in. × 8 in. **19.** No **21.** No
23. Yes, $\frac{21}{2}$, 17
29. c) 13, the number of reflections form a Fibonacci sequence.

Review Exercises

1. 2, 3, 4, 6, 8 **2.** 2, 3, 4, 6, 9 **3.** $2^6 \cdot 3$ **4.** $2^4 \cdot 3 \cdot 5$
5. $2^2 \cdot 3^2 \cdot 5$ **6.** $2^2 \cdot 3^2 \cdot 5 \cdot 7$ **7.** $2^6 \cdot 3 \cdot 5$
8. GCD = 4, LCM = 160 **9.** GCD = 5, LCM = 240
10. GCD = 4, LCM = 7992
11. GCD = 40, LCM = 6720
12. GCD = 4, LCM = 480 **13.** GCD = 36, LCM = 432
14. −1 **15.** 2 **16.** −2 **17.** −5 **18.** −7 **19.** 1
20. 4 **21.** 4 **22.** −6 **23.** 12 **24.** −15 **25.** −24
26. 2 **27.** −4 **28.** 6 **29.** 6 **30.** 3 **31.** 0.625
32. 0.8 **33.** 0.75 **34.** 3.75 **35.** $0.\overline{428571}$ **36.** $0.41\overline{6}$
37. 0.375 **38.** 0.875 **39.** $0.\overline{285714}$ **40.** $\frac{624}{1000} = \frac{78}{125}$
41. $\frac{2}{3}$ **42.** $\frac{243}{100}$ **43.** $\frac{61}{33}$ **44.** $\frac{12,083}{1000}$ **45.** $\frac{21}{5000}$ **46.** $\frac{97}{45}$
47. $\frac{232}{99}$ **48.** $\frac{2506}{495}$

In Ex. 49–54 other answers are possible.

49. 0.00425 **50.** $\frac{27}{40}$ **51.** 2.4065 **52.** $\frac{19}{24}$ **53.** 0.505
54. 0.25 **55.** $\frac{13}{15}$ **56.** $\frac{1}{10}$ **57.** $\frac{41}{36}$ **58.** $\frac{28}{45}$ **59.** $\frac{35}{54}$
60. $\frac{53}{28}$ **61.** $\frac{1}{6}$ **62.** $\frac{13}{40}$ **63.** $\frac{8}{15}$ **64.** $2\sqrt{3}$ **65.** $6\sqrt{2}$
66. $4\sqrt{2}$ **67.** $-3\sqrt{3}$ **68.** $8\sqrt{2}$ **69.** $-20\sqrt{3}$
70. $8\sqrt{3}$ **71.** $3\sqrt{2}$ **72.** $4\sqrt{3}$ **73.** 3 **74.** $2\sqrt{7}$
75. $\frac{3\sqrt{2}}{2}$ **76.** $\frac{\sqrt{15}}{5}$ **77.** $15 + 5\sqrt{5}$ **78.** $4\sqrt{3} + 3\sqrt{2}$
79. Commutative property of addition
80. Commutative property of multiplication
81. Associative property of addition
82. Distributive property of multiplication over addition
83. Commutative property of addition
84. Commutative property of addition
85. Associative property of multiplication
86. Commutative property of multiplication
87. Distributive property of multiplication over addition

88. Commutative property of multiplication
89. Closed **90.** Closed **91.** Not closed **92.** Closed
93. Not closed **94.** Not closed **95.** 8 **96.** $\frac{1}{9}$ **97.** 5
98. 125 **99.** 1 **100.** $\frac{1}{125}$ **101.** 64 **102.** 81
103. 3.2×10^3 **104.** 4.23×10^{-5} **105.** 1.68×10^{-3}
106. 4.95×10^6 **107.** 420 **108.** 0.0000387
109. 0.000175 **110.** 100,000 **111.** 6.4×10^2
112. 1.38×10^5 **113.** 2.1×10^1 **114.** 3.0×10^0
115. 33,600,000,000 **116.** 22.5 **117.** 3200 **118.** 5.0
119. 25 **120.** Arithmetic, 23, 28
121. Geometric, −324, 972 **122.** Arithmetic, −16, −20
123. Geometric, $\frac{1}{16}, \frac{1}{32}$ **124.** Arithmetic, 16, 19
125. Geometric, −2, 2 **126.** 4, 9, 14, 19, 24
127. −6, −9, −12, −15, −18
128. $-\frac{1}{2}, -\frac{5}{2}, -\frac{9}{2}, -\frac{13}{2}, -\frac{17}{2}$ **129.** 4, 8, 16, 32, 64
130. 16, 8, 4, 2, 1 **131.** $\frac{1}{2}, -\frac{1}{4}, \frac{1}{8}, -\frac{1}{16}, \frac{1}{32}$ **132.** 14
133. −36 **134.** 92 **135.** 216 **136.** $\frac{1}{4}$ **137.** −48
138. −84 **139.** −25 **140.** 632 **141.** 52 **142.** 170
143. 33 **144.** Arithmetic, $a_n = -3n + 10$
145. Arithmetic, $a_n = 5n - 5$
146. Arithmetic, $a_n = -\frac{3}{2}n + \frac{11}{2}$
147. Geometric, $a_n = 3(2)^{n-1}$
148. Geometric, $a_n = 4(-1)^{n-1}$
149. Geometric, $a_n = 5\left(\frac{1}{3}\right)^{n-1}$ **150.** Yes, 21, 34 **151.** No
152. No **153.** Yes, 26, 42

Chapter 5 Test

1. 2, 3, 4, 6, 8, 9 **2.** $2^3 \cdot 3^2 \cdot 5$ **3.** −6 **4.** −22 **5.** −50
6. 0.4355 **7.** $\frac{17}{144}$ **8.** 0.375 **9.** $\frac{49}{20}$ **10.** $\frac{121}{240}$ **11.** $\frac{19}{30}$
12. $8\sqrt{2}$ **13.** $\frac{\sqrt{15}}{3}$
14. Closed; the product of any two integers is an integer.
15. Associative property of addition
16. Distributive property of multiplication over addition
17. 64 **18.** 32 **19.** $\frac{1}{49}$ **20.** 8.0×10^6 **21.** $a_n = -4n$
22. −187 **23.** 243 **24.** 1023 **25.** $3(2)^{n-1}$
26. 1, 1, 2, 3, 5, 8, 13, 21

Chapter 6

Section 6.1 Exercises

1. 16 **3.** -64 **5.** 686 **7.** 11 **9.** 28 **11.** 15
13. $-\frac{83}{18}$ **15.** $\frac{5}{2}$ **17.** 0 **19.** 0 **21.** No **23.** Yes
25. No **27.** Yes **29.** Yes **31. a)** 240 mi **b)** 336 mi
33. $4280 **35.** 88.72 min

Section 6.2 Exercises

1. $-2x$ **3.** $5x - 7$ **5.** $5x - 2y + 3$ **7.** $-8x + 2$
9. $-5x + 3$ **11.** $-1.1x + 7.4$ **13.** $\frac{1}{12}x - 2$
15. $10x - 5y + 4$ **17.** $7r - 20$ **19.** $-2.2x + 5$
21. $-\frac{7}{20}x + \frac{17}{15}$ **23.** $4.52x - 13.5$ **25.** 6 **27.** 1
29. 44/9 **31.** 2/3 **33.** -2 **35.** 3 **37.** 21 **39.** -15
41. No solution **43.** All real numbers **45.** 56/11 **47.** 3
49. 12 **51.** $50.38 **53.** $1770.93 **55.** $247.06
57. $199.26 **59.** 0.3 cc **61. b)** -1
63. a) An equation that has no solution
   b) You will get a false statement like $0 = 3$.

Section 6.3 Exercises

1. 240 **3.** 32 **5.** 4 **7.** 45 **9.** 54 **11.** 2 **13.** 2000
15. 6 **17.** 5 **19.** 2 **21.** 500 **23.** 233 **25.** $-3/2$
27. 4 **29.** 0.5 **31.** 14
33. $y = \dfrac{-3x + 5}{2}$ or $y = -\frac{3}{2}x + \frac{5}{2}$
35. $y = \dfrac{-4x + 14}{7}$ or $y = -\frac{4}{7}x + 2$
37. $y = \dfrac{2x + 6}{3}$ or $y = \frac{2}{3}x + 2$
39. $y = \dfrac{-4x - 6}{3}$ or $y = -\frac{4}{3}x - 2$
41. $y = \dfrac{5x - 3z - 2}{7}$ or $y = \frac{5}{7}x - \frac{3}{7}z - \frac{2}{7}$
43. $t = d/r$ **45.** $c = p - a - b$ **47.** $l = \dfrac{v}{wh}$ **49.** $d = \dfrac{c}{\pi}$
51. $b = y - mx$ **53.** $l = \dfrac{p - 2w}{2}$ or $l = \frac{1}{2}p - w$
55. $c = 3A - a - b$ **57.** $h = \dfrac{3V}{B}$ **59.** $C = \frac{5}{9}(F - 32)$
61. $m = \dfrac{y_2 - y_1}{x_2 - x_1}$ **63.** $1500 **65.** 3052.08 in.^{3}
67. 256 gametes **69.** 18.8 mg **71.** $\approx$$11,325,227,000,000
73. 59.3%

Section 6.4 Exercises

1. $7 - 3x$ **3.** $6y + 8$ **5.** $3r - 10$ **7.** $2(5 - x)$
9. $\dfrac{6 + x}{7}$ **11.** $(7y - 5) + 11$ **13.** $x + 8 = 24, x = 16$
15. $\dfrac{x}{7} = 5, x = 35$ **17.** $11 + 2x = 9, x = -1$
19. $3x + 4 = 16, x = 4$ **21.** $x + 11 = 3x + 1, x = 5$
23. $x + 3 = 5(x + 7), x = -8$
25. $6200 + 500y = 12{,}200$; 12 yr
27. $c + 0.06c = 380$, $358.49
29. $x + x + 2x = 84$, two younger: 21 hr, older 42 hr
31. $20 = 2l + 2(\frac{2}{3}l)$, $l = 6$ ft, $w = 4$ ft
33. $2x + x + (x + 50) = 6170$, $B = 1530$, $A = 3060$,
   $C = 1580$
35. $c - 0.2c = 18$, $22.50
37. $186 = 2(w + 9) + 2w$, $w = 42$ ft, $l = 51$ ft
39. $65m = 760$, ≈ 12 months
41. $22 = 6h$, 4 or more hours
43. $\dfrac{r}{2} + 0.07r = 227$, $426.13
45. $0.75 + 0.50(2)(m - 1) = 6.25$, where m = minutes;
   $6\frac{1}{2}$ min; or $0.75 + 0.5h = 6.25$, where h = half minutes,
   then add 1 min to get $6\frac{1}{2}$ min

Section 6.5 Exercises

1. When both sides of an inequality are being multiplied
   or divided by a negative number, the order of the
   inequality must be reversed.
3. Yes
5.
7.
9.
11.
13.
15.
17.
19.
21.
23.

25.

2 3 4 5 6 7 8

27.

−9 −8 −7 −6 −5 −4

29.

−16 −15 −14 −13 −12 −11

31.

−21 −20 −19 −18 −17 −16

33.

−2 −1 0 1 2 3

35.

−3 −2 −1 0 1 2

37.

−2 −1 0 1 2

39.

−1 0 1 2 3 4 5

41.

2 3 4 5 6

43. $94 \le x \le 100$, assuming 100 is the highest grade possible
45. $x + 3x < 21$, $B = 5$ hr, $A = 15$ hr
47. a) $180 + 60x \le 1200$ **b)** 17 boxes **49.** $x \le \$15.57$
51. $81.6 \le x \le 100$, assuming 100 is the highest grade possible

Section 6.6 Exercises

1. Plotting points, using intercepts, and using the slope and the y intercept

Exercises 3, 5, 7, 9

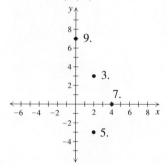

Exercises 11, 13, 15, 17

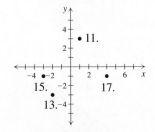

19. $(4, 3)$ **21.** $(-2, 4)$ **23.** $(-3, 1)$ **25.** $(-5, -3)$
27. $(2, -3)$ **29. a)** $(2, 5)$ **b)** 10 sq. units
31. $(7, 2)$ or $(-1, 2)$ **33.** $b = -4$ **35.** $b = 3$
37. $(2, -10), (-1, 5)$ **39.** $(0, 5), (-\frac{5}{2}, 0)$ **41.** $(0, \frac{4}{3}), (8, 0)$
43. $(0, 1), (-4, -2)$

45.
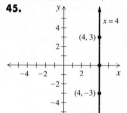

47.
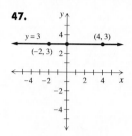

49.
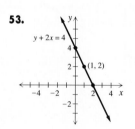

51.
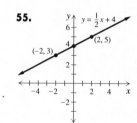

53.
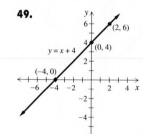

55.
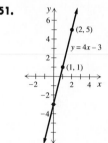

57.
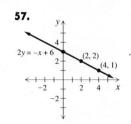

59.
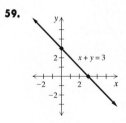

61.
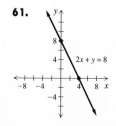

63.
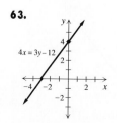

65.

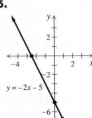

67.

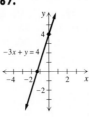

69. $-\frac{1}{3}$ **71.** $\frac{6}{5}$ **73.** 0 **75.** $-\frac{3}{8}$ **77.** $-\frac{4}{3}$

79.

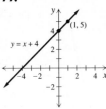

81.

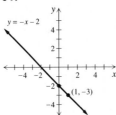

83.

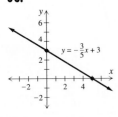

85.

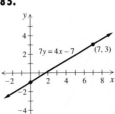

87.

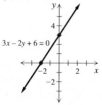

89. $y = \frac{1}{2}x - 1$ **91.** $y = -\frac{2}{3}x - 2$

93. a)

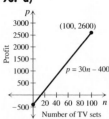

b) About $1700
c) About 13

95. a)

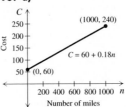

b) About $170
c) About 400 mi

97. a) Solve the equations for y to put them in slope intercept form. Then compare the slopes and y intercepts. If the slopes are equal but the y intercepts are different, then the lines are parallel.

b) The lines are parallel.

Section 6.7 Exercises

1.

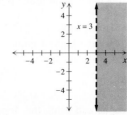

3.

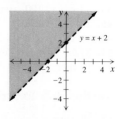

5.

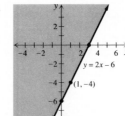

7.

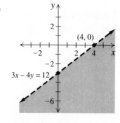

9.

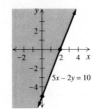

11.

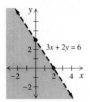

13.

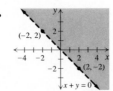

15.

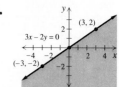

17.

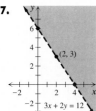

$3x + 2y = 12$

19.

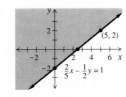

$(5, 2)$

$\frac{2}{5}x - \frac{1}{2}y = 1$

21.

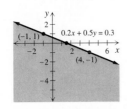

$0.2x + 0.5y = 0.3$

$(-1, 1)$

$(4, -1)$

23. a) $2l + 2w \le 30,\ 0 \le l \le 15,\ 0 \le w \le 15$

b)

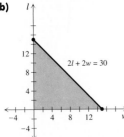

$2l + 2w = 30$

5. If a vertical line touches more than one point on the graph, then for each value of x there is not a unique value for y, and the graph is not a function.

7. Function, domain: $x = -2, 1, 2, 3$; range: $y = -3, -1, 1, 2$

9. Function, domain: all real numbers, $\mathbb{R}$; range: all real numbers, $\mathbb{R}$

11. Function, domain: $\mathbb{R}$, range: $y = 2$

13. Function, domain: $\mathbb{R}$, range $y \ge -4$

15. Function, domain: $0 \le x \le 8$, range: $-1 \le y \le 1$

17. Function, domain: $0 \le x \le 12$, range: $y = 1, 2, 3$

19. Not a function **21.** Function, domain: $\mathbb{R}$, range: $y > 0$

23. Not a function **25.** 9 **27.** 18 **29.** -14 **31.** 52

33. 4 **35.** -8 **37.** -48

39. a) Upward

b) $x = 0$

c) $(0, 1)$

d) $(0, 1)$

e) No x intercepts

g) Domain: $\mathbb{R}$, range: $y \ge 1$

f)

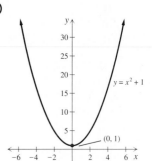

$y = x^2 + 1$

$(0, 1)$

41. a) Downward

b) $x = 0$

c) $(0, -9)$

d) $(0, -9)$

e) No x intercepts

g) Domain: $\mathbb{R}$, range: $y \le -9$

f)

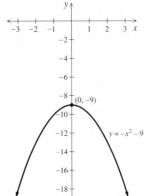

$(0, -9)$

$y = -x^2 - 9$

Section 6.8 Exercises

3. $(x + 2)(x + 6)$ **5.** $(x - 4)(x + 3)$ **7.** $(x + 6)(x - 4)$

9. $(x + 5)(x - 2)$ **11.** $(x - 7)(x - 3)$ **13.** $(x - 8)(x + 1)$

15. $(x + 7)(x - 4)$ **17.** $(x + 9)(x + 7)$ **19.** $(2x - 1)(x + 2)$

21. $(3x - 1)(x + 1)$ **23.** $(5x + 2)(x + 2)$

25. $(5x + 3)(x - 2)$ **27.** $(5x - 4)(x - 2)$

29. $(3x + 4)(x - 6)$ **31.** 6, -3 **33.** $-\frac{5}{3}, \frac{3}{4}$ **35.** $-1, -2$

37. 3, 5 **39.** $-3, 5$ **41.** 1, 3 **43.** $-2, 9$ **45.** 4, -9

47. 1, $-\frac{3}{2}$ **49.** $-\frac{1}{5}, -2$ **51.** $\frac{1}{3}, 1$ **53.** $\frac{1}{4}, 2$ **55.** 4, 5

57. 1, 3 **59.** $-1, 9$ **61.** No real solution

63. $\dfrac{5 \pm \sqrt{41}}{2}$ **65.** No real solution **67.** $\dfrac{1 \pm \sqrt{17}}{8}$

69. $-\frac{5}{2}, -1$ **71.** 1, $\frac{7}{3}$ **73.** $\dfrac{-3 \pm \sqrt{15}}{2}$

75. a) Two **b)** One **c)** No real solution

Section 6.9 Exercises

1. Any set of ordered pairs

3. The set of values that can be used for the independent variable

43. a) Downward

b) $x = 0$

c) $(0, 2)$

d) $(0, 2)$

e) $(-\sqrt{2}, 0), (\sqrt{2}, 0)$

g) Domain: $\mathbb{R}$, range: $y \le 2$

f)

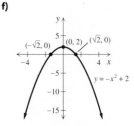

$(-\sqrt{2}, 0)$ $(0, 2)$ $(\sqrt{2}, 0)$

$y = -x^2 + 2$

45. a) Upward   **f)**
b) $x = 0$
c) $(0, -3)$
d) $(0, -3)$
e) $\left(\dfrac{-\sqrt{6}}{2}, 0\right), \left(\dfrac{\sqrt{6}}{2}, 0\right)$
g) Domain: $\mathbb{R}$,
range: $y \geq -3$

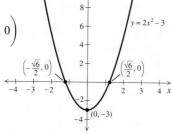

47. a) Upward
b) $x = 1$
c) $(1, -7)$
d) $(0, -6)$
e) $(1 - \sqrt{7}, 0), (1 + \sqrt{7}, 0)$
f)

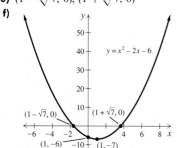

g) Domain: $\mathbb{R}$,
range: $y \geq -7$

49. a) Upward   **f)**
b) $x = -\dfrac{5}{2}$
c) $(-2.5, -0.25)$
d) $(0, 6)$
e) $(-3, 0), (-2, 0)$
g) Domain: $\mathbb{R}$,
range: $y \geq -0.25$

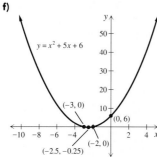

51. a) Downward   **f)**
b) $x = 2$
c) $(2, -2)$
d) $(0, -6)$
e) No x intercepts
g) Domain: $\mathbb{R}$,
range: $y \leq -2$

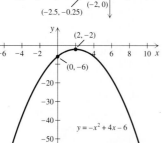

53. a) Upward   **f)**
b) $x = \dfrac{1}{2}$
c) $\left(\dfrac{1}{2}, -\dfrac{25}{2}\right)$
d) $(0, -12)$
e) $(-2, 0), (3, 0)$
g) Domain: $\mathbb{R}$,
range: $y \geq -\dfrac{25}{2}$

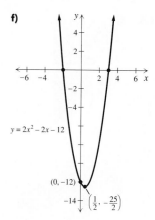

55. Domain: $\mathbb{R}$
range: $y > 0$

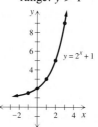

57. Domain: $\mathbb{R}$
range: $y > 0$

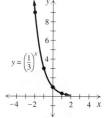

59. Domain: $\mathbb{R}$
range: $y > 1$

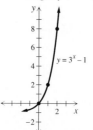

61. Domain: $\mathbb{R}$
range: $y > 0$

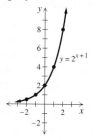

63. Domain: $\mathbb{R}$
range: $y > -1$

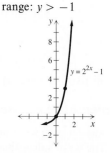

65. Domain: $\mathbb{R}$
range: $y > -1$

67. \$2252   **69. a)** 244 ft   **b)** 100 ft   **71.** 12.5 psi

Review Exercises

1. 31 **2.** -21 **3.** 69 **4.** -47 **5.** $\frac{33}{8}$ **6.** 23

7. $2x - 2$ **8.** $-x - 6$ **9.** $3x - \frac{13}{2}$ **10.** 5

11. 6 **12.** 0 **13.** -31 **14.** $\frac{54}{5}$ **15.** $0.12\overline{6}$

16. 375 min or 6 hr 15 min **17.** 36 **18.** 13 **19.** 101.5

20. 6 **21.** $y = 2x - 6$ **22.** $y = \dfrac{-3x + 8}{4}$ or $y = -\frac{3}{4}x + 2$

23. $y = \dfrac{2x + 22}{3}$ or $y = \frac{2}{3}x + \frac{22}{3}$

24. $y = -x + 2z$ **25.** $l = \dfrac{A}{w}$ **26.** $w = \dfrac{P - 2l}{2}$

27. $x_2 = 3A - x_1 - x_3$ **28.** $d = \dfrac{a_n - a_1}{n - 1}$ **29.** $8 - 7x$

30. $(9 + x) + 4$ **31.** $6r + 7$ **32.** $8/q - 11$

33. $12 - 3x = 21$; $x = -3$ **34.** $2x - 11 = x + 4$; $x = 15$

35. $9(x + 4) = 72$; $x = 4$

36. $10x + 14 = 8(x + 12)$; $x = 41$

37. $x + x + x + x + 2(x + 3) = 24$; width 3 ft, length 6 ft

38. $x + x + 2x = 60$; 15 hr for two workers, 30 hr for third worker

39. $x + 2x + 300 = 1200$; first \$900, second \$300

40. $30x = 480$, 16 hr

41. **42.**

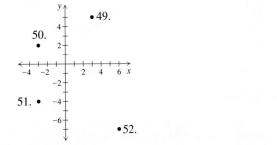

43. **44.**

45. **46.**

47. **48.**

Exercises 49–52

49. **50.** **51.** **52.**

53. $D(3, -3)$; $A = 25$ sq. units **54.** $D(4, 1)$; $A = 21$ sq. units

55.

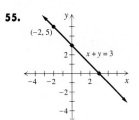

56.

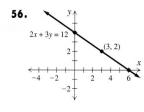

57.

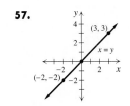

58.

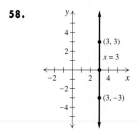

59.

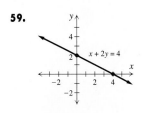

60.

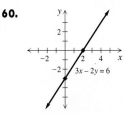

61.

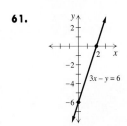

62.

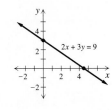

63. $-\frac{1}{4}$ **64.** $-\frac{5}{2}$ **65.** $\frac{7}{6}$ **66.** $\frac{1}{3}$

67.

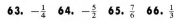

68.

69.

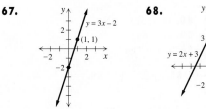

70.

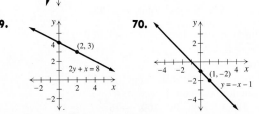

71. $y = 2x + 4$ **72.** $y = -x + 1$

73. a)

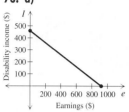

b) About $160
c) About $160

74. a)

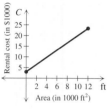

b) About $6400
c) About 4120 ft^2

75.

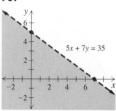

76.

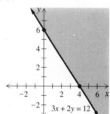

77.

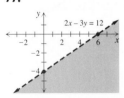

78.

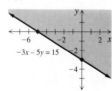

79. $(x + 4)(x + 5)$ **80.** $(x + 3)(x - 2)$
81. $(x - 6)(x - 4)$ **82.** $(x - 8)(x - 3)$
83. $(2x + 3)(x - 5)$ **84.** $(3x - 1)(x + 2)$ **85.** $-1, -2$

86. $1, 5$ **87.** $6, -\frac{3}{2}$ **88.** $-2, -\frac{1}{3}$ **89.** $\dfrac{5 \pm \sqrt{21}}{2}$

90. $1, 2$ **91.** No real solution **92.** $\dfrac{1 \pm \sqrt{13}}{2}$

93. Function, domain: $x = -2, -1, 2, 3$; range: $y = -1, 0, 2$
94. Not a function
95. Function, domain: $\mathbb{R}$, range: $y \leq 12$
96. Function, domain: $\mathbb{R}$, range: $\mathbb{R}$
97. 4 **98.** 17 **99.** 39 **100.** -16

101. a) Downward
b) $x = 0$
c) $(0, 7)$
d) $(0, 7)$
e) $(-\sqrt{7}, 0), (\sqrt{7}, 0)$
g) Domain: $\mathbb{R}$, range: $y \leq 7$

f)

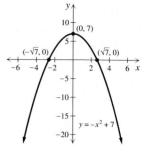

102. a) Upward
b) $x = 4$
c) $(4, -78)$
d) $(0, -30)$
e) $(4 - \sqrt{26}, 0), (4 + \sqrt{26}, 0)$

f)

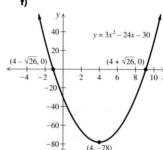

g) Domain: $\mathbb{R}$, range: $y \geq -78$

103. Domain: $\mathbb{R}$
range: $y > 0$

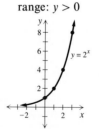

104. Domain: $\mathbb{R}$
range: $y > 0$

105. 22.8 mpg **106. a)** 4208 **b)** 4250 **107.** 68.7%

Chapter 6 Test

1. -58 **2.** -2 **3.** -6 **4.** $4x - 5 = 19, 6$
5. $100 + 0.06s = 300 + 0.04s$, $10,000 **6.** 648
7. $y = \dfrac{3x - 6}{2}$ or $y = \frac{3}{2}x - 3$

8.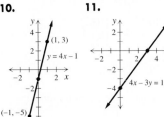

9. $-\frac{1}{3}$

10.
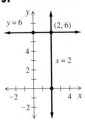

11.
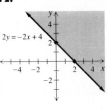

12.
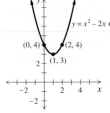

13.
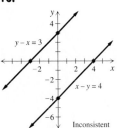
Inconsistent

15.
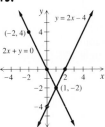

13. 2, 4 **14.** $\dfrac{-3 \pm \sqrt{41}}{4}$ **15.** No **16.** -62

17. a) Upward
b) $x = 1$
c) $(1, 3)$
d) $(0, 4)$
e) No x intercepts
g) Domain: $\mathbb{R}$,
range: $y \geq 3$

f)
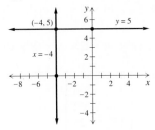

17.
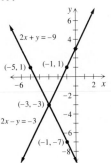

19.
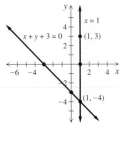

Chapter 7

Section 7.1 Exercises

1. A system of equations that has a solution
3. A dependent system of equations is one that has an infinite number of solutions.

5.
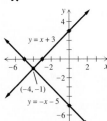

7.
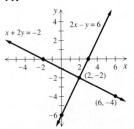

21.
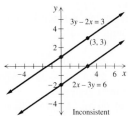
Inconsistent

23.
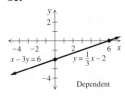
Dependent

25. a) One unique solution, the lines intersect at one and only one point.
b) No solution, the lines do not intersect.
c) Infinitely many solutions, the lines coincide.
27. No solution **29.** One solution
31. Infinitely many solutions **33.** No solution
35. Infinitely many solutions **37.** Not perpendicular
39. Perpendicular
41. a) $c = 10,000 + 3000n$ **b)**
$c = 5,000 + 4000n$
c) 5 terminals

9.

11.

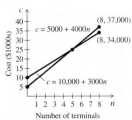

43. a) $s = 0.08d$
$s = 200 + 0.04d$
c) \$5000
d) Option 1,
straight 8% of sales

b)

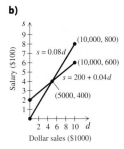

45. a) $R = 165x$
$C = 8400 + 95x$
c) 120 units
d) Loss of \$1400
e) About 138 units

b)

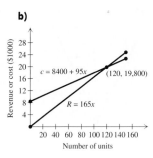

Section 7.2 Exercises

1. The system is dependent if the same value is obtained on both sides of the equal sign.
3. $(2, 4)$ **5.** $(0, -3)$ **7.** No solution, inconsistent system
9. $(4, -2)$ **11.** Dependent, no solution **13.** $(-5, 2)$
15. $(\frac{11}{2}, 19)$ **17.** $(\frac{2}{5}, -\frac{1}{5})$
19. No solution, inconsistent system **21.** $(2, 0)$
23. $(\frac{2}{3}, 2)$ **25.** $(-2, 0)$ **27.** $(-1, 8)$ **29.** $(-1, -2)$
31. $(2, 7)$ **33.** No solution, inconsistent system
35. $(-1, 2)$ **37.** $c = 14 + 0.15m$ **a)** 60 mi **b)** No frills
$c = 20 + 0.05m$
39. $s = 200 + 0.05x$, \$2000
$s = 0.15x$
41. $x + y = 100$, 70 gal 5% butterfat, 30 gal skim
$0.05x + 0.00y = 100(0.035)$
43. $0.01A + 0.20B = 20$, 80 g of Mix A, 60 g of Mix B
$0.06A + 0.02B = 6$
45. $c = 0.10x + 6$, **a)** 40 checks **b)** Bank B
$c = 0.20x + 2$
47. \$66.67

Section 7.3 Exercises

3. b) $\begin{bmatrix} -1 & 2 & 7 \\ -1 & 1 & -2 \end{bmatrix}$ **5. b)** $\begin{bmatrix} 11 & -14 \\ 10 & -15 \end{bmatrix}$

7. $\begin{bmatrix} 4 & 1 \\ 6 & 5 \end{bmatrix}$ **9.** $\begin{bmatrix} 2 & 4 \\ -6 & 9 \\ 4 & 8 \end{bmatrix}$ **11.** $\begin{bmatrix} 3 & 6 \\ 6 & -6 \end{bmatrix}$

13. $\begin{bmatrix} 1 & 0 & -7 \\ 9 & 8 & -7 \\ 6 & 1 & -9 \end{bmatrix}$ **15.** $\begin{bmatrix} 8 & 14 \\ 0 & 4 \end{bmatrix}$ **17.** $\begin{bmatrix} 2 & 23 \\ 12 & 4 \end{bmatrix}$

19. $\begin{bmatrix} 16 & 15 \\ -8 & 6 \end{bmatrix}$ **21.** $\begin{bmatrix} 20 & 19 \\ 24 & 18 \end{bmatrix}$ **23.** $\begin{bmatrix} 15 \\ 22 \end{bmatrix}$

25. $\begin{bmatrix} 1 & 0 \\ 0 & 1 \end{bmatrix}$

27. Matrices A and B do not have the same dimensions and cannot be added, $A \times B = \begin{bmatrix} 6 & -6 & 9 \\ 10 & -4 & 18 \end{bmatrix}$

29. Cannot be added, $A \times B = \begin{bmatrix} 26 & 38 \\ 24 & 24 \end{bmatrix}$

31. Cannot be added, $A \times B = \begin{bmatrix} 17 \\ 39 \end{bmatrix}$

33. $A + B = B + A = \begin{bmatrix} 14 & 8 \\ 5 & -1 \end{bmatrix}$

35. $A + B = B + A = \begin{bmatrix} 8 & 0 \\ 6 & -8 \end{bmatrix}$

37. $(A + B) + C = A + (B + C) = \begin{bmatrix} 9 & 0 \\ 7 & 10 \end{bmatrix}$

39. $(A + B) + C = A + (B + C) = \begin{bmatrix} 5 & 5 \\ 7 & -37 \end{bmatrix}$

41. No **43.** Yes **45.** Yes

47. $(A \times B) \times C = A \times (B \times C) = \begin{bmatrix} 25 & 11 \\ 68 & 28 \end{bmatrix}$

49. $(A \times B) \times C = A \times (B \times C) = \begin{bmatrix} 16 & -10 \\ -24 & 2 \end{bmatrix}$

51. $(A \times B) \times C = A \times (B \times C) = \begin{bmatrix} 17 & 0 \\ -7 & 0 \end{bmatrix}$

53.
Large	Small
38	50
56	72
17	26
10	14

55. $[2546, 3358]$ **57.** No

59. Yes, $A \times B = B \times A = \begin{bmatrix} 1 & 0 \\ 0 & 1 \end{bmatrix}$ **61.** No

Section 7.4 Exercises

1. $(3, 4)$ **3.** $(-2, -1)$ **5.** $(3, 3)$ **7.** $(-2, 1)$ **9.** $(1, 1)$
11. $(1, -2)$
13. Cherries: \$4 per pound, mints: \$5 per pound
15. Pizza: \$15, beer: \$4

Section 7.5 Exercises

1.

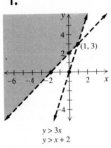

$y > 3x$
$y > x + 2$

3.

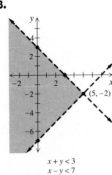

$x + y < 3$
$x - y < 7$

5.

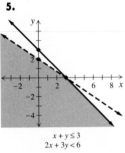

$x + y \le 3$
$2x + 3y < 6$

7.

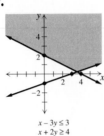

$x - 3y \le 3$
$x + 2y \ge 4$

9.

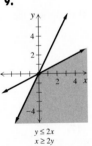

$y \le 2x$
$x \ge 2y$

11.

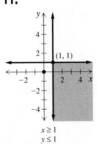

$x \ge 1$
$y \le 1$

13.

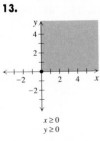

$x \ge 0$
$y \ge 0$

15.

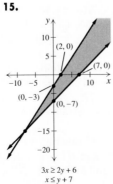

$3x \ge 2y + 6$
$x \le y + 7$

17. Yes

Section 7.6 Exercises

1. Max (3, 2) = 14
Min (0, 0) = 0

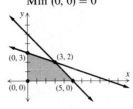

3. Max (3, 1) = 25
Min (0, 0) = 0

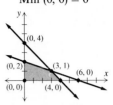

5. Max (4, 3) = 16.5
Min (2, 0) = 3

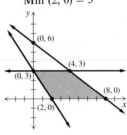

7. a) $x + y \le 18$
$x \ge 2$
$x \le 5$
$y \ge 2$
b) $P = 20x + 25y$
c)

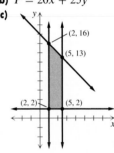

d) (2, 2), (2, 16), (5, 13),
(5, 2)
e) $2 - A$, $16 - B$
f) $440

9. 300 packs of regular hot dogs and 50 packs of all beef hot dogs. Profit is $110.

Review Exercises

1.

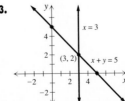

2.

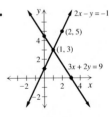

3.

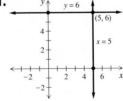

4.

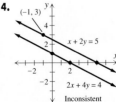

Inconsistent

5. One solution   **6.** Infinite number of solutions
7. No solution   **8.** One solution   **9.** (2, 1)   **10.** (0, −2)
11. (−2, −8)   **12.** No solution, inconsistent   **13.** (9, 3)
14. (3, 4)   **15.** (−1, −1)   **16.** (2, 0)   **17.** (30, 15)
18. Infinite number of solutions, dependent

19. $\begin{bmatrix} 3 & -3 \\ -3 & 13 \end{bmatrix}$ **20.** $\begin{bmatrix} 1 & 9 \\ 11 & -3 \end{bmatrix}$ **21.** $\begin{bmatrix} 6 & 9 \\ 12 & 15 \end{bmatrix}$

22. $\begin{bmatrix} 1 & 24 \\ 29 & -14 \end{bmatrix}$ **23.** $\begin{bmatrix} -19 & 12 \\ -31 & 16 \end{bmatrix}$ **24.** $\begin{bmatrix} -22 & -27 \\ 18 & 19 \end{bmatrix}$

25. (0, 2)   **26.** (−2, 2)   **27.** (1, 2)   **28.** (1, 0)
29. Fixed fee: $7, mileage fee: $1.50
30. 3 gal 30%, 2 gal 80%   **31.** 50
32. nickels: 5, dimes: 55   **33.**

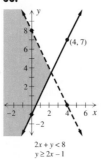

$$2x + y < 8$$
$$y \geq 2x - 1$$

34.

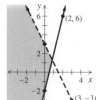

$$y < -2x + 3$$
$$y \geq 4x - 2$$

35.

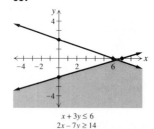

$$x + 3y \leq 6$$
$$2x - 7y \geq 14$$

36.

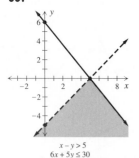

$$x - y > 5$$
$$6x + 5y \leq 30$$

37. Max (9, 0) = 54

Chapter 7 Test

1.

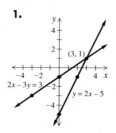

2. One solution   **3.** (4, 3)   **4.** (4, 10)   **5.** (−1, 5)
6. (−1, 3)   **7.** No solution, inconsistent system   **8.** (2, 2)
9. $\begin{bmatrix} -7 & 7 \\ 7 & -7 \end{bmatrix}$ **10.** $\begin{bmatrix} -17 & 1 \\ 5 & -13 \end{bmatrix}$ **11.** $\begin{bmatrix} 14 & -34 \\ -23 & 25 \end{bmatrix}$

12. $y \geq -2x + 4$
    $2x - y < 8$

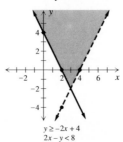

$$y \geq -2x + 4$$
$$2x - y < 8$$

13. 5 lb of each   **14.** $15 per month, $300 initiation fee

Chapter 8

Section 8.1 Exercises

1. Ray, $\overrightarrow{AB}$ **3.** Ray, $\overrightarrow{BA}$ **5.** Line, $\overleftrightarrow{AB}$ **7.** {C}
9. $\angle BFC$ **11.** $\overline{BD}$ **13.** $\overline{BC}$ **15.** $\overline{BC}$ **17.** { }
19. $\angle ABE$ **21.** { }   **23.** $\overrightarrow{BC}$ **25.** $\angle EBC$ **27.** $\overrightarrow{AC}$
29. $\overline{BE}$ **31.** Obtuse   **33.** Straight   **35.** Acute   **37.** 60°
39. $22\frac{1}{2}°$ **41.** 3°   **43.** 85°   **45.** 89°   **47.** 75°   **49.** $166\frac{1}{4}°$
51. 65.3°   **53.** c   **55.** d   **57.** a
59. $\angle 1 = 72°$, $\angle 2 = 18°$ **61.** $\angle 1 = 144°$, $\angle 2 = 36°$
63. Two lines in the same plane that do not intersect
65. An infinite number   **67. a)** $\angle 1 = 135.5°$, $\angle 2 = 44.5°$
69. $\angle 1 = 26°$. $\angle 2 = 154°$ **71.** $\angle 1 = 22°$, $\angle 2 = 68°$
73. $\angle 1 = 77°$, $\angle 2 = 13°$ **75.** A line
77. a) Yes   **b)** No   **c)** An infinite number

In Ex. 79–87 other answers are possible.
79. GDC and HIJ **81.** HGB, GDC, and FGD
83. $\overline{GD}$ and FGH **85.** $\overline{GD}$ and $\overline{BC}$ **87.** $\overline{FG}, \overline{GD},$ and $\overline{GB}$

89. It may be more stable on an unleveled surface.

91. a) Undefined terms, definitions, axioms (postulates), theorems

93.

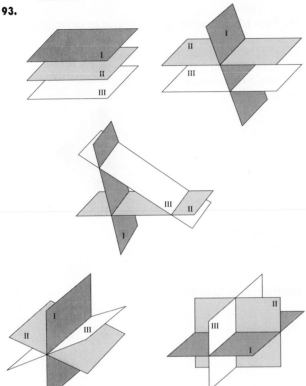

Section 8.2 Exercises

1. Two figures are similar if their corresponding angles have the same measure and their corresponding sides are in proportion.
3. Pentagon **5.** Octagon **7.** Heptagon **9.** Right
11. Obtuse **13.** Equilateral **15.** Scalene **17.** Isosceles
19. Parallelogram **21.** Square **23.** 35° **25.** 170°
27. a)

b) $180° \times 6 = 1080°$ **c)** $(8 - 2)180° = 1080°$

29. 1260° **31.** 720° **33.** 1440° **35.** 60°; 120°
37. 128.6°; 51.4° **39.** 108°; 72° **41.** 55° **43.** 35°
45. 80° **47.** 100° **49.** $\overline{B'C'} = 3$, $\overline{A'C'} = 5$
51. $\overline{BC} = 6$, $\overline{DF} = 3.2$ **53.** 6 **55.** $1\frac{2}{3}$ **57.** 3 **59.** 130°
61. 50° **63.** 5 **65.** 12 **67.** 30° **69.** 4 **71.** 8
73. 70°

Section 8.3 Exercises

1. 6 in.² **3.** 18 ft² **5.** 18 in.², 20 in. **7.** 25 ft², 20 ft
9. 6000 cm² or 0.6 m², 6.54 m or 654 cm
11. 12.56 in.², 12.56 in. **13.** 7.065 ft², 9.42 ft **15.** 13 ft
17. $\sqrt{24} \approx 4.9$ m **19.** 13.76 ft² **21.** 8 in.²
23. 3.8125 m² **25.** 6 mm² **27.** 0.77 yd²
29. 164.7 ft² **31.** 126,000 cm²
33. 0.0608 m² **35.** $100,200 **37.** $837.20 **39.** 53.3 gal
41. a) 17,517.84 ft² **b)** 3.5 bags
   c) $79.80 assuming they purchase 4 bags
43. It becomes 9 times as large. **45.** Its area is doubled.
47. The area remains the same.
49. The area remains the same. **51.** 0.02211 ha
53. The round burger from Burger King is about 0.62 in.² larger.
55. a) 2410 ft² **b)** 9.64 gal
   c) $189.50 + tax, assuming 10 gal were purchased
57. 52 ft² **59.** $lw - 2s^2$

Section 8.4 Exercises

1. 4 ft³ **3.** 75.36 in.³ **5.** 2400 in.³ **7.** 292.5 ft³
9. 900 in.³ **11.** 30.51 ft³ **13.** 31.4 m³ **15.** 24 ft³
17. 108 ft³ **19.** 7.85 yd³ **21.** 1,500,000 cm³ **23.** 4 m³
25. a) 28,750 in.³ **b)** 16.64 ft³
27. a) 54,000 cm³ **b)** 54,000 mℓ **c)** 54 ℓ
29. a) 108 ft³ **b)** 4 yd³ **31.** 82,944,000 ft³
33. 1.77 mm³
35. a) $A_{\text{round}} = 63.6$ in.², $A_{\text{rect}} = 63$ in.²
   b) $V_{\text{round}} = 127.2$ in.³, $V_{\text{rect}} = 126$ in.³
37. 6 cuts **39.** 6 edges **41.** 9 faces **43.** 8 vertices
45. 4 times the volume
47. No, a half-inch cube has a volume of $\frac{1}{8}$ in.³. A half cubic inch has a volume of $\frac{1}{2}$ in.³.
49. a) 6.75 ft³ **b)** 50.6 gal **51.** 18.4 in.

Section 8.5 Exercises

1. Vertices: 4; arcs: 5 **3.** Vertices: 7; arcs: 11
5. Each graph has the same number of arcs from the corresponding vertices.
7. Yes; start at C and end at D, or start at D and end at C.
9. Yes; start at any point and end where you started.
11. No; 4 odd vertices

13. Yes; start at D and end at E, or start at E and end at D.
15. a) 2 odd, 4 even   **b)** Yes
   c) Start at B and end at E, or start at E and end at D.
17. a) 2 odd, 4 even   **b)** Yes
   c) Start at B and end at F, or start at F and end at B.
19. a) 4 odd, 1 even   **b)** No   **21. a)** 5 odd, 1 even   **b)** No
23. Door must be placed in room C.
25. Yes. However, you must start from either the island on the right or the land below the islands.
27. No, it is not possible assuming your starting and ending points are considered vertices.
29. a) KY, MO, AR, MS, AL, GA, NC, VA
   b) VA, TN, GA
31.

33. Arcs $=$ Vertices $+$ Regions $- 2$

Section 8.6 Exercises

1. One edge   **3.** One strip
5. a) No   **b)** 2   **c)** 2
   d) Two strips, one inside the other
7. No, it does not.
9. a)

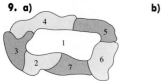

   b)

11. a)

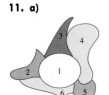

   b)

13. a)

   b)

15. Inside of curve

Section 8.7 Exercises

7. a) The sum of the measures of the angles of a triangle is 180°.

b) The sum of the measures of a triangle is less than 180°.
c) The sum of the measures of a triangle is greater than 180°.
9. A sphere
11. Each type of geometry can be used in its own frame of reference.
13. Positive curvature: Riemannian geometry
   Zero curvature: Euclidean geometry
   Negative curvature: hyperbolic geometry
15. a)

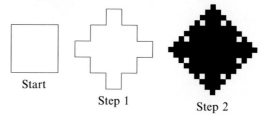

Start   Step 1   Step 2

   b) Infinite
   c) Finite

Section 8.8 Exercises

The answers for Ex. 1–9 are not drawn to scale.

1.   **3.**   **5.**

7.   **9.**

11. RT 90 FD 40 LT 90 FD 20 LT 90 FD 20
   RT 90 FD 20
13. RT 90 FD 30 RT 180 FD 60 RT 135 FD 43
   RT 135 FD 30 BK 30 LT 90 FD 30
15. LT 45 FD 40 RT 90 FD 40 RT 90 FD 40
   RT 90 FD 40 RT 135
17. REPEAT 4[FD 20 FD 10 RT 90 FD 10 RT 90
   FD 10 RT 90 FD 10 RT 90 BK 20 RT 90]
19. LT 90 FD 20 RT 90 FD 30 RT 90 FD 40
   RT 90 FD 30 RT 90 FD 20 RT 90

The answers for Ex. 21–37 are not drawn to scale.

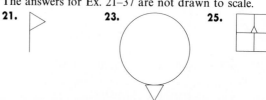

21.   **23.**   **25.**

27. **29.**

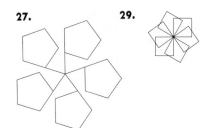

31. SETX 50 SETY 30 SETX −40 SETY −10

33. SETX 40 SETY 40 SETX 0
SETY 10 SETX 20

35. **37.**

39. LT 45 FD 20 RT 90 FD 40 RT 90 FD 20
RT 90 FD 20 RT 90 FD 40 LT 90 FD 20
BK 40 LT 90 FD 40 RT 90 FD 40 RT 90
FD 40 HOME

41. REPEAT 4[LT 30 FD 20 RT 120 FD 20
RT 120 FD 20 LT 120]

43. LT 90 PENTAGON REPEAT 2[RT 108 FD 40
LT 72 PENTAGON]

Review Exercises

1. $\overline{BF}$ **2.** $\overrightarrow{BH}$ **3.** $\{C\}$ **4.** $\triangle BFC$ **5.** $\overleftrightarrow{DC}$ **6.** { }
7. 3 in. **8.** 7 in. **9.** 100° **10.** 50°
11. ⊀1 = 50°; ⊀2 = 60°; ⊀3 = 120°; ⊀4 = 130°;
⊀5 = 50°; ⊀6 = 50°
12. 20 in.² **13.** 30 in.² **14.** 13 in.² **15.** 72 in.²
16. 50.24 cm² **17.** 120 in.³ **18.** 401.92 in.³ **19.** 60 mm²
20. 28 ft³ **21.** 540 m³ **22.** 3140 mm³ **23.** 960 ft³
24. $647.50
25. a) $48\sqrt{2}$ or 67.9 ft³ **b)** 4618.8 lb **c)** 511.3 gal
26. Yes; start at A or E. **27.** Yes; start at B or D.
28. Yes; start at D or E. **29.** Yes; start at any point.
30. No
32. Euclidean: Given a line and a point not on the line,
one and only one line can be drawn parallel to the
given line through the given point.
Elliptical: Given a line and a point not on the line, no
line can be drawn through the given point parallel to
the given line.
Hyperbolic: Given a line and a point not on the line,
two or more lines can be drawn through the given
point parallel to the given line.

The answers for Ex. 33–37 are not drawn to scale.

33. **34.** H **35.**

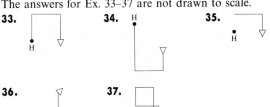

36. **37.**

Chapter 8 Test

1. $\overrightarrow{AE}$ **2.** $\overline{BC}$ **3.** $\overrightarrow{BF}$ **4.** $\overleftrightarrow{AE}$ **5.** 52.6° **6.** 86.3°
7. 37° **8.** 540° **9.** ≈ 5.38 cm
10. a) 16 in. **b)** 48 in. **c)** 96 in.² **11.** 904.32 in.³
12. 252.8 ft³ **13.** 96 ft³ **14.** Yes, C or D **15.** No

Chapter 9

Section 9.1 Exercises

1. An operation that can be performed on only two
elements
3. $1 - 2 \neq 2 - 1$ **5.** $(1 - 2) - 3 \neq 1 - (2 - 3)$
$-1 \neq 1$ $\qquad -4 \neq 2$
7. Closed; identity element; inverses; associative property
9. No; no identity element **11.** Yes
13. No, not associative
15. No, not closed. For example, $\sqrt{2} \cdot \sqrt{2} = 2$, and 2 is a
rational number.
17. No, 0 has no multiplicative inverse.

Section 9.2 Exercises

1. 11 **3.** 8 **5.** 7 **7.** 5 **9.** 2 **11.** 2 **13.** 3 **15.** 9
17. 7
19. If the first number is greater than the second, subtract
to obtain the answer. Otherwise add 12 to the first
number and then subtract to obtain the answer. For
example, $6 - 9 = (6 + 12) - 9 = 18 - 9 = 9$
21. 1 **23.** 1 **25.** 5 **27.** 2 **29.** 3 **31.** 2
33. Yes, satisfies the five required properties
35. a) {0, 1, 2, 3} **b)** * **c)** Yes **d)** Yes; 0
e) Yes: 0–0, 1–3, 2–2, 3–1 **f)** $(2*3)*1 = 2*(3*1)$
g) Yes; $2*3 = 3*2$ **h)** Yes
37. a) {19, 23, 27} **b)** % **c)** Yes **d)** Yes, 19
e) Yes, 19–19, 23–27, 27–23
f) $(19\%23)\%27 = 19\%(23\%27)$ **g)** Yes;
$23\%27 = 27\%23$ **h)** Yes

39. a) No, no identity element
   b) $(1 \odot 2) \odot 3 \neq 1 \odot (2 \odot 3)$
41. Not associative: $(\triangle 3 \square) 3 \triangle \neq \triangle 3 (\square 3 \triangle)$, not commutative: $\square 3 \triangle \neq \triangle 3 \square$
43. No inverse for $\sim$ or for *, not associative
45. No identity element, no inverses, not associative, not commutative
47. a)

+	E	O
E	E	O
O	O	E

   b) Yes, satisfies the five properties
51. a) Is closed; identity element is 6; inverses: 1–5, 2–2, 3–3, 4–4, 5–1, 6–6; is associative—for example,
$$(2 \infty 5) \infty 3 = 2 \infty (5 \infty 3)$$
$$3 \infty 3 = 2 \infty 2$$
$$6 = 6$$
   b) $3 \infty 1 \neq 1 \infty 3$
$$2 \neq 4$$
53. $4^3 = 64$

Section 9.3 Exercises

1. Sunday **3.** Thursday **5.** Monday **13.** 3 **15.** 1
17. 0 **19.** 2 **21.** 2 **23.** 3 **25.** 1 **27.** 12 **29.** 3
31. 1 **33.** 1 **35.** 9 **37.** 2 **39.** 4 **41.** 5 **43.** 8
45. { } **47.** 6 **49.** 7
51. a) 1996, 2000, 2004, 2008, 2012 **b)** 3004
   c) 2504, 2508, 2512, 2516, 2520, 2524
53. a) 3 P.M.–11 P.M. **b)** 7 A.M.–3 P.M. **c)** 7 A.M.–3 P.M.
55. a)

+	0	1	2	3
0	0	1	2	3
1	1	2	3	0
2	2	3	0	1
3	3	0	1	2

   b) Yes **c)** Yes, 0 **d)** Yes, 0–0, 1–3, 2–2, 3–1
   e) $(1 + 2) + 3 = 1 + (2 + 3)$
   f) Yes, $2 + 3 = 3 + 2$ **g)** Yes **h)** Yes
57. a)

×	0	1	2	3
0	0	0	0	0
1	0	1	2	3
2	0	2	0	2
3	0	3	2	1

   b) Yes **c)** Yes, 1
   d) No, no inverse for 0 or for 2, inverse of 1 is 1, inverse of 3 is 3
   e) $(1 \times 2) \times 3 = 1 \times (2 \times 3)$
   f) Yes, $2 \times 3 = 3 \times 2$ **g)** No

59. 14 calendars **61.** ? = 7 **63.** ? = 3 **65.** $x = 0$
67. $x = 1$ **69. b)**
$$
\begin{array}{r}
4236 \equiv 6 \ (\text{mod } 9) \\
+ \ 3784 \equiv +4 \ (\text{mod } 9) \\
\hline
8020 \equiv 1 \ (\text{mod } 9)
\end{array}
$$

Review Exercises

1. Set of elements, binary operation **2.** 5 **3.** 1 **4.** 8
5. 9 **6.** 9 **7.** 10 **8.** 6
9. Closure, identity element, inverses, associative property
10. A commutative group **11.** Yes
12. No, no inverse for any integer except 1 **13.** Yes
14. No, no inverse for 0
15. Not every element has an inverse, and not associative.
16. No identity element
17. No identity element and not associative
18. a) $\{1, 0, ?, \triangle\}$ **b)** $\otimes$ **c)** Yes **d)** Yes, 1
   e) Yes; 1–1, $\bigcirc$–$\triangle$, ?–?, $\triangle$–$\bigcirc$
   f) $(\bigcirc \otimes ?) \otimes \triangle = \bigcirc \otimes (? \otimes \triangle)$
   g) Yes; $\bigcirc \otimes ? = ? \otimes \bigcirc$ **h)** Yes
19. 2 **20.** 4 **21.** 3 **22.** 5 **23.** 2 **24.** 1 **25.** 2
26. 13 **27.** 6 **28.** 4 **29.** 2 **30.** No solution **31.** 6
32. 0, 2, 4, 6 **33.** 5 **34.** 6 **35.** 3 **36.** 5 **37.** 7
38.

+	0	1	2	3	4	5
0	0	1	2	3	4	5
1	1	2	3	4	5	0
2	2	3	4	5	0	1
3	3	4	5	0	1	2
4	4	5	0	1	2	3
5	5	0	1	2	3	4

Yes, a commutative group
39.

×	0	1	2	3
0	0	0	0	0
1	0	1	2	3
2	0	2	0	2
3	0	3	2	1

No, no inverse for 0
40. a) No, she will be off.
   b) Yes, she will have the evening off.

Chapter 9 Test

1. A set of elements and a binary operation
2. Closure, identity element, inverses, associative property, commutative property
3. No, not all elements have inverses.

4.

+	1	2	3
1	2	3	1
2	3	1	2
3	1	2	3

5. Yes, it is a commutative group
6. No, not closed   **7.** No, no identity element
8. Yes, it is a commutative group   **9.** 1   **10.** 3   **11.** 6
12. 1   **13.** 5   **14.** 3   **15.** 2, 5   **16.** 2
17. a)

×	0	1	2
0	0	0	0
1	0	1	2
2	0	2	1

  b) No, no inverse for 0

Chapter 10

Section 10.1 Exercises

1. 25.0%   **3.** 41.7%   **5.** 45.7%   **7.** 1367.8%
9. 34,267.0%   **11.** 0.254   **13.** 0.00004   **15.** 0.073468
17. 0.0075   **19.** 16.4%
21. a) Approximately 0.07%   **b)** Approximately 84.5%
  c) Approximately 0.85%
23. 10.0%   **25.** 43.8%   **27. a)** 13.6%   **b)** 15.1%
29. a) 125   **b)** 41.4% decrease
31. a) 374   **b)** 35.0% decrease   **33.** $0.01x(70) = 14$, 20%
35. $0.17(400) = x$, 68   **37.** $0.20x = 7$, 35
39. $0.05(32.50) = x$, $1.63   **41.** $0.3x = 57$, 190
43. $400(0.17) = x$, 68   **45.** 7.5% increase
47. 118.4% markup   **49.** 62.5% increase
51. 40.3% markup
53. Sales price is not consistent with the ad. The sales price was $2.25 higher.
55. He will have a loss of $10.

Section 10.2 Exercises

1. $2.40   **3.** $15.17   **5.** $15.85   **7.** $38.26   **9.** $1451.25
11. $600   **13.** 1.5 years   **15.** 3.2%   **17.** 13%
19. a) $22.50   **b)** $1522.50
21. a) $135.42   **b)** $2364.58   **c)** 13.7%
23. a) $437.50   **b)** $7\frac{1}{4}$%   **c)** $362.69   **25.** 138 days
27. 136 days   **29.** 134 days   **31.** October 6
33. October 31   **35.** $1253.61   **37.** $2163.02
39. $712.45   **41.** $424.97   **43.** $1256.14
45. a) $206.50, $204.50, $202.33   **b)** 13.33

Section 10.3 Exercises

1. a) $2205.00   **b)** $205.00   **3. a)** $2811.98   **b)** $411.98
5. a) $3062.61   **b)** $562.61   **7. a)** $8541.33   **b)** 3041.33
9. a) $2247.20   **b)** $247.20   **11. a)** $2252.99   **b)** $252.99
13. a) $2533.54   **b)** $533.54   **15. a)** $1935.63   **b)** $435.63
17. a) $3172.17   **b)** $3243.40   **19.** $3514.98
21. $8554.57
23. a) $1220.19, $220.19   **b)** $1485.95, $485.95
  c) $2191.12, 1191.12   **d)** No
27. 6.67%   **29.** $7209.56   **31.** $6356.34   **33.** 12%
35. 8.84% simple rate

Section 10.4 Exercises

1. a) $46.80   **b)** 12.00%   **3. a)** $10.14   **b)** 11.50%
5. $3.89   **7. a)** $9.38   **b)** $18.68   **c)** 12.75%   **d)** 13.50%
9. 10.75%   **11. a)** 11.25%   **b)** $83.70   **c)** $1469.41
13. a) $24.64   **b)** $918.33   **15. a)** $.78   **b)** $286.76
17. a) $1585.37   **b)** $20.61   **c)** $1862.97
19. a) The average daily balance is $1504.72 or $80.65 less.
  b) The finance charge is $19.56, or $1.05 less.
21. a) 6 months   **b)** $9.75   **c)** Save $.39 with credit card.

Section 10.5 Exercises

1. a) $32,500   **b)** $764.40
3. a) $9800   **b)** $88,200   **c)** $1764.00
5. a) $601.25   **b)** Yes
7. a) $174,471.60   **b)** $99,471.60   **c)** $34.12
9. a) $27,000.00   **b)** $2160.00   **c)** $887.76
  d) $402,019.20   **e)** $267,019.20   **f)** $32.76
11. a) $17,000.00   **b)** $841.25   **c)** $558.96   **d)** $661.96
  e) Yes   **f)** $20.63   **g)** $251,763.20   **h)** $166,763.20
13. a) $630.58   **b)** 8.9%   **c)** 8.1%
15. a) $716.45   **c)** 9.38%
  b) and **d)**

Payment Number	Interest	Balance of Principal	Loan
1	$667.50	$48.95	$88,951.05
2	$667.13	$49.32	$88,901.73
3	$666.76	$49.69	$88,852.04
4	$694.53	$21.92	$88,830.12
5	$694.36	$22.09	$88,808.03
6	$694.18	$22.27	$88,785.76

  e) 9.46%

Review Exercises

1. 12.5% **2.** 60.0% **3.** 50.0% **4.** 4.5% **5.** 0.8%
6. 125.0% **7.** 0.26 **8.** 0.115 **9.** 0.00056 **10.** 0.004
11. 0.008$\overline{3}$ **12.** 0.0000045 **13.** 50%
14. 10.31% decrease **15.** $0.01x(80) = 20$, 25%
16. $0.20x = 44$, 220 **17.** $0.18(540) = x$, 97.2
18. \$7.34 **19.** 35 **20.** 29.4% **21.** \$7.50 **22.** 10.0%
23. \$450.00 **24.** 6 months **25.** \$5017.50
26. a) \$138.00 **b)** \$2438.00
27. a) \$690.00 **b)** \$3310.00 **c)** 13.9%
28. a) $7\frac{1}{2}$% **b)** \$830.00 **c)** \$941.18 **29.** \$8508.92
30. a) \$595.51, \$95.51 **b)** \$597.03, \$97.03
  c) \$597.81, \$97.81 **31.** \$7589.04 **32.** 5.76%
33. a) 11.00% **b)** \$493.02 **c)** \$4350.73
34. a) \$109.18 **b)** \$2014.11
35. a) 11.25% **b)** \$83.70 **c)** \$1469.41
36. a) \$6.31 **b)** \$847.61 **37. a)** \$3.94 **b)** \$448.46
38. a) \$3125.00 **b)** \$9375 **c)** \$2812.50 **d)** 13.50%
39. a) \$7.40 **b)** 13.5%
40. a) \$47,495 **b)** \$3972.00 **c)** \$993.00 **d)** \$715.34
  e) \$940.34 **f)** Yes
41. a) \$22,475.00 **b)** \$772.02 **c)** \$13.49 **d)** \$300,402.20
  e) \$210,502.20
42. a) \$662.29 **b)** 9.15% **c)** 9.35%

Chapter 10 Test

1. \$23.33 **2.** 7 years **3.** \$313.50 **4.** \$3313.50
5. \$3697.50 **6.** \$442.50 **7.** 11.00%
8. a) \$38.79 **b)** \$1503.59 **9. a)** \$9.15 **b)** \$794.92
10. a) \$8.10 **b)** \$802.37 **11.** \$2507.32 **12.** \$107.32
13. \$5776.94, \$2176.94 **14.** \$6530.35, \$1730.35
15. \$38,625.00 **16.** \$5770.00 **17.** \$1442.50
18. \$1156.43 **19.** \$1433.51 **20.** Yes **21.** \$316,168.20
22. \$161,668.20

Chapter 11

Section 11.1 Exercises

1. The official definition by the Weather Service is **(d)**.
  The specific location is the place where the rain gauge
  is. The Weather Service uses sophisticated
  mathematical equations to calculate these probabilities.
3. No, it means that the average person with traits
  similar to Mr. Bennett's will live another 42.34 years.
5. $\frac{1}{2}$
7. Roll a die many times and find the relative frequency
  of 4's to the total number of rolls.
13. a) $\frac{1}{2}$ **b)** $\frac{7}{20}$ **c)** $\frac{3}{20}$ **15. a)** 0.92 **b)** 400

17. a) $\frac{25}{40}$ or 0.625 **b)** $\frac{33}{40}$ or 0.825
19. a) 0 **b)** $\frac{50}{250} = 0.2$ **c)** 1 **21. a)** 0.24 **b)** 0.76
23. a) 0.51 **b)** 0.49

Section 11.2 Exercises

1. $\frac{1}{5}$ **3.** $\frac{1}{13}$ **5.** $\frac{12}{13}$ **7.** $\frac{1}{2}$ **9.** 0 **11.** $\frac{1}{52}$
13. a) $\frac{1}{3}$ **b)** $\frac{1}{3}$ **c)** $\frac{1}{3}$ **15. a)** $\frac{1}{3}$ **b)** $\frac{1}{3}$ **c)** $\frac{1}{3}$ **17.** $\frac{1}{4}$ **19.** 1
21. $\frac{1}{6}$ **23.** $\frac{1}{2}$ **25.** $\frac{2}{5}$ **27.** $\frac{8}{15}$ **29.** $\frac{2}{3}$ **31.** $\frac{3}{5}$ **33.** $\frac{3}{13}$
35. $\frac{5}{13}$ **37.** $\frac{9}{13}$ **39.** $\frac{3}{4}$ **41.** $\frac{2}{3}$ **43.** $\frac{2}{51}$ **45.** $\frac{7}{17}$ **47.** $\frac{49}{51}$
49. $\frac{1}{2}$ **51.** $\frac{2}{5}$ **53.** $\frac{4}{7}$ **55.** $\frac{6}{35}$ **57.** $\frac{14}{25}$ **59.** $\frac{16}{25}$ **61.** $\frac{4}{25}$
63. $\frac{11}{36}$ **65.** $\frac{2}{3}$

Section 11.3 Exercises

1. 25:1 **3. a)** $\frac{2}{5}$ **b)** $\frac{3}{5}$ **c)** 3:2 **d)** 2:3 **5.** 5:1 **7.** 2:1
9. 12:1, 1:12 **11.** 3:1, 1:3 **13.** 1:1 **15.** 3:1
17. a) 999,999:1 **b)** 1:999,999 **19.** 13:15 **21.** 1:6
23. 27:1 **25. a)** $\frac{1}{9}$ **b)** $\frac{8}{9}$ **27. a)** $\frac{2}{7}$ **b)** $\frac{5}{7}$ **29.** 3:5
31. a) 20:1 **b)** $\frac{20}{21}$

Section 11.4 Exercises

3. \$250 **5. a)** $\frac{7}{16}$ in. or 0.4375 in. **b)** $\frac{217}{16}$ or 13.56 in.
7. a) No, a disadvantage **b)** Yes, it is an advantage.
9. a) $-$\$.50 **b)** \$.50 **c)** \$500 **11. a)** \$3 **b)** \$1
13. a) \$4.25 **b)** \$2.25 **15.** \$40.00 gain **17.** \$150,000
19. 325 **21.** 170.41 calls **23.** \$10,000 **25. a)** -5.3¢
27. -25.4¢ **29.** Favors the player

Section 11.5 Exercises

1. a)   **b)** $\frac{1}{4}$ **c)** $\frac{1}{2}$ **d)** $\frac{1}{4}$

3. a)   **b)** 0 **c)** $\frac{1}{6}$ **d)** $\frac{2}{3}$

5. a)   **b)** $\frac{1}{4}$ **c)** $\frac{3}{4}$

7. a)

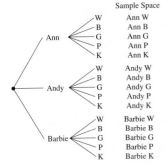

Sample Space

Ann	W	Ann W
	B	Ann B
	G	Ann G
	P	Ann P
	K	Ann K
Andy	W	Andy W
	B	Andy B
	G	Andy G
	P	Andy P
	K	Andy K
Barbie	W	Barbie W
	B	Barbie B
	G	Barbie G
	P	Barbie P
	K	Barbie K

b) $\frac{2}{3}$ **c)** $\frac{7}{15}$ **d)** $\frac{1}{15}$ **e)** $\frac{2}{15}$

9. a)

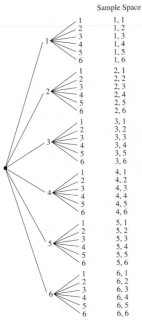

Sample Space

b) $\frac{1}{6}$ **c)** $\frac{1}{6}$ **d)** $\frac{1}{36}$ **e)** No

11. a)

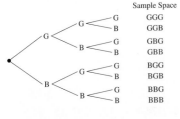

Sample Space

	G	G	GGG
		B	GGB
G	G		
	B	G	GBG
		B	GBB
	G	G	BGG
B		B	BGB
	B	G	BBG
		B	BBB

b) $\frac{1}{8}$ **c)** $\frac{7}{8}$ **d)** $\frac{3}{4}$

13. a)

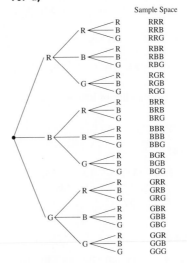

Sample Space

b) $\frac{1}{27}$ **c)** $\frac{1}{27}$ **d)** $\frac{26}{27}$

15. a) 24

b)

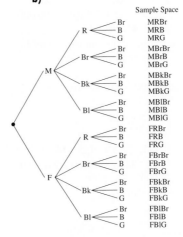

Sample Space

c) $\frac{1}{24}$ **d)** $\frac{1}{8}$

Section 11.6 Exercises

1. a) No, both events cannot occur at the same time.
 b) No, being healthy can affect your happiness.

3. Events A and B cannot occur at the same time.
 Therefore, $P(A \text{ and } B) = 0$.

5. $\frac{1}{2}$ **7.** $\frac{1}{2}$ **9.** $\frac{2}{13}$ **11.** $\frac{8}{13}$ **13.** $\frac{17}{26}$ **15. a)** $\frac{1}{169}$ **b)** $\frac{1}{221}$

17. a) $\frac{1}{16}$ **b)** $\frac{1}{17}$ **19. a)** $\frac{1}{8}$ **b)** $\frac{13}{102}$ **21. a)** $\frac{100}{169}$ **b)** $\frac{10}{17}$

23. $\frac{1}{4}$ **25.** $\frac{1}{8}$ **27.** $\frac{9}{64}$ **29.** $\frac{1}{8}$ **31.** $\frac{3}{8}$ **33.** $\frac{1}{32}$ **35.** $\frac{1}{8}$

37. $\frac{1}{8}$ **39. a)** $\frac{1}{16}$ **b)** 0 **41. a)** $\frac{9}{16}$ **b)** $\frac{1}{2}$

43. a) $\frac{1000}{2197}$ **b)** $\frac{60}{143}$ **45. a)** $\frac{48}{2197}$ **b)** $\frac{3}{143}$ **47.** $\frac{325}{16,539}$
49. $\frac{196}{5513}$ **51.** 0.7 **53.** 0.343 **55.** $\frac{1}{4}$ **57.** $\frac{27}{1024}$ **59.** $\frac{243}{1024}$
61. $\frac{1}{80}$ **63.** $\frac{6859}{8000}$
65. a) No **b)** 0.001 **c1)** 0.00004 **c2)** 0.00096
   c3) 0.998001
67. 0.04 **69.** 0.24 **71.** 0.024 **73.** 0.023424 **75.** $\frac{1}{1000}$
77. $\frac{8}{15}$
79. Favors the dealer: The probability of losing is
   $\frac{39}{52} \cdot \frac{38}{51} = 0.56$.

Section 11.7 Exercises

1. $\frac{1}{3}$ **3.** $\frac{1}{2}$ **5.** $\frac{1}{2}$ **7.** $\frac{1}{3}$ **9.** $\frac{1}{13}$ **11.** $\frac{1}{16}$ **13.** $\frac{1}{7}$ **15.** $\frac{1}{8}$
17. $\frac{1}{2}$ **19.** $\frac{1}{6}$ **21.** $\frac{1}{6}$ **23.** $\frac{2}{3}$ **25.** $\frac{19}{25}$ **27.** $\frac{4}{15}$ **29.** $\frac{10}{11}$
31. $\frac{3}{19}$ **33.** $\frac{44}{47}$ **35.** $\frac{11}{21}$ **37.** $\frac{10}{23}$ **39.** $\frac{18}{23}$

Section 11.8 Exercises

3. 720 **5.** 5040 **7.** 20 **9.** 1 **11.** 40,320 **13.** 40,320
15. 56 **17. a)** 1000 **b)** $\frac{1}{1000}$ **19.** 1,000,000,000
21. 120 **23.** 120 **25.** 56 **27.** 10,000,000,000
29. 67,600 **31.** 10,400 **33.** 6,760,000 **35.** 728,000
37. a) 8,000,000 **b)** 6,400,000,000 **c)** 64,000,000,000,000
39. 427,518,000 **41.** 3,628,800 **43.** 40,320
45. a) The order is important. Since the numbers may be
   repeated it is not a true permutation lock.
   b) 64,000 **c)** 59,280
47. 420 **49.** 2520 **51.** 3,628,800 **53.** 40,320 **55.** 211

Section 11.9 Exercises

3. 15 **5.** 1 **7.** 495 **9.** 1 **11.** $\frac{1}{2}$ **13.** 455 **15.** 5
17. 120 **19.** 15,504 **21.** 1287 **23.** 2,598,960 **25.** 150
27. 24 **29.** 22,308 **31.** 294,000 **33.** 1200
35. a) 45 **b)** 56 **37. a)** 6 **b)** 10 **c)** $_nC_2$

Section 11.10 Exercises

1. $\frac{1}{5}$ **3.** $\frac{3}{14}$ **5.** $\frac{1}{12}$ **7.** $\frac{5}{28}$ **9.** $\frac{253}{9996}$ **11.** $\frac{1}{10}$ **13.** $\frac{7}{10}$
15. $\frac{3}{20}$ **17.** $\frac{3}{44}$ **19.** $\frac{34}{55}$ **21.** $\frac{10}{91}$ **23.** $\frac{59}{65}$ **25.** $\frac{1}{77}$ **27.** $\frac{5}{77}$
29. $\frac{1}{3,838,380}$ **31.** $\frac{1}{137,085}$ **33. a)** $\frac{1}{123,760}$ **b)** $\frac{1}{30,940}$
35. $\frac{5}{729}$ **37.** $\frac{821}{1458}$ **39. a)** $\frac{1}{2,162,160}$ **b)** $\frac{1}{6435}$ **41.** $\frac{1}{24}$

Review Exercises

2. 0.80 **4.** 0.62 **5.** $\frac{1}{2}$ **6.** $\frac{4}{5}$ **7.** 1 **8.** $\frac{1}{5}$ **9.** $\frac{3}{10}$
10. $\frac{51}{100}$ **11.** $\frac{23}{50}$ **12.** $\frac{49}{100}$ **13. a)** 2:1 **b)** 1:2 **14.** 23:2
15. $\frac{2}{3}$ **16.** 2:3 **17. a)** $-\$1.20$ **b)** $-\$3.60$ **c)** \$0.80
18. a) $-15¢$ **b)** $15¢$ **c)** $-\$15.38$ **19.** 660 people

20. a)   **b)** Sample Space   **c)** $\frac{1}{12}$

	Sample Space
T—J	TJ
T—G	TG
T—C	TC
J—T	JT
J—G	JG
J—C	JC
G—T	GT
G—J	GJ
G—C	GC
C—T	CT
C—J	CJ
C—G	CG

21. a)   **b)** Sample Space   **c)** $\frac{1}{4}$

	Sample Space
H—R	HR
H—B	HB
H—G	HG
H—P	HP
T—R	TR
T—B	TB
T—G	TG
T—P	TP

22. a) $\frac{1}{4}$ **b)** $\frac{2}{9}$ **23. a)** $\frac{1}{100}$ **b)** 0 **24. a)** $\frac{4}{25}$ **b)** $\frac{2}{15}$
25. a) $\frac{9}{100}$ **b)** $\frac{1}{15}$ **26.** $\frac{1}{15}$ **27.** 0 **28.** $\frac{1}{5}$ **29.** $\frac{1}{4}$
30. Against, 3:1; in favor, 1:3 **31.** \$1.25 **32.** $\frac{1}{8}$ **33.** $\frac{5}{8}$
34. In favor, 3:5; against, 5:3 **35.** \$3.75 **36.** $\frac{7}{8}$ **37.** $\frac{7}{8}$
38. $\frac{25}{32}$ **39.** $\frac{7}{32}$ **40.** $\frac{1}{8}$ **41.** $\frac{6}{17}$ **42.** $\frac{10}{17}$ **43.** $\frac{10}{17}$ **44.** $\frac{12}{17}$
45. a) 120 **b)** \$222,220 **46.** 60 **47.** 60 **48.** 504
49. 20 **50. a)** 3003 **b)** 3,628,800 **51.** $\frac{1}{25,827,165}$
52. 1120 **53.** 560 **54.** $\frac{1}{221}$ **55.** $\frac{1}{12}$ **56.** $\frac{1}{18}$ **57.** $\frac{1}{24}$
58. $\frac{11}{12}$ **59.** $\frac{5}{91}$ **60.** $\frac{15}{91}$ **61.** $\frac{3}{13}$ **62.** $\frac{10}{13}$

Chapter 11 Test

1. $\frac{17}{36}$ **2.** $\frac{3}{4}$ **3.** $\frac{1}{2}$ **4.** $\frac{3}{4}$ **5.** $\frac{1}{4}$ **6.** $\frac{3}{14}$ **7.** $\frac{3}{14}$ **8.** $\frac{2}{7}$
9. $\frac{15}{28}$ **10.** $\frac{8}{13}$ **11.** 18
12.

	Sample Space
1—a	1a
1—b	1b
1—c	1c
2—a	2a
2—b	2b
2—c	2c
3—a	3a
3—b	3b
3—c	3c
4—a	4a
4—b	4b
4—c	4c
5—a	5a
5—b	5b
5—c	5c
6—a	6a
6—b	6b
6—c	6c

13. $\frac{1}{18}$ **14.** $\frac{4}{9}$ **15.** $\frac{2}{3}$ **16.** 374,400 **17.** 5:9 **18.** $\frac{2}{7}$
19. 25¢ **20.** $\frac{4}{7}$ **21.** 120 **22.** $\frac{1}{7}$ **23.** $\frac{6}{7}$ **24.** $\frac{175}{396}$

Chapter 12

Section 12.1 Exercises

7. a) The entire group of items being studied
   b) A subset of the population
9. a) A sample obtained by selecting every *n*th item on a list or production line
   b) Use the random number table to select the first item, then select every *n*th item after that.
11. a) and **b)** Divide the population into parts (strata), then draw random samples from each stratum.

Section 12.2 Exercises

3. Being the largest department store does not imply that it is inexpensive.
5. Most driving is done close to home.
7. Many elderly people retire to Florida.
9. Because four out of six dentists prefer a toothpaste, it does not mean it is the best. Also, the total number of dentists surveyed is not given, and what does "best" mean in this context?
11. Because it has fewer calories one cannot conclude that one will stay healthy by eating it.
13. Because most prefer it does not mean that they buy it. Other things such as cost must be considered.
15. a)

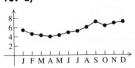

b)

17. a)

TV Ratings

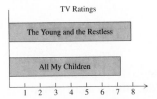

b)

TV Ratings

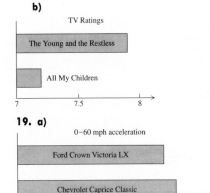

19. a)

0–60 mph acceleration

b) The new graph gives the impression of a smaller difference between the two cars.

Section 12.3 Exercises

1. a) 19 **b)** 7 **c)** 17 **d)** 42–48 **e)** 49–55

3.

Grade	Number of Students
0	4
1	6
2	3
3	4
4	3
5	3
6	7
7	3
8	12
9	3
10	2

5.

IQ	Number of Students
80– 84	2
85– 89	5
90– 94	8
95– 99	8
100–104	12
105–109	5
110–114	4
115–119	1
120–124	3
125–129	1
130–134	0
135–139	1

7.

IQ	Number of Students
80– 90	8
91–101	22
102–112	11
113–123	7
124–134	1
135–145	1

9.

Age	Number of Presidents
40–45	2
46–51	11
52–57	16
58–63	6
64–69	5

11.

Age	Number of Presidents
42–46	3
47–51	10
52–56	12
57–61	9
62–66	4
67–71	2

13.

Electoral Votes	Number of States and D.C.
1– 8	29
9–16	14
17–24	4
25–32	2
33–40	1
41–48	0
49–56	1

15.

Electoral Votes	Number of States and D.C.
0–10	33
11–21	12
22–32	4
33–43	1
44–54	1

17.

Population (millions)	Number of Urban Areas
2.0– 3.9	10
4.0– 5.9	11
6.0– 7.9	9
8.0– 9.9	2
10.0–11.9	2
12.0–13.9	0
14.0–15.9	0
16.0–17.9	1

19.

Population (millions)	Number of Urban Areas
2.9– 5.8	21
5.9– 8.8	10
8.9–11.8	3
11.9–14.8	0
14.9–17.8	1

21.

Population Density per Square Mile	Number of States
0.5– 100.5	29
100.6– 200.6	10
200.7– 300.7	4
300.8– 400.8	2
400.9– 500.9	1
501.0– 601.0	0
601.1– 701.1	1
701.2– 801.2	1
801.3– 901.3	0
901.4–1001.4	1
1001.5–1101.5	1

23.

Population Density per Square Mile	Number of States
0.5– 150.5	34
150.6– 300.6	9
300.7– 450.7	2
450.8– 600.8	1
600.9– 750.9	1
751.0– 901.0	1
901.1–1051.1	2

Marginal note on page 584: There are 6 *F*'s in the statement.

Section 12.4 Exercises

1. Psychology—486; Business—551; Engineering—385; Sciences—374; English—580; History and Political Science—400; Other—824.

3.

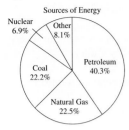

Sources of Energy

Nuclear 6.9%
Other 8.1%
Petroleum 40.3%
Coal 22.2%
Natural Gas 22.5%

d)

Visits	Number of Families
1	4
2	2
3	8
4	8
5	6
6	11
7	9
8	3
9	0
10	1

5. a) and **b)**

Years employees have worked for ABC Co.

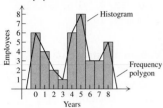

7. a) and **b)**

Annual salaries of management at ABC Co.

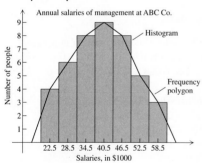

9. a) 29 **b)** 6 **c)** 8 **d)** 237

e)

Number of Plants	Number of Houses
6	3
7	7
8	8
9	6
10	4
11	0
12	1

11. a) 8 **b)** 24 **c)** 52

e)

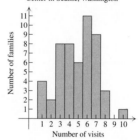

Number of visits selected families have made to the Pacific Science Center in Seattle, Washington

13.

Histogram

Frequency polygon

15. a)

Salaries	Number of Companies
25	1
26	7
27	4
28	3
29	2
30	3
31	3
32	2

b) and **c)**

Starting salaries of employees at 25 different companies

17. a)

Circulation (millions)	Number of Magazines
1.9– 6.3	25
6.4–10.8	2
10.9–15.3	0
15.4–19.8	2
19.9–24.3	1

b) and **c)**

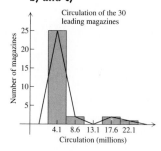

Circulation of the 30 leading magazines

Section 12.5 Exercises

1. 5.2, 5, 8, 4.5   **3.** 74.7, 80, none, 70.5   **5.** 8, 8, none, 8
7. 72.9, 60, none, 85   **9.** 10.7, 11, 12, 10.5
11. 477.2, 463, none, 524.5   **13.** 80.2, 80.5, none, 78
17. Mean: $95.62; median: $100.55; mode: none; midrange: $87.07
19. a) 523.89, 190.5, none, 2141.15
   b) 10.13, 9.3, none, 11.3
21. 498
23. a) Mean = 43,400   **b)** Cannot find median
   c) Cannot find mode   **d)** Midrange = 93,300
25. One example: 50, 60, 70, 80, 90   **27.** 91 or better
29. Mode   **31.** Mean, mode, and midrange will change

33. The data must be ranked.
35. Kevin was taller than approximately 43% of all kindergarten children.
37. About 75% of the students travel no more than 47.6 minutes one way to Central Valley College.
39. a) No
   b) Yes, Kendra was in the better relative position.

Section 12.6 Exercises

1. First set, because the scores have a greater spread from the mean.
3. All the data are the same.   **7.** 9, $\sqrt{11.2} = 3.35$
9. 12, $\sqrt{12.67} = 3.56$ **11.** 0, 0   **13.** 2, $\sqrt{0.67} = 0.82$
15. 4, $\sqrt{2.86} = 1.69$ **17.** 28, $\sqrt{120} = 10.95$
19. 18, $\sqrt{42} = 6.48$
21. d) When every piece of data in a distribution is multiplied by some number n, the mean and standard deviation of the new set of data will be n times the mean and standard deviation of the original set of data.
   e) $\bar{x} = 20$, $s = 10$

Section 12.7 Exercises

5. The mode is the lowest value, the median is greater than the mode, and the mean is greater than the median
9. 0.50   **11.** 0.818   **13.** 0.088   **15.** 0.041   **17.** 0.015
19. 0.113   **21.** 44.5%   **23.** 90.5%   **25.** 97.1%
27. 97.5%   **29.** 17.5%   **31.** 34.1%   **33.** 68.2%
35. 81.8%   **37.** 30.8%   **39.** 36.1%   **41.** 30.8%
43. 99.4%   **45.** 50%   **47.** 17.4%   **49.** 77.3%
51. 29.1%   **53.** 59.9%   **55.** 69.2%   **57.** 42.1%
59. 54,560   **61.** 11.1%
63. 3.6%—A; 10%—B; 74.9%—C; 8.6%—D; 2.9%—F
65. a) $\bar{x} = 5.33$ **b)** 3.00
   c) $\bar{x} + 1.1s = 8.63$
      $\bar{x} - 1.1s = 2.03$
      $\bar{x} + 1.5s = 9.83$
      $\bar{x} - 1.5s = 0.83$
      $\bar{x} + 2s = 11.33$
      $\bar{x} - 2s = -0.67$
      $\bar{x} + 2.5s = 12.83$
      $\bar{x} - 2.5s = -2.17$
   d) $\frac{17}{30} = 56.7\%$, $\frac{28}{30} = 93.3\%$, $\frac{30}{30} = 100\%$, $\frac{30}{30} = 100\%$

Review Exercises

1. a) The entire group of items being studied
   b) A subset of the population

2. A sample where every item in the population has the same chance of being selected

3. The candy bars may have lots of calories, or fat, or salt. Therefore it may not be healthy to eat them.

4. Sales may not necessarily be a good indicator of profit, and expenses must also be considered.

5. a)

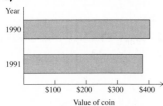

b)

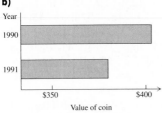

6. a)

Class	Frequency
20	4
21	6
22	1
23	5
24	2
25	1
26	4
27	1
28	2
29	1
30	8

b) and **c)**

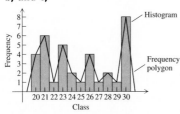

7. a)

Class	Frequency
30– 38	4
39– 47	4
48– 56	6
57– 65	7
66– 74	7
75– 83	7
84– 92	4
93–101	1

b) and **c)**

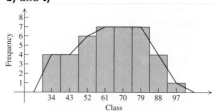

8. 77   **9.** 78   **10.** None   **11.** 77.5   **12.** 27

13. $\sqrt{116.8} = 10.81$ **14.** 13   **15.** 13   **16.** None

17. 13.5   **18.** 19   **19.** $\sqrt{40} = 6.32$ **20.** 68.2%

21. 95.4%   **22.** 94.5%   **23.** 5.5%   **24.** 72.6%   **25.** 50%

26. 24.9%   **27.** 56.8%   **28.** 81.8%   **29.** 175 lb

30. 180 lb   **31.** 25%   **32.** 25%   **33.** 14%   **34.** 18,700 lb

35. 233 lb   **36.** 145.6 lb   **37.** 3.675   **38.** 2   **39.** 3

40. 7   **41.** 14   **42.** $\sqrt{8.28} = 2.88$

43.

Class	Frequency
0– 1	7
2– 3	14
4– 5	10
6– 7	6
8– 9	1
10–11	1
12–13	0
14–15	1

44. and **45.**

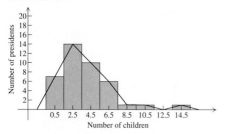

46. No **47.** No **48.** No

Chapter 12 Test

1. 20 **2.** 21 **3.** 24 **4.** 18 **5.** 12 **6.** $\sqrt{24.5} = 4.95$

7.

Class	Frequency
5–10	7
11–16	5
17–22	1
23–28	7
29–34	5
35–40	3
41–46	2

8.

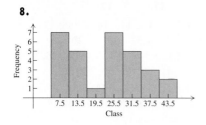

9.

10. \$475 **11.** \$450 **12.** 75% **13.** 79% **14.** \$48,000
15. \$520 **16.** \$420 **17.** 34.1% **18.** 34.5% **19.** 33.7%
20. 27.4%

Appendix A

1. 36 kg **3.** 1.26 m **5.** 12 m² **7.** 333.3 lb
9. 1687.5 acres **11.** 55.1 ℓ **13.** 1.52 fl oz **15.** 54 kg
17. See Table A in Appendix A **19.** 2 **21.** 30° Celsius
23. c **25.** a **27.** f **29.** mg, 0.001 g **31.** dg, 0.1 g
33. dag, 10 g **35.** kg, 1 000 g **37.** 30 m **39.** 2.4 ℓ
41. 0.001 34 ℓ **43.** 1 427 ℓ **45.** 0.000 076 m **47.** 830 cm
49. 94 500 g **51.** 895 000 mℓ **53.** 0.002 4 km
55. 620 cm, 5.6 dam, 0.47 km **57.** 2.2 kg, 2 400 g, 24 300 dg
59. a) 1.3 m **b)** 130 cm **61.** 310 cm **63.** 1 hr 23.3 min
65. 281 g **67.** 114.75 mg **69.** 400.5 mg **71.** 184.5 mg
73. a) 472 mg **b)** 3 776 mg or 3.776 g **75.** 82.4°F
77. −21.1°C **79. b)** 1.0 cc

PHOTO CREDITS

INDEX

A BRIEF LOOK AT THE HISTORY OF MATHEMATICS

Mathematical Periods	CHAPTER 7 Systems of Equations and Inequalities	CHAPTER 8 Geometry	CHAPTER 9 Mathematical Systems
Egyptian and Babylonian period **3000** B.C.–**600** B.C.		**3000** B.C. Great pyramids constructed. Although the mathematics involved was accurate, the procedures used were trial and error, rather than formal mathematical techniques. **2000** B.C. Volume of frustrum of pyramid, formulas for land area and granary volumes, rough formulas for area of circle and cylinder in Moscow Papyrus.	
Greek period **600** B.C.–A.D. **500**		**600** B.C. Thales, said to be the first mathematician to prove a theorem, is considered the first true mathematician. Pythagoras made numerous contributions to geometry, including the Pythagorean theorem. **300** B.C. Euclid's *Elements* laid the foundation for present-day high school geometry. The *Elements* emphasized the deductive method. **240** B.C. Archimedes computed π, conic sections, Archimedes spiral.	
Hindu and Arabian period (the dark ages of mathematics) A.D. **500**–A.D. **1200**		**980** Abul-Wefa—geometric constructions and trigonometric tables. **1100** Omar Khayyam—geometry, solutions of cubic equations, calendar.	
Period of transition A.D. **1200**–A.D. **1550**		**1634** Herigone first used (to represent angle. **1636** Descartes—formula relating edges, faces, and vertices of convex polyhedron.	
Modern period A.D. **1550–present** **(Century of geniuses 1600–1700)**	**1800** Gauss and Legendre used the method of least squares to solve systems of linear equations. **1857** Arthur Cayley solved systems of linear equations using matrices. **1950** George Dantzig introduced the simplex algorithm to solve linear programming problems.	**1640** Desargues introduced projective geometry, conic sections. **1650** Pascal—projective geometry, conic sections, Pascal's triangle. **1733** Saccheri, forerunner of non-Euclidean geometry. **1750** Euler contributed much to geometry and topology, including Euler's formulas. **1805** Legendre reorganized Euclid's *Elements* in his *Elements de Geometrie*. Also contributed to calculus and number theory. **1825** Gauss, Bolyai, and Lobachevski contributed to the development of non-Euclidean geometry. Gauss—point set topology. **1854** Riemann—Riemannian geometry, Riemann surfaces, Riemann integral in calculus. **1895** Poincaré—systematic development of combinatorial topology. **1899** Hilbert—formalization of geometry, contributions to number theory and differential equations.	**1820** Gauss introduced modulo arithmetic. **1825** Abel's work on elliptical functions led to Galois's development (1830) of the theory of groups. **1857** Cayley invented matrices.